Universitext

Universitext

Universitext is a series of textbooks that presents material from a wide variety of mathematical disciplines at master's level and beyond. The books, often well class-tested by their author, may have an informal, personal even experimental approach to their subject matter. Some of the most successful and established books in the series have evolved through several editions, always following the evolution of teaching curricula, into very polished texts.

Thus as research topics trickle down into graduate-level teaching, first textbooks written for new, cutting-edge courses may make their way into *Universitext*.

More information about this series at http://www.springer.com/series/223

Vladimir Kadets

A Course in Functional
Analysis and Measure Theory

Translated from the Russian by Andrei Iacob

Springer

Vladimir Kadets
School of Mathematics
  and Computer Science
V. N. Karazin Kharkiv
  National University
Kharkiv
Ukraine

ISSN 0172-5939   ISSN 2191-6675   (electronic)
Universitext
ISBN 978-3-319-92003-0   ISBN 978-3-319-92004-7   (eBook)
https://doi.org/10.1007/978-3-319-92004-7

Library of Congress Control Number: 2018943248

Mathematics Subject Classification (2010): 46-01, 47-01, 28-01

Printed on acid-free paper

This Springer imprint is published by the registered company Springer International Publishing AG
part of Springer Nature
The registered company address is: Gewerbestrasse 11, 6330 Cham, Switzerland

*To my family, teachers,
friends and students.*

Preface to the English Translation

The original Russian version of this textbook appeared in electronic preprint form in 2004 and was then published officially in 2006 under the title "A Course in Functional Analysis". At that time, the Functional Analysis course in my University was obligatory for pure math majors. It ran over three semesters (each for 4 hours weekly) and included the Measure Theory course as an integral part. The textbook was intended to cover those three semesters, as well as the optional one-semester course in Topological Vector Spaces. Later, Measure Theory was separated from the Functional Analysis course, which explains why the 2012 Ukrainian translation of the book appeared with the changed title "A Course in Functional Analysis and Measure Theory". We live in an epoch of permanent changes. In 2015, the former School of Mechanics and Mathematics was renamed the School of Mathematics and Computer Science, some new courses were introduced, and others were modified; in particular, the Functional Analysis course was divided into two, entitled "Functional Analysis" (for Bachelors) and "Advanced Functional Analysis" (for Masters), so an appropriate title for the book should be something long and unpleasant like "Courses in Measure Theory, Functional Analysis, and Advanced Functional Analysis in one book, with an introduction to Topological Vector spaces".☺

The Functional Analysis universe is immense, so one book cannot cover everything. There is some basic material, such as orthonormal systems in Hilbert spaces, or the Hahn–Banach theorem, that every decent Functional Analysis book ought to include, but advanced topics, to a great extent, reflect the mathematical tastes and research interests of the author. The present book is not an exception.

The book consists of 18 chapters, the structure of which is reflected in detail in the "Contents" below. The reader is assumed to be familiar with the basics of Mathematical Analysis and Linear Algebra and to have some elementary knowledge of the language of General Topology. In Chap. 1 "Metric and Topological Spaces", we briefly recall the well-known basic definitions and facts, discuss the terminology and notation adopted in the book, and present in more detail results that are important for our course, such as a compactness criterion in $C(K)$ or the Baire theorem. In Chaps. 2–4, we present the key results about the Lebesgue measure and integral, starting from basic definitions and reaching the most often applied results, such as the Lebesgue dominated convergence theorem and the Fubini theorem on the integration of functions of two variables. In the next two Chaps. 5 and 6, we already pass to elements of Functional Analysis, which are supposed to help the reader understand the more advanced questions of Measure Theory collected in Chap. 7, "Absolute Continuity of Measures and Functions. The Connection Between Derivative and Integral". The main subject of Chap. 8 is the description of continuous linear functionals on $C(K)$ as integrals with respect to charges (signed measures). In Chap. 9, we talk about various forms of the Hahn–Banach theorem and its applications. The closed graph theorem, uniform boundedness principle, and their multiple applications can be found in Chap. 10. Chapter 11 is devoted to spectral properties of bounded linear operators in Banach spaces; in particular, it treats the spectral theory of compact operators. Chapters 12 and 13 cover basic results about Hilbert spaces and self-adjoint operators, up to the construction of Borel functions of self-adjoint operators and of spectral measures. In Chap. 14, "Operators in L_p", we start with the duality between L_p and $L_{p'}$, and then pass to the Fourier transform theory in L_1, L_2, and, through an operator interpolation theorem, in L_p. The title of Chap. 15, "Fixed Point Theorems and Applications", is self-explanatory. Chapters 16–18, "Topological Vector Spaces", "Elements of Duality Theory", and "The Krein–Milman Theorem and Applications", present material for the optional course in Topological Vector Spaces mentioned above.

Acknowledgments. All those mentioned in the dedication: my family, teachers, friends and students, should be mentioned here as well. In my student years, I learned a lot from my teachers, who influenced my tastes in mathematics (especially Boris Yakovlevich Levin, who taught me Mathematical Analysis and Complex Analysis, and Victor Dmitrievich Golovin, my teacher in Measure Theory and Functional Analysis). But undoubtedly the major influence on me as a mathematician came from my father, Mikhail Kadets, and the members of the Kharkiv Banach space theory seminar. My wife Anna Vishnyakova works in the same department. She was the very first reader of the Russian version of this textbook and used it in her teaching practice; her encouragement and advice helped me a lot. Communication with my co-authors, friends, and students (my apologies for not mentioning everyone by name: I am rather lucky to have long lists in all three mutually intersecting categories) broadened my horizons and influenced, indirectly or directly, the selection of material and the exposition style. It is perhaps interesting and encouraging to mention here that my nephew Boris Shumyatsky, niece Asya Shumyatska, and children Lucy and Borys survived my lectures in Measure Theory and Functional Analysis without complaints.

I am grateful to Igor Chyzhykov from the Ivan Franko National University of Lviv, who did the Ukrainian translation of this book (together with Ya. Magola) and organized its publication; special thanks to him for providing me with the LaTeX source files of the Ukrainian translation, which facilitated the preparation of the English version.

It was a nice surprise for me when, in August 2016, Rémi Lodh, mathematics editor at Springer Verlag, wrote to me that "a senior mathematician in Germany" recommended my book for translation into the English. So, my special thanks to that senior mathematician, unknown to me, to Rémi Lodh, and to Springer Verlag for following his/her advice, and for all the cooperation and help in all stages of the book's preparation. I wish to express my gratitude to Andrei Iacob for his excellent translation, not only of this book, but also of our joint book with my father "Series in Banach Spaces", published by Birkhäuser in 1997.

Kharkiv, Ukraine   Vladimir Kadets
January 2018

Contents

Introduction

Functional Analysis is devoted to the study of various structures that are defined on infinite-dimensional linear spaces. Normed, Banach, and topological vector spaces, Hilbert spaces, function spaces, Banach algebras, spaces of operators—this is a rather incomplete list of the basic objects of Functional Analysis. Although some results that belong to this direction of mathematics also emerged earlier, Functional Analysis established itself as an independent field in the 1920s. Banach's famous monograph [1], published in its first version in 1931, in Polish, provided an overview of the period of establishment of this new branch of mathematics, period in which a leading role was played by the members of the Lviv mathematics school under the leadership of Stefan Banach. Through the efforts of many mathematicians, Functional Analysis has grown into one of the most interesting fields of contemporary mathematics, a field whose active development continues intensively to the present day.

The textbook you are holding in your hands was elaborated on the basis of a Functional Analysis course which has been delivered by the author since 1990 in the Mathematics Department of the School of Mechanics and Mathematics of the Kharkiv National University. The course is divided into three semesters, with the first semester devoted first and foremost to the study of Measure Theory and the Lebesgue integral, and the second and third dealing with the fundamental structures of Functional Analysis and Operator Theory. In addition, for students interested in making a deeper acquaintance with the subject, special courses, such as "Topological Vector Spaces" and "Introduction to the Theory of Banach spaces", are offered. Some of the material (and the closer to the end of the textbook one gets, the more material of this kind is included in the text) can be regarded not as part of the main course, but as a bridge connecting the standard course to the specialized ones. For the reader's convenience, the textbook includes additional sections covering material from preliminarily delivered courses. Thus, we recall the necessary terminology from Linear Algebra and reproduce the introductory chapters of the theory of metric spaces and compact sets, which belong more to Mathematical Analysis and Topology.

To strengthen the analogy with the Riemann integral, the theory of the Lebesgue integral is treated on the basis of Fréchet's definition, namely by means of the convergent integral sums analogous to the Riemann integral sums. One of the advantages of this approach is the simplicity with which this definition extends to vector-valued functions.

In mathematics, nothing can be learned without solving problems. This text includes many exercises: simple ones, designed to facilitate the mastering of the new notions, as well as more complicated ones, which help to move deeper into the subject. The material that goes beyond the standard course is often presented in survey form. Proofs are frequently replaced by chains of exercises, which, once solved, allow the reader to independently obtain the stated results. Exercises include also some of the assertions used in the main text. This is done in those cases where the assertions seem too elementary to us to require writing their proofs explicitly, or are not too obvious, yet fully accessible to the students and serve as good training tests. Some exercises are supplied with comments, placed at the end of the corresponding chapter.

Functional Analysis is founded on a geometric approach to the study of objects that are actually analytic in nature: functions, equations, series, sequences. This approach, which enables one to use geometric intuition in complicated analytic problems, proved to be highly productive. Thanks to it, powerful methods were developed in the framework of Functional Analysis that found application in a variety of fields of mathematics. The language of Functional Analysis found its way into branches of pure and applied mathematics such as Harmonic Analysis, Differential and Integral Equations, Approximate Computation Methods, Linear Programming, Optimization Methods, and this list of applications could go on. The present textbook is also an attempt to represent the main ideas and directions of such applications, first and foremost to questions in Harmonic Analysis.

The preparation of this book extended over a period of several years, and at different stages of its writing, many teachers and students participated in the discussion of the text and provided advice. I want to thank all the listeners of this course, and especially my former students Yu. Zabelyshinskii and I. Rud', who provided access to their course notes; to my colleagues A. Vishnyakova and L. Bezuglaia, who used the draft of the textbook to teach a Functional Analysis course, and whose remarks enabled me to improve the text, and also to the many students who used the textbook to study Functional Analysis and provided a list of misprints. I am deeply grateful to V. Maslyuchenko, A. Plichko, M. Popov, and V. Romanov who provided reviews of this textbook, and to T. Banakh, who made a series of useful remarks that were taken into account in the work on the manuscript.

Chapter 1
Metric and Topological Spaces

Topological, and especially metric spaces, are frequently mentioned and used in courses on mathematical analysis, linear algebra (in which one of the most important examples of metric spaces, the finite-dimensional Euclidean space, is examined), differential geometry (in which geodesic curves and the intrinsic metric of a surface are studied), and also, it goes without saying, in topology courses. For this reason, we only briefly recall the well-known definitions and facts, discuss the terminology and notation adopted in this book, while dwelling in more detail on issues that possibly are not treated in other courses.

1.1 Sets and Maps

In the exposition of functional analysis it is assumed that the reader is familiar with the notion of set and the simplest operations on sets: union and intersection of a finite or infinite number of sets, difference, complement, symmetric difference, Cartesian product, as well as with the notions of relation, function, graph of a relation or of a function, equivalence classes; terms like countable or uncountable sets, and so on. In the main part of the this course we will not use the technique of transfinite numbers and transfinite induction; but the reader will undoubtedly benefit from making acquaintance with the elements of the theory of transfinite numbers in, say, Kelley's textbook [23], the "Appendix" of which provides a strict formal exposition of the theory, or in Natanson's book [31], where the exposition is less formal, but readily accessible. Some finer questions of measure theory and functional analysis require a command of the method of transfinite induction. We shall touch upon such questions only in exercises and comments to them (though not often).

As a rule, sets will be denoted by uppercase Latin letters, and their elements by lowercase letters. The terms "collection (of elements)", "system (of elements)", or

© Springer International Publishing AG, part of Springer Nature 2018
V. Kadets, *A Course in Functional Analysis and Measure Theory*,
Universitext, https://doi.org/10.1007/978-3-319-92004-7_1

"family (of elements)" will often be used with the same meaning as "set". Now let us explain some of the terms and notations used in the text.

— $A \backslash B$ — the (set-theoretic) difference of the sets A and B: $A \setminus B$ consists of all elements of A that do not belong to B.
— $A \triangle B$ — the symmetric difference of the sets A and B: $A \triangle B = (A \backslash B) \cup (B \backslash A)$. Another, equivalent definition: $A \triangle B = (A \cup B) \backslash (A \cap B)$.
— $A \times B$ — Cartesian product of the sets A and B: $A \times B = \{(a, b) : a \in A, b \in B\}$. In other words, the Cartesian product of the sets A and B is the set of all ordered pairs, where the first coordinate belongs to A and the second to B.
— $\prod_{k=1}^{n} A_k$ — the Cartesian product of the sets $A_1, \ldots, A_n$:

$$\prod_{k=1}^{n} A_k = \{(a_1, \ldots, a_n) : a_k \in A_k\}.$$

Formally, the Cartesian product operation is not associative. For example, the elements of the set $(A \times B) \times C$ are of the form $((a, b), c)$, whereas those of $A \times (B \times C)$ are of the form $(a, (b, c))$. At the same time, both $((a, b), c)$ and $(a, (b, c))$ are naturally identified with the triple (a, b, c). If one agrees to use such an identification, then the Cartesian product operation becomes associative, and one has that $\left(\prod_{k=1}^{n} A_k\right) \times \left(\prod_{k=n+1}^{m} A_k\right) = \prod_{k=1}^{m} A_k$.

— 2^A — the collection of all subsets of A.
— $\mathbb{R}$ — the set of all real numbers (alternative name — the real line or axis).
— $\mathbb{Q}$ — the set of all rational numbers.
— $\mathbb{Z}$ — the set of all integers.
— $\mathbb{N}$ — the set of all natural numbers.
— $\mathbb{C}$ — the set of all complex numbers.
— $\mathbb{R}^n$ — the n-dimensional coordinate space, i.e., the Cartesian product of n copies of the real line.
— $\mathbb{R}^+ = \{t \in \mathbb{R} : t \geqslant 0\}$ (the non-negative half-line).

In the exercises given below we have collected some relations among sets that will be used later in various arguments. Generally such relations will be used without proof; their verification is purely technical and requires only a few routine manipulations with logic expressions and enumeration of possible cases.

Exercises

1. Let $A = \bigcup_{k=1}^{\infty} A_k$, $B = \bigcup_{k=1}^{\infty} B_k$. Then $A \cap B = \bigcup_{k,j=1}^{\infty} (A_k \cap B_j)$.

2. Let $A, B \subset \Omega$. Then $(\Omega \setminus A) \triangle (\Omega \setminus B) = A \triangle B$.

3. For any sets A_1, A_2, A_3 one has the inclusion $A_1 \triangle A_3 \subset (A_1 \triangle A_2) \cup (A_2 \triangle A_3)$.

4. Let $\{A_n\}_{n \in M}$ and $\{B_n\}_{n \in M}$ be two collections of sets. Then

$$\left(\bigcap_{n \in M} A_n \right) \triangle \left(\bigcap_{n \in M} B_n \right) \subset \bigcup_{n \in M} (A_n \triangle B_n)$$

and

$$\left(\bigcup_{n \in M} A_n \right) \triangle \left(\bigcup_{n \in M} B_n \right) \subset \bigcup_{n \in M} (A_n \triangle B_n).$$

5. For any map $f : X \to Y$ and any subsets $A, B \subset X$, it holds that $f(A \cup B) = f(A) \cup f(B)$; also,

6. $f(A \cap B) \subset f(A) \cap f(B)$.

7. Give an example where $f(A \cap B) \neq f(A) \cap f(B)$.

8. The map $f : X \to Y$ is injective if and only if, for any subsets $A, B \subset X$, one has that $f(A \cap B) = f(A) \cap f(B)$.

9. Let $f_1 : X \to Y_1$, $f_2 : X \to Y_2$, and let the map $f : X \to Y_1 \times Y_2$ be given by the rule $f(x) = (f_1(x), f_2(x))$. Then $f^{-1}(A_1 \times A_2) = f_1^{-1}(A_1) \cap f_2^{-1}(A_2)$ for any subsets $A_1 \subset Y_1$, $A_2 \subset Y_2$.

10. Let $\{A_n\}_{n \in M}$ be some collection of subsets of a set Ω. Then the *de Morgan formulas* hold:

$$\Omega \setminus \bigcap_{n \in M} A_n = \bigcup_{n \in M} (\Omega \setminus A_n) \quad \text{and} \quad \Omega \setminus \bigcup_{n \in M} A_n = \bigcap_{n \in M} (\Omega \setminus A_n).$$

1.2 Topological Spaces

1.2.1 Terminology

A family τ of subsets of the set X is called a *topology* if it satisfies the following axioms:

1. The empty set and the set X itself belong to τ.
2. The union of any collection of sets of the family τ belongs to τ.
3. The intersection of any finite number of sets of the family τ belongs to τ.

A set equipped with a topology is called a *topological space*. If on a set one considers only one topology, then the corresponding topological space will be denoted by the same letter as the set itself. If the topology needs to be specified, then we will use the notation (X, τ). The sets belonging to the family τ are said to be *open in the topology* τ (or simply *open*, if it is clear which topology one is talking about).

The simplest example of a topology on an arbitrary set X is the *discrete topology* 2^X, where as the open sets one takes all subsets of X. Another standard example of a topological space is the real line $\mathbb{R}$, where the open sets are the finite or countable unions of open intervals.

Let X be a topological space and $x \in X$. A subset $U \subset X$ is called an *open neighborhood* of the point x if U is open and $x \in U$. A set U is called a *neighborhood* of the point x if it contains an open neighborhood of x. A topological space X is said to be *separated in the sense of Hausdorff*, or a *Hausdorff space* (or simply *separated*), if it satisfies the following *separation axiom*:

4. For any points $x, y \in X$, $x \neq y$, there exist neighborhoods U and V of x and y, respectively, such that $U \cap V = \emptyset$.

Henceforth we will, as a rule, consider Hausdorff spaces.

Let X be a topological space and $x \in X$. A family $\mathfrak{U}$ of subsets of X is called a *neighborhood basis* of the point x if all the elements of $\mathfrak{U}$ are neighborhoods of x, and for any neighborhood U of x there exists a neighborhood $V \in \mathfrak{U}$ such that $V \subset U$.

A topology can be defined locally, i.e., starting not with the entire family of open sets, but with bases of open neighborhoods. Thus, suppose that for each point x of the set X there is given a non-empty family $\mathfrak{U}_x$ of sets with the following properties:

— if $U \in \mathfrak{U}_x$, then $x \in U$;
— if $U_1, U_2 \in \mathfrak{U}_x$, then there exists a set $U_3 \in \mathfrak{U}_x$ such that $U_3 \subset U_1 \cap U_2$;
— if $U \in \mathfrak{U}_x$ and $y \in U$, then there exists a set $V \in \mathfrak{U}_y$ such that $V \subset U$.

Then there exists a unique topology on X for which the families $\mathfrak{U}_x$ are neighborhood bases of the corresponding points. This topology is given as follows: a point x is said to be an *interior point of the set* A if some neighborhood $U \in \mathfrak{U}_x$ of the point x is contained in A; the set A is declared as *open* if all its points are interior points. In other words, a set is open if and only if together with any of its points it also contains a neighborhood of that point.

Let A be a subset of the topological space X. The set A is said to be *closed* if its complement $X \setminus A$ is open. The union of any finite number of closed sets is closed, while the intersection of any number of closed sets is closed. The *closure* of a set A is defined as the intersection of all closed sets that contain A, and is denoted by $\overline{A}$. The set $\overline{A}$ is the smallest, with respect to inclusion, closed set containing A. A point $x \in X$ is called a *limit point* for A if every neighborhood of x contains a point of A different from x. The closure of the set A consists of the points of A itself and all its limit points. The set A is said to be *dense* in the set B if $\overline{A} \supset B$. The set A is said to be *dense* if it is dense in the entire space X. A topological space X is called *separable* if X contains a countable dense subset.

Let (x_n) be a sequence of elements of the topological space X. The point $x \in X$ is called the *limit* of the sequence (x_n) if for any neighborhood U of the point x all the terms of the sequence, starting with one of them, belong to U. In this case we say that the sequence (x_n) *converges* to x. A point x is called a *limit point* for the sequence (x_n) if any neighborhood U of x contains infinitely many terms of the sequence.

A map f acting from a topological space X to a topological space Y is said to be *continuous* if for any open set A in Y its preimage $f^{-1}(A)$ is an open set in X. Continuity can be reformulated in terms of neighborhoods: the map f is continuous if for any point $x \in X$ and any neighborhood U of the point $f(x)$ there exists a neighborhood V of x such that $f(V) \subset U$. The map $f : X \to Y$ is called a *homeomorphism* if it is bijective, continuous, and its inverse map $f^{-1} : Y \to X$ is continuous. Two topological spaces are said to be *homeomorphic* if there exists a homeomorphism between them. Suppose that on the set X there are given two topologies, τ_1 and τ_2. By definition, the topology τ_1 is *stronger* than the topology τ_2 (or, equivalently, τ_2 is *weaker* than τ_1) if any set open in the topology τ_2 is also open in the topology τ_1. In other words, the topology τ_1 is stronger than the topology τ_2 if the identity map $x \mapsto x$, acting from the topological space (X, τ_1) to the topological space (X, τ_2), is continuous. The relation "τ_1 is stronger than τ_2" is denoted $\tau_1 \succ \tau_2$.

If a set is closed, then it is also closed in any stronger topology. Correspondingly, the closure of any set in a weaker topology contains the closure of that set in the stronger topology. A limit point of a set remains a limit point in any stronger topology. If a sequence (x_n) converges to x in the topology τ_1 and $\tau_1 \succ \tau_2$, then (x_n) converges to x in the topology τ_2, too.

Let A be a subset of the topological space X. A set $B \subset A$ is said to be *open in* A if B can be represented as the intersection of some open subset of the space X with A. The subsets open in A define on A a topology, called the *induced topology*. A subset of a topological space X, equipped with the induced topology, is called a *subspace of the topological space* X. For example, the set $\mathbb{Z}$ of integers, equipped with the discrete topology, is a subspace of $\mathbb{R}$, while $\mathbb{R}$, in turn, is a subspace of the space $\mathbb{C}$ of all complex numbers. The induced topology is also referred to as the *restriction of the topology of the space X to the subset A*.

Exercises

1. In a Hausdorff topological space every subset consisting of a single point (singleton) is closed.

2. Suppose the topological space X satisfies the following separation axiom: each single-point set is a closed subset of X. Suppose further that $x \in X$ is a limit point for A. Then every neighborhood of x contains infinitely many points of A.

3. Suppose the topological space X contains an uncountable collection of pairwise disjoint open subsets. Then X is not separable.

4. Let A, B, C be subsets of the topological space X such that A is dense in B and B is dense in C. Then A is dense in C.

5. If the system of neighborhoods of a point x has a countable basis, then there exists a decreasing — with respect to inclusion — sequence of neighborhoods that constitutes a neighborhood basis of x.

6. Let A be a subset of the topological space X, and x a limit point of A. Suppose x has a countable neighborhood basis. Then there exists a sequence of points of A that converges to x.

7. Let $f : X \to Y$ be a continuous map. If the subset A is dense in X, then $f(A)$ is dense in $f(X)$.

8. Let $f : X \to Y$ be a continuous map. If A is a dense subset of X and $f(X)$ is dense in Y, then $f(A)$ is dense in Y.

9. Let $f : X \to Y$ be a continuous map. Let A be a dense subset of X and B a closed subset of Y. If $f(A) \subset B$, then also $f(X) \subset B$.

10. Give an example of a continuous function $f : [0, 1] \to [0, 1]$ and a dense set $A \subset [0, 1]$ such that $f^{-1}(A)$ is not dense in $[0, 1]$.

11. Can two dense subsets of a topological space be disjoint?

12. Two continuous maps from a topological space X into a Hausdorff topological space Y that coincide on a dense subset $X_1 \subset X$ coincide everywhere on X.

13. The *interior* of a set A is defined as the set of all interior points of A. Show that the interior is an open set.

14. Suppose the set $A \subset X$ intersects all dense subsets of the space X. Then A has a non-empty interior.

15. The composition of two continuous maps is continuous.

16. Consider the following topology τ on $\mathbb{R}$: for each number x take as the neighborhood basis of x the family of all sets of the form $\{x\} \cup ((x - a, x + a) \cap \mathbb{Q}), a > 0$. Show that the topological space $(\mathbb{R}, \tau)$ thus constructed is separable, but contains a non-separable subspace.

1.2.2   The Product of Two Topological Spaces

Let X_1 and X_2 be topological spaces. We define on the Cartesian product $X_1 \times X_2$ a topology by taking for each point $x = (x_1, x_2) \in X_1 \times X_2$ the neighborhood basis $\mathfrak{U}_x$ consisting of all sets of the form $U_1 \times U_2$, where U_1 is a neighborhood of the point x_1 in X_1 and U_2 is a neighborhood of the point x_2 in X_2. The topology thus described is called the *product topology*, and the set $X_1 \times X_2$ equipped with the product topology is called the *product of the topological spaces* X_1 and X_2.

Consider the mappings $P_j : X_1 \times X_2 \to X_j$, $j = 1, 2$, which send the point $x = (x_1, x_2)$ to its jth coordinates: $P_1(x) = x_1$, $P_2(x) = x_2$. These maps are called *coordinate projectors* or *projections*.

Exercises

1. The coordinate projectors are continuous.

2. Among all topologies on $X_1 \times X_2$ in which the coordinate projectors are continuous, the product topology is the weakest.

3. The usual topology on $\mathbb{R}^2 = \mathbb{R} \times \mathbb{R}$ coincides with the corresponding product topology.

4. Let $x, x_n \in X_1 \times X_2$. Show that the convergence of the sequence (x_n) to a point x in the product topology is equivalent to the simultaneous convergence of the sequences $(P_1(x_n))$ to $P_1(x)$ and $(P_2(x_n))$ to $P_2(x)$. This justifies calling the product topology also the "*topology of componentwise convergence*" (also called the *topology of coordinatewise* or *pointwise convergence*).

5. Let X, Y_1, Y_2 be topological spaces, $f_1 : X \to Y_1$, $f_2 : X \to Y_2$, and let the map $f : X \to Y_1 \times Y_2$ be defined by the rule $f(x) = (f_1(x), f_2(x))$. Then f is continuous if and only if the two maps f_1 and f_2 are continuous.

6. The functions $(x, y) \mapsto x + y$ and $(x, y) \mapsto x \cdot y$ are continuous as functions from $\mathbb{R}^2$ to $\mathbb{R}$.

7. From the two preceding exercises and the theorem on the composition of continuous maps derive the theorem asserting the continuity of the sum and of the product of functions acting from a topological space into $\mathbb{R}$.

8. Deduce the theorem on the limit of the sum and product of numerical sequences from Exercises 3, 4, and 6.

9. Denote by $[0, 1]$ the unit interval equipped with the usual topology, and by $[0, 1]_d$ the same interval, equipped with the discrete topology. Describe the product topology on $X_1 \times X_2$, when: a) $X_1 = X_2 = [0, 1]$; b) $X_1 = X_2 = [0, 1]_d$; c) $X_1 = [0, 1]$, $X_2 = [0, 1]_d$.

10. Any product of Hausdorff spaces is Hausdorff.

11. The coordinate projectors are *open mappings*: the image of an open set under a coordinate projector is again an open set.

1.2.3 Compact Spaces

A Hausdorff topological space X is said to be *compact*, if it is non-empty and from any open cover of X one can extract a finite subcover. In more detail: X is compact if for every family $\mathfrak{U}$ of open sets whose union is the entire space X there exists a finite number of elements $U_1, \ldots, U_n \in \mathfrak{U}$ whose union is, as before, the entire X. A subset A of a topological space X is called a *compact set* if A is a compact set in the induced topology. In other words, A is a compact set if any two distinct points of A can be separated by a pair of neighborhoods and for any family $\mathfrak{U}$ of open subsets of X whose union contains A, there exist finitely many elements $U_1, \ldots, U_n$ of the family $\mathfrak{U}$, whose union also contains A. Any compact subset of a Hausdorff topological space is closed; any closed subset of a compact space is itself compact.

Let X, Y be Hausdorff spaces with X compact, and let $f : X \to Y$ be continuous. Then $f(X)$ is a compact subset of Y (this assertion is an easy consequence of the definitions). In particular, if X is a compact space and Y is a Hausdorff space, then the image of any closed subset K of X under f is closed. Consequently, if X is compact, Y is Hausdorff, and the map $f : X \to Y$ is not only continuous, but also bijective, then the map $f^{-1} : Y \to X$ is also continuous, i.e., f is a homeomorphism. The last assertion can be reformulated as follows: suppose on X there are given two separated topologies, $\tau_1 \succ \tau_2$, and X is compact in the topology τ_1. Then $\tau_1 = \tau_2$.

A family of sets $\mathfrak{W}$ is said to be *centered* if the intersection of any finite collection of sets from $\mathfrak{W}$ is not empty.

Theorem 1. *A Hausdorff topological space K is compact if and only if any centered family of closed subsets of K has a common point.*

Proof. Let K be a compact space and $\mathfrak{W}$ be a centered family of subsets of K. Suppose that the sets of $\mathfrak{W}$ have no common point, i.e., the intersection $\bigcap_{W \in \mathfrak{W}} W$ is empty. Passing to complements, we see that $\bigcup_{W \in \mathfrak{W}} (K \setminus W) = K$. Hence, the open sets of the form $K \setminus W$ form a cover of the compact space K. Extract a finite subcover $K \setminus W_1, \ldots, K \setminus W_n$, $W_i \in \mathfrak{W}$, $\bigcup_{i=1}^{n} (K \setminus W_i) = K$. But this last condition means that the set $\bigcap_{i=1}^{n} W_i$ is empty. This contradicts the assumption that the family $\mathfrak{W}$ is centered.

Conversely, suppose that every centered family of closed subsets of the space K has a common point. We claim that the space K is compact. Indeed, let the family $\mathfrak{U}$ of open sets cover the space K. Then the complements of the elements of $\mathfrak{U}$ form a family $\mathfrak{W}$ of closed sets with empty intersection. By assumption, the family $\mathfrak{W}$ cannot be centered, hence there exists a finite collection of sets $W_1, \ldots, W_n \in \mathfrak{W}$ that have an empty intersection. Then $\bigcup_{i=1}^{n} (K \setminus W_i) = K$, $K \setminus W_i \in \mathfrak{U}$, i.e., from the cover $\mathfrak{U}$ one can extract a finite subcover. The theorem is proved. □

Theorem 2. *Any infinite subset of a compact space has a limit point.*

Proof. Let A be an infinite subset of the compact space K. Consider the family $\mathfrak{W}$ of all closed subsets $W \subset K$ with the property that the difference $A \setminus W$ consists of finitely many points. The family $\mathfrak{W}$ is centered, hence there exists a point x which belongs to all the elements of $\mathfrak{W}$. We claim that the point x is a limit point of A. Indeed, let U be an arbitrary open neighborhood of x. Then the complement $K \setminus U$ does not contain x, and hence does not belong to $\mathfrak{W}$. Therefore, $A \setminus (K \setminus U) = A \cap U$ contains infinitely many points. □

Let us state, without proof, the Urysohn lemma on the functional separation of sets and the Tietze extension theorem. The proofs of these well-known results (even in a somewhat more general formulation) can be found, for instance, in K. Kuratowski's textbook [25, v. 1].

Lemma 1 (**Urysohn's lemma**). *Let A and B be disjoint closed subsets of a compact space K. Then there exists a continuous function $f : K \to [0, 1]$ that is equal to 0 on A and to 1 on B.*

Theorem 3 (**Tietze's theorem**). *Every continuous real-valued function defined on a closed subset of a compact space extends to a continuous function defined on the entire space.*

Exercises

1. Let K be a compact space, $x \in K$, and A a closed subset of K such that $x \notin A$. Then in K there exist two open disjoint subsets U and V such that $x \in U$ and $A \subset V$.

2. Let A and B be disjoint closed subsets of the compact space K. Then in K there exist two open subsets U and V such that $A \subset U$ and $B \subset V$.

The properties of compact spaces formulated in the previous two exercises can be regarded as strengthenings of the Hausdorff separation axiom (Subsection 1.2.1, Axiom 4). Topological spaces in which any two disjoint closed subsets can be separated by disjoint neighborhoods (as in Exercise 2 above) are called *normal spaces*. Urysohn's lemma is valid not only for compact spaces, but also for arbitrary normal spaces. A sketch of the proof of this fact is given in Exercises 4–6 below.

3. Let A and B be disjoint closed subsets of K, and $f : K \to [0, 1]$ a continuous function equal to 0 on A and 1 on B. Let D denote the set $\{n/2^m : m \in \mathbb{N}; \ 1 \leqslant n < 2^m\}$ of all dyadic rational points of the interval $(0, 1)$; and finally, for any $r \in D$, let $F_r = f^{-1}([r, 1])$. Then the sets F_r have the following properties: (1) all F_r are closed; (2) for any $r_1 < r_2$ there exists an open set $G = G_{r_1, r_2}$ satisfying $F_{r_1} \supset G \supset F_{r_2}$ (in particular, $F_{r_1} \supset F_{r_2}$); (3) $B \subset F_r \subset K \setminus A$ for any $r \in D$.

4. Suppose some family F_r of sets has the properties (1)–(3) listed in the preceding exercise. Define the function $f : K \to [0, 1]$ by $f(x) = \sup\{r \in D : x \in F_r\}$ (if in this equality the set is empty, then its supremum is taken to be 0). Then the function f is continuous, $F_r = f^{-1}([r, 1])$ for all $r \in D$, $f(x) = 0$ on A, and $f(x) = 1$ on B.

5. Let K be a normal topological space, F a closed set in K, G an open set in K, and $F \subset G$. Then there exist sets $\widetilde{F}$ and $\widetilde{G}$ in K, such that $\widetilde{F}$ is closed, $\widetilde{G}$ is open, and $F \subset \widetilde{G} \subset \widetilde{F} \subset G$.

6. Let A and B be disjoint closed subsets of the normal space K. The there exists a family of sets $F_r, r \in D$, with the properties (1)–(3). (The sets F_r can be constructed sequentially: first $F_{1/2}$, then $F_{1/4}$ and $F_{3/4}$, and so on, making sure that at each step properties (1)–(3) are satisfied.)

1.2.4   Semicontinuous Functions

Let X be a topological space. A function $f : X \to \mathbb{R}$ is said to be *lower semicontinuous* if for any $a \in \mathbb{R}$ the set $f^{-1}((a, +\infty))$ is open. In other words, the function f is lower semicontinuous if for any point $x \in X$ and any $a \in \mathbb{R}$, the condition $f(x) > a$ implies the existence of an entire neighborhood of the point x on which all values of f are also bigger than a. A function $f : X \to \mathbb{R}$ is said to be *upper semicontinuous* if the function $-f$ is lower semicontinuous. The function f is upper semicontinuous if and only if for any $a \in \mathbb{R}$ the set $f^{-1}((-\infty, a))$ is open. A function $f : X \to \mathbb{R}$ is continuous if and only if it is both lower and upper semicontinuous. The set of lower semicontinuous (respectively upper semicontinuous, continuous) real-valued functions on X will be denoted by $LSC(X)$ (respectively, by $USC(X)$ and $C(X)$), the notation being self-explanatory.

Example 1. Let $A \subset X$ be an arbitrary subset, and put

$$\mathbb{1}_A(x) = \begin{cases} 1, & \text{if } x \in A, \\ 0, & \text{if } x \in X \setminus A \end{cases}$$

($\mathbb{1}_A$ is called the *characteristic function* of the set A). The function $\mathbb{1}_A$ is lower semicontinuous if and only if the set A is open, and upper semicontinuous if and only if A is closed.

Theorem 1. *The class $LSC(X)$ has the following properties:*

1. *If $f, g \in LSC(X)$, then $f + g \in LSC(X)$.*
2. *If $f \in LSC(X)$ and $g \in C(X)$, then $f - g \in LSC(X)$.*
3. *If $f \in LSC(X)$ and $\lambda \in [0, +\infty)$, then $\lambda f \in LSC(X)$.*
4. *The supremum of any pointwise bounded collection of lower semicontinuous functions belongs again to $LSC(X)$. In detail: let $S \subset LSC(X)$ and let the function $f : X \to \mathbb{R}$ be given by $f(x) = \sup\{g(x) : g \in S\}$. Then $f \in LSC(X)$.*
5. *If $f, g \in LSC(X)$, then $\min\{f, g\} \in LSC(X)$.*

Proof. 1. For any $a \in \mathbb{R}$, the set $(f + g)^{-1}((a, +\infty))$ can be represented as a union of open sets:

$$(f + g)^{-1}((a, +\infty)) = \bigcup_{t \in \mathbb{R}} \left(f^{-1}((t, +\infty)) \cap g^{-1}((a - t, +\infty)) \right).$$

As such, it is itself open.

2. This follows from the previous assertion, since $-g \in C(X) \subset LSC(X)$.

3. $(\lambda f)^{-1}((a, +\infty)) = f^{-1}((a/\lambda, +\infty))$.

4. The supremum of a set of numbers is bigger than a if and only if at least one of the numbers in the set is bigger than a. Hence, $f^{-1}((a, +\infty))$ can be written as the union of open sets $\bigcup_{g \in S} g^{-1}((a, +\infty))$, and consequently is open.

5. $(\min\{f, g\})^{-1}((a, +\infty)) = f^{-1}((a, +\infty)) \cap g^{-1}((a, +\infty))$, and now recall that the intersection of two open sets is open. $\square$

Theorem 2. *Any lower semicontinuous function on a compact space is bounded from below.*

Proof. Let $f \in LSC(X)$, where X is a compact space. Set $A_n = f^{-1}((-n, +\infty))$, $n \in \mathbb{N}$. The sets A_n increase with the growth of n and form an open cover of the compact space X. Hence, there exists an $n_0 \in \mathbb{N}$ such that $A_{n_0} = X$, and consequently $f(t) > -n_0$ for all points $t \in X$. $\square$

Theorem 3. *Let X be a compact space and $f \in LSC(X)$. Then the function f coincides with the supremum of the family of all continuous functions that majorize f. In other words, for any $x \in X$ and any $\varepsilon > 0$ there exists a function $g \in C(X)$ such that $g \leqslant f$ at all points, and $g(x) \geqslant f(x) - \varepsilon$.*

Proof. With no loss of generality, one can assume that f is non-negative: by the preceding theorem, this can be achieved by adding to f a sufficiently large constant. Fix a point $x \in X$ and denote $f(x) - \varepsilon$ by a. If $a \leqslant 0$, then the function $g \equiv 0$ satisfies all the conditions of the theorem. Hence, we can assume that $a > 0$. Applying Urysohn's lemma to the pair of disjoint closed sets $A = f^{-1}((-\infty, a])$ and $B = \{x\}$, we deduce that there exists a continuous function $h : X \to [0, 1]$ that is equal to 0 on A and 1 on B. We claim that $g = ah$ is the sought-for function. Indeed, at the points $t \in X$ where $g(t) = 0$, the inequality $0 \leqslant g(t) \leqslant f(t)$ is obvious. As for the points where $g(t) \neq 0$, they lie in $X \setminus A$, i.e., at these points we have $f(t) > a \geqslant g(t)$. $\square$

Exercises

1. Let X be a compact space and $f \in LSC(X)$. Then there exists a point $x \in X$ for which $f(x) = \min_{t \in X} f(t)$.

Let X be a topological space, $f : X \to \mathbb{R}$, and let $\mathfrak{U}_x$ be the system of neighborhoods of $x \in X$. The *lower limit* of the function f at the point x is the number $\underline{\lim}_{t \to x} f(t) \in \mathbb{R} \cup \{-\infty\}$, defined by the formula

$$\varliminf_{t \to x} f(t) = \sup_{V \in \mathfrak{U}_x} \inf_{t \in V \setminus \{x\}} f(t).$$

The *upper limit*, $\varlimsup_{t \to x}$ is defined in a similar manner:

$$\varlimsup_{t \to x} f(t) = \inf_{V \in \mathfrak{U}_x} \sup_{t \in V \setminus \{x\}} f(t).$$

Other commonly used names for these quantities are *limit inferior* and *limit superior*, with the corresponding notations $\liminf_{t \to x} f(t)$ and $\limsup_{t \to x} f(t)$.

2. A function $f : X \to \mathbb{R}$ on a topological space X is lower semicontinuous if and only if the inequality $f(x) \leqslant \varliminf_{t \to x} f(t)$ holds for all $x \in X$.

3. Let $f : X \to \mathbb{R}$ be an arbitrary bounded function on the topological space X. The function $\underline{f}(x) = \min \left\{ f(x), \varliminf_{t \to x} f(t) \right\}$ is called the *lower envelope* of the function f, while $\overline{f}(x) = \max \left\{ f(x), \varlimsup_{t \to x} f(t) \right\}$ is called the *upper envelope* of f. Prove that $\underline{f}$ is lower semicontinuous and $\overline{f}$ is upper semicontinuous.

4. Let $f : X \to \mathbb{R}$ be an arbitrary function, $g \in LSC(X)$, and assume that $g \leqslant f$. Then $g \leqslant \underline{f}$.

1.3   Metric Spaces

1.3.1   The Axioms of Metric. Sequences and Topology

A function of two variables $\rho : X \times X \to \mathbb{R}^+$ is called a *metric* on the set X if it has the following properties:

1. $\rho(x, x) = 0$;
2. if $\rho(x, y) = 0$, then $x = y$ (non-degeneracy);
3. $\rho(x, y) = \rho(y, x)$ (symmetry);
4. $\rho(x, z) \leqslant \rho(x, y) + \rho(y, z)$ (triangle inequality).

The properties listed above are called the *metric axioms*. The quantity $\rho(x, y)$ is referred to as the *distance* between the elements x and y. A set equipped with a metric is called a *metric space*.

A subset of a metric space X, endowed with the metric of X, is called a *subspace of the metric space* X.

Let X be a metric space, $x_0 \in X$, and $r > 0$. By $B_X(x_0, r)$ (or $B(x_0, r)$, when it is clear what space one is talking about) one denotes the *open ball* of radius r centered at x_0: $B_X(x_0, r) = \{ x \in X : \rho(x, x_0) < r \}$. The *topology of a metric space* is given by means of balls: the balls centered at x_0 form a neighborhood basis of the point x_0. In other words, a subset A of the metric space X is declared to be open if together with any of its points, A contains some ball centered at that point: for any $x \in A$, there exists an $r > 0$ such that $B_X(x, r) \subset A$. A sequence (x_n) of elements of the metric space X converges to an element x if $\rho(x_n, x) \to 0$ as $n \to \infty$. Since a metric space is simultaneously a topological space, all basic topological notions remain meaningful in metric spaces. The peculiarity of the metric spaces is that in them topological concepts can be equivalently defined in terms of convergence of sequences (*sequential definitions*). Some of these definitions are given below.

Let A be a subset of the metric space X. A point $x \in X$ is said to be a *limit point* for A if there exists a sequence of elements $x_n \in A \setminus \{x\}$ which converges to x. A subset

$A \subset X$ is said to be *closed* if it contains all its limit points. A subset $A \subset X$ is said to be *open* if its complement $X \setminus A$ is closed.

Thus, in metric spaces the convergence of sequences uniquely determines the topology. This fact can be alternatively explained by providing sequential definitions of the concepts of continuity and homeomorphism. Thus, let X and Y be metric spaces. A map $f : X \to Y$ is said to be *continuous* if it takes convergent sequences into convergent ones: for any $(x_n), x \in X$, if $x_n \to x$, then $f(x_n) \to f(x)$. As for the notions of homeomorphism and homeomorphic spaces, they are defined in terms of continuity (see Subsection 1.2.1). Here is one more definition: a map $f : X \to Y$ is called an *isometry* (or *bijective isometry*) if it is bijective and preserves the metric, i.e., $\rho(x_1, x_2) = \rho(f(x_1), f(x_2))$ for all $x_1, x_2 \in X$. Two metric spaces X and Y are said to be *isometric* if there exists an isometry between them.

Exercises

1. In a metric space the neighborhood system of any point has a countable basis.

2. Show that for metric spaces the sequential definitions given above are equivalent to the topological ones.

3. For two subsets A and B of a metric space, the distance between them is defined as the infimum of the distances between their elements: $\rho(A, B) = \inf_{a \in A, b \in B} \rho(a, b)$. Prove that this "distance" obeys neither the non-degeneracy axiom, nor the triangle inequality.

4. Let X and Y be metric spaces. Define a metric on their Cartesian product $X \times Y$ by the rule $\rho((x_1, y_1), (x_2, y_2)) = \rho(x_1, x_2) + \rho(y_1, y_2)$. Verify the metric axioms for this expression. Show that the topology generated by this metric on $X \times Y$ coincides with the usual product topology. In particular, the convergence in this metric coincides with the coordinatewise (componentwise) convergence: $(x_n, y_n) \to (x, y)$ in $X \times Y$ if and only if $x_n \to x$ in X and $y_n \to y$ in Y.

5. Let X and Y be metric spaces and $f : X \to Y$ be a continuous map. Then the *graph*$\Gamma(f) = \{(x, f(x)) : x \in X\}$ of the map f is closed in $X \times Y$. The statement of this exercise remains valid for separated topological spaces. The separation of which of the two spaces X, Y is important here, and of which not?

6. Give an example of discontinuous function $f : \mathbb{R} \to \mathbb{R}$ with closed graph.

7. Show that the graph of a continuous map $f : X \to Y$ is homeomorphic to the space X.

8. Let Y be a subspace of the metric space X. Then on Y one has the topology induced by the topology of X, and also the topology given by the metric of the space Y. Show that these topologies coincide.

9. The *closed ball* of radius r centered at the point x_0 of the metric space X is defined as the set $\overline{B}_X(x_0, r) = \{x \in X : \rho(x, x_0) \leqslant r\}$. Show that every open ball is an open set, and every closed ball is a closed set.

10. Using the example of the metric space consisting of two points, $X = \{0, 1\}$, $\rho(0, 1) = 1$, show that the closure of an open ball does not necessarily coincide with the corresponding closed ball. Using the example of the metric space consisting of three points, $X = \{0, 1, 2\}$, equipped with the natural metric, show that a closed ball of larger radius can be strictly included in a ball of smaller radius[1] (of course, in this situation the centers of the balls cannot coincide). What values can the ratio of radii for closed balls strictly included in one another take?

11. Equip the space $\mathbb{R}^\omega$ of all numerical sequences with the metric

$$\rho(x, y) = \sum_{n=1}^{\infty} \frac{1}{2^n} \frac{|x_n - y_n|}{1 + |x_n - y_n|},$$

where x_n and y_n are the components of the elements x and y, respectively. Verify the metric axioms. Prove that convergence in this metric coincides with the componentwise (coordinatewise) convergence.

12. Another metric on $\mathbb{R}^\omega$ that induces the same topology is the Fréchet metric, defined by

$$\rho_1(x, y) = \inf_{n \in \mathbb{N}} \left\{ \frac{1}{n} + \max_{1 \leqslant k \leqslant n} |x_k - y_k| \right\}.$$

13. Any subspace of a separable metric space is separable. (For general topological spaces this is not the case: see Exercise 16 in Subsection 1.2.1.)

14. Which of the metric spaces you know about are separable, and which not?

1.3.2   *Distance of a Point to a Set. Continuity of Distance*

The distance of a point x of a metric space X to a non-empty subset $A \subset X$ is defined to be the infimum of the distances of x to the elements of A: $\rho(x, A) = \inf_{a \in A} \rho(x, a)$. We note that the point x belongs to the closure of the set A if and only if $\rho(x, A) = 0$.

Proposition 1. *The function $x \mapsto \rho(x, A)$ is continuous.*

Proof. Let $x, y \in X$. By the triangle inequality,

$$\rho(x, A) = \inf_{a \in A} \rho(x, a) \leqslant \inf_{a \in A} \rho(y, a) + \rho(x, y) = \rho(y, A) + \rho(x, y).$$

---

[1] "... and the other explained to me that inside the globe there was another globe much bigger than the outer one" (Jaroslav Hašek, "The good solider Švejk", Chapter "Švejk thrown out of the lunatic asylum").

Hence, $\rho(x, A) - \rho(y, A) \leqslant \rho(x, y)$. Since x and y play symmetric roles, $\rho(y, A) - \rho(x, A) \leqslant \rho(x, y)$, that is

$$|\rho(x, A) - \rho(y, A)| \leqslant \rho(x, y) \tag{1}$$

for all $x, y \in X$. This inequality obviously yields the required continuity (in fact, we actually established not only the continuity, but also that the Lipschitz condition with constant equal to 1 holds; see the corresponding definition in Exercise 4 of Subsection 1.3.4). $\qquad\square$

If $A = \{z\}$ is a singleton, we obtain a useful particular case of inequality (1):

$$|\rho(x, z) - \rho(y, z)| \leqslant \rho(x, y),$$

which leads to the following important property of the distance.

Proposition 2. *The distance is a continuous function of two variables, i.e., if $x_n \to x$, $y_n \to y$, then $\rho(x_n, y_n) \to \rho(x, y)$.*

Proof. Indeed, $|\rho(x_n, y_n) - \rho(x, y)| \leqslant |\rho(x_n, y_n) - \rho(x_n, y)| + |\rho(x_n, y) - \rho(x, y)| \leqslant \rho(y_n, y) + \rho(x_n, x) \to 0$ as $n \to \infty$. $\qquad\square$

Exercises

1. The *Hausdorff distance* between two closed subsets A and B of a metric space X is defined as

$$\rho_H(A, B) = \max \left\{ \sup_{b \in B} \rho(b, A), \ \sup_{a \in A} \rho(a, B) \right\}.$$

Show that the Hausdorff distance ρ_H is indeed a metric on the family of all non-empty bounded closed subsets of the metric space X (for comparison, see Exercise 3 in Subsection 1.3.1).

2. Let A and B be disjoint closed subsets of the metric space X. Consider the sets $A_1 = \{x \in X : \rho(x, A) < \rho(x, B)\}$ and $B_1 = \{x \in X : \rho(x, A) > \rho(x, B)\}$. Verify that A_1 and B_1 are disjoint open neighborhoods of the sets A and B, respectively. This will show that every metric space is a normal topological space (see the exercises in Subsection 1.2.3).

3. Let A and B be disjoint closed subsets of the metric space X. For each $x \in X$, put

$$f(x) = \frac{\rho(x, A)}{\rho(x, A) + \rho(x, B)}.$$

Then f is a continuous function, equal to 0 on A and to 1 on B, and taking all its values in the interval $[0, 1]$. This will provide a simple proof of Urysohn's lemma in metric spaces (see Subsection 1.2.3). Moreover, in contrast to the general Urysohn lemma, the function f constructed here is equal to 0 *only* at the points of the set A and equal to 1 *only* at the points of the set B.

4. Let A be a closed subset of a metric space X and $f : A \to [0, 1]$ a continuous function. Extend the function f to $X \setminus A$ by means of the Haudsorff formula

$$f(x) = \inf_{t \in A} \left\{ f(t) + \frac{\rho(t, x)}{\rho(x, A)} - 1 \right\}.$$

Verify that the thus extended function f is continuous on the entire space X. Deduce from this Tietze's theorem (Subsection 1.2.3) in the case of metric spaces.

1.3.3   Completeness

A sequence (x_n) of elements of the metric space X is called a *Cauchy sequence* (or a *fundamental sequence*) if $\rho(x_n, x_m) \to 0$ as $n, m \to \infty$. In detail, (x_n) is a Cauchy sequence if for any $\varepsilon > 0$ there exists a number N, beginning with which all distances between pairs of elements x_n are smaller than ε. If a sequence $x_n \in X$ has a limit $x \in X$, then it is a Cauchy sequence: indeed, $\rho(x_n, x_m) \leqslant \rho(x_n, x) + \rho(x, x_m) \to 0$ as $n, m \to \infty$. A metric space X is said to be *complete* if every Cauchy sequence in X has a limit. As known from calculus, the spaces $\mathbb{R}, \mathbb{C}$, as well as all finite-dimensional Euclidean spaces, are complete.

Let us recall a number of facts.

Theorem 1. *Any closed subspace of a complete metric space is itself complete; a complete subspace of any metric space is closed.*

Proof. Let A be a closed subset of the metric space X, and $x_n \in A$ form a Cauchy sequence. Since X is complete, the sequence (x_n) has a limit $x \in X$. Since A is closed, this limit belongs to A. This establishes the completeness of A.

Conversely, suppose A is complete and the sequence of $x_n \in A$ has a limit $x \in X$. Then (x_n) is a Cauchy sequence. Thanks to completeness, (x_n) has a limit in A, and by the uniqueness of the limit, this limit coincides with x. Thus, $x \in A$. This shows that A is closed and completes the proof. $\qquad\square$

Let A be a non-empty set of the metric space X. The *diameter* of the set A is defined as $\operatorname{diam}(A) = \sup_{x, y \in A} \rho(x, y)$.

Theorem 2 (Nested sets theorem). *Let $A_1 \supset A_2 \supset \cdots$ be a decreasing chain of non-empty closed subsets of a complete metric space X and let $\operatorname{diam}(A_n) \to 0$ as $n \to \infty$. Then the intersection $\bigcap_{n=1}^{\infty} A_n$ is non-empty and consists of exactly one point.*

Proof. Pick in each set A_n a point a_n. Let N be some natural number, and let $k, j > N$. Then since the sequence A_n is decreasing, the points a_k and a_j belong to the set A_N. Therefore, $\rho(a_j, a_k) \leqslant \operatorname{diam}(A_N) \to 0$ as $N \to \infty$, i.e., (a_n) is a Cauchy sequence. Denote the limit of this sequence by a. For any N and any $k > N$, the point a_k lies in A_N. Consequently, $a = \lim_{n \to \infty} a_k$ also lies in A_N. We have shown that $a \in A_N$ for all N, i.e., the intersection of the sets A_n is not empty. Now note that $\bigcap_{n=1}^{\infty} A_n \subset A_N$ for all N, and so

$$\operatorname{diam}\left(\bigcap_{n=1}^{\infty} A_n\right) \leqslant \operatorname{diam} A_N \to 0, \quad N \to \infty.$$

But a set of diameter zero necessarily reduces to a single point. This completes the proof of the theorem. □

Exercises

1. Suppose the Cauchy sequence (x_n) in the metric space X contains a convergent subsequence. Then the sequence (x_n) itself also converges.

2. The metric space X is complete if and only if any sequence (x_n) satisfying $\sum_{n=1}^{\infty} \rho(x_n, x_{n+1}) < \infty$ converges.

3. Consider a cube of unit side length and a ball of unit radius in the three-dimensional Euclidean space $\mathbb{R}^3$. Which of these objects has a larger diameter? Does the answer change if the objects are considered in the four-dimensional space? The five-dimensional one? (The unit cube in $\mathbb{R}^n$ is the set of all vectors whose components lie between 0 and 1, while the unit ball is the set of all vectors for which the sum of the squares of their components is not larger than 1.)

4. Show that in an incomplete space the nested sets theorem fails.

5. Give an example of a decreasing chain $A_1 \supset A_2 \supset A_3 \supset \cdots$ of closed subsets of the real line with void intersection.

6. Construct a homeomorphism between the open interval $(0, 1)$ and the real line $\mathbb{R}$. This will show that completeness is a metric property, and not a topological one: an incomplete space and a complete space can be homeomorphic.

7. Show that if X, Y are complete metric spaces, then the Cartesian product $X \times Y$, equipped with the metric introduced in Exercise 4 of Subsection 1.3.1, is also complete.

8. Show that the space $\mathbb{R}^\omega$ considered in Exercise 11 of Subsection 1.3.1 is complete.

9. In a Euclidean space the diameter of a ball is twice its radius. Is this true in an arbitrary metric space?

10. Show that in the Euclidean plane any set of unit diameter can be included in a disc of radius $\frac{1}{\sqrt{3}}$ (Jung's theorem).

11. Show that in the Euclidean plane any set of unit diameter can be decomposed into three sets, each of diameter less than 1 (Borsuk's theorem).

Exercises 10 and 11 belong to a direction in mathematics called *combinatorial geometry*. Combinatorial geometry studies problems concerned with the mutual disposition of geometric figures, optimal covers, decomposition into smaller parts, and so on. In spite of the seeming simplicity of their formulations, such problems often turn out to be highly non-trivial; many naturally arising problems in the field remain unsolved at this time. An example is provided by *Borsuk's Problem*: *Can any set of diameter 1 in the 4-dimensional Euclidean space be divided into 5 pieces, each of diameter smaller than* 1? For details about this problem and other questions of combinatorial geometry, refer to the monographs [7, 16, 17].

12. A closed subset of a topological space is called *perfect* if it does not have isolated points (in other words, if each point of the set is a limit point of the set itself). Prove that in a complete metric space the cardinality of any perfect set is not smaller than the cardinality of the continuum.

13. Prove that a metric space is separable if and only if for each $\varepsilon > 0$ the space can be covered by a countable number of balls or radius ε.

14. Let A be a uncountable subset of the complete separable metric space X. A point $x \in X$ is called a *condensation point* of the set A if the intersection of A with any neighborhood of the point x is uncountable. Prove that the set A_c of condensation points of A is not empty, is perfect, and such that the difference $A \setminus A_c$ is at most countable.

1.3.4   Uniform Continuity. The Extension Theorem

Definition 1. Let X and Y be metric spaces. A map $f : X \rightarrow Y$ is said to be *uniformly continuous* if for any $\varepsilon > 0$ there exists a $\delta = \delta(\varepsilon) > 0$ such that for any two elements $x_1, x_2 \in X$ satisfying $\rho(x_1, x_2) < \delta$, the distance between their images is smaller than ε: $\rho(f(x_1), f(x_2)) \leqslant \varepsilon$.

We note that every uniformly continuous map is continuous, but in general continuity does not imply uniform continuity: as an example, consider the function $f(x) = 1/x$ on the open interval $(0, 1)$.

Lemma 1. *Let $f : X \rightarrow Y$ be a uniformly continuous map of the metric space X into the metric space Y. Then for any Cauchy sequence (x_n) of points of X its image $(f(x_n))$ is a Cauchy sequence in Y.*

Proof. Given any $\varepsilon > 0$, take $\delta(\varepsilon)$ as in the definition of uniform continuity. By the definition of a Cauchy sequence, there exists a number $N = N(\varepsilon)$ such that beginning with $N = N(\varepsilon)$ the distances between all the pairs of elements x_n become smaller than $\delta(\varepsilon)$, i.e., for any $n, m > N$ it holds that $\rho(x_n, x_m) < \delta(\varepsilon)$. But then also $\rho(f(x_n), f(x_m)) \leqslant \varepsilon$ for all $n, m > N$.   □

Theorem 1 (Extension theorem). *Let X_1 be a subspace of the metric space X, $\overline{X}_1$ the closure of the set X_1 in X, and Y a complete metric space. Then any uniformly continuous map $f : X_1 \rightarrow Y$ extends uniquely to a uniformly continuous map $\bar{f} : \overline{X}_1 \rightarrow Y$.*

Proof. For each point $x \in \overline{X}_1$ there exists a sequence of elements $x_n \in X_1$ which converges to x. Since the space Y is complete, the preceding lemma shows that the sequence $(f(x_n))$ has a limit. Furthermore, this limit does not depend on the choice of the sequence (x_n), but only on the point x. Indeed, if $x_n, y_n \in X_1$ are two different sequences that converge to x, then the "mixed" sequence $x_1, y_1, x_2, y_2, \ldots$ also converges to x. Hence, the sequence of images $f(x_1), f(y_1), f(x_2), f(y_2), \ldots$ converges to a limit. It follows that the sequences $(f(x_n))$ and $(f(y_n))$ must have the same limit. Denote by $\bar{f}(x)$ the common limit of all sequences of the form $(f(x_n))$, where $x_n \in X_1$ and $x_n \rightarrow x$.

   If $x \in X_1$, then for (x_n) one can take the sequence $(x, x, x, \ldots)$. In this case $\bar{f}(x) = f(x)$, so we have shown that the map $\bar{f}$ is an extension of f. It remains to verify the uniform continuity of $\bar{f}$. We take an arbitrary $\varepsilon > 0$ and show that the $\delta = \delta(\varepsilon)$ given by the definition of the uniform continuity of the map f also works for $\bar{f}$. Let $x, y \in \overline{X}_1$ be arbitrary elements satisfying $\rho(x, y) < \delta$, and let $x_n, y_n \in X_1$ be such that $x_n \rightarrow x$ and $y_n \rightarrow y$ as $n \rightarrow \infty$. Since $\rho(x_n, y_n) \leqslant \rho(x_n, x) + \rho(x, y) + \rho(y, y_n)$, it follows that $\rho(x_n, y_n) < \delta$ for sufficiently large n. Therefore, for n large we have the inequality $\rho(f(x_n), f(y_n)) \leqslant \varepsilon$. Letting $n \rightarrow \infty$, we obtain the required inequality $\rho(\bar{f}(x), \bar{f}(y)) \leqslant \varepsilon$.   □

Exercises

A quantitative characteristic of the uniform continuity of a map is provided by the quantity

$$\omega(f, \varepsilon) = \sup\{\delta > 0 : (\rho(x_1, x_2) < \delta) \implies (\rho(f(x_1), f(x_2)) \leqslant \varepsilon)\}$$

(here we adopt the convention that the supremum of the empty set is 0).

1. The map f is uniformly continuous if and only $\omega(f, \varepsilon) > 0$ for all $\varepsilon > 0$.

2. For a function f on a segment or on the line, the *modulus of continuity* is *semi-additive*: $\omega(f, \varepsilon_1 + \varepsilon_2) \geqslant \omega(f, \varepsilon_1) + \omega(f, \varepsilon_2)$ for all $\varepsilon_1, \varepsilon_2 > 0$.

3. Give an example of a metric space X and a real-valued function f on X whose modulus of continuity is not semi-additive.

4. Suppose the map $f : X \to Y$ satisfies the Lipschitz condition (there exists $C > 0$ such that $\rho(f(x_1), f(x_2)) \leqslant C\rho(x_1, x_2)$ for all $x_1, x_2 \in X$). Then f is uniformly continuous. Estimate from below the modulus of continuity of f.

5. Calculate the modulus of continuity of an isometry.

1.3.5   Pseudometric Spaces and the Associated Metric Spaces. The Completion of a Metric Space

A function of two variables $\rho : X \times X \to \mathbb{R}^+$ is called a *pseudometric* on the set X if it satisfies the metric Axioms 1, 3, and 4 ($\rho(x, x) = 0$, $\rho(x, y) = \rho(y, x)$, and $\rho(x, z) \leqslant \rho(x, y) + \rho(y, z)$), but not necessarily Axiom 2 (non-degeneracy axiom). A set equipped with a pseudometric is called a *pseudometric space*. The topology on a pseudometric space is given in the same way as on a metric space, by means of balls. The main difference between pseudometric spaces and metric spaces is that the topology defined by a pseudometric is not separated. Let us show that "gluing together" those points of a pseudometric space that cannot be separated from one another naturally yields a metric space.

Thus, let (X, ρ) be a pseudometric space. Two elements $x, y \in X$ are said to be *ρ-equivalent* (written $x \approx y$) if $\rho(x, y) = 0$.

Theorem 1. *The relation $\approx$ is an equivalence relation on X. If A, B are equivalence classes, and $a \in A$, $b \in B$ arbitrary representatives, then the quantity $\rho(A, B) = \rho(a, b)$ does not depend on the choice of these representatives and gives a metric on the space $X/\approx$ of all equivalence classes generated by the relation $\approx$.*

Proof. The symmetry property of the relation $\approx$ is obvious. Next, note that

$$\text{if } x, y, z \in X \text{ and } z \approx y, \text{ then } \rho(x, y) = \rho(x, z). \tag{i}$$

Indeed, by the triangle inequality, $\rho(x, y) \leqslant \rho(x, z) + \rho(z, y) = \rho(x, z)$ and $\rho(x, z) \leqslant \rho(x, y) + \rho(y, z) = \rho(x, y)$. This immediately yields the transitivity of the relation $\approx$. The fact that the number $\rho(a, b)$ does not depend on the choice of the representatives $a \in A$, $b \in B$ of the equivalence classes A, B is also an obvious consequence of (i). The symmetry and triangle inequality for the function ρ on $X/\approx$ follow from the corresponding properties of the pseudometric ρ on X. Finally, the non-degeneracy of the metric ρ on $X/\approx$ is a result of the performed "gluing": if $A, B \in X/\approx$ are equivalence classes for which $\rho(A, B) = 0$, then there exist representatives $a \in A, b \in B$, such that $\rho(a, b) = 0$. Thus, $a \approx b$, and so the equivalence classes A and B coincide.   □

The space $X/\approx$ just described is called the *metric space associated with the pseudometric space* X.

As in a metric space, a sequence (x_n) of elements of the pseudometric space X is called a *Cauchy*, or *fundamental* sequence, if $\rho(x_n, x_m) \to 0$ as $n, m \to \infty$. A pseudometric space X is said to be complete if any Cauchy sequence in X has a limit. The mapping $F : X \to X/\approx$ that associates to each element its equivalence class preserves distances, and hence preserves the Cauchy property and the convergence of sequences. Consequently, the pseudometric space X is complete if and only if the metric space $X/\approx$ is complete.

Definition 1. Let X be an incomplete metric space. A metric space $Y \supset X$ is called a *completion* of the space X if Y is a complete space, the restriction of the metric of Y to X coincides with the original metric of the space X (that is, X is a subspace of Y), and X is a dense subset of Y.

By solving the chain of exercises given below, the reader will establish the existence of a completion for every incomplete space and the uniqueness of this completion up to an isometry.

Exercises

1. Let X be a metric space. Define $\widetilde{X}$ to be the space of all Cauchy sequences in X. Let $x, y \in \widetilde{X}$, $x = (x_n)_{n\in\mathbb{N}}$, $y = (y_n)_{n\in\mathbb{N}}$. Put $\rho(x, y) = \lim_{n\to\infty} \rho(x_n, y_n)$. Verify that the quantity $\rho(x, y)$ is well defined for any $x, y \in \widetilde{X}$ and gives a pseudometric on $\widetilde{X}$.

2. Prove that $\widetilde{X}$ is a complete pseudometric space.

3. Denote by $\widetilde{\widetilde{X}}$ the metric space associated with the pseudometric space $\widetilde{X}$. Identify each element x of the space X with the equivalence class of the Cauchy sequence $(x, x, x, \ldots)$. Verify that under this identification X is a subspace of the space $\widetilde{\widetilde{X}}$.

4. Show that $\widetilde{\widetilde{X}}$ is a completion of the space X.

5. Uniqueness of the completion: let Y_1, Y_2 be two completions of the space X. Then there exists a bijective isometry $S : Y_1 \to Y_2$ which keeps fixed the elements of the space X ($S(x) = x$ for all $x \in X$). That is to say, from the point of view of their metric structure, the spaces Y_1 and Y_2 are indistinguishable.

6. Use the existence of a completion to extend the result of Exercise 14 in Subsection 1.3.3 to an incomplete separable space.

1.3.6 Sets of First Category and Baire's Theorem

A subset A of a topological space X is said to be *nowhere dense* if A is not dense in any non-empty open subset of X. In other words, the set A is nowhere dense if its closure contains no open sets. Since in a metric space for any point the closed balls of non-zero radii centered at that point constitute a neighborhood basis, for metric spaces the above definition can be reformulated as follows: the subset A is nowhere dense if any ball $\overline{B}_X(x_0, r)$, $r > 0$, contains a smaller closed ball of non-zero radius in which there are no points of A.

Typical examples of nowhere dense sets are the Cantor set in the interval (see Subsection 1.4.4), and rectifiable curves in the plane. We should emphasize that when we speak about a nowhere dense set we need to specify its ambient space. For instance, a segment is a nowhere dense set in the plane, but not on the line; $A = \{0\}$ is nowhere dense on the real line, but in the set of natural numbers the same set A is open.

Theorem 1 (Baire's theorem). *A complete metric space cannot be covered by a countable collection of nowhere dense subsets.*

Proof. Let X be a complete metric space and $A_1, A_2, \ldots$ be nowhere dense subsets of X. We need to show that $\bigcup_{n=1}^{\infty} A_n$ does not coincide with X. Since A_1 is nowhere dense in X, there exists a closed ball $B_1 = \overline{B}_X(x_1, r_1)$ with $0 < r_1 < 1/2$ which does not intersect A_1. Since A_2, in its turn, is nowhere dense (in particular, A_2 is not dense in B_1), there exists a closed ball $B_2 = \overline{B}_X(x_2, r_2)$ with $0 < r_2 < 1/4$, which is contained in B_1 and does not intersect A_2. Continuing this argument, we obtain a decreasing chain $B_1 \supset B_2 \supset B_3 \supset \cdots$ of closed balls whose radii tend to zero, and such that each B_n does not intersect the corresponding set A_n. By the nested sets theorem, the sets B_n have a common point, which we denote by x. Since $x \in B_n$ for all n, and $B_n \cap A_n = \emptyset$, we conclude that x does not belong to any of the sets A_n. Thus, we have shown that there exists a point $x \in X \setminus \bigcup_{n=1}^{\infty} A_n$, i.e., the sets A_n do not cover the entire space X. $\square$

In connection with Baire's theorem just proved, the following terminology was introduced. A subset A of a topological space X is called a *set of first category* (or a *meagre set* in X) if A can be written as a countable union of nowhere dense subsets in X. A subset of X that is not of first category is called a *set of second category* in X. In these terms Baire's theorem asserts that any complete metric space is a set of second category in itself.

Exercises

1. Verify that the complement of a dense open set is a nowhere dense set.

2. Check that in the proof of Baire's theorem the ball B_1 can be chosen to lie inside a given open subset of X. Deduce from this that every open subset of a complete metric space X is a set of second category in X.

3. Show that in an incomplete metric space Baire's theorem may not hold.

4. Verify the following properties: a subset of a set of first category is also a set of first category; a finite or countable union of sets of first category is also a set of first category; if a set contains a subset of second category, then it itself is a set of second category.

5. Is it true that the intersection of two sets of second category is always of second category?

6. Use Baire's theorem to prove Cantor's theorem asserting that the interval $[0, 1]$ is not countable.

7. Show that a single-point subset $A = \{x\}$ of a topological space X is nowhere dense if and only if x is a limit point in X. In conjunction with Baire's theorem, this easily yields the following weakened version of Exercise 12 of Subsection 1.3.3: any perfect set in a complete metric space is uncountable.

8. Suppose the infinitely differentiable function f on the interval $[0, 1]$ has the following property: for every point $t \in [0, 1]$ there exists a number $n = n(t)$ such that the n-th derivative of f at t is equal to zero. Using the sets $A_n = \{t \in [0, 1] : f^{(n)}(t) = 0\}$ and Baire's theorem, show that on some interval $[a, b] \subset [0, 1]$ the function f is a polynomial.

9. Under the assumptions of the preceding exercise, show that the function f is actually a polynomial on the entire interval $[0, 1]$. The proof of this theorem by Ernest Corominas and Ferran Sunyer i Balaguer (1954), as well as several other non-trivial applications of Baire's theorem, can be found in [6, Chap. 1, Sect. 10].

10. Prove the following analogue of Baire's theorem: any compact topological space is a set of second category in itself.

1.4 Compact Sets in Metric Spaces

1.4.1 Precompact Sets

Let X be a metric space, $A, C \subset X$, and $\varepsilon \geqslant 0$. The set C is called an ε-*net* for A if $\bigcup_{x \in C} B(x, \varepsilon) \supset A$; in other words, for each $a \in A$ there exists an $x \in C$ such that $\rho(x, a) < \varepsilon$. Yet another reformulation: the set C is an ε-net for A if and only if $\rho(a, C) < \varepsilon$ for every $a \in A$. From the triangle inequality it follows that if C is an

ε-net for A and D is an ε-net for C, then D is an 2ε-net for A. For example, the center of an open ball of radius r is an r-net for that ball, while the set $C = \{\frac{1}{3}, \frac{2}{3}\}$ is a $\frac{1}{3}$-net for the interval $(0, 1)$. The set C is called a *finite ε-net for* A if C is an ε-net for A and has a finite number of elements.

Lemma 1. *If the set A admits a finite ε-net, then A also admits a finite 2ε-net consisting of elements of A.*

Proof. Let C be a finite ε-set for A. In each ball $B(c, \varepsilon)$ with $c \in C$, if $B(c, \varepsilon)$ intersects A, pick a point $x \in B(c, \varepsilon) \cap A$. The resulting finite set of elements is the sought-for 2ε-net. $\square$

A subset A of a metric space X is said to be *precompact* if for any $\varepsilon > 0$ there exists a finite ε-net for A.

We note the following obvious properties of precompact sets: if $A \supset B$ and A is precompact, then B is precompact; any finite union of precompact sets is precompact. Every precompact set is bounded, i.e., is contained in some ball of finite radius (for this it even suffices that there exists a finite ε-net for some fixed value of ε). A set in $\mathbb{R}^n$ is precompact if and only if it is bounded.

Lemma 2. *Suppose the set A admits a precompact ε-net for every $\varepsilon > 0$. Then A is precompact.*

Proof. Choose a precompact $(\varepsilon/2)$-net B for A, and a finite $(\varepsilon/2)$-net C for B. Then C will be a finite ε-net for A. $\square$

Theorem 1. *Let A be a subset of the metric space X. Then the following conditions are equivalent:*

1. *A is precompact.*
2. *For any $\varepsilon > 0$, from any sequence of elements of A one can extract a subsequence in which all distances between pairs of terms are no larger than ε.*
3. *From any sequence of elements of A one can extract a Cauchy subsequence.*

Proof. 1. $\Longrightarrow$ 2. Let A be precompact, and $\{a_n\}_{n\in\mathbb{N}} \subset A$. Cover A by a finite number of balls of radius $\varepsilon/2$. Then at least one of these balls contains an infinite subsequence of the sequence (a_n).

2. $\Longrightarrow$ 3. Let $\{a_n\}_{n\in\mathbb{N}} \subset A$. Applying condition 2. successively with $\varepsilon = 1$, $\varepsilon = 1/2$, $\varepsilon = 1/3$, ..., we obtain infinite sets of indices $N_1 \supset N_2 \supset N_3 \supset \cdots$, for which $\mathrm{diam}\{a_n\}_{n\in N_k} \leqslant 1/k$. We construct an increasing set of indices M, taking the first element in N_1, the second in N_2, the third in N_3, and so on. The subsequence $(a_n)_{n\in M}$ is a Cauchy sequence because for every $k \in \mathbb{N}$ all distances between pairs of its terms of starting with the k-th, are not larger than $1/k$.

3. $\Longrightarrow$ 1. Suppose the set A is not precompact. Then there exists an $\varepsilon > 0$ such that no finite set is an ε-net for A. Let us show that there exists a sequence $\{a_n\}_{n\in\mathbb{N}} \subset A$ for which all distances between pairs of elements are larger than or equal to ε.

Such a sequence cannot have Cauchy subsequences. We proceed as follows. For a_1 take an arbitrary element of A. The set $C_1 = \{a_1\}$ does not constitute an ε-net, so there exists an $a_2 \in A$ such that $\rho(a_2, C_1) \geqslant \varepsilon$. In the set $C_2 = \{a_1, a_2\}$ the distances between pairs of elements are larger than or equal to ε. The set C_2 does not constitute an ε-net, hence there exists an $a_3 \in A$ such that $\rho(a_3, C_2) \geqslant \varepsilon$. Suppose we have already constructed the elements $a_1, \ldots, a_n$ of the sought-for sequence with distances between pairs not smaller than ε. The set $C_n = \{a_1, \ldots, a_n\}$ is finite, and so it is not an ε-net for A. Choose the point $a_{n+1} \in A$ so that $\rho(a_{n+1}, C_n) \geqslant \varepsilon$. Continuing the described process indefinitely, we obtain the required sequence. $\qquad\square$

In complete spaces this result can be strengthened.

Theorem 2. *Let A be a closed subset of a complete metric space X. Then the following conditions are equivalent:*

1. *A is compact.*
2. *A is precompact.*
3. *From any sequence of elements of A one can extract a convergent subsequence.*

Proof. The equivalence 2. $\Longleftrightarrow$ 3. follows from the preceding theorem; the implication 1. $\Longrightarrow$ 3. follows from the fact that any subset of a compact set, in particular any subsequence, has a limit point. It remains to establish the implication 2. $\Longrightarrow$ 1. To this end we remark first that for any centered family $\mathfrak{W}$ of subsets of a closed precompact set D and any $\varepsilon > 0$ there exists a closed subset B of D such that the family $\mathfrak{W}_1 = \{V \cap B : V \in \mathfrak{W}\}$ is again centered and $\mathrm{diam}(B) < \varepsilon$. Indeed, it suffices to cover the precompact set D by a finite number of closed subsets of diameter smaller than ε; then at least one of these subset can be taken as B. Now let us show that any centered family $\mathfrak{W}$ of closed subsets of our precompact set A has a common element. By Theorem 1 of Subsection 1.2.3, this will mean that the set A is compact.

Thus, suppose A is precompact and $\mathfrak{W}$ is a centered family; then there exists a closed subset $B_1 \subset A$ with $\mathrm{diam}(B_1) < 1$ such that the family $\mathfrak{W}_1 = \{V \cap B_1 : V \in \mathfrak{W}\}$ is again centered. Since B_1 is also a precompact set, there exists a closed subset $B_2 \subset B_1$ with $\mathrm{diam}(B_2) < 1/2$ such that the family $\mathfrak{W}_2 = \{V \cap B_2 : V \in \mathfrak{W}\}$ is centered. Continuing in this way, we produce a decreasing chain $B_1 \supset B_2 \supset B_3 \supset \cdots$ of closed subsets with $\mathrm{diam}(B_n) \to 0$ such that for each n the family $\{V \cap B_n : V \in \mathfrak{W}\}$ is centered. In particular, all intersections $V \cap B_n$, with $V \in \mathfrak{W}$ and $n \in \mathbb{N}$, are not empty. By the nested sets theorem (Subsection 1.3.3, Theorem 2), $\bigcap_{n=1}^{\infty} B_n$ is not empty and consists of exactly one point, which we denote by x. Now let us consider an arbitrary element $V \in \mathfrak{W}$ and show that $x \in V$, i.e., x is the required common point of all sets in the family $\mathfrak{W}$. Indeed, since for any $n \in \mathbb{N}$ the intersection $V \cap B_n$ is not empty and $x \in B_n$, we have $\rho(x, V) < \mathrm{diam}(B_n)$ for all n, i.e., $\rho(x, V) = 0$, as needed. $\qquad\square$

Exercises

1. Every compact metric space is separable.

2. A Cartesian product of precompact (compact) sets in the metric of Exercise 4 in Subsection 1.3.1 is also precompact (respectively, compact).

3. Let K, X be metric spaces, $f : K \to X$ a continuous map, and suppose K is compact. Then f is uniformly continuous.

For a subset A of the metric space X denote by $n_A(r)$ the largest possible number of pairwise disjoint balls of radius r centered at points of A. Show that:

4. A is precompact if and only if $n_A(r) < \infty$ for any r.

5. The function $r \mapsto n_A(r)$ does not increase with the growth of r.

6. The function $n_A(r)$ is bounded (in a neighborhood of zero) if and only if the set A is finite.

7. Let A be a bounded set in $\mathbb{R}^m$ with non-empty interior. Then $n_A(r)$ has the same order of growth at zero as r^{-m}. This shows that $n_A(r)$ can be used to define a dimension of the set A.

Let us remark that the exact values of $n_A(r)$ are not easy to calculate even for relatively simple sets, like, for instance, a ball in $\mathbb{R}^3$. The classical problem of the densest packing of balls (spheres) in $\mathbb{R}^3$ was solved only in 1998! Finding a possibly exact estimate of the numbers $n_A(r)$ for sets in $\mathbb{R}^m$ is of great practical value. For instance, if one identifies a signal consisting of m numerical components with a point in $\mathbb{R}^m$, then the distance measures how easy it is to recognize these signals. Accordingly, the task of determining the possible number of recognizable signals of a given power reduces to the search for the possibly largest numbers of pairwise disjoint balls of radius r in a fixed ball.

1.4.2   Spaces of Continuous Maps and Functions. Arzelà's Theorem

Let Γ be a set, and X a metric space. We endow the set of all bounded X-valued functions on Γ (i.e., bounded maps from Γ to X) with the metric ρ defined by $\rho(f, g) = \sup_{t \in \Gamma} \rho(f(t), g(t))$.[2] The resulting metric space of bounded X-valued functions is denoted by $\ell_\infty(\Gamma, X)$. The metric of this space is called the *uniform metric*, and the convergence in $\ell_\infty(\Gamma, X)$ coincides with the uniform convergence.

---

[2]In this definition the symbol ρ is used with two distinct meanings: on the left — as the distance in $\ell_\infty(\Gamma, X)$, and on the right — as the distance in X. This ambiguity can be removed by denoting the metric on the space X by ρ_X.

Theorem 1. *If X is a complete metric space, then the space $\ell_\infty(\Gamma, X)$ is also complete.*

Proof. Let (f_n) be an arbitrary Cauchy sequence in $\ell_\infty(\Gamma, X)$. Then for any $t \in \Gamma$ the values $f_n(t)$ form a Cauchy sequence in X: $\rho(f_n(t), f_m(t)) \leqslant \rho(f_n, f_m) \to 0$ as $n, m \to \infty$.

Since the space X is complete, the sequence $(f_n(t))$ has a limit, which we denote by $f(t)$. To show that (f_n) converges to f uniformly, let us write in more detail the definition of a Cauchy sequence: for any $\varepsilon > 0$ there exists an $N \in \mathbb{N}$, such that $\rho(f_n, f_m) \leqslant \varepsilon$ for all $n, m > N$. Making the definition of the metric in $\ell_\infty(\Gamma, X)$ explicit, we see that for any $\varepsilon > 0$ there exists an $N \in \mathbb{N}$ such that for any $t \in \Gamma$ and any $n, m > N$ one has $\rho(f_n(t), f_m(t)) \leqslant \varepsilon$. Letting here $m \to \infty$, we conclude that $\rho(f_n(t), f(t)) \leqslant \varepsilon$ for all $t \in \Gamma$. In conjunction with the boundedness of the functions f_n, this shows that the function f is bounded, i.e., $f \in \ell_\infty(\Gamma, X)$. Further, for $n > N$ we take the supremum over $t \in \Gamma$ in the inequality $\rho(f_n(t), f(t)) \leqslant \varepsilon$ to conclude that $\rho(f_n, f) \leqslant \varepsilon$ for all $n > N$. Therefore, $f_n \to f$ in the metric of the space $\ell_\infty(\Gamma, X)$, which establishes the completeness of this space. $\square$

Let K be a compact topological space and X be a metric space. The set of continuous maps from K to X, equipped with the uniform metric, is called the *space of continuous X-valued functions* and is denoted by $C(K, X)$. The distance in $C(K, X)$ can be expressed by the formula $\rho(f, g) = \max_{t \in K} \rho(f(t), g(t))$. By the well-known theorem of calculus asserting the continuity of the limit of a uniformly convergent sequence of continuous functions, $C(K, X)$ is a closed subspace of the space $\ell_\infty(K, X)$. Hence, if X is a complete metric space, then $C(K, X)$ is also complete.

We recall also that any function $f \in C(K, X)$ is uniformly continuous (being a continuous map on a compact metric space).

Lemma 1. *Let X be a precompact space and Γ a finite set. Then the space $\ell_\infty(\Gamma, X)$ is precompact.*

Proof. Let A be a finite ε-net for X. Then the set $\ell_\infty(\Gamma, A)$ of all maps from Γ to A is a finite ε-net for $\ell_\infty(\Gamma, X)$. $\square$

Lemma 2. *Suppose the family G of continuous functions constitutes a precompact set in $C(K, X)$. Then the set $G(K) = \bigcup_{f \in G} f(K)$ is precompact in X.*

Proof. Let $G_1 \subset G$ be a finite ε-net for G. Put $G_1(K) = \bigcup_{f \in G_1} f(K)$. Since each of the sets $f(K)$ is compact (as the image of a compact set under a continuous map), $G_1(K)$ is compact, being a finite union of compact sets. At the same time, $G_1(K)$ is an ε-net for $G(K)$. By Lemma 2 of the preceding Subsection 1.4.1, $G(K)$ is precompact in X. $\square$

Definition 1. Let K and X be metric spaces. A family G of maps from K to X is said to be *equicontinuous* if for every $\varepsilon > 0$ there exists a $\delta > 0$ such that for any map $f \in G$ and any points $t_1, t_2 \in K$ satisfying $\rho(t_1, t_2) < \delta$, the distance between the images of these points is not larger than ε: $\rho(t_1, t_2) < \delta \implies \rho(f(t_1), f(t_2)) \leqslant \varepsilon$.

Lemma 3. *Let K and X be metric spaces with K compact, and let the family G of continuous maps be precompact in $C(K, X)$. Then the family G is equicontinuous.*

Proof. Fix $\varepsilon > 0$ and choose a finite ε-net $G_1 \subset G$ for G. Since every map $g \in G_1$ is uniformly continuous and the number of these maps is finite, there exists a $\delta > 0$ such that for any map $g \in G_1$ and any points $t_1, t_2 \in K$ such that $\rho(t_1, t_2) < \delta$ one has $\rho(g(t_1), g(t_2)) \leqslant \varepsilon$. Now let $f \in G$. By the definition of an ε-net, there exists a $g \in G_1$ such that $\rho(f, g) < \varepsilon$. By the triangle inequality, for any $t_1, t_2 \in K$ satisfying $\rho(t_1, t_2) < \delta$ we have

$$\rho(f(t_1), f(t_2)) \leqslant \rho(f(t_1), g(t_1)) + \rho(g(t_1), g(t_2)) + \rho(g(t_2), f(t_2)) \leqslant 3\varepsilon.$$

Since ε is arbitrary, the equicontinuity of the family G is established. □

The following theorem was obtained in 1895 by Arzelà for subsets of $C[a, b]$. A weaker form was discovered earlier by Ascoli, therefore the result is often referred to as the "Arzelà–Ascoli theorem". The extension to $C(K)$ for an arbitrary metric compact space K was made by Fréchet in 1906.

Theorem 2 (Arzelà's theorem). *Let K and X be metric spaces with K compact, and let $G \subset C(K, X)$. In order for the family G to be precompact it is necessary and sufficient that the following two conditions be satisfied: (1) G is equicontinuous, and (2) the images of all the maps in the family G are contained in one and the same precompact set $Y \subset X$.*

Proof. The necessity of the two conditions was already established in Lemmas 2 and 3 above. Let us prove their sufficiency. Fix an $\varepsilon > 0$ and the corresponding $\delta = \delta(\varepsilon)$ from the definition of the uniform continuity of the family G. Pick a finite δ-net Γ in K. Now consider the restriction mapping $F : G \to \ell_\infty(\Gamma, Y)$, which associates to each map $f \in G$ its restriction to Γ. By Lemma 1, the subspace $\ell_\infty(\Gamma, Y)$ is precompact, and hence $F(G)$ is precompact as well. Therefore, there exists a finite set $G_1 \subset G$ such that $F(G_1)$ is an ε-net in $F(G)$. We claim that G_1 is a 3ε-net for G.

Indeed, let $f \in G$ be an arbitrary map. By the definition of the set G_1, there exists a $g \in G_1$ such that $\rho(F(f), F(g)) < \varepsilon$. Interpreting the definition of the set F and the metric in $\ell_\infty(\Gamma, Y)$, we see that $\rho(f(t), g(t)) < \varepsilon$ for all $t \in \Gamma$. Further, for each $x \in K$ there exists a $t \in \Gamma$ such that $\rho(x, t) < \delta$ (because Γ is a δ-net in K). Finally, recalling that δ was taken from the definition of uniform continuity, we have

$$\rho(f(x), g(x)) \leqslant \rho(f(x), f(t)) + \rho(f(t), g(t)) + \rho(g(t), g(x)) < 3\varepsilon.$$

Since this holds for all $x \in K$, it follows that

$$\rho(f, g) = \max_{x \in K} \rho(f(x), g(x)) < 3\varepsilon.$$

Hence, G has a finite 3ε-net for any $\varepsilon > 0$, so the proof is complete. □

Corollary 1. *If under the assumptions of Arzelà's theorem the space X is complete, then for a set $G \subset C(K, X)$ to be compact it is necessary and sufficient that the following three conditions be satisfied:* (1) G *is equicontinuous,* (2) *the images of all the maps in the family G are contained in one and the same precompact set $Y \subset X$, and* (3) G *is a closed subset of the space $C(K, X)$.* □

In the most important particular cases, when the range space X is $\mathbb{R}$, $\mathbb{C}$, or $\mathbb{R}^n$, the precompact sets in X are simply the bounded sets. Condition 2) in Arzelà's theorem can be restated in a simpler way: the family G is uniformly bounded, i.e., $\sup_{f \in G, t \in K} \rho(0, f(t)) < \infty$. Concerning the uniform continuity, we provide a sufficient condition which is rather convenient in practice: if all the functions of the family G satisfy the Lipschitz condition with a common constant (i.e., if there exists a $c > 0$ such that for any $f \in G$ one has $\rho(f(t), f(\tau)) \leqslant c\rho(t, \tau)$ for all $t, \tau \in K$), then G is uniformly continuous.

Exercises

1. Why is the distance between any two elements of the space $C(K, X)$ finite?

2. Why in the definition of the uniform metric on $C(K, X)$ are we allowed to write "max" instead of "sup"?

3. Verify that the metric axioms hold for the uniform metric.

4. Show that if $C(K, X)$ is a complete metric space, then the space X is also complete.

Denote by $C[0, 1]$ the metric space $C([0, 1], \mathbb{R})$.

5. No non-empty open set in $C[0, 1]$ can be equicontinuous. In particular, in $C[0, 1]$ there are bounded, but at the same time not precompact, sets.

For the sets in $C[0, 1]$ listed below, check whether they are (a) bounded, (b) open, (c) closed, (d) equicontinuous, (e) precompact, (f) compact:

6. $A_1 = \{f : 0 \leqslant f(t) \leqslant 1 \text{ for all } t \in [0, 1]\}$.

7. $A_2 = \{f : f(t) > 0 \text{ for all } t \in [0, 1]\}$.

8. The set A_3 of those functions from A_2 for which $\int_0^1 f(t)dt < 1$.

9. The set A_4 of all continuously differentiable functions that satisfy the condition $\max_{t \in [0,1]} |f'(t)| \leqslant 1$.

10. The set A_5 of all continuously differentiable functions that satisfy the condition $\int_0^1 |f'(t)|^2 dt \leqslant 1$.

11. The set A_6 of all continuously differentiable functions that satisfy the condition $\int_0^1 |f'(t)|dt \leqslant 1$.

12. $A_7 = A_1 \cap A_4$.

13. The set A_8 of all convex functions belonging to A_1.

1.4.3 Application: The Isoperimetric Problem

The *isoperimetric problem* in the plane is the problem of finding, among all closed convex curves of a given length, the curve that bounds the maximal possible area. This classical problem, already considered in ancient Greece,[2] has numerous generalizations which play an important role in the geometry of convex bodies (see W. Blaschke's wonderful book [5]) and functional analysis (see the monograph of V. Milman and G. Schechtman [30], which is small, yet very rich in ideas and results).

Assuming that the isoperimetric problem has a solution, one can prove by elementary methods that the sought-for optimal curve is necessarily a circle. Some of these elementary proofs, like, say, Steiner's four-hinge method (§1 in Blaschke's book), are so simple and elegant, that they are often included in the syllabus of extracurricular mathematics school clubs. Establishing the existence of a solution itself turned out to be rather challenging; the first proof was provided by Weierstrass in the 1870s. Since then mathematics, in its development, has traveled a long path and now, armed with such strong tools as the theory of compact spaces — in particular Arzelà's theorem — we are able to prove the aforementioned theorem of Weierstrass without major efforts.

Let us denote by G the family of all functions $f : [0, 2\pi] \to \mathbb{R}^2$ that satisfy $f(0) = f(2\pi) = 0$ and the Lipschitz condition with constant 1, and for which $f([0, 2\pi])$ is a convex curve (in other words, G is the family of parametrically-given convex curves). Every convex curve of length no bigger than 2π, and which starts and ends at zero, can be identified with a function from G. To do this it suffices to consider the natural parametrization of the curve, i.e., take as parameter the arc length of the curve, measured from zero to the current point. For each function $g \in G$ we denote by $s(g)$ the area bounded by the curve $f([0, 2\pi])$.

Theorem 1. *The family of functions G is a compact set in $C[0, 2\pi]$. Moreover, s is a continuous function on G, and hence s attains its supremum on G.*

---

[2]The isoperimetric problem is associated with the legend of queen Dido, the founder of Carthage (in modern-day Tunisia). When the colonists arrived in the new place, the locals did not receive them kindly. The request of allotting a piece of land for the construction of the new city was de facto rejected by the following statement: "You are allowed to use as much land as can be encompassed by the skin (hide) of an ox." However, Dido did not allow herself to get confused and faced the challenge. She ordered her people to cut the skin into very thin strips and, connecting them, mark the border of the new settlement. Needless to say, in doing so it was desirable to obtain the largest area, i.e., to solve the isoperimetric problem.

Proof. As we mentioned at the end of the preceding subsection, the existence of a common Lipschitz constant means that the family we are dealing with is equicontinuous. Further, for any function $g \in G$ it holds that $\rho(0, g(t)) = \rho(g(0), g(t)) \leqslant |t| \leqslant 2\pi$, hence the family G is uniformly bounded. A uniform (or even pointwise) limit of a sequence of functions satisfying the Lipschitz condition with constant 1 itself satisfies the Lipschitz condition with constant 1. Convexity is also not affected by passing to such a limit. Hence, the family G is closed. Thus, compactness of the family G is established. It remains to verify that the function s is continuous. Given f_1 and f_2 in G, denote the figures bounded by these curves by F_1 and F_2, respectively, and denote $\rho(f_1, f_2)$ by ε. Further, denote by $F_{1,\varepsilon}$ the set of all points lying at distance at most ε from F_1. Now choose on the interval $[0, 2\pi]$ an ε-net $\{t_1, \ldots, t_n\}$ with $n < 2\pi/\varepsilon$. Thanks to the Lipschitz condition, the set $\{f_1(t_1), \ldots, f_1(t_n)\}$ is an ε-net on the curve $f_1([0, 2\pi])$. For each k take a disc of radius 2ε centered at the point $f_1(t_k)$. The union of these n discs and the set F_1 covers the whole set $F_{1,\varepsilon}$, and hence also covers the set F_2. We have $s(f_2) \leqslant s(f_1) + 4n\pi\varepsilon^2 \leqslant s(f_1) + 8\pi^2\varepsilon$. Since in this argument the roles of the functions f_1 and f_2 can be switched, we conclude that $|s(f_2) - s(f_1)| \leqslant 8\pi^2\varepsilon = 8\pi^2\rho(f_2, f_1)$, i.e., the function s is not just continuous, it actually satisfies the Lipschitz condition. $\qquad\square$

Exercises

1. Show that the set $F_{1,\varepsilon}$ appearing in the proof of the preceding theorem is convex.

2. Show that $F_{1,\varepsilon} \supset F_2$.

3. Provide the details of the proof that the family G is closed in $C[0, 2\pi]$.

4. Prove that the supremum of the areas of all convex figures of a given perimeter l coincides with the supremum of the areas of all figures bounded by rectifiable curves of length l. In other words, in the isoperimetric problem the convexity condition is not essential.

1.4.4 The Cantor Set

The *ternary expansion* of a number $x \in [0, 1]$ is the representation of the number in the form $x = \frac{x_1}{3} + \frac{x_2}{3^2} + \frac{x_3}{3^3} + \cdots$, where the *expansion digits* x_k are 0, 1, or 2. In abridged form, one writes $x = (0.x_1x_2...)_3$. Some numbers have two ternary expansions, for instance $(0.10000...)_3 = (0.02222...)_3$. The *Cantor set* is defined as the subset $\mathcal{K} \subset [0, 1]$ consisting of the numbers having at least one ternary expansion that does not contain the digit 1. The structure of the Cantor set can be better understood by looking at its complement. Specifically, the numbers whose ternary expansion necessarily has 1 as the first digit form the interval $\Delta_1^1 = (\frac{1}{3}, \frac{2}{3})$. The numbers for which the first digit is not 1, but the second is necessarily 1, form together two intervals, $\Delta_1^2 = (\frac{1}{9}, \frac{2}{9})$ and $\Delta_2^2 = (\frac{7}{9}, \frac{8}{9})$. This reasoning provides a description of the entire complement of $\mathcal{K}$. Accordingly, $\mathcal{K}$ can be seen as the result of the following construction: in the first step one discards from the interval $[0, 1]$ its middle third, i.e., the interval $(\frac{1}{3}, \frac{2}{3})$. Two intervals remain, $[0, \frac{1}{3}]$ and $[\frac{2}{3}, 1]$. Then in each of the remaining intervals one discards the middle third. Now one is left with four intervals. Again one discards from of each of these remaining intervals the middle third. What is left at the end of this infinite process is precisely the Cantor set $\mathcal{K}$.

Exercises

1. Show that the Cantor set $\mathcal{K}$ is closed.

2. Show that $\mathcal{K}$ is a perfect set, i.e., it has no isolated points.

3. Shows that the cardinality of the Cantor set is equal to the cardinality of the continuum.

4. Determine the rate of growth at zero of the quantity $n_{\mathcal{K}}(r)$ for the Cantor set (for the definition, see the exercises in Subsection 1.4.1).

5. Show that the Cantor set is nowhere dense in the interval $[0, 1]$.

6. Show that for every compact metric space X there exists a surjective continuous mapping $f : \mathcal{K} \to X$.

7. Consider the set $2^{\mathbb{N}}$ of all subsets of the natural numbers $\mathbb{N}$ and endow it with the following topology: for any subset $A \subset \mathbb{N}$, a neighborhood basis is given by the family of sets $U_n(A) = \{B \subset \mathbb{N} : B \cap \{1, 2, \ldots, n\} = A \cap \{1, 2, \ldots, n\}\}$. Verify that with this topology $2^{\mathbb{N}}$ is homeomorphic to the Cantor set.

Chapter 2
Measure Theory

2.1 Systems of Sets and Measures

2.1.1 Algebras of Sets

Let Ω be a fixed set and $\mathbb{A}$ a family of subsets of Ω. The family $\mathbb{A}$ is called an *algebra of sets* on Ω (or simply an *algebra* on Ω) if it obeys the following axioms:

1. $\Omega \in \mathcal{A}$.
2. If $A \in \mathbb{A}$, then also $\Omega \setminus A \in \mathbb{A}$.
3. If $A_1, A_2 \in \mathbb{A}$, then also $A_1 \cap A_2 \in \mathbb{A}$.

The reader will easily be able to verify that if $\mathbb{A}$ is an algebra of sets on Ω, then:

— $\emptyset \in \mathbb{A}$;

— the intersection of any finite collection of sets from $\mathbb{A}$ again lies in $\mathbb{A}$;

— the union of any two sets in $\mathbb{A}$ again lies in $\mathbb{A}$ (here it is helpful to observe that the complement of a union is the intersection of the individual complements);

— the union of any finite collection of sets from $\mathbb{A}$ again lies in $\mathbb{A}$;

— the set-theoretic difference and symmetric difference of any two sets from $\mathbb{A}$ again lies in $\mathbb{A}$.

Let us introduce a useful notation. Let the sets $A_1, A_2, \ldots$ be disjoint (i.e., pairwise disjoint — no two of them intersect). Then their union will be denoted by the *disjoint union* symbol $\bigsqcup_{k=1}^{\infty} A_k$. Accordingly, whenever we use the disjoint union symbol $\sqcup$, we will assume that the sets involved in the union are pairwise disjoint. For instance, the notation $C = A \sqcup B$ means that A and B are disjoint and $A \cup B = C$.

Proposition 1. *For any sequence $A_1, A_2, \ldots$ of elements of the algebra $\mathbb{A}$ there exist elements $\widetilde{A}_1, \widetilde{A}_2, \ldots \in \mathbb{A}$ such that $\widetilde{A}_k \subset A_k$ for all k, and $\bigcup_{k=1}^{\infty} A_k = \bigsqcup_{k=1}^{\infty} \widetilde{A}_k$.*

© Springer International Publishing AG, part of Springer Nature 2018
V. Kadets, *A Course in Functional Analysis and Measure Theory*,
Universitext, https://doi.org/10.1007/978-3-319-92004-7_2

Proof. The required pairwise disjoint sets $\widetilde{A}_k$ can be constructed in different ways. The simplest approach is to remove from each set the points belonging to the preceding sets, i.e., take $\widetilde{A}_1 = A_1$ and $\widetilde{A}_k = A_k \setminus \left(\bigcup_{j=1}^{k-1} A_j \right)$ for $k > 1$.   □

Clearly, Proposition 1 holds for countable as well as for finite sequences of sets.

An example of an algebra of sets is provided by the family 2^Ω of all subsets of a set Ω. Other, less trivial examples are provided in the exercises below.

Theorem 1. *Let Φ be a family of subsets of the set Ω. Then among all algebras on Ω that contain Φ as a subfamily there exists a smallest one with respect to inclusion.*

Proof. Define $\mathbb{A}$ to be the intersection of all algebras of sets on Ω that contain Φ. In other words, a set A belongs to $\mathbb{A}$ if and only if A belongs to all algebras that contain Φ as a subfamily. Obviously, any algebra of sets that contains Φ also contains $\mathbb{A}$. At the same time, it is easy to verify that $\mathbb{A}$ satisfies the axioms of algebras of sets:

1. Ω belongs to all algebras on Ω that contain Φ, hence $\Omega \in \mathbb{A}$.

2. If $A \in \mathbb{A}$, then A belongs to all algebras on Ω that contain Φ. Hence, $\Omega \setminus A$ belongs to all algebras of sets on Ω that contain Φ, and so $\Omega \setminus A \in \mathbb{A}$.

3. If $A_1, A_2 \in \mathbb{A}$, then both sets A_1 and A_2 belong to all algebras of sets that contain Φ. Hence, $A_1 \cap A_2$ belongs to all algebras of sets that contain Φ, and so $A_1 \cap A_2 \in \mathbb{A}$.   □

The smallest algebra that contains Φ is denoted by $\mathbb{A} = \mathbb{A}(\Phi)$ and is called the *algebra generated by the family* Φ. In this case we also say that Φ *generates the algebra* $\mathbb{A}$. A constructive description of the algebra generated by a family of sets is given in Exercise 6 below.

Exercises

1. Which of the axioms of an algebra of sets are not satisfied for the collection of all finite subsets of the interval $[0, 1]$? And for the collection of all infinite subsets of $[0, 1]$?

2. Describe the smallest algebra of sets on $[0, 1]$ that contains all single-point subsets.

3. Verify that the family of sets $\{\emptyset, [0, 1/2],]1/2, 1], [0, 1]\}$ is an algebra on $[0, 1]$.

4. A *subinterval* of the interval $[0, 1]$ is any open, closed, or semi-open interval contained in $[0, 1]$. Verify that the sets that are finite unions of subintervals constitute an algebra on $[0, 1]$. Is this algebra generated by the family of all open subintervals of $[0, 1]$? Or by the family of all semi-open subintervals?

5. Verify that the intersection of any collection of algebras on a set Ω is again an algebra.

6. Let Φ be a family of subsets of Ω that includes Ω as an element. Show that the sets obtained from the elements of the family Φ by a finite number of operations of intersection and passage to the complement constitute an algebra. In fact, this algebra coincides with the algebra generated by the family Φ.

7. Let $\mathbb{A}$ be a family of subsets of Ω that satisfies Axioms 1 and 2 of an algebra of sets and is closed under taking the union of two sets. Show that $\mathbb{A}$ is an algebra.

2.1.2 σ-Algebras of Sets. Borel Sets

A family Σ of subsets of a set Ω is called a *σ-algebra* if it is an algebra of sets and is stable under the countable union operation: for any sequence A_n, $n \in \mathbb{N}$, of elements of the algebra Σ their union is an element of Σ. Passing to complements we immediately see that a σ-algebra is also stable under taking countable intersections (de Morgan formulas: Sect. 1.1, Exercise 10). As follows from Proposition 1 of the preceding subsection, if the family Σ is an algebra of sets, then in order to verify that Σ is a σ-algebra, it suffices to check the stability under union property not for all countable unions, but only for unions of pairwise disjoint sets. To verify the correctness of the following definition, one proceeds as in the proof of Theorem 1 of Subsection 2.1.1.

Definition 1. Let Φ be a family of subsets of a set Ω. The smallest σ-algebra Σ that contains Φ is called the *σ-algebra generated by the family* Φ. Σ coincides with the intersection of all σ-algebras on Ω that contain Φ.

Let us rephrase the definition as the following statement:

Proposition 1. *If a σ-algebra Σ_0 contains the family Φ, then Σ_0 also contains the entire σ-algebra generated by the family Φ.* $\qquad\square$

Let Ω be a topological space. The σ-algebra $\mathfrak{B}$ generated by the family of all open subsets of Ω is called the *σ-algebra of Borel sets* on Ω. The elements of the σ-algebra $\mathfrak{B}$ are called *Borel sets*.

Unfortunately, in general the σ-algebra generated by a family of sets, in particular, the family of Borel subsets of a topological space, does not admit a nice constructive description analogous to that given in Exercise 6 of the preceding Subsection 2.1.1. Nevertheless, one can get a feeling for the meaning of Borel sets from the following considerations. The family $\mathfrak{B}$ contains all open subsets of the space Ω. Since $\mathfrak{B}$ is an algebra, it also contains the complements of all open sets, that is, all the closed sets. Being a σ-algebra, $\mathfrak{B}$ contains all countable unions of closed sets (the latter are called *sets of class F_σ*, or simply *F_σ-sets*). Moreover, $\mathfrak{B}$ contains all countable intersections of open sets (the latter are called *sets of class* G_δ, or simply G_δ*-sets*). Countable unions of G_δ-sets are called *sets of class* $G_{\delta\sigma}$, or simply $G_{\delta\sigma}$*-sets*). Similarly, countable intersections of F_σ-sets are called *sets of class* $F_{\sigma\delta}$, or simply $F_{\sigma\delta}$*-sets*). Continuing, countable unions of $F_{\sigma\delta}$-sets form the class $F_{\sigma\delta\sigma}$. In a similar manner one introduces the Borel classes $G_{\delta\sigma\delta}$, $F_{\sigma\delta\sigma\delta}$, and so on, to infinity. All these classes of Borel sets are contained in the σ-algebra of Borel sets, but even on $\Omega = [a, b]$ they do not exhaust the whole class of Borel sets.[1] For more details on Borel sets we refer the reader to Kuratowski's book [25, Chap. 2, § 30]. The importance of studying Borel sets comes from the fact that sets that arise naturally in problems of analysis, such as sets of points of continuity, smoothness, or convergence, etc., are usually Borel sets, and in fact belong to Borel classes of not too high index.

The next useful statement illustrates the fact that one and the same σ-algebra can be generated by different systems of sets.

Proposition 2. *The collection of the sets* $(a, +\infty)$ *with* $a \in \mathbb{R}$ *generates the* σ-*algebra* $\mathfrak{B}$ *of Borel sets on the real line.*

Proof. Denote the σ-algebra generated by the sets $(a, +\infty)$, $a \in \mathbb{R}$, by B_1. We need to show that $B_1 = \mathfrak{B}$. Since $\mathfrak{B}$ contains all the open sets, it contains in particular all the sets $(a, +\infty)$. By Proposition 1, this means that $B_1 \subset \mathfrak{B}$. Again by Proposition 1, to establish the opposite inclusion it suffices to show that all open sets lie in B_1. So let $b \in \mathbb{R}$ be arbitrary. The closed half-line $[b, +\infty)$ can be represented as a countable intersection of sets of the form $(a, +\infty)$, namely, $[b, +\infty) = \bigcap_{n=1}^{\infty} (b - \frac{1}{n}, +\infty)$. Consequently, $[b, +\infty) \in B_1$. The σ-algebra B_1 contains all open intervals, because $(a, b) = (a, +\infty) \setminus [b, +\infty)$. Since every open set on the line is the union of an at most countable collection of open intervals, all open sets are elements of the σ-algebra B_1. The proof is complete. $\square$

Definition 2. The *restriction* of the family of subsets Φ to a subset $A \subset \Omega$ is the collection Φ_A of all intersections of the elements of the family Φ with A: $\Phi_A = \{A \cap B : B \in \Phi\}$.

Exercises

1. Define the family Σ of subsets of the interval $[0, 1]$ as follows: a set belongs to Σ if either itself or its complement is at most countable. Is the family Σ a σ-algebra?

2. Is the family of countable unions of subintervals of the interval $[0, 1]$ a σ-algebra?

---

[1]To obtain all the Borel sets, one needs to define the classes $G_{\delta\sigma\delta...}$ and $F_{\sigma\delta\sigma...}$ not only in the case where the index $\sigma\delta\sigma \ldots$ is a finite sequence, but also for arbitrary countable ordinals. Here we run into one of the problems of measure theory that require familiarity with ordinal numbers and transfinite induction.

3. Let Ω be a set, Σ a σ-algebra on Ω, and $A \subset \Omega$. Then Σ_A is a σ-algebra on A.

4. Let X be a topological space and A be a Borel subset of X. Regard A as a subspace of X. Show that every subset $B \subset A$ that is a Borel set in the subspace A is also a Borel set in the original space X.

5. Do the sets of first category in the interval $[0, 1]$ form a σ-algebra? Describe the smallest σ-algebra of sets on $[0, 1]$ that contains all the subsets of first category.

6. Describe the smallest σ-algebra on the interval $[0, 1]$ which contains all the subsets of second category.

7. Let A be a dense G_δ-set in the complete metric space X. Show that $X \setminus A$ is a set of first category in X.

8. The intersection of a finite or countable number of dense G_δ-sets in a complete metric space is again a dense G_δ-set.

9. Let A be a G_δ-set in a complete metric space X and let $\overline{A}$ be the closure of A. Then $\overline{A} \setminus A$ is a set of first category in X.

10. Give an example of a decreasing chain of countable dense subsets of an interval which has an empty intersection.

11. A countable dense subset of an interval cannot be a G_δ-set.

12. Let f be a real-valued function on an interval. Show that the set $\mathrm{dc}(f)$ of all discontinuity points of f is an F_σ-set.

13. Write the collection of all open intervals with rational endpoints as a sequence (a_n, b_n), $n = 1, 2, \ldots$, and consider the sets $A_n = (-\infty, a_n] \cup [b_n, +\infty)$. In these notations, $\mathrm{dc}(f) = \bigcup_{n=1}^{\infty} \left(\overline{f^{-1}(A_n)} \setminus f^{-1}(A_n) \right)$.

Definition. A function $f : [0, 1] \to \mathbb{R}$ is said to be of the *first (Baire) class* if it can be represented as the pointwise limit of a sequence of continuous functions $f_n \in C[0, 1]$. For details on functions of the first class, see [25, Chap. 2, §31].

14. Let $f : [0, 1] \to \mathbb{R}$ be a function of the first class. Then $f^{-1}([a, +\infty)) \in G_\delta$ for any $a \in \mathbb{R}$.

15. For functions f of the first class the set $\mathrm{dc}(f)$ is of first category; consequently, f necessarily has continuity points.

16. Show that the set of all differentiability points of a continuous function on an interval is a Borel set. To which Borel class does it belong?

17. Let (f_n) be a sequence of continuous functions on an interval. Show that the set of all convergence points of the sequence (f_n) is a Borel set. To which Borel class does it belong?

18. Prove that every open (respectively closed) subset of a metric space is an F_σ-set (respectively, G_δ-set). In arbitrary topological spaces this assertion is in general not true.

19. Show that the classes F_σ and G_δ on an interval do not coincide.

20. Show that in a separable metric space the σ-algebra generated by the family of all open balls coincides with the σ-algebra of Borel sets.

21. Does the preceding assertion remain true if one discards the separability assumption?

22. Show that the σ-algebra of Borel sets on the real line is generated by a countable collection of sets (σ-algebras with this property are said to be *countably generated*).

23. Show that the cardinality of any countably-generated σ-algebra is not larger than the cardinality of the continuum. Show that, in particular, the σ-algebra of Borel sets on the real line has the cardinality of the continuum.

2.1.3 Products of σ-Algebras

Let (Ω_1, Σ_1) and (Ω_2, Σ_2) be sets endowed with σ-algebras. A *rectangle* in $\Omega_1 \times \Omega_2$ is any set of the form $A_1 \times A_2$, where $A_1 \in \Sigma_1$ and $A_2 \in \Sigma_2$. We define the σ-algebra $\Sigma_1 \otimes \Sigma_2$ on the Cartesian product $\Omega_1 \times \Omega_2$ as the smallest σ-algebra which contains all rectangles.

Exercises

1. Let $\mathfrak{B}_1$ and $\mathfrak{B}_2$ be the Borel σ-algebras on the topological spaces X_1 and X_2, respectively, and $\mathfrak{B}$ be the σ-algebra of Borel sets on $X_1 \times X_2$. Then $\mathfrak{B}_1 \otimes \mathfrak{B}_2 \subset \mathfrak{B}$.

2. The product of the σ-algebras of Borel sets on two separable metric spaces X_1 and X_2 coincides with the σ-algebra of Borel sets on $X_1 \times X_2$. In particular, the product of the σ-algebras of Borel sets on the (coordinate) lines coincides with the σ-algebra of Borel sets on the plane.

3. Does the preceding assertion remain valid if one drops the separability assumption?

4. Show that $2^{\mathbb{N}} \otimes 2^{\mathbb{N}} = 2^{\mathbb{N} \times \mathbb{N}}$.

5. Is it true that $2^{[0,1]} \otimes 2^{[0,1]} = 2^{[0,1] \times [0,1]}$?

6. Let $A \in \Sigma_1 \otimes \Sigma_2$, $t_1 \in \Omega_1$. Put $A_{t_1} = \{t_2 \in \Omega_2 : (t_1, t_2) \in A\}$. Show that $A_{t_1} \in \Sigma_2$.

2.1.4 Measures: Finite and Countable Additivity

The reader has undoubtedly already encountered the notion of measure, though perhaps under a different name. For instance, the number of elements in a set is a measure on the family $\mathbb{N}_f$ of all finite subsets of the set of natural numbers; area is a measure on the family of plane figures that have an area; the length of a rectifiable curve, the volume, the mass, are all examples of measures. In Subsection 2.3.1 we will construct a central example in measure theory, the Lebesgue measure on an interval.

Before giving the formal definition, let us remark that in the initial stage of our exposition all values of a measure are assumed to be finite non-negative numbers, i.e., the value $+\infty$ is not permitted. This simplifies the exposition and brings in some convenient additional properties of the measures under consideration, like for example the property from Exercise 2 of this subsection. Measures of this kind are often called *finite positive measures*. A more general kind of measure, called σ-*finite*, which may also take infinite values, will appear in Subsection 2.3.7. Later on, in Sect. 7.1, we will study a generalization to the case of values of arbitrary sign, in Subsection 8.4.5 to the case when the values are complex numbers, and in Sect. 13.4 even more general measures whose values are elements of a vector space. All these generalizations are widely used in Functional Analysis and Operator Theory, but before they appear in our textbook, the word "measure" will be used exclusively for finite positive measures.

Definition 1. Let Ω be a set with a family of subsets Φ given on it. A set function $\mu : \Phi \to \mathbb{R}$ is called a *finitely additive measure* if it satisfies the following conditions:

1. $\mu(A) \geqslant 0$ for any $A \in \Phi$.

2. If $A_1, A_2, \ldots, A_n \in \Phi$, the sets A_k are pairwise disjoint, and $\bigcup_{k=1}^{n} A_k \in \Phi$, then $\mu\left(\bigcup_{k=1}^{n} A_k\right) = \sum_{k=1}^{n} \mu(A_k)$.

Assume that $\emptyset \in \Phi$. Then from Condition 2 it follows that $\mu(\emptyset) + \mu(\emptyset) = \mu(\emptyset \cup \emptyset) = \mu(\emptyset)$, i.e., $\mu(\emptyset) = 0$. Note, however, that there can be non-empty sets of measure zero.

If the domain Φ of a finitely additive measure is an algebra of sets, then Condition 2 can be restated in a simpler way:

$2'$. For any pair of disjoint sets $A_1, A_2 \in \Phi$ the measure of their union is equal to the sum of their individual measures: $\mu(A_1 \sqcup A_2) = \mu(A_1) + \mu(A_2)$.

Let us list a number of properties of finitely additive measures.

Proposition 1. *Let μ be a finitely additive measure on an algebra $\mathbb{A}$ of subsets of the set Ω. Then:*

(a) *If $A_1, A_2 \in \mathbb{A}$, then $\mu(A_1 \setminus A_2) = \mu(A_1) - \mu(A_1 \cap A_2)$. If moreover $A_2 \subset A_1$, then $\mu(A_1 \setminus A_2) = \mu(A_1) - \mu(A_2)$.*

(b) *If* $A_1, A_2 \in \mathbb{A}$ *and* $A_2 \subset A_1$, *then* $\mu(A_2) \leqslant \mu(A_1)$. *In particular, if* $\mu(A_1) = 0$, *then also* $\mu(A_2) = 0$.

(c) *If* $\mu(A_2) = 0$, *then* $\mu(A_1 \setminus A_2) = \mu(A_1)$.

(d) $\mu(A_1 \cup A_2) = \mu(A_1) + \mu(A_2) - \mu(A_1 \cap A_2) \leqslant \mu(A_1) + \mu(A_2)$.

(e) $\mu\left(\bigcup_{k=1}^{n} A_k\right) \leqslant \sum_{k=1}^{n} \mu(A_k)$ *for any sets* $A_1, A_2, \ldots, A_n \in \mathbb{A}$.

Proof. (a) $A_1 = (A_1 \setminus A_2) \sqcup (A_1 \cap A_2)$. Hence, $\mu(A_1) = \mu(A_1 \setminus A_2) + \mu(A_1 \cap A_2)$.

(b) is a direct consequence of (a): $\mu(A_1) - \mu(A_2) = \mu(A_1 \setminus A_2) \geqslant 0$.

(c) If $\mu(A_2) = 0$, then also $\mu(A_2 \cap A_1) = 0$. It remains to apply assertion (a).

(d) Write $A_1 \cup A_2$ as the union $A_1 \cup A_2 = (A_1 \setminus A_2) \sqcup (A_2 \setminus A_1) \sqcup (A_2 \cap A_1)$ of three disjoint sets. Then we have

$$\mu(A_1 \cup A_2) = \mu(A_1 \setminus A_2) + \mu(A_2 \setminus A_1) + \mu(A_1 \cap A_2)$$
$$= \left(\mu(A_1 \setminus A_2) + \mu(A_1 \cap A_2)\right) + \left(\mu(A_2 \setminus A_1) + \mu(A_1 \cap A_2)\right) - \mu(A_1 \cap A_2)$$
$$= \mu(A_1) + \mu(A_2) - \mu(A_1 \cap A_2).$$

(e) is derived from (d) by induction on n. □

The most studied and most useful finitely additive measures are the countably additive measures, i.e., the measures that, in addition to conditions 1 and 2 of Definition 1, obey the *countable additivity axiom*:

$$\mu\left(\bigsqcup_{k=1}^{\infty} A_k\right) = \sum_{k=1}^{\infty} \mu(A_k)$$

for every disjoint collection of sets $A_n \in \Phi$, $n = 1, 2, \ldots$, with $\bigsqcup_{k=1}^{\infty} A_k \in \Phi$.

Countably-additive measures are also called σ-*additive*.

For a measure given on a σ-algebra Σ, the verification of countable additivity is somewhat simpler: if $A_n \in \Sigma$, $n = 1, 2, \ldots$, then automatically also $\bigcup_{k=1}^{\infty} A_k \in \Sigma$.

A countably additive measure μ given on a σ-algebra Σ of subsets of a set Ω is called a *probability measure* if $\mu(\Omega) = 1$.

Proposition 2. *Let* μ *be a countably additive measure given on a* σ-*algebra* Σ *of subsets of a set* Ω. *Then:*

1. *If* $A_n \in \Sigma$, $n = 1, 2, \ldots$, *is an increasing chain of sets (i.e.,* $A_1 \subset A_2 \subset \cdots \subset A_n \subset \cdots$), *then* $\mu\left(\bigcup_{k=1}^{\infty} A_k\right) = \lim_{k \to \infty} \mu(A_k)$.

2. *If* $A_n \in \Sigma$, $n = 1, 2, \ldots$, *is a decreasing chain of sets (i.e.,* $A_1 \supset A_2 \supset \cdots \supset A_n \supset \cdots$), *then* $\mu\left(\bigcap_{k=1}^{\infty} A_k\right) = \lim_{k \to \infty} \mu(A_k)$.

Proof. The proofs of assertions 1 and 2 of the proposition may be carried out in a similar way; moreover, one follows from the other by passing to complements. So let us prove, for example, the first assertion. Suppose the sets A_n form an increasing chain.

Put $A_\infty := \bigcup_{k=1}^\infty A_k$ and $B_n = A_{n+1} \setminus A_n$. The sequence of sets $A_1, B_1, B_2, B_3, \ldots$ is disjoint (i.e., its elements are pairwise disjoint), and $A_1 \sqcup \left(\bigsqcup_{k=1}^n B_k\right) = A_{n+1}$, $A_1 \sqcup \left(\bigsqcup_{k=1}^\infty B_k\right) = A_\infty$. Using the countable additivity assumption and the definition of the sum of a series, we have

$$\mu(A_\infty) = \mu(A_1) + \sum_{j=1}^\infty \mu(B_j) = \lim_{k \to \infty} \left(\mu(A_1) + \sum_{j=1}^k \mu(B_j)\right) = \lim_{k \to \infty} \mu(A_k).$$

The assertion is proved. $\qquad\square$

Here are two very simple yet useful remarks.

Proposition 3. *Let μ be a countably additive measure given on a σ-algebra Σ of subsets of a set Ω, $A_n \in \Sigma$, $n = 1, 2, \ldots$. Then:*

1. $\mu\left(\bigcup_{k=1}^\infty A_k\right) \leqslant \sum_{k=1}^\infty \mu(A_k)$. *In particular, if $\mu(A_k) = 0$ for all k, then also* $\mu\left(\bigcup_{k=1}^\infty A_k\right) = 0$.

2. *If $\mu(A_i \cap A_j) = 0$ for any $i, j \in \mathbb{N}$, $i \neq j$, then $\mu\left(\bigcup_{k=1}^\infty A_k\right) = \sum_{k=1}^\infty \mu(A_k)$.*

Proof. 1. Since the sets $\bigcup_{k=1}^n A_k$ form an increasing (with respect to n) chain, then by Assertion 1 of Proposition 2, in the inequality $\mu\left(\bigcup_{k=1}^n A_k\right) \leqslant \sum_{k=1}^n \mu(A_k)$ proved in Proposition 1 we are allowed to take the limit $n \to +\infty$.

2. Consider the sets $D = \bigcup_{i,j \in \mathbb{N}} (A_i \cap A_j)$ and $A_k' = A_k \setminus D$. The auxiliary sets A_k' are already pairwise disjoint. Since $\mu(D) = 0$, we have $\mu(A_k') = \mu(A_k)$ and $\mu\left(\bigcup_{k=1}^\infty A_k\right) = \mu\left(\bigcup_{k=1}^\infty A_k'\right)$. It remains to use the countable additivity. $\qquad\square$

Exercises

1. Prove the Assertion 2 of Proposition 2.

2. If the countably additive measure μ is given on a σ-algebra Σ, then $\mu(A_n) \to 0$ as $n \to \infty$, for any disjoint sequence of sets $A_n \in \Sigma$, $n = 1, 2, \ldots$.

3. Let μ be a finitely additive measure given on a σ-algebra Σ of subsets of a set Ω. Suppose that for any increasing chain of sets one has that $\mu\left(\bigcup_{k=1}^\infty A_k\right) = \lim_{k \to \infty} \mu(A_k)$. Then the measure μ is countably additive.

4. For a finitely additive measure μ defined on a σ-algebra, countable additivity is equivalent to the following condition: for any sets $A_n \in \Sigma$, $n = 1, 2, \ldots$, that form a decreasing chain with empty intersection, $\lim_{k \to \infty} \mu(A_k) = 0$.

5. Let (b_m) be a sequence of positive numbers such that $\sum_{m=1}^\infty b_m < \infty$. On the set $\mathbb{N}$ of all natural numbers consider the σ-algebra $2^{\mathbb{N}}$ of all subsets. For any $A \in 2^{\mathbb{N}}$ define its measure $\mu(A)$ as $\mu(A) = \sum_{m \in A} b_m$. Verify that μ is a countably additive measure.

6. Show that what the preceding exercise describes is the general form of a countably additive measure on $2^{\mathbb{N}}$.

7. Give an example of a finitely additive, but not countably additive measure on some algebra of subsets of $\mathbb{N}$.

8. Give an example of a finitely additive, but not countably additive measure on the σ-algebra $2^{\mathbb{N}}$ of all subsets of the set $\mathbb{N}$ of natural numbers.

9. Prove assertion (e) of Proposition 1 and also Proposition 3 by using Proposition 1 of Subsection 2.1.1.

10. Let μ be a finitely additive measure on an algebra Σ of subsets of a set Ω. Let $A, A_j \in \Sigma$, $A \subset \bigcup_{j=1}^{n} A_j$. Then $\mu(A) \leqslant \sum_{j=1}^{n} \mu(A_j)$. Does this assertion remain true if we replace n by $+\infty$?

11. In the setting of the preceding exercise, suppose that for some $k \in \mathbb{N}$ each point of the set A belongs to at least k distinct sets A_j, $1 \leqslant j \leqslant n$ (a so-called k-fold cover). Then $\mu(A) \leqslant \frac{1}{k} \sum_{j=1}^{n} \mu(A_j)$.

2.1.5 Measure Spaces. Completeness. Completion of a σ-Algebra with Respect to a Measure

A triple (Ω, Σ, μ), where Ω is a set endowed with a σ-algebra Σ of its subsets and μ is a countably additive measure on Σ, is called a *measure space*. Recall that the measures we are considering at this stage take only finite non-negative values. If one needs to stress this feature one may use the name *"finite measure space"*. If, in addition, μ is a probability measure (i.e., $\mu(\Omega) = 1$), then (Ω, Σ, μ) is called a *probability space*. In probability theory the set Ω is referred to as the space of elementary events, the elements of the σ-algebra Σ as events, and $\mu(A)$ as the probability of the event A taking place. The measurable functions, which will be treated in the next chapter, are referred to in the setting of probability theory as random variables, while the integral of a random variable is referred to as its mathematical expectation. We will not use the probabilistic terminology, but many of the problems that will be studied in the next chapter also play a role in probability theory.

Definition 1. A measure space (Ω, Σ, μ) is said to be *complete* (alternatively, one says that Σ is *complete with respect to the measure* μ) if the following condition is satisfied: for any $A \in \Sigma$ such that $\mu(A) = 0$, if $B \subset A$, then $B \in \Sigma$.

If the σ-algebra Σ is not complete with respect to the measure μ, then one can naturally extend the domain of definition of μ to a wider σ-algebra Σ' that will already be complete with respect to the extended μ. The extension procedure for achieving this, described below, is called the *completion* of the σ-algebra Σ with respect to the measure μ.[2]

Thus, let (Ω, Σ, μ) be a measure space (which may be incomplete). A subset $B \subset \Omega$ is said to be *negligible* if there exists a set $A \in \Sigma$ such that $\mu(A) = 0$ and $B \subset A$. We list the following obvious properties of negligible sets:

— if the set B is negligible and $B \in \Sigma$, then $\mu(B) = 0$;
— if the set B is negligible, then all its subsets are negligible;
— the union of any finite or countable family of negligible subsets is negligible (this follows from Proposition 3 in Sect. 2.1.4).

Two sets $A_1, A_2 \subset \Omega$ are said to be *equivalent* (and one writes $A_1 \sim A_2$), if their symmetric difference $A_1 \triangle A_2$ is negligible. The relation $\sim$ is symmetric and reflexive (obviously), as well as transitive: if $A_1 \sim A_2$ and $A_2 \sim A_3$, then the symmetric difference $A_1 \triangle A_3 \subset (A_1 \triangle A_2) \cup (A_2 \triangle A_3)$ is negligible, i.e., $A_1 \sim A_3$. Let us mention some further properties.

Lemma 1. 1. *If $A \sim B$, then $(\Omega \setminus A) \sim (\Omega \setminus B)$.*

2. *If $A_n \sim B_n$, $n \in M$, where M is a finite or countable index set, then $\bigcup_{n \in M} A_n \sim \bigcup_{n \in M} B_n$ and $\bigcap_{n \in M} A_n \sim \bigcap_{n \in M} B_n$.*

3. *If $B_1 \sim B_2$, $B_1, B_2 \in \Sigma$, then $\mu(B_1) = \mu(B_2)$.*

Proof. Assertion 1 follows from the relation $(\Omega \setminus A) \triangle (\Omega \setminus B) = A \triangle B$, and assertion 2 from the relations

$$\left(\bigcap_{n \in M} A_n \right) \triangle \left(\bigcap_{n \in M} B_n \right) \subset \bigcup_{n \in M} (A_n \triangle B_n)$$

and

$$\left(\bigcup_{n \in M} A_n \right) \triangle \left(\bigcup_{n \in M} B_n \right) \subset \bigcup_{n \in M} (A_n \triangle B_n).$$

Let us prove assertion 3. Since $\mu(B_1 \triangle B_2) = 0$, the sets $B_1 \setminus B_2$ and $B_2 \setminus B_1$ have measure zero. We have $\mu(B_1) = \mu(B_1 \cap B_2) + \mu(B_1 \setminus B_2) = \mu(B_1 \cap B_2) = \mu(B_1 \cap B_2) + \mu(B_2 \setminus B_1) = \mu(B_2)$. $\qquad\square$

Let us introduce the promised new collection of sets Σ' as follows: $A \in \Sigma'$ if there exists a set $B \in \Sigma$ such that $A \sim B$.

Theorem 1. *The family of sets Σ' contains the σ-algebra Σ and is itself a σ-algebra on Ω.*

Proof. If $A \in \Sigma$, then $A \in \Sigma'$: it suffices to take $B = A$ in the definition. Now let us verify that Σ' satisfies the axioms of a σ-algebra.

---

[2] We advise the reader to regard the assertions proved in this subsection as exercises and attempt to provide her/his own proofs.

1. $\Omega \in \Sigma'$.

2. If $A \in \Sigma'$, then also $\Omega \setminus A \in \Sigma'$. Indeed, by the definition, there exists a $B \in \Sigma$ such that $A \sim B$. But then $\Omega \setminus B \in \Sigma$ and $(\Omega \setminus A) \sim (\Omega \setminus B)$.

3. Suppose the set A_n belongs to Σ', $B_n \in \Sigma$, and $A_n \sim B_n$ for all $n = 1, 2, \ldots$ Then $\bigcup_{n=1}^{\infty} B_n \in \Sigma$ and $\bigcup_{n=1}^{\infty} A_n \sim \bigcup_{n=1}^{\infty} B_n$. Therefore, $\bigcup_{n=1}^{\infty} A_n \in \Sigma'$. $\square$

Let us extend the measure μ to a measure μ', defined now on Σ'. Let $A \in \Sigma'$ and $B \in \Sigma$ be such that $A \sim B$. Set $\mu'(A) = \mu(B)$. This definition is correct thanks to assertion 3 of the Lemma 1, that is, $\mu'(A)$ depends only on A and not on the choice of B.

Theorem 2. *The measure μ' is countably additive.*

Proof. Let $A_n \in \Sigma'$ be a disjoint sequence of sets and let $B_n \in \Sigma$ such that $A_n \sim B_n$. Since $A_i \cap A_j = \emptyset$ for any pair $i, j \in \mathbb{N}$ with $i \neq j$, and since $B_i \cap B_j \sim A_i \cap A_j$, one has $\mu(B_i \cap B_j) = 0$. Using assertion 2 of Proposition 3 in Subsection 2.1.4 and the relations $A_n \sim B_n$ and $\bigcup_{n=1}^{\infty} A_n \sim \bigcup_{n=1}^{\infty} B_n$, we conclude that

$$\mu'\left(\bigcup_{k=1}^{\infty} A_k\right) = \mu\left(\bigcup_{k=1}^{\infty} B_k\right) = \sum_{k=1}^{\infty} \mu(B_k) = \sum_{k=1}^{\infty} \mu'(A_k). \qquad \square$$

The measure space (Ω, Σ', μ') so constructed is called the *completion* of the measure space (Ω, Σ, μ). The measure μ' is usually denoted by the same letter μ as the original measure. This does not lead to confusion, since $\mu' = \mu$ on Σ.

Note that the completion procedure of a measure space considered above can be viewed as a particular case of Lebesgue's procedure for extending a measure, which will be addressed in Subsection 2.2.3 below. This fact will be formulated in Exercise 5 of that subsection.

Exercises

1. The completion of a measure space is a complete space.

2. A measure space is complete if and only if it coincides with its completion.

3. Let (Ω, Σ', μ') be the completion of the measure space (Ω, Σ, μ), $A \subset \Omega$. Show that

— $A \in \Sigma'$ if and only there exist sets $B, C \in \Sigma$ such that $B \subset A \subset C$ and $\mu(B) = \mu(C)$;

— $A \in \Sigma'$ if and only there exist a set $B \in \Sigma$ and a negligible set C such that $A = B \cup C$.

4. Let (Ω, Σ, μ) be a measure space. Show that the expression $\rho(A, B) = \mu(A \triangle B)$ gives a pseudometric on Σ.

5. Suppose $A_n \in \Sigma$ and the series $\sum_{n=1}^{\infty} \rho(A_n, A_{n+1})$ converges. Then the sequence (A_n) converges to $A_{\infty} = \bigcap_{n=1}^{\infty} \left(\bigcup_{k=n}^{\infty} A_k \right)$ in the pseudometric ρ.

6. Show that (Σ, ρ) is a complete pseudometric space.

2.1.6 Operations on Measures. δ-Measure. Atoms, Purely Atomic and Non-atomic Measures

Let Ω be a set endowed with a σ-algebra Σ. For measures on Σ there are natural operations of addition and of multiplication by positive real numbers: $(\mu_1 + \mu_2)(A) = \mu_1(A) + \mu_2(A)$ and $(a\mu)(A) = a\mu(A)$. We leave to the reader to verify that the operations thus introduced take the class of countably additive measures into itself.

Definition 1. An *atom* of the measure μ is a set $A \in \Sigma$ with the property that $\mu(A) > 0$ and for any $B \in \Sigma_A$ either $\mu(B) = 0$, or $\mu(A \setminus B) = 0$. If a measure has atoms it is called *atomic*; in the opposite case, the measure is called *non-atomic* (or *atomless*). A measure is called *purely atomic* if Ω can be written as the union of a finite or countable number of atoms.

A typical example of a purely atomic measure is the *δ-measure*. Specifically, let x be an arbitrary point of Ω. The δ-measure concentrated (or supported) in the point x is the measure δ_x defined by $\delta_x(A) = 1$, if $x \in A$, and $\delta_x(A) = 0$, if $x \notin A$.

Recall that two sets $A_1, A_2 \in \Sigma$ are said to be equivalent (written $A_1 \sim A_2$) if $\mu(A_1 \triangle A_2) = 0$. For example, for the measure δ_x its atom Ω is equivalent to the singleton $\{x\}$. The equivalence class of a set $A \in \Sigma$ is denoted by $[A]$.

By solving the exercises proposed below, the reader will obtain, in particular, the proofs of the following theorems:

Theorem 1. *Every countably additive measure on a σ-algebra can be written as the sum of a purely atomic measure and a non-atomic measure; moreover, this representation is unique.*

Theorem 2. *Suppose Ω is a separable metric space, the σ-algebra Σ contains all the Borel sets, and μ is a countably additive measure on Σ. Then each atom of the measure μ is equivalent to a singleton.*

Exercises

1. Any set equivalent to an atom is itself an atom.

2. If the atoms A_1, A_2 of the measure μ are not equivalent, then $\mu(A_1 \cap A_2) = 0$.

3. Let $A_n \in \Sigma$, $n = 1, 2, \ldots$, be a finite or countable sequence of pairwise non-equivalent atoms of the measure μ. Then there exists a disjoint sequence $A'_n \in \Sigma$, $n = 1, 2, \ldots$, of atoms of the measure μ, such that $A'_k \sim A_k$, $k = 1, 2, \ldots$ (use Proposition 1 of Subsection 2.1.1).

4. All representatives of an equivalence class of sets have the same measure.

As suggested by the previous exercise, let us define the measure of an equivalence class to be the measure of any representative of the class, that is, $\mu([A]) := \mu(A)$.

5. The equivalence class of an atom will be called an *atomic class*. The sum of the measures of any finite number of pairwise distinct atomic classes is not larger than $\mu(\Omega)$.

6. There are at most countably many distinct atomic classes.

7. There exists a finite or countable disjoint sequence $A_1, A_2, \ldots$ of atoms of the measure μ such that any atom of μ is equivalent to one of the atoms A_n.

8. In the setting of the preceding exercise, put $A_\infty := \bigcup_{k=1}^\infty A_k$ and define measures μ_1 and μ_2 on Σ by the rules $\mu_1(A) = \mu(A \cap A_\infty)$ and $\mu_2(A) = \mu(A \setminus A_\infty)$, respectively. Verify that μ_1 and μ_2 are countably additive measures, $\mu = \mu_1 + \mu_2$, the measure μ_1 is purely atomic, and the measure μ_2 is non-atomic. This establishes the representation in Theorem 1.

9. Let $\mu = \mu'_1 + \mu'_2$ be a decomposition into a purely atomic and a non-atomic measure, and B be an atom of the measure μ. Then B is an atom for μ'_1 and $\mu(B) = \mu'_1(B)$. Conversely, any atom of the measure μ'_1 is μ'_1-equivalent to an atom of the measure μ.

10. In the setting of the preceding exercise, the measure μ'_1 coincides with the measure μ_1 from Exercise 8 above. This establishes the uniqueness in Theorem 1.

11. Let Ω and Σ be as in the formulation of Theorem 2, and let $A \in \Sigma$. Then the set A can be partitioned into at most countably many pairwise disjoint sets from Σ, with diameters not larger than ε.

12. Under the conditions of Theorem 2, for any atom A of the measure μ and any $\varepsilon > 0$, there exists an atom $A_1 \subset A$ (automatically equivalent to the atom A) such that $\mathrm{diam}(A_1) < \varepsilon$.

13. Under the conditions of Theorem 2, let A be an atom of the measure μ. Using the preceding exercise, construct a chain $A \supset A_1 \supset A_2 \supset \cdots \supset A_n \supset \cdots$ of atoms with $\mathrm{diam}(A_n) < 1/n$. Show that the intersection of the sets in this chain consists of a single point, and that the resulting singleton is an atom. This completes the proof of Theorem 2.

14. Let Σ be the σ-algebra on $[0, 1]$ considered in Exercise 1 of Subsection 2.1.2. Put $\mu(A) = 0$, if A is at most countable, and $\mu(A) = 1$, if the complement of A is at most countable. Verify that the interval $[0, 1]$ is an atom of the measure μ, but it is not equivalent to any singleton.

15. Let μ be a countably additive non-atomic measure on the σ-algebra Σ. Then for any $A \in \Sigma$ and any $\alpha \in (0, 1)$, there exists a subset $B \in \Sigma_A$ such that $\mu(B) = \alpha\mu(A)$.

2.2 Extension of Measures

Often a measure is initially defined naturally on some relatively narrow class of sets, and before one starts using this measure, one needs to extend it to a wider class of sets. This situation is encountered even in school textbooks: area is defined first for rectangles, then for triangles, and then, via decomposition into smaller parts, for arbitrary polygons. Further, approximating a disc by polygons, one can define the area of a disc. One proceeds similarly to define the volume of figures in space. In this section we study a general scheme for extension of measures and apply it to construct the example of measure of highest importance for us, the Lebesgue measure on an interval.

2.2.1 Extension of a Measure from a Semiring of Sets to the Algebra Generated by the Semiring

Definition 1. A family Φ of subsets of a set Ω is called a *unital semiring* if:

1. $\Omega \in \Phi$.

2. If $A, B \in \Phi$, then $A \cap B \in \Phi$.

3. For any set $A \in \Phi$, its complement $\Omega \setminus A$ can be written as the union of a finite number of pairwise disjoint elements of the family Φ.

For a set $A \subset \Omega$ a *basic representation* is a representation $A = \bigsqcup_{k=1}^{n} A_k$, where $A_k \in \Phi$. Needless to say, there may exist sets that admit no basic representation.

Theorem 1. *Let Φ be a unital semiring. Then the family $\mathbb{A}$ of all sets that admit a basic representation constitutes the smallest algebra of sets $\mathbb{A}(\Phi)$ containing Φ.*

Proof. Let us show that $\mathbb{A}$ is an algebra of sets. Let $A, B \in \mathbb{A}$, and let $A = \bigsqcup_{k=1}^{n} A_k$ and $B = \bigsqcup_{j=1}^{m} B_k$ be corresponding basic representations. Then $A \cap B = \bigsqcup_{k=1}^{n} \bigsqcup_{j=1}^{m} A_k \cap B_j$ is a basic representation for $A \cap B$. Hence, $\mathbb{A}$ is stable under taking the intersection of a finite number of sets.

Now let us show that $\mathbb{A}$ is stable under taking complements. Thus, let $A = \bigsqcup_{k=1}^{n} A_k$ be an arbitrary element of $\mathbb{A}$, $\{A_k\}_{k=1}^{n} \subset \Phi$. By Axiom 3 in the definition of a unital semiring, all the sets $\Omega \setminus A_k$ lie in $\mathbb{A}$. Therefore, by what was proved above, $\Omega \setminus A = \bigcap_{k=1}^{n} (\Omega \setminus A_k)$ also lies in $\mathbb{A}$. Hence, $\mathbb{A}$ is an algebra. It remains to observe that any algebra of sets that contains all the elements of Φ must also contain their finite unions, i.e., all elements of $\mathbb{A}$. This shows that $\mathbb{A} = \mathbb{A}(\Phi)$. $\square$

Theorem 2. *Any finitely additive measure μ given on a unital semiring Φ extends uniquely to a finitely additive measure on the algebra $\mathbb{A}(\Phi)$ generated by Φ.*

Proof. Let us start with uniqueness. Let μ' be some extension to $\mathbb{A}(\Phi)$ of the measure μ, and $A = \bigsqcup_{k=1}^{n} A_k$ be a basic representation of an element A of the algebra $\mathbb{A}(\Phi)$. Then $\mu'(A) = \sum_{k=1}^{n} \mu'(A_k) = \sum_{k=1}^{n} \mu(A_k)$. Hence, $\mu'(A)$ is uniquely determined by the measure μ.

Now let us show that the expression $\mu'(A) = \sum_{k=1}^{n} \mu(A_k)$ obtained above does indeed give a finitely additive measure on $\mathbb{A}(\Phi)$. We begin by verifying the correctness of our definition, namely, by showing that $\mu'(A)$ is determined by the set A, i.e., it does not depend on the choice of a basic representation of A. Let $A = \bigsqcup_{k=1}^{n} A_k$ and $A = \bigsqcup_{j=1}^{m} B_j$ be two different basic representations of the set $A \in \mathbb{A}(\Phi)$. Consider the sets $C_{i,j} = A_i \cap B_j$. They are pairwise disjoint, and $A_i = \bigsqcup_{j=1}^{m} C_{ij}$, $B_j = \bigsqcup_{i=1}^{n} C_{ij}$. We have

$$\sum_{i=1}^{n} \mu(A_i) = \sum_{i=1}^{n} \left(\sum_{j=1}^{m} \mu(C_{ij}) \right) = \sum_{j=1}^{m} \left(\sum_{i=1}^{n} \mu(C_{ij}) \right) = \sum_{j=1}^{m} \mu(B_j).$$

The correctness of the definition is thus established. The finite additivity of the measure μ' is rather simply to verify. Indeed, let $A, B \in \mathbb{A}(\Phi)$ be disjoint sets, and let $A = \bigsqcup_{i=1}^{n} A_i$ and $B = \bigsqcup_{j=1}^{m} B_j$ be basic representations. Taken together, the sets A_i and B_j, $i = 1, \ldots, n$, $j = 1, \ldots, m$, provide a basic representation for $A \sqcup B$. Therefore, $\mu'(A \sqcup B) = \sum_{i=1}^{n} \mu(A_i) + \sum_{j=1}^{m} \mu(B_j) = \mu'(A) + \mu'(B)$. $\square$

Theorem 3. *Let μ be a countably additive measure on the unital semiring Φ, and μ' be its extension to the algebra $\mathbb{A} = \mathbb{A}(\Phi)$ constructed in Theorem 2. Then the measure μ' is also countably additive.*

Proof. Let $A_n \in \mathbb{A}$, $n = 1, 2, \ldots$, be a disjoint sequence of sets and suppose their union $B = \bigsqcup_{n=1}^{\infty} A_n$ also belongs to the algebra $\mathbb{A}$. Further, let $B = \bigsqcup_{j=1}^{m} B_j$ be a basic representation for B, and $A_k = \bigsqcup_{i=1}^{m_k} A_{ki}$ be basic representations for A_k. Then $B_j = \bigsqcup_{k=1}^{\infty} (A_k \cap B_j) = \bigsqcup_{k=1}^{\infty} \bigsqcup_{i=1}^{m_k} (A_{ki} \cap B_j)$, and all the sets figuring in the last representation belong to the semiring Φ. Since the measure μ on Φ is countably additive, $\mu(B_j) = \sum_{k=1}^{\infty} \sum_{i=1}^{m_k} \mu(A_{ki} \cap B_j)$. Therefore,

$$\mu'(B) = \sum_{j=1}^{m} \mu(B_j) = \sum_{j=1}^{m} \left(\sum_{k=1}^{\infty} \sum_{i=1}^{m_k} \mu(A_{ki} \cap B_j) \right)$$

$$= \sum_{k=1}^{\infty} \left(\sum_{j=1}^{m} \sum_{i=1}^{m_k} \mu(A_{ki} \cap B_j) \right) = \sum_{k=1}^{\infty} \mu' \left(\bigcup_{i=1}^{m_k} \bigcup_{j=1}^{m} A_{ki} \cap B_j \right) = \sum_{k=1}^{\infty} \mu'(A_k).   \square$$

Definition 2. Let Φ be a family of sets, $\mu \colon \Phi \to \mathbb{R}^+$. The set function μ is said to be *countably semiadditive* if for any sets $A, B_k \in \Phi$ the inclusion $A \subset \bigcup_{k=1}^\infty B_k$ implies $\mu(A) \leqslant \sum_{k=1}^\infty \mu(B_k)$.

Theorem 4 (Countable additivity test). *Let μ be a finitely additive measure on the unital semiring Φ which satisfies the countable semiadditivity condition. Then μ is countably additive.*

Proof. Let $A, B_k \in \Phi$, $A = \bigsqcup_{k=1}^\infty B_k$. We need to show that $\mu(A) = \sum_{k=1}^\infty \mu(B_k)$. Thanks to the countable semiadditivity condition, in order to do this it suffices to prove the inequality $\sum_{k=1}^\infty \mu(B_k) \leqslant \mu(A)$. Let μ' be the extension of the measure μ to the algebra $\mathbb{A}(\Phi)$ constructed in Theorem 2. From the inclusion $A \supset \bigcup_{k=1}^n B_k$ and the already established finite additivity of the measure μ' it follows that $\mu(A) = \mu'(A) \geqslant \mu'\left(\bigcup_{k=1}^n B_k\right) = \sum_{k=1}^n \mu'(B_k) = \sum_{k=1}^n \mu(B_k)$. It remains to let $n \to \infty$. $\square$

Exercises

1. What allowed us, in the proof of Theorem 3, to regroup the terms in the infinite sum? In general the sum of a series may change as a result of such an action. Why this did not happen in our case here?

2. Give an example of a family of sets Φ on the interval $[0, 1]$ and of a finitely additive measure μ on Φ such that no extension of μ to the algebra generated by Φ will be a finitely additive measure.

3. Let Φ be a family of sets on $[0, 1]$, μ a finitely additive measure on Φ. Can it happen that the measure μ has more than one extension to the algebra generated by Φ, with preservation of finite additivity?

4. Justify the equality $A_i = \bigcup_{j=1}^m C_{ij}$ in the proof of Theorem 2. Where was an analogous relation used in Theorem 3?

5. Let Φ be a unital semiring. Show that $\emptyset \in \Phi$.

6. Let Φ denote the family of all triangles in the plane (here triangles are considered together with their interior). For each $A \in \Phi$, let $r(A)$ denote the radius of the disc inscribed in the triangle A. Verify that the set function r is countably semiadditive on Φ. Is r a finitely additive measure on Φ?

2.2.2 Outer Measure

In this subsection Ω will be a set endowed with an algebra of subsets $\mathbb{A}$ and a countably additive measure μ. As already mentioned, the most natural domain of definition for a countably additive measure is not an algebra, but a σ-algebra of sets. Hence, it is highly desirable to know how to extend a countably additive measure to the σ-algebra generated by the algebra $\mathbb{A}$. The first idea that comes to mind is to proceed by analogy with Theorem 2 of the preceding subsection. Namely, consider disjoint countable unions of sets from $\mathbb{A}$. If all these sets again lie in $\mathbb{A}$, it means that we were actually dealing from the very beginning with a σ-algebra. In the opposite case, we define the measure of any such union as the sum of the measures of its components. The correctness of this definition can be justified, but in contrast to Theorem 2, the class of sets to which the measure is extended in the indicated way will now not be a Σ-algebra. Moreover, that class will not even be invariant under passage to complements, i.e., it will not even be an algebra! Hence, it is further necessary to define the measure in some way on the complements of the obtained sets. And then what should one do with the unions of such complements? Although in principle the idea just described can actually be implemented (some remarks on this theme are made below in Subsection 2.2.4), many technical difficulties can be avoided by using a different approach, based on the notion of outer measure. It is to H. Lebesgue that we owe this last approach.

Definition 1. Let $A \subset \Omega$ be an arbitrary set. The *outer measure* of the set A is the quantity

$$\mu^*(A) = \inf \left\{ \sum_{k=1}^{\infty} \mu(A_k) : \ A_k \in \mathbb{A}, \ A \subset \bigcup_{k=1}^{\infty} A_k \right\}.$$

The outer measure is defined already on all subsets of the set Ω, but on such a wide class of sets it does not even enjoy the finite additivity property. In the next subsection we will construct a σ-algebra of sets $\Sigma \supset \mathbb{A}$ on which μ^* will be countably additive, thus solving the problem of extending the measure μ. In the present subsection we do some preparatory work.

Properties of the Outer Measure:

1. *Monotonicity*: if $A \subset B$, then $\mu^*(A) \leqslant \mu^*(B)$.
2. *Semiadditivity*: for any sets $A, B \subset \Omega$ it holds that $\mu^*(A \cup B) \leqslant \mu^*(A) + \mu^*(B)$.
3. *Countable semiadditivity*: if $A \subset \bigcup_{k=1}^{\infty} B_k$, then $\mu^*(A) \leqslant \sum_{k=1}^{\infty} \mu^*(B_k)$.
4. $\mu^*(A) = \inf \left\{ \sum_{k=1}^{\infty} \mu(B_k) : \ B_k \in \mathbb{A}, \ A \subset \bigsqcup_{k=1}^{\infty} B_k \right\}$.
5. If $A \in \mathbb{A}$, then $\mu^*(A) = \mu(A)$, that is, μ^* is an extension of the measure μ.

Proof. 1. For $\mu^*(A)$ the infimum in the definition of μ^* is taken for a wider family of sets $\{A_k\}_1^{\infty}$ than for $\mu^*(B)$: if $B \subset \bigcup_{k=1}^{\infty} A_k$, then also $A \subset \bigcup_{k=1}^{\infty} A_k$. The infimum over a wider family does not exceed the infimum over a narrower one.

2. Again, instead of the infimum over all covers we are taking the infimum over a narrower class:

$\mu^*(A \cup B) \leqslant$

$$\inf \left\{ \sum_{k=1}^{\infty} \mu(A_k) + \sum_{k=1}^{\infty} \mu(B_k) : A_k, B_k \in \mathbb{A}, \ A \subset \bigcup_{k=1}^{\infty} A_k, B \subset \bigcup_{k=1}^{\infty} B_k \right\}$$
$$= \mu^*(A) + \mu^*(B).$$

3. Here one argues as in item 2.

4. Let $A \subset \Omega$ be an arbitrary set. Denote by

$$\nu(A) = \inf \left\{ \sum_{k=1}^{\infty} \mu(B_k) : \ B_k \in \mathbb{A}, \ A \subset \bigcup_{n=1}^{\infty} B_k \right\}$$

the right-hand side of the equality in question. By Proposition 1 in Subsection 2.1.1, for any family of sets $A_k \in \mathbb{A}$, $A \subset \bigcup_{k=1}^{\infty} A_k$, there exists a disjoint family of sets $B_k \in \mathbb{A}, A \subset \bigsqcup_{k=1}^{\infty} B_k$, such that $B_k \subset A_k$ for all k. For this family we have $\mu(B_k) \leqslant \mu(A_k)$, and therefore $\sum_{k=1}^{\infty} \mu(B_k) \leqslant \sum_{k=1}^{\infty} \mu(A_k)$. Consequently, $\nu(A) \leqslant \mu^*(A)$. The opposite inequality follows from the fact that the infimum in the definition of $\nu(A)$ is taken over a narrower class of sets than in the definition of the outer measure.

5. Let $A \in \mathbb{A}$, $A_k \in \mathbb{A}$, and $A \subset \bigsqcup_{k=1}^{\infty} A_k$. Set $B_k = A \cap A_k$. Then $B_k \in \mathbb{A}$, $A = \bigsqcup_{k=1}^{\infty} B_k$, and $\mu(B_k) \leqslant \mu(A_k)$. Now we use the countable additivity of the measure μ: $\mu(A) = \sum_{k=1}^{\infty} \mu(B_k) \leqslant \sum_{k=1}^{\infty} \mu(A_k)$. Taking the infimum over all such families $\{A_k\}_1^{\infty}$ yields $\mu(A) \leqslant \mu^*(A)$. The opposite inequality is obtained if in the definition of the outer measure we take the specific family $\{A_k\}_1^{\infty}$ given by $A_1 = A$, $A_k = \emptyset$ for $k \geqslant 2$. $\qquad \square$

By analogy with Exercises 4–6 in Subsection 2.1.5, we introduce on the collection of all subsets of the set Ω the *pseudometric ρ generated by the outer measure* by the rule $\rho(A, B) = \mu^*(A \triangle B)$.

Properties of the Pseudometric ρ:

1. For any sets $A, B, C \subset \Omega$ the *triangle inequality* $\rho(A, C) \leqslant \rho(A, B) + \rho(B, C)$ holds (this justifies using the term "pseudometric" for ρ).

2. $|\mu^*(A) - \mu^*(B)| \leqslant \rho(A, B)$ for any $A, B \subset \Omega$. In particular, the outer measure μ^* is continuous with respect to ρ.

3. $\rho(A, B) = \rho(\Omega \setminus A, \Omega \setminus B)$, i.e., passage to the complement is an isometry.

4. $\rho(A_1 \cap A_2, B_1 \cap B_2) \leqslant \rho(A_1, B_1) + \rho(A_2, B_2)$.

5. $\rho \left(\bigcup_{n \in M} A_n, \bigcup_{n \in M} B_n \right) \leqslant \sum_{n \in M} \rho(A_n, B_n)$, where the index set M is finite or countable and $A_n, B_n \subset \Omega$.

Proof. For each of the listed properties we provide the relations that yield it (together with properties of outer measure):

   1. $A \triangle C \subset (A \triangle B) \cup (B \triangle C)$.

   2. $A \subset B \cup (A \triangle B)$, $B \subset A \cup (A \triangle B)$.

3. $(\Omega \setminus A) \triangle (\Omega \setminus B) = A \triangle B$.

4. $\left(\bigcap_{n \in M} A_n \right) \triangle \left(\bigcap_{n \in M} B_n \right) \subset \bigcup_{n \in M} (A_n \triangle B_n)$ (applied to $M = \{1, 2\}$).

5. $\left(\bigcup_{n \in M} A_n \right) \triangle \left(\bigcup_{n \in M} B_n \right) \subset \bigcup_{n \in M} (A_n \triangle B_n)$. $\qquad\qquad\qquad$ $\square$

Recall that in Exercises 2–4 proposed at the very beginning of Sect. 1.1 the reader was asked to check the set-theoretical relations used in the above proof. Now is a good occasion to look at those exercises again. ☺

Exercises

1. Fill in the details of the proof that the outer measure is countably semiadditive.

2. Show that $\rho \left(\bigcap_{n \in M} A_n, \bigcap_{n \in M} B_n \right) \leqslant \sum_{n \in M} \rho(A_n, B_n)$ for all finite or countable collections of sets $A_n, B_n \subset \Omega$.

3. Show that the mapping $(A, B) \mapsto A \cap B$ is continuous as a function of two variables with respect to the pseudometric ρ.

A set A is called ρ-*negligible* if $\rho(A, \varnothing) = 0$.
Derive the following properties of ρ-negligible sets:

4. A is ρ-negligible if and only if $\mu^*(A) = 0$.

5. If $A \subset B$ and B is ρ-negligible, then so is A.

6. Any finite or countable union of ρ-negligible sets is ρ-negligible.

7. If our algebra $\mathbb{A}$ is a σ-algebra, then the definition of negligible sets from Subsection 2.1.5 applied to the measure space $(\Omega, \mathbb{A}, \mu)$ is equivalent to the definition of ρ-negligible sets.

2.2.3 Extension of a Measure from an Algebra to a σ-Algebra

Let, as in the preceding subsection, Ω be a set endowed with an algebra $\mathbb{A}$ of its subsets and a countably additive measure μ. A set $A \subset \Omega$ is called *measurable* if it belongs to the closure of the family $\mathbb{A}$ with respect to the pseudometric ρ. We let Σ denote the collection of all measurable subsets of Ω. In detail, $A \in \Sigma$ if and only if for any $\varepsilon > 0$ there exists a $B \in \mathbb{A}$ such that $\rho(A, B) < \varepsilon$. Obviously, $\mathbb{A} \subset \Sigma$. Note that the class of measurable sets depends not only on the algebra $\mathbb{A}$, but also on the measure μ. Whenever we need to emphasize that the measurable sets considered are generated specifically by the measure μ, we will refer to them not simply as measurable, but as μ-*measurable* sets.

Example 1. If the set A is ρ-negligible (i.e., $\rho(A, \emptyset) = 0$ or, equivalently, $\mu^*(A) = 0$), then A is measurable.

Lemma 1. *Let $A \subset \Omega$ and suppose that for any $\varepsilon > 0$ there exists a set $B \in \Sigma$ such that $\rho(A, B) < \varepsilon$. Then $A \in \Sigma$.*

Proof. Here we can simply refer to the fact that the closure of a set is a closed set. But we can also fill in the details: Pick a set $B \in \Sigma$ such that $\rho(A, B) < \varepsilon/2$. By the definition of measurable sets, for this B there exists a set $C \in \mathbb{A}$ such that $\rho(B, C) < \varepsilon/2$. By the triangle inequality, $\rho(A, C) < \varepsilon$. $\qquad \square$

Lemma 2. *Any countable union of elements of the algebra $\mathbb{A}$ is measurable.*

Proof. Let $A_n \in \mathbb{A}$ be a disjoint sequence of sets, $A = \bigsqcup_{n=1}^{\infty} A_n$. Since $\sum_{n=1}^{\infty} \mu(A_n) \leqslant \mu(\Omega)$, the series converges, and for every $\varepsilon > 0$ there exists an index n such that $\sum_{k=n+1}^{\infty} \mu(A_k) < \varepsilon$. Denote $\bigsqcup_{k=1}^{n} A_k$ by B. Then we have that $B \in \mathbb{A}$ and $\rho(A, B) = \mu^* \left(\bigsqcup_{k=n+1}^{\infty} A_k \right) \leqslant \sum_{k=n+1}^{\infty} \mu(A_k) < \varepsilon$. To reduce the case of an arbitrary sequence of sets $A_n \in \mathbb{A}$ to the case just treated it suffices to apply Proposition 1 of Subsection 2.1.1. $\qquad \square$

Theorem 1. *The family Σ of all measurable subsets of Ω is a σ-algebra.*

Proof. First, $\Omega \in \Sigma$, because $\Omega \in \mathbb{A}$. Next, let $A, B \in \Sigma$, and let $A_n, B_n \in \mathbb{A}$ be sequences that approximate A and B, respectively, in the sense that $\rho(A, A_n) \to 0$ and $\rho(B, B_n) \to 0$ as $n \to \infty$. Then $A_n \cap B_n \in \mathbb{A}$ and

$$\rho(A \cap B, A_n \cap B_n) \leqslant \rho(A, A_n) + \rho(B, B_n) \to 0 \quad \text{as } n \to \infty.$$

Hence, $A \cap B \in \Sigma$. Analogously, $\Omega \setminus A_n \in \mathbb{A}$ and $\rho(\Omega \setminus A, \Omega \setminus A_n) = \rho(A, A_n) \to 0$ as $n \to \infty$, which means that $\Omega \setminus A \in \Sigma$. It remains to establish the stability under the operation of taking countable unions. So, let $A_n \in \Sigma$, $n = 1, 2, \ldots$. Pick sets $B_n \in \mathbb{A}$ such that $\rho(A_n, B_n) \leqslant \varepsilon 2^{-n}$. Then $\bigcup_{n=1}^{\infty} B_n$ is a measurable set which ε-approximates $\bigcup_{n=1}^{\infty} A_n$: $\rho \left(\bigcup_{n=1}^{\infty} A_n, \bigcup_{n=1}^{\infty} B_n \right) \leqslant \sum_{n \in \mathbb{N}} \rho(A_n, B_n) = \varepsilon$. $\qquad \square$

Theorem 2. *The restriction of the outer measure μ^* to the σ-algebra Σ is countably additive.*

Proof. First let us show that the outer measure is finitely additive on Σ. Let $A_1, A_2 \in \Sigma$ be a disjoint pair and let $\varepsilon > 0$. By the definition of measurable sets, there exist sets $B_1, B_2 \in \mathbb{A}$ such that $\rho(A_1, B_1) + \rho(A_2, B_2) < \varepsilon$. The sets B_j may intersect, but this intersection cannot be large: by Property 4 of the pseudometric ρ,

$$\mu(B_1 \cap B_2) = \mu^*(B_1 \cap B_2) = \rho(\emptyset, B_1 \cap B_2)$$
$$= \rho(A_1 \cap A_2, B_1 \cap B_2) \leqslant \rho(A_1, B_1) + \rho(A_2, B_2) \leqslant \varepsilon.$$

Now let us use properties 2 and 5 of the function ρ. We get

$$\left|\mu^*(A_1 \cup A_2) - \left(\mu^*(A_1) + \mu^*(A_2)\right)\right| \leqslant |\mu(B_1 \cup B_2) - (\mu(B_1) + \mu(B_2))| + 2\varepsilon$$
$$= \mu(B_1 \cap B_2) + 2\varepsilon \leqslant 3\varepsilon.$$

Since ε is arbitrary, it follows that $\mu^*(A_1 \cup A_2) = \mu^*(A_1) + \mu^*(A_2)$, which establishes the finite additivity.

To complete the proof we use Theorem 4 of Subsection 2.2.1: any finitely additive measure on a unital semiring (and hence, also on a σ-algebra, since σ-algebras are also semirings) that is countably semiadditive is automatically countably additive.[3]
                                                                                              $\square$

Thus, we managed to extend the measure μ to a countably additive measure given on the σ-algebra $\Sigma \supset \mathbb{A}$. Hence, we simultaneously established the existence of such an extension to the σ-algebra generated by the algebra $\mathbb{A}$. Combining this with the results of Subsection 2.2.1, we obtain the following assertion.

Theorem 3. *Every countably additive measure given on a unital semiring can be extended to a countably additive measure on the σ-algebra generated by the semiring.*
                                                                                              $\square$

The resulting extension of the measure to the σ-algebra of μ-measurable sets will be denoted by the same letter μ as the original measure. That is, by definition, $\mu(A) = \mu^*(A)$ for all $A \in \Sigma$. The uniqueness of the extension and other useful properties of the construction described above are given as exercises below. Before passing to the exercises, let us emphasize that the measurability of all sets A for which $\mu^*(A) = 0$ (Example 1) is an important addition to Theorems 1 and 2.

Proposition 1. *The measure space (Ω, Σ, μ) obtained by the extension procedure described in this subsection is complete.*
                                                                                              $\square$

Exercises

1. Let Ω be a set endowed with an algebra $\mathbb{A}$ of subsets and a countably additive measure μ; let Σ_1 be the σ-algebra generated by $\mathbb{A}$, and μ_1 some extension of the measure μ to Σ_1. Then $\mu_1(A) \leqslant \mu^*(A)$ for all $A \in \Sigma_1$.

2. Suppose that on some algebra of sets $\widetilde{\Sigma}$ there are given two finitely additive measures μ_1 and μ_2 which satisfy $\mu_1(\Omega) = \mu_2(\Omega)$ and $\mu_1(A) \leqslant \mu_2(A)$ for all $A \in \widetilde{\Sigma}$. Then the measures μ_1 and μ_2 coincide.

---

[3]If our mathematician found his tea kettle empty on Monday, filled it with water, and boiled it to prepare tea, then on Tuesday, in order to prepare tea, he must first empty completely the kettle to reduce the task to the previous one. We have proceeded in a similar way. Our measure was already given on a σ-algebra. To achieve countable additivity, we reduced the problem to the criterion where the measure is given on a semiring and where in the proof we need to extend the measure from the semiring to the algebra.

3. Use the preceding two exercises to establish the uniqueness of the countably additive extension from an algebra to the σ-algebra it generates. Deduce from this the uniqueness of the extension in Theorem 3.

4. Establish the uniqueness of the extension to any σ-algebra that lies between Σ_1 and Σ.

5. Show that the completion of the measure space $(\Omega, \Sigma_1, \mu^*)$ coincides with the space (Ω, Σ, μ^*). In particular, if our algebra $\mathbb{A}$ is a σ-algebra, then the completion of the measure space $(\Omega, \mathbb{A}, \mu)$ is equal to (Ω, Σ, μ^*).

6. Let $(\Omega, \mathbb{A}, \mu)$ be a complete measure space. Show that the family Σ of measurable sets constructed for $(\Omega, \mathbb{A}, \mu)$ by the recipe described in the present subsection coincides with $\mathbb{A}$.

2.2.4 A Monotone Class Theorem for Sets

In this subsection we go deeper into the structure of measurable sets and we prove a theorem that will be needed in Subsection 4.4.4.

Let (Ω, Σ, μ) be the measure space obtained, as described above, by extending the measure μ from some unital semiring $\Phi \subset \Sigma$. More precisely, from the semiring we constructed the algebra $\mathbb{A}(\Phi)$ it generates, from this algebra the outer measure μ^*, and then for the outer measure the class of measurable sets (which is just our Σ) and the measure on Σ defined by the rule $\mu(A) = \mu^*(A)$. Let Φ_1 denote the family of all sets that can be represented as the union of a disjoint finite or countable collection of elements of the semiring Φ. Further, denote by Φ_2 the family of all sets that can be represented as the intersection of a decreasing sequence of sets from the family Φ_1. Since the family of measurable sets Σ is a σ-algebra, Φ_1 and Φ_2 consist of measurable sets.

Proposition 1. *The class of sets Φ_1 is stable with respect to taking the intersection of finitely many sets and under taking the union of disjoint finite or countable collections of sets.*

Proof. Let $A = \bigsqcup_{k=1}^{\infty} A_k$ and $B = \bigsqcup_{k=1}^{\infty} B_k$ be two arbitrary elements of the family Φ_1, written as corresponding countable unions of disjoint collections of elements of the semiring Φ (to avoid treating finite unions separately, we note that here some of the sets A_k, B_k may be empty). Then $A \cap B$ can also be written as a disjoint countable union of elements of Φ: $A \cap B = \bigsqcup_{k,j=1}^{\infty} (A_k \cap B_j) \in \Phi_1$.

Further, let the sets A_n be written as the corresponding disjoint unions $A_n = \bigsqcup_{k=1}^{\infty} A_{n,k} \in \Phi_1$ and be themselves disjoint. Then $\bigsqcup_{n=1}^{\infty} A_n = \bigsqcup_{n,k=1}^{\infty} A_{n,k} \in \Phi_1$. $\qquad\square$

Proposition 2. *For any set $A \subset \Omega$,*

$$\mu^*(A) = \inf\{\mu(B) : B \in \Phi_1, \ B \supset A\}.$$

Proof. Each element of the algebra $\mathbb{A}$, being a disjoint union of elements of the semiring Φ, lies in Φ_1. Hence, any countable disjoint union of elements of the algebra $\mathbb{A}$ also lies in Φ_1. It remains to use Property 4 of the outer measure (see Subsection 2.2.2), where we replace $\bigsqcup_{k=1}^{\infty} B_k$ by B.   □

Proposition 3. *For any set $A \subset \Omega$ there exists a set $B \in \Phi_2$ such that $A \subset B$ and* $\mu^*(A) = \mu(B)$.

Proof. By the preceding proposition, for any $n \in \mathbb{N}$ there exists a set $B_n \in \Phi_1$, $B_n \supset A$, such that $\mu(B_n) < \mu^*(A) + 1/n$. With no loss of generality, we may assume that the sets B_n form a decreasing chain (otherwise, we replace B_n by $B'_n = \bigcap_{k=1}^{n} B_k$). The intersection of the sets of this decreasing chain is the sought-for set B.   □

Definition 1. Let (Ω, Σ, μ) be a measure space. A family $\mathcal{M} \subset \Sigma$ is called a *monotone class of sets* if it obeys the following axioms:

A. If $A, B \in \mathcal{M}$ and $A \cap B = \varnothing$, then $A \cup B \in \mathcal{M}$.

B. If $A, B \in \mathcal{M}$ and $A \subset B$, then $B \setminus A \in \mathcal{M}$.

C. If $B \in \mathcal{M}$, $A \in \Sigma$, $A \subset B$, and $\mu(B) = 0$, then $A \in \mathcal{M}$.

D. If $A_n \in \mathcal{M}$, $n = 1, 2, \ldots$, is an increasing chain of sets, then $\bigcup_{n=1}^{\infty} A_n \in \mathcal{M}$.

We note that from axiom A it follows that a monotone class is stable under taking the union of finitely many disjoint sets. Hence, applying axiom D, we deduce that a monotone class is stable under taking the union of countably many disjoint sets.

Theorem 1 (Monotone class theorem for sets). *Let (Ω, Σ, μ) be a measure space obtained, as described in Sect. 2.2, by extending the measure μ from some unital semiring $\Phi \subset \Sigma$. Further, let $\mathcal{M} \subset \Sigma$ be a monotone class which contains all the elements of the semiring Φ. Then $\mathcal{M} = \Sigma$.*

Proof. Since $\Omega \in \mathcal{M} \subset \Sigma$, axiom B of a monotone class implies that for any element $A \in \mathcal{M}$ the complement $\Omega \setminus A$ also belongs to $\mathcal{M}$. Passing in axiom D to complements, we deduce that the class $\mathcal{M}$ is stable under taking the intersection of decreasing chains of sets.

The families Φ_1 and Φ_2 introduced in the begining of this subsection lie in $\mathcal{M}$. By Proposition 3, for any set $A \in \Sigma$ of measure 0, there exists a set $B \in \Phi_2 \subset \mathcal{M}$ such that $A \subset B$ and $\mu(B) = 0$. Hence, by axiom C, every set A of measure 0 lies in the class $\mathcal{M}$. Finally, consider an arbitrary set $A \in \Sigma$. We use again Proposition 3 and choose a $B \in \mathcal{M}$ such that $A \subset B$ and $\mu(A) = \mu(B)$. Then $C = B \setminus A$ is a set of measure 0, hence $C \in \mathcal{M}$. It remains to apply axiom B to deduce that $A = B \setminus C \in \mathcal{M}$.   □

Exercises

1. Show that the class Φ_1 is the family of all sets that can be written as a (not necessarily disjoint) finite or countable union of elements of the algebra $\mathbb{A}$.

2. Show that the class Φ_1 is the family of all sets that can be written as a (not necessarily disjoint) finite or countable union of elements of the semiring Φ.

3. Show that the class Φ_1 is stable under taking the union of finitely or countably many sets (possibly not disjoint).

4. Show that the class of sets Φ_2 is stable under taking the intersection of finitely or countably many sets.

5. Justify the inclusion $\Phi_2 \subset \mathcal{M}$ in the proof of the last theorem.

6. Where in the proof of the theorem was the condition $A \in \Sigma$ (i.e., measurability) used? Is it not possible to prove in the same way that every subset $A \subset \Omega$ belongs to $\mathcal{M}$?

7. On the interval $[0, 1]$ consider the semiring Φ (which in fact is an algebra) consisting of the finite sets and their complements. Show that in this case the class Φ_1 is not an algebra of sets.

8. In the setting of the preceding exercise, describe the class Φ_2. Show that in the present case Φ_2 is a σ-algebra of sets.

9. Let Ω be a set consisting of four points, $\Sigma = 2^\Omega$, and define the measure of an element $A \in \Sigma$ as the number of points in the set A (the so-called *counting measure*). Show that the family $\mathcal{M}$ of all subsets consisting of an even number of elements is a monotone class that does not coincide with Σ. Doesn't this example contradict the preceding theorem?

10. Prove the following version of the monotone class theorem: let (Ω, Σ_0, μ) be a measure space, $\Phi \subset \Sigma_0$ a unital semiring that generates the σ-algebra Σ_0, and (Ω, Σ_0', μ) the completion of the measure space (Ω, Σ_0, μ). Further, let Σ be a σ-algebra that lies between Σ_0 and Σ_0' (i.e., $\Sigma_0 \subset \Sigma \subset \Sigma_0'$), and $\mathcal{M} \subset \Sigma$ be a monotone class of sets that contains all elements of the semiring Φ. Show that $\mathcal{M} \supset \Sigma$.

11. Use Proposition 3 in order to obtain an easy solution of Exercise 7 from Subsection 2.2.2.

2.3 Measures on an Interval and on the Real Line

2.3.1 The Lebesgue Measure on the Interval

As we already mentioned, length is a measure on the family of intervals. In this subsection we apply the general theory of extension of measures to this example, historically the first and fundamental for the entire theory of measures.

Let $\Omega = (\omega_1, \omega_2)$ be a non-degenerate bounded open interval. A *subinterval* of Ω is any open, closed, or semi-open interval lying in Ω, i.e., any subset of the form $[a, b]$, $[a, b)$, (a, b), or $(a, b]$ that is contained in Ω. In particular, the empty set as well as the one-point sets (singletons) are subintervals.

The family of all subintervals of Ω is a semiring of sets, which we denote by Φ. For any subinterval $\Delta \in \Phi$ we let $\lambda(\Delta)$ denote its length, i.e., $\lambda(\Delta) = b - a$, where a and b are the left and right endpoints of Δ.

Theorem 1. *The length λ is a countably additive measure on the semiring Φ.*

Proof. By the countable additivity test (Theorem 4 in Subsection 2.2.1), we need to verify that the measure λ is finitely additive and countably semiadditive.

Finite additivity. Let Δ_k, $k = 1, 2, \ldots, n$, be disjoint subintervals, written in increasing order, let a_k and b_k be the endpoints of Δ_k, and let $\bigcup_{k=1}^n \Delta_k = \Delta \in \Phi$. Then a_1 and b_n coincide with the endpoints of the interval Δ, and $a_{k+1} = b_k$, $k = 1, 2, \ldots, n - 1$. We have $\sum_{k=1}^n \lambda(\Delta_k) = \sum_{k=1}^n (b_k - a_k) = b_n - a_1 = \lambda(\Delta)$.

Countable semiadditivity. Let $\Delta \subset \bigcup_{k=1}^\infty \Delta_k$, $\Delta_k, \Delta \in \Phi$, and let a and b be the endpoints of the interval Δ, and a_k and b_k the endpoints of Δ_k. Take an arbitrary $\varepsilon > 0$ and, slightly moving away from the original endpoints, introduce auxiliary intervals $\Delta' \subset \Delta$ and $\Delta'_k \supset \Delta_k$, such that Δ' is closed, Δ'_k are open subsets of the interval Ω, and

$$\lambda(\Delta) - \lambda(\Delta') + \sum_{k=1}^\infty \left(\lambda(\Delta'_k) - \lambda(\Delta_k) \right) < \varepsilon, \tag{1}$$

i.e., such that the endpoints are not displaced too much. For the new intervals, as before, we have the inclusion $\Delta' \subset \bigcup_{k=1}^\infty \Delta'_k$, but now this inclusion has a different meaning, namely, we obtained a cover of a compact set, something we are familiar with! Let us choose a finite subcover, i.e., take a finite set of indices $N \subset \mathbb{N}$ such that $\Delta' \subset \bigcup_{k \in N} \Delta'_k$. Thanks to the finite additivity we may apply assertion (e) of Proposition 1 from Subsection 2.1.4: $\lambda(\Delta') \leqslant \sum_{k \in N} \lambda(\Delta'_k) \leqslant \sum_{k=1}^\infty \lambda(\Delta'_k)$. Using condition (1), we see that $\lambda(\Delta) \leqslant \sum_{k=1}^\infty \lambda(\Delta_k) + \varepsilon$. In view of the arbitrariness of ε, the countable semiadditivity is established. $\square$

Now let us extend the measure λ by means of the extension recipe described in Sect. 2.2. This yields a countably additive measure (denoted again by λ), given on a σ-algebra $\Sigma \supset \Phi$, such that $\lambda((a, b)) = b - a$ for any interval (a, b), and the resulting measure space $(\Omega, \Sigma, \lambda)$ is complete. The elements of the σ-algebra Σ are called *Lebesgue-measurable* sets, and the constructed measure λ on Σ is called the *Lebesgue measure*.

For the moment, the Lebesgue measure is given in a somewhat encoded way, by referring to the general scheme of measure extension. The purpose of the remarks below is to recast this definition in maximal detail and accessibility.

Remarks

1. Let $A = \bigsqcup_{k=1}^{\infty} (a_k, b_k) \subset \Omega$ (recall that this is the general form of an open subset of the interval $\Omega = (\omega_1, \omega_2)$). Then A is Lebesgue measurable, and $\lambda(A) = \sum_{k=1}^{\infty} |b_k - a_k|$.

2. Any one-point set (singleton) is measurable and its Lebesgue measure is equal to zero. Consequently, the Lebesgue measure of any finite or countable set is also equal to zero.

3. The outer measure of any set $A \subset \Omega$ can be calculated by the rule $\lambda^*(A) = \inf \left\{ \sum_{k=1}^{\infty} |b_k - a_k| : \bigcup_{k=1}^{\infty} [a_k, b_k] \supset A \right\}$.

4. If $\bigcup_{k=1}^{\infty} [a_k, b_k] \supset A$, then every interval $[a_k, b_k]$ can be replaced by a slightly larger open interval such that the sum of lengths changes by an arbitrarily small amount. This means that in the formula given above one can equally well use open intervals instead of closed intervals:

$$
\begin{aligned}
\lambda^*(A) &= \inf \left\{ \sum_{k=1}^{\infty} |c_k - d_k| : \bigcup_{k=1}^{\infty} (c_k, d_k) \supset A \right\} \\
&= \inf \{ \lambda(B) : \ B \text{ open}, \ B \supset A \} \\
&= \inf \left\{ \sum_{k=1}^{\infty} |b_k - a_k| : \bigsqcup_{k=1}^{\infty} (a_k, b_k) \supset A \right\}.
\end{aligned}
\tag{2}
$$

5. By the definition, a subset A of Ω is Lebesgue measurable if for any $\varepsilon > 0$ there exists a set B which is a finite union of intervals such that $\lambda^*(A \triangle B) < \varepsilon$.

6. By the definition of the extension of a measure, $\lambda(A) = \lambda^*(A)$ for any Lebesgue-measurable set A.

7. The outer measure and hence the Lebesgue measure of a set A do not depend on the interval $\Omega \supset A$ with which the construction starts. Accordingly, henceforth we do not have to specify in which particular interval all the sets considered lie.

8. The Lebesgue measure of a set does not change under translations $A \mapsto A + t$.

9. If $\lambda^*(A) = 0$, then A is Lebesgue measurable and $\lambda(A) = 0$ (see the Example 1 in Subsection 2.2.3). Such sets are called *negligible*, or *sets of measure zero*.

10. Since any open subset of an interval is Lebesgue measurable, so is every Borel subset of an interval (Σ is a *sigma*-algebra containing all the open sets, and $\mathfrak{B}$ is, by definition, the smallest σ-algebra that contains all the open sets). In particular, all closed sets, all G_δ-sets, all F_σ-sets, etc, are Lebesgue measurable.

Theorem 2. *A set $A \subset \Omega$ is Lebesgue measurable if and only if it can be written as the difference $B \setminus C$ of a G_δ-set B and a set $C \subset B$ with $\lambda^*(C) = 0$.*

Proof. Since the G_δ-sets, as well as the negligible sets, are measurable, any difference of such sets is also measurable. Hence, we only have to prove the converse statement. Suppose the subset $A \subset \Omega$ is Lebesgue measurable. By definition, $\lambda(A) = \lambda^*(A)$. By formula (2) for the outer measure, for any $n \in \mathbb{N}$ there exists an open set $B_n \supset A$ such that $\lambda(B_n) < \lambda(A) + 1/n$. Put $B = \bigcap_{n=1}^{\infty} B_n$. The set B, as required, belongs to the class G_δ and contains A. Further, for any $n \in \mathbb{N}$, $\lambda(A) \leqslant \lambda(B) \leqslant \lambda(B_n) < \lambda(A) + 1/n$. Hence, $\lambda(A) = \lambda(B)$. It remains to put $C = B \setminus A$.   □

Passing to complements we obtain the following:

Corollary 1. *The set $A \subset \Omega$ is Lebesgue measurable if and only if it can be written as the disjoint union of an F_σ-set and a negligible set.*   □

Since the Borel sets on an interval are Lebesgue measurable, the preceding theorem and its corollary apply to them. We deduce that, although the Borel subsets of the interval do not reduce to the F_σ- and G_δ-sets, they do not differ too much from the sets of the latter two classes. Moreover, the F_σ-sets and G_δ-sets can be obtained from one another by adding or removing negligible sets.

The aforementioned way of expressing measurable sets in terms of Borel classes and negligible sets provides useful information on the structure of the Lebesgue measure and the Lebesgue-measurable sets. However, one should not let ourselves be excessively seduced by the beauty of the picture just painted: while negligible sets can indeed be neglected from the point of view of measure theory, in many other settings they may have a rather complicated structure.

Example 1 (A set of measure zero with the cardinality of the continuum). Recall that the Cantor set is the closed subset $\mathcal{K} \subset [0, 1]$ consisting of the numbers whose triadic expansion either does not contain the digit 1, or contains it only as the last digit in the expansion (Subsection 1.4.4). The Cantor set can be also constructed by means of the following procedure of step-by-step removal of parts from the interval $[0, 1]$. In the first step, remove the set $\Delta_1^1 = (1/3, 2/3)$. Let $K_1 = [0, 1] \setminus \Delta_1^1$. The set K_1 consists of two intervals of length $1/3$. In each of these intervals recede from the endpoints by $1/3$ of its length and remove the resulting middle subintervals, $\Delta_1^2 = (1/9, 2/9)$ and $\Delta_2^2 = (7/9, 8/9)$. Let $K_2 = K_1 \setminus (\Delta_1^2 \cup \Delta_2^2)$. In the n-th step, the set K_n will consist of 2^n intervals of length $1/3^n$, and to obtain K_{n+1} remove from the middle of each component of K_n its third. The Cantor set coincides with $\bigcap_{n=1}^{\infty} K_n$. The measure of the set K_n is the sum of the measures of its components, i.e., $\lambda(K_n) = 2^n/3^n$, which tends to zero when $n \to \infty$. Since $\lambda(\mathcal{K}) \leqslant \lambda(K_n)$ for all n, we conclude that $\lambda(\mathcal{K}) = 0$.

To prove that the Cantor set has continuum cardinality we can proceed in different ways. Here is one of the simplest. $\mathcal{K}$ is a subset of $[0, 1]$, so that the cardinality of card $\mathcal{K}$ does not exceed that of the continuum. To prove the opposite inequality, we construct an injective map of a set having the cardinality of the continuum into $\mathcal{K}$.

Namely, to each dyadic fraction $x \in (0, 1)$ we assign the triadic fraction $f(x)$ by keeping all the zeros of the fraction x unchanged and replacing every 1 by a 2. Then $x \mapsto f(x)$ is the required injective map. Another proof of the fact that $\mathcal{K}$ has the cardinality of the continuum is given in Exercise 12 of Subsection 1.3.3.

Exercises

1. Calculate the Lebesgue measure of the following sets:

A. $[1, 3] \cup [5, 6]$;

B. $(2, 4) \setminus ([1, 3] \cup [5, 6])$;

C. $((2, 4) \setminus [1, 3]) \cup [5, 6]$;

D. $[1, 4] \triangle [2, 6]$;

E. $\bigcup_{n=1}^{\infty} \left[\frac{1}{2^n}, \frac{1}{2^{n+1}} \right]$;

F. $\bigcup_{n=1}^{\infty} \left[\frac{1}{n}, \frac{1}{2n} \right]$;

G. the set of rational numbers in the interval $[0, 1]$;

H. the set of irrational numbers in the interval $[0, 1]$.

2. Let $A \subset [0, 1]$ be such that the complement of A has Lebesgue measure zero. Then A is dense in $[0, 1]$.

3. For every $A \subset [0, 1]$ consider on the interval $[0, 1]$ the function f given by $f(x) = \lambda^*(A \cap [0, x])$. Show that f is continuous.

4. Construct in the interval $[0, 1]$ a nowhere dense set of positive Lebesgue measure.

5. Construct in the interval $[0, 1]$ a set A of the second category with $\lambda(A) = 0$.

6. Is the Lebesgue measure atomic or non-atomic?

Show that:

7. If a measurable set has non-zero Lebesgue measure, then it has the cardinality of the continuum.

8. The cardinality of the family of negligible sets on an interval is equal to the cardinality of the family of all subsets of the interval. Therefore (see Exercise 23 in Subsection 2.1.2), on an interval there exist negligible sets that are not Borel sets.

9. Every negligible set is contained in a negligible G_δ-set.

The *inner measure* of a set $A \subset \Omega$ is defined as

$$\lambda_*(A) = \sup\{\lambda(B) : B \subset A, \ B \text{ closed}\}.$$

Prove that:

10. $\lambda_* \leqslant \lambda^*$.

11. A set $A \subset \Omega$ is Lebesgue measurable if and only if $\lambda_*(A) = \lambda^*(A)$.

2.3.2   A Bit More Terminology. The Meaning of the Term "Almost Everywhere"

Let (Ω, Σ, μ) be a measure space. The elements of the σ-algebra Σ are called *measurable sets*. If on Ω we need to consider simultaneously several σ-algebras, and we need to specify which σ-algebra we have in mind, then we will refer to the elements of the σ-algebra Σ as Σ-measurable sets. For example, suppose that on an interval one considers, along with the Lebesgue-measurable sets, the σ-algebra $\mathfrak{B}$ of Borel sets. Then in accordance with the above terminology, the Borel sets can be alternatively referred to as $\mathfrak{B}$-measurable sets, or Borel-measurable sets.

Recall that a set $A \subset \Omega$ is called negligible (see Subsection 2.1.5) if A is contained in a measurable set of measure zero. If (Ω, Σ, μ) is a complete measure space (say, an interval with the Lebesgue measure), then the definition simplifies: "negligible set" and "set of measure zero" are synonyms. A set is said to be *of full measure* if its complement is negligible.

An assertion P concerning points of the set Ω is said to hold *for almost all* points $t \in \Omega$, or to hold *almost everywhere* (abbreviated a.e.) if the set of all points t for which P is not true is negligible. For instance, a function $f \colon \Omega \to \mathbb{R}$ is said to be zero (or vanish) almost everywhere (abbreviated as $f \overset{\text{a.e.}}{=} 0$) if the set of all points t where $f(t) \neq 0$ is negligible. Further, we write $f \overset{\text{a.e.}}{\geqslant} g$ if the set of all points t where $f(t) < g(t)$ is negligible; and so on. Considerations and estimates that are made almost everywhere are considerably more convenient than the usual pointwise considerations. Thus, on an interval (tacitly assumed to be endowed with the Lebesgue measure), if a function has a finite or countable number of discontinuity points, then it is often not necessary to define the function at those points or one can define it in the way most suitable for the task at hand, since for almost all values of the argument this does not affect the function.

We mention two important properties that are immediate consequences of the properties of negligible sets (Subsection 2.1.5).

— Suppose that statement P_1 implies statement P_2, and that P_1 holds almost everywhere. Then P_2 also holds almost everywhere.

— Let P_j, $j \in M$, be a finite or countable set of statements, and P be the statement that all the statements P_j hold simultaneously. If all P_j hold almost everywhere, then P also holds almost everywhere.

Exercises

1. Prove the last two assertions.

2. Explain what the negation of the assertion $f \overset{\text{a.e.}}{\geqslant} g$ means. Does it coincide with the assertion $f \overset{\text{a.e.}}{<} g$?

3. Can the assertions $f \overset{\text{a.e.}}{\geqslant} g$ and $f \overset{\text{a.e.}}{<} g$ hold simultaneously?

4. Show that if $f \overset{\text{a.e.}}{\geqslant} g$ and $g \overset{\text{a.e.}}{\geqslant} h$, then $f \overset{\text{a.e.}}{\geqslant} h$.

5. Show that if simultaneously $f \overset{\text{a.e.}}{\geqslant} g$ and $f \overset{\text{a.e.}}{\leqslant} g$, then $f \overset{\text{a.e.}}{=} g$.

6. Suppose that two continuous functions on an interval coincide a.e. with respect to the Lebesgue measure. Show that then the functions coincide at all points.

7. Does the preceding statement remain true if we replace the Lebesgue measure by an arbitrary countably additive measure given on the Borel subsets of the interval?

2.3.3   Lebesgue's Theorem on the Differentiability of Monotone Functions

The proof of existence theorems often make use of the following idea: instead of constructing the required object explicitly, one shows that in some or another sense there are "many" such objects. But if already there are many, then they of course exist. For instance, the simplest proof of the existence of transcendental numbers is based on considerations of cardinality: the algebraic numbers form a countable set, therefore the transcendental number not only exist, but constitute the "main mass" of all numbers. In Exercise 15 of Subsection 2.1.2 it is shown how, in exactly the same manner, to prove existence theorems (specifically, to prove the existence of continuity points for the pointwise limit of a sequence of continuous functions) one can use sets of the first and second category. In each such argument the most important point is to appropriately choose what notion of "smallness" should be used. In the present subsection we will, as a first non-obvious application of measure theory, prove that every monotone function has points of differentiability. Actually, we will prove more:

Theorem 1. *Every monotone function on an interval is differentiable almost everywhere, i.e., the set of points at which the function is not differentiable has Lebesgue measure zero.*

For the reader to truly appreciate the depth and elegance of this result, I insistently advise her/him to set aside the book for the moment and reflect on it for at least a couple of days. I have to sincerely confess that although in my time I got "hooked" on this problem, I failed to solve it by myself. On the other hand, this experience provided a good stimulus for studying measure theory, and subsequently my teacher had no need to convince me that this field is important.

Before we can prove the theorem, we need some notation and lemmas. Let g be an upper-continuous function given on the interval $\Omega = [\omega_1, \omega_2]$ (see Subsection 1.2.4 for the corresponding definition). An interior point x of Ω is said to be *invisible from the right* for the function g if there exists a $t > x, t \in \Omega$, such that $g(x) < g(t)$.

Lemma 1 (F. Riesz's "rising sun" lemma). *Let g be an upper-semicontinuous function on Ω. Then the set A of all points that are invisible from the right for the function g is open. Moreover, if one writes A canonically as a union of disjoint subintervals $\Delta_k = (a_k, b_k)$, then $g(a_k + 0) \leqslant g(b_k)$. (Here $g(a_k + 0)$ is understood as the upper limit of the function $g(t)$ as $t \to a_k + 0$.)*

Proof. Let x_0 be a point that is invisible from the right and let $t_0 > x_0$ such that $g(x_0) < g(t_0)$. Then thanks to the semi-continuity of the function g, the point x_0 has an entire neighborhood where $g(x) < g(t_0)$. All the points of this neighborhood are invisible from the right. Hence, A is an open set. Now let $\Delta = (a, b)$ be one of the intervals composing A, i.e., $(a, b) \subset A, a, b \notin A$. Suppose that the conclusion of the lemma is false. Then there exists a point $x_0 \in \Delta$ such that $g(x_0) > g(b)$. Consider the set D of all points $x \in [x_0, b)$ at which $g(x) \geqslant g(x_0)$. Then D is a non-empty closed and bounded set. Denote the extreme right point of D by x_1. Since x_1 is invisible from the right, there is in Ω a point $t_0 > x_1$ such that $g(t_0) > g(x_1)$. Clearly, t_0 cannot lie to the right of the point b, as otherwise b will also be invisible from the right: $g(t_0) > g(x_1) \geqslant g(x_0) > g(b)$. It follows that $t_0 \in (x_1, b)$. But then $t_0 \in D$, that is, x_1 is not the extreme right point of the set D: contradiction.   □

Note that, by symmetry, an analogous statement holds true for points invisible from the left (the point x is *invisible from the left* if there exists a $t < x, t \in \Omega$, for which $g(x) < g(t)$); only at the endpoints of the intervals composing the set of points invisible from the left will the opposite condition $g(a_k) \geqslant g(b_k - 0)$ be satisfied.

Lemma 2 (negligibility test). *Suppose the set $A \subset (\omega_1, \omega_2)$ has the following property: there exists a $\theta \in (0, 1)$ such that $\lambda^*(A \cap (a, b)) \leqslant \theta(b - a)$ for any subinterval $(a, b) \subset (\omega_1, \omega_2)$. Then A is negligible.*

Proof. Let $B = \bigsqcup_{k \in M} \Delta_k$ be an arbitrary open set containing A and Δ_k be the (finitely or countably many) open subintervals that constitute A. By the hypothesis, $\lambda^*(A) = \sum_{k \in M} \lambda^*(A \cap \Delta_k) \leqslant \theta \sum_{k \in M} \lambda(\Delta_k) = \theta\lambda(B)$. Taking the infimum over all such sets B, we get the inequality $\lambda^*(A) \leqslant \theta\lambda^*(A)$, which can hold only if $\lambda^*(A) = 0$.   □

Before we address the proof of the main theorem, we need to make several preparatory remarks. The term "increasing function" will be used here as meaning "non-decreasing function", i.e., we will not require that the function be strictly increasing.

It suffices to prove the theorem for increasing functions: the case of decreasing functions is obtained via multiplication by -1. Let $\Omega = [\omega_1, \omega_2]$ and let $f : \Omega \to \mathbb{R}$ be an increasing function. For any interior point x of the interval Ω we define the following quantities, finite or equal to $+\infty$:

- right upper derivative number $R(x) = \overline{\lim\limits_{t \to x+0}} \dfrac{f(t) - f(x)}{t - x}$;

- right lower derivative number $r(x) = \varliminf\limits_{t \to x+0} \dfrac{f(t) - f(x)}{t - x}$;

- left upper derivative number $L(x) = \overline{\lim\limits_{t \to x-0}} \dfrac{f(x) - f(t)}{x - t}$;

- the left lower derivative number $l(x) = \varliminf\limits_{t \to x-0} \dfrac{f(x) - f(t)}{x - t}$.

To prove the theorem, we need to show that all the derivative numbers listed above are almost everywhere equal to one another and finite. To do this, it in turn suffices to show that for any increasing function f on Ω the following two relations hold:

$$R(x) \overset{\text{a.e.}}{<} \infty, \tag{1}$$

$$R(x) \overset{\text{a.e.}}{\leqslant} l(x). \tag{2}$$

Indeed, applying (2) to the auxiliary function $g(x) = -f(-x)$ and returning to the original function, we get the condition $L(x) \overset{\text{a.e.}}{\leqslant} r(x)$. Combining these conditions with the obvious inequalities $r(x) \overset{\text{a.e.}}{\leqslant} R(x)$ and $l(x) \overset{\text{a.e.}}{\leqslant} L(x)$, we conclude that $R(x) \overset{\text{a.e.}}{\leqslant} l(x) \overset{\text{a.e.}}{\leqslant} L(x) \overset{\text{a.e.}}{\leqslant} r(x) \overset{\text{a.e.}}{\leqslant} R(x)$, i.e., all the inequalities in this chain are in fact equalities.

Proof of Lebesgue's theorem. Let $f : [\omega_1, \omega_2] \to \mathbb{R}$ be an increasing function. Since a monotone function may have only discontinuities of the first kind, we can assume for the sake of convenience that the function f is upper-semicontinuous: it suffices to redefine the function at the points of discontinuity, putting $f(t) = \overline{\lim\limits_{x \to t}} f(x)$. The reader should verify that under this modification the set of differentiability points does not change, and the derivative numbers change in at most countably many points (the discontinuity points of f), i.e., they remain unchanged almost everywhere.

For any $C > 0$ consider the set $R_{>C} = \{x \in (\omega_1, \omega_2) : R(x) > C\}$. To prove the relation $R(x) \overset{\text{a.e.}}{<} \infty$ we need to estimate from above by zero the outer measure of the set R_∞ of all points x of (ω_1, ω_2) for which $R(x) = \infty$. Since $R_\infty \subset R_{>C}$, it suffices to verify that $\lambda^*(R_{>C}) \to 0$ as $C \to \infty$.

Let $x \in R_{>C}$. Then there exists a point $t > x$ such that $(f(t) - f(x))/(t - x) > C$, i.e., $f(t) - Ct > f(x) - Cx$. Hence, the set $R_{>C}$ consists of the points that are invisible from the right for the function $g(x) = f(x) - Cx$. By the rising sun lemma, $R_{>C}$ is contained in the open set $B = \bigsqcup_{k \in M} (a_k, b_k)$, with $g(a_k + 0) \leqslant g(b_k)$. That is,

$$b_k - a_k \leqslant \frac{1}{C}(f(b_k) - f(a_k + 0)) \leqslant \frac{1}{C}(f(b_k) - f(a_k)).$$

The intervals $(f(a_k), f(b_k))$ are disjoint subintervals of $(f(\omega_1), f(\omega_2 - 0))$ (here we use the monotonicity of f). Consequently,

$$\lambda^*(R_{>C}) \leqslant \sum_{k \in M} (b_k - a_k) \leqslant \frac{1}{C} \sum_{k \in M} (f(b_k) - f(a_k)) \leqslant \frac{1}{C}(f(\omega_2 - 0) - f(\omega_1)),$$

(3)

which tends to 0 as $C \to \infty$.

Now let us turn to the proof of the relation $R(x) \overset{\text{a.e.}}{\leqslant} l(x)$. Denote by D the set of all points $x \in (\omega_1, \omega_2)$ in which $R(x) > l(x)$. Further, for any pair of rational numbers (C, c) with $0 < c < C$, denote by $D(C, c)$ the set of all points $x \in (\omega_1, \omega_2)$ in which simultaneously $l(x) < c$ and $R(x) > C$. Since the set of pairs of rational numbers is countable, there are countably many sets $D(C, c)$. Now D is the union of our collection of sets $D(C, c)$, so to show that D is negligible, it suffices to show that all sets $D(C, c)$ have measure zero. To this end we will use the negligibility test provided by Lemma 2, with $\theta = c/C$.

Let $(a, b) \subset \Omega$ be an arbitrary interval, and let $x \in D(C, c) \cap (a, b)$. Since $l(x) < c$, there exists a $t \in (a, x)$ such that $(f(x) - f(t))/(x - t) < c$. Then $f(x) - cx < f(t) - ct$, i.e., the point x is invisible from the left for the function $g(y) = f(y) - cy$ on the interval (a, b). Applying again the rising sun lemma, we see that the set $D(C, c) \cap (a, b)$ is contained in a finite or countable union of disjoint intervals $(\alpha_k, \beta_k) \subset (a, b)$, $k \in N$, and at the endpoints of these intervals $f(\beta_k - 0) - f(\alpha_k) \leqslant c(\beta_k - \alpha_k)$.

Inequality (3) was proved for an increasing function on an arbitrary interval. Recall that $D(C, c) \subset R_{>C}$, and apply condition (3) to the function f on the interval (α_k, β_k):

$$\lambda^*(D(C, c) \cap (\alpha_k, \beta_k)) \leqslant \lambda^*(R_{>C} \cap (\alpha_k, \beta_k))$$
$$\leqslant \frac{1}{C}(f(\beta_k - 0) - f(\alpha_k)) \leqslant \frac{c}{C}(\beta_k - \alpha_k).$$

It remains to use the fact that, by construction,

$$D(C, c) \cap (a, b) = \bigcup_{k \in N} (D(C, c) \cap (\alpha_k, \beta_k)),$$

and the countable semiadditivity of the outer measure:

$$\lambda^*(D(C, c) \cap (a, b)) \leqslant \sum_{k \in N} \lambda^*(D(C, c) \cap (\alpha_k, \beta_k)) \leqslant \frac{c}{C} \sum_{k \in N} (\alpha_k - \beta_k) \leqslant \frac{c}{C}(b - a).$$

Thus, the condition of the negligibility test is satisfied, hence $\lambda^*(D(C, c)) = 0$. $\square$

Exercises

1. Give an example of a continuous monotone function with a dense set of points of non-differentiability.

2. Prove the Borel measurability of all sets that appear in the theorem on the differentiability of monotone functions (the set $R_{>c}$, the set D of all points $x \in (\alpha, \beta)$ such that $R(x) > l(x)$, and so on).

3. Prove **Fubini's theorem on the differentiation of series**: *If the series $\sum_{n=1}^{\infty} f_n$ of increasing functions on an interval converges at every point to the function f, then the series of derivatives $\sum_{n=1}^{\infty} f'_n$ converges almost everywhere to f'.*

4. Let A be a Lebesgue-measurable subset of the interval $[a, b]$. The *density of the set A at the point $x \in [a, b]$* is the limit as $\alpha, \beta \to +0$ (if it exists) of the expression $\lambda([x - \alpha, x + \beta] \cap A)/(\alpha + \beta)$. A point $x \in [a, b]$ is called a *density point of A* if the density at x exists and is equal to 1. Prove that almost every point $x \in A$ is a density point of A (**Lebesgue's density theorem**).

2.3.4 The Difficult Problem of Measure Theory. Existence of Sets that are not Lebesgue Measurable

The *difficult problem of measure theory* we are going to address here is to construct a σ-additive measure μ with $\mu([0, 1]) = 1$, defined on the family of **all** subsets of the interval $[0, 1]$ and invariant under translations, i.e., if both the set A and its translate $A + t$ lie in the interval, then $\mu(A) = \mu(A + t)$. In this subsection we will show that this problem is not solvable, i.e., there is no measure with the required properties. The construction presented here is due to Vitali.

The reasoning plan is natural. Namely, we will assume that a measure μ with the required properties exists, study its properties, and eventually reach a contradiction. First let us note that the measure of any singleton is equal to zero. Indeed, points are obtained from one another by translations, and consequently they all have the same measure, call it α. Assuming that $\alpha > 0$, the measure of the entire interval $[0, 1]$ would be infinite, since the interval has infinitely many points. Hence, $\alpha = 0$. Therefore, we can consider $[0, 1)$ instead of $[0, 1]$ without affecting the measure of subsets. The interval $[0, 1)$ can be regarded as being rolled into a circle. Let us introduce on $[0, 1)$ the operation $+_1$ of addition modulo 1: $a +_1 b$ equals the fractional part of the number $a + b$. On the circle, this operation corresponds to the counter-clockwise rotation of the point $2\pi a$ by the angle $2\pi b$. If $A \subset [0, 1)$ and $t \in [0, 1)$, then instead of the usual translate $A + t$ it is more convenient to consider the translate $A +_1 t$, corresponding to the rotation on the circle, since here one does not need to track whether or not part of the set falls outside the interval. Obviously, $\mu(A) = \mu(A +_1 t)$, since $A +_1 t = (A \cap [0, 1 - t) + t) \bigsqcup (A \cap [1 - t, 1) + t - 1)$, i.e., the set A splits into two parts, one that is translated to the right of the interval $[0, 1)$, and the second to the left of it.

Now let us introduce on $[0, 1)$ the following equivalence relation: $a \sim b$ if $a - b \in \mathbb{Q}$ ($\mathbb{Q}$ denotes, as usual, the set of rational numbers). Pick a single element in each equivalence class and denote the set of selected elements by E. Note that the sets $E +_1 t$ with $t \in \mathbb{Q} \cap [0, 1)$ are pairwise disjoint. Indeed, if $E +_1 t$ intersects $E +_1 \tau$ at the point x, where $t, \tau \in \mathbb{Q} \cap [0, 1)$, then the elements $x -_1 t$ and $x -_1 \tau$, which belong to the same equivalence class, both belong to E, which is impossible by construction. There are infinitely many sets $E +_1 t$; they are obtained from one another by translations, and are disjoint. Hence, their measures must be all equal, and equal to zero. But $[0, 1) = \bigcup_{t \in \mathbb{Q} \cap [0,1)} (E +_1 t)$, whence $\mu([0, 1)) = 0$: contradiction.

Theorem 1. *There exist subsets of the interval $[0, 1]$ that are not Lebesgue measurable.*

Proof. Assuming that all the subsets of $[0, 1]$ are Lebesgue measurable, the Lebesgue measure would be a translation-invariant σ-additive probability measure defined on the family of all subsets of $[0, 1]$. But we have just established that there is no measure with such properties. $\square$

Exercises

1. In the proof of the non-solvability of the difficult problem of measure theory considered above, find the place where the Axiom of Choice was used.

2. Give an example of a countably additive probability measure defined on all the subsets of an interval. (Needless to say, such a measure will not be invariant under translations.)

3. Show that the interval $[0, 1]$ can be written as the union of two disjoint sets A and B such that $\lambda^*(A) = \lambda^*(B) = 1$. Show that both sets A and B must be non-measurable. This will provide another proof of the existence of non-measurable sets.

4. In the setting of the preceding exercise, show that $\lambda^*(A \cap C) = \lambda^*(B \cap C) = \lambda(C)$ for any Lebesgue-measurable set C.

5. If one adopts the Continuum Hypothesis, the interval $[0, 1]$ can be expressed as the disjoint union of a continuum cardinality family of sets that have outer measure equal to 1.

2.3.5 *Distribution Functions and the General form of a Borel Measure on the Interval*

A *Borel measure* on a topological space X is a countably additive measure given on the σ-algebra of all Borel subsets of X. For instance, the Lebesgue measure on the family of Borel subset of the interval $\Omega = [\omega_1, \omega_2]$ is a Borel measure on Ω. In this subsection we will establish a bijective correspondence between the Borel measures on an interval and the increasing right-continuous functions on that interval.

Definition 1. Let μ be a Borel measure on the interval $\Omega = [\omega_1, \omega_2]$. The *distribution function* of the measure μ is the function $F : \Omega \to \mathbb{R}^+$ given by $F(t) = \mu([\omega_1, t])$.

Theorem 1. *The distribution function of a Borel measure on an interval is a (not strictly) increasing right-continuous function.*

Proof. If $\omega_1 \leqslant a < b \leqslant \omega_2$, then $[\omega_1, a] \subset [\omega_1, b]$, and consequently $F(a) = \mu([\omega_1, a]) \leqslant \mu([\omega_1, b]) = F(b)$. Let us prove the right continuity. Let $t_n \in \Omega$ be a decreasing sequence that converges to t. Then $[\omega_1, t_n]$ is a decreasing sequence of sets, and $[\omega_1, t] = \bigcap_{n=1}^{\infty} [\omega_1, t_n]$. It follows that

$$F(t) = \mu([\omega_1, t]) = \lim_{n \to \infty} \mu([\omega_1, t_n]) = \lim_{n \to \infty} F(t_n). \qquad \square$$

Theorem 2. *Let $F : \Omega \to \mathbb{R}^+$ be an increasing right-continuous function on the interval $\Omega = [\omega_1, \omega_2]$. Then there exists a unique Borel measure μ on $[\alpha, \beta]$ that has F as distribution function.*

Proof. We argue by analogy with the construction of the Lebesgue measure. Let Φ be the semiring of all subintervals of the interval Ω. Define the measure μ on Φ as follows: $\mu([\omega_1, a]) = F(a)$, $\mu([a, b]) = F(b) - F(a - 0)$ for $a > \omega_1$ (these two formulas can be combined into one if we adopt the convention that $F(\omega_1 - 0) = 0$), $\mu((a, b]) = F(b) - F(a)$, $\mu((a, b)) = F(b - 0) - F(a)$, and $\mu([a, b)) = F(b - 0) - F(a - 0)$.

These formulas were not chosen arbitrarily: this is precisely the way in which a Borel measure and its distribution function must be related. The finite additivity of the measure μ defined above is readily established. To prove its countable semiadditivity (which in turn will imply its countable additivity), we first note that the measure of any subinterval coincides with the supremum of the measures of all closed subintervals contained in it, as well as with the infimum of the measures of all open subintervals that contain it. Then we must use the finite cover lemma in the same way we did in the proof of the countable semiadditivity of the Lebesgue measure on Φ (Theorem 1, Subsection 2.3.1). To complete the proof, it remains to use the theorem asserting the existence and uniqueness of the extension of a measure from a unital semiring to the σ-algebra it generates (Theorem 3 and Exercises 1–3 in Subsection 2.2.3). $\square$

We provide below a chain of simple exercises, which, once solved, will allow the reader to obtain an important theorem on the structure of monotone functions, namely, the representation as a sum of a continuous function and a step function. (See Subsection 2.1.6 for the necessary definitions and results. In particular, we recall Theorem 2 of that subsection asserting that any atom of a Borel measure reduces to a singleton.)

Exercises

1. The mapping that assigns to each Borel measure on an interval its distribution function is additive, i.e., it takes a sum of measures into the sum of the corresponding distribution functions.

Let M be a finite or countable subset of the interval $[\alpha, \beta]$ and $h: M \to \mathbb{R}^+$ be a function satisfying $\sum_{x \in M} h(x) < \infty$. The *step function* associated to the set M and the function h is defined by the formula $f_{M,h}(t) = \sum_{x \in M \cap [\alpha, t]} h(x)$.

2. To understand why one uses the term "step function", draw the graph of the function $f_{M,h}$ on the interval $[0, 3]$ for $M = \{1, 2\}$ and $h(1) = h(2) = 1$.

3. What is the distribution function of the δ-measure δ_x concentrated in a point $x \in [\alpha, \beta]$?

4. Let μ be a purely atomic Borel measure on the interval $[\alpha, \beta]$. Then the distribution function of μ is a step function.

5. Let μ be a Borel measure on the interval $[\alpha, \beta]$, F its distribution function, and $\alpha < t \leqslant \beta$. Then $\mu(\{t\}) = F(t) - F(t - 0)$, $\mu(\{\alpha\}) = F(\alpha)$.

6. A Borel measure on the interval $[\alpha, \beta]$ is non-atomic if and only its distribution function is continuous and vanishes at the point α.

7. From the representability of a measure as the sum of a non-atomic measure and a purely atomic one it follows that every non-negative, right-continuous, increasing function on the interval $[\alpha, \beta]$ can be uniquely represented at the sum of a continuous increasing function that vanishes at the point α and a step function.

8. Every right-continuous increasing function on the interval $[\alpha, \beta]$ can be represented as the sum of a continuous increasing function and a step function. This representation is unique up to a constant term, meaning that to one of the terms one can add, and from the other subtract, the same number, without affecting the sum. There are no other sources of non-uniqueness.

9. Every increasing function on the interval $[\alpha, \beta]$ is uniquely representable as the sum of a right-continuous function and a function that differs from zero in at most countably many points.

10. Every increasing function f on the interval $[\alpha, \beta]$ is uniquely (up to an additive constant) representable as a sum of three terms, $f_1 + f_2 + f_3$, where f_1 is a continuous function, f_2 a step function, and f_3 a function that differs from zero in at most countably many points.

2.3.6 The Cantor Staircase and a Measure Uniformly Distributed on the Cantor Set

The description of Borel measures on an interval obtained in the preceding subsection may give the impression that all these measures are similar to the Lebesgue measure (at least after one removes atoms). While this impression is in some sense correct, the picture is not as simple as a first glance might suggest. Below we construct a non-atomic probability Borel measure on the interval $[0, 1]$ that is concentrated on the Cantor set: the complement of the Cantor set is negligible from the point of view of that measure. Thus, in some sense this measure has properties opposite to those of the Lebesgue measure: the Lebesgue measure of the Cantor set is equal to 0, while that of the complement of the Cantor set is equal to 1 (Subsection 2.3.1 Example 1). The constructed measure and its distribution function — the Cantor staircase — will provide a source of important examples in the sequel.

As usual, we denote the Cantor set by $\mathcal{K}$, and the intervals of length $1/3^n$ that are removed from $[0, 1]$ at the nth step of the construction of $\mathcal{K}$ by Δ_j^n, $j = 1, 2, \ldots, 2^{n-1}$. For fixed n we label the intervals Δ_j^n in increasing order: $\Delta_1^1 = (1/3, 2/3)$, $\Delta_1^2 = (1/9, 2/9)$, $\Delta_2^2 = (7/9, 8/9)$, and so on.

The main idea of the construction of the sought-for measure is to define its distribution function F so that the measures of all intervals Δ_j^n will be equal to zero, while the measures of the symmetric parts of the set $\mathcal{K}$ will be equal to one another. Thus, $\mathcal{K} = (K \cap [0, 1/3]) \cup (\mathcal{K} \cap [2/3, 1])$, and the parts $\mathcal{K} \cap [0, 1/3]$ and $\mathcal{K} \cap [2/3, 1]$ are symmetric, so it is natural to set their measures equal to $1/2$. By the same symmetry considerations, the measures of the sets $\mathcal{K} \cap [0, 1/9]$, $\mathcal{K} \cap [2/9, 3/9]$, $\mathcal{K} \cap [6/9, 7/9]$, and $\mathcal{K} \cap [8/9, 1]$ must be equal to $1/4$, and so on. Accordingly, we define the distribution function F as follows: we put $F(t) = \frac{1}{2}$ on Δ_1^1, $F(t) = 1/4$ on Δ_1^2, $F(t) = 3/4$ on $\Delta_2^2, \ldots$, and on Δ_j^n we put $F(t) = (2j - 1)/2^n, \ldots$. We have thus defined the distribution function on a dense set, namely, the complement of $\mathcal{K}$. It is readily seen that this function F is uniformly continuous on $[0, 1] \setminus \mathcal{K}$: if $|x - y| < 1/3^n$, then $|F(x) - F(y)| \leqslant 1/2^n$. It follows (Subsection 1.3.4) that F extends uniquely to a continuous function on the whole interval. The resulting monotone continuous function is called the *Cantor staircase* and is denoted by $F_{\mathcal{K}}$. The Borel measure $\mu_{\mathcal{K}}$ on $[0, 1]$ with distribution function $F_{\mathcal{K}}$ is referred to as a *measure, uniformly distributed on the Cantor set*.

Exercises

1. To understand the origin of the term "Cantor staircase", draw the graph of the function $F_{\mathcal{K}}$.

2. Provide the explicit expression of the value of $F_{\mathcal{K}}(t)$ in terms of the triadic expansion of the number t.

3. Prove that the image of the Cantor set under the function $F_\mathcal{K}(t)$ is the whole interval $[0, 1]$.

4. Let Ω_1 be a set, (Ω, Σ, μ) a measure space, $f : \Omega_1 \to \Omega$ a surjective map. Then the family of sets $\Sigma_1 = \{ f^{-1}(A) : A \in \Sigma \}$ is a σ-algebra on Ω, and the rule $\mu_1 \left(f^{-1}(A) \right) = \mu(A)$ defines a countably additive measure μ_1 on Σ_1. If the measure space (Ω, Σ, μ) is complete, is the measure space $(\Omega_1, \Sigma_1, \mu_1)$ necessarily complete?

5. For a function $f : [0, 1] \to [0, 1]$, call the point $x \in [0, 1]$ a *sticking point* if its preimage $f^{-1}(x)$ consists of more than one point. Show that the set of sticking points of a monotone function is at most countable.

6. Let $f : [0, 1] \to [0, 1]$ be an increasing function. Then for any collection of subsets $A_n \subset [0, 1]$, $n \in M$, the symmetric difference of the sets $\bigcup_{n \in M} f(A_n)$ and $f \left(\bigcup_{n \in M} A_n \right)$ is at most countable.

7. Let $f : [0, 1] \to [0, 1]$ be an increasing function. Then the family of all subsets $A \subset [0, 1]$ for which $f(A)$ is a Borel set is a σ-algebra containing all the intervals.

8. From the preceding exercise it follows that the image of a Borel set under a monotone function is again a Borel set.

9. Let μ be a non-atomic Borel measure on $[0, 1]$, F its distribution function, and λ the Lebesgue measure. Then $\mu(A) = \lambda(F(A))$ for any Borel subset $A \subset [0, 1]$.

2.3.7 σ-Finite Measures and the Lebesgue Measure on the Real Line

In many problems it is reasonable to allow a measure to take not only finite positive values, but on some sets also the value $+\infty$. One of these generalizations is the notion of a σ-finite measure.

Definition 1. Let Σ be a σ-algebra of subsets of a set Ω. A map $\mu : \Sigma \to [0, +\infty]$ is called a σ-*finite measure* if it satisfies the following axioms:

1. Countable additivity: $\mu \left(\bigsqcup_{k=1}^{\infty} A_k \right) = \sum_{k=1}^{\infty} \mu(A_k)$ for all $A_k \in \Sigma$.

2. σ-finiteness: the whole set Ω can be written as $\Omega = \bigsqcup_{k=1}^{\infty} A_k$, where $A_k \in \Sigma$ and $\mu(A_k) < \infty$.

A typical example of a σ-finite measure is the *Lebesgue measure on the real line*.

Definition 2. A set $A \subset \mathbb{R}$ is said to be Lebesgue measurable if its intersection with any finite interval is a Lebesgue-measurable subset of that interval. The Lebesgue measure of a set A is defined by means of the measures of the intersections of A with finite intervals: $\lambda(A) = \sum_{n=-\infty}^{+\infty} \lambda(A \cap [n, n+1))$.

A triple (Ω, Σ, μ), where Ω is a set endowed with a σ-algebra Σ of subsets and μ is a σ-finite measure on Σ, is called a *space with σ-finite measure* or a *σ-finite measure space*. The ordinary measure spaces are also called *spaces with finite measure*, or *finite measure spaces* whenever one needs to emphasize that the measure is finite.

Exercises

1. The Lebesgue-measurable subsets of the real line form a σ-algebra, and the Lebesgue measure on the line is a σ-finite measure.

2. Any Borel subset of the real line is Lebesgue measurable.

3. For measurable subsets A of the real line the Lebesgue measure can be calculated by the formulas

$$\lambda(A) = \lim_{n,m \to \infty} \lambda(A \cap [-n, m]) = \lim_{n \to \infty} \lambda(A \cap [-n, n]).$$

4. The Lebesgue measure of an open subset of the real line is equal to the sum of the lengths of the intervals composing the subset.

5. The Lebesgue measure of a Lebesgue-measurable subset A of the real line coincides with its outer measure $\lambda^*(A) = \inf\{\lambda(B) : B$ open, $B \supset A\}$.

6. Every Lebesgue-measurable subset of the real line can be written as the union of a Borel set and a set of measure zero.

7. Let M be an index set and $(\Omega_n, \Sigma_n, \mu_n)$, $n \in M$, be finite measure spaces such that the sets Ω_n are pairwise disjoint. Put $\Omega = \bigcup_{n \in M} \Omega_n$, define the σ-algebra Σ as the family of all sets of the form $A = \bigcup_{n \in M} A_n$ with $A_n \in \Sigma_n$, and put $\mu(A) = \sum_{n \in M} \mu(A_n)$. Under what conditions is (Ω, Σ, μ) a σ-finite measure space? A finite measure space?

8. Let (Ω, Σ, μ) be a σ-finite measure space. Then for any increasing sequence A_n of measurable sets it holds that $\mu\left(\bigcup_{k=1}^{\infty} A_k\right) = \lim_{k \to \infty} \mu(A_k)$.

9. For finite measures we noted earlier the following property (Proposition 2 in Subsection 2.1.4): if $A_n \in \Sigma$, $n = 1, 2, \ldots$, is a decreasing chain of sets (i.e., $A_1 \supset A_2 \supset \cdots \supset A_n \supset \cdots$), then $\mu\left(\bigcap_{k=1}^{\infty} A_k\right) = \lim_{k \to \infty} \mu(A_k)$. On the example of the Lebesgue measure show that for σ-finite measures this property does not hold: there exists a decreasing chain of measurable sets A_n such that $\lambda(A_n) = +\infty$ and $\lambda\left(\bigcap_{k=1}^{\infty} A_k\right) = 0$ (i.e., $\neq \lim_{k \to \infty} \mu(A_k)$). Nevertheless, if one assumes that the measures of the sets A_n are finite, the property also holds in σ-finite measure spaces.

Comments on the Exercises

Subsection 2.1.1

Exercise 2. The smallest algebra of sets on $[0, 1]$ that contains all single-point subsets (singletons) consists of all finite subsets and all subsets (called *cofinite*) whose complement is finite.

Subsection 2.1.2

Exercise 7. Write A as $\bigcap_{n=1}^{\infty} A_n$, where the A_n are open. Then $A_n \supset A$, and hence are dense sets. Accordingly, $X \setminus A = \bigcup_{n=1}^{\infty} (X \setminus A_n)$ and all the sets $X \setminus A_n$ are closed and nowhere dense (see Exercise 1 in Subsection 1.3.6).

Exercise 8. Pass to complements and use the preceding exercise.

Exercise 9. By Exercise 7 in Subsection 2.1.2, $\overline{A} \setminus A$ is a set of first category in $\overline{A}$.

Exercise 11. Any countable subset of an interval is a set of first category, and any dense G_δ-subset of an interval is of second category.

Exercise 12. Let $\underline{f}$ and $\overline{f}$ be the lower and respectively upper envelope of the function f (Exercise 3 in Subsection 1.2.4). Then $\mathrm{dc}(f)$ can be expressed as $\bigcup_{n=1}^{\infty} \left\{ t : \overline{f}(t) - \underline{f}(t) \geqslant 1/n \right\}$, where each of the sets $\left\{ t : \overline{f}(t) - \underline{f}(t) \geqslant 1/n \right\}$ is closed.

Exercise 14. Suppose the functions f_n are continuous and converge to f at all points as $n \to \infty$. Then $f^{-1}([a, +\infty)) = \bigcap_{m=1}^{\infty} \bigcap_{k=1}^{\infty} \bigcup_{n=k}^{\infty} f_k^{-1}((a - \frac{1}{m}, +\infty)) \in G_\delta$.

Exercise 15. Use Exercises 13, 9 and 14.

Exercise 18. See Lemma 1 in Subsection 8.1.1.

Exercise 19. Use Exercise 11.

Subsection 2.1.3

Exercise 1. Fix an open set $U \subset X_1$ and consider the family Σ of all subsets $V \subset X_2$ such that $U \times V \in \mathfrak{B}$. Then Σ is a σ-algebra that contains all the open sets, and so $\Sigma \supset \mathfrak{B}_2$. That is, the Borel sets are precisely the sets of the form $U \times V$, where $U \subset X_1$ is open and $V \in \mathfrak{B}_2$. Now fix $V \in \mathfrak{B}_2$ and consider the family Ψ of all subsets $U \subset X_1$ with the property that $U \times V \in \mathfrak{B}$. Then Ψ is a σ-algebra on X_1 that contains all open sets, and so $\Psi \supset \mathfrak{B}_1$. Hence, all the "rectangles" $A_1 \times A_2$, where $A_1 \in \mathfrak{B}_1$ and $A_2 \in \mathfrak{B}_2$, belong to $\mathfrak{B}$. It follows that $\mathfrak{B}_1 \otimes \mathfrak{B}_2 \subset \mathfrak{B}$.

Exercise 2. For any point $x = (x_1, x_2) \in X_1 \times X_2$, the product $\mathfrak{B}_1 \otimes \mathfrak{B}_2$ contains all neighborhoods of x of the form $B(x_1, r_1) \times B(x_2, r_2)$, i.e., a neighborhood basis of x. In view of the separability, every open set in $X_1 \times X_2$ can be written as a union of sets of the form $B(x_1, r_1) \times B(x_2, r_2)$. Therefore, $\mathfrak{B}_1 \otimes \mathfrak{B}_2$ contains all the open sets, and hence also all the Borel sets in $X_1 \times X_2$.

Exercise 3. In general, the answer may be negative for $X_1 = X_2 = [0, 1]$ endowed with the discrete metric. See Exercise 5 and the comments to it given below.

Exercise 5. The answer relies on axiomatic set theory (say, some version of the Continuum Hypothesis is assumed to hold). In some of the axiomatic systems these families of sets are not identical, see [50]. I am grateful to Taras Banakh for communicating this reference to me.

Exercise 6. Fix $t_1 \in \Omega_1$ and consider the family Σ of all sets $A \subset \Omega_1 \times \Omega_2$ for which $A_{t_1} \in \Sigma_2$. Then Σ is a σ-algebra that contains all the "rectangles". Consequently, $\Sigma \supset \Sigma_1 \otimes \Sigma_2$.

Subsection 2.1.5

Exercises 4–6. Although these problems can be solved directly, the most elegant way of doing this relies on the correspondence between sets and their characteristic functions (Subsection 3.1.3), and the utilization of the normed space $L_1(\Omega, \Sigma, \mu)$ and its completeness (Subsections 6.1.3 and 6.3.2). Namely, the map $A \mapsto \mathbb{1}_A$ reduces all these problems to the analogous problems for the subset $\{\mathbb{1}_A : A \in \Sigma\} \subset L_1(\Omega, \Sigma, \mu)$. The solution of Exercise 4 reduces to proving that $\mu(A \triangle B) = \|\mathbb{1}_A - \mathbb{1}_B\|$, and that of Exercise 6 — to the closedness of $\{\mathbb{1}_A : A \in \Sigma\}$ in the complete metric space $L_1(\Omega, \Sigma, \mu)$.

Subsection 2.1.6

Exercise 2. Suppose $\mu(A_1 \cap A_2) \neq 0$. Since A_1 is an atom, for any subset of nonzero measure of the set A_1 its complement in A_1 has measure 0. Consequently, we have $\mu(A_1 \setminus (A_1 \cap A_2)) = 0$. For the same reason, $\mu(A_2 \setminus (A_1 \cap A_2)) = 0$. Therefore, also $\mu(A_1 \triangle A_2) = 0$.

Exercise 9. The innocent looking "conversely" part is far from easy. Let us sketch the shortest proof known to us, which relies on Hahn's decomposition for charges (signed measures) (Subsection 7.1.2). Let B be an atom for μ'_1. On the σ-algebra Σ_B of subsets of B define the charge ν by $\nu(A) = \mu'_1(A) - \mu'_2(A)$. Let $B = B^+ \sqcup B^-$ be Hahn's decomposition of B for ν, that is, $\nu(A) \geqslant 0$ for all $A \subset B^+$ and $\nu(D) \leqslant 0$ for all $D \subset B^-$. Then for every $A \in \Sigma_{B^+}$ with $\mu(A) > 0$ we have $\mu'_1(A) = \frac{1}{2}(\nu(A) + \mu(A)) > 0$, so $\mu'_1(A) = \mu'_1(B)$ because B was an atom of μ'_1. This applies also to $A = B^+$, and consequently $\mu'_1(B^+) = \mu'_1(B)$. This means that for every $A \in \Sigma_{B^+}$ with $\mu(A) > 0$ we have $\mu'_1(A) = \mu'_1(B^+)$ and so $\mu'_1(B^+ \setminus A) = 0$. Then,

$$\mu_2'(B^+ \setminus A) = \mu_1'(B^+ \setminus A) - \nu(B^+ \setminus A) = -\nu(B^+ \setminus A) \leqslant 0,$$

so also $\mu_2'(B^+ \setminus A) = 0$. Thus, we have shown that $\mu(B^+ \setminus A) = 0$ for every $A \in \Sigma_{B^+}$ with $\mu(A) > 0$, i.e. B^+ is an atom for μ. μ_1'-equivalence of B and B^+ follows from the already established equality $\mu_1'(B^+) = \mu_1'(B)$.

Exercise 15. A much more general result, known as the *Lyapunov convexity theorem*, can be found in Subsection 18.2.4.

Subsection 2.2.1

Exercise 2. Take $\Phi = \left\{ [0, 1], \left[\frac{1}{2}, 1 \right] \right\}$ and put $\mu([0, 1]) = 1$, $\mu \left(\left[\frac{1}{2}, 1 \right] \right) = 2$.

Exercise 3. Take $\Phi = \left\{ \left[0, \frac{2}{3} \right], \left[\frac{1}{3}, 1 \right] \right\}$ and put $\mu \left(\left[0, \frac{2}{3} \right] \right) = \mu \left(\left[\frac{1}{3}, 1 \right] \right) = 1$. Then $\mathbb{A}(\Phi) = \left\{ \emptyset, [0, 1], \left[0, \frac{2}{3} \right], \left[\frac{1}{3}, 1 \right], \left[\frac{1}{3}, \frac{2}{3} \right], \left[0, \frac{1}{3} \right), \left(\frac{2}{3}, 1 \right], \left[0, \frac{1}{3} \right) \sqcup \left(\frac{2}{3}, 1 \right] \right\}$, and one can define the extended measure to have $\mu \left(\left[\frac{1}{3}, \frac{2}{3} \right] \right) = 0$ (and then $\mu \left(\left[0, \frac{1}{3} \right) \right) = \mu \left(\left(\frac{2}{3}, 1 \right] \right) = 1$), or alternatively to have $\mu \left(\left[\frac{1}{3}, \frac{2}{3} \right] \right) = \mu \left(\left[0, \frac{1}{3} \right) \right) = \mu \left(\left(\frac{2}{3}, 1 \right] \right) = \frac{1}{2}$.

Exercise 6. According to a theorem of the author [58], the assertion in question remains valid in a considerably more general setting: for convex sets in Hilbert space and inscribed balls, instead of triangles in the plane and discs. Later a proof in the two-dimensional case was independently obtained by András Bezdek [49].

Subsection 2.2.4

Exercise 10. Use the fact (established in the exercises at the end of the previous Subsection 2.2.3) that the measure space obtained, as described in Sects. 2.2.1–2.2.3, by extending the measure μ from the semiring $\Phi \subset \Sigma_0$, is equal to (Ω, Σ_0', μ). After that check that the proof of the monotone class theorem for sets works in this case as well.

Subsection 2.3.1

Exercise 4. The required set A can be constructed analogously to the Cantor set, with the difference that the removed intervals must be "small", namely, with total length smaller than 1. Moreover, one can ensure that $\lambda(A)$ is arbitrarily close to 1.

Exercise 5. Taking $A = \bigcup_{n=1}^{\infty} A_n$, where A_n are nowhere dense sets with $\lambda(A_n) \geqslant 1 - 1/n$, we obtain a set of first category with $\lambda(A) = 1$. The complement of A is the required set.

Subsection 2.3.3

Exercise 3. See [36, Chap. 1, §2]. Another solution, based on the theory of Lebesgue integration, is outlined in Exercise 6 of Subsection 7.2.4.

Exercise 4. See [36, Chap. 1, §2]. Another, more natural solution based on the derivative of a Lebesgue integral, regarded as a function of the upper integration limit, is outlined in Exercise 1 of Subsection 7.2.4.

Subsection 2.3.4

Exercise 3. In order that $\lambda^*(A) = \lambda^*(B) = 1$, it is necessary and sufficient that both A and B intersect all closed sets of non-zero measure. Since there exist only a continuum of closed subsets of the interval, one can write the closed sets of positive measure as a transfinite sequence K_γ, $\gamma < c$, where c is the smallest ordinal of continuum cardinality. Now for each $\gamma < c$ we choose two distinct points $a_\gamma, b_\gamma \in K_\gamma \setminus (\{a_\beta\}_{\beta<\gamma} \cup \{b_\beta\}_{\beta<\gamma})$. That such a choice is possible is justified by the fact that at each step the set K_γ is a continuum, while the set $\{a_\beta\}_{\beta<\gamma} \cup \{b_\beta\}_{\beta<\gamma}$ of already chosen points has cardinality smaller than that of the continuum. It remains to put $A = \{a_\gamma\}_{\gamma<c}$, $B = [0,1] \setminus A \supset \{b_\gamma\}_{\gamma<c}$.

Exercise 5. One of the possible constructions is provided in the remarks at the end of the proof of [63, Theorem 2.16].

Subsection 2.3.7

Exercise 9. The promised example is $A_n = [n, +\infty)$, $n = 1, 2, \ldots$

Chapter 3
Measurable Functions

Measure and integration theory studies above all real-valued functions. To avoid unnecessary repetition, let us agree that, unless otherwise stipulated, the term "function" will be used for real-valued functions. Thus, when we say "function f on Ω", we mean that f is a function from Ω to $\mathbb{R}$. For functions whose range does not lie in $\mathbb{R}$ we will use the term "map" or "mapping".

The operations on functions will be understood pointwise. For example, $f_1 + f_2$ is the function on Ω given by the rule $(f_1 + f_2)(t) = f_1(t) + f_2(t)$, the function $\max\{f, g\}$ is defined as $\max\{f, g\}(t) = \max\{f(t), g(t)\}$, and so on. The limit of a sequence of functions will also be understood as the pointwise limit.

3.1 Measurable Functions and Operations on Them

In this section (Ω, Σ) will be a set endowed with a σ-algebra of its subsets. All functions, unless otherwise stipulated, will be assumed to be defined on Ω; the elements of the σ-algebra Σ will be referred to as measurable sets.

3.1.1 Measurability Criterion

Definition 1. Let (Ω_1, Σ_1) and (Ω_2, Σ_2) be sets endowed with σ-algebras of subsets. A map $f : \Omega_1 \to \Omega_2$ is said to be *measurable* if $f^{-1}(A) \in \Sigma_1$ for all $A \in \Sigma_2$.

As the definition indicates, measurable maps play in measure theory the same role that continuous maps do in the theory of topological spaces. Particular examples of measurable maps are the *measurable functions* introduced below.

© Springer International Publishing AG, part of Springer Nature 2018
V. Kadets, *A Course in Functional Analysis and Measure Theory*,
Universitext, https://doi.org/10.1007/978-3-319-92004-7_3

Definition 2. A function f on Ω is said to be *measurable* (more specifically, *measurable with respect to the σ-algebra Σ, or Σ-measurable*), if for any Borel subset $A \subset \mathbb{R}$ the set $f^{-1}(A)$ is measurable.

Theorem 1. *Let (Ω_1, Σ_1) and (Ω_2, Σ_2) be sets endowed with σ-algebras of their subsets, and let Λ be a family of subsets of Ω_2 that generates the σ-algebra Σ_2. In order for the map $f : \Omega_1 \to \Omega_2$ to be measurable it is necessary and sufficient that for any set $A \in \Lambda$ its preimage $f^{-1}(A)$ lies in the σ-algebra Σ_1.*

Proof. If f is measurable, then the preimage of any set $A \in \Sigma_2$ lies in Σ_1. In particular, Σ_1 contains the preimages of all sets $A \in \Lambda$.

Conversely, suppose that Σ_1 contains all sets of the form $f^{-1}(A)$ with $A \in \Lambda$. We need to show that the preimages of all elements of the family Σ_2 lie in Σ_1. To do this, we introduce the following family Λ_1 of subsets of the set Ω_2: a set A belongs to Λ_1 if $f^{-1}(A) \in \Sigma_1$. It is readily verified that Λ_1 is a σ-algebra and contains all elements of the family Λ. Since Σ_2 is the smallest σ-algebra containing Λ, it follows that $\Sigma_2 \subset \Lambda_1$, as we needed to show. $\qquad\square$

Let $f : \Omega \to \mathbb{R}$ be a function and $a \in \mathbb{R}$. Denote $f^{-1}((a, +\infty))$ by $f_{>a}$, i.e., $f_{>a}$ is the set of all $t \in \Omega$ at which $f(t) > a$. Since (see Subsection 2.1.2, Proposition 2) the sets $(a, +\infty)$ with $a \in \mathbb{R}$ generate the σ-algebra $\mathfrak{B}$ of Borel set on $\mathbb{R}$, we obtain the following simple measurability criterion:

Corollary 1. *The function $f : \Omega \to \mathbb{R}$ is measurable if and only if all the sets $f_{>a}$ with $a \in \mathbb{R}$ are measurable.* $\qquad\square$

Corollary 2. *Let (Ω, Σ), (Ω_1, Σ_1), and (Ω_2, Σ_2) be sets endowed with σ-algebras of subsets. Endow, as usual, the Cartesian product $\Omega_1 \times \Omega_2$ with the σ-algebra $\Sigma_1 \otimes \Sigma_2$ (see Subsection 2.1.3). Then for any measurable maps $f_1 : \Omega \to \Omega_2$ and $f_2 : \Omega \to \Omega_2$, the map $f : \Omega \to \Omega_1 \times \Omega_2$ given by the rule $f(t) = (f_1(t), f_2(t))$ is also measurable.*

Proof. By definition, the σ-algebra $\Sigma_1 \otimes \Sigma_2$ is generated by the sets $A_1 \times A_2$ with $A_1 \in \Sigma_1$ and $A_2 \in \Sigma_2$. We have $f^{-1}(A_1 \times A_2) = f_1^{-1}(A_1) \cap f_2^{-1}(A_2) \in \Sigma$. $\qquad\square$

If we take for Ω a topological space and for Σ the σ-algebra $\mathfrak{B}$ of Borel sets on Ω, we obtain a particular case of measurability, Borel measurability:

Definition 3. A function f on the topological space Ω is said to be *Borel measurable* if the preimage $f^{-1}(A)$ of any Borel subset A of the real line is a Borel subset of Ω.

As an example of a Borel-measurable function one can take any continuous function. Indeed, for a continuous function f all the sets $f_{>a}$ are open, and hence belong to the σ-algebra $\mathfrak{B}$ of Borel sets, i.e., the above measurability criterion applies.

For an arbitrary set $A \in \Sigma$ we can consider the σ-algebra Σ_A of all measurable subsets of A. If the restriction of the function f to A is measurable with respect to the σ-algebra Σ_A, then f is said to be *measurable on the subset A*.

Exercises

1. If the function f is measurable, then for any $a \in \mathbb{R}$ the sets $f_{\neq a} = \{t \in \Omega : f(t) \neq a\}$, $f_{=a} = \{t \in \Omega : f(t) = a\}$, $f_{\leqslant a} = \{t \in \Omega : f(t) \leqslant a\}$, $f_{<a} = \{t \in \Omega : f(t) < a\}$, and $f_{\geqslant a} = \{t \in \Omega : f(t) \geqslant a\}$ are measurable.

2. Let f be a Borel-measurable function on the interval $[a, b]$. Then the set of maximum points of f is a Borel set.

3. The set of local maximum points of a Borel-measurable function on the real line is a Borel set.

4. Let (Ω_1, Σ_1) and (Ω_2, Σ_2) be sets endowed with σ-algebras of subsets, and let $\Omega_1 \times \Omega_2$ be endowed with the σ-algebra $\Sigma_1 \otimes \Sigma_2$. Prove that the projection maps P_1 and P_2, which send each element $(t_1, t_2) \in \Omega_1 \times \Omega_2$ into its coordinates t_1 and t_2, respectively, are measurable.

5. Prove the converse of Corollary 2: if the map $f : \Omega \to \Omega_1 \times \Omega_2$ given by $f(t) = (f_1(t), f_2(t))$ is measurable, then the maps f_1 and f_2 are also measurable.

6. Show that every monotone function on the real line is Borel measurable.

7. Let f be a Borel-measurable function on the interval $[a, b]$. Then the set of maximum points of f is a Borel set.

8. Let f be a measurable function on Ω. Prove that the functions $|f|$, sign f, f^+, and f^- are measurable.

9. If the function f is measurable, then λf is measurable for any $\lambda \in \mathbb{R}$.

10. Let the function f be measurable on Ω. Then f is measurable on any subset $A \in \Sigma$.

11. Suppose that Ω can be written as the union of two measurable subsets A and B, and the function f is measurable on both A and B. Then f is measurable on Ω.

12. Give an example of a bijective measurable map $f : \Omega_1 \to \Omega_2$ whose inverse is not measurable.

13. Let $g : \mathbb{R} \to \mathbb{R}$ be a continuous function and A be a Lebesgue-measurable set in $\mathbb{R}$.

(a) Is the set $g(A)$ necessarily Borel measurable?

(b) Lebesgue measurable?

(c) Can the set $g^{-1}(A)$ be not Lebesgue measurable?

14. Let $g : \mathbb{R} \to \mathbb{R}$ be a continuous function and A be an open subset of $\mathbb{R}$. Then $g(A)$ is a Borel set. Moreover, $g(A)$ is an F_σ-set.

15. Let $g : \mathbb{R} \to \mathbb{R}$ be a continuous function and A be a Borel set in $\mathbb{R}$. Can the set $g(A)$ be not Borel?

16. Let (Ω, Σ, μ) be a measure space. Two measurable functions f and g on Ω are said to be *equimeasurable*, if $\mu(f_{>a}) = \mu(g_{>a})$ for all $a \in \mathbb{R}$. Show that if f and g are equimeasurable then $\mu(f^{-1}(A)) = \mu(g^{-1}(A))$ for any Borel set A of real numbers.

3.1.2 Elementary Properties of Measurable Functions

Theorem 1. *Let* (Ω_1, Σ_1), (Ω_2, Σ_2), *and* (Ω_3, Σ_3) *be sets endowed with σ-algebras of subsets, and let* $f : \Omega_1 \to \Omega_2$ *and* $g : \Omega_2 \to \Omega_3$ *be measurable maps. Then the composition* $g \circ f : \Omega_1 \to \Omega_3$ *is also a measurable map.*

Proof. Let $A \in \Sigma_3$. Then $g^{-1}(A) \in \Sigma_2$, and so $(g \circ f)^{-1}(A) = f^{-1}(g^{-1}(A)) \in \Sigma_1$, as needed.   □

Corollary 1.

1. *Suppose the function* $f : \Omega \to \mathbb{R}$ *is measurable and the function* $g : \mathbb{R} \to \mathbb{R}$ *is Borel measurable. Then the composition* $g \circ f$ *is also measurable.*

2. *In particular, if* $f : \Omega \to \mathbb{R}$ *is measurable and* $g : \mathbb{R} \to \mathbb{R}$ *is continuous, then* $g \circ f$ *is measurable.*

3. *Suppose the functions* $f_1, f_2 : \Omega \to \mathbb{R}$ *are measurable, and the function* $g : \mathbb{R}^2 \to \mathbb{R}$ *of two variables is continuous. Then the function* $f(t) = g(f_1(t), f_2(t))$ *is measurable.*

Proof. Only item 3 requires a proof. Consider the plane $\mathbb{R}^2 = \mathbb{R} \times \mathbb{R}$, endowed with the σ-algebra of Borel sets, or, which is the same, with the product of the σ-algebras of Borel sets on the line $\mathbb{R}$. By Corollary 2 in the preceding subsection, the function $F : \Omega \to \mathbb{R}^2$ defined by the rule $F(t) = (f_1(t), f_2(t))$ is measurable. It remains to note that $f = g \circ F$ and apply the preceding theorem.   □

Theorem 2. *The class of measurable functions on* (Ω, Σ) *enjoys the following properties: if the functions* f *and* g *are measurable, then so are the functions* $f + g$, fg, $\max\{f, g\}$, *and* $\min\{f, g\}$. *Moreover, the functions* $|f|$, $\operatorname{sign} f$, $f^+ = \max\{f, 0\}$, $f^- = (-f)^+$, *and* λf *with any* $\lambda \in \mathbb{R}$ *are measurable. If* f *does not vanish at any point, then the function* $1/f$ *is measurable.*

Proof. The functions $g_1(x, y) = x + y$ and $g_2(x, y) = xy$ of two variables are continuous, and so are the functions $\max\{x, y\}$ and $\min\{x, y\}$. By item 3 of the last corollary, this implies that the functions $f + g$, fg, $\max\{f, g\}$, and $\min\{f, g\}$ are measurable. The continuity of the functions $|t|$, t^+, t^-, and λt, in conjunction with item 2 of the preceding corollary, guarantee the measurability of the functions $|f|$,

f^+, f^-, and λf. The measurability of the function sign f follows from item 1 of the same corollary and the Borel measurability of the function sign t. Finally, if f does not vanish at any point, then the function $1/f$ can be represented as the composition of the measurable function $f : \Omega \to \mathbb{R} \setminus \{0\}$ (where $\mathbb{R} \setminus \{0\}$ is endowed with the σ-algebra of Borel sets), and the continuous — and hence Borel-measurable — function $1/t : \mathbb{R} \setminus \{0\} \to \mathbb{R}$. $\qquad\square$

Theorem 3. *Suppose the sequence (f_n) of measurable functions converges pointwise to a function f, i.e., for any $t \in \Omega$, $f_n(t) \to f(t)$ as $n \to \infty$. Then f is a measurable function.*

Proof. Fix a number $a \in \mathbb{R}$. The value of the function f at the point $t \in \Omega$ is larger than a if and only if there exist a rational number $r \in \mathbb{Q}$ and a number $n \in \mathbb{N}$ such that for any $m > n$ it holds that $f_m(t) > a + r$. Translating this statement into the language of measure theory, we conclude that $f_{>a} = \bigcup_{r \in \mathbb{Q}} (\bigcup_{n=1}^{\infty} \bigcap_{m=n+1}^{\infty} (f_m)_{>a+r} \in \Sigma$. $\qquad\square$

Applying this theorem to the sequence of partial sums of a series we obtain the following statement.

Corollary 2. *If a series of measurable functions converges pointwise, then its sum is a measurable function.* $\qquad\square$

Exercises

1. Prove directly that if the functions f and g are measurable, then for any $a \in \mathbb{R}$ the set $(f + g)_{>a}$ belongs to Σ. According to the criterion in the preceding section, this will provide another proof of the measurability of the sum of two measurable functions.

2. Express the sets $(\max\{f, g\})_{>a}$ and $(\min\{f, g\})_{>a}$ in terms of the analogous sets for the functions f and g.

3. If the functions f and g are measurable, then the sets of points $t \in \Omega$ in which $f = g$, $f \neq g$, $f > g$, and $f < g$, respectively, are measurable.

4. Let (f_n) be a pointwise bounded sequence of measurable functions. Then the functions $f = \sup_n f_n$ and $g = \overline{\lim}_{n \to \infty} f_n$ are also measurable.

5. Let A denote the set of all differentiability points of the function f on the line (see Exercise 13 in Subsection 2.1.2). Show that the function f' is Borel-measurable on A.

6. Identify in the standard way the field $\mathbb{C}$ of complex numbers with the plane $\mathbb{R}^2$, and endow $\mathbb{C}$ with the σ-algebra of Borel subsets of the plane. A measurable map $f : \Omega \to \mathbb{C}$ is called a *measurable complex-valued function*. Prove that $f : \Omega \to \mathbb{C}$ is measurable if and only if the real-valued functions $\operatorname{Re} f$ and $\operatorname{Im} f$ are measurable.

7. Prove the following properties of complex-valued functions:

(1) if the functions f and g are measurable, then so is their sum $f + g$;

(2) if the function f is measurable, then so is λf for any $\lambda \in \mathbb{C}$;

(3) if the functions f and g are measurable, then so is their product fg;

(4) if the function f is measurable, then $|f|$ is a measurable real-valued function.

3.1.3 The Characteristic Function of a Set

Let Ω be a set and A be a subset of Ω. The *characteristic function* of the set A is the function $\mathbb{1}_A$ on Ω equal to 1 on A and equal to zero on the complement $\Omega \setminus A$ of A. Alternative notations found in the literature are χ_A and I_A. We note that the last notation is most frequently encountered in probability theory, where the characteristic function of a set is called the *indicator* of that set, and the term "characteristic function" is used for a completely different object. Of course, it would be reasonable, in the notation for the characteristic function, to account not only for the set A, but also for the ambient set Ω. For instance, one and the same set A of real numbers can be regarded as a subset of an interval in one situation, and as a subset of the real line in another. In the first case the function $\mathbb{1}_A$ is defined on the interval, and in the second on the real line, and the same symbol is used in both situations. This slight ambiguity does not have unpleasant consequences: here, like in many other situations, a function defined on a subset is tacitly extended to the ambient set by zero.

The properties listed in Exercises 1–5 below will be used in the sequel, and for this reason the reader is advised to pay close attention to them.

Exercises

1. Let (Ω, Σ) be a set endowed with a σ-algebra of subsets, and $A \subset \Omega$. The function $\mathbb{1}_A$ is measurable if and only if the set A is measurable.

2. $\mathbb{1}_{A \cup B} = \max\{\mathbb{1}_A, \mathbb{1}_B\}$.

3. $\mathbb{1}_{A \cap B} = \min\{\mathbb{1}_A, \mathbb{1}_B\} = \mathbb{1}_A \cdot \mathbb{1}_B$.

4. If the sets A and B are disjoint, then $\mathbb{1}_{A \cup B} = \mathbb{1}_A + \mathbb{1}_B$.

5. Let $A = \bigsqcup_{n=1}^{\infty} A_n$. Then $\mathbb{1}_A = \sum_{n=1}^{\infty} \mathbb{1}_{A_n}$.

6. Let (A_n) be a sequence of sets. Then $\varlimsup_{n \to \infty} \mathbb{1}_{A_n}$ is the characteristic function of a set A, called the *upper limit* of the sequence (A_n). Express the set A in terms of the sets A_n by means of the usual operations of union and intersection.

7. Consider the set $2^\mathbb{N}$ of all subsets of the natural numbers with the topology described in Exercise 7 of Subsection 1.4.4. Verify that a sequence of sets converges in this topology if and only if the characteristic functions of the sets converge pointwise to the corresponding characteristic function.

3.1.4 Simple Functions. Lebesgue Approximation of Measurable Functions by Simple Ones. Measurability on the Completion of a Measure Space

Let (Ω, Σ) be a set endowed with a σ-algebra. A function f on Ω is called *simple* if it can be represented as $f = \sum_{n=1}^{\infty} a_n \mathbb{1}_{A_n}$, where $A_n \in \Sigma$ is a disjoint sequence of sets and a_n are numbers. Since the sets A_n are disjoint, the series $\sum_{n=1}^{\infty} a_n \mathbb{1}_{A_n}$ does not merely converge pointwise: for any point $t \in \Omega$ all the terms of the series, except possibly for one (with the index n for which $t \in A_n$), vanish at t. On each of the sets A_n the function f is equal to the constant a_n, and $f(t) = 0$ in the complement of the union of all A_n. Simple functions are also called *countably-valued functions* or, in more detail, *countably-valued measurable functions*. This terminology is justified by the following assertion.

Theorem 1. *The function f is simple if and only if it is measurable and the set of its values (i.e., its image, or range) is at most countable.*

Proof. The measurability of a simple function $f = \sum_{n=1}^{\infty} a_n \mathbb{1}_{A_n}$ can be verified directly (the preimage of any set under f is a finite or countable union of some of the sets A_n); alternatively, one can refer to the measurability of the sum of a series of measurable functions. Further, $f(\Omega) \subset \{a_n\}_{n=1}^{\infty} \cup \{0\}$, which shows that the set of all values of f is at most countab le. Conversely, suppose f is measurable and the set M of its values is at most countable. Then for any $t \in M$, the set $f^{-1}(t)$ is measurable and $f = \sum_{t \in M} t \mathbb{1}_{f^{-1}(t)}$. $\qquad\square$

If the set of values of a simple function is finite, then the function is said to be *finitely-valued*.

Theorem 2. *The classes of finitely-valued and countably-valued functions are stable under taking sums and products, as well as the maximum and the minimum of two functions.*

Proof. We already know that the listed operations preserve measurability. Now let f and g be two functions on Ω, and let $f(\Omega)$ and $g(\Omega)$ be their images. If $f(\Omega)$ and $g(\Omega)$ are finite (countable), then the sets

$$f(\Omega) + g(\Omega) = \{t + r : t \in f(\Omega), \, r \in g(\Omega)\}$$

and

$$f(\Omega) \cdot g(\Omega) = \{t \cdot r : t \in f(\Omega), \, r \in g(\Omega)\}$$

are finite (respectively, countable). The assertion of the theorem follows from the fact that the images of the functions $f + g$, fg, $\max\{f, g\}$, and $\min\{f, g\}$ lie in $f(\Omega) + g(\Omega)$, $f(\Omega) \cdot g(\Omega)$, $f(\Omega) \cup g(\Omega)$, and $f(\Omega) \cup g(\Omega)$, respectively.   □

Measurable functions can have a rather complicated structure. For this reason, to facilitate the study of their structure one uses approximations of measurable functions by simple functions.

Theorem 3. *Let f be a measurable function on Ω. Then for any $\varepsilon > 0$ there exists a simple function $f_\varepsilon \leqslant f$ which at all points differs from f by at most ε. Moreover, if $f \geqslant 0$, then f_ε can also be chosen to be non-negative, and if f is bounded, then for f_ε one can take a finitely-valued function.*

Proof. For each integer n introduce the number $t_n = n\varepsilon$ and the intervals $\Delta_n = [t_n, t_{n+1})$. Denote the set $f^{-1}(\Delta_n)$ by A_n. Some of the sets A_n may be empty. In particular, if $f \geqslant 0$, then all the A_n with index $n < 0$ are empty. Further, if f is bounded in modulus by some constant C, then all the A_n with $|n| > (C/\epsilon) + 1$ are empty. The sets A_n are pairwise disjoint, their union is the whole Ω, and on A_n the values of the function f satisfy the inequalities $t_n \leqslant f(t) < t_{n+1}$. We define the function f_ε so that its value on A_n is equal to the corresponding t_n: $f_\varepsilon = \sum_{n=1}^{\infty} t_n \mathbb{1}_{A_n}$.

The function f_ε defined in this way enjoys all the properties stated in the theorem. Indeed, on each A_n we have $t_n = f_\varepsilon(t) \leqslant f(t) < t_{n+1}$, i.e., $f(t) - \varepsilon < f_\varepsilon(t) \leqslant f(t)$ at all points $t \in \Omega$. If $f \geqslant 0$, then f_ε cannot take negative values t_n: the sets A_n that correspond to negative t_n will be empty. If f is bounded, then all the A_n, except for a finite number of them, will be empty, and so f_ε will be finitely-valued.   □

Corollary 1. *For any measurable function f there exists a non-decreasing sequence $f_1 \leqslant f_2 \leqslant \cdots$ of simple functions which converges uniformly to f. If, in addition, f is non-negative (bounded), then the functions f_n can be chosen to be non-negative (respectively, finitely-valued).*

Proof. We use the preceding theorem and chose a simple function f_1 such that $0 \leqslant f - f_1 \leqslant 1$. The function $f - f_1$ is measurable and non-negative, so by the preceding theorem there exists a simple non-negative function g_1 which satisfies the inequalities $0 \leqslant f - f_1 - g_1 \leqslant 1/2$. Put $f_2 = f_1 + g_1$. Then $f_1 \leqslant f_2$ and $0 \leqslant f - f_2 \leqslant 1/2$. The function $f - f_2$ is again measurable and non-negative, and so one can approximate it by a simple function g_2: $0 \leqslant f - f_2 - g_2 \leqslant 1/3$. Naturally, we define the function f_3 as $f_2 + g_2$. Continuing this process, we obtain an increasing sequence of simple functions satisfying the conditions $0 \leqslant f - f_n \leqslant 1/n$, which ensures that the sequence converges uniformly. Ensuring that the additional non-negativity or finite-valuedness requirements in the statement of the corollary are satisfied presents no difficulty.   □

The proof of the next result is based on the fact that measurable functions can be approximated by simple ones.

Theorem 4. *Let (Ω, Σ, μ) be a measure space and (Ω, Σ', μ) be its completion. Then for any Σ'-measurable function f on Ω, there exists a Σ-measurable function g that coincides with f almost everywhere.*

Proof. First we will prove this assertion for simple functions. Let $f = \sum_{n=1}^{\infty} a_n \mathbb{1}_{A_n}$, where the sets A_n belong to the σ-algebra Σ' and are disjoint. In each of the sets A_n we choose a subset $B_n \in \Sigma$ for which $\mu(A_n \setminus B_n) = 0$ (see Exercise 3 in Subsection 2.1.5). Then $g = \sum_{n=1}^{\infty} a_n \mathbb{1}_{B_n}$ is the sought-for function. Now let f be an arbitrary Σ'-measurable function, let (f_n) be a sequence of simple Σ'-measurable functions that converges pointwise to f, and finally let g_n be Σ-measurable functions that coincide almost everywhere with the corresponding f_n. Denote by $A \subset \Omega$ the negligible set in the complement of which $f_n = g_n$ for all $n = 1, 2, \ldots$. By the definition of negligible sets, there exists a Σ-measurable set B of null measure such that $B \supset A$. Consider the full-measure set $C = \Omega \setminus A$. The functions $g_n \cdot \mathbb{1}_C$ are Σ-measurable, converge on C to f, and vanish in the complement of C. That is, the functions $g_n \cdot \mathbb{1}_C$ converge pointwise to $g = f \cdot \mathbb{1}_C$, and, by Theorem 3 of Subsection 3.1.2, this limit function is Σ-measurable. It remains to observe that $g = f$ almost everywhere, since the set B where this equality can fail is negligible. $\square$

Exercises

1. The function f_ε figuring in the statement of Theorem 3 can be chosen so that $f_\varepsilon(\Omega) \subset f(\Omega)$.

2. Let X be a metric space endowed with the σ-algebra of Borel sets, and let $f: \Omega \to X$ be a measurable map. Then the following conditions are equivalent:

— for every $\varepsilon > 0$ there exists a countably-valued map $f_\varepsilon: \Omega \to X$ such that $\rho(f(t), f_\varepsilon(t)) \leqslant \varepsilon$ for all $t \in \Omega$;

— the set $f(\Omega)$ is separable.

3. In the setting of the preceding exercise, the following conditions are equivalent:

— for every $\varepsilon > 0$, there exist a finitely-valued measurable map $f_\varepsilon: \Omega \to X$ such that $\rho(f(t), f_\varepsilon(t)) \leqslant \varepsilon$ for all $t \in \Omega$;

— the set $f(\Omega)$ is precompact.

4. The map f_ε in the two preceding exercises can be chosen so that it will satisfy $f_\varepsilon(\Omega) \subset f(\Omega)$.

5. Show that for every Lebesgue-measurable function f on the interval one can find an equimeasurable decreasing function $\widetilde{f}$ (for the definition of equimeasurability, see Exercise 16 in Subsection 3.1.1). This function $\widetilde{f}$ is called a *decreasing rearrangement* of the function f.

3.2 Main Types of Convergence

In this section (Ω, Σ, μ) will be a fixed finite measure space, and the functions f, f_n, and all the others will be assumed, unless otherwise stipulated, to be defined on Ω, measurable, and real-valued.

3.2.1 Almost Everywhere Convergence

The sequence of functions (f_n) is said to *converge almost everywhere* to the function f (written $f_n \xrightarrow{\text{a.e.}} f$) if the set of all points $t \in \Omega$ at which the numerical sequence $f_n(t)$ does not converge to $f(t)$ as $n \to \infty$ is negligible.

We note the following elementary properties of almost everywhere convergence, the verification of which is left to the reader.

A. If $f_n \xrightarrow{\text{a.e.}} f$ and $f_n \xrightarrow{\text{a.e.}} g$, then $f \overset{\text{a.e.}}{=} g$.

B. If $f_n \xrightarrow{\text{a.e.}} f$ and $f_n \overset{\text{a.e.}}{=} g_n$, then $g_n \xrightarrow{\text{a.e.}} f$.

C. If $f_n \xrightarrow{\text{a.e.}} f$, $g_n \xrightarrow{\text{a.e.}} g$, and $f_n \overset{\text{a.e.}}{\leqslant} g_n$, then $f \overset{\text{a.e.}}{\leqslant} g$.

D. If $G: \mathbb{R}^2 \to \mathbb{R}$ is a continuous function, $f_n \xrightarrow{\text{a.e.}} f$ and $g_n \xrightarrow{\text{a.e.}} g$, then $G(f_n, g_n) \xrightarrow{\text{a.e.}} G(f, g)$. This implies, in particular, the theorems on the limit of a sum and of a product.

Almost everywhere convergence plays an important role in the theory of the Lebesgue integral. Under relatively mild additional assumptions (see Subsection 4.4) the integral of the limit function can be calculated as the limit of the integrals of the terms of the sequence. Moreover, almost everywhere convergence is in many respects far more convenient to work with than the usual pointwise convergence. First of all, it is a more general type of convergence, so it is easier to verify. Next, here, as in general when one deals with properties that hold almost everywhere, we can ignore the behavior of functions on negligible sets. For example, for a piecewise-continuous or for a monotone function it is not at all necessary to define the values in discontinuity points, as they have no influence whatsoever on almost everywhere convergence! On the other hand, almost everywhere convergence has an essential drawback: this convergence is not generated by a metric or topology, so there is no natural way of defining a "rate of convergence" for it. Let us give an example of a problem where this drawback shows up.

Definition 1. Let X and Y be two families of measurable functions on Ω. We say that X is a.e. *dense in* Y (dense in the sense of almost everywhere convergence) if for any $f \in Y$ there exists a sequence (f_n) of elements of the family X such that $f_n \xrightarrow{\text{a.e.}} f$.

Theorem 1. *Suppose that X is a.e. dense in Y and Y is a.e. dense in Z. Then X is a.e. dense in Z.* □

This natural property is important not only from the point of view of the inner harmony of the theory of almost everywhere convergence, but also from the point of view of applications. For instance, it enables one to show that the family of continuous functions on an interval is a.e. dense in the family of all Lebesgue-measurable functions on that interval. Although these results can be established using only the definition of almost everywhere convergence, devising such proofs is far from simple (we invite the reader to have a try at it!). If, on the contrary, the convergence had been given by some topology, the problem would have been rather trivial (see Exercise 4 in Subsection 1.2.1). Fortunately, here the following subtle idea comes to the rescue. As it turns out, the space of measurable functions carries a topology for which the notion of denseness of a subset coincides precisely with a.e. denseness, though the convergence in it (the so-called convergence in measure) is not equivalent to almost everywhere convergence. The study of this topology and the corresponding type of convergence is addressed next.

3.2.2 Convergence in Measure. Examples

Let a and ε be strictly positive numbers, f a measurable function. We denote by $U_{a,\varepsilon}(f)$ the set of all measurable functions g for which $\mu\left(|g - f|_{>a}\right) < \varepsilon$. (Here, as earlier, the symbol $h_{>a}$ stands for the set of all points $t \in \Omega$ at which $h(t) > a$). The *topology of convergence in measure* on the space of all measurable functions on Ω is the topology in which a neighborhood basis of f is provided by the sets $U_{a,\varepsilon}(f)$ with $a, \varepsilon > 0$. Accordingly, a sequence of functions (f_n) is said to *converge in measure* to the function f (written $f_n \xrightarrow{\mu} f$) if for any $a > 0$,

$$\mu\left(|f_n - f|_{>a}\right) \to 0 \quad \text{as } n \to +\infty.$$

Theorem 1. *Convergence in measure enjoys the following properties:*

A. $f_n \xrightarrow{\mu} f$ *if and only if* $f_n - f \xrightarrow{\mu} 0$.

B. *If* $f_n \xrightarrow{\mu} f$ *and* $f_n \xrightarrow{\mu} g$, *then* $f \overset{a.e.}{=} g$.

C. *If* $f_n \xrightarrow{\mu} f$ *and* $f_n \overset{a.e.}{=} g_n$, *then* $g_n \xrightarrow{\mu} f$.

Proof. Properties A and C are obvious. We prove property B. Let A be the set of all points $t \in \Omega$ at which $f(t) \neq g(t)$, and A_n the set of all points $t \in \Omega$ at which $|f(t) - g(t)| > 1/n$. Since $A = \bigcup_{n \in \mathbb{N}} A_n$, it suffices to show that $\mu(A_n) = 0$ for all n. For any $k \in \mathbb{N}$, at each point $t \in A_n$ either $|f(t) - f_k(t)| > 1/(2n)$, or $|g(t) - f_k(t)| > 1/(2n)$. Hence, if we denote by $B_{n,k}$ the set of all points at which $|f(t) - f_k(t)| > 1/(2n)$, and by $C_{n,k}$ the set of all points where $|g(t) - f_k(t)| > 1/(2n)$, then $A_n \subset B_{n,k} \cup C_{n,k}$. By the definition of convergence in measure, for fixed n and $k \to \infty$, the measures of the sets $B_{n,k}$ and $C_{n,k}$ tend to 0. Hence, $\mu(A_n)$ can only be 0. $\square$

Theorem 2. *Let X be a family of measurable functions on Ω. Then every point in the closure of X in the topology of convergence in measure is the limit of a sequence of elements of X that converges in measure.*

Proof. We use here the idea of Exercise 6 of Subsection 1.2.1. Let f be a point in the closure of the set X. Note that the neighborhood $U_{a,\varepsilon}(f)$ increases with the growth of a, as well as with the growth of ε. Consider the neighborhoods $U_n = U_{1/n,1/n}(f)$. Clearly, $U_1 \supset U_2 \supset \cdots$ and together the sets U_n constitute a neighborhood basis of f (if $U_{a,\varepsilon}(f)$ is an arbitrary neighborhood of f, then $U_{a,\varepsilon}(f) \supset U_n$ for $n >$ max $\{1/a, 1/\varepsilon\}$). By the definition of the closure, all sets $X \cap U_n$ are non-empty. Pick in each set $X \cap U_n$ an element f_n. Then (f_n) is the sought-for sequence of elements of the set X that converges in measure to f. $\square$

Example 1 (sliding hump). In the interval $[0, 1]$ consider the subintervals $I_{n,k} = [\frac{k-1}{2^n}, \frac{k}{2^n}], n = 0, 1, 2, \ldots, k = 1, \ldots, 2^n$. For fixed n, the intervals $I_{n,k}, k = 1, \ldots, 2^n$, cover the whole interval $[0, 1]$. Now consider the sequence of functions $f_1 = \mathbb{1}_{[0,1]}$, $f_2 = \mathbb{1}_{[0,1/2]}$, $f_3 = \mathbb{1}_{[1/2,1]}$, $\ldots, f_{2^n+k} = \mathbb{1}_{I_{n,k}}, \ldots$. For each $a > 0$, the set of points $x \in [0, 1]$ where $|f_{2^n+k}(x)| > a$ is either empty (if $a \geqslant 1$), or coincides with $I_{n,k}$. Since the lengths of the intervals $I_{n,k}$ tend to zero when $n \to \infty$, the sequence (f_n) tends to zero in measure (with respect to the Lebesgue measure). At the same time, the sequence (f_n) does not tend to zero at *any point*, since every point of the interval $[0, 1]$ belongs to infinitely many intervals $I_{n,k}$. This example allows one to get a feeling for the meaning of the convergence in measure, and at the same time shows that convergence in measure is not equivalent to almost everywhere convergence.

Exercises

1. In the preceding example, find a subsequence of the sequence (f_n) that tends to 0 at every point.

2. Why are the sets $|f_n - f|_{>a}$ in the definition of convergence in measure measurable?

3. Verify that our definition of convergence in measure is correct, i.e., that convergence in the topology of convergence in measure is indeed equivalent to the condition appearing in the definition.

4. If $f_n \xrightarrow{\mu} f, g_n \xrightarrow{\mu} g$, and $f_n \overset{\text{a.e.}}{\leqslant} g_n$, then $f \overset{\text{a.e.}}{\leqslant} g$.

5. On the segment $[0, 1]$ consider the sequence of functions $g_n(x) = x^n$. Show that $g_n \xrightarrow{\mu} 0$ (in the sense of the Lebesgue measure). Does this sequence converge to zero pointwise? Almost everywhere?

6. Flesh out the proof of Theorem 2.

7. $\mu\left(|f - h|_{>a}\right) \leqslant \mu\left(|f - g|_{>\frac{a}{2}}\right) + \mu\left(|g - h|_{>\frac{a}{2}}\right)$ for any measurable functions f, g, h and any $a > 0$.

8. Let $f_n \xrightarrow{\mu} f$ and $g_n \xrightarrow{\mu} g$. Then $f_n + g_n \xrightarrow{\mu} f + g$.

9. By definition, (f_n) is a Cauchy sequence in the sense of convergence in measure if $\mu(|f_n - f_m|_{>a}) \to 0$ as $n, m \to \infty$. Prove that any sequence that converges in measure is a Cauchy sequence in the above sense.

10. The sequence of functions $\sin(\pi n x)$ on $[0, 1]$ does not tend in measure to any function; moreover, it does not contain a subsequence that converges in measure.

11. Let f_n be an increasing sequence of functions and let $f_n \xrightarrow{\mu} f$. Then $f_n \xrightarrow{\text{a.e.}} f$.

12. The expression $\rho(f, g) = \inf_{a \in (0, +\infty)} \left\{a + \mu(|f - g|_{>a})\right\}$ is a pseudometric that generates the topology of convergence in measure.

13. Another example: the pseudometric $d(f, g) = \inf\left\{a > 0 : \mu(|f - g|_{>a}) \leqslant a\right\}$ also gives the topology of convergence in measure.

14. Let (Ω, Σ, μ) be a finite measure space and let the measure μ be purely atomic. Then for functions on Ω convergence in measure is equivalent to convergence almost everywhere. If μ is not purely atomic, then these two types of convergence are not equivalent.

3.2.3 Theorems Connecting Convergence in Measure to Convergence Almost Everywhere

Definition 1. The *upper limit of a sequence of sets* (A_n) is the set $\overline{\lim}\, A_n = \bigcap_{n=1}^{\infty} \bigcup_{k=n}^{\infty} A_k$.

Another commonly used name for the same object is the *limit superior*, with the corresponding notation $\lim \sup_{n \to \infty} A_n$. That our use of the terms "upper limit" or "limit superior" is natural will become clear once Exercise 6 in Subsection 3.1.3 is solved.

Lemma 1 (on the upper limit of a sequence of sets). *Let $A_n \in \Sigma$ and $A_\infty = \overline{\lim}\, A_n$. Then*

(i) $\mu(A_\infty) \geqslant \overline{\lim}\, \mu(A_n)$. *In particular, if $\mu(A_\infty) = 0$, then $\mu(A_n) \to 0$ as $n \to \infty$.*

(ii) *If $\sum_{n=1}^{\infty} \mu(A_n) < \infty$, then $\mu(A_\infty) = 0$.*

Proof. Consider the sets $B_n = \bigcup_{k=n}^{\infty} A_k$. Then $A_\infty = \bigcap_{n=1}^{\infty} B_n$. Since the sets B_n form a decreasing chain,

$$\lim_{n\to\infty} \mu(B_n) = \mu(A_\infty). \tag{3.1}$$

To prove assertion (i), it remains to note that $B_n \supset A_n$, and $\mu(B_n) \geqslant \mu(A_n)$. Further, if $\sum_{n=1}^{\infty} \mu(A_n) < \infty$, then $\mu(B_n) \leqslant \sum_{k=n}^{\infty} \mu(A_k) \to 0$ as $n \to \infty$, which in view of (3.1) yields assertion (ii). $\square$

We note that in probability theory the assertion (ii) of the preceding lemma is known as the "Borel–Cantelli lemma".

Theorem 1 (Lebesgue). *Convergence almost everywhere implies convergence in measure. Precisely, if f, f_n are measurable functions on Ω and $f_n \to f$ almost everywhere, then $f_n \xrightarrow{\mu} f$.*

Proof. By hypothesis, the set D of all points at which f_n does not converge to f is negligible (of measure zero). Fix $a > 0$. Consider the sets $A_n = |f_n - f|_{>a}$ and $A_\infty = \overline{\lim} A_n$. By the definition of the upper limit, $A_\infty = \bigcap_{n=1}^{\infty} \bigcup_{k=n}^{\infty} A_k$, i.e., A_∞ is the set of all points $t \in \Omega$ with the property that for any $n \in \mathbb{N}$ there exists a $k > n$ such that $|f_n(t) - f(t)| > a$. Hence, $A_\infty \subset D$ and $\mu(A_\infty) = 0$. By the preceding lemma, $\mu(A_n) \to 0$ as $n \to \infty$, i.e., $\mu(|f_n - f|_{>a}) \to 0$ as $n \to \infty$. $\square$

Lemma 2. *Let f_n be measurable functions, and a_n and ε_n be positive numbers such that $a_n \to 0$ as $n \to \infty$ and $\sum_{n=1}^{\infty} \varepsilon_n < \infty$. Moreover, suppose that f_n satisfy the condition $\mu\left(|f_n|_{>a_n}\right) < \varepsilon_n$. Then $f_n \xrightarrow{a.e.} 0$.*

Proof. Denote by D the set of all points where f_n does not tend 0, and set $A_n = |f_n|_{>a_n}$, $B_n = \bigcup_{k=n}^{\infty} A_k$, $A_\infty = \overline{\lim} A_n = \bigcap_{n=1}^{\infty} B_n$. Let $t \in \Omega$ be an arbitrary point such that $f_n(t)$ does not tend to zero. For any $n \in \mathbb{N}$, there exists $k \geqslant n$ such $f_k(t) > a_k$, that is, $t \in B_n$. Hence, $D \subset B_n$ for all n, and $D \subset A_\infty$. At the same time, $\sum_{n=1}^{\infty} \mu(A_n) < \sum_{n=1}^{\infty} \varepsilon_n < \infty$ by hypothesis. Applying assertion (ii) of the lemma on the upper limit of sequence of sets, we conclude that $\mu(D) \leqslant \mu(A_\infty) = 0$. $\square$

Theorem 2 (F. Riesz). *Any sequence of measurable functions that converges in measure contains a subsequence that converges almost everywhere.*

Proof. Suppose that $f_n \xrightarrow{\mu} f$. Fix $a_n, \varepsilon_n > 0$, such that the conditions of the preceding lemma are satisfied, and choose an increasing sequence of indices m_n such that $\mu\left(|f_{m_n} - f|_{>a_n}\right) < \varepsilon_n$. By Lemma 2, $f_{m_n} - f \xrightarrow{a.e.} 0$, hence $f_{m_n} \xrightarrow{a.e.} f$. $\square$

Theorem 3 (convergence in measure criterion). *The sequence of measurable functions* (f_n) *converges in measure to the function* f *if and only if any subsequence of the sequence* (f_n), *in its turn, contains a subsequence that converges to* f *almost everywhere.*

Proof. Suppose $f_n \xrightarrow{\mu} f$. Then each subsequence of the sequence (f_n) also converges in measure, so by the preceding theorem, it contains a subsequence that converges to f almost everywhere. Conversely, suppose that f_n does not converge in measure to f. Then there exist a, $\varepsilon > 0$ and a subsequence (g_n) of (f_n) such that none of the functions g_n lies in the neighborhood $U_{a,\varepsilon}(f)$. It follows that the subsequence (g_n) does not contain subsequences that converge in measure to f, and hence, by Theorem 1, neither does it contain subsequences that converge almost everywhere to f.   □

Corollary 1. *If* $G : \mathbb{R}^2 \to \mathbb{R}$ *is a continuous function,* $f_n \xrightarrow{\mu} f$ *and* $g_n \xrightarrow{\mu} g$, *then* $G(f_n, g_n) \xrightarrow{\mu} G(f, g)$. *In particular, it follows that* $f_n + g_n \xrightarrow{\mu} f + g$ *and* $f_n g_n \xrightarrow{\mu} fg$.

Proof. Use the preceding criterion and the corresponding properties of convergence almost everywhere.   □

Corollary 2 (**Theorem 1 in Subsection** 3.2.1). *Let X, Y and Z be sets of measurable functions on* Ω. *If X is a.e. dense in Y and Y is a.e. dense in Z, then X is a.e. dense in Z.*

Proof. By Theorem 1, in the topology of convergence in measure X is dense in Y and Y is dense in Z. Therefore (Exercise 4 of Subsection 1.2.1), X is dense in Z in the topology of convergence in measure. Hence, by Theorem 2 of Subsection 3.2.2, the set X is *sequentially dense* in Z in the sense of convergence in measure, i.e., for any $f \in Z$ there exists a sequence (f_n) of elements of the set X such that $f_n \xrightarrow{\mu} f$. It remains to apply Theorem 2.   □

Exercises

1. Solve Exercise 4 in Subsection 3.2.2 based on the results obtained in the current subsection.

2. Let (f_n) be a Cauchy sequence in the sense of convergence in measure (see Exercise 9 in Subsection 3.2.2). Then (f_n) contains a subsequence that converges almost everywhere.

3. If a sequence of measurable functions is Cauchy in the sense of convergence in measure, then it has a limit in the same sense.

4. Suppose that in some space X of measurable functions on a finite measure space almost everywhere convergence coincides with convergence in some topology τ on X. Then in X almost everywhere convergence coincides with convergence in measure.

5. Almost everywhere convergence in the space of all measurable functions on an interval cannot be given by a topology.

6. The subset of all continuous functions is a.e. dense in the space of all measurable functions on an interval.

7. Let (A_n) be a decreasing chain of sets. Then $\overline{\lim} A_n = \bigcap_{n=1}^{\infty} A_n$ and $\mu(\overline{\lim} A_n) = \lim_{n \to \infty} \mu(A_n)$.

8. For any increasing chain of sets A_n it also holds that $\mu(\overline{\lim} A_n) = \lim_{n \to \infty} \mu(A_n)$, because in this case $\overline{\lim} A_n = \bigcup_{n=1}^{\infty} A_n$.

9. Give an example in which $\mu(\overline{\lim} A_n) \neq \overline{\lim}_{n \to \infty} \mu(A_n)$.

10. A point $t \in \Omega$ belongs to $\overline{\lim} A_n$ if and only if t belongs to infinitely many of the sets A_n.

11. Consider the functions $f_n = \mathbb{1}_{(n,\infty)}$ on $\mathbb{R}$. Verify that f_n converge almost everywhere on $\mathbb{R}$ to 0 but does not converge in measure. This example shows that Theorem 1 does not extend to σ-finite measure spaces.

12. Let (Ω, Σ, μ) be a σ-finite measure space, then any sequence (f_n) of measurable functions on Ω that converges in measure to a measurable function f contains a subsequence that converges to f almost everywhere. In other words, Theorem 2 remains valid in σ-finite measure spaces.

3.2.4   Egorov's Theorem

The functions $g_n(x) = x^n$ on the interval $[0, 1]$ provide a typical example of a sequence that converges at each point, but does not converge uniformly. At the same time, the convergence can be improved if one removes an arbitrarily small neighborhood of the point 1: on the remaining interval $[0, 1 - \varepsilon]$ the convergence will already be uniform. A similar situation arises in the theory of power series: a series converges to its sum uniformly not in the entire disc of convergence, but in any disc of a slightly smaller radius. These facts are particular cases of the following very general result.

Theorem 1 (Egorov's theorem). *Suppose that* $f_n \xrightarrow{\text{a.e.}} f$ *on* Ω. *Then for every* $\varepsilon > 0$ *there exists a set* $A = A_\varepsilon \in \Sigma$ *with* $\mu(A) < \varepsilon$, *on the complement of which* (f_n) *converges uniformly to* f.

Proof. Fix $a_n, \varepsilon_n > 0$ such that $a_n \to 0$ as $n \to \infty$ and $\sum_{n=1}^{\infty} \varepsilon_n < \varepsilon$. Consider the sets $A_{m,n} = |f_m - f|_{>a_n}$ and $B_{m,n} = \bigcup_{k=m}^{\infty} A_{k,n}$. For fixed n, the sets $B_{m,n}$ form a chain decreasing with m, and $\mu \left(\bigcap_{m=1}^{\infty} B_{m,n} \right) = 0$ (since $\bigcap_{m=1}^{\infty} B_{m,n}$ is included in the negligible set D consisting of all points at which f_n does not tend to f). Consequently, $\mu(B_{m,n}) \to 0$ as $m \to \infty$. Now for each n pick an index m_n such that $\mu(B_{m_n,n}) < \varepsilon_n$. Let us prove that $A = \bigcup_{n=1}^{\infty} B_{m_n,n}$ is the required set. First, $\mu(A) \leqslant \sum_{n=1}^{\infty} \varepsilon_n < \varepsilon$. Further, $\Omega \setminus A \subset \Omega \setminus B_{m_n,n}$, that is, for every $k > m_n$ the set $A_{k,n} = |f_k - f|_{>a_n}$ does not contain points of $\Omega \setminus A$. It follows that $\sup_{t \in \Omega \setminus A} |f_k(t) - f(t)| \leqslant a_n$ for $k > m_n$, which establishes the uniform convergence on $\Omega \setminus A$. $\qquad\square$

Exercises

1. Use Exercise 6 in Subsection 3.2.3 and Egorov's theorem to obtain the following result: **Luzin's theorem.** *For any Lebesgue-measurable function f on the interval $[a,b]$ and any $\varepsilon > 0$ there exists a measurable set A with $\mu(A) < \varepsilon$, such that the restriction of f to $[a, b] \setminus A$ is continuous.*

2. Show that in the statement of Luzin's theorem the set A can be chosen to be open.

3. In the statement of Egorov's theorem, can the condition $\mu(A) < \varepsilon$ be replaced by $\mu(A) = 0$? What about the analogous question for Luzin's theorem?

4. In the statement of Egorov's theorem, can the sequence f_n, which converges almost everywhere, be replaced by a sequence which converges in measure?

5. Where in Egorov's theorem did the measurability of the involved functions play a role?

Comments on the Exercises

Subsection 3.1.1

Exercise 2. Denote the supremum of the values of the function f on $[a, b]$ by m. Then the set of maximum points of f coincides with $f_{=m}$.

Exercise 3. Write all intervals with rational endpoints as a sequence (a_n, b_n), $n \in \mathbb{N}$, and denote the set of points of "true" maximum of the function f on (a_n, b_n) by M_n. The sought-for set of local maxima of f coincides with $\bigcup_{n=1}^{\infty} M_n$.

Exercise 4. Take as (Ω_1, Σ_1) the interval $[0, 1]$ endowed with the σ-algebra of Lebesgue-measurable sets, and take for (Ω_2, Σ_2) the same interval with the σ-algebra of Borel sets, and for f the identity map.

Exercise 13. (a) No (even for the function $g(x) = x$).

(b) No. Let g be the Cantor staircase (Subsection 2.3.6), extended to $(-\infty, 0)$ by 0, and to $(1, +\infty)$ by 1. Let $B \subset [0, 1]$ be a set that is not Lebesgue measurable. With no loss of generality, we may assume that B consists only of irrational points (otherwise one can replace B by $B \setminus \mathbb{Q}$). As the required A take $g^{-1}(B)$. Then A is a subset of the Cantor set, whence $\lambda(A) = 0$, so A is Lebesgue measurable. However, $f(A) = B$ is not measurable.

(c) It can. To produce an example, one needs to come up with a continuous strictly monotone function which maps some set of positive measure into a set of measure 0.

Exercise 14. One needs to represent A as the union of a sequence of compact sets and recall that the image of a compact set under a continuous map is again compact.

Exercise 15. It can. The author is not aware of a simple example. A set that is the image of a Borel set under a continuous map is called an *analytic set*, or a *projective set of class* 1. The existence of an analytic set that is not Borel is a particular case of Theorem VI in §38 of the monograph [25, vol. 1].

Subsection 3.2.3

Exercise 6. Continuous functions can be used to approximate characteristic functions of intervals; linear combinations of characteristic functions of intervals can in turn be used to approximate characteristic functions of open sets; then characteristic functions of open sets to approximate characteristic functions of arbitrary Lebesgue-measurable sets; then linear combinations of characteristic functions of measurable sets (i.e., finitely-valued functions) to approximate simple functions; and finally, simple functions to approximate arbitrary measurable functions. An analogous statement will be proved in a considerably more general situation in Subsection 8.3.3.

Exercise 12. Write Ω as a disjoint union of sets Ω_m, $m = 1, 2, \ldots$, of finite measure. Successively applying on each set Ω_m the theorem asserting that from any sequence that converges in measure one can extract an almost-everywhere convergent subsequence, we construct an infinite sequence of sets of indices $\mathbb{N} \supset N_1 \supset N_2 \supset N_3 \supset \cdots$ such that on each of the sets Ω_m the sequence $\{f_n\}_{n \in N_m}$ converges almost everywhere. Picking a diagonal subsequence n_m (i.e., one for which $n_1 \in N_1$, $n_2 \in N_2$ and $n_2 > n_1$, $n_3 \in N_3$ and $n_3 > n_2$, and so on), we obtain a subsequence f_{n_m} which converges almost everywhere on each set Ω_j, i.e., converges almost everywhere on $\Omega = \bigsqcup_{j=1}^{\infty} \Omega_j$.

Subsection 3.2.4

Exercise 1. In a more general situation Luzin's theorem will be proved in Subsection 8.3.3.

Chapter 4
The Lebesgue Integral

4.1 Convergence Along a Directed Set. Partitions

4.1.1 Directed Sets

We recall that a relation $\succ$ on a set G is called an *order relation*, or simply an *order*, if it satisfies the following conditions:

1. $g \succ g$ for any $g \in G$ (reflexivity).
2. If $g_2 \succ g_1$ and $g_1 \succ g_2$, then $g_1 = g_2$ (antisymmetry).
3. If $g_2 \succ g_1$ and $g_3 \succ g_2$, then $g_3 \succ g_1$ (transitivity).

A set G endowed with a binary relation $\succ$ is called a *directed set* if the following axioms are satisfied:

(a) $g \succ g$ for any $g \in G$;
(b) if $g_2 \succ g_1$ and $g_3 \succ g_2$, then $g_3 \succ g_1$;
(c) for any two elements $g_1, g_2 \in G$ there exists an element $g_3 \in G$ such that $g_3 \succ g_1$ and $g_3 \succ g_2$.

We note that often when the notion of a directed set is defined one requires that the relation $\succ$ is an order relation, whereas in our definition a directed set does not have to satisfy Condition 2 of order relations.

Exercise

1. In which of the examples below is the relation $\succ$ on the set $\mathbb{Z}$ of integers an order relation? In which of these examples is $(\mathbb{Z}, \succ)$ a directed set?

© Springer International Publishing AG, part of Springer Nature 2018
V. Kadets, *A Course in Functional Analysis and Measure Theory*,
Universitext, https://doi.org/10.1007/978-3-319-92004-7_4

1. $n_1 \succ n_2$, if $n_1 > n_2$.

2. $n_1 \succ n_2$, if $n_1 \geqslant n_2$.

3. $n_1 \succ n_2$, if $n_1 \leqslant n_2$.

4. $n_1 \succ n_2$, if $|n_1| \geqslant |n_2|$.

5. $n_1 \succ n_2$, if $n_1 \geqslant n_2$ and n_1 is divisible by n_2.

6. $n_1 \succ n_2$, if $n_1 \geqslant n_2$ and $n_1 - n_2$ is divisible by 2.

Let $(G, \succ)$ be a directed set. Two elements $g_1, g_2 \in G$ are said to be equivalent (written $g_1 \sim g_2$), if $g_2 \succ g_1$ and $g_1 \succ g_2$.

7. Verify that "$\sim$" is an equivalence relation.

4.1.2   Limit Along a Directed Set. Cauchy's Criterion

Let $(G, \succ)$ be a directed set and $f : G \to \mathbb{R}$ a function. The number $a \in \mathbb{R}$ is called the *limit of the function f along the directed set* $(G, \succ)$ if for any $\varepsilon > 0$ there exists a $g \in G$ such that $|f(g_1) - a| < \varepsilon$ for any $g_1 \succ g$. In this case one writes $a = \lim_{(G,\succ)} f$, or, if the directed set is clear from the context, $a = \lim_g f(g)$. The function $f : G \to \mathbb{R}$ is said to *converge along the directed set* $(G, \succ)$ if $\lim_{(G,\succ)} f$ exists.

Let us list a number of simple properties of the limit along a directed set.

1. If $a = \lim_{(G,\succ)} f$, then for any $g \in G$ and any $\varepsilon > 0$ there exists an element $g_1 \succ g$ such that $|f(h) - a| < \varepsilon$ for all $h \succ g_1$.

2. If $a = \lim_{(G,\succ)} f$ and $b = \lim_{(G,\succ)} f$, then $a = b$ (uniqueness of the limit).

3. Suppose that for the functions f_1 and f_2 there exists a $g \in G$ such that $f_1(h) = f_2(h)$ for all $h \succ g$. If one of these functions converges along the directed set $(G, \succ)$, then so does the other, and $\lim_{(G,\succ)} f_1 = \lim_{(G,\succ)} f_2$. It follows that for the existence of the limit it is not necessary that the function be defined on the entire set G: it suffices that it is defined for all h that succeed some fixed element $g \in G$.

4. Suppose that $f_1 \leqslant f_2$ and the limits of the functions f_1 and f_2 along the directed set G exists. Then $\lim_{(G,\succ)} f_1 \leqslant \lim_{(G,\succ)} f_2$.

5. Let $a_1 = \lim_{(G,\succ)} f_1$, $a_2 = \lim_{(G,\succ)} f_2$, and suppose the function of two variables $F : \mathbb{R}^2 \to \mathbb{R}$ is continuous at the point (a_1, a_2). Then $\lim_g F(f_1(g), f_2(g))$ exists and is equal to $F(a_1, a_2)$.

6. If $\lim_{(G,\succ)} f$ exists, then for any scalar $t \in \mathbb{R}$ the limit $\lim_{(G,\succ)} tf$ exists, and $\lim_{(G,\succ)} tf = t \lim_{(G,\succ)} f$.

7. If $a_1 = \lim_{(G,\succ)} f_1$ and $a_2 = \lim_{(G,\succ)} f_2$, then $a_1 + a_2 = \lim_{(G,\succ)}(f_1 + f_2)$.

For example, let us prove property 5 (incidentally, properties 6 and 7 are consequences of property 5). Fix $\varepsilon > 0$ and choose $\delta > 0$ such that for any point $(b_1, b_2) \in \mathbb{R}^2$, if $\max\{|a_1 - b_1|, |a_2 - b_2|\} < \delta$, then $|F(a_1, a_2) - F(b_1, b_2)| < \varepsilon$. Since $a_1 = \lim_{(G,\succ)} f_1$, there exists a $g \in G$ such that $|f_1(h) - a_1| < \delta$ for all $h \succ g$.

Further, since $a_2 = \lim_{(G,\succ)} f_2$, then by the first of the properties listed above, there exists a $g_1 \succ g$ such that $|f_2(h) - a_2| < \delta$ for all $h \succ g_1$. Then for any $h \succ g_1$ one simultaneously has that $|f_1(h) - a_1| < \delta$ and $|f_2(h) - a_2| < \delta$. Hence, for any $h \succ g_1$ one has $|F(a_1, a_2) - F(f_1(h), f_2(h))| < \varepsilon$, as we needed to prove.

Theorem 1 (Cauchy's criterion for convergence along a directed set). *For the function $f: G \to \mathbb{R}$ to converge along the directed set $(G, \succ)$ it is necessary and sufficient that for every $\varepsilon > 0$ there exist an element $g \in G$ such that $|f(g) - f(h)| < \varepsilon$ for all $h \succ g$.*

Proof. Necessity. Suppose f converges along the directed set $(G, \succ)$ and $\lim_g f(g) = a$. By the definition of the limit, for every $\varepsilon > 0$ there exists a $g \in G$ such that $|f(h) - a| < \varepsilon/2$ for all $h \succ g$. Then for any h that succeeds g one has $|f(g) - f(h)| < |f(g) - a| + |a - f(h)| < \varepsilon$.

Sufficiency. Let us first use the condition of the theorem with $\varepsilon = 1$. Let $g_1 \in G$ be such that $|f(g_1) - f(h)| < 1$ for all $h \succ g_1$. Now we use the condition with $\varepsilon = 1/2$. Denote by g_2 an element such that $g_2 \succ g_1$ and $|f(g_2) - f(h)| < 1/2$ for all h succeeding g_2. Continuing this reasoning, we obtain a sequence $g_1 \prec g_2 \prec g_3 \prec \cdots$ such that $|f(g_n) - f(h)| < 1/n$ for all $h \succ g_n$. In particular, $|f(g_n) - f(g_m)| < 1/n$ for all $m, n \in \mathbb{N}$, $m > n$. Therefore, the numerical sequence $(f(g_n))$ satisfies the Cauchy condition, and so it converges. Denote $\lim_{n \to \infty} f(g_n)$ by a. We claim that $\lim_g f(g) = a$. Indeed, for any $\varepsilon > 0$ there exists an $n_0 \in \mathbb{N}$ such that $2/n_0 < \varepsilon$. By construction, for every $h \succ g_{n_0}$ we have $|f(g_{n_0}) - f(h)| < 1/n_0$. In particular, since $g_n \succ g_{n_0}$ for all $n > n_0$, we see that for any $n > n_0$ and any $h \succ g_{n_0}$,

$$|f(g_n) - f(h)| \leqslant |f(g_n) - f(g_{n_0})| + |f(g_{n_0}) - f(h)| < \frac{1}{n_0} + \frac{1}{n_0} < \varepsilon.$$

Letting $n \to \infty$ in the obtained inequality $|f(g_n) - f(h)| < \varepsilon$, we conclude that $|a - f(h)| \leqslant \varepsilon$ for all $h \succ g_{n_0}$. □

Exercises

1. Consider $\mathbb{R}$ with the natural directed set structure: $a \succ b$ if $a \geqslant b$. Verify that the limit of a function with respect to this directed set coincides with the usual limit as $t \to +\infty$.

2. Describe other examples of limits known from calculus, such as $\lim_{t \to -\infty}$, $\lim_{t \to \infty}$, $\lim_{t \to a}$, and $\lim_{t \to a-0}$, as limits along the corresponding directed sets.

3. The Riemann integral is defined as a limit of integral sums. Write this kind of limit as a limit along a directed set.

4. Let $\mathbb{N}_f$ denote the family of all finite subsets of the set $\mathbb{N}$ of natural numbers. We say that the finite set A succeeds the finite set B if $A \supset B$. Show that, equipped with this order relation, $\mathbb{N}_f$ is a directed set.

5. Let (a_n) be an arbitrary sequence of numbers. Define the function $s : \mathbb{N}_f \to \mathbb{R}$ by the formula $s(A) = \sum_{n \in A} a_n$. Show that the function s has a limit along the directed set $\mathbb{N}_f$ if and only if the series $\sum_{n=1}^{\infty} a_n$ is absolutely convergent. In this case $\lim_A s(A) = \sum_{n=1}^{\infty} a_n$.

6. Define the limit along a directed set for functions taking values in an arbitrary topological space. Then show that the Cauchy criterion holds for convergence along a directed set for functions with values in complete metric spaces.

4.1.3 Partitions

From this point on till the end of Sect. 4.5, (Ω, Σ, μ) will be a finite measure space and A a measurable subset of Ω (i.e., $A \in \Sigma$). Unless otherwise stipulated, the functions f, f_n, will be defined on A and take real values.

Let $A \in \Sigma$ be an arbitrary non-empty set. A *partition* of the set A is a finite or countable collection D of pairwise disjoint non-empty measurable subsets $\Delta_k \subset A$, $k = 1, 2, \ldots$, such that $\bigcup_k \Delta_k = A$. To avoid treating each time separately the cases when the number of elements in a partition is finite or countable, henceforth we will write partitions as countable collections of measurable sets, with the understanding that the collection may also be finite.

A partition $D = \{\Delta_k\}_{k=1}^{\infty}$ of the set A is said to be *admissible* for the function f if for every element $\Delta_k \in D$ of non-zero measure, $\sup_{t \in \Delta_k} |f(t)| < \infty$ and $\sum_{k=1}^{\infty} \sup_{t \in \Delta_k} (|f(t)| \mu(\Delta_k)) < \infty$.

By definition, the partition $D_1 = \{\Delta_k^1\}_{k=1}^{\infty}$ *succeeds* the partition $D_2 = \{\Delta_k^2\}_{k=1}^{\infty}$ if D_1 is a *refinement* of the partition D_2. In other words, $D_1 \succ D_2$ if for any $k, j \in \mathbb{N}$, if Δ_k^1 intersect Δ_j^2, then $\Delta_k^1 \subset \Delta_j^2$. One also says that the partition D_1 is *finer* than D_2, or *refines* D_2.

Theorem 1. *If the partition $D = \{\Delta_k\}_{k=1}^{\infty}$ of the set A is admissible for the function f, then so is any finer partition $D_1 = \{\Delta_k^1\}_{k=1}^{\infty} \succ D$.*

Proof. Group together the sets Δ_k^1 that lie in the same element of the partition D:

$$\sum_{k=1}^{\infty} \sup_{t \in \Delta_k^1} |f(t)| \mu(\Delta_k^1) = \sum_{j=1}^{\infty} \sum_{\Delta_k^1 \subset \Delta_j} \sup_{t \in \Delta_k^1} |f(t)| \mu(\Delta_k^1)$$

$$\leq \sum_{j=1}^{\infty} \sup_{t \in \Delta_j} |f(t)| \sum_{\Delta_k^1 \subset \Delta_j} \mu(\Delta_k^1) = \sum_{j=1}^{\infty} \sup_{t \in \Delta_j} |f(t)| \mu(\Delta_j) < \infty.$$

The theorem is proved.   □

The verification of the following properties of admissible partitions is left to the reader.

1. Suppose the partition D is admissible for the function f, and $a \in \mathbb{R}$ is an arbitrary scalar. Then the partition D is admissible for the function af.

2. Suppose the partition D is simultaneously admissible for the functions f and g. Then D is admissible for the function $f + g$.

Let $D = \{\Delta_k\}_{k=1}^{\infty}$ be a partition of the set A. A sequence $T = \{t_k\}_1^{\infty} \subset \Omega$ is called a *selection of marked points* for D if $t_k \in \Delta_k$ for every $k \in \mathbb{N}$. Let (D_1, T_1) and (D_2, T_2) be partitions with respective collections of marked points. By definition, we say that the pair (D_1, T_1) *succeeds* the pair (D_2, T_2) if D_1 succeeds D_2.

Exercises

1. Show that for two partitions $D_1 = \{\Delta_k^1\}_{k=1}^{\infty}$ and $D_2 = \{\Delta_k^2\}_{k=1}^{\infty}$ of the set A the following conditions are equivalent:

(a) $D_1 \succ D_2$;

(b)   for every $k \in \mathbb{N}$ there exists a $j \in \mathbb{N}$ such that $\Delta_k^1 \subset \Delta_j^2$;

(c)   for every $j \in \mathbb{N}$ there exists a subset of indices $J \subset \mathbb{N}$ such that $\bigcup_{k \in J} \Delta_k^1 = \Delta_j^2$.

2. Show that the relation $\succ$ introduced on the set of partitions of the set A is an order relation.

3. Let $D_1 = \{\Delta_k^1\}_{k=1}^{\infty}$ and $D_2 = \{\Delta_k^2\}_{k=1}^{\infty}$ be partitions of the set A. Define a new partition D_3 by arranging in a sequence all the non-empty sets of the form $\Delta_k^1 \cap \Delta_j^2$, $k, j \in \mathbb{N}$. Show that D_3 refines both D_1 and D_2; hence, the family of all partitions of the set A is a directed set.

4. Let D_1, D_2, and D_3 be the partitions in the preceding exercise. Show that if the partition D refines both D_1 and D_2, then $D \succ D_3$.

5. Show that the family of all pairs (D, T) of partitions with marked points is a directed set.

4.2 Integrable Functions

4.2.1 Integral Sums

Definition 1. Let $A \in \Sigma$, $f : A \to \mathbb{R}$ be a function, $D = \{\Delta_k\}_{k=1}^{\infty}$ be an admissible partition of A, and $T = \{t_k\}_1^{\infty}$ be a collection of marked points. The *integral sum* of the function f on the set A, associated with the pair (D, T), is the number

$$S_A(f, D, T) = \sum_{k=1}^{\infty} f(t_k)\mu(\Delta_k).$$

Note that the admissibility of the partition D guarantees the absolute convergence of the series $\sum_{k=1}^{\infty} f(t_k)\mu(\Delta_k)$ in the definition of the integral sum. This absolute convergence is necessary for the integral sum to depend on the partition with marked points itself, and not on the order in which the elements of the partition are written.

The verification of the following properties of integral sums is again left to the reader.

1. $S_A(af, D, T) = aS_A(f, D, T)$.

2. $S_A(f + g, D, T) = S_A(f, D, T) + S_A(g, D, T)$.

3. If $f \geqslant 0$ on the set A, then $S_A(f, D, T) \geqslant 0$.

4. If $f \geqslant g$ on the set A, then $S_A(f, D, T) \geqslant S_A(g, D, T)$.

5. If the function f is identically equal to a constant a on the set A, then any partition D is admissible for f and $S_A(f, D, T) = a\mu(A)$.

6. If on the set A it holds that $f \geqslant a$, then $S_A(f, D, T) \geqslant a\mu(A)$.

7. If on the set A it holds that $f \leqslant b$, then $S_A(f, D, T) \leqslant b\mu(A)$.

By analogy with the Riemann integral sums one can introduce the upper and lower integral sums for partitions of general form.

Definition 2. Let $f : A \to \mathbb{R}$ be a function on the measurable set A, and $D = \{\Delta_k\}_{k=1}^{\infty}$ be an admissible partition of A. The *upper integral sum* of the function f associated to the partition D is the number

$$\overline{S}_A(f, D) = \sum_{k=1}^{\infty} \sup_{t \in \Delta_k} (f(t)\mu(\Delta_k)),$$

and the *lower integral sum* is the number

$$\underline{S}_A(f, D) = \sum_{k=1}^{\infty} \inf_{t \in \Delta_k} (f(t)\mu(\Delta_k)).$$

Remark 1. By the definition of an admissible partition, for each $\Delta_k \in D$ of non-zero measure, $\sup_{t \in \Delta_k} |f(t)| < \infty$. Consequently, all the terms $\sup_{t \in \Delta_k} (f(t)\mu(\Delta_k))$ and $\inf_{t \in \Delta_k} (f(t)\mu(\Delta_k))$ in the definition of the upper and lower integral sums are finite. Their sums will also be finite, thanks to the condition $\sum_{k=1}^{\infty} \sup_{t \in \Delta_k} (|f(t)|\mu(\Delta_k)) < \infty$. Henceforth, whenever we write upper and lower integral sums, we will remember that the terms corresponding to sets $\Delta_k \in D$ with $\mu(\Delta_k) = 0$ are themselves equal to zero. The remaining terms can be written without additional parentheses as $\sup_{t \in \Delta_k} f(t)\mu(\Delta_k)$ and $\inf_{t \in \Delta_k} f(t)\mu(\Delta_k)$, without risk of running into the indeterminacy $\infty \cdot 0$.

Lemma 1. *Suppose the partition* $D = \{\Delta_k\}_{k=1}^{\infty}$ *of the set* A *is admissible for the function* f. *Then:*

(1) *for any choice T of marked points,*

$$\underline{S}_A(f, D) \leqslant S_A(f, D, T) \leqslant \overline{S}_A(f, D).$$

(2) *Further, suppose $D_1 \succ D$. Then*

$$\underline{S}_A(f, D) \leqslant \underline{S}_A(f, D_1) \leqslant \overline{S}_A(f, D_1) \leqslant \overline{S}_A(f, D).$$

(3) *Finally,*

$$\underline{S}_A(f, D) = \inf_T S_A(f, D, T)$$

and

$$\overline{S}_A(f, D) = \sup_T S_A(f, D, T).$$

Proof. (1) Since $t_k \in \Delta_k$, $\inf_{t \in \Delta_k} f(t) \leqslant f(t_k) \leqslant \sup_{t \in \Delta_k} f(t)$ for all $k \in \mathbb{N}$, which implies the needed inequalities.

(2) Let $D_1 = \{\Delta_k^1\}_{k=1}^\infty$. Grouping together the sets Δ_k^1 that are included in one and the same element of the partition D, we have

$$\overline{S}_A(f, D_1) = \sum_{j=1}^\infty \sum_{\Delta_k^1 \subset \Delta_j} \sup_{t \in \Delta_k^1} f(t) \mu(\Delta_k^1)$$

$$\leqslant \sum_{j=1}^\infty \sup_{t \in \Delta_j} f(t) \sum_{\Delta_k^1 \subset \Delta_j} \mu(\Delta_k^1) = \overline{S}_A(f, D).$$

One similarly verifies that $\underline{S}_A(f, D) \leqslant \underline{S}_A(f, D_1)$.

(3) To show the equality $\overline{S}_A(f, D) = \sup_T S_A(f, D, T)$, we construct for each $\delta > 0$ a collection $T_\delta = \{t_k\}_1^\infty$ of marked points such that $f(t_k) \geqslant \sup_{t \in \Delta_k} f(t) - \delta$. Then

$$S_A(f, D, T_\delta) \geqslant \sum_{k=1}^\infty \sup_{t \in \Delta_k} f(t) \mu(\Delta_k) - \sum_{k=1}^\infty \delta \mu(\Delta_k) = \overline{S}_A(f, D) - \delta \mu(A),$$

which in view of the arbitrariness of δ yields the required relation. The equality $\underline{S}_A(f, D) = \inf_T S_A(f, D, T)$ is established in the same manner. $\square$

Exercises

1. In general, the sum of a series may change when its terms are permuted. Why were we allowed to regroup the terms in the estimates carried out in the proof of the lemma?

2. Let $D_1 = \{\Delta_k^1\}_{k=1}^\infty$ be a partition of the set A which is admissible for f, and let $D_2 = \{\Delta_k^2\}_{k=1}^\infty$ be such that $D_2 \succ D_1$. Define the countable-valued functions $\overline{f}_i$ and $\underline{f}_i$, $i = 1, 2$, by the rules $\overline{f}_i = \sum_{k\in\mathbb{N}:\ \mu(\Delta_k^i)\neq 0} \left(\sup_{t\in\Delta_k^i} f(t)\right) \mathbb{1}_{\Delta_k^i}$ and $\underline{f}_i = \sum_{k\in\mathbb{N}:\ \mu(\Delta_k^i)\neq 0} \left(\inf_{t\in\Delta_k^i} f(t)\right) \mathbb{1}_{\Delta_k^i}$, respectively. Show that at almost all points of the set A it holds that $\underline{f}_1 \leqslant \underline{f}_2 \leqslant \overline{f}_2 \leqslant \overline{f}_1$. (We note that in this exercise the unpleasantly looking sums $\sum_{k\in\mathbb{N}:\ \mu(\Delta_k^i)\neq 0}$ appear instead of the pretty ones $\sum_{k=1}^\infty$ only in order to avoid values $\pm\infty$ for some of the terms $\sup_{t\in\Delta_k^i} f(t)$ and $\inf_{t\in\Delta_k^i} f(t)$. For, say, bounded f, we could use the ordinary summation $\sum_{k=1}^\infty$.)

4.2.2   Definition and Simplest Properties of the Lebesgue Integral

Definition 1. Let $A \in \Sigma$ be a measurable set and $f : A \to \mathbb{R}$ a function on A. The number $a \in \mathbb{R}$ is called the *integral* (specifically, *Lebesgue integral*) of the function f on the set A with respect to the measure μ (notation: $a = \int_A f\,d\mu$) if for every $\varepsilon > 0$ there exists an admissible partition D_ε of A such that for any partition D that refines D_ε, and any choice of marked points T for D, one has $|a - S_A(f, D, T)| \leqslant \varepsilon$. The function $f : A \to \mathbb{R}$ is said to be *integrable* on the set A with respect to the measure μ, or μ-*integrable on A*, if the corresponding integral exists.

In other words, the function f is integrable on A if, starting with some partition, the integral sums are defined and the limit of the integral sums along the directed set of partitions with marked points, described in Subsection 4.1.3, exists. This limit is called the *Lebesgue integral* of f and is denoted by $f = \int_A f\,d\mu$.

The assertions about the Lebesgue integral listed below are straightforward consequences of the corresponding properties of integral sums and of the limit along a directed set.

1. Let $f : A \to \mathbb{R}$ be an integrable function and $\lambda \in \mathbb{R}$. Then the function λf is also integrable and $\int_A \lambda f\,d\mu = \lambda \int_A f\,d\mu$.

2. If the functions f and g are integrable on A, then so is the function $f + g$, and $\int_A (f + g)d\mu = \int_A f\,d\mu + \int_A g\,d\mu$.

3. If the integrable function f is greater than or equal to zero on the set A, then $\int_A f\,d\mu \geqslant 0$.

4. If $f \geqslant g$ on the set A and both f and g are integrable on A, then $\int_A f\,d\mu \geqslant \int_A g\,d\mu$.

5. If $f : A \to \mathbb{R}$ is an integrable function, $f \geqslant 0$, and $\int_A f\,d\mu = 0$, then any function g satisfying the inequality $0 \leqslant g \leqslant f$ is also integrable on A, and $\int_A g\,d\mu = 0$.

6. Let $a \in \mathbb{R}$ be a constant. Then $\int_A a\,d\mu = a\mu(A)$.

7. Let $f : A \to \mathbb{R}$ be an integrable function, $a \in \mathbb{R}$, and $f \leqslant a$ on A. Then $\int_A f \, d\mu \leqslant a\mu(A)$. Similarly, if $f \geqslant b$, then $\int_A f \, d\mu \geqslant b\mu(A)$.

Theorem 1. *For a function $f : A \to \mathbb{R}$, where $A \in \Sigma$, the following conditions are equivalent:*

(1) *f is integrable and $\int_A f \, d\mu = a$;*

(2) *for any $\varepsilon > 0$ there exists an admissible partition $D_\varepsilon = \{\Delta_j\}_{j=1}^\infty$ of the set A such that for any choice T of marked points, $|a - S_A(f, D_\varepsilon, T)| < \varepsilon$;*

(3) *for any $\varepsilon > 0$ there exists an admissible partition D_ε of the set A such that the corresponding upper and lower integral sums of the function f approximate a to within ε: $|a - \overline{S}_A(f, D_\varepsilon)| \leqslant \varepsilon$ and $|a - \underline{S}_A(f, D_\varepsilon)| \leqslant \varepsilon$.*

Proof. The implication (1) $\Longrightarrow$ (2) is obvious. The implication (2) $\Longrightarrow$ (3) follows from Lemma 1 proved in the preceding subsection (item (3) of that lemma). Indeed, all the integral sums $S_A(f, D_\varepsilon, T)$ lie in the interval $[a - \varepsilon, a + \varepsilon]$ by assumption; consequently, $\underline{S}_A(f, D) = \inf_T S_A(f, D, T)$ and $\overline{S}_A(f, D) = \sup_T S_A(f, D, T)$ also lie in that interval. The same lemma yields the implication (3) $\Longrightarrow$ (1). Namely, let D_ε be the partition in item (3). By the indicated lemma, for any partition D that refines D_ε,

$$a - \varepsilon \leqslant \underline{S}_A(f, D_\varepsilon) \leqslant \underline{S}_A(f, D) \leqslant \overline{S}_A(f, D) \leqslant \overline{S}_A(f, D_\varepsilon) \leqslant a + \varepsilon.$$

Further, for any choice $T = \{t_k\}_1^\infty$ of marked points for D,

$$\underline{S}_A(f, D) \leqslant S_A(f, D, T) \leqslant \overline{S}_A(f, D).$$

Hence, $a - \varepsilon \leqslant S_A(f, D, T) \leqslant a + \varepsilon$ and $|a - S_A(f, D, T)| \leqslant \varepsilon$. $\square$

Example 1. Let $\{A_k\}_1^\infty$ be a partition of the set $A \in \Sigma$ into measurable sets, $f = \sum_{k=1}^\infty a_k \mathbb{1}_{A_k}$ be a countably-valued measurable function, and suppose the series $\sum_{k=1}^\infty a_k \mu(A_k)$ converges absolutely. Then the function f is integrable on A and $\int_A f \, d\mu = \sum_{k=1}^\infty a_k \mu(A_k)$. Indeed, if for D we take the partition of A into the sets $\{A_k\}_1^\infty$, then

$$\underline{S}_A(f, D) = \overline{S}_A(f, D) = \sum_{k=1}^\infty a_k \mu(A_k).$$

It remains to apply condition (3) in Theorem 1 with $D_\varepsilon = D$.

By the Cauchy criterion for convergence along a directed set, a function $f : A \to \mathbb{R}$ is integrable on the set A if and only if for every $\varepsilon > 0$ there exist an admissible partition D_ε of A and a choice T of marked points such that $|S_A(f, D_\varepsilon, T) - S_A(f, D, \widetilde{T})| < \varepsilon$ for any $D \succ D_\varepsilon$ and any choice $\widetilde{T}$ of marked points of the partition D.

Since according to Lemma 1 in the preceding subsection $[\underline{S}_A(f, D_\varepsilon), \overline{S}_A(f, D_\varepsilon)]$ is the smallest closed interval containing all possible values of the sums of the form $S_A(f, D, \widetilde{T})$, we obtain the following useful reformulation of the Cauchy criterion.

Theorem 2. *A function* $f: A \to \mathbb{R}$ *is integrable on the set* A *if and only if for any* $\varepsilon > 0$ *there exists an admissible partition* D_ε *of* A *such that the corresponding upper and lower integral sums of the function* f *differ by less than* ε: $\left| \overline{S}_A(f, D_\varepsilon) - \underline{S}_A(f, D_\varepsilon) \right| < \varepsilon$. $\qquad\square$

Theorem 3. *Let* $f: A \to \mathbb{R}$ *be an integrable function. Then the function* $|f|$ *is also integrable.*

Proof. Let $\varepsilon > 0$ and let $D_\varepsilon = \{\Delta_j\}_{j=1}^\infty$ be the partition provided for the function f by the preceding theorem. Then

$$\left| \overline{S}_A(|f|, D_\varepsilon) - \underline{S}_A(|f|, D_\varepsilon) \right| = \sum_{k=1}^\infty \left(\sup_{t \in \Delta_k} (|f(t)|\mu(\Delta_k)) - \inf_{t \in \Delta_k} (|f(t)|\mu(\Delta_k)) \right)$$

$$\leqslant \sum_{k=1}^\infty \left(\sup_{t \in \Delta_k} (f(t)\mu(\Delta_k)) - \inf_{t \in \Delta_k} (f(t)\mu(\Delta_k)) \right) = \overline{S}_A(f, D_\varepsilon) - \underline{S}_A(f, D_\varepsilon) < \varepsilon.$$

Hence, for every $\varepsilon > 0$ we proved the existence of a partition D_ε for which $\left| \overline{S}_A(|f|, D_\varepsilon) - \underline{S}_A(|f|, D_\varepsilon) \right| < \varepsilon$. By Theorem 2, this establishes the integrability of the function $|f|$. $\qquad\square$

Corollary 1. *Let* $f: A \to \mathbb{R}$ *be an integrable function. Then the functions* f^+ *and* f^- *are also integrable.*

Proof. Recall that, by definition, $f^+(t)$ coincides with $f(t)$ for all t where $f(t) > 0$; for those t where $f(t) \leqslant 0$, $f^+(t) = 0$. Similarly, $f^-(t) = |f(t)|$ at the points where $f(t) \leqslant 0$, and at the remaining points $f^-(t) = 0$. Since $f^+ = \frac{1}{2}(f + |f|)$ and $f^- = \frac{1}{2}(|f| - f)$, the assertion of the corollary follows from the preceding theorem and the properties of the integrals listed earlier. $\qquad\square$

Corollary 2. *Let* f *and* g *be two integrable functions. Then the functions* $\max\{f, g\}$ *and* $\min\{f, g\}$ *are also integrable.*

Proof. The assertion is a straightforward consequence of the formulas $\max\{f, g\} = \frac{1}{2}(f + g + |f - g|)$ and $\min\{f, g\} = \frac{1}{2}(f + g - |f - g|)$. $\qquad\square$

Exercises

1. Prove the implication (1) $\implies$ (2) in Theorem 1 of Subsection 4.2.2.

2. Prove Theorem 2 of Subsection 4.2.2.

3. Why in Theorem 3 of Subsection 4.2.2 is D_ε an admissible partition for the function $|f|$?

4. Verify the formulas $f^+ = \frac{1}{2}(f + |f|), f^- = \frac{1}{2}(|f| - f), \max\{f, g\} = \frac{1}{2}(f + g + |f - g|)$, and $\min\{f, g\} = \frac{1}{2}(f + g - |f - g|)$ that figure in the last two corollaries.

5. Let A be a set of measure zero. Show that any function f on A is integrable and $\int_A f \, d\mu = 0$.

6. Let f and g be two functions defined on the measurable set A and such that f and g coincide almost everywhere. If f is integrable, then so is g, and $\int_A g \, d\mu = \int_A f \, d\mu$.

7. Suppose the functions f and g are integrable on A and $f \leqslant g$ almost everywhere. Then $\int_A f \, d\mu \leqslant \int_A g \, d\mu$.

8. Let Σ be the σ-algebra of Lebesgue-measurable subsets of the interval $[a, b]$, λ be the Lebesgue measure on $[a, b]$, and $f : [a, b] \to \mathbb{R}$ be a Riemann-integrable function. Use Theorem 1 of Subsection 4.2.2 to prove that the function f is Lebesgue integrable on $[a, b]$ and $\int_{[a,b]} f \, d\lambda = \int_a^b f(t) dt$.

9. Here is a more general result. Let $F : [a, b] \to \mathbb{R}$ be a monotone non-decreasing Stieltjes function, and μ be the corresponding Borel measure, i.e., F is the distribution of μ (see Subsection 2.3.5). Then any Stieltjes-integrable function $f : [a, b] \to \mathbb{R}$ is μ-integrable on $[a, b]$ and $\int_{[a,b]} f \, d\mu = \int_a^b f(t) dF(t)$.

10. Let $A \subset [a, b]$ be a dense subset of Lebesgue measure zero. Show that the characteristic function $\mathbb{1}_A$ is not Riemann integrable, but is Lebesgue integrable on $[a, b]$. What is $\int_{[a,b]} \mathbb{1}_A d\lambda$ equal to?

11. Show that the function $f(x) = 1/x$ is not Lebesgue integrable on the interval $(0, 1]$.

12. Prove the following reformulation of Theorem 2 in Subsection 4.2.2: a function $f : A \to \mathbb{R}$ is integrable on the set A if and only if for any $\varepsilon > 0$ and any partition D of A, there exists an admissible partition $D_\varepsilon \succ D$, such that the corresponding upper and lower integral sums of f differ by less than ε: $|\overline{S}_A(f, D_\varepsilon) - \underline{S}_A(f, D_\varepsilon)| < \varepsilon$.

13. Give an example of two countably-valued integrable functions whose product is not integrable.

14. Let μ be the measure on $\mathbb{N}$ described in Exercise 5 of Subsection 2.1.4. Then a function $f : \mathbb{N} \to \mathbb{R}$ is integrable on $\mathbb{N}$ with respect to the measure μ if and only if the series $\sum_{n=1}^\infty f(n) b_n$ is absolutely convergent. In this case $\int_{\mathbb{N}} f \, d\mu = \sum_{n=1}^\infty f(n) b_n$.

15. The definition of the integral remains valid for complex-valued functions. Verify the properties

$$\int_A \lambda f \, d\mu = \lambda \int_A f \, d\mu, \quad \int_A (f + g) d\mu = \int_A f \, d\mu + \int_A g \, d\mu$$

and the inequality

$$\left| \int_A f \, d\mu \right| \leqslant \int_A |f| \, d\mu$$

for complex-valued functions and complex scalars.

16. Let f be a complex-valued function on A, and denote by f_1 and f_2 the real and imaginary parts of f. Show that f is integrable if and only if f_1 and f_2 are integrable, and in this case $\int\limits_A f \, d\mu = \int\limits_A f_1 d\mu + i \int\limits_A f_2 d\mu$.

17. Verify for complex-valued functions the validity of the equivalence (1) $\Longleftrightarrow$ (2) in Theorem 1, of the integrability criterion for countably-valued functions (Example 1), and of the assertion of Theorem 3, all from Subsection 4.2.2.

4.2.3   The Integral as a Set Function

Theorem 1. *Let* $f : A \to \mathbb{R}$ *be a function integrable on* A, *and* B *a measurable subset of* A. *Then* f *is integrable on* B.

Proof. By Theorem 2 in Subsection 4.2.2, for any $\varepsilon > 0$ there exists an admissible partition $D_\varepsilon = \{\Delta_j\}_{j=1}^\infty$ of the set A such that the corresponding upper and lower integral sums of the function f differ by less than ε: $|\overline{S}_A(f, D_\varepsilon) - \underline{S}_A(f, D_\varepsilon)| < \varepsilon$. Consider the set K of those indices k for which Δ_k intersects B. Then the sets $\Delta_k^1 = B \cap \Delta_k, k \in K$, constitute an admissible partition of B. Denote this partition by D_ε^1. Note that $\sup_{t \in \Delta_k^1} f(t) \leqslant \sup_{t \in \Delta_k} f(t)$, $\inf_{t \in \Delta_k^1} f(t) \geqslant \inf_{t \in \Delta_k} f(t)$, and $\mu(\Delta_k^1) \leqslant \mu(\Delta_k)$. Let us estimate the quantity $|\overline{S}_A(f, D_\varepsilon^1) - \underline{S}_A(f, D_\varepsilon^1)|$. We have

$$
\begin{aligned}
|\overline{S}_B(f, D_\varepsilon^1) - \underline{S}_B(f, D_\varepsilon^1)| &= \sum_{k \in K} \left(\sup_{t \in \Delta_k^1} \left[f(t)\mu(\Delta_k^1) \right] - \inf_{t \in \Delta_k^1} \left[f(t)\mu(\Delta_k^1) \right] \right) \\
&\leqslant \sum_{k=1}^\infty \left(\sup_{t \in \Delta_k} \left[f(t)\mu(\Delta_k) \right] - \inf_{t \in \Delta_k} \left[f(t)\mu(\Delta_k) \right] \right) \\
&= |\overline{S}_A(f, D_\varepsilon) - \underline{S}_A(f, D_\varepsilon)| < \varepsilon.
\end{aligned}
$$

Thus, we have shown that for any $\varepsilon > 0$ there exists a partition of the set B such that the corresponding upper and lower integral sums differ by less than ε. By Theorem 2 of Subsection 4.2.2, this establishes the integrability of the function f on B. $\square$

Theorem 2. *Let* $A_1, A_2 \in \Sigma$ *be disjoint sets and let the function* f *be integrable on both* A_1 *and* A_2. *Then* f *is integrable on the union* $A_1 \cup A_2$, *and*

$$\int\limits_{A_1 \cup A_2} f \, d\mu = \int\limits_{A_1} f \, d\mu + \int\limits_{A_2} f \, d\mu.$$

Proof. Denote $\int_{A_i} f \, d\mu$ by a_i, $i = 1, 2$. We use condition (2) in Theorem 1 of Subsection 4.2.2. Fix $\varepsilon > 0$ and choose admissible partitions D_1 and D_2 of the sets A_1 and A_2, respectively, with the property that for any choice T_1 and T_2 of marked points, one has $|a_i - S_{A_i}(f, D_i, T_i)| < \varepsilon$, $i = 1, 2$. Construct a partition D of the union $A_1 \cup A_2$ by taking all the elements of the partitions D_1 and D_2. Now let T be an arbitrary collection of marked points for D and denote by T_i the part of T that falls in the set A_i, $i = 1, 2$. Then

$$S_{A_1 \cup A_2}(f, D, T) = S_{A_1}(f, D_1, T_1) + S_{A_2}(f, D_2, T_2),$$

and consequently

$$|a_1 + a_2 - S_{A_1 \cup A_2}(f, D, T)| \leqslant |a_1 - S_{A_1}(f, D_1, T_1)| + |a_2 - S_{A_2}(f, D_2, T_2)| < 2\varepsilon.$$

Since ε was arbitrary, the aforementioned integrability criterion applies. $\qquad\square$

Corollary 1. *If the function f is integrable and non-negative on A, then the set function $G(B) = \int_B f \, d\mu$ is a finitely-additive measure on the family $\Sigma_A = \{B \in \Sigma : B \subset A\}$.* $\qquad\square$

Theorem 3. *Suppose the function f takes on the set A only non-negative values. Let $\{A_k\}_1^\infty$ be a partition of the set A into measurable subsets. Suppose also that f is integrable on each A_k and the series $\sum_{k=1}^\infty \int_{A_k} f \, d\mu$ converges. Then f is integrable on the whole set A and $\int_A f \, d\mu = \sum_{k=1}^\infty \int_{A_k} f \, d\mu$.*

Proof. We argue by analogy with the preceding proof. Denote $\int_{A_k} f \, d\mu$ by a_k, $k = 1, 2, \ldots$. Fix an $\varepsilon > 0$ and admissible partitions D_k of the sets A_k, $k = 1, 2, \ldots$, such that for any choice T_k of marked points for the D_k, $k = 1, 2, \ldots$, one has $|a_k - S_{A_k}(f, D_k, T_k)| < \varepsilon/2^k$. Now construct a partition $D = \{\Delta_j\}_{j=1}^\infty$ of the set A by taking all the elements of the partitions D_k, $k = 1, 2, \ldots$. For any choice T of marked points for D, denote by T_k the part of T consisting of the points that fall in the corresponding set A_k, $k = 1, 2, \ldots$. Then

$$\sum_{j=1}^\infty \sup_{t \in \Delta_j} \left[|f(t)| \mu(\Delta_j) \right] = \sup_T \sum_{j=1}^\infty \left[f(t_j) \mu(\Delta_j) \right]$$

$$= \sup_T \sum_{k=1}^\infty S_{A_k}(f, D_k, T_k) \leqslant \sum_{k=1}^\infty \left(a_k + \frac{\varepsilon}{2^k} \right) < \infty.$$

Hence, we have proved that the partition D is admissible for f. Further,

$$\left| \sum_{k=1}^\infty a_k - S_A(f, D, T) \right| \leqslant \sum_{k=1}^\infty |a_k - S_{A_k}(f, D_k, T_k)| < \sum_{k=1}^\infty \frac{\varepsilon}{2^k} = \varepsilon.$$

By item (2) in Theorem 1 in Subsection 4.2.2, the proof is complete. $\qquad\square$

Corollary 2. *Under the assumptions of Corollary 1 (non-negativity and integrability of f on A), the set function $G(B) = \int_B f \, d\mu$ is not merely finitely-additive: in fact, it is a countably-additive measure on Σ_A.*

Proof. Let $\{B_k\}_1^\infty$ be a partition of some set $B \in \Sigma$ into measurable subsets. In view of the already proved finite-additivity of the set function G, we have

$$\sum_{k=1}^n \int_{B_k} f \, d\mu = \int_{\bigcup_{k=1}^n B_k} f \, d\mu \leqslant \int_B f \, d\mu$$

for all $n \in \mathbb{N}$. Therefore,

$$\sum_{k=1}^\infty \int_{B_k} f \, d\mu \leqslant \int_B f \, d\mu < \infty,$$

and now we are under the conditions of the theorem. Applying the theorem, we conclude that

$$\int_{\bigcup_{k=1}^\infty B_k} f \, d\mu = \int_B f \, d\mu = \sum_{k=1}^\infty \int_{B_k} f \, d\mu. \qquad \square$$

We are now ready to prove the main result of this subsection.

Theorem 4. *Let $(B_k)_1^\infty$ be a sequence of pairwise-disjoint measurable sets, and $A = \bigcup_{k=1}^\infty B_k$. Then the function f is integrable on A if and only if f is integrable on each B_k and the series $\sum_{k=1}^\infty \int_{B_k} |f| \, d\mu$ converges. Moreover, in this case*

$$\int_B f \, d\mu = \sum_{k=1}^\infty \int_{B_k} f \, d\mu.$$

Proof. In the case where $f \geqslant 0$ the result follows from Theorem 3 and Corollary 2. Hence, the assertion holds true for the functions f^+ and f^-. To complete the proof, it suffices to apply the relations $|f| = f^+ + f^-$ and $f = f^+ - f^-$. $\qquad \square$

Corollary 3. *Let $\{B_k\}_1^\infty$ be a partition of the set $A \in \Sigma$ into measurable sets and $f = \sum_{k=1}^\infty b_k \mathbb{1}_{B_k}$ be a countably-valued function. Then for f to be integrable on A it is necessary and sufficient that the series $\sum_{k=1}^\infty b_k \mu(B_k)$ be absolutely convergent. In this case,*

$$\int_A f \, d\mu = \sum_{k=1}^\infty b_k \mu(B_k). \qquad \square$$

Example 1. Let $(A_k)_{k=1}^\infty$ be a sequence of pairwise disjoint subsets of non-zero measure of the interval $[0, 1]$ such that $\bigcup_{k=1}^\infty A_k = [0, 1]$. Consider the countably-valued function $f = \sum_{k=1}^\infty \frac{(-1)^k}{\mu(A_k)} \mathbb{1}_{A_k}$. By Corollary 3, the function f is not Lebesgue on $[0, 1]$. Put $B_n = A_{2n-1} \cup A_{2n}$. Then $(B_n)_1^\infty$ is again a sequence of pairwise-disjoint subsets of measure zero of the interval $[0, 1]$ and $\bigcup_{n=1}^\infty B_n = [0, 1]$. Now note that the function f is integrable on each set B_n and $\int_{B_n} f \, d\lambda = 0$. Consequently, the series $\sum_{n=1}^\infty \int_{B_n} f \, d\lambda$ is absolutely convergent. Thus, in general the convergence (and even absolute convergence) of the series of integrals over subsets does not imply the integrability of the function on the union of these subsets.

Remark 1. Suppose the function f is defined almost everywhere on the set A, i.e., there exists a set $B \subset A$ of measure zero such that f is defined on $A \setminus B$. As readily follows from Exercise 5 in Subsection 4.2.2, the following assertions are equivalent:

(a) f is integrable on $A \setminus B$;

(b) f can be extended to the entire set A so that it becomes integrable on A;

(c) any extension of f to the whole set A is integrable on A.

It is also obvious that the value of the integral does not change if the values of the function are changed on a negligible set. Therefore, in the framework of integration theory one can consider functions that are defined almost everywhere. This proves very convenient when one deals with functions such as $\frac{1}{\sqrt{x}}$, $\frac{x}{|x|}$, and so on: we don't have to worry about how to redefine the function at points of discontinuity.

Exercises

1. Doesn't Example 1 above contradict the assertion of Theorem 4?

2. Let $A \in \Sigma$. Denote by Σ_A the collection of all elements of the σ-algebra Σ that are subsets of A, and let $\mu_1 : \Sigma_A \to \mathbb{R}$ be the restriction of the measure μ to Σ_A (i.e., $\mu_1(B) = \mu(B)$ for all $B \in \Sigma_A$). Verify that (A, Σ_A, μ_1) is again a measure space.

3. Let $A \in \Sigma$ and let f be defined and integrable on A. Extend the function f to $\Omega \setminus A$ by zero. Show that the function f redefined in this way is integrable on Ω.

4. As we already mentioned (Exercises 15–17 in Subsection 4.2.2), the definition of the integral can be extended to complex-valued functions. Verify that Theorems 1, 2 and 4 hold for complex-valued functions.

4.3 Measurability and Integrability

4.3.1 Measurability of Integrable Functions

The following simple estimate proves very useful when dealing with the Lebesgue integral. In Analysis its habitual name is *Chebyshev's inequality*, or *the first Chebyshev inequality*. In the setting of Probability Theory, the same estimate, written in the language of random variables and probability, is usually called *Markov's inequality*.

Lemma 1 (Chebyshev's inequality). *Let $a > 0$ be a constant, g an integrable function on $A \in \Sigma$, $g \geqslant 0$, and $B \subset A$ a measurable set such that $g(t) \geqslant a$ for all $t \in B$. Then*

$$\mu(B) \leqslant \frac{1}{a} \int_A g \, d\mu.$$

Proof. We have

$$\int_A g \, d\mu \geqslant \int_B g \, d\mu \geqslant \int_B a \, d\mu = a\mu(B). \qquad \square$$

Theorem 1. *If a measure space is complete, then every function integrable on a set is measurable on that set.*

Proof. Let the function f be integrable on the set $A \in \Sigma$. Choose a sequence of increasingly finer admissible partitions $D_j = \{\Delta_k^j\}_{k=1}^\infty$, $D_1 \prec D_2 \prec D_3 \prec \cdots$, for each of which $|\overline{S}_A(f, D_j) - \underline{S}_A(f, D_j)| < 1/j$. By analogy with Exercise 2 in Subsection 4.2.1, define two sequences of countably-valued functions,

$$\overline{f}_j = \sum_{k\in\mathbb{N}:\ \mu(\Delta_k^j)\neq 0} \sup_{t\in\Delta_k^j} f(t)\, \mathbb{1}_{\Delta_k^j} \quad \text{and} \quad \underline{f}_j = \sum_{k\in\mathbb{N}:\ \mu(\Delta_k^j)\neq 0} \inf_{t\in\Delta_k^j} f(t)\, \mathbb{1}_{\Delta_k^j}.$$

These functions are integrable on A,

$$\int_A \overline{f}_j \, d\mu = \overline{S}_A(f, D_j), \quad \text{and} \quad \int_A \underline{f}_j \, d\mu = \underline{S}_A(f, D_j).$$

Outside of the negligible set Δ formed by the union of all those Δ_k^j that have zero measure the sequence $(\overline{f}_j)$ is pointwise non-increasing and bounded below by the function f. Consequently, on $A \setminus \Delta$ the functions $\overline{f}_j$ have a pointwise limit when $j \to \infty$, which we denote by $\overline{f}$. Similarly, we denote by $\underline{f}$ the pointwise limit of the functions $\underline{f}_j$ when $j \to \infty$. The functions $\underline{f}$ and $\overline{f}$ are measurable on $A \setminus \Delta$ (as limits of sequences of measurable functions), and satisfy $\underline{f} \leqslant f \leqslant \overline{f}$. If we can prove that $\underline{f} = \overline{f}$ almost everywhere, then it will follow that $\underline{f} = f = \overline{f}$ almost everywhere, and hence that f is measurable (here we use the assumption that the measure is complete). So, denote by B the set of all points of $A \setminus \Delta$ where $\underline{f} \neq \overline{f}$, and by B_n the set of all points where $\overline{f} - \underline{f} > \frac{1}{n}$. Since $B = \bigcup_{n=1}^\infty B_n$, it suffices to prove that $\mu(B_n) = 0$ for all n. Observe that $\overline{f}_j - \underline{f}_j \geqslant \overline{f} - \underline{f}$ for all $j \in \mathbb{N}$, and so on B_n one has $\overline{f}_j - \underline{f}_j > \frac{1}{n}$. By the Chebyshev inequality,

$$\mu(B_n) \leqslant n \int_A (\overline{f}_j - \underline{f}_j)\, d\mu = n(\overline{S}_A(f, D_j) - \underline{S}_A(f, D_j)) < \frac{n}{j}.$$

Letting $j \to \infty$, we get the desired equality $\mu(B_n) = 0$. $\qquad \square$

Remark 1. Analogously to Exercise 2 from Subsection 4.2.1, the summation over only those k for which the sets Δ_k^j have non-zero measure, which we used in the definition of the functions $\overline{f}_j$ and $\underline{f}_j$, instead of the ordinary summation $\sum_{k=1}^\infty$, is done in order to avoid possible infinite values at some points. There is another way to fix this inconvenience, proceeding in the spirit of Remark 1 at the end of the Subsection 4.2.3. Namely, we could use the formulas

$$\overline{f}_j = \sum_{k=1}^{\infty} \sup_{t \in \Delta_k^j} f(t) \, \mathbb{1}_{\Delta_k^j} \quad \text{and} \quad \underline{f}_j = \sum_{k=1}^{\infty} \inf_{t \in \Delta_k^j} f(t) \, \mathbb{1}_{\Delta_k^j},$$

but agree that these functions are not defined at all points, but only almost everywhere.

Remark 2. If (Ω, Σ, μ) is an incomplete measure space and (Ω, Σ', μ) is its completion, then a function that is integrable on (Ω, Σ, μ) may be not Σ-measurable, but is necessarily Σ'-measurable. As we know, these two types of measurability differ only slightly: for every Σ'-measurable function f there exists a Σ-measurable function that coincides with f almost everywhere. To avoid being hindered each time we encounter this inessential difference, in the setting of integration theory we will assume that, unless otherwise stipulated, the measure spaces under consideration are complete. Accordingly, henceforth, *all integrable functions will be assumed to be measurable*. Another convention most frequently encountered in the literature is that on incomplete measure spaces one considers only measurable integrable functions, i.e., measurability is treated as a necessary part of the definition of integrability.

Exercises

1. Justify the existence of the sequence of partitions D_j in the proof of the last theorem.

2. The proof of Theorem 1 used the completeness of the measure μ and the assertion that if $f \overset{\text{a.e.}}{=} g$ and f is measurable, then g is also measurable. Verify this assertion! Can we manage without completeness here?

3. The proof of Theorem 1 used implicitly the fact that $\sup_{t \in \Delta_k^j}$ and $\inf_{t \in \Delta_k^j}$ are finite whenever $\mu(\Delta_k^j) \neq 0$. Why is this fact true?

From the measurability of an integrable function and Chebyshev's inequality proved at the beginning of the present subsection one derives the following useful assertion.

4. Let $\int_A |f| \, d\mu = 0$. Then on the set A the function f is equal to zero almost everywhere.

4.3.2   The Uniform Limit Theorem

Theorem 1. *Suppose the sequence of functions (f_n) converges uniformly on the set A to a function f. If all f_n are integrable on A, then f is also integrable on A, and $\int_A f \, d\mu = \lim_{n \to \infty} \int_A f_n \, d\mu$.*

Proof. Let $a_n = \int_A f_n \, d\mu$. The sequence (a_n) is Cauchy:

$$|a_n - a_m| \leqslant \int_A |f_n - f_m| \, d\mu \leqslant \sup_{t \in A} |f_n(t) - f_m(t)| \mu(A) \to 0, \quad n, m \to \infty.$$

Denote the limit of (a_n) by $\dot{a}$. Let ε be an arbitrary positive number. Thanks to the uniform convergence of the sequence (f_n) tò f, there exists a number $N = N(\varepsilon)$ such that for every $n > N$ and any $t \in A$, one has $|f_n(t) - f(t)| < \varepsilon/(4\mu(A))$. Now fix an $n > N$ such that $|a_n - a| < \varepsilon/4$. Using the integrability of the function f_n, construct an admissible partition $D = \{\Delta_k\}_{k=1}^{\infty}$ of the set A with the following property: for any choice $T = \{t_k\}_1^{\infty}$ of marked points it holds that $|a_n - S_A(f_n, D, T)| < \varepsilon/2$.

Since $|f_n(t) - f(t)| < \varepsilon/(4\mu(A))$, the partition D is also admissible for the function f:

$$\sum_{k=1}^{\infty} \sup_{t \in \Delta_k} [|f(t)| \mu(\Delta_k)] \leqslant \sum_{k=1}^{\infty} \sup_{t \in \Delta_k} [|f_n(t)| \mu(\Delta_k)] + \frac{\varepsilon}{4} < \infty.$$

Further, for any choice $T = \{t_k\}_1^{\infty}$ of marked points we have

$$|a - S_A(f, D, T)| = \left| a - \sum_{k=1}^{\infty} f(t_k)(\Delta_k) \right|$$

$$\leqslant |a - a_n| + \left| a_n - \sum_{k=1}^{\infty} f_n(t_k) \mu(\Delta_k) \right| + \left| \sum_{k=1}^{\infty} (f_n(t_k) - f(t_k)) \mu(\Delta_k) \right|$$

$$< \frac{\varepsilon}{4} + \frac{\varepsilon}{2} + \frac{\varepsilon}{4} = \varepsilon.$$

By criterion (2) in Theorem 1 of Subsection 4.2.2, the function f is integrable and $\int_A f \, d\mu = a$. To complete the proof, it remains to recall that a denotes $\lim_{n \to \infty} \int_A f_n \, d\mu$. $\qquad \square$

4.3.3 An Integrability Condition for Measurable Functions

Theorem 1. *If the measurable function f admits an integrable majorant, then it is itself integrable. More precisely: suppose that on the set A the function f is integrable, $|f| \leqslant g$, and the function g is integrable. Then f is also integrable.*

Proof. First we treat the special case where f is a countably-valued function, i.e. $f = \sum_{k=1}^{\infty} a_k \mathbb{1}_{A_k}$, where (A_k) is a sequence of pairwise disjoint measurable sets. In this case the inequality $|f| \leqslant g$ means that $|a_k| \leqslant g(t)$ for all $t \in A_k$. Consequently, the series $\sum_{k=1}^{\infty} a_k \mu(A_k)$ is absolutely convergent:

$$\sum_{k=1}^{\infty} |a_k| \, \mu(A_k) \leqslant \sum_{k=1}^{\infty} \int_{A_k} g \, d\mu \leqslant \int_A g \, d\mu < \infty.$$

By Example 1 in Subsection 4.2.2 (or by Corollary 3 of Subsection 4.2.3), the function f is integrable.

The general case may be deduced from two already known results: the theorem on approximation of a measurable function by countably-valued functions (Theorem 3 of Subsection 3.1.4), and the uniform limit theorem. Indeed, suppose f is measurable, $|f| \leqslant g$, and g is integrable. Construct a sequence (f_n) of measurable countably-valued functions such that $\sup_A |f_n(t) - f(t)| < 1/n$. Then $|f_n| \leqslant g + 1/n$, hence, by the particular case treated above, the functions f_n are integrable. Hence, we were able to represent the function f as the limit of a uniformly convergent sequence of integrable functions. This establishes the integrability of f. □

The integrability condition for measurable functions established above will prove useful in many situations. The reason is that measurability is preserved by the usual operations on functions: sum, product, passing to the limit, etc. Thanks to this, the verification of the measurability of some given function is usually not too difficult. Finding an integrable majorant is easier than verifying integrability based directly on the definition.

Exercises

1. Let f be a measurable function and let $|f|$ be integrable. Then f is integrable.

2. Suppose that for the measurable function f there exist an admissible partition. Then f is integrable.

3. Any bounded measurable function is integrable.

4. The product of a bounded measurable function with an integrable function is again integrable.

5. Describe the measure spaces on which every measurable function is integrable.

6. Let (Ω, Σ, μ) be a finite measure space, E the space of all measurable scalar-valued functions on Ω, and F be the subspace of E consisting of all functions that vanish almost everywhere. Denote by $L_0(\Omega, \Sigma, \mu)$ the quotient space E/F. To simplify the terminology, it is customary to say that the elements of the space $L_0(\Omega, \Sigma, \mu)$ are measurable functions of Ω, but with the understanding that functions that are equal almost everywhere are identified. Let $f, g \in L_0(\Omega, \Sigma, \mu)$. Put $\rho(f, g) = \int_\Omega \frac{|f-g|}{1+|f-g|} d\mu$. Show that ρ gives a metric on $L_0(\Omega, \Sigma, \mu)$, and moreover that convergence in this metric coincides with the convergence in measure.

4.4 Passage to the Limit Under the Integral Sign

In Sect. 4.2 we made acquaintance with the Lebesgue integral and we saw that although the Lebesgue integral is a more general notion than the Riemann integral, it retains all the convenient properties of the integral familiar from calculus. We now turn to exhibiting the advantages of the Lebesgue integral over the Riemann integral. We will show that for the Lebesgue integral not only is the uniform limit theorem valid, but several considerably more general theorems on the passage to the limit under the integral sign also hold, which are particularly convenient in applications.

Recall that in this chapter, till the end of Sect. 4.5, we are dealing with a finite measure space (Ω, Σ, μ), $A \in \Sigma$, and unless otherwise stipulated, the functions f, f_n, are defined on A and take real values. The extension of the obtained results to σ-finite measure spaces will carried out in Subsection 4.6.2.

4.4.1 Fatou's Lemma

Theorem 1 (Fatou's lemma). *Suppose that on the set A there is given a sequence (f_n) of non-negative integrable functions such that (f_n) converges almost everywhere to some function f, and the integrals of the functions (f_n) are jointly bounded, i.e., $\int_A f_n \, d\mu \leqslant C < \infty$ for all n. Then f is integrable and*

$$\int_A f \, d\mu \leqslant \varliminf_{n \to \infty} \int_A f_n \, d\mu.$$

Proof. We use Egorov's theorem (Subsection 3.2.4). Pick a measurable subset $A_1 \subset A$ with $\mu(A \backslash A_1) \leqslant 1/2$, on which the sequence (f_n) converges uniformly to f. Denote $A \setminus A_1$ by B_1. Applying again Egorov's theorem, pick in B_1 a measurable subset A_2 with $\mu(B_1 \setminus A_2) \leqslant 1/4$, on which (f_n) also converges uniformly to f. Denote $B_1 \setminus A_2$ by B_2. Continuing this process, we produce a sequence (A_j) of pairwise-disjoint measurable sets and a decreasing sequence of sets B_j, $A_{j+1} \subset B_j$, $B_{j+1} = B_j \setminus A_{j+1}$, $\mu(B_j) \leqslant 1/2^j$, with the property that on each A_j the sequence (f_n) converges to f uniformly.

By the uniform limit theorem, the function f is integrable on each set A_j. Further, for any $N \in \mathbb{N}$ we have the estimate

$$\sum_{k=1}^{N} \int_{A_k} f \, d\mu = \lim_{n \to \infty} \sum_{k=1}^{N} \int_{A_k} f_n \, d\mu = \lim_{n \to \infty} \int_{\bigcup_{k=1}^{N} A_k} f_n \, d\mu \leqslant \varliminf_{n \to \infty} \int_A f_n \, d\mu,$$

and so $\sum_{k=1}^{\infty} \int_{A_k} f \, d\mu \leqslant \varliminf_{n \to \infty} \int_A f_n \, d\mu$. By Theorem 3 in Subsection 4.2.3, the function f is integrable on $D = \bigcup_{k=1}^{\infty} A_k$, and $\int_D f \, d\mu \leqslant \varliminf_{n \to \infty} \int_A f_n \, d\mu$. It remains to observe that, by construction, the complement of D in A has measure zero: $\mu(A \setminus \bigcup_{k=1}^{\infty} A_k) = \mu(\bigcap_{k=1}^{\infty} B_k) = \lim_{n \to \infty} \mu(B_n) = 0$. Hence, the function f is integrable on the whole set A and $\int_A f \, d\mu \leqslant \varliminf_{n \to \infty} \int_A f_n \, d\mu$. $\qquad \square$

Remark 1. The assumption in the formulation of Fatou's lemma that the functions f_n are non-negative can be slightly relaxed: it suffices to require that all f_n are greater than or equal to some integrable function g. Indeed, in this case the functions $f_n - g$ are non-negative, and Fatou's lemma applies to them in the original formulation. That is, the function $f - g$ is integrable (and hence so is $f = g + (f - g)$), and $\int_A (f - g)d\mu \leqslant \varliminf\limits_{n\to\infty} \int_A (f_n - g)d\mu$. It remains to add $\int_A g d\mu$ to both sides of the last inequality to obtain the desired estimate $\int_A f\, d\mu \leqslant \varliminf\limits_{n\to\infty} \int_A f_n d\mu$.

Exercises

1. If the measurable function f is positive and the integrals of all integrable functions that are smaller than f are jointly bounded, then f is integrable.

2. Verify that the example of the step functions $f_n = n \, \mathbb{1}_{(0,1/n)}$, given on $A = [0, 1]$ equipped with the Lebesgue measure, shows that under the assumptions of Fatou's lemma $\int_A f d\mu$ is not necessarily equal to $\varliminf\limits_{n\to\infty} \int_A f_n d\mu$.

3. Give an example showing that under the assumptions of Fatou's lemma the limit $\lim\limits_{n\to\infty} \int_A f_n d\mu$ does not necessarily exist.

4. Show that in Fatou's lemma the assumption that the functions f_n are non-negative can be replaced by the assumption that $f_n \geqslant 0$ almost everywhere.

5. Let $(A_k)_1^\infty$ be a sequence of pairwise disjoint subsets of non-zero measure of the interval $[0, 1]$. Consider the sequence of integrable functions $f_n = \sum_{k=1}^{2n} \frac{(-1)^k}{\lambda(A_k)} \mathbb{1}_{A_k}$. Show that the integrals of the functions f_n are equal to 0 (and consequently are jointly bounded), that the functions f_n converge at each point to the function $f = \sum_{k=1}^{\infty} \frac{(-1)^k}{\lambda(A_k)} \mathbb{1}_{A_k}$, but the limit function f is not integrable with respect to the Lebesgue measure λ. Which of the hypotheses of Fatou's lemma is not satisfied here?

4.4.2 Lebesgue's Dominated Convergence Theorem

Theorem 1. *Suppose that on the set A there is given a sequence (f_n) of integrable functions which converges almost everywhere to a function f. Suppose further that the sequence (f_n) has an integrable majorant g (i.e., g is integrable and $|f_n| \leqslant g$ for all n). Then the limit function f is integrable and*

$$\int_A f\, d\mu = \lim_{n\to\infty} \int_A f_n d\mu.$$

Proof. All the functions f_n are bounded from below by the integrable function $-g$, and the integrals of the f_n's are jointly bounded: $\int_A f_n d\mu \leqslant \int_A g \, d\mu < \infty$. By the remark made after the proof of the Fatou lemma, the function f is integrable and $\int_A f \, d\mu \leqslant \varliminf_{n \to \infty} \int_A f_n d\mu$. Applying the same reasoning to the functions $-f_n$ yields

$$\int_A (-f)d\mu \leqslant \varliminf_{n\to\infty} \int_A (-f_n)d\mu = -\varlimsup_{n\to\infty} \int_A f_n d\mu,$$

i.e.,

$$\varlimsup_{n\to\infty} \int_A f_n d\mu \leqslant \int_A f \, d\mu \leqslant \varliminf_{n\to\infty} \int_A f_n d\mu. \tag{$*$}$$

But when can the upper limit of a sequence be bounded from above by the lower limit of the same sequence? Only when the sequence has a true limit. Hence, from the double inequality $(*)$ it follows that the limit $\lim_{n\to\infty} \int_A f_n d\mu$ exists and $\int_A f \, d\mu$ is equal to this limit. $\square$

Exercises

1. Based on Lebesgue's theorem and Exercise 14 in Subsection 4.2.2 on the relationship between series and integrals for functions defined on $\mathbb{N}$ (see also Exercise 5 of Subsection 2.1.4), prove the following *dominated convergence theorem for series*: Suppose given an infinite matrix $(a_{n,m})_{n,m=1}^{\infty}$ in which every column converges to a corresponding limit: $\lim_{n\to\infty} a_{n,m} = a_m$. Further, suppose that there exists a sequence (b_m) of positive numbers, $\sum_{m=1}^{\infty} b_m < \infty$, which dominates all the rows of the matrix, i.e., $|a_{n,m}| \leqslant b_m$ for all $n, m \in \mathbb{N}$. Then the series $\sum_{m=1}^{\infty} a_m$ is absolutely convergent and $\sum_{m=1}^{\infty} a_m = \lim_{n\to\infty} \sum_{m=1}^{\infty} a_{n,m}$.

2. Formulate and prove the analogue of Fatou's lemma for series. ·

3. Show that the assumption $|f_n| \leqslant g$ in the statement of the Lebesgue dominated convergence theorem can be replaced by the assumption that $|f_n| \overset{\text{a.e.}}{\leqslant} g$.

4.4.3 Levi's Theorems on Sequences and Series

Theorem 1 (Levi's theorem on monotone sequences). *Suppose* $f_1 \leqslant f_2 \leqslant f_3 \leqslant \cdots$ *is a non-decreasing sequence of functions that are integrable on* A, *and the integrals of the functions* f_n *are jointly bounded by a constant* $C < \infty$. *Then the sequence* (f_n) *converges almost everywhere to an integrable function* f, *and* $\int_A f \, d\mu = \lim_{n\to\infty} \int_A f_n d\mu$.

Proof. With no loss of generality we may assume that all $f_n \geqslant 0$ (the general case reduces to this particular one by introducing the auxiliary functions $f_n - f_1$). Thanks to monotonicity, at each point $t \in A$ the sequence $(f_n(t))$ converges to either a finite limit, or to $+\infty$. Denote by B the set of those points $t \in A$ where $f_n(t) \to +\infty$ as $n \to \infty$. We claim that B is a set of measure zero. Indeed, consider for each pair $n, m \in \mathbb{N}$ the set $B_{n,m} = \{t \in A : f_n(t) > m\}$, and put $B_m = \bigcup_{n=1}^{\infty} B_{n,m}$. In other words, B_m is the set of those points $t \in A$ where, starting with some index n, the values of f_n are larger than m. Clearly, $B = \bigcap_{m=1}^{\infty} B_m$. By the Chebyshev inequality (Lemma 1 in Subsection 4.3.1),

$$\mu(B_{n,m}) \leqslant \frac{1}{m} \int_A f_n d\mu \leqslant \frac{C}{m}.$$

Since for fixed m the sets $B_{n,m}$ grow as n grows,

$$\mu(B_m) = \lim_{n \to \infty} \mu(B_{n,m}) \leqslant C/m.$$

In its turn, the set B_m decreases with the growth of m, i.e.,

$$\mu(B) = \lim_{m \to \infty} \mu(B_m) \leqslant \lim_{m \to \infty} C/m = 0,$$

as claimed.

Now define the function f on B in an arbitrary way (for instance, put $f = 0$ on B), and on $A \setminus B$, where, by construction, for each t the sequence $(f_n(t))$ has a finite limit, put $f(t) = \lim_{n \to \infty} f_n(t)$. With this definition, the sequence f_n converges almost everywhere to f, and so, by Fatou's lemma, f is integrable. Moreover, f is an integrable majorant for all functions f_n, so to complete the proof it remains to apply the Lebesgue dominated convergence theorem.   □

Theorem 2 (Levi's theorem on series). *Suppose that on the set A there is given a sequence (f_n) of non-negative integrable functions satisfying $\sum_{n=1}^{\infty} \int_A f_n d\mu < \infty$. Then the series $\sum_{n=1}^{\infty} f_n$ converges almost everywhere to an integrable function f, and $\int_A f \, d\mu = \sum_{n=1}^{\infty} \int_A f_n d\mu$.*

Proof. It suffices to note that the sequence of partial sums of the series $\sum_{n=1}^{\infty} f_n$ satisfies the conditions of Levy's theorem on monotone sequences.   □

Exercises

1. Show that the condition $f_n \leqslant f_{n+1}$, $n \in \mathbb{N}$, in the statement of Levi's theorem on monotone sequences can be replaced by the condition "for all $n \in \mathbb{N}$, $f_n \leqslant f_{n+1}$ almost everywhere".

2. Provide the details of the proof of Levi's theorem on series.

3. Using the representation of a function as the difference of its positive and negative parts, prove the following strengthening of Levi's theorem on series: Suppose the functions f_n are integrable on the set A and $\sum_{n=1}^{\infty} \int_A |f_n| d\mu < \infty$. Then the series $\sum_{n=1}^{\infty} f_n$ converges almost everywhere to an integrable function f, and $\int_A f \, d\mu = \sum_{n=1}^{\infty} \int_A f_n d\mu$. This result will be used later, in Subsection 6.3.2, to prove the completeness of the space L_1.

4. Derive the solution of Exercise 4 of Subsection 4.3.1 by applying Levi's theorem on monotone sequences to the sequence $f_n = n|f|$.

4.4.4 A Monotone Class Theorem for Functions

Definition 1. Let (Ω, Σ, μ) be a finite measure space. A family $\mathcal{E}$ of functions integrable on Ω is called a *monotone class of functions* if it obeys the following axioms:

(1) if $f_1, f_2 \in \mathcal{E}$, then $a_1 f_1 + a_2 f_2 \in \mathcal{E}$ for any scalars a_1, a_2 (linearity);

(2) if $f_1, f_2, \ldots, f_n, \ldots \in \mathcal{E}$, the f_n form a non-decreasing sequence which converges at each point to a function f, and $\sup_n \int_\Omega f_n d\mu = C < \infty$, then $f \in \mathcal{E}$ (analogue of Levi's theorem);

(3) if $f \in \mathcal{E}$, $f \geqslant 0$, and $\int_\Omega f \, d\mu = 0$, then all the measurable functions g which satisfy the double inequality $0 \leqslant g \leqslant f$ also belong to $\mathcal{E}$ (analogue of completeness).

Note that by passing from f_n to $-f_n$ one readily obtains another property of a monotone class:

(2′) if $f_1, f_2, \ldots f_n, \ldots \in \mathcal{E}$, the functions f_n form a non-increasing sequence which converges at each point to a function f, and $\inf_n \int_\Omega f_n d\mu > -\infty$, then $f \in \mathcal{E}$.

Theorem 1. *Let (Ω, Σ, μ) be a finite measure space obtained, as described in Sect. 2.2, by extending the measure μ from some unital semiring $\Phi \subset \Sigma$ to Σ. Let $\mathcal{E}$ be a monotone class of functions which contains the characteristic functions of all elements of the semiring Φ. Then $\mathcal{E}$ coincides with the set of all integrable functions on Ω.*

Proof. Denote by $\mathcal{M}$ the family of all sets whose characteristic functions belong to $\mathcal{E}$. Then $\mathcal{M}$ is a monotone class that contains Φ as a subclass. By the monotone class theorem for sets (Subsection 2.2.4), $\mathcal{M} = \Sigma$. Hence, the class $\mathcal{E}$ contains the characteristic functions of all measurable sets.

Every finitely-valued integrable function has the form $\sum_{k=1}^{n} a_k \mathbb{1}_{A_k}$, with $A_k \in \Sigma$, and thanks to linearity all such functions lie in $\mathcal{E}$. Every non-negative countably-valued integrable function $f = \sum_{k=1}^{\infty} a_k \mathbb{1}_{A_k}, a_k \geqslant 0$, is the limit of a non-decreasing sequence of finitely-valued functions $f_n = \sum_{k=1}^{n} a_k \mathbb{1}_{A_k}$, and consequently also lies in $\mathcal{E}$. Further, every non-negative integrable function can be represented as the limit of a non-decreasing sequence of non-negative countably-valued integrable functions, and

finally, every integrable function f is representable as the difference $f = f^+ - f^-$
of two integrable functions. $\square$

Exercises

1. Let $\mathcal{E}$ be a monotone class, $f \in \mathcal{E}$, $f \geqslant 0$ and $\int_\Omega f \, d\mu = 0$. Then any measurable
function g that satisfies the inequality $\|g\| \leqslant f$ also lies in $\mathcal{E}$. Moreover, if the measure
space (Ω, Σ, μ) is complete (which we tacitly assume in this chapter), then every
function g satisfying $|g| \leqslant f$ is automatically measurable and belongs to $\mathcal{E}$.

2. Show the independence of the axioms of a monotone class. In other words, give
examples of families of integrable functions that satisfy two of the monotone class
axioms, but not the remaining axiom. For instance, an example of a family that
satisfies axioms (1) and (3), but not axiom (2), and so on.

3. Verify that Theorem 1 can be proved in the slightly more general situation of a
monotone class of measurable integrable functions defined on the (possibly incomplete) measure space (Ω, Σ, μ) described in Exercise 10 of Subsection 2.2.4.

4.5   Multiple Integrals

4.5.1   Products of Measure Spaces

Let $(\Omega_1, \Sigma_1, \mu_1)$ and $(\Omega_2, \Sigma_2, \mu_2)$ be finite measure spaces and $\Omega = \Omega_1 \times \Omega_2$. Following Subsection 2.1.3, a *rectangle* in Ω is any set of the form $A_1 \times A_2$, where
$A_1 \in \Sigma_1$ and $A_2 \in \Sigma_2$. We let Φ denote the family of all rectangles in Ω.

Theorem 1. *The family Φ is a unital semiring.*

Proof. Let $A = A_1 \times A_2$ and $B = B_1 \times B_2$ be arbitrary rectangles. Then $A \cap B = (A_1 \cap B_1) \times (A_2 \cap B_2)$ is again a rectangle. Next, $\Omega \setminus A = ((\Omega_1 \setminus A_1) \times \Omega_2) \sqcup (A_1 \times (\Omega_2 \setminus A_2))$, that is, the complement of a rectangle can be written as the disjoint
union of two rectangles. Finally, the space Ω is itself a rectangle. $\square$

Define the measure μ on Φ by the rule $\mu(A_1 \times A_2) = \mu_1(A_1) \cdot \mu_2(A_2)$.

Theorem 2. *The measure μ on Φ is countably additive.*

Proof. Let $A = A_1 \times A_2$, $B_n = A_{1,n} \times A_{2,n}$, and suppose the rectangles B_n are pairwise disjoint and their union is the rectangle A. For any $t \in A_1$, denote by $N(t)$ the set
of all indices n for which $t \in A_{1,n}$. Then the family of sets $A_{2,n}$, $n \in N(t)$, is disjoint,
and $\bigcup_{n \in N(t)} A_{2,n} = A_2$. Consider on A_1 the auxiliary functions $f_n = \mu_2(A_{2,n}) \mathbb{1}_{A_{1,n}}$.

These functions are μ_1-integrable, and their integrals are equal to $\mu(B_n)$. We note that for each $t \in A_1$ it holds that

$$\sum_{n \in \mathbb{N}} f_n(t) = \sum_{n \in N(t)} \mu_2(A_{2,n}) = \mu_2 \left(\bigcup_{n \in N(t)} A_{2,n} \right) = \mu_2(A_2).$$

It remains to integrate both sides of this equality, which is allowed by Levi's theorem on series, to conclude that

$$\sum_{n \in \mathbb{N}} \mu(B_n) = \int_{A_1} \sum_{n \in \mathbb{N}} f_n(t) d\mu_1 = \int_{A_1} \mu_2(A_2) d\mu_1 = \mu_1(A_1) \cdot \mu_2(A_2) = \mu(A),$$

as needed.   □

Let us apply the extension of measure recipe described in Sect. 2.2 to the measure μ on Φ. The resulting measure space (Ω, Σ, μ) is called the *product* of the measure spaces $(\Omega_1, \Sigma_1, \mu_1)$ and $(\Omega_2, \Sigma_2, \mu_2)$. The measure μ is denoted by $\mu_1 \times \mu_2$, and the elements of the σ-algebra Σ are also referred to as $\mu_1 \times \mu_2$-measurable sets. It is clear that the σ-algebra Σ includes as a subsystem the smallest σ-algebra that contains all rectangles (the latter was denoted by $\Sigma_1 \otimes \Sigma_2$).

Remark 1. Let $(\Omega_k, \Sigma_k, \mu_k)$, $k \in \{1, 2, \ldots, n\}$ be a finite collection of measure spaces. A *parallelepiped* in $\prod_{k=1}^{n} \Omega_k = \Omega_1 \times \Omega_2 \times \cdots \times \Omega_n$ is a set of the form $\prod_{k=1}^{n} A_k$ with $A_k \in \Sigma_k$. Put $\mu \left(\prod_{k=1}^{n} A_k \right) = \prod_{k=1}^{n} \mu(A_k)$. One can show that the parallelipipeds form a unital semiring and that the measure μ is countably additive on this semiring. Applying again the extension of measure recipe, we obtain a measure space, called the *product* of the spaces $(\Omega_k, \Sigma_k, \mu_k)$. However, to avoid repeating in the case of arbitrary n the reasoning used in the case $n = 2$, it is more convenient to define the product of a finite number of measure spaces by induction on the number of factors. That is, we first take the product of two spaces, then take the product of the resulting space by the third space, then the product by the next space, and so on. This second approch is also more convenient because the theorems obtained for the product of a pair of measure spaces can be subsequently extended by induction to an arbitrary finite number of factors.

Remark 2. For the moment, we only have a formal definition of the product of measure spaces. The reader will get a better feeling about what this notion means after studying the next subsection. In particular, Exercise 1 in Subsection 4.5.2, where an explicit formula for the calculation of the measure analogous to the formula for the area of a curvilinear trapezoid will prove helpful. However, one can successfully work with the product of measures without resorting to this formula, using only the countable additivity and completeness of the measure, together with the formula for the measure of a rectangle.

Exercises

Regard the unit square K as the product of intervals $[0, 1] \times [0, 1]$ and define the two-dimensional Lebesgue measure $\lambda \times \lambda$ on K as the product of the usual one-dimensional Lebesgue measures on the corresponding factor intervals.

1. Verify that the diagonal of the square K is a set of measure zero.

2. Let the set $A \subset [0, 1]$ have one-dimensional measure zero. Show that the set of all points $x = (x_1, x_2) \in K$ for which $x_1 - x_2 \in A$ has two-dimensional measure zero.

3. Let A be a subset of K that has an area. Show that $(\lambda \times \lambda)(A)$ coincides with the area of A.

4. Let f be an arbitrary measurable function on $[0, 1]$. Show that the function $g : K \to \mathbb{R}$ defined by the rule $g(x_1, x_2) = f(x_1)$ is measurable on K.

5. Show that the function g given by $g(x_1, x_2) = x_1 - x_2$ is measurable on K.

6. Show that the function from the preceding exercise has the following property: for any Lebesgue measurable set $A \subset [0, 1]$, the preimage $f^{-1}(A)$ is Lebesgue measurable in the square.

7. Let f be a measurable function on $[0, 1]$. Show that the function $g : K \to \mathbb{R}$ defined by the rule $g(x_1, x_2) = f(x_1 - x_2)$ is measurable on K.

8. Consider the function $g : K \to \mathbb{R}$ defined by the rule $g(x_1, x_2) = x_1$. Let D denote the main diagonal of the square K. Now define the measure μ on the σ-algebra $\mathfrak{B}$ of Borel subsets of the square K by the rule $\mu(A) = \lambda(g(A \cap D))$. Verify that μ is a countably-additive measure on $\mathfrak{B}$ and that the values of the measures μ and $\lambda \times \lambda$ coincide on the rectangles of the form $[a, b] \times [0, 1]$ and $[0, 1] \times [a, b]$, but on squares of the form $[a, b] \times [a, b]$ they differ.

9. Show that $\mathfrak{B}$ is the smallest σ-algebra of subsets of the square K that contains all rectangles of the form $[a, b] \times [0, 1]$ and $[0, 1] \times [a, b]$. Combined with the preceding exercise, this fact provides a nice (although not the simplest possible, see the comment to Exercise 3 of Subsection 2.2.1) example of two countably-additive measures which coincide on a family of subsets that generates a given σ-algebra, but do not coincide on the whole σ-algebra.

4.5.2 Double Integrals and Fubini's Theorem

Throughout Subsections 4.5.2 and 4.5.3, $(\Omega_1, \Sigma_1, \mu_1)$ and $(\Omega_2, \Sigma_2, \mu_2)$ will be finite measure spaces, and (Ω, Σ, μ) will denote their product. Each element of the set $\Omega = \Omega_1 \times \Omega_2$ has the form (t_1, t_2), where $t_1 \in \Omega_1$ and $t_2 \in \Omega_2$. Accordingly, it is natural to regard any function f defined on Ω as a function $f(t_1, t_2)$ of two variables and, by analogy with what is done in calculus, call the integral $\int_\Omega f d\mu$ a *double integral*. When we consider the integral with respect to one of the variables with the other kept fixed, we will use the expression $\int_{\Omega_1} f(t_1, t_2) d\mu_1(t_1)$, where the notation $d\mu_1(t_1)$ emphasizes with respect to which variable one is integrating.

Definition 1. We say that for the function $f : \Omega \to \mathbb{R}$ there exists the *iterated integral* $\int_{\Omega_2} \left[\int_{\Omega_1} f(t_1, t_2) d\mu_1(t_1) \right] d\mu_2(t_2)$ if, for almost every fixed value of the variable $t_2 \in \Omega_2$, the function $t_1 \mapsto f(t_1, t_2)$ is μ_1-integrable on Ω_1, and the function $g(t_2) = \int_{\Omega_1} f(t_1, t_2) d\mu_1(t_1)$ of the variable t_2 is μ_2-integrable on Ω_2.

Theorem 1 (Fubini's Theorem). *If the function $f : \Omega \to \mathbb{R}$ is integrable as a function of two variables (i.e., the double integral $\int_\Omega f d\mu$ exists), then the iterated integral of f exists, and the double integral is equal to the iterated integral:*

$$\int_\Omega f d\mu = \int_{\Omega_2} \left[\int_{\Omega_1} f(t_1, t_2) d\mu_1(t_1) \right] d\mu_2(t_2).$$

Proof. We say that the function $f : \Omega \to \mathbb{R}$ belongs to the *Fubini class* (and write $f \in \text{Fub}(\mu)$), if f is μ-integrable on Ω, the iterated integral of f exists, and the double integral is equal to the iterated integral: $\int_\Omega f d\mu = \int_{\Omega_2} \left[\int_{\Omega_1} f(t_1, t_2) d\mu_1(t_1) \right] d\mu_2(t_2)$. We need to show that the Fubini class coincides with the class of functions that are integrable as functions of two variables. Since the Fubini class contains the characteristic functions of all rectangles, it suffices to show (see Theorem 1 in Subsection 4.4.4) that the Fubini class is a monotone class of functions on (Ω, Σ, μ). The first of the monotone class axioms, linearity, is readily verified. The verification of the second and third axioms requires some efforts.

We need to prove two statements.

A. If $f_n \in \text{Fub}(\mu)$, $n \geqslant 1$, the functions f_n form a non-decreasing sequence that converges at each point to some function f, and $\sup_n \int_\Omega f_n d\mu = C < \infty$, then $f \in \text{Fub}(\mu)$.

B. If $f \in \text{Fub}(\mu)$, $f \geqslant 0$, $\int_\Omega f d\mu = 0$, and the measurable function g satisfies the inequality $0 \leqslant g \leqslant f$, then $g \in \text{Fub}(\mu)$.

Let us start with assertion A. Denote $\int_{\Omega_1} f_n(t_1, t_2) d\mu_1(t_1)$ by $g_n(t_2)$. By hypothesis, the functions g_n are defined almost everywhere and are integrable on Ω_2; moreover, $\int_{\Omega_2} g_n d\mu_2 = \int_\Omega f_n d\mu$. Further, g_n form almost everywhere a non-decreasing sequence of functions, and $\sup_n \int_{\Omega_2} g_n d\mu_2 = \sup_n \int_\Omega f_n d\mu = C < \infty$. By Levi's theorem, the sequence (g_n) converges almost everywhere on Ω_2 to an integrable function g, and

$$\int_{\Omega_2} g d\mu_2 = \lim_{n \to \infty} \int_{\Omega_2} g_n d\mu_2 = \lim_{n \to \infty} \int_\Omega f_n d\mu. \tag{1}$$

Denote by D the set of all points $t_2 \in \Omega_2$ for which the values $g_n(t_2)$ and $g(t_2)$ are defined, $\int_{\Omega_1} f_n(t_1, t_2)d\mu_1(t_1) = g_n(t_2)$, $g_n(t_2)$ do not decrease with the growth of n, and converge to $g(t_2)$. By construction, $\mu_2(\Omega_2 \setminus D) = 0$. For each point $t_2 \in D$, the functions $f_n(\cdot, t_2)$ are integrable with respect to the first variable, do not decrease with the growth of n, and converge to the function $f(\cdot, t_2)$. Moreover, we have $\int_{\Omega_1} f_n(t_1, t_2)d\mu_1(t_1) = g_n(t_2) \leqslant g(t_2) < \infty$. Applying again Levi's theorem, but now with respect to the variable t_1, we conclude that for every $t_2 \in D$ (i.e., for almost every value of the variable t_2) the function f is integrable with respect to t_1, and

$$\int_{\Omega_1} f(t_1, t_2)d\mu_1(t_1) = \lim_{n \to \infty} \int_{\Omega_1} f_n(t_1, t_2)d\mu_1(t_1) = \lim_{n \to \infty} g_n(t_2) = g(t_2). \quad (2)$$

Finally, Levi's theorem, applied to the functions f_n on Ω (i.e., with respect to both variables), yields the equality $\lim_{n \to \infty} \int_\Omega f_n d\mu = \int_\Omega f \, d\mu$. Combining this equality with relations (1) and (2), we see that

$$\int_{\Omega_2} \left[\int_{\Omega_1} f(t_1, t_2)d\mu_1(t_1) \right] d\mu_2(t_2) = \int_{\Omega_2} g \, d\mu_2$$

$$= \lim_{n \to \infty} \int_{\Omega_2} g_n d\mu_2 = \lim_{n \to \infty} \int_\Omega f_n d\mu = \int_\Omega f \, d\mu,$$

that is, $f \in \mathrm{Fub}(\mu)$.

Now let us prove assertion B. First of all, the relations $0 \leqslant g \leqslant f$ and $\int_\Omega f \, d\mu = 0$ imply that g is integrable on Ω and

$$\int_\Omega g \, d\mu = 0. \quad (3)$$

Denote $\int_{\Omega_1} f(t_1, t_2)d\mu_1(t_1)$ by $h(t_2)$. Since $f \in \mathrm{Fub}(\mu)$, the function h is defined almost everywhere on Ω_2, is integrable, and $\int_{\Omega_2} h \, d\mu_2 = \int_\Omega f \, d\mu = 0$. Thanks to non-negativity, h vanishes almost everywhere on Ω_2 (surely the respected reader has already managed to solve Exercise 4 in Subsection 4.3.1). Denote by Δ the set of all points $t_2 \in \Omega_2$ in which $h(t_2) = 0$. The complement of Δ in Ω_2 has measure zero, and for each fixed $t_2 \in \Delta$ the function $f(t_1, t_2)$ is integrable on Ω_1 with respect to the variable t_1, and $\int_{\Omega_1} f(t_1, t_2)d\mu_1(t_1) = 0$. Again thanks to positivity, for each fixed $t_2 \in \Delta$, we have $f(t_1, t_2) = 0$ for almost every $t_1 \in \Omega_1$. But due to the inequality $0 \leqslant g \leqslant f$, at the points where $f(t_1, t_2) = 0$ one also has $g(t_1, t_2) = 0$. Hence, for each fixed $t_2 \in \Delta$ we have $g(t_1, t_2) = 0$ for almost all $t_1 \in \Omega_1$. Therefore, for $t_2 \in \Delta$ (i.e., for almost every value of t_2) the function g is integrable with respect to t_1, and $\int_{\Omega_1} g(t_1, t_2)d\mu_1(t_1) = 0$. In conjunction with equality (3), this shows that the function g belongs to the Fubini class. $\qquad \square$

Remark 1. Since in the assumptions of Fubini's theorem the variables t_1 and t_2 play equivalent roles, one can also exchange their roles in the conclusion of the theorem. Hence, if the double integral exists, then the iterated integrals are defined in the two possible orders of integration, and both these integrals are equal to the double integral. Therefore, if the double integral exists, one can change the order of integration in the iterated integral without affecting the result:

$$\int_{\Omega_2}\left[\int_{\Omega_1} f(t_1, t_2)d\mu_1(t_1)\right] d\mu_2(t_2) = \int_{\Omega_1}\left[\int_{\Omega_2} f(t_1, t_2)d\mu_2(t_2)\right] d\mu_1(t_1).$$

It is in this form that Fubini's theorem is most often applied.

Exercises

1. Let A be a measurable subset of $\Omega = \Omega_1 \times \Omega_2$. For any $t_1 \in \Omega_1$ denote by A_{t_1} the set of all points $t_2 \in \Omega_2$ such that $(t_1, t_2) \in A$. Using Fubini's theorem, show that $A_{t_1} \in \Sigma_2$ for almost every $t_1 \in \Omega_1$ and $\mu(A) = \int_{\Omega_1} \mu_2(A_{t_1})d\mu_1(t_1)$.

2. In the setting of the preceding exercise, suppose that the function f is μ-integrable on A. Show that $\int_A f\, d\mu = \int_{\Omega_1}\left[\int_{A_{t_1}} f(t_1, t_2)d\mu_2(t_2)\right]d\mu_1(t_1)$.

3. Suppose the set $A_1 \subset [0, 1]$ is not Lebesgue measurable. Define the set $A \subset [0, 1] \times [0, 1]$ as the union of the sets $A_1 \times [0, 1/2]$ and $([0, 1] \setminus A_1) \times (1/2, 1]$. Show that for the function $f = \mathbb{1}_A$ the iterated integral $\int_{[0,1]}\left[\int_{[0,1]} f(t, \tau)d\lambda(\tau)\right]d\lambda(t)$ exists. That is this integral equal to? Show that the integral $\int_{[0,1]}\left[\int_{[0,1]} f(t, \tau)d\lambda(t)\right]d\lambda(\tau)$ does not exist. Is f integrable as a function of two variables? Measurable?

4. Give an example of a function on the square for which the two iterated integrals exist, but are not equal to one another.

5. Give an example of a function on the square for which the two iterated integrals exist and are equal, but the function is not integrable as a function of two variables.

4.5.3   A Converse to Fubini's Theorem

As the exercises above show, changing the order of integration in an iterated integral is not always possible. One condition under which this can be done, namely the joint integrability of the function, does indeed sound pleasant, but it is too abstract. Indeed, how does one determine for a concrete function, say, of two variables, that it is integrable as a function of two variables? It would have been much simpler to be able to deal with the iterated integral if, after verifying that it exists for some function in one order, we could be sure that this function is also integrable in the other order, as well as a function of two variables. Unfortunately, not everything in life is so simple. Yet, as the next theorem shows, there are no special grounds for complaining about life (at least concerning this subject). This "inverse Fubini theorem" is due to L. Tonelli.

Theorem 1. *Let f be a non-negative measurable function on $\Omega = \Omega_1 \times \Omega_2$ for which the iterated integral $\int_{\Omega_2} \left[\int_{\Omega_1} f(t_1, t_2) d\mu_1(t_1) \right] d\mu_2(t_2)$ exists. Then for f the double integral also exists, and consequently*

$$\int_{\Omega_2} \left[\int_{\Omega_1} f(t_1, t_2) d\mu_1(t_1) \right] d\mu_2(t_2) = \int_{\Omega_1} \left[\int_{\Omega_2} f(t_1, t_2) d\mu_2(t_2) \right] d\mu_1(t_1).$$

Proof. Consider the sets $A_n = \{t \in \Omega : f(t) \leqslant n\}$ and the functions $f_n = f \cdot \mathbb{1}_{A_n}$. Each f_n is bounded and measurable on Ω, and hence μ-integrable (see Theorem 1 and Exercise 3 in Subsection 4.3.3). Further,

$$\int_{\Omega} f_n d\mu = \int_{\Omega_2} \left[\int_{\Omega_1} f_n(t_1, t_2) d\mu_1(t_1) \right] d\mu_2(t_2) \leqslant \int_{\Omega_2} \left[\int_{\Omega_1} f(t_1, t_2) d\mu_1(t_1) \right] d\mu_2(t_2),$$

i.e., the integrals $\int_{\Omega} f_n d\mu$ are bounded from above by a constant that does not depend on n. Finally, the functions f_n form a non-decreasing sequence and converge pointwise to f. To complete the proof, it remains to apply Levi's theorem. $\qquad\square$

Recall that if a function is measurable, then its integrability is equivalent to the integrability of its modulus.

Corollary 1. *For a measurable function f on Ω the following conditions are equivalent:*

(1) *f is integrable on Ω as a function of two variables;*

(2) *for the function $|f|$ there exists the iterated integral*

$$\int_{\Omega_2} \left[\int_{\Omega_1} |f(t_1, t_2)| d\mu_1(t_1) \right] d\mu_2(t_2). \qquad\square$$

Remark 1. Since the product $\prod_{k=1}^{n} (\Omega_k, \Sigma_k, \mu_k)$ of a finite number of measure spaces is constructed inductively, as $\left(\prod_{k=1}^{n-1} (\Omega_k, \Sigma_k, \mu_k) \right) \times (\Omega_n, \Sigma_n, \mu_n)$, the results of the last two subsections can be generalized with no difficulty to the case of multiple integrals.

Remark 2. As already mentioned, for incomplete measure spaces the integrability of
a function does not necessarily imply its measurability: to obtain measurability one
still has to remove a set of measure zero. A product of measure spaces is complete by
construction, but in principle the factors themselves may also be incomplete spaces.
In such a case, if for some reason or another we need that a function of two variables
be measurable as a function of the first or the second variable, then we need to restrict
ourselves to $\Sigma_1 \otimes \Sigma_2$-measurable functions. We leave to the reader to verify that in
this case, for each fixed value of one of the variables, the function is measurable
in the other variable. To show this, it is reasonable to consider first characteristic
functions of sets (see Exercise 6 in Subsection 2.1.3) and then use the approximation
of measurable functions by countably-valued functions. When one speaks about the
version of the Fubini theorem for a product of incomplete measure spaces and one
also needs the measurability of the inner integral as a function of the other variable,
then one runs into an unavoidable difficulty: for some values of t_2, the inner integral
$\int_{\Omega_1} f(t_1, t_2)d\mu_1(t_1)$ may be not defined (for example, it may be infinite). One can
overcome this difficulty by either considering bounded functions, or considering
non-negative functions and extending the definition of measurability to functions
that are allowed to take the value $+\infty$ at some points. In this way the Fubini theorem
for characteristic functions of sets can be deduced from the version of the monotone
class theorem for sets given in Exercise 10 of Subsection 2.2.4, with Φ being the
semiring of rectangles and $\Sigma_1 \otimes \Sigma_2$ being the σ-algebra in question, and then one
can appeal to the proof of the monotone class theorem for functions, together with
the fact that in Levi's theorem on monotone sequences the exceptional set on which
the sequence of functions goes to infinity is measurable.

Exercises

1. Show that if the function f on $[0, 1] \times [0, 1]$ is Riemann integrable as a function
of two variables, then it is also Lebesgue integrable as a function of two variables.

2. Prove the formula for passing to polar coordinates in the Lebesgue integral.

4.6 The Lebesgue Integral on an Interval and on the Real
Line

4.6.1 The Lebesgue Integral and the Improper Integral
on an Interval

As we already remarked in Exercise 8 of Subsection 4.2.2, from the condition (2)
of Theorem 1 in Subsection 4.2.2 it obviously follows that any Riemann integrable
function $f : [a, b] \to \mathbb{R}$ is also Lebesgue integrable. Moreover, by Theorem 1 in

Subsection 4.3.3, all bounded measurable functions on an interval are Lebesgue integrable.

If a function is Riemann integrable, then it necessarily is bounded. For this reason in calculus one studies in detail the improper integral as a means of defining the integral for some unbounded functions on an interval. To get a better feeling about the nature of the Lebesgue integral, we discuss below the connection between the improper integral and the Lebesgue integral.

Theorem 1. *Suppose the function $f : [a, b] \to \mathbb{R}$ is continuous everywhere except at the point a and is Lebesgue integrable on $[a, b]$. Then for f there exists the improper integral $\int_a^b f(t)dt$, and this integral is equal to the corresponding Lebesgue integral $\int_{[a,b]} f \, d\lambda$.*

Proof. Let $a_n \in (a, b]$, $a_n \to a$ as $n \to \infty$. Consider the auxiliary functions $f_n = f \cdot \mathbb{1}_{[a_n,b]}$. The f_n form a sequence of (both Riemann and Lebesgue) integrable functions that converge almost everywhere to the function f; moreover, $|f|$ serves as a majorant for all the functions f_n. By Lebesgue's theorem, $\int_{[a,b]} f_n d\lambda = \int_{[a_n,b]} f \, d\lambda$ tends to $\int_{[a,b]} f \, d\lambda$ as $n \to \infty$. But by the definition of the improper integral, for f there exists the improper integral, which is equal to $\int_{[a,b]} f \, d\lambda$. $\qquad\square$

Theorem 2. *Suppose the function $f : [a, b] \to \mathbb{R}$ is continuous everywhere except at the point a, and is non-negative. If the improper integral of f exists, then f is Lebesgue integrable on $[a, b]$.*

Proof. Let a_n and (f_n) be as in the proof of the preceding theorem. The sequence (f_n) is non-decreasing and tends almost everywhere to the function f. Next, by the definition of the improper integral, $\int_{[a,b]} f_n d\lambda = \int_{[a_n,b]} f \, d\lambda$ tends to $\int_a^b f(t)dt$ as $n \to \infty$. It remains to apply Levi's theorem on monotone sequences. $\qquad\square$

From calculus the reader is certainly familiar with examples of functions for which the improper integral exists, but the modulus of which is not integrable even in the improper sense. Such functions are not Lebesgue integrable, because the modulus of a Lebesgue-integrable function is itself Lebesgue integrable.

In what follows, if a function f is Lebesgue integrable on a segment, then for its Lebesgue integral we will use the notation $\int_{[a,b]} f(t)d\lambda$, as well as the notation $\int_a^b f(t)dt$, which is more usual in calculus courses.

Exercises

Which of the functions f listed below is Lebesgue integrable on the interval $[a, b]$?

1. $f(t) = t^2$, $[a, b] = [0, 1]$.

2. $f(t) = t^{-2}$, $[a, b] = [0, 1]$.

3. $f(t) = t^{-2}, [a, b] = [1, 2]$.

4. $f(t) = \sin(t^{-2}), [a, b] = [0, 1]$.

5. $f(t) = (\sin t)^{-2}, [a, b] = [0, 1]$.

6. $f(t) = t^{-1/2}, [a, b] = [0, 1]$.

7. $f(t) = 1/t, [a, b] = [-1, 1]$.

Which of the functions f on the square $[0, 1] \times [0, 1]$ listed below are integrable with respect to the Lebesgue measure in the plane, and which not?

8. $f(x, y) = x + y$.

9. $f(x, y) = 1/(x + y)$.

10. $f(x, y) = x - y$.

11. $f(x, y)) = 1/(x - y)$.

12. $f(x, y) = \sin \left(1/(x - y)\right)$.

13. For which values of the parameter α is the function $f(x, y) = (x^2 + y^2)^{-\alpha}$ integrable on $[0, 1] \times [0, 1]$?

4.6.2   The Integral with Respect to a σ-finite Measure

In Sects. 4.2–4.5 we studied the Lebesgue integral on a finite measure space (Ω, Σ, μ). To successfully define the Lebesgue integral on the real line, or, say, on an unbounded subset of the plane, we also have to consider the case of countably-additive measures that are allowed to take the value $+\infty$ on some elements of the σ-algebra Σ. We refer to such measures as *infinite*.

Thus, let (Ω, Σ, μ) be an infinite measure space. A subset $A \in \Sigma$ is called a *set of σ-finite measure* (alternatively, one says that the measure μ *is σ-finite on A*), if A can be written as a countable union of sets of finite measure. σ-finite measures have already been mentioned in Subsection 2.3.7. If the measure μ is σ-finite on A, then A can be represented as $A = \bigsqcup_{n=1}^{\infty} A_n, 0 < \mu(A_n) < \infty$, where the sets A_j are pairwise disjoint. We note also that any countable union of sets of σ-finite measure is itself a set of σ-finite measure.

For functions defined on a set A of σ-finite measure one can introduce partitions of the set A into subsets of finite measure. One can also introduce integral sums and the Lebesgue integral in the same way we proceeded for sets of finite measure. The reader can verify independently that the proofs of the main properties of the integral remain valid in this case, too. The only difficulty that one must overcome in extending the properties of the integral from the case of a finite measure to that of a σ-finite measure is that functions that are constant on A are not integrable. We insistently advise the reader to go over again the entire scheme for the construction of the Lebesgue integral presented above with the goal of independently constructing, following the already available model, the theory of the Lebesgue integral on a set of σ-finite measure. In the present subsection we will propose a roundabout path whereby, using an artificial device, we can reduce integration with respect to a σ-finite measure to the already familiar case of integration with respect to a finite measure. This allows one to reduce the properties of the integral with respect to a σ-finite measure to the corresponding already known results.

Let A be a set of σ-finite measure. Fix some representation of A in the form $A = \bigsqcup_{n=1}^{\infty} A_n$, with $A_n \in \Sigma$ and $0 < \mu(A_n) < \infty$. We are already familiar with the notion of integral on any set of finite measure, in particular on each of the sets A_n. Relying on this, we can give the following

Definition 1. A function f is said to be *integrable on A with respect to the σ-finite measure* μ if f is μ-integrable on each of the sets A_k and $\sum_{k=1}^{\infty} \int_{A_k} |f| d\mu < \infty$. In this case we put $\int_A f \, d\mu = \sum_{k=1}^{\infty} \int_{A_k} f \, d\mu$.

Let $a_n = 2^n \mu(A_n)$. Define on the family Σ_A of all measurable subsets of the set A a new measure μ_1 by the formula

$$\mu_1(B) = \sum_{n=1}^{\infty} \frac{\mu(B \cap A_n)}{a_n}.$$

The triplet (A, Σ_A, μ_1) so defined is a finite measure space. Consider on A the function $g = \sum_{n=1}^{\infty} a_n \mathbb{1}_{A_n}$.

Lemma 1. *Let $B \in \Sigma_A$ and $\mu(B) < \infty$. Then the function $h : B \to \mathbb{R}$ is integrable on B with respect to the measure μ if and only if the function $f \cdot g$ is integrable on the set B with respect to the measure μ_1. In this case, $\int_B h \, d\mu = \int_B hg \, d\mu_1$.*

Proof. On Σ_{A_n} we have $\mu_1 = \frac{1}{a_n} \mu$, and on A_n the function g is equal to the constant a_n. Hence, for $B \subset A_n$ the assertion is obvious:

$$\int_B h \, d\mu = \int_B a_n h \, d \left(\frac{1}{a_n} \mu \right) = \int_B hg \, d\mu_1.$$

An arbitrary set B can be written as $B = \bigsqcup_{n=1}^{\infty}(B \cap A_n)$, where the sets $B \cap A_n$ are disjoint and each of them is contained in the corresponding A_n. Since on B not only μ_1, but also the measure μ is finite, we can apply Theorem 4 of Subsection 4.2.3 on the countable additivity of the integral as a set function to the integral on B with respect to μ_1, as well as with respect to μ, and deduce our assertion by combining the already proved assertions on the sets $B \cap A_n$. $\square$

Lemma 2. *The function f is integrable on the set A with respect of the measure μ if and only if the function $f \cdot g$ is integrable on A with respect to the measure μ_1. In this case, $\int_A f \, d\mu = \int_A fg \, d\mu_1$.*

Proof. We apply Lemma 1 to each of the sets A_k, using Definition 1 and applying Theorem 4 of Subsection 4.2.3 to the measure μ_1.   ☐

In view of Lemma 2, the linearity of the integral, the fact that one can integrate inequalities, the measurability of integrable functions, the integrability criterion for measurable functions, Fatou's lemma, Lebesgue's dominated convergence theorem, Levi's theorem — all these properties for the integral with respect to the measure μ are obvious consequences of the corresponding properties for the measure μ_1.

The following property of the integral over a set of σ-finite measure means that the integral does not depend on the representation of the set A in the form $A = \bigsqcup_{n=1}^{\infty} A_n$ (the reader undoubtedly wondered whether this is true when Definition 1 was stated).

Theorem 1. *Let A be a set of σ-finite measure, $A = \bigsqcup_{n=1}^{\infty} B_n$, where $B_n \in \Sigma$ and $\mu(B_n) < \infty$. The integral $\int_A f \, d\mu$ exists if and only if the function f is integrable on each of the sets B_n and $\sum_{k=1}^{\infty} \int_{B_k} |f| \, d\mu < \infty$. In this case, $\int_A f \, d\mu = \sum_{k=1}^{\infty} \int_{B_k} f \, d\mu$.*

Proof. Let μ_1 be the finite measure figuring in Lemmas 1 and 2. Using Lemma 2 and applying Theorem 4 of Subsection 4.2.3 to μ_1, we see that the function f is μ-integrable on A if and only if the function $f \cdot g$ is μ_1-integrable on each B_n and the series $\sum_{k=1}^{\infty} \int_{B_k} |f| g \, d\mu_1$ converges. Then one has $\int_A f \, d\mu = \sum_{k=1}^{\infty} \int_{B_k} f g \, d\mu_1$. To complete the proof, it remains to apply Lemma 1 to the sets B_k, which is possible because the measure μ is finite on each B_k.   ☐

Exercises

In all the exercises below A_n and μ_1 are as in Definition 1 and Lemma 1.

1. Suppose $B \subset A_n$ for some n. Then $\mu_1(B) = \mu(B)/a_n$.

2. Let D be a partition of the set A into subsets A_n. Then for any partition $D_1 = \{\Delta_k\}_{k=1}^{\infty} \succ D$ and any choice T of marked points the integral sum $S_A(f, D, T) = \sum_{k=1}^{\infty} f(t_k) \mu(\Delta_k)$ with respect to the measure μ coincides with the integral sum $S_A(fg, D, T) = \sum_{k=1}^{\infty} (fg)(t_k) \mu_1(\Delta_k)$ with respect to the measure μ_1. Therefore, $\int_A f \, d\mu$ can be defined as a limit of integral sums, and by Lemma 2 this definition is equivalent to the preceding one.

3. Show that for a set $B \subset A$ the conditions $\mu(B) = 0$ and $\mu_1(B) = 0$ are equivalent.

4. For a sequence of functions (f_n) on A the following conditions are equivalent: $f_n \to f$ a.e. with respect to the measure μ; $f_n \to f$ a.e. with respect to the measure μ_1; $f_n g \to fg$ a.e. with respect to the measure μ_1.

5. Let λ be the Lebesgue measure on the real line. The measurable function f on the line is λ-integrable if and only if

$$\sum_{k=-\infty}^{\infty} \int_{[k,k+1)} |f| d\lambda < \infty,$$

and then

$$\int_{\mathbb{R}} f \, d\lambda = \sum_{k=-\infty}^{\infty} \int_{[k,k+1)} f \, d\lambda = \lim_{n \to \infty} \int_{[-n,n)} f \, d\lambda.$$

6. Based on the already proved theorems about passage to the limit under the integral sign with respect to a finite measure, prove the Fatou lemma, the Lebesgue dominated convergence theorem, and Levi's theorems on sequences and series for the case of a σ-finite measure.

7. (Caution!) The uniform limit theorem **does not hold** for the integral with respect to a σ-finite measure. Show this by means of the following example: the sequence $f_n = \frac{1}{2n} \mathbb{1}_{[-n,n]}$ of functions integrable on the whole real line converges uniformly to zero, but the integrals of these functions are all equal to 1.

8. Prove Theorem 1 without the assumption $\mu(B_n) < \infty$.

9. Let $(\Omega_1, \Sigma_1, \mu_1)$ and $(\Omega_2, \Sigma_2, \mu_2)$ be σ-finite measure spaces, and let $\{A_k\}_1^\infty$ and $\{B_k\}_1^\infty$ be corresponding partitions of the sets Ω_1 and Ω_2 into subsets of finite measure. Then the rectangles $\{A_k \times B_j\}_{k,j=1}^\infty$ form a partition of the Cartesian product $\Omega_1 \times \Omega_2$ into subsets of finite measure. Using this partition, prove the Fubini theorem and the inverse Fubini theorem for products of σ-finite measure spaces.

10. One can also define the integral on sets whose measure is infinite without assuming their σ-finiteness, but this definition does not extend the given one too much. Recall that the *support* of a function $f : A \to \mathbb{R}$ is defined as the set supp $f = \{t \in A : f(t) \neq 0\}$. Call a function f *integrable* on A, if supp f is a set of σ-finite measure and f is integrable on supp f. By definition, $\int_A f \, d\mu = \int_{\text{supp } f} f \, d\mu$. Verify that the main properties of the integral with respect to a σ-finite measure remain valid for the more general integral thus introduced.

11. Show that it is not possible to extend the notion of an integral to functions whose supports are not sets of σ-finite measure with preservation of the main properties of the integral, specifically, of properties (a) if $f \geqslant g$ on the set A and both f and g are integrable on A, then $\int_A f \, d\mu \geqslant \int_A g \, d\mu$, and (b) $\int_A a \, d\mu = a\mu(A)$ for any constant $a \in \mathbb{R}$ and any measurable set A of finite measure.

4.6.3 Convolution

Definition 1. Given functions f and g on the real line, we say that their *convolution* is defined if for almost every $t \in \mathbb{R}$ the function $f(\tau)g(t - \tau)$ is Lebesgue integrable on $\mathbb{R}$ as a function of the variable τ.

If this is the case, then the *convolution* of the functions f and g is the function $f * g$ defined for almost every $t \in \mathbb{R}$ by the formula

$$(f * g)(t) = \int_{\mathbb{R}} f(\tau)g(t - \tau)d\lambda(\tau).$$

The notion of convolution proves important in probability theory (the density of the distribution of the sum of two independent random variables is the convolution of the densities of the distributions of the summands), as well as in Fourier transform theory (the Fourier transform of the convolution of two functions is equal to the product of the Fourier transforms of the factors). In this subsection we establish the following useful result.

Theorem 1. *If the functions f and g are Lebesgue integrable on the real line, then their convolution is defined. Moreover, the function $f * g$ is integrable on the line and*

$$\int_{\mathbb{R}} |f * g| d\lambda \leqslant \int_{\mathbb{R}} |f| d\lambda \int_{\mathbb{R}} |g| d\lambda.$$

Proof. First of all, we remind the reader that a product of integrable functions is not necessarily integrable, that is, for some values of the parameter t the function $f(\tau) g(t - \tau)$ may not be integrable with respect to the variable τ. But where have we already encountered an assertion of the kind "for almost all values of the first variable, the function is integrable with respect to the second variable"? In Fubini's theorem, of course! So it is precisely to Fubini's theorem that we will reduce our assertion.

Note that the functions $f(\tau)$, $g(t - \tau)$, and so also their product $f(\tau)g(t - \tau)$ are measurable as functions of two variables (Exercises 4–7 in Subsection 4.5.1). According to the integrability criterion for measurable functions (Theorem 1 in Subsection 4.3.3), to show that $f(\tau)g(t - \tau)$ is integrable on $\mathbb{R} \times \mathbb{R}$, it suffices to establish the integrability of the positive function $|f(\tau)g(t - \tau)|$. For this, in turn, it suffices (see Subsection 4.5.3) to verify that the iterated integral exists.

By assumption, the function g is integrable, and consequently the function $g(t - \tau)$ is integrable with respect to t for any fixed value of τ. We have

$$\int_{\mathbb{R}} |f(\tau)g(t - \tau)| d\lambda(t) = |f(\tau)| \int_{\mathbb{R}} |g(t - \tau)| d\lambda(t) = |f(\tau)| \int_{\mathbb{R}} |g(t)| d\lambda(t).$$

Now we can easily calculate the iterated integral:

$$\int_{\mathbb{R}} \left[\int_{\mathbb{R}} |f(\tau)g(t - \tau)| d\lambda(t) \right] d\lambda(\tau) = \int_{\mathbb{R}} |f(\tau)| d\lambda(\tau) \int_{\mathbb{R}} |g(t)| d\lambda(t). \quad (1)$$

Hence, the product $f(\tau)g(t - \tau)$ is integrable as a function of two variables. Applying Fubini's theorem, we see that for almost every $t \in \mathbb{R}$ the function $f(\tau)g(t - \tau)$ is Lebesgue integrable on $\mathbb{R}$ as a function of the variable τ and the function $(f * g)(t) = \int_{\mathbb{R}} f(\tau)g(t - \tau)d\lambda(\tau)$ is integrable with respect to the variable t, i.e., the convolution exists and is integrable. The claimed inequality $\int_{\mathbb{R}} |f * g| d\lambda \leq \int_{\mathbb{R}} |f| d\lambda \int_{\mathbb{R}} |g| d\lambda$ is a direct consequence of relation (1). □

Exercises

1. The convolution operation is commutative, i.e., $f * g = g * f$ for any integrable functions f and g.

As we observed above, for complex-valued functions the definition and properties of the integral do not differ essentially from the corresponding definitions and properties for real-valued functions. One of the branches of mathematics where integration of complex-valued functions is ubiquitous is harmonic analysis: the theory of Fourier series and integrals and of problems concerned with them. In this book we will often address various problems of harmonic analysis to demonstrate applications of the material we are treating.

2. Let f be a complex-valued function on $\mathbb{R}$. Show that for every $t \in \mathbb{R}$ the function $f(\tau)e^{it\tau}$ is integrable on $\mathbb{R}$ as a function of the variable τ.

3. The *Fourier transform* of the integrable function f on the real line is the function $\widehat{f}$ on $\mathbb{R}$ (alternative customary notations are $F(f)$ and $\mathcal{F}(f)$) defined by the formula $\widehat{f}(t) = \int_{\mathbb{R}} f(\tau)e^{it\tau}d\lambda(\tau)$. Show that the function $\widehat{f}$ is bounded on $\mathbb{R}$.

4. Show that the function $\widehat{f}$ is continuous and tends to 0 at infinity.

5. Suppose the functions f and g are integrable on the line. Then $\widehat{f * g} = \widehat{f} \cdot \widehat{g}$.

6. Let f and g be 2π-periodic functions on the real line that are integrable on the interval $[0, 2\pi]$. The convolution of f and g on the interval $[0, 2\pi]$ is the function $f * g$ defined for $t \in [0, 2\pi]$ by the rule $(f * g)(t) = \int_{[0,2\pi]} f(\tau)g(t - \tau)d\lambda_1(\tau)$, where $\lambda_1 = \frac{1}{2\pi}\lambda$ is the normalized Lebesgue measure on the interval. Show that, as in the case of the convolution on the line, the convolution of two integrable functions on the interval is well defined, the function $f * g$ is itself integrable on the interval, and

$$\int_{[0,2\pi]} |f * g| d\lambda_1 \leq \int_{[0,2\pi]} |f| d\lambda_1 \int_{[0,2\pi]} |g| d\lambda_1.$$

7. Recall that the *Fourier coefficients* of the integrable function f on the interval $[0, 2\pi]$ are the numbers $\widehat{f}_n = \int_{[0,2\pi]} f(t)e^{int}d\lambda_1, n \in \mathbb{Z}$. In the setting of the preceding exercise, show that the Fourier coefficients of the function $f * g$ on the interval $[0, 2\pi]$ are the products of the corresponding Fourier coefficients of the functions f and g.

8. Show that $\widehat{f}_n \to 0$ as $n \to \infty$.

Comments on the Exercises

Subsection 4.2.2

Exercise 8. Any Riemann-integrable function satisfies condition (2) of Theorem 1 of Subsection 4.2.2; moreover, in the definition of the Riemann integral, one can take as a partition any partition into a finite number of sufficiently small intervals.

Subsection 4.3.1

Exercise 3. By the definition of an admissible partition!

Exercise 4. By the Chebyshev inequality (Lemma 1 of Subsection 4.3.1), all the sets $|f|_{>1/n} = \{t \in A : |f(t)| > 1/n\}$ have measure zero. Hence, their union, i.e., the set of all points t in which $f(t) \neq 0$, has measure zero.

Subsection 4.6.3

Fourier Analysis is one of the subjects where the methods of Functional Analysis are widely applicable. In particular, the most natural solution for Exercise 4 will appear in Theorem 1 of Subsection 14.2.2, and for Exercise 8 — in Corollary 1 of Subsection 10.4.3. We formulate these facts now just in order to advertise the powerful methods that will appear later in our book.

Chapter 5
Linear Spaces, Linear Functionals, and the Hahn–Banach Theorem

5.1 Linear Spaces

5.1.1 Main Definitions

Since the reader is assumed to be familiar with elementary linear algebra, this short subsection does not pretend to be a linear algebra manual. Our goal here is modest: just fixing notations that will be used later.

Let $\mathbb{K}$ be a field. A set X with operations of addition of its elements and multiplication of its elements by elements of $\mathbb{K}$ is called a *linear space over the field* $\mathbb{K}$ (or a $\mathbb{K}$-*linear space, vector space over the field* $\mathbb{K}$, or $\mathbb{K}$-*vector space*) if X is an abelian group with respect to addition and for any $\lambda, \mu \in \mathbb{K}$ and any $x, y \in X$ the following relations connecting multiplication by elements of $\mathbb{K}$ with addition hold:

— $1 \cdot x = x$;

— $(\lambda\mu)x = \lambda(\mu x)$;

— $(\lambda + \mu)x = \lambda x + \mu x$;

— $\lambda(x + y) = \lambda x + \lambda y$.

Functional analysis courses usually treat linear spaces over the field of real or complex numbers. Henceforth, the symbol $\mathbb{K}$ will be used whenever an argument is equally valid for the real and the complex numbers. The elements of the field $\mathbb{K}$ will be also referred to as numbers or scalars, while the elements of the linear space itself will be called "vectors".

Let A be a subset of the linear space X. An element $x \in X$ is said to be a *linear combination* of elements of the set A if it can be written as

$$x = \sum_{k=1}^{n} \lambda_k x_k,$$

© Springer International Publishing AG, part of Springer Nature 2018
V. Kadets, *A Course in Functional Analysis and Measure Theory*,
Universitext, https://doi.org/10.1007/978-3-319-92004-7_5

where $\lambda_k \in \mathbb{K}$ and $x_k \in A$. The set of all linear combinations of elements of the set A is called the *linear span* (alternatively, *linear hull* or *linear envelope*) of the set A; we denote it by Lin A. We note that even if the set A is infinite, when we speak about linear combinations we consider only finite (though arbitrarily large) collections of elements of A.

A subset Y of the linear space X is called a *linear subspace* (we will simply say "subspace" if the context is clear) if for any $x, y \in Y$ and any $\lambda, \mu \in \mathbb{K}$, the linear combination $\lambda x + \mu y$ also lies in Y. The linear span of the set A is the smallest linear subspace containing A; accordingly, it is also called the *linear subspace spanned by A*.

A subset A of the linear space X is said to be *complete* if Lin $A = X$; further, A is said to be a *linearly independent set* if for any finite subset $\{x_k\}_{k=1}^n \subset A$, the equality $\sum_{k=1}^n \lambda_k x_k = 0$ can hold only if all the coefficients λ_k are equal to 0. A complete linearly independent set is called a *Hamel basis*. If the linear space X contains a finite Hamel basis, then it is said to be *finite-dimensional*; in the opposite case the space is said to be *infinite-dimensional*. In contrast to linear algebra, functional analysis studies mainly infinite-dimensional spaces.

Exercises

1. Recall how one proves that in a finite-dimensional space X all Hamel bases have the same number of elements (this number is called the *dimension of the space X* and is denoted by dim X). Show that:

— if a linear space contains an infinite linearly independent set, then the space is infinite-dimensional;

— if a linear space contains a finite complete set of vectors, then the space is finite-dimensional;

— if a linear space contains a countable complete set of vectors, then any Hamel basis is at most countable;

— if a linear space contains a countable linearly independent set of vectors, then any Hamel basis contains at least a countable number of elements;

— if a linear space contains a countable Hamel basis, then any other Hamel basis of this space is countable.

Prove the last three statements above with "countable" replaced by any other "fixed cardinality".

2. Show that the following spaces are infinite-dimensional:

— the space $\mathcal{P}$ of all polynomials in one variable;

— the space $C[a, b]$ of all continuous scalar-valued functions on the interval $[a, b]$;

— the space of all Lebesgue-integrable scalar-valued functions on the interval $[a, b]$.

3. Describe the compact spaces K for which the space $C(K)$ is finite-dimensional. (Recall that in this book the terms "compact space" and "Hausdorff compact space" are synonymous.) For which measure spaces (Ω, Σ, μ) is the space of all Lebesgue-integrable scalar functions on (Ω, Σ, μ) finite-dimensional?

5.1.2 Ordered Sets and Zorn's Lemma

Let $(\Gamma, \prec)$ be an ordered set, i.e., a set endowed with an order relation $\prec$. If two elements $a, b \in \Gamma$ satisfy $b \prec a$, we say that the element a *majorizes* the element b. If, in addition, $a \neq b$, then we say that a *strictly majorizes* b. A subset $A \subset \Gamma$ is said to be *bounded* if Γ contains an element that majorizes all elements of A. Such an element is called an *upper bound* of the subset A. A subset $A \subset \Gamma$ is called a *chain*, or a *linearly ordered subset*, if any two elements $a, b \in A$ are comparable, i.e., either $a \prec b$, or $b \prec a$. An element $a \in \Gamma$ is called a *maximal element* of the set Γ if Γ contains no element that strictly majorizes a. The ordered set Γ is called *inductive* if $\Gamma \neq \emptyset$ and every chain in Γ bounded.

Although historically the following statement retained the name "lemma", nowadays it is often taken as one of the axioms of set theory. Essentially, this statement replaces for uncountable sets the principle of mathematical induction. If one assumes the most common axiomatic system in modern set theory ZFC (Zermelo–Fraenkel + Axiom of Choice), then Zorn's lemma is not an axiom, but is a theorem. We don't provide a proof here because its nature lies too far from our subject. The proof can be found in books devoted to the foundations of set theory, or in the already cited set-theoretic appendix in [23] or set-theoretic chapters of [31].

Lemma 1 (Zorn's Lemma). *Every inductive ordered set has a maximal element.*
□

Exercises

1. Consider the following order on the coordinate plane $\mathbb{R}^2$: $(x_1, x_2) \prec (y_1, y_2)$ if either $x_1 + x_2 < y_1 + y_2$, or $(x_1, x_2) = (y_1, y_2)$. Give in each case an example of a subset of $\mathbb{R}^2$ that with respect to the indicated order has the property listed below:

— has no maximal element;

— has two maximal elements;

— has infinitely many maximal elements.

2. Show that in every ordered set any finite chain Γ contains a largest element: there exists an $a \in \Gamma$ such that $b \prec a$ for all $a \in \Gamma$. Is this assertion true for infinite chains?

5.1.3   Existence of Hamel Bases

Theorem 1. *Every linear space contains a Hamel basis.*

Proof. Let X be a linear space. Denote by Γ the family of all linearly independent subsets of the space X, equipped with the natural order relation (by inclusion): the subset A majorizes the subset B if $A \supset B$. Let us prove that the ordered set Γ is inductive. To this end, we pick an arbitrary chain $\Gamma_1 \subset \Gamma$ and show that the set $M = \bigcup_{A \in \Gamma_1} A$, i.e., the union of all sets that are elements of Γ_1, is an upper bound for the chain Γ_1 in Γ. Since M majorizes all elements of Γ_1, we only need to show that $M \in \Gamma$. In other words, we have to verify that M is a linearly independent set. Let $A = \{a_1, a_2, \ldots, a_n\}$ be an arbitrary finite subset of M; let $B_k, k = 1, 2, \ldots, n$ be the corresponding elements of the chain Γ_1 such that $a_k \in B_k$. Since the sets B_k are pairwise comparable, one of them (say, B_j), contains all the others. Thus, $A \subset B_j$, B_j is linearly independent, and so A is also linearly independent. Thus, we have shown that every finite subset of the set M is linearly independent, and consequently so is M. By Zorn's lemma, Γ has a maximal element A. We claim that A is the sought-for Hamel basis. Indeed, any element of the family Γ is linearly independent, and hence, in particular, so is A. Let us show that the set A is complete. Suppose $\mathrm{Lin}\, A \neq X$. Pick an arbitrary element $x \in X \setminus \mathrm{Lin}\, A$. Then $A \cup \{x\}$ is a linearly independent set that strictly majorizes A, which contradicts the maximality of the set A. $\square$

Exercises

1. Show that any linearly independent set in a linear space can be completed to a Hamel basis.

2. Let X_1 be a subspace of the linear space X. Show that there exists a subspace $X_2 \subset X$ with the following properties: $\mathrm{Lin}\,(X_1 \cup X_2) = X$, $X_1 \cap X_2 = \{0\}$.

3. Exhibit a Hamel basis in the space $\mathcal{P}$ of all polynomials in one variable.[1]

5.1.4   Linear Operations on Subsets

Let A_1, A_2 be subsets of the linear space X. By $A_1 + A_2$ we denote the set of all elements of the form $a_1 + a_2$ with $a_1 \in A_1$ and $a_2 \in A_2$. In geometry, $A_1 + A_2$ is called the *Minkowski sum* of the sets A_1 and A_2. Analogously, for $A \subset X$ and $t \in \mathbb{K}$, we denote by tA the set of all elements of the form tx with $x \in A$. The set $(-1)A$ is naturally denoted by $-A$, and for $A_1 + (-A_2)$ we simply write $A_1 - A_2$.

---

[1] It is interesting that, despite the existence theorem proved above, no explicit example of a Hamel basis in more complicated infinite-dimensional spaces (say, in $C[0, 1]$) is known.

Exercises

1. $0 \in A_1 - A_2$ if and only if the sets A_1 and A_2 intersect.

2. If $A \subset X$ is a subspace, then $A + A = A$ and $tA = A$ for every $t \neq 0$.

3. If A_1 and A_2 are subspaces, then $A_1 + A_2$ is also a subspace.

4. If A_1 and A_2 are subspaces, then $\text{Lin}\,(A_1 \cup A_2) = A_1 + A_2$.

5. If A_1 and A_2 are subspaces and $A_1 \cap A_2 = \{0\}$, then every element $x \in A_1 + A_2$ has a **unique** decomposition as $x = a_1 + a_2$, with $a_1 \in A_1$ and $a_2 \in A_2$. In this case, to emphasize the uniqueness of the decomposition, one uses the *direct sum* symbol $\oplus$: instead of $A_1 + A_2$ one writes $A_1 \oplus A_2$.

6. A subset Y of a linear space X is a subspace if and only if $\lambda_1 Y + \lambda_2 Y \subset Y$ for all scalars λ_1 and λ_2.

7. Let A_1 and A_2 be closed segments in the plane $\mathbb{R}^2$. What geometric figure is $A_1 + A_2$? When is $A_1 + A_2$ a segment?

8. Let A be a closed set in the plane. Show that $A + A = 2A$ if and only if A is convex.

5.2 Linear Operators

5.2.1 Injectivity and Surjectivity

Let X, Y be linear spaces over the field $\mathbb{K}$. A mapping $T : X \to Y$ is called a *linear operator* if for any $x_1, x_2 \in X$ and any $\lambda_1, \lambda_2 \in \mathbb{K}$ it holds that $T(\lambda_1 x_1 + \lambda_2 x_2) = \lambda_1 T(x_1) + \lambda_2 T(x_2)$.

In particular, when $Y = \mathbb{K}$, a linear operator $T : X \to \mathbb{K}$ is called a *linear functional*.

A linear operator $T : X \to Y$ is said to be *injective* if its *kernel* $\text{Ker}\, T = T^{-1}(0)$ reduces to zero. An operator T is said to be *surjective* if its *image* (or *range*) $T(X)$ coincides with the whole space Y. Finally, an operator is said to be *bijective*, or *invertible*, if it is both injective and surjective. In other words, if the equation $Tx = b$ has a solution for any right-hand side $b \in Y$, then the operator T is surjective; if from the solvability of the equation $Tx = b$ for a given right-hand side it necessarily follows that its solution is unique, then the operator T is injective. Hence, bijectivity means existence and uniqueness of the solution for any right-hand side.

Exercises

1. Verify that the kernel and image of an operator are linear subspaces.

2. Let X and Y be linear spaces and X_1 be a subspace of X. Prove that any linear operator $T: X_1 \to Y$ can be extended to a linear operator acting from X to Y.

3. Let X, Y, and Z be linear spaces, $U: X \to Y$ and $V: Y \to Z$ be linear operators, and $T = V \circ U$. Show that: (a) if T is injective, then U is also injective; (b) if T is surjective, then V is also surjective. Are the converse assertions true?

4. Let $T: X \to Y$ be a linear operator and $A_1, A_2 \subset X$. Then $T(A_1 + A_2) = T(A_1) + T(A_2)$.

5. Let $T_1, T_2: X \to Y$ be linear operators and $A \subset X$. Then $(T_1 + T_2)(A) \subset T_1(A) + T_2(A)$. Give an example where $(T_1 + T_2)(A) \neq T_1(A) + T_2(A)$.

6. Let $T: X \to Y$ be a linear operator and $A_1, A_2 \subset Y$. Then $T^{-1}(A_1 + A_2) \supset T^{-1}(A_1) + T^{-1}(A_2)$. Give an example where $T^{-1}(A_1 + A_2) \neq T^{-1}(A_1) + T^{-1}(A_2)$.

5.2.2   Quotient Spaces

Let X be a linear space and X_1 be a subspace of X. We introduce on X the following equivalence relation: $x \sim y$ if $x - y \in X_1$. The equivalence class of an element x is readily seen to be the set $[x] = x + X_1 = \{x + y : y \in X_1\}$. The set of all such equivalence classes, equipped with the operations described in Subsection 5.1.4, is called the *quotient space* (or *factor space*) of the space X by the subspace X_1, and is denoted[2] by X/X_1. Let us list the simplest properties of the linear operations on a quotient space, which imply, in particular, that every quotient space is a linear space.

1. The equivalence class of zero is the zero element of the quotient space:
$$[0] + [x] = X_1 + (x + X_1) = x + (X_1 + X_1) = x + X_1 = [x].$$

2. $\lambda_1[x_1] + \lambda_2[x_2] = [\lambda_1 x_1 + \lambda_2 x_2]$:
$\lambda_1[x_1] + \lambda_2[x_2] = \lambda_1(x_1 + X_1) + \lambda_2(x_2 + X_1) = \lambda_1 x_1 + \lambda_2 x_2 + (\lambda_1 X_1 + \lambda_2 X_1) = \lambda_1 x_1 + \lambda_2 x_2 + X_1 = [\lambda_1 x_1 + \lambda_2 x_2]$.

A quotient space comes with a closely related operator q, the *quotient map* of the space X onto X/X_1, given by $q(x) = [x]$. The linearity of the quotient map follows from the property 2. above. The quotient map is surjective.

---

[2]Like in computer programming languages, in mathematics slash and backslash have rather different meanings. Caution: the notation X/X_1 stands for the quotient space, whereas $X \setminus X_1$ stands for the set-theoretic difference.

An important example of a quotient space arises naturally in integration theory (see Problem 6 in Subsection 4.3.3). Let (Ω, Σ, μ) be a (finite or σ-finite) measure space. One denotes by $L_0(\Omega, \Sigma, \mu)$ the quotient space of the space X of all measurable scalar-valued functions on Ω by the subspace X_1 of functions with negligible support. In this example the corresponding equivalence relation is already well-known to the reader: $f \sim g$ if and only if $f = g$ almost everywhere.

5.2.3 Injectivization of a Linear Operator

Let X and Y be linear spaces and $T: X \to Y$ be a linear, not necessarily injective, operator. The *injectivization* of T is defined to be the operator $\widetilde{T}: X/\text{Ker } T \to Y$ which associates to the equivalence class $[x]$ of the element $x \in X$ the element Tx: $\widetilde{T}([x]) = Tx$.

Exercises

1. Show that the operator $\widetilde{T}$ is well defined, namely, that if the equivalence classes of the elements x_1 and x_2 coincide, then $Tx_1 = Tx_2$. In other words, $\widetilde{T}([x])$ does not depend on the choice of the representative in the equivalence class $[x]$.

2. Verify the linearity and injectivity of the operator $\widetilde{T}$.

3. Verify that $T = \widetilde{T} \circ q$, where $q: X \to X/\text{Ker } T$ is the quotient map operator. Hence, every operator can be written as the composition of a surjective operator and an injective one. Compare to Exercise 3 in Subsection 5.2.1.

Let X be a linear space, X_1 a subspace of X, and X' the space of all linear functionals on X. The *annihilator* in X' of the subspace X_1 is the set $X_1^\perp$ of all functionals $f \in X'$ for which $\text{Ker } f \supset X_1$. For each linear functional g on X/X_1 we define the functional $Ug \in X'$ by the rule $(Ug)(x) = g([x])$. In other words, Ug is the composition of the functional g and the quotient map.

4. Show that $U: (X/X_1)' \to X'$ is a linear injective operator.

5. Show that $U((X/X_1)') = X_1^\perp$, i.e., $X_1^\perp$ is a linear subspace isomorphic to $(X/X_1)'$.

5.3   Convexity

Functional analysis deals mainly with analytical objects — functions, sequences, limits, and so on, but the approach to these objects differs essentially from the approach of mathematical analysis (calculus). Instead of studying individual functions or sequences with certain properties, one considers the corresponding spaces, subspaces, or subsets. Thanks to such an approach, many problems of analysis can be reduced to problems on the mutual disposition or properties of sets in appropriate spaces. For example, the question of the approximability of a continuous function by polynomials reduces to the question of whether the space of polynomials is dense in the space of continuous functions with respect to some metric; the problem of defining the integral for functions that are not Riemann integrable can be formulated as the problem of extending a linear functional to a larger space; the Cauchy problem for a differential equation becomes the problem of searching for the fixed point of an appropriate mapping. This approach allows one, in the search for a solution, to use geometric considerations, draw sketches, examine examples where as models for infinite-dimensional sets one can take figures in the plane or three-dimensional space. However, in order to be able to successfully use our geometric intuition in problems of functional analysis it is necessary to accumulate some experience. One has to learn how to distinguish the essential properties of the model from the specifics of the planar figure, and learn to translate the ideas inspired by the figure into rigorous mathematical reasoning. To this end one needs, in particular, to define correctly, whenever possible, infinite-dimensional analogues of notions and constructions that are ordinarily used in geometric arguments. In this section we introduce such analogues for the notions of line, segment, convex set, as well as that of the partition of three-dimensional space into two half-spaces by a plane.

5.3.1   Definitions and Properties

Let X be a linear space, and let $x, y \in X$, $x \neq y$. The *straight line passing through the points x and y* is defined as the set of all elements of the form $\lambda x + (1 - \lambda)y$, where λ runs through the whole real line. The *segment joining x and y* is defined to be the set of all elements of the form $\lambda x + (1 - \lambda)y$ with $\lambda \in [0, 1]$. A subset $A \subset X$ is said to be *convex* if together with any pair of its points it contains the segment joining them.

Exercises

1. Lines and segments are convex sets.

2. The intersection of any number of convex sets is convex.

3. The union of two convex sets is convex.

4. Let X, Y be linear spaces, $U: X \to Y$ a linear operator. Show that for any convex subset A of X the set $U(A)$ is also convex. Show that if B is a convex subset of Y, then the set $U^{-1}(B)$ is convex.

Which of the sets in the space s of all real numerical sequences listed below are convex?

5. $\{a = (a_n)_1^\infty : \inf_n a_n > 1 \}$.

6. $\{a = (a_n)_1^\infty : \inf_n a_n < 1 \}$.

7. $\{a = (a_n)_1^\infty : \sup_n a_n = +\infty \}$.

8. $\{a = (a_n)_1^\infty : \lim_{n \to \infty} a_n = +\infty\}$.

Which of the sets in the space of all real-valued functions on the interval $[0, 1]$ sequences listed below are convex?

9. The set of all continuous functions.

10. The set of all discontinuous functions.

11. The set of all infinitely differentiable functions.

12. The set of all nowhere differentiable functions.

Let A be a subset of the linear space X. Show that

13. $A + A \supset 2A$.

14. If A is convex, then $A + A = 2A$.

5.3.2 Convex Hull

Let $\{x_k\}_{k=1}^n$ be an arbitrary finite collection of vectors of the linear space X. A vector of the form $x = \sum_{k=1}^n \lambda_k x_k$ is called a *convex combination* of the vectors x_k if the coefficients $\lambda_1, \ldots, \lambda_n$ are non-negative numbers and satisfy $\sum_{k=1}^n \lambda_k = 1$.

Proposition 1. *Let A be a convex set in a linear space X, and let $\{x_k\}_{k=1}^n \subset A$. Then any convex combination of the vectors x_k lies in A.*

Proof. We proceed by induction on the number n of vectors figuring in the convex combination. The induction base, i.e., the case $n = 2$, follows directly from the definition of a convex set. Let us pass from $n - 1$ to n. Let $\{\lambda_k\}_{k=1}^n$ be nonnegative numbers, and let $\sum_{k=1}^{n} \lambda_k = 1$ and $x = \sum_{k=1}^{n} \lambda_k x_k$. If among the scalars λ_k there is one equal to zero, then x is actually a combination of $n-1$ vectors, so that $x \in A$ by the induction hypothesis. Now assume that all λ_k are different from zero. Put $\mu = \sum_{k=1}^{n-1} \lambda_k$ and $\mu_k = \lambda_k/\mu, k = 1, \ldots, n-1$. Then $y = \sum_{k=1}^{n-1} \mu_k x_k$, being a convex combination of $n-1$ vectors from A, belongs to A. Since $x = \mu y + \lambda_n x_n$, the vector x is a convex combination of two vectors from A, and so $x \in A$.   □

Let A be an arbitrary subset of the linear space X. The set of all convex combinations of vectors from A is called the *convex hull* (or *convex envelope*) of the set A and is denoted by $\mathrm{conv}(A)$ or $\mathrm{conv}\, A$.

Exercises

Adopting the preceding definition, show that:

1. $\mathrm{conv}(A)$ is a convex set.

2. $\mathrm{conv}(A)$ is the smallest convex set that contains A.

3. If A consists of only two points, then $\mathrm{conv}(A)$ is the segment joining those points.

Show that

4. The convex hull of a set of three points in the plane is the triangle with vertices at those points.

5. The convex hull of any subset A of the plane is the union of all triangles with vertices at points of A.

Is this last assertion valid for subsets of three-dimensional space?

5.3.3   Hypersubspaces and Hyperplanes

A subspace Y of the linear space X is called a *hypersubspace* if there exists a vector $e \in X \setminus Y$ such that $\mathrm{Lin}\{e, Y\} = X$.

Proposition 1. *For a subspace Y of the linear space X the following conditions are equivalent:*

1. *Y is a hypersubspace in X.*
2. *$\dim(X/Y) = 1$.*
3. *There exists a non-zero linear functional f on X such that $\mathrm{Ker}\, f = Y$.*

Proof. 1. $\Longrightarrow$ 2. Let Y be a hypersubspace in X, and let $e \in X \setminus Y$ be such that Lin$\{e, Y\} = X$. Consider the equivalence class of e, $[e] \in X/Y$. Then $[e] \neq 0$, since $e \notin Y$. At the same time, Lin $[e] = X/Y$, i.e., in X/Y there exists a basis consisting of a single element.

2. $\Longrightarrow$ 3. Since dim$(X/Y) = 1$, there exists a bijective linear operator $U : X/Y \to \mathbb{K}$ between our quotient space and the field $\mathbb{K}$. Consider the functional $f = U \circ q$, where $q : X \to X/Y$ is the quotient map. For this functional, Ker $f = Y$.

3. $\Longrightarrow$ 1. Let f the functional figuring in assertion 3. Pick a vector $e \in X$ for which $f(e) = 1$. Clearly, $e \in X \setminus Y$. Let us show that Lin$\{e, Y\} = X$. To this end we take an arbitrary vector $x \in X$ and decompose it as $x = f(x)e + (x - f(x)e)$. In this decomposition the second term $x - f(x)e$ lies in Ker $f = Y$, hence we managed to write the vector x as $ae + y$, with $a \in \mathbb{K}$ and $y \in Y$. $\qquad\square$

A subset A of the linear space X is called a *hyperplane* if it is the translate of a hypersubspace. In other words, a hyperplane in X is any set of the form $x + Y$, where $x \in X$ and Y is a hypersubspace of X.

Exercises

1. Let f be a nonzero linear functional on the space X, and a an arbitrary number. Show that the *level set* $f_a = \{x \in X : f(x) = a\}$ of the functional f is a hyperplane in X.

2. Show that every hyperplane in X is the level set of some non-zero linear functional.

3. Let Y be a hypersubspace in X. Show that Lin$\{e, Y\} = X$ for every vector $e \in X \setminus Y$.

4. Suppose the kernels of two linear functionals coincide. Show that these functionals are linearly dependent.

5. Let Y be a hyperplane in the real linear space X. Introduce on $X \setminus Y$ the following equivalence relation: $x_1 \sim x_2$ if the segment joining the points x_1 and x_2 does not intersect the hyperplane Y. Verify that this is indeed an equivalence relation. The set $X \setminus Y$ splits into two equivalence classes, called the *half-spaces determined by the hyperplane* Y.

6. Suppose a hyperplane is the level set f_a of the functional f in the real linear space X. Verify that the half-spaces determined by the hyperplane f_a are the sets $f_{<a} = \{x \in X : f(x) < a\}$ and $f_{>a} = \{x \in X : f(x) > a\}$.

7. Why is the assumption that the space is real important in the two preceding exercises?

Definition 1. We say that the subspace Y of the linear space X has (is of) *codimension* $n \in \mathbb{N}$ (and write $\mathrm{codim}_X Y = n$) if the dimension of the quotient space X/Y is equal to n. If X/Y is infinite-dimensional, we put $\mathrm{codim}_X Y = +\infty$.

In the exercises below n stands, as usual, for a (finite) natural number.

8. $\mathrm{codim}_X Y \leq n$ if and only if there exist n vectors $\{x_k\}_1^n \subset X$ such that $\mathrm{Lin}\{x_1, \ldots, x_n, Y\} = X$.

9. $\mathrm{codim}_X Y \leq n$ if and only if Y has nonzero intersection with any subspace Z of X of dimension greater than or equal to $n + 1$.

10. $\mathrm{codim}_X Y = n$ if and only if there exists a subspace Z of X such that $\dim Z = n$, $Z \cap Y = \{0\}$, and $Z + Y = X$.

11. Let $Y \subset Z$ be two subspaces of the linear space X. Then $\mathrm{codim}_X Y = \mathrm{codim}_Z Y + \mathrm{codim}_X Z$.

12. For any two subspaces Y, Z of the space X it holds that

$$\max\{\mathrm{codim}_X Y, \mathrm{codim}_X Z\} \leq \mathrm{codim}_X(Y \cap Z) \leq \mathrm{codim}_X Y + \mathrm{codim}_X Z.$$

13. $\mathrm{codim}_X Y = \dim Y^\perp$.

14. $\mathrm{codim}_X Y \leq n$ if and only if there exists a collection of n linear functionals in X such that the intersection of their kernels is contained in Y.

15. $\mathrm{codim}_X Y = n$ if and only if there exists a collection of n linearly independent linear functionals on X such that the intersection of their kernels coincides with Y.

16. Let f and $\{f_k\}_{k=1}^n$ be linear functionals on X and suppose $\mathrm{Ker}\, f \supset \bigcap_{k=1}^n \mathrm{Ker}\, f_k$. Then $f \in \mathrm{Lin}\{f_k\}_{k=1}^n$.

5.4 The Hahn–Banach Theorem on the Extension of Linear Functionals

5.4.1 Convex Functionals

A real-valued function p given on the linear space X is called a *convex functional* (a precise, but too long name, should be *convex positively-homogeneous functional*) if it satisfies the following conditions:

— $p(\lambda x) = \lambda p(x)$ for any vector $x \in X$ and any non-negative number λ (positive homogeneity);

— $p(x + y) \leq p(x) + p(y)$ for any $x, y \in X$ (triangle inequality).

Exercises

1. Let p be a convex functional. Then $p(0) = 0$.

2. Let $t > 0$ and let p be a convex functional. Then tp is a convex functional.

3. The modulus of a convex functional is also a convex functional.

4. Let p_1 and p_2 be convex functionals. Then $p_1 + p_2$ and $\max\{p_1, p_2\}$ are also convex functionals.

5. Every linear functional on a real linear space is a convex functional. Verify that if p and $-p$ are convex functionals, then p is a linear functional.

6. Let p be a convex functional and $x \in X$ be a fixed vector such that $p(x) \le 0$ and $p(-x) \le 0$. Then $p(x) = p(-x) = 0$.

Which of the expressions listed below define convex functionals on the space $C[0, 1]$ of continuous functions on the interval $[0, 1]$?

7. $p_1(f) = \max\{f(t) : t \in [0, 1]\}$.

8. $p_2(f) = \min\{f(t) : t \in [0, 1]\}$.

9. $p_3(f) = p_1(f) - p_2(f)$.

10. $p_4(f) = |f|$.

Which of the expressions listed below define convex functionals on the space ℓ_∞ of all bounded numerical sequences?

11. $p_5(x) = x_5$.

12. $p_6(x) = x_5 \cdot x_6$.

13. $p_7(x) = \sum_{n=1}^{\infty} |x_n|$.

14. $p_8(x) = \varlimsup_{n \to \infty} x_n$.

15. $p_9(x) = \sum_{n=1}^{\infty} \frac{1}{2^n} |x_n|$.

(In the last five exercises x stands for the sequence $(x_1, x_2, \ldots, x_n, \ldots)$.)

Convex functionals and convex sets are closely connected. The next subsection is devoted to the description of this connection.

5.4.2   The Minkowski Functional

A subset A of the linear space X is said to be *absorbing* if for any $x \in X$ there exists an $n \in \mathbb{N}$ such that $\frac{1}{t}x \in A$ for every $t > n$. Note that an absorbing set A necessarily contains the zero element of the space and satisfies $\bigcup_{n=1}^{\infty} nA = X$.

Let A be a convex absorbing set in X. The *Minkowski functional* of the set A is the real-valued function on X defined as

$$\varphi_A(x) = \inf \left\{ t > 0 : \frac{1}{t}x \in A \right\}.$$

The functional φ_A is connected with the convex set A by the following obvious relations:

— if $x \in A$, then $\varphi_A(x) \leq 1$;
— if $\varphi_A(x) < 1$, then $x \in A$.

Proposition 1. *Let A be a convex absorbing set in X. Then $\varphi_A(x)$ is a convex functional that takes only non-negative values.*

Proof. Let us show first that the functional $\varphi_A(x)$ is positively homogeneous. Take $\lambda > 0$ and $x \in X$. Then

$$\varphi_A(\lambda x) = \inf \left\{ t > 0 : \frac{1}{t}\lambda x \in A \right\} = \inf \left\{ \lambda s > 0 : \frac{1}{s}x \in A \right\}$$

$$= \lambda \inf \left\{ s > 0 : \frac{1}{s}x \in A \right\} = \lambda \varphi_A(x)$$

(in the last chain of equalities we used the change of notation $t \to \lambda s$).

Now let us verify the triangle inequality. Take $x, y \in X$. We need to show that $\varphi_A(x + y) \leq \varphi_A(x) + \varphi_A(y)$. Obviously, it suffices to show that for any $a > \varphi_A(x)$ and $b > \varphi_A(y)$ one has the inequality $\varphi_A(x + y) \leq a + b$. So, fix numbers $a > \varphi_A(x)$ and $b > \varphi_A(y)$. Then $\varphi_A \left(\frac{x}{a} \right) < 1$ and $\varphi_A \left(\frac{y}{b} \right) < 1$, i.e., $\frac{x}{a} \in A$ and $\frac{y}{b} \in A$. In view of the convexity of A, the vector $\frac{x+y}{a+b} = \frac{a}{a+b}\frac{x}{a} + \frac{b}{a+b}\frac{y}{b}$ also belongs to A. Hence, $\varphi_A \left(\frac{x+y}{a+b} \right) \leq 1$, and the needed inequality $\varphi_A(x + y) \leq a + b$ is proved. □

Exercises

A set A in the linear space X is said to be *balanced* if $\lambda A \subset A$ for any scalar λ with $|\lambda| \leq 1$.

1. Let A be a convex balanced set. Then $Y = \bigcup_{n=1}^{\infty} nA$ is a linear subspace of X and A is an absorbing set in Y.

Let p be a convex functional that takes only non-negative values. Put $A = \{x \in X : p(x) < 1\}$. Verify that

2. A is a convex absorbing set.

3. For any $x \in A$ there exists an $\varepsilon > 0$ such that $(1 + \varepsilon)x \in A$.

4. $\varphi_A = p$.

Let A and B be convex absorbing sets. Verify that:

5. $\varphi_{A \cap B} = \max(\varphi_A, \varphi_B)$.

6. $\varphi_{-A}(x) = \varphi_A(-x)$.

5.4.3   The Hahn–Banach Theorem—Analytic Form

The Hahn–Banach theorem on the extension of linear functionals that will be proved in the present subsection (alternatively known as the analytic form of the Hahn–Banach theorem) is one of the most important theorems in functional analysis. It is frequently used, both in the subject itself and in applications of functional analysis to a wide circle of related fields. Some of these applications will be treated in this book. The Hahn–Banach theorem is traditionally regarded as one of the "fundamental principles of functional analysis". Such "fundamental principles" also include the geometric form of the Hahn–Banach theorem (Subsection 9.3.2), Banach's inverse operator theorem, the open mapping and the closed graph theorems, as well as the Banach–Steinhaus theorem (see Chap. 10).

Theorem 1. *Suppose that on the real linear space X there is given a convex functional p; let Y be a subspace of X and f be a linear functional on Y such that $f(y) \leq p(y)$ for all $y \in Y$. Then f can be extended to a linear functional g defined on the whole X, with preservation of the majorization condition: $g(x) \leq p(x)$ for all $x \in X$.*

Proof. We first treat the special case where Y is a hypersubspace of X. Let $e \in X \setminus Y$ be a vector such that $\mathrm{Lin}\{e, Y\} = X$. Any element of X is uniquely representable as $x = \lambda e + y$, where $y \in Y$ and λ is a real number. Hence, the sought-for functional g, the extension of the functional f from Y to X, is uniquely determined by its value on the vector e: $g(\lambda e + y) = \lambda g(e) + f(y)$. In order for the majorization condition to be preserved, the number $g(e)$ must satisfy the requirement

$$\lambda g(e) + f(y) \leq p(\lambda e + y) \text{ for all } \lambda \in \mathbb{R} \text{ and all } y \in Y. \qquad (*)$$

For $\lambda = 0$, condition $(*)$ is satisfied thanks to the assumption of the theorem. For positive λ, $(*)$ can be recast as

$$g(e) \leq \frac{1}{\lambda}(p(\lambda e + y) - f(y)) \quad \text{for all } \lambda > 0 \text{ and all } y \in Y,$$

whereas for negative $\lambda = -\mu$ (∗) becomes

$$g(e) \geq -\frac{1}{\mu}(p(-\mu e + v) - f(v)) \quad \text{for all } \mu > 0 \text{ and all } v \in Y.$$

Hence, for the sought-for extension to exist it is necessary and sufficient that the following inequality be satisfied:

$$\sup\left\{-\frac{1}{\mu}(p(-\mu e + v) - f(v)) : \mu > 0,\ v \in Y\right\}$$
$$\leq \inf\left\{\frac{1}{\lambda}(p(\lambda e + y) - f(y)) : \lambda > 0,\ y \in Y\right\}.$$

Let us verify this inequality. Take $\lambda,\ \mu > 0$ and $y,\ v \in Y$. Moving the terms involving f to the left-hand side and those involving p to the right-hand side, we reduce the inequality

$$-\frac{1}{\mu}(p(-\mu e + v) - f(v)) \leq \frac{1}{\lambda}(p(\lambda e + y) - f(y))$$

to

$$\frac{1}{\mu}f(v) + \frac{1}{\lambda}f(y) \leq \frac{1}{\mu}p(-\mu e + v) + \frac{1}{\lambda}p(\lambda e + y).$$

The latter follows from the majorization condition for f; indeed,

$$\frac{1}{\mu}f(v) + \frac{1}{\lambda}f(y) = f\left(\frac{1}{\mu}v + \frac{1}{\lambda}y\right) \leq p\left(\frac{1}{\mu}v + \frac{1}{\lambda}y\right)$$
$$\leq \frac{1}{\mu}p(-\mu e + y) + \frac{1}{\lambda}p(\lambda e + y).$$

Thus, the particular case of a subspace of codimension 1 is dealt with. What we proved can be restated as follows: a linear functional f given on Y and satisfying the majorization condition can be extended to the linear span of the subspace Y and an arbitrary vector, with preservation of the majorization condition. Iterating this statement, we can produce an extension of the functional f to the linear span of the subspace Y and an arbitrary finite number of vectors. Unfortunately, since the ambient space X is infinite-dimensional, this argument does not always yield an extension to the whole X. For this reason, to complete the argument we need to use Zorn's lemma — a standard recipe for organizing induction arguments in the case of an infinite number of steps.

So, let Γ denote the family of all pairs (Z, h), where Z is a subspace of X containing Y, and h is a linear functional on Z that coincides with f on Y and satisfies the majorization condition on Z. Essentially, the elements of Γ are extensions of the functional f. We need to show that among the elements $(Z, h) \in \Gamma$ there is one with $Z = X$. We introduce on Γ an order relation by setting $(Z_1, h_1) \succ (Z_2, h_2)$ if $Z_1 \supset Z_2$ and the restriction of the functional h_1 to Z_2 coincides with h_2. It is readily seen that the ordered set Γ is inductive. Hence, by Zorn's lemma, Γ contains a maximal element (Z_0, h_0). Suppose Z_0 does not coincide with the whole space X. Then by the already established particular case of the Hahn–Banach theorem, the functional h_0 can be extended, with preservation of the majorization, to a subspace of the form $\mathrm{Lin}\{e, Z_0\}$ that includes Z_0 strictly. The existence of such an extension contradicts the maximality of the pair (Z_0, h_0). Therefore, $Z_0 = X$, which completes the proof. $\qquad\square$

Remark 1. The Hahn–Banach theorem establishes the existence of an extension, but this extension is not necessarily unique (give an example with a two-dimensional space X and a one-dimensional subspace Y). Accordingly, in all applications of the Hahn–Banach theorem that will be encountered in the sequel we will assert the existence of some or another object, but uniqueness will not necessarily hold.

Exercises

1. Where in the above formulation of the Hahn–Banach theorem did the assumption that the space considered, and hence the functionals involved, are real play a role?

2. Try to formulate and prove a generalization of the Hahn–Banach theorem to the complex case.

3. In the case where the quotient space X/Y has a finite or countable Hamel basis, the Hahn–Banach theorem can be proved without resorting to Zorn's lemma. How?

5.5 Some Applications of the Hahn–Banach Theorem

5.5.1 Invariant Means on a Commutative Semigroup

Let G be a commutative (Abelian) semigroup; we denote the semigroup operation on G by the symbol '+'. Consider the space $\ell_\infty(G)$ of all bounded real-valued functions on G. Each element $g \in G$ generates a *shift* or *translation operator* $S_g : \ell_\infty(G) \to \ell_\infty(G)$ by the rule $\left(S_g F\right)(h) = F\left(g + h\right)$. A linear functional I on $\ell_\infty(G)$ is called an *invariant mean* on G if it satisfies the following conditions:

— $\inf_{g \in G} F(g) \le I(F) \le \sup_{g \in G} F(g)$ for any function $F \in \ell_\infty(G)$ (i.e., $I(F)$ is a mean value for F);

— $I(S_g F) = I(F)$ for any function $F \in \ell_\infty(G)$ and any $g \in G$ (shift or translation invariance).

In this subsection we show that on any commutative semigroup G there exists an invariant mean.

Given a function $F \in \ell_\infty(G)$, set

$$p(F) = \inf\left\{ \sup_{h\in G} \frac{1}{n}\sum_{k=1}^{n} F(g_k + h) : n \in \mathbb{N},\ (g_k)_{k=1}^{n} \in G^n \right\},$$

where the infimum is taken over all finite collections of elements $g_k \in G$, possibly with repetitions.

Proposition 1. *p is a convex functional on $\ell_\infty(G)$ which satisfies for any function $F \in \ell_\infty(G)$ and any $g \in G$ the conditions*

(1) $p(S_g F - F) \le 0$;

(2) $p(F - S_g F) \le 0$

(*according to Exercise 5 in Subsection 5.4.1, conditions* (1) *and* (2) *mean that the estimated quantities are in fact equal to zero*).

Proof. The positive homogeneity is here obvious, so let us verify the triangle inequality. Let $F_1, F_2 \in \ell_\infty(G)$, $\varepsilon > 0$. Pick elements $g_k^1, k = 1, 2, \ldots, n_1$, and g_k^2, $k = 1, 2, \ldots, n_2$, of the semigroup G such that

$$\sup_{h\in G}\left\{ \frac{1}{n_i}\sum_{k=1}^{n_i} F_i(g_k^i + h) \right\} < p(F_i) - \varepsilon, \quad i = 1, 2.$$

Then

$$p(F_1 + F_2) \le \sup_{h\in G}\left\{ \frac{1}{n_1 n_2}\sum_{j=1}^{n_2}\sum_{k=1}^{n_1} (F_1 + F_2)(g_k^1 + g_j^2 + h) \right\}.$$

Using the fact that the supremum of a sum is not larger than the sum of the suprema of its summands, we continue the estimate as

$$\le \frac{1}{n_2}\sum_{j=1}^{n_2} \sup_{h\in G}\left\{ \frac{1}{n_1}\sum_{k=1}^{n_1} F_1(g_k^1 + g_j^2 + h) \right\} + \frac{1}{n_1}\sum_{k=1}^{n_1} \sup_{h\in G}\left\{ \frac{1}{n_2}\sum_{j=1}^{n_2} F_2(g_k^1 + g_j^2 + h) \right\}$$

$$\le p(F_1) + p(F_2) - 2\varepsilon,$$

which in view of the arbitrariness of ε establishes the triangle inequality.

Now let us verify condition (1). We have

$$p(S_g F - F) \le \sup_{h\in G}\left\{ \frac{1}{n}\sum_{k=1}^{n} (S_g F - F)(h + kg) \right\}$$

$$= \sup_{h\in G} \frac{1}{n}\big(F(h + (n+1)g) - F(h + g)\big) \le \frac{2}{n}\sup_{h\in G}|F(h)|.$$

Letting here $n \to \infty$, we obtain the needed estimate. Condition (2) is verified analogously. □

Denote the function identically equal to 1 on G by $\mathbb{1}$. We mention two more obvious properties of the functional p:

— $p(c\mathbb{1}) = c$ for every $c \in \mathbb{R}$;

— if the function F is everywhere smaller than or equal to 0, then $p(F) \le 0$.

Theorem 1. *On any commutative semigroup G there exists an invariant mean.*

Proof. Consider in $\ell_\infty(G)$ the subspace $Y = \mathrm{Lin}\{\mathbb{1}\}$. Define the functional f on Y by $f(c \cdot \mathbb{1}) = c$. Clearly, f is linear and $f \le p$ on Y. Using the Hahn–Banach theorem, we extend f to a linear functional I on the whole space $\ell_\infty(G)$, preserving the majorization. Let us show that I is an invariant mean. First, note that the functional I is *monotone*, meaning that if $F_1, F_2 \in \ell_\infty(G)$ and $F_1 \le F_2$ at all points, then $I(F_1) \le I(F_2)$. Indeed, since $F_1 - F_2 \le 0$, we have $I(F_1) - I(F_2) = I(F_1 - F_2) \le p(F_1 - F_2) \le 0$. Next, the monotonicity of the functional I implies that if the function F is bounded from above and from below by constants, i.e., $c_1\mathbb{1} \le F \le c_2\mathbb{1}$, then $I(F)$ is bounded by the same constants: $c_1 \le I(F) \le c_2$. This establishes the first condition in the definition of an invariant mean. The second condition — shift invariance — is an immediate consequence of the majorization condition and properties (1) and (2) of the functional p:

$$I(S_g F) - I(F) = I(S_g F - F) \le p(S_g F - F) \le 0,$$

$$I(F) - I(S_g F) = I(F - S_g F) \le p(F - S_g F) \le 0.$$

Remark 1. The commutativity of the semigroup is not a necesary condition for the existence of an invariant mean. For details on groups that admit invariant means we refer to the monograph [32].

5.5.2 The "easy" Problem of Measure Theory

Recall that we proved earlier (Subsection 2.3.4) the insolvability of the what we called the *difficult problem of measure theory*: the construction of a shift-invariant countably-additive probability measure μ defined on all subsets of the interval $[0, 1)$. From this we then deduced the existence of sets that are not Lebesgue measurable: if every subset of the interval were Lebesgue measurable, then the Lebesgue measure would be a solution of the difficult problem of measure theory. At the same time, the analogous problem where countable additivity is replaced by finite additivity (what we call here the *"easy" problem of measure theory*) is already solved.

Theorem 1 (Banach). *There exists a finitely-additive measure μ, defined on all subsets of the interval $[0, 1)$, such that $\mu([0, 1)) = 1$ and μ is shift invariant (i.e., $\mu(A + t) = \mu(A)$ for all subsets $A \subset [0, 1)$ and all $t \in \mathbb{R}$ such that $A + t \subset [0, 1)$).*

Proof. We equip the interval $[0, 1)$ with the operation of addition modulo 1: the sum of the numbers a and b modulo 1 is the fractional part of the number $a + b$. Let I be an invariant mean on this group. The measure μ defined by the rule $\mu(A) = I(\mathbb{1}_A)$, where $\mathbb{1}_A$ is the characteristic function of the set A, does the job. □

It is interesting that to construct a similar measure on the sphere of the three-dimensional Euclidean space (i.e., a finitely-additive probability measure defined on all subsets of the sphere and invariant under the isometries of the sphere) is already not possible (Hausdorff [54, 1914]). The reason is the complicated structure of the group of isometries of the sphere. The reader interested in questions concerned with the existence of invariant measures is referred to the monograph [43] and the survey [65], where one can read about facts in the spirit of the famous Banach–Tarski paradox, where the sphere is cut into a finite number of "pieces", from which one can then assemble two new spheres of exactly the same size. The possibility of such a partition would of course lead to a contradiction if the "pieces" could be "measured" by means of a finitely-additive measure invariant under isometries.

Exercises

4. Let μ be a measure and I be the invariant mean appearing in the preceding theorem. Although we have proved that these objects exists, no computation rules were provided. Moreover, the invariant mean (and hence the invariant measure) on the interval is not unique. Nevertheless, for some functions the invariant mean can be calculated, based on the definition of this object. Show that $\mu([0, 1/n)) = 1/n$.

5. Let $m < n$. Show that $\mu([0, m/n)) = m/n$.

6. Let $[a, b) \subset [0, 1)$. Show that $\mu([a, b)) = b - a$.

7. Let f be a piecewise constant function on $[0, 1)$. Then $I(f) = \int_0^1 f(t)dt$.

8. Show that $I(f) = \int_0^1 f(t)dt$ for any Riemann-integrable function on the interval.

For bounded Lebesgue-integrable functions $I(f)$ may not coincide with the Lebesgue integral. Nevertheless, if we carry out the proof of the theorem asserting the existence of an invariant mean (Subsection 5.5.1) for the case of the interval, taking for Y not the subspace of constants, but the subspace of bounded Lebesgue-integrable functions, then one can show that:

9. There exists an invariant mean I on the interval such that $I(f) = \int_{[0,1)} f(t)d\lambda$ for every bounded Lebesgue-integrable function.

10. Based on the existence of an invariant measure on the interval, show that there exists a finitely-additive shift-invariant measure on the real line, defined on all subsets, with the property that the measure of any interval is equal to its length. Needless to say, this measure is also allowed to take infinite values.

Consider the semigroup $\mathbb{N}$ of natural numbers with the addition operation. The functions on $\mathbb{N}$ are sequences; the invariant mean on $\mathbb{N}$ is called the *generalized Banach limit* and denoted by the symbol Lim. Verify that:

11. The generalized Banach limit of any bounded sequence lies between the upper and lower limits of the sequence.

12. If the sequence $x = (x_1, x_2, \ldots, x_n, \ldots)$ has a limit, then $\text{Lim } x = \lim_{n \to \infty} x_n$.

13. If the sequence $x = (x_1, x_2, \ldots, x_n, \ldots)$ is *uniformly Cesàro convergent* to the number s (i.e., the sequence $(x_{k+1} + x_{k+2} + \cdots + x_{k+n})/n$ converges uniformly in k to s as $n \to \infty$), then $\text{Lim } x = s$.

14. Using the example of the sequence $x = (1, 0, 1, 0, \ldots)$ convince yourself that the generalized Banach limit of a sequence is not necessarily a limit point of the sequence.

15. Using the example of the sequences $x = (1, 0, 1, 0, \ldots)$ and $y = (0, 1, 0, 1, \ldots)$ convince yourself that the generalized Banach limit is not a multiplicative functional: $\text{Lim }(xy)$ is not necessarily equal to the product of $\text{Lim } x$ and $\text{Lim } y$.

Comments on the Exercises

Subsection 5.1.3

Exercise 1. Let G be a linearly independent subset of X. Denote by Γ the family of all linearly independent subsets of X that contain G. After that proceed as in the proof of Theorem 1.

Subsection 5.3.3

Exercise 4. Denote the functionals considered by f and g, and $\text{Ker } f = \text{Ker } g$ by Y. Then Y is a hypersubspace of X. Hence, there exists a vector $e \in X \setminus Y$ such that $\text{Lin}\{e, Y\} = X$. The numbers $a = f(e)$ and $b = g(e)$ are different from 0, because $e \notin Y$. The functional $ag - bf$, being a linear combination of the functionals f and g, is equal to zero both on Y and on the vector e. It follows that $ag - bf$ vanishes on the entire space $X = \text{Lin}\{e, Y\}$, i.e., f and g are linearly dependent.

Exercises 13–16. Here the duality between quotient spaces and annihilators described in Exercises 4 and 5 of Subsection 5.2.3 will be of help. A direct (without resorting to the codimension argument) solution of Exercise 16 will be provided in Lemma 1 of Subsection 16.3.2.

Subsection 5.4.3

Exercise 2. For complex spaces substitute in the statement of Theorem 1 the conditions $f(y) \leq p(y)$ and $g(y) \leq p(y)$ by $\mathrm{Re}\, f(y) \leq p(y)$ and $\mathrm{Re}\, g(y) \leq p(y)$, respectively.

Subsection 5.5.2

Exercise 10. This assertion admits in a certain sense a converse, proved in 1948 by Lorentz [69]: if all generalized Banach limits of a bounded numerical sequence $x = (x_1, x_2, \ldots, x_n, \ldots)$ are equal to one and the same number s, then the sequence x is uniformly Cesàro convergent to s.

Chapter 6
Normed Spaces

6.1 Normed Spaces, Subspaces, and Quotient Spaces

6.1.1 Norms. Examples

Let X be a linear space. A mapping $x \mapsto \|x\|$ that associates to each element of the space X a nonnegative number is called a *norm* if it obeys the following axioms:

(1) if $\|x\| = 0$, then $x = 0$ (non-degeneracy);

(2) $\|\lambda x\| = |\lambda| \|x\|$ for all $x \in X$ and all scalars λ;

(3) $\|x + y\| \leqslant \|x\| + \|y\|$ (triangle inequality).

Conditions (2) and (3) show that a norm is a particular case of a convex functional. In connection with this we suggest that the reader return to the exercises in Subsection 5.4.1 and examine which of the properties 1–5 of convex functionals hold in the case of a norm, and also which of the functionals p_i in Exercises 7–15 are norms.

Definition 1. A linear space X endowed with a norm is called a *normed space*.

Let us note that if a linear space X is endowed with some norm, then one has a normed space, but if the same linear space is endowed with another norm, then it already becomes a different normed space. Below we provide examples of normed spaces that will be repeatedly encountered in the sequel. The verification of the norm axioms in these examples is left to the reader.

Examples

1. Let K be a compact topological space. We let $C(K)$ denote the normed space of continuous scalar-valued functions on K with the norm $\|f\| = \max\{|f(t)| : t \in K\}$. An important particular case of the space $C(K)$ is the space $C[a, b]$ of continuous functions on the interval $[a, b]$.

© Springer International Publishing AG, part of Springer Nature 2018
V. Kadets, *A Course in Functional Analysis and Measure Theory*,
Universitext, https://doi.org/10.1007/978-3-319-92004-7_6

2. ℓ_1 is the space of numerical sequences $x = (x_1, x_2, \ldots, x_n, \ldots)$ that satisfy
the condition $\sum_{n=1}^{\infty} |x_n| < \infty$, equipped with the norm $\|x\| = \sum_{n=1}^{\infty} |x_n|$. Since any
sequence can be regarded as a function defined on the set $\mathbb{N}$ of natural numbers,
the space ℓ_1 is a particular case of the space $L_1(\Omega, \Sigma, \mu)$ studied below in Subsection 6.1.3: $\Omega = \mathbb{N}$, Σ is the family of all subsets of $\mathbb{N}$, and μ is the counting
measure.

3. ℓ_∞ denotes the space of all bounded numerical sequences with the norm $\|x\| =$
$\sup_n |x_n|$.

4. c_0 is the space of all numerical sequences that tend to zero. The norm on c_0 is
given in the same way as on ℓ_∞.

Definition 2. A mapping $x \mapsto p(x)$ that associates to each element of the linear
space X a non-negative number is called a *seminorm* if it obeys the norm axioms (2)
and (3).

Exercises

1. Give an example of a seminorm on $\mathbb{R}^2$ which is not a norm.

2. Give an example of a convex functional on $\mathbb{R}^2$ that is not a seminorm.

3. Let B be a convex absorbing set in the linear space X. Suppose, in addition, that
B is a *balanced set*, i.e., $\lambda B \subset B$ for any scalar λ with $|\lambda| \leqslant 1$. Then the Minkowski
functional of B (see Subsection 5.4.2) is a seminorm.

6.1.2 The Metric of a Normed Space and Convergence. Isometries

Let X be a normed space. The *distance between the elements* $x_1, x_2 \in X$ is defined
by $\rho(x_1, x_2) = \|x_2 - x_1\|$. From the norm axioms is follows that ρ is indeed a metric
on X. Hence, every normed space is simultaneously a metric space, and so all the
notions defined in metric spaces — open and closed sets, compact sets, limit points,
completeness, etc., — also make sense in normed spaces. In particular, a sequence
(x_n) of elements of the normed space X converges to the element x if $\|x_n - x\| \to 0$
as $n \to \infty$. An essential difference in terminology between normed and metric spaces
shows up in the definition of isometries: in a normed space one additionally requires
that the map in question is linear.

A linear operator T acting from a normed space X to a normed space Y is called
an *isometric embedding* if $\|Tx\| = \|x\|$ for all $x \in X$.

A bijective isometric embedding is called an *isometry*. The normed spaces X and
Y are said to be *isometric* if there exists an isometry between them.

Exercises

1. Suppose the sequence (x_n) of elements of a normed space converges to the element x. Show that $\|x_n\| \to \|x\|$ as $n \to \infty$.

2. Consider in the space ℓ_1 the elements $x_n = (n^k/(n+1)^{k+1})_{k=1}^{\infty}$. Write in explicit form the coordinates of x_1 and x_2. What are the norms of x_1 and x_2? Calculate the norms $\|x_n\|$ for arbitrary n.

3. Show that convergence in $C(K)$ is the uniform convergence on K. In particular, convergence in $C[a, b]$ is uniform convergence on $[a, b]$, a type of convergence well known from calculus.

4. Show that for any $a < b$ the space $C[a, b]$ is isometric to the space $C[0, 1]$.

5. If the compact spaces K_1 and K_2 are homeomorphic, then the space $C(K_1)$ is isometric to $C(K_2)$. Conversely, if $C(K_1)$ is isometric to $C(K_2)$, then K_1 and K_2 are homeomorphic (this converse is far from trivial).

6. Show that in the space ℓ_1 the convergence of a sequence of vectors $x_n = (x_n^k)_{k=1}^{\infty}$ to a vector $x = (x^k)_{k=1}^{\infty}$ implies the coordinatewise convergence: $x_n^k \to x^k$ as $n \to \infty$, for all $k = 1, 2, \ldots$. On the other hand, coordinatewise convergence does not imply convergence in ℓ_1.

7. The sequence (x_n) in Exercise 2 above can be regarded as a sequence in ℓ_1, and also as one in c_0. What are the norms $\|x_n\|$ in c_0 equal to? Show that the sequence (x_n) converges coordinatewise to 0 and converges to 0 in c_0, but does not converge in ℓ_1.

8. Let X be some sequence space. The *positive cone* in X is defined to be the set of vectors of X all the coordinates of which are non-negative. Consider the three cases $X = c_0$, $X = \ell_1$, and $X = \ell_{\infty}$. In each of them prove that the positive cone is closed and convex, and describe its interior and boundary.

6.1.3 The Space L_1

Let (Ω, Σ, μ) be a (finite or infinite) measure space, E the linear space of all μ-integrable scalar-valued functions on Ω, and F the subspace of E consisting of all functions that vanish almost everywhere. By $L_1(\Omega, \Sigma, \mu)$ we denote the quotient space E/F. The analogous quotient space $L_0(\Omega, \Sigma, \mu)$ was mentioned in Subsection 5.2.2. To simplify the terminology, one usually says that the elements of the space $L_1(\Omega, \Sigma, \mu)$ are functions integrable on Ω, with the understanding that two functions that coincide almost everywhere are identified. The norm in $L_1(\Omega, \Sigma, \mu)$ is given by the formula $\|f\| = \int_{\Omega} |f(t)| d\mu$. An important particular case of the space

$L_1(\Omega, \Sigma, \mu)$ is the space $L_1[a, b]$ of Lebesgue-integrable functions on an interval $[a, b]$. In this case $\Omega = [a, b]$, Σ is the family of all Lebesgue-measurable subsets of the interval, and μ is the Lebesgue measure.

Exercises

1. Show that $L_1(\Omega, \Sigma, \mu)$ is a normed space.

2. Show that for any $a < b$ the space $L_1[a, b]$ is isometric to the space $L_1[0, 1]$.

3. Show that the space $L_1[0, 1]$ is isometric to the space $L_1(-\infty, +\infty)$.

4. Show that the space $L_1[0, 1]$ is isometric to the space $L_1([0, 1] \times [0, 1])$.

5. The convergence of a sequence of functions in $L_1(\Omega, \Sigma, \mu)$ implies its convergence in measure, but if the measure is not purely atomic (a typical example is the space $L_1[a, b]$), then convergence in $L_1(\Omega, \Sigma, \mu)$ does not imply convergence almost everywhere.

6. If (Ω, Σ, μ) is a finite measure space and a sequence of integrable functions converges uniformly on Ω, then this sequence also converges in $L_1(\Omega, \Sigma, \mu)$.

7. Show that regardless of what norm the space $L_1[a, b]$ is endowed with, convergence in this norm cannot coincide with convergence in measure. (Compare with Exercise 6 in Subsection 4.3.3.)

8. Consider the positive cone in $L_1(\Omega, \Sigma, \mu)$, that is, the set G of all functions from $L_1(\Omega, \Sigma, \mu)$ that are almost everywhere greater than or equal to zero. Show that G is a closed convex set that has no interior points.

9. By analogy with the preceding exercise, consider the positive cone in $C(K)$. Show that this set is convex and closed, and describe its interior and boundary.

Let p be a seminorm on the space X. The *kernel* of the seminorm p is the set Ker p of all points $x \in X$ such that $p(x) = 0$.

10. Ker p is a linear subspace of X.

11. The expression $\rho(x_1, x_2) = p(x_2 - x_1)$ defines a pseudometric on X.

12. Show that for any $x \in X$ and any $y \in$ Ker p, we have $p(x + y) = p(x)$.

13. The expression $\|[x]\| = p(x)$ defines a norm on the quotient space $X/\text{Ker } p$.

Since the expression $p(f) = \int_\Omega |f(t)| d\mu$ gives a seminorm on the linear space E of all scalar-valued μ-integrable functions on Ω, $F = $ Ker p is the subspace of E consisting of all functions that vanish almost everywhere, the definition given above for the space $L_1(\Omega, \Sigma, \mu)$ is a particular case of the construction described in Exercises 10–13 of this subsection.

6.1.4 Subspaces and Quotient Spaces

A linear subspace Y of the normed space X, equipped with the norm of X, is called *subspace of the normed space X*. Hence, any subspace of a normed space is itself a normed space.

Let Y be a closed subspace of the normed space X, $x \in X$ an arbitrary element, and $[x] = x + Y$ the corresponding element of the quotient space X/Y. Define $\|[x]\| = \inf_{y \in Y} \|x + y\|$. In other words, $\|[x]\|$ is the distance in X from 0 to the set $x + Y$. Since Y is a subspace, and hence $Y = -Y$, the original definition is equivalent to the following one: $\|[x]\| = \inf_{y \in Y} \|x - y\|$. The geometric meaning of the latter is that $\|[x]\|$ is the distance in X from x to the subspace Y.

Proposition 1. *The expression $\|[x]\|$ introduced above gives a norm on the space X/Y.*

Proof. Let us verify the norm axioms.

1. Suppose $\|[x]\| = 0$. Then $\inf_{y \in Y} \|x - y\| = 0$, and so x is a limit point of the subset Y. Since Y is closed, $x \in Y$ and $[x] = Y = [0]$.

2. Since Y is a subspace, $\lambda Y = Y$ for any nonzero scalar λ. We have $\|[\lambda x]\| = \inf_{y \in Y} \|\lambda x + y\| = \inf_{y \in Y} \|\lambda x + \lambda y\| = |\lambda| \inf_{y \in Y} \|x + y\| = |\lambda| \cdot \|[x]\|$.

3. Let $x_1, x_2 \in X$ and $\varepsilon > 0$. By the definition of the infimum, there exist y_1, $y_2 \in Y$, such $\|x_1 + y_1\| < \|[x_1]\| + \varepsilon$ and $\|x_2 + y_2\| < \|[x_2]\| + \varepsilon$. It follows that

$$\|[x_1 + x_2]\| = \inf_{y \in Y} \|x_1 + x_2 + y\| \leqslant \|x_1 + x_2 + y_1 + y_2\|$$
$$\leqslant \|x_1 + y_1\| + \|x_2 + y_2\| \leqslant \|[x_1]\| + \|[x_2]\| + 2\varepsilon,$$

which in view of the arbitrariness of ε means that the needed triangle inequality holds.

Henceforth we will always assume that the quotient space of a normed space is equipped with the norm described above.

Example Let (Ω, Σ, μ) be a measure space, X the space of all bounded measurable functions on Ω, endowed with the norm $\|f\| = \sup_{t \in \Omega} |f(t)|$, and Y the subspace of X consisting of the functions that vanish almost everywhere. The corresponding quotient space X/Y is denoted by $L_\infty(\Omega, \Sigma, \mu)$.

Exercises

1. Prove the following formula for the norm in $L_\infty(\Omega, \Sigma, \mu)$:

$$\|f\|_\infty = \inf_{A \in \Sigma,\, \mu(A)=0} \left\{ \sup_{t \in \Omega \setminus A} |f(t)| \right\}.$$

2. Prove the inequality $|f| \overset{\text{a.e.}}{\leqslant} \|f\|_\infty$.

3. Show that $\|f\|_\infty$ is equal to the infimum of the set of all constants c such that $|f| \overset{\text{a.e.}}{\leqslant} c$.

4. In the space $C[a, b]$ consider the subspace Y consisting of the constants (i.e., constant functions). Show that the norm of the element $[f]$ of the quotient space $C[a, b]/Y$ can be calculated by the formula

$$\|[f]\| = \frac{1}{2}\Big(\max\{f(t) : t \in [a, b]\} - \min\{f(t) : t \in [a, b]\}\Big).$$

5. The space ℓ_1 can be regarded as a linear subspace of c_0, though it will not be a normed subspace of c_0: the norm given on ℓ_1 does not coincide with the norm on c_0. Show that ℓ_1 is not closed and is dense in c_0. Show that as a subset of c_0, ℓ_1 belongs to the class F_σ.

6. Show that the space c_0 of all sequences that converge to zero is closed in ℓ_∞.

7. Show that the norm of the element $[a]$ in the space ℓ_∞/c_0 is calculated by the formula $\|[a]\| = \varlimsup_{n \to \infty} |a_n|$, where a_n are the coordinates of the element $a \in \ell_\infty$.[1]

6.2   Connection Between the Unit Ball and the Norm. L_p Spaces

6.2.1   Properties of Balls in a Normed Space

Let X be a normed space, $x_0 \in X$, and $r > 0$. As usual, we denote by $B_X(x_0, r)$ the open ball of radius r centered at x_0:

$$B_X(x_0, r) = \{x \in X : \|x - x_0\| < r\}.$$

The *unit ball* B_X in the space X is the open ball of unit radius centered at zero: $B_X = \{x \in X : \|x\| < 1\}$. The *unit sphere* S_X and *closed unit ball* $\overline{B}_X$ are similarly defined as

$$S_X = \{x \in X : \|x\| = 1\}, \quad \text{and} \quad \overline{B}_X = \{x \in X : \|x\| \leqslant 1\}.$$

Let us list some of the simplest properties of the objects just introduced.

---

[1] In Soviet times, one of the Kharkiv newspapers published a paper on the fulfillment of the production plan by highly productive workers ("peredoviks"), entitled "The [production] norm is not a limit!". The last assertion above can be considered as a counterexample to this assertion.

— The unit ball is an open set, while the unit sphere and the closed unit ball are closed sets.

— $B_X(x_0, r) = x_0 + r B_X$.

— B_X is a convex absorbing set (see Exercise 2 in Subsection 5.4.2).

— B_X is a balanced set, i.e., for any scalar λ such that $|\lambda| \leqslant 1$, we have $\lambda B_X \subset B_X$.

— For any $x_0 \in X$ and $r > 0$, the linear span of the ball $B_X(x_0, r)$ coincides with the whole space X.

Exercises

1. Prove that the closure of the open ball $B_X(x_0, r)$ in a normed space is the closed ball $\overline{B}_X(x_0, r)$. Compare with Exercise 10 in Subsection 1.3.1.

2. The space of numerical rows $x = (x_1, x_2, \ldots, x_n)$ with the norm $\|x\| = \sum_{k=1}^n |x_n|$ is denoted by ℓ_1^n; the analogous space of rows with the norm $\|x\| = \sup_n |x_n|$ is denoted by ℓ_∞^n. The spaces ℓ_1^n and ℓ_∞^n are finite-dimensional analogues of the spaces ℓ_1 and ℓ_∞. Construct in the coordinate plane the unit balls of the spaces ℓ_1^2 and ℓ_∞^2. Exhibit an isometry between these two spaces.

3. Construct in the three-dimensional coordinate space the unit balls of the spaces ℓ_1^3 and ℓ_∞^3. Show that these normed spaces are not isometric.

4. Nested balls principle. Let X be a complete normed space, and $B_n = \overline{B}_X(x_n, r_n)$ be a decreasing (with respect to inclusion) sequence of closed balls. Show that the intersection $\bigcap_{n=1}^\infty B_n$ is not empty. (In contrast to the nested sets principle, here one does not assume that the diameters of the balls tend to zero, but neither does one assert that the intersection consists of a single point.)

5. Give an example of a complete metric space in which the assertion of the preceding exercise is not true.

6.2.2 Definition of the Norm by Means of a Ball. The Spaces L_p

Let B be a convex absorbing set in the linear space X. Recall (see Subsection 5.4.2) that the Minkowski functional of the set B is the function on X given by the formula $\varphi_B(x) = \inf \{ t > 0 : t^{-1} x \in B \}$.

Theorem 1. *Let B be a convex, absorbing, balanced set in the space X which also has the following algebraic boundedness property: for each $x \in X \setminus \{0\}$ there exists an $a > 0$ such that $ax \notin B$. Then the Minkowski functional φ_B gives a norm on X.*

Proof. The fact that φ_B is a convex functional was already established in Subsection 5.4.2. Since the set B is balanced, $\varphi_B(\lambda x) = \varphi_B(|\lambda|x) = |\lambda|\varphi_B(x)$ for all $x \in X$ and all scalars λ, i.e., φ_B is a seminorm. Finally, if $x \in X \setminus \{0\}$, then thanks to the algebraic boundedness there exists an $a > 0$ such that $ax \notin B$. Hence, $\varphi_B(x) \geqslant \frac{1}{a} > 0$, which establishes the non-degeneracy of the Minkowski functional. $\square$

Let (Ω, Σ, μ) be a (finite or not) measure space, and $p \in [1, \infty)$ a fixed number. We denote by $L_p(\Omega, \Sigma, \mu)$ the subset of the space $L_0(\Omega, \Sigma, \mu)$ of all measurable scalar-valued functions on Ω consisting of the functions for which the integral $\int_\Omega |f(t)|^p d\mu$ exists. Here, as in the case of the space $L_0(\Omega, \Sigma, \mu)$, functions in $L_p(\Omega, \Sigma, \mu)$ that are equal almost everywhere are regarded as one and the same element. For $f \in L_p(\Omega, \Sigma, \mu)$, we put $\|f\|_p = \left(\int_\Omega |f(t)|^p d\mu \right)^{1/p}$.

Theorem 2. $L_p(\Omega, \Sigma, \mu)$ *is a linear space, and* $\|\cdot\|_p$ *is a norm on* $L_p(\Omega, \Sigma, \mu)$.

Proof. Consider the set $B_p \subset L_p(\Omega, \Sigma, \mu)$ consisting of all functions for which $\int_\Omega |f(t)|^p d\mu < 1$. Let $f, g \in B_p$ and $\lambda \in [0, 1]$. Since the function $|x|^p$ is convex on $\mathbb{R}$, for any $t \in \Omega$ we have the numerical inequality

$$|\lambda f(t) + (1 - \lambda)g(t)|^p \leqslant \lambda |f(t)|^p + (1 - \lambda)|g(t)|^p.$$

Integrating this inequality we conclude that $\lambda f + (1 - \lambda)g \in B_p$, and so B_p is a convex set. It is readily verified that the set B_p is balanced and algebraically bounded. From the fact that B_p is convex and balanced and the obvious equality $L_p(\Omega, \Sigma, \mu) = \bigcup_{n=1}^\infty nB_p$ it follows that $L_p(\Omega, \Sigma, \mu)$ is a linear space and B_p is an absorbing set in $L_p(\Omega, \Sigma, \mu)$ (Exercise 1 in Subsection 5.4.2). Consequently, the Minkowski functional of the set B_p is defined on $L_p(\Omega, \Sigma, \mu)$ and gives a norm on this linear space. It remains to observe that $\|\cdot\|_p$ coincides with φ_{B_p}. Indeed, for any $f \in L_p(\Omega, \Sigma, \mu)$, $\frac{1}{t}f \in B_p$ if and only if $t > \|f\|_p$, i.e., $\|f\|_p = \varphi_{B_p}(f)$. $\square$

In what follows, $L_p(\Omega, \Sigma, \mu)$ will be regarded as a normed space equipped with the norm $\|\cdot\|_p$. Important particular cases are the spaces $L_p[a, b]$ (i.e., the case $\Omega = [a, b]$ with the Lebesgue measure) and the spaces ℓ_p, where the role of Ω is played by $\mathbb{N}$, $\Sigma = 2^{\mathbb{N}}$, and μ is the counting measure (the measure of a set is the number of its elements). Since every function defined on the set $\mathbb{N}$ of natural numbers can be regarded as a sequence, ℓ_p is usually defined as the space of numerical sequences $x = (x_k)_{k \in \mathbb{N}}$ that satisfy the condition $\sum_{k=1}^\infty |x_k|^p < \infty$, equipped with the norm $\|x\|_p = \left(\sum_{k=1}^\infty |x_k|^p \right)^{1/p}$.

Exercises

1. Suppose the linear space X is endowed with two norms, $\|\cdot\|_1$ and $\|\cdot\|_2$, and let B_1 and B_2 be the corresponding unit balls. Then $B_1 \subset B_2$ if and only if the inequality $\|\cdot\|_1 \geqslant \|\cdot\|_2$ holds in X.

2. Suppose the linear space X is endowed with three norms, $\|\cdot\|_1$, $\|\cdot\|_2$, $\|\cdot\|_3$, and let B_1, B_2, and and B_3 be the corresponding unit balls. Suppose $\|\cdot\|_3$ is expressed in terms of $\|\cdot\|_1$ and $\|\cdot\|_2$ as $\|x\|_3 \doteq \max\{\|x\|_1, \|x\|_2\}$. Then $B_3 = B_1 \cap B_2$.

3. ℓ_p, regarded as a set, increases with the growth of p, while for fixed x the norm $\|x\|_p$ decreases with the growth of p.

4. The set ℓ_0 of terminating (finitely supported) sequences (i.e., sequences in which, starting with some index, all terms are equal to 0) is dense in ℓ_p for any $p \in [1, \infty)$.

5. If $p_1 < p$, then the set ℓ_{p_1} is dense in the space ℓ_p.

6. Let B be a convex, absorbing, balanced, and algebraically bounded set in the normed space X. Endow X with the norm defined by the Minkowski functional of B. In order for the unit ball of this norm to coincide with B it is necessary and sufficient that B have the following property: for every $x \in B$, there exists an $\varepsilon > 0$, such that $(1 + \varepsilon)x \in B$.

7. For $1 \leqslant p < \infty$, the set of bounded functions is dense in $L_p[a, b]$.

8. For $1 \leqslant p < \infty$, the set of continuous functions is dense in $L_p[a, b]$.

9. For $1 \leqslant p < \infty$, the set of all polynomials is dense in $L_p[a, b]$.

10. For $1 \leqslant p < \infty$, the set of continuous functions satisfying the condition $f(0) = 0$ is dense in $L_p[0, b]$.

11. The set of continuous functions is not dense $L_\infty[a, b]$.

6.3 Banach Spaces and Absolutely Convergent Series

A *Banach space* is a complete normed space, i.e., a normed space in which every Cauchy sequence converges. Banach spaces constitute the most important class of normed spaces: they are the spaces most often encountered in applications, and many of the most important results of functional analysis revolve around the notion of Banach space.[2]

6.3.1 Series. A Completeness Criterion in Terms of Absolute Convergence

Let (x_n) be a sequence of elements of the normed space X. The *partial sums* of the series $\sum_{n=1}^{\infty} x_n$ are the vectors $s_n = \sum_{k=1}^{n} x_k$. If the partial sums of the series

---

[2]At least in the opinion of the author of these lines, who specializes in the theory of Banach spaces.

$\sum_{n=1}^{\infty} x_n$ converge to an element x, the series is said to *converge* and the element x is called the *sum* of the series. The equality $\sum_{n=1}^{\infty} x_n = x$ is the generally adopted short way of writing that "the series $\sum_{n=1}^{\infty} x_n$ converges and its sum is equal to x". The series $\sum_{n=1}^{\infty} x_n$ is called *absolutely convergent* if $\sum_{n=1}^{\infty} \|x_n\| < \infty$.

Proposition 1 (**Cauchy convergence criterion for series**). *For the series $\sum_{n=1}^{\infty} x_n$ of elements of a Banach space X to converge it is necessary and sufficient that $\left\| \sum_{k=n}^{m} x_k \right\| \to 0$ as $n, m \to \infty$.*

Proof. Convergence of a series is equivalent to convergence of the sequence of its partial sums s_n. In turn, in a complete space convergence of a sequence is equivalent to the sequence being Cauchy. It remains to note that $s_m - s_n = \sum_{k=n+1}^{m} x_k$. $\square$

Proposition 2. *Suppose the series $\sum_{n=1}^{\infty} x_n$ of elements of the Banach space X converges absolutely. Then the series $\sum_{n=1}^{\infty} x_n$ converges.*

Proof. Since the numerical series $\sum_{n=1}^{\infty} \|x_n\|$ converges, $\sum_{k=n}^{m} \|x_k\| \to 0$ as $n, m \to \infty$. Consequently, $\|\sum_{k=n}^{m} x_k\| \leqslant \sum_{k=n}^{m} \|x_k\| \to 0$ as $n, m \to \infty$. To complete the proof, it remains to apply Proposition 1. $\square$

Proposition 3. *Let X be normed space that is not complete. Then in X there exists an absolutely convergent, but not convergent series.*

Proof. Since X is not complete, there exists a Cauchy sequence $v_n \in X$ which does not have a limit. By the definition of a Cauchy sequence, $\|v_n - v_m\| \to 0$ as $n, m \to \infty$. It follows that there exists an $n_1 \in \mathbb{N}$ such that $\|v_n - v_m\| < \frac{1}{2}$ for all $n, m \geqslant n_1$. Analogously, pick an $n_2 \geqslant n_1$ such that $\|v_n - v_m\| < \frac{1}{4}$ for all $n, m \geqslant n_2$. Continuing this argument, we obtain an increasing sequence of indices n_j such that $\|v_n - v_m\| < \frac{1}{2^j}$ for all $n, m \geqslant n_j$. Then for the sequence v_{n_j} it holds that

$$\|v_{n_2} - v_{n_1}\| < \frac{1}{2}, \quad \|v_{n_3} - v_{n_2}\| < \frac{1}{4}, \quad \dots, \quad \|v_{n_{j+1}} - v_{n_j}\| < \frac{1}{2^j}, \dots.$$

Now we define the sought-for series $\sum_{j=1}^{\infty} x_j$ by $x_1 = v_{n_1}$, $x_2 = v_{n_2} - v_{n_1}$, ..., $x_j = v_{n_j} - v_{n_{j-1}}$, and so on. The constructed series is absolutely convergent: $\sum_{j=2}^{\infty} \|x_j\| < \frac{1}{2} + \frac{1}{4} + \dots = 1$. At the same time, its partial sums are equal to v_{n_j}, and so they form (see Exercise 1 in Subsection 1.3.3) a divergent sequence. $\square$

Propositions 2 and 3 provide the following characterization of complete normed spaces.

Theorem 1. *For the normed space X to be complete it is necessary and sufficient that every absolutely convergent series in X be convergent.* $\square$

6.3.2 Completeness of the Space L_1

Sobriety is a life norm ...
True, but is life complete with this norm?[3]

We begin by proving a reformulation of Levi's theorem, essentially stated above in Exercise 3 of Subsection 4.4.3.

Lemma 1. *Suppose the series $\sum_{n=1}^{\infty} f_n$ of functions from $L_1 = L_1(\Omega, \Sigma, \mu)$ converges absolutely in the norm of this space. Then the series $\sum_{n=1}^{\infty} f_n$ converges almost everywhere to an integrable function f and $\|f\| \leqslant \sum_{n=1}^{\infty} \|f_n\|$.*

Proof. By the definition of the norm in L_1, the absolute convergence means that $\sum_{n=1}^{\infty} \int_{\Omega} |f_n| d\mu < \infty$. By Levi's theorem, the series $\sum_{n=1}^{\infty} |f_n|$ converges almost everywhere to an integrable function g, and $\int_{\Omega} g\, d\mu = \sum_{n=1}^{\infty} \int_{\Omega} |f_n| d\mu$. Denote the set of measure 0 in the complement of which $\sum_{n=1}^{\infty} |f_n|$ converges by A. For each point $t \in \Omega \setminus A$, the numerical series $\sum_{n=1}^{\infty} f_n(t)$ converges absolutely to some number $f(t)$. Thus, we defined on $\Omega \setminus A$ (i.e., almost everywhere on Ω) a function f, and the series $\sum_{n=1}^{\infty} f_n$ converges to f at all points of $\Omega \setminus A$. Extend f to the set A by 0. The function f is measurable on $\Omega \setminus A$, being the pointwise limit of a sequence of measurable functions; moreover, f has an integrable majorant, namely, the function g. Hence, f is integrable and

$$\int_{\Omega} |f| d\mu \leqslant \int_{\Omega} g\, d\mu = \sum_{n=1}^{\infty} \int_{\Omega} |f_n| d\mu. \qquad \square$$

Theorem 1. *L_1 is a Banach space.*

Proof. We use the theorem of the preceding subsection, i.e., the completeness criterion in terms of absolute convergence. Suppose the series $\sum_{n=1}^{\infty} f_n$ of L_1-functions converges absolutely. By the preceding lemma, $\sum_{n=1}^{\infty} f_n$ converges almost everywhere to an integrable function f. We claim that the series $\sum_{n=1}^{\infty} f_n$ converges to f in the norm of the space L_1. Indeed, $\|f - \sum_{n=1}^{k} f_n\| = \|\sum_{n=k}^{\infty} f_n\| \leqslant \sum_{n=k}^{\infty} \|f_n\| \to 0$ as $k \to \infty$. $\qquad \square$

Exercise

Prove the completeness of the space L_p.

The completeness of the space L_p will be established later, in Chap. 14, by an indirect argument. Nevertheless, the reader will profit from finding a direct proof of this fact.

---

[3] A joke from the times of the Gorbachev anti-alcoholism campaign in Soviet Union, 1985–1990. Quoted from a toast given by Ya.G. Prytula at the banquet for the International Conference on Functional Analysis and its Applications dedicated to the 110th anniversary of Stefan Banach, May 28–31, 2002, Lviv, Ukraine.

6.3.3 Subspaces and Quotient Spaces of Banach Spaces

Let X be a Banach space. A linear subspace $Y \subset X$, equipped with the norm of X, is called a *subspace of the Banach space X* if Y is closed in X. Hence, a subspace of a Banach space is itself a Banach space. As the reader had undoubtedly noticed, the meaning of the term "subspace" depends on where this subspace is considered. Since a Banach space is simultaneously a metric as well as a linear and normed space, the term "subspace" is somewhat overloaded. For this reason we emphasize once again that in Banach spaces subspaces will be tacitly understood to be closed linear subspaces. If for some reason we need to consider a non-closed linear subspace, we will state explicitly that the subspace is not closed.

Theorem 1. *Let X be a Banach space and Y be a subspace of X. Then the quotient space X/Y is also a Banach space.*

Proof. Let $x_n \in X$ be such that the corresponding equivalence classes form an absolutely convergent series: $\sum_n \|[x_n]\| < \infty$. By the completeness criterion, we need to prove that the series $\sum_n [x_n]$ converges to some element of the quotient space. To this end we pick in each class $[x_n]$ a representative y_n such that $\|y_n\| \leqslant \|[x_n]\| + \frac{1}{2^n}$. Then $\sum_n y_n$ is an absolutely convergent series in X, which in view of the completeness of X means that the series $\sum_n y_n$ converges in X to some element x. We claim that $\sum_n [y_n] = [x]$. Indeed,

$$\left\| [x] - \sum_{k=1}^{n} [x_k] \right\| = \left\| [x] - \sum_{k=1}^{n} [y_k] \right\| = \left\| [x - \sum_{k=1}^{n} y_k] \right\| \leqslant \left\| x - \sum_{k=1}^{n} y_k \right\| \to 0$$

as $n \to \infty$. $\qquad\qquad\qquad\qquad\qquad\qquad\qquad\qquad\qquad\qquad\qquad\qquad\qquad\qquad\square$

Exercises

12. Let X be a Banach space, and let $x_n \in X$ be a fixed sequence of nonzero vectors. We introduce the space E of all numerical sequences $a = (a_n)_1^\infty$ for which the series $\sum_{n=1}^{\infty} a_n x_n$ converges. We endow the space E with the norm $\|a\| = \sup\{\|\sum_{n=1}^{N} a_n x_n\| : N = 1, 2, \ldots\}$. Verify that E is a Banach space.

13. Let X be a Banach space, Y a nontrivial subspace of X (i.e., Y is closed and $Y \neq X$). Prove that Y is nowhere dense in X.

14. Show that a Banach space cannot be represented as a countable union of non-trivial subspaces.

15. Show that a Hamel basis of an infinite-dimensional Banach space is not countable.

16. Let $\mathcal{P}$ be the space of all polynomials (of arbitrarily large degree) with real coefficients, equipped with the norm $\|a_0 + a_1 t + \cdots + a_n t^n\| = |a_0| + |a_1| \cdots + |a_n|$. Is $\mathcal{P}$ complete?

17. Denote by $\{e_n\}_1^\infty$ the *standard basis* of the space ℓ_1: $e_1 = (1, 0, 0, \ldots)$, $e_2 = (0, 1, 0, \ldots), \ldots$. Show that for every $a = (a_n)_1^\infty \in \ell_1$ the series $\sum_{n=1}^\infty a_n e_n$ converges to a. Is the convergence absolute?

18. Consider in ℓ_∞ the sequence $\{e_n\}_1^\infty$ from the previous exercise. What are the partial sums of the series $\sum_{n=1}^\infty e_n$ equal to? Does this series converge to the element $x = (1, 1, \ldots) \in \ell_\infty$? Describe the elements $a = (a_n)_1^\infty \in \ell_\infty$ for which the series $\sum_{n=1}^\infty a_n e_n$ converges to a. For which a will the convergence be absolute?

19. Prove that the space $L_\infty(\Omega, \Sigma, \mu)$ is complete.

20. Prove that in each of the spaces $L_p(\Omega, \Sigma, \mu)$ with $1 \leqslant p \leqslant \infty$, the subspace of finite-valued measurable functions is dense.

21. The space ℓ_p with $1 \leqslant p < \infty$ is separable, whereas ℓ_∞ is not separable.

6.4 Spaces of Continuous Linear Operators

6.4.1 A Continuity Criterion for Linear Operators

Definition 1. Let X and Y be normed spaces. A linear operator $T: X \to Y$ is said to be *bounded* if it maps bounded sequences into bounded sequences. In other words, if $x_n \in X$ and $\sup_n \|x_n\| < \infty$ imply $\sup_n \|Tx_n\| < \infty$.

The main purpose of this subsection is to prove that for a linear operator continuity and boundedness are equivalent.

Theorem 1. *Let X and Y be normed spaces. For a linear operator $T: X \to Y$ the following conditions are equivalent:*

(1) *T is continuous;*

(2) *T maps sequences that converge to zero into sequences that converge to zero;*

(3) *T maps sequences that converge to zero into bounded sequences;*

(4) *T is bounded.*

Proof. The implications (1) $\Longrightarrow$ (2) $\Longrightarrow$ (3) $\Longleftarrow$ (4) are obvious: indeed, condition (2), i.e., the continuity of the operator at zero, is a particular case of condition (1); condition (3) follows from (2) as well as from (4), because sequences that converge to zero are bounded. Now let us prove the converse implications.

$(2) \implies (1)$. Suppose the sequence of vectors $x_n \in X$ converges to the vector $x \in X$. Then $x_n - x \to 0$ as $n \to \infty$, so by condition (2), $T x_n - T x = T(x_n - x) \to 0$ as $n \to \infty$. That is, convergence of x_n to x implies convergence of $T x_n$ to $T x$.

$(3) \implies (2)$. We proceed by reductio ad absurdum. Suppose condition (2) is not satisfied: there exists a sequence (x_n) in X which converges to zero, but such that $T x_n$ does not converge to zero. Then one can extract from (x_n) a subsequence, denoted (v_n), for which $\inf_n \|T v_n\| = \varepsilon > 0$. Consider the vectors $w_n = \frac{1}{\sqrt{\|v_n\|}} v_n$. The sequence w_n still converges to 0, but $\|T w_n\| \geqslant \frac{\varepsilon}{\sqrt{\|v_n\|}} \to \infty$, which contradicts condition (3).

$(3) \implies (4)$. Suppose condition (4) is not satisfied: there exists a bounded sequence (x_n) in X such that $\sup_n \|T x_n\| = \infty$. Then one can extract from (x_n) a subsequence, denoted (v_n), for which $\|T v_n\| \to \infty$. Consider the vectors $w_n = \frac{1}{\sqrt{\|T v_n\|}} v_n$. The sequence (w_n) already converges to 0, but $\|T w_n\| = \sqrt{\|T v_n\|} \to \infty$, which contradicts condition (3). $\square$

Exercises

1. Let X and Y be normed spaces, $T: X \to Y$ a continuous linear operator. Then $\operatorname{Ker} T = T^{-1}(0)$ is a closed linear subspace in X. (**N.B.!**) This is a simple yet important fact, and in the sequel will be used without further clarifications.

2. The image (range) of a continuous operator is not necessarily closed. Examine this in the case of the integration operator $T: C[0, 1] \to C[0, 1]$, $(Tf)(t) = \int_0^t f(\tau)d\tau$.

6.4.2   The Norm of an Operator

The *norm* of the linear operator T, acting from the normed space X into the normed space Y, is defined as

$$\|T\| = \sup_{x \in S_X} \|T x\|.$$

Proposition 1. *Let $\|T\| < \infty$. Then $\|T x\| \leqslant \|T\| \cdot \|x\|$ for any $x \in X$.*

Proof. For $x = 0$ the inequality holds trivially. Now let $x \neq 0$. Since $x/\|x\| \in S_X$, we have $\|T(x/\|x\|)\| \leqslant \|T\|$. Therefore, $\|T x\| = \|x\| \cdot \|T(x/\|x\|)\| \leqslant \|T\| \cdot \|x\|$, as claimed. $\square$

Proposition 2. *Let X and Y be normed spaces. For a linear operator $T: X \to Y$ the following conditions are equivalent:*

(1) T *is bounded*;

(2) $\|T\| < \infty$;

(3) *there exists a constant* $C > 0$ *such that* $\|Tx\| \leqslant C\|x\|$ *for all* $x \in X$.

Proof. (1) $\Longrightarrow$ (2). Suppose $\|T\| = \infty$. Then for any positive integer n there exists a vector $x_n \in S_X$ such that $\|Tx_n\| > n$. The sequence (x_n) is bounded, and the images of its terms tend in norm to infinity. This contradicts condition (1). The implication (2) $\Longrightarrow$ (3) was proved in Proposition 1 (with $C = \|T\|$). It remains to show that (3) $\Longrightarrow$ (1). Let $x_n \in X$ be a bounded sequence: $\|x_n\| \leqslant K$ for some constant K. Then, by condition (3), $\|Tx_n\| \leqslant CK$ for all n. Hence, the operator T maps bounded sequences into bounded sequences, as we needed to prove. $\square$

Remark 1. If condition (3) of the preceding theorem is satisfied, then

$$\|T\| = \sup_{x \in S_X} \|Tx\| \leqslant \sup_{x \in S_X} C\|x\| = C.$$

That is, if $\|Tx\| \leqslant C\|x\|$ for all $x \in X$, then $\|T\| \leqslant C$. This observation is often used in the estimation of norms of operators.

Remark 2. In the literature one encounters quite a few equivalent definitions of the norm of an operator:

— $\|T\| = \sup_{x \in B_X} \|Tx\|$;

— $\|T\| = \sup_{x \in \overline{B}_X} \|Tx\|$;

— $\|T\| = \sup_{x \in X \setminus \{0\}} \dfrac{\|Tx\|}{\|x\|}$;

— $\|T\|$ is the infimum of all constants $C \geqslant 0$ such that the inequality $\|Tx\| \leqslant C\|x\|$ is satisfied for all $x \in X$.

The verification of the equivalence of these definitions is left to the reader.

We let $L(X, Y)$ denote the space of all continuous linear operators acting from the normed space X into the normed space Y. $L(X, Y)$ is naturally endowed with linear operations: if $T_1, T_2 \in L(X, Y)$ are operators and λ_1, λ_2 are scalars, then the operator $\lambda_1 T_1 + \lambda_2 T_2 \in L(X, Y)$ acts according to the rule $(\lambda_1 T_1 + \lambda_2 T_2) x = \lambda_1 T_1 x + \lambda_2 T_2 x$. We described above how to introduce a norm on $L(X, Y)$ — the norm of the operator, but it remains to verify that the norm axioms are indeed satisfied.

Proposition 3. *The space* $L(X, Y)$ *of continuous operators is a normed space.*

Proof. Let us verify the norm axioms (Subsection 6.1.1).

1. Suppose $\|T\| = 0$. Then the operator T is equal to 0 on all elements of the unit sphere of the space X, which in view of its linearity means that T is equal to zero on the entire space X.

2. $\|\lambda T\| = \sup_{x \in S_X} \|\lambda Tx\| = |\lambda| \sup_{x \in S_X} \|Tx\| = |\lambda| \, \|T\|.$

3. Let $T_1, T_2 \in L(X, Y)$ and $x \in X$. By Proposition 1,

$$\|(T_1 + T_2)x\| \leqslant \|T_1 x\| + \|T_2 x\| \leqslant \|T_1\| \cdot \|x\| + \|T_2\| \cdot \|x\| = (\|T_1\| + \|T_2\|) \cdot \|x\|.$$

By Remark 1, this yields the needed triangle inequality: $\|T_1 + T_2\| \leqslant \|T_1\| + \|T_2\|$.
□

The norm of an operator is an important concept that will be frequently used in our text. For this reason the reader who has no experience working with norms is strongly advised to seriously pay attention to the exercises given below.

Exercises

1. Let $T \in L(X, Y)$ and $x_1, x_2 \in X$. Then $\|Tx_1 - Tx_2\| \leqslant \|T\| \cdot \|x_1 - x_2\|$.

2. Let $T_1, T_2 \in L(X, Y)$ and $x \in X$. Then $\|T_1 x - T_2 x\| \leqslant \|T_1 - T_2\| \cdot \|x\|$.

3. Let X, Y, Z be normed spaces, $T_1 \in L(X, Y)$, $T_2 \in L(Y, Z)$. Prove the *multiplicative triangle inequality* for the composition of operators: $\|T_2 \circ T_1\| \leqslant \|T_2\| \cdot \|T_1\|$.

4. Let X be a Banach space, $(x_n)_1^\infty$ a bounded sequence in X, and $\{e_n\}_1^\infty$ the standard basis in the space ℓ_1 (see Exercise 6 in Subsection 6.3.3). Define the operator T: $\ell_1 \to X$ by the formula $Ta = \sum_{n=1}^\infty a_n x_n$, for any element $a = (a_n)_1^\infty$ of the space ℓ_1. Show that T is a continuous linear operator, $Te_n = x_n$, and $\|T\| = \sup_n \|x_n\|$. Show that any continuous linear operator $T: \ell_1 \to X$ can be described as indicated above.

5. Let X be a normed space and X_1 a closed subspace of X. Show that the quotient map q of the space X onto the space X/X_1 (see Subsection 5.2.2) is a continuous linear operator. Calculate the norm $\|q\|$. Show that $q(B_X) = B_{X/X_1}$.

6. Let X and Y be normed spaces, and $T: X \to Y$ a linear operator. Prove that the injectivization $\widetilde{T}$ of the operator T (see Subsection 5.2.3) is a continuous linear operator and $\|\widetilde{T}\| = \|T\|$.

7. In the setting of the preceding exercise, suppose that $T(B_X) = B_Y$. Show that in this case $\widetilde{T}$ is a bijective isometry of the spaces $X/\mathrm{Ker}\, T$ and Y.

8. Let $\mathcal{P}$ be the space of all polynomials, as in Exercise 5 in Subsection 6.3.3, and let $D_m: \mathcal{P} \to \mathcal{P}$ be the m-th derivative operator. Verify that D_m is a linear operator and calculate its norm. Is D_m a continuous operator?

9. Equip the linear space $\mathcal{P}$ of polynomials with the norm $\|a_0 + a_1 t + \cdots + a_n t^n\|_1 = \sum_{k=0}^n k! \, |a_k|$. Denote the resulting normed space by $\mathcal{P}_1$. Is the m-th derivative operator $D_m: \mathcal{P}_1 \to \mathcal{P}_1$ continuous? What is its norm?

10. Let X and Y be normed spaces and $T\colon X \to Y$ be a bijective linear operator. Show that the operator T is an isometry if and only if $\|T\| = \|T^{-1}\| = 1$.

6.4.3 Pointwise Convergence

Theorem 1. *Suppose X and Y are normed spaces, $T_n\colon X \to Y$ is a linear operator, and the limit $\lim_{n\to\infty} T_n x$ exists for all $x \in X$. Then the map $T\colon X \to Y$ given by the recipe $T(x) = \lim_{n\to\infty} T_n x$ is a linear operator.*

Proof. Indeed,

$$T(ax_1 + bx_2) = \lim_{n\to\infty} T_n(ax_1 + bx_2)$$

$$= a \lim_{n\to\infty} T_n(x_1) + b \lim_{n\to\infty} T_n(x_2) = aT(x_1) + bT(x_2). \qquad \square$$

Definition 1. A sequence of operators $T_n\colon X \to Y$ is said to *converge pointwise* to the operator $T\colon X \to Y$ if $Tx = \lim_{n\to\infty} T_n x$ for all $x \in X$.

Theorem 2. *Suppose the sequence of operators $T_n \in L(X, Y)$ converges pointwise to the operator $T\colon X \to Y$ and $\sup_n \|T_n\| = C < \infty$. Then $T \in L(X, Y)$ and $\|T\| \leqslant C$.*

Proof. The estimate $\|Tx\| = \lim_{n\to\infty} \|T_n x\| \leqslant C\|x\|$ holds for all $x \in X$. $\qquad \square$

Theorem 3. *If the sequence of operators $T_n \in L(X, Y)$ converges to the operator T in the norm of the space $L(X, Y)$, then it also converges pointwise to T.*

Proof. Indeed,

$$\|T_n x - Tx\| = \|(T_n - T)x\| \leqslant \|T_n - T\| \cdot \|x\| \to 0 \quad \text{as } n \to \infty. \qquad \square$$

Exercises

1. Let $X = C[0, 1]$, $Y = \mathbb{R}$, and let the operators $T_n \in L(X, Y)$ act as $T_n(f) = f(0) - f(1/n)$. Calculate the norms of T_n.

2. Pointwise convergence does not imply convergence in norm. Example: the sequence of operators from the preceding exercise tends to 0 pointwise, but not in norm.

3. The following general fact is known (Josefson and Nissenzweig, [55, 72], see also [47]): *On any infinite-dimensional normed space there exists a sequence of linear functionals that converges to 0 pointwise, but not in norm.* Give corresponding examples in all infinite-dimensional normed spaces you know.

4. Under the assumptions of Theorem 2, show that $\|T\| \leqslant \underline{\lim}_{n \to \infty} \|T_n\|$. In other words, the norm on $L(X, Y)$ is lower semicontinuous with respect to pointwise convergence.

5. Introduce on $L(X, Y)$ a topology in which convergence coincides with pointwise convergence.

6.4.4 Completeness of the Space of Operators. Dual Space

Theorem 1. *Let X be a normed space and Y a Banach space. Then $L(X, Y)$ is a Banach space.*

Proof. We use the definition. Suppose the operators $T_n \in L(X, Y)$ form a Cauchy sequence: $\|T_n - T_m\| \to 0$ as $n, m \to \infty$. Then for any point $x \in X$ the values $T_n x$ form a Cauchy sequence in the complete space Y, because $\|T_n x - T_m x\| \leqslant \|T_n - T_m\| \cdot \|x\| \to 0$ as $n, m \to \infty$. Hence, for any $x \in X$ the sequence $(T_n x)$ has a limit. Define the operator $T : X \to Y$ by the rule $Tx = \lim_{n \to \infty} T_n x$. By Theorem 1 of the preceding subsection, the operator T is linear. Since every Cauchy sequence is bounded, Theorem 2 of the preceding subsection shows that $T \in L(X, Y)$. It remains to verify that $T = \lim_{n \to \infty} T_n$ in the norm of the space $L(X, Y)$. Since the sequence T_n is Cauchy, for any $\varepsilon > 0$ there exists a number $N(\varepsilon)$ such that $\|T_N - T_M\| < \varepsilon$ for all $M > N > N(\varepsilon)$. Then for any point x of the unit sphere S_X of X it also holds that $\|T_N x - T_M x\| < \varepsilon$ for $M > N > N(\varepsilon)$. Letting here $M \to \infty$ in the last inequality, we obtain $\|T_N x - Tx\| < \varepsilon$. Now if in the left-hand side of this inequality we take the supremum over $x \in S_X$, we get $\|T_N - T\| \leqslant \varepsilon$ for $N > N(\varepsilon)$, i.e., $T = \lim_{n \to \infty} T_n$, as needed. $\square$

The *dual* (or *conjugate*) *space* of the normed space X is the space X^* of all continuous linear functionals on X, equipped with the norm $\|f\| = \sup_{x \in S_X} |f(x)|$. In other words, if X is a real space, then $X^* = L(X, \mathbb{R})$, while if X is a complex space, $X^* = L(X, \mathbb{C})$. Since $\mathbb{R}$ and $\mathbb{C}$ are complete spaces, the theorem above shows that the space X^* is complete, regardless of whether the space X itself is complete or not. The space X^* will also be referred to simply as the *dual* of X.

As was the case with the norm of an operator (see Remark 2 in Subsection 6.4.2), there are other standard definitions for the norm of a functional. We provide one of them that is specific for functionals rather than for general operators.

Remark 1. Let X be a real normed space, and let $f \in X^*$. Then $\|f\| = \sup_{x \in S_X} f(x)$.

Proof. We use the symmetry of the sphere: $x \in S_X$ if and only if $-x \in S_X$. Hence, $\sup_{x \in S_X} f(x) = \sup_{x \in S_X} f(-x)$. Consequently,

$$\|f\| = \sup_{x \in S_X} |f(x)| = \sup_{x \in S_X} \max\{f(x), -f(x)\}$$

$$= \max\left\{ \sup_{x \in S_X} f(x), \sup_{x \in S_X} f(-x) \right\} = \sup_{x \in S_X} f(x). \qquad \square$$

Exercises

1. Let X be a real normed space, and $f \in X^*$. Then $\|f\| = \sup_{x \in \overline{B}_X} f(x)$.

2. Let X be a complex normed space, and $f \in X^*$. Then $\|f\| = \sup_{x \in S_X} \operatorname{Re} f(x)$.

3. On the space ℓ_∞ of all bounded numerical sequences $x = (x_1, x_2, \ldots)$, equipped with the norm $\|x\| = \sup_n |x_n|$, define the functional f by the formula $f(x) = \sum_{n=1}^\infty a_n x_n$, where $a = (a_1, a_2, \ldots)$ is a fixed element of the space ℓ_1. Show that $\|f\| = \sum_{n=1}^\infty |a_n|$.

4. On the space $C[0, 1]$ consider the linear functional F defined by the rule $F(x) = \int_0^{1/2} x(t)\,dt - \int_{1/2}^1 x(t)\,dt$. Show that $\|F\| = 1$ and that $|F(x)| < 1$ for all $x \in S_{C[0,1]}$. This example shows that the supremum in the definition of the norm of a functional (or operator) is not necessarily attained.

6.5 Extension of Operators

In this section we consider several simple yet useful conditions under which a continuous operator can be extended from a subspace of a normed space to the entire ambient space.

6.5.1 Extension by Continuity

Theorem 1. *Let X_1 be a dense subspace of the normed space X, Y a Banach space, and $T_1 \in L(X_1, Y)$. Then the operator T_1 admits a unique extension $T \in L(X, Y)$.*

Proof. Since the subspace X_1 is dense, for any $x \in X$ there exists a sequence of vectors $x_n \in X_1$ which converges to x. Then $(T_1 x_n)$ is a Cauchy sequence in Y:

$$\|T_1 x_n - T_1 x_m\| \leqslant \|T_1\| \cdot \|x_n - x_m\| \to \infty \quad \text{as } n, m \to \infty.$$

Denote the limit of this sequence by Tx. Then

$$\|Tx\| = \lim_{n \to \infty} \|T_1 x_n\| \leqslant \|T_1\| \lim_{n \to \infty} \|x_n\| = \|T_1\| \cdot \|x\|.$$

Note that Tx does indeed depend only on x, and not on the choice of the sequence x_n: if $x_n' \in X_1$ is some other sequence that converges to x, then $\|T_1 x_n - T_1 x_n'\| \leqslant \|T_1\| \cdot \|x_n - x_n'\| \to 0$ as $n \to \infty$, and so the sequences $(T_1 x_n)$ and $(T_1 x_n')$ have the same limit. Hence, for every $x \in X$ we defined a map $T: X \to Y$ by the rule $Tx = \lim_{n \to \infty} T_1 x_n$, where $x_n \in X_1$ form a sequence that converges to x. It remains to show that T is the sought-for operator. Let us verify that the operator T is linear. Let $x_1, x_2 \in X$, $x_1^n, x_2^n \in X_1$, $x_2^n \to x_2$, $x_1^n \to x_1$ as $n \to \infty$. Then

$$T(a_1 x_1 + a_2 x_2) = \lim_{n\to\infty} T_1(a_1 x_1^n + a_2 x_2^n)$$
$$= a_1 \lim_{n\to\infty} T_1 x_1^n + a_2 \lim_{n\to\infty} T_1 x_2^n = a_1 T x_1 + a_2 T x_2$$

for all scalars a_1, a_2. Thanks to the already established inequality $\|Tx\| \leqslant \|T_1\| \cdot \|x\|$, the operator T is continuous, i.e., $T \in L(X, Y)$. This proves the existence of the extension. Its uniqueness follows from the fact that two continuous functions which coincide on a dense set coincide everywhere. $\square$

Exercises

1. In the argument above we skipped the verification of the fact that the operator T is an extension of the operator T_1. Complete this step.

2. Show that under the conditions of the preceding theorem $\|T\| \leqslant \|T_1\|$.

3. Let X and Y be normed spaces, $X_1 \subset X$ be an arbitrary subspace, and $T \in L(X, Y)$ be an extension of the operator $T_1 \in L(X_1, Y)$. Show that $\|T\| \geqslant \|T_1\|$.

4. Combining Exercises 2 and 3 above, show that under the assumptions of Theorem 1, $\|T\| = \|T_1\|$.

5. Give an example of a continuous function which is defined on a dense subset of the interval $[0, 1]$, but which cannot be extended to a continuous function on the whole interval.

6. Show that every continuous linear operator is a uniformly continuous mapping. Deduce the main theorem of the present subsection from the theorem, given in Subsection 1.3.4, on the extension of uniformly continuous mappings. Moreover, the linearity of the extended operator can be deduced from the uniqueness of the extension.

6.5.2 Projectors; Extension from a Closed Subspace

Let X_1 be a subspace of the normed space X. The operator $P \in L(X, X)$ is called a *projector* onto X_1 if $P(X) \subset X_1$ and $Px = x$ for all $x \in X_1$.

Theorem 1. *For a subspace X_1 of the normed space X, the following conditions are equivalent:*

(1) *in X there exists a projector onto X_1;*

(2) *for any normed space Y, any operator $T_1 \in L(X_1, Y)$ extends to an operator $T \in L(X, Y)$.*

Proof. (1) $\implies$ (2). Define $T \in L(X, Y)$ by the rule $Tx = T_1(Px)$.

(2) $\implies$ (1). Take $Y = X_1$ and define $T_1 \in L(X_1, Y)$ by the rule $T_1x = x$. Let $T \in L(X, Y)$ be an extension of the operator T_1. Since in our case $Y \subset X$, we can regard T as an operator from X to X. We have $T(X) \subset Y = X_1$, and $Tx = T_1x = x$ for all $x \in X_1$. Hence, T is the required projector onto X_1. $\qquad\square$

Exercises

1. Provide the details of the proof of the implication (1) $\implies$ (2) in the preceding theorem.

2. Let X_1 be a subspáce of the normed space X and $P \in L(X, X)$ be a projector onto X_1. Then $P(X) = X_1 = \text{Ker}(I - P)$ and the subspace X_1 is closed in X.

3. Suppose that under the conditions of the preceding exercise $X_1 \neq \{0\}$. Then $\|P\| \geqslant 1$.

4. For a subspace X_1 of the normed space X the following conditions are equivalent:

— in X there exists a projector P onto X_1 with $\|P\| = 1$;

— for any normed space Y, any operator $T_1 \in L(X_1, Y)$ extends to an operator $T \in L(X, Y)$ with $\|T\| = \|T_1\|$.

5. Let $X = \ell_1^3$ (see Exercise 2 in Subsection 6.2.1 for the definition), and let X_1 be the subspace consisting of all elements for which the sum of their coordinates is equal to zero. Show that in X there is no projector P onto X_1 with $\|P\| = 1$.

Comments on the Exercises

Subsection 6.1.2

Exercise 1. Since $\|x\| = \rho(0, x)$, the result follows from the continuity of the distance (Subsection 1.3.2).

Exercise 5. See Subsection 18.2.1.

Subsection 6.2.2

Exercise 3. See Theorem 2 in Subsection 14.1.2.

Exercise 7. Let $g \in L_p[a, b]$. Consider the sequence of truncations

$$g_n = \min\{n, \max\{g, -n\}\}.$$

The sequence of functions $|g_n - g|^p$ converges almost everywhere to zero and admits the integrable majorant $|g|^p$. Hence, by the Lebesgue dominated convergence theorem, $\|g_n - g\|_p \to 0$ as $n \to \infty$.

Exercise 8. By the preceding exercise, it suffices to show that any bounded function $f \in L_p[a, b]$ can be approximated in the metric of L_p by continuous functions. By Exercise 6 in Subsection 3.2.3, there exists a sequence of continuous functions (f_n) that converges to f a.e. With no loss of generality we can assume that all f_n are bounded in modulus by the same constant C as f (otherwise we replace f_n by the truncations $\widetilde{f}_n = \min\{C, \max\{f_n, -C\}\}$). The convergence of $\|f_n - f\|_p$ to 0 follows from the Lebesgue dominated convergence theorem.

Subsection 6.4.2

Exercise 5. $[x] \in q(B_X) \iff \exists y \in B_X : [y] = [x] \iff \|[x]\| < 1 \iff [x] \in B_{X/X_1}$.

Exercise 6.

$$\|\widetilde{T}\| = \sup_{[x] \in B_{X/X_1}} \|\widetilde{T}[x]\| = \sup_{[x] \in q(B_X)} \|\widetilde{T}[x]\|$$

$$= \sup_{x \in B_X} \|\widetilde{T}[x]\| = \sup_{x \in B_X} \|Tx\| = \|T\|.$$

Chapter 7
Absolute Continuity of Measures and Functions. The Connection Between Derivative and Integral

7.1 Charges. The Hahn and Radon–Nikodým Theorems

The family of all finite measures on a fixed σ-algebra does not constitute a linear space: such measures can be added, but already the difference of two of them may take negative values, and consequently not be a measure. This creates some inconveniences, so to avoid them one introduces a generalized notion of measure that is allowed to take negative values. Such generalized measures are called charges, or signed measures. \

In this section (Ω, Σ) is a set with a given σ-algebra of its subsets. Unless otherwise stipulated, all functions will be assumed to be defined on Ω, and all measures and charges to be defined on Σ.

7.1.1 The Boundedness of Charges Theorem

A mapping $\nu \colon \Sigma \to \mathbb{R}$ is called a *charge*, or a *signed measure*, if it satisfies the *countable additivity* condition: for any sequence of pairwise disjoint sets $A_n \in \Sigma$, $n = 1, 2, \ldots$, the series $\sum_{k=1}^{\infty} \nu(A_k)$ converges and $\nu\left(\bigcup_{k=1}^{\infty} A_k\right) = \sum_{k=1}^{\infty} \nu(A_k)$.

We immediately note that this definition implies that the series $\sum_{k=1}^{\infty} \nu(A_k)$ converges for all permutations of its terms, i.e., it converges absolutely. Many of the properties of measures carry over to charges, with the same proofs. Thus, the charge of the empty set is equal to zero (in the definition of countable additivity take all $A_n = \emptyset$); charges are finitely additive (take $A_n = \emptyset$ for all $n > N$). In particular, we shall use the following assertions:

— if $A_n \in \Sigma$, $n = 1, 2, \ldots$, is an increasing chain of sets, then $\nu\left(\bigcup_{k=1}^{\infty} A_k\right) = \lim_{k \to \infty} \nu(A_k)$;

— if $A_n \in \Sigma$, $n \in \mathbb{N}$, form a decreasing chain of sets, then $\nu\left(\bigcap_{k=1}^{\infty} A_k\right) = \lim_{k \to \infty} \nu(A_k)$.

© Springer International Publishing AG, part of Springer Nature 2018
V. Kadets, *A Course in Functional Analysis and Measure Theory*,
Universitext, https://doi.org/10.1007/978-3-319-92004-7_7

Nevertheless, some care should be taken when working with charges: for instance, it may well happen that $\nu(A) > \nu(B)$ for some sets $A \subset B \in \Sigma$, something that is not possible for measures.

For charges there are defined natural operations of addition and multiplication by scalars: $(a_1\nu_1 + a_2\nu_2)(A) = a_1\nu_1(A) + a_2\nu_2(A)$, as well as inequalities: $\mu \geqslant \nu$, if $\mu(A) \geqslant \nu(A)$ for all $A \in \Sigma$.

For $A \in \Sigma$ we define $\nu^+(A) = \sup\{\nu(B) : B \in \Sigma_A\}$. Since for B one can take, in particular, the empty set, $\nu^+(A) \geqslant 0$.

Lemma 1. *Let* $A = \bigsqcup_{k=1}^{\infty} A_k$. *Then* $\nu^+(A) = \sum_{k=1}^{\infty} \nu^+(A_k)$. *In other words,* ν^+ *satisfies the countable additivity condition.*

Proof. Every set $B \in \Sigma_A$ can be written as $B = \bigsqcup_{n=1}^{\infty} B_n$ with $B_n \subset A_n$: it suffices to put $B_n = B \cap A_n$. We have

$$\nu^+(A) = \sup\{\nu(B) : B \in \Sigma_A\} = \sup\left\{\sum_{n=1}^{\infty} \nu(B_n) : B_n \in \Sigma_{A_n}, n \in \mathbb{N}\right\}$$

$$= \sum_{n=1}^{\infty} \sup\{\nu(B_n) : B_n \in \Sigma_{A_n}\} = \sum_{n=1}^{\infty} \nu^+(A_n). \qquad \square$$

Lemma 2. *Suppose* $\nu^+(A) = +\infty$. *Then for any* $n \in \mathbb{N}$ *there exists a measurable set* $B \subset A$ *such that* $\nu^+(B) = +\infty$ *and* $|\nu(B)| > n$.

Proof. Take $B_1 \in \Sigma_A$ with $\nu(B_1) > n + |\nu(A)|$ and put $B_2 = A \setminus B_1$. Then $|\nu(B_2)| \geqslant |\nu(B_2)| - |\nu(A)| > n$. Since $\nu^+(B_1) + \nu^+(B_2) = \nu^+(A) = +\infty$, at least one of the numbers $\nu^+(B_1)$ and $\nu^+(B_2)$ is infinite. The corresponding B_i is the sought-for set B. $\qquad \square$

Theorem 1. ν^+ *is a finite countably-additive measure on* Σ.

Proof. The countable additivity was already established in Lemma 1. Let us prove the finiteness. Suppose there exists a set $A \in \Sigma$ such that $\nu^+(A) = +\infty$. By Lemma 2, there exists a set $A_1 \in \Sigma_A$ such that $\nu^+(A_1) = +\infty$ and $|\nu(A_1)| > 1$. Applying Lemma 2 again to the set A_1 with $n = 2$, we produce a set $A_2 \subset A_1$ such that $\nu^+(A_2) = +\infty$ and $|\nu(A_2)| > 2$. Continuing this process, we obtain an increasing sequence of sets $A_1, A_2, \ldots, A_n, \ldots \in \Sigma$ such that $|\nu(A_n)| > n$, which contradicts the condition $\nu\left(\bigcap_{k=1}^{\infty} A_k\right) = \lim_{k \to \infty} \nu(A_k)$ mentioned at the beginning of this subsection. $\qquad \square$

Corollary 1 (boundedness of charges). *For every charge* ν *there exist constants* $C_1, C_2 \in \mathbb{R}$ *such that* $C_1 \leqslant \nu(A) \leqslant C_2$ *for all* $A \in \Sigma$.

Proof. It suffices to take $C_2 = \nu^+(\Omega)$ and $C_1 = -(-\nu)^+(\Omega)$. $\qquad \square$

Definition 1. The measure ν^+ is called the *positive part* of the charge ν; the measure $\nu^- = (-\nu)^+$ is called the *negative part* of the charge ν; finally, the measure $|\nu| = \nu^+ + \nu^-$ is called the *variation* of the charge ν.

Theorem 2. *Every charge v admits the* Jordan decomposition $v = v^+ - v^-$. *Hence, every charge can be written as the difference of two measures.*

Proof. For any $A \in \Sigma$ it holds that

$$v^+(A) - v(A) = \sup\{v(B) - v(A) : B \in \Sigma_A\} = \sup\{-v(A \setminus B) : B \in \Sigma_A\}.$$

Denote $A \setminus B$ by C. When B runs over Σ_A, so does C. Consequently,

$$v^+(A) - v(A) = \sup\{-v(C) : C \in \Sigma_A\} = (-v)^+(A) = v^-(A). \qquad \square$$

Exercises

1. Prove the formulas $-v^-(A) = \inf\{v(B) : B \in \Sigma_A\}$ and $v(A) = \sup\{v(B) : B \in \Sigma_A\} + \inf\{v(B) : B \in \Sigma_A\}$.

2. Show that $|v(A)| \leqslant |v|(A)$ for all $A \in \Sigma$.

3. Let μ be a measure and v a charge. Suppose that $\mu \geqslant v$. Then $\mu \geqslant v^+$.

4. Let μ be a measure and v a charge. Suppose that $|v(A)| \leqslant \mu(A)$ for all $A \in \Sigma$. Then $|v| \leqslant \mu$.

5. Verify that the assertion of Exercise 4 in Subsection 2.1.4 remains valid for charges and even for more general set functions: Let v be a finitely-additive set function given on a σ-algebra Σ and taking values in a normed space X. Suppose that for any decreasing chain of sets $A_1, A_2, \ldots, A_n, \ldots \in \Sigma$ with empty intersection one has $\lim_{k \to \infty} v(A_k) = 0$. Then v is a countably-additive set function.

6. Verify that the formula $\|v\| := |v|(\Omega)$ defines a norm on the space $M(\Omega, \Sigma)$ of all charges on Σ. Show that the normed space $M(\Omega, \Sigma)$ is complete.

7.1.2 The Hahn Decomposition Theorem

Lemma 1. *For every charge v on Σ there exists a set $\Omega^+ \in \Sigma$ such that $v^+(\Omega^+) = v^+(\Omega)$ and $v^-(\Omega^+) = 0$.*

Proof. Fix a sequence $\varepsilon_n > 0$ satisfying $\sum_{n=1}^\infty \varepsilon_n < \infty$. For each $n \in \mathbb{N}$ choose a set $A_n \in \Sigma$ with $v(A_n) > v^+(\Omega) - \varepsilon_n$. Then

$$v^+(A_n) > v^+(\Omega) - \varepsilon_n, \tag{1}$$

$$v^-(A_n) = v^+(A_n) - v(A_n) \leqslant v^+(\Omega) - v(A_n) \leqslant \varepsilon_n. \tag{2}$$

Now put $\Omega^+ = \overline{\lim} \, A_n$. By assertion (i) of the lemma on the upper limit of a sequence of sets (Lemma 1 in Subsection 3.2.3), applied to the measure ν^+, estimate (1) implies that $\nu^+(\Omega^+) \geqslant \overline{\lim} \, \nu^+(A_n) = \nu^+(\Omega)$. Assertion (ii) of the same lemma, now applied to ν^-, yields the equality $\nu^-(\Omega^+) = 0$: here estimate (2) helped. $\square$

Theorem 1 (Hahn's theorem). *For every charge ν there exists a decomposition of Ω as the union of two disjoint measurable sets Ω^+ and Ω^- with the property that any subset of Ω^+ has charge greater than or equal to zero, while any subset of Ω^- has charge smaller than or equal to zero. This decomposition is unique up to $|\nu|$-equivalence.*

Proof. Take for Ω^+ the corresponding set from Lemma 1. Since $\nu^-(\Omega^+) = 0$, no subset of Ω^+ can have a negative charge. Now put $\Omega^- = \Omega \setminus \Omega^+$. Since $\nu^+(\Omega^-) = \nu^+(\Omega) - \nu^+(\Omega^+) = 0$, no subset of Ω^- can have a positive charge. This establishes the existence of the decomposition. Now let us show the uniqueness. Let $\Omega_1^+ \sqcup \Omega_1^-$ be another decomposition with the same properties. Then $\nu^-(\Omega_1^+) = 0$, $\nu^-(\Omega^+) = 0$, and so $\nu^-(\Omega_1^+ \bigtriangleup \Omega^+) = 0$. Analogously, $\nu^+(\Omega_1^- \bigtriangleup \Omega^-) = 0$. But $\Omega_1^+ \bigtriangleup \Omega^+ = \Omega_1^- \bigtriangleup \Omega^-$ (the symmetric difference of two sets coincides with the symmetric difference of their complements), so we also have $\nu^+(\Omega_1^+ \bigtriangleup \Omega^+) = 0$ and $\nu^-(\Omega_1^- \bigtriangleup \Omega^-) = 0$. Therefore, $|\nu|\,(\Omega_1^+ \bigtriangleup \Omega^+) = |\nu|\,(\Omega_1^- \bigtriangleup \Omega^-) = 0$. $\square$

The sets Ω^+ and Ω^- are called the *positivity* and *negativity* sets of the charge ν, and the decomposition $\Omega = \Omega^+ \sqcup \Omega^-$ is called the *Hahn decomposition*.

Exercises

1. Show that $\nu^+(A) = \nu(A \cap \Omega^+)$ and $\nu^-(A) = -\nu(A \cap \Omega^-)$.

2. Solve Exercise 3 of Subsection 7.1.1 based on the preceding exercise.

By analogy with Subsection 2.1.6, an *atom* of the charge ν is a subset $A \in \Sigma$ such that $\nu(A) \neq 0$ and for any $B \in \Sigma_A$ either $\nu(B) = 0$, or $\nu(A \setminus B) = 0$. If a charge has atoms, it is called *atomic*; if there are no atoms, the charge is called *non-atomic* or *atomless*. The charge ν is called *purely atomic* if Ω can be written as the union of a finite or countable number of disjoint atoms of ν.

3. The atoms of the charge ν coincide with the atoms of the measure $|\nu|$. The charge ν is non-atomic if and only if the measure $|\nu|$ is non-atomic.

4. Any charge can be uniquely represented as the sum of a purely atomic charge and a non-atomic charge.

In the normed space $M(\Omega, \Sigma)$ (see Exercise 6 of Subsection 7.1.1) consider the subsets $M_{\mathrm{at}}(\Omega, \Sigma)$ of purely atomic charges and $M_{\mathrm{nonat}}(\Omega, \Sigma)$ of non-atomic charges. Prove that:

5. $M_{\mathrm{at}}(\Omega, \Sigma)$ and $M_{\mathrm{nonat}}(\Omega, \Sigma)$ are closed subspaces in $M(\Omega, \Sigma)$ and $M(\Omega, \Sigma) = M_{\mathrm{at}}(\Omega, \Sigma) \oplus M_{\mathrm{nonat}}(\Omega, \Sigma)$.

6. Show that if $\nu_1 \in M_{\mathrm{at}}(\Omega, \Sigma)$ and $\nu_2 \in M_{\mathrm{nonat}}(\Omega, \Sigma)$, then $\|\nu_1 + \nu_2\| = \|\nu_1\| + \|\nu_2\|$.

As in the case of measures, we introduce admissible partitions (i.e., such that $\sum_{k=1}^{\infty} \lim_{t \in \Delta_k} [|f(t)| \cdot |\nu|(\Delta_k)] < \infty$) and integral sums of a function f with respect to the charge ν: $S_A(f, D, T, \nu) = \sum_{k=1}^{\infty} f(t_k)\nu(\Delta_k)$. The integrability and integral of a function with respect to a charge are also defined by means of integral sums.

7. A measurable function f is integrable with respect to the charge ν on the set $A \in \Sigma$ if and only if f is integrable on A with respect to ν on both $A \cap \Omega^+$ and $A \cap \Omega^-$.

8. A measurable function f is integrable with respect to the charge ν on the set $A \in \Sigma$ if and only if f is integrable on A with respect to the measure ν^+, as well as with respect to the measure ν^-. Moreover, $\int_A f\, d\nu = \int_A f\, d\nu^+ - \int_A f\, d\nu^-$.

9. A measurable function f is integrable with respect to the charge ν on the set $A \in \Sigma$ if and only if f is integrable on A with respect to the measure $|\nu|$.

10. Show that the expression $\int_A f\, d\nu$ is linear in f, as well as in ν, and is also countably additive with respect to A.

11. Prove the inequality $\left|\int_A f\, d\nu\right| \leqslant \int_A |f|\, d|\nu|$.

7.1.3 Absolutely Continuous Measures and Charges

Let ν be a charge and μ be a countably-additive measure on Σ. We say that the charge ν is *absolutely continuous* with respect to the measure μ (and write $\nu \ll \mu$)[1], if for any $A \in \Sigma$ with $\mu(A) = 0$ one also has $\nu(A) = 0$.

At a first glance, the notion of absolute continuity just introduced does not elicit any associations with the usual notion of continuity. However, such associations become evident once we provide the following equivalent formulation:

Theorem 1. *For a charge ν and a measure μ the following conditions are equivalent:*

(1) $\nu \ll \mu$;

(2) $|\nu| \ll \mu$;

(3) *for every $\varepsilon > 0$ there exists a $\delta > 0$ such that for any $A \in \Sigma$, if $\mu(A) < \delta$, then $|\nu|(A) < \varepsilon$.*

---

[1] Do not confuse with the inequality $\nu \leqslant \mu$!

Proof. (1) $\implies$ (2). Suppose $\mu(A) = 0$. Then $\mu(B) = 0$ for all $B \in \Sigma_A$. Hence, $\nu(B) = 0$ for all $B \in \Sigma_A$, i.e., all quantities $\nu^+(A)$, $\nu^-(A)$, and $|\nu|(A)$ are equal to zero.

(2) $\implies$ (3). Suppose that condition (3) is not satisfied. Then there is an $\varepsilon > 0$ such that for any $\delta > 0$ there exists a set $A \in \Sigma$ such that $\mu(A) < \delta$ and $|\nu|(A) \geqslant \varepsilon$. Applying this with $\delta = 2^{-n}$, we produce a sequence of sets $A_n \in \Sigma$ such that $\mu(A_n) < 2^{-n}$ and $|\nu|(A_n) \geqslant \varepsilon$. Then for the set $B = \overline{\lim} A_n$ it holds that $|\nu|(B) \geqslant \varepsilon$ and $\mu(B) = 0$, which contradicts the assumption that $|\nu| \ll \mu$.

(3) $\implies$ (1). Suppose $\mu(A) = 0$. Then $\mu(A) < \delta$ for any $\delta > 0$, and so $|\nu|(A) < \varepsilon$ for all $\varepsilon > 0$. Therefore, $|\nu|(A) = 0$, and consequently $\nu(A) = 0$. $\qquad\square$

Exercises

1. Let $\nu \ll \mu$. Then $\nu^+ \ll \mu$ and $\nu^- \ll \mu$.

2. Denote by $M_{abs}(\Omega, \Sigma, \mu)$ the subset of $M(\Omega, \Sigma)$ consisting of all charges that are continuous with respect to the measure μ. Show that $M_{abs}(\Omega, \Sigma, \mu)$ is a closed subset of $M(\Omega, \Sigma)$.

3. Endow Σ with the pseudometric $\rho(A, B) = \mu(A \bigtriangleup B)$. Show that the charge ν defines a continuous mapping of the pesudometric space (Σ, ρ) into $\mathbb{R}$ if and only if $\nu \ll \mu$.

4. Let μ be a finite countably-additive measure and ν a finitely-additive measure on Σ. Suppose that for every $\varepsilon > 0$ there exists a $\delta > 0$ such that for any $A \in \Sigma$, if $\mu(A) < \delta$, then $\nu(A) < \varepsilon$. Then the measure ν is also countably additive.

5. Extend the results of the present subsection to the case of a σ-finite measure μ.

7.1.4 The Charge Induced by a Function

Let μ be a fixed measure. Given an arbitrary function $f \in L_1(\Omega, \Sigma, \mu)$, we define the charge μ_f by the formula

$$\mu_f(A) = \int_A f \, d\mu.$$

If $\mu(A) = 0$, then also $\int_A f \, d\mu = 0$; therefore, $\mu_f \ll \mu$. Below, in Subsection 7.1.6, we will see that μ_f is a typical example of an absolutely continuous charge.

Theorem 1. *Let $f \in L_1(\Omega, \Sigma, \mu)$. Then for every $\varepsilon > 0$ there exists a $\delta > 0$ such that for any $A \in \Sigma$, if $\mu(A) < \delta$, then $\int_A |f| d\mu < \varepsilon$.*

Proof. Apply the main theorem of the preceding subsection to the measure $\mu_{|f|}$. □

Theorem 2. *For $f, g \in L_1(\Omega, \Sigma, \mu)$ the following conditions are equivalent:*

(1) *$f \leqslant g$ almost everywhere with respect to the measure μ;*

(2) *$\mu_f \leqslant \mu_g$.*

In particular, $f = g$ almost everywhere with respect to the measure μ if and only if $\mu_f = \mu_g$.

Proof. The implication (1) $\Longrightarrow$ (2) is obvious. Let us prove the converse implication. Denote by A the set of all $t \in \Omega$ for which $f(t) > g(t)$. Then, on the one hand, $f - g > 0$ on A, while on the other hand, $\int_A (f - g)d\mu = \mu_f(A) - \mu_g(A) \leqslant 0$. That is, $\mu(A) = 0$. □

Let us mention another simple yet useful corollary of Levi's theorem.

Theorem 3. *Suppose $f_n \in L_1(\Omega, \Sigma, \mu)$ is an increasing sequence, and $\mu_{f_n} \leqslant \nu$ for all n. Then the sequence (f_n) converges μ-almost everywhere to a function $f \in L_1(\Omega, \Sigma, \mu)$, and $\mu_f \leqslant \nu$.*

Exercises

1. Find explicit expressions for $(\mu_f)^+$, $(\mu_f)^-$, and $|\mu_f|$.

2. Verify that the mapping $f \mapsto \mu_f$ is a linear isometric embedding of the space $L_1(\Omega, \Sigma, \mu)$ in $M(\Omega, \Sigma)$.

3. Find Ω^+ and Ω^- for the charge μ_f.

Recall that in Exercises 7–11 of Subsection 7.1.2 we defined the integral of a function with respect to a charge.

4. A measurable function g on Ω is integrable with respect to the charge μ_f if and only if the function gf is integrable with respect to the measure μ. If this is the case, then $\int_\Omega gf \, d\mu = \int_\Omega g \, d\mu_f$.

The statement of the last exercise is conventionally written as the equality $d\mu_f = f \, d\mu$, which means that under the integral sign one of these expressions can be replaced by the other.

5. Verify that Theorems 1 and 2 remain valid for a σ-finite μ.

7.1.5 Strong Singularity

Let ν_1 and ν_2 be charges on Σ. The charge ν_1 is said to be *strongly singular* with respect to the charge ν_2 (notation: $\nu_1 \perp \nu_2$) if there exists a decomposition of the set Ω as a union of disjoint sets $A_1, A_2 \in \Sigma$ such that $|\nu_1|(A_2) = |\nu_2|(A_1) = 0$. In other words, the charges ν_1 and ν_2 are concentrated on two distinct and disjoint sets: ν_1 on A_1, and ν_2 on A_2. As one can see from the definition, the strong singularity relation is symmetric: $(\nu_1 \perp \nu_2) \Longleftrightarrow (\nu_2 \perp \nu_1)$. For example, $\nu^+ \perp \nu^-$ for any charge ν. For a pair of measures μ, ν on Σ, we introduce the family $F(\mu, \nu)$ of all nonnegative μ-integrable measurable functions f for which $\mu_f \leqslant \nu$. Further, we introduce the quantity

$$
m(\mu, \nu) = \sup \left\{ \int_\Omega f \, d\mu : f \in F(\mu, \nu) \right\}.
$$

We note here one obvious property of the notions just introduced: if $0 \leqslant \nu_1 \leqslant \nu_2$, then $F(\mu, \nu_1) \subset F(\mu, \nu_2)$, and hence $m(\mu, \nu_1) \leqslant m(\mu, \nu_2)$.

Lemma 1. *The following conditions on the pair of measures μ, ν are equivalent:*

— $\mu \perp \nu$;

— $m(\mu, \nu) = 0$.

Proof. If $\mu \perp \nu$, then there exists a partition $\Omega = A_1 \sqcup A_2$ for which $\mu(A_2) = \nu(A_1) = 0$. Let $f \in F(\mu, \nu)$. Then $\int_{A_1} f \, d\mu = \mu_f(A_1) \leqslant \nu(A_1) = 0$, and hence $\int_\Omega f \, d\mu = \int_{A_1} f \, d\mu + \int_{A_2} f \, d\mu = 0$. Therefore, $m(\mu, \nu) = 0$.

   Conversely, suppose that $m(\mu, \nu) = 0$. Consider the auxiliary charges $\mu - n\nu$ and the corresponding Hahn decompositions $\Omega = \Omega_n^+ \sqcup \Omega_n^-$. Set $A_1 = \bigcap_{n=1}^\infty \Omega_n^+$, $A_2 = \bigcup_{n=1}^\infty \Omega_n^-$. We claim that these sets form the sought-for partition $\Omega = A_1 \sqcup A_2$, with $\mu(A_2) = \nu(A_1) = 0$. Indeed, since

$$
\int_A \frac{1}{n} \mathbb{1}_{\Omega_n^-} d\mu = \frac{1}{n} \mu \left(A \cap \Omega_n^- \right) \leqslant \nu \left(A \cap \Omega_n^- \right) \leqslant \nu(A) \quad \text{for all } A \in \Sigma,
$$

we infer that $\frac{1}{n} \mathbb{1}_{\Omega_n^-} \in F(\mu, \nu)$. Hence, $\int_\Omega \frac{1}{n} \mathbb{1}_{\Omega_n^-} d\mu = 0$, i.e., $\mu(\Omega_n^-) = 0$ for all n. Thus, we have shown that $\mu(A_2) = 0$. Further, $A_1 \subset \Omega_n^+$, that is, $(\mu - n\nu)(A_1) \geqslant 0$ for all n. Consequently, $\nu(A_1) = 0$, too.   □

Exercises

1. If $\mu \perp \nu$ and $\tilde{\mu} \ll \mu$, then $\tilde{\mu} \perp \nu$.

2. If $\mu \perp \nu$ and $\nu \ll \mu$, then $\nu = 0$.

3. Let ν_1 and ν_2 be charges on Σ and $\nu_1 \perp \nu_2$. Then $|\nu_1 + \nu_2| = |\nu_1| + |\nu_2|$. Is the converse statement true?

4. Suppose $|\nu_1 + \nu_2| = |\nu_1 - \nu_2| = |\nu_1| + |\nu_2|$. Then $\nu_1 \perp \nu_2$.

5. Denote by $M_{ss}(\Omega, \Sigma, \mu)$ the subset of $M(\Omega, \Sigma)$ consisting of all charges that are strongly singular with respect to the measure μ. Show that $M_{ss}(\Omega, \Sigma, \mu)$ is a closed subspace of $M(\Omega, \Sigma)$.

6. Let $\nu_1 \in M_{abs}(\Omega, \Sigma, \mu)$ and $\nu_2 \in M_{ss}(\Omega, \Sigma, \mu)$. Then $\|\nu_1 + \nu_2\| = \|\nu_1\| + \|\nu_2\|$.

7. Let ν_1 be a non-atomic charge and ν_2 be a purely atomic charge. Then $\nu_1 \perp \nu_2$.

8. Verify that the quantity $m(\mu, \nu)$ is lower semiadditive in the second variable, i.e., $m(\mu, \nu_1 + \nu_2) \geqslant m(\mu, \nu_1) + m(\mu, \nu_2)$.

9. If $\nu_1 \geqslant \nu_2 \geqslant 0$, then $m(\mu, \nu_1 - \nu_2) \leqslant m(\mu, \nu_1) - m(\mu, \nu_2)$.

7.1.6 The Radon–Nikodým Theorem

In this subsection μ will be a fixed measure on Σ. We continue to use the notations $F(\mu, \nu)$ and $m(\mu, \nu)$ introduced in the preceding subsection.

Theorem 1. *Any charge ν on Σ admits a unique decomposition $\nu = \eta_1 + \eta_2$ as the sum of a charge η_1 that is absolutely continuous with respect to the measure μ and a charge $\eta_2 \perp \mu$. Moreover, the charge η_1 has the form μ_f with $f \in L_1(\Omega, \Sigma, \mu)$.*

Proof. We begin with the uniqueness of the representation. Thus, suppose that in addition to the representation $\nu = \eta_1 + \eta_2$, there exists another similar representation $\nu = \widetilde{\eta}_1 + \widetilde{\eta}_2$. Then $\eta_1 - \widetilde{\eta}_1 \ll \mu$ and $\eta_1 - \widetilde{\eta}_1 = \widetilde{\eta}_2 - \eta_2 \perp \mu$. That is to say (see Exercise 2 in Subsection 7.1.5), $\eta_1 - \widetilde{\eta}_1 = 0$, and consequently $\widetilde{\eta}_2 - \eta_2 = 0$.

To establish the existence of the needed decomposition, it suffices to consider the case where $\nu \geqslant 0$: indeed, the general case can be dealt with by considering separately the problem on the positivity and negativity sets of the charge ν. The proof relies on the so-called *exhaustion method*: the sought-for function f will be constructed as the sum of a series $\sum_{n=1}^{\infty} f_n$, each term of which "absorbs" a part of the quantity $m(\mu, \nu)$ that was not absorbed by the preceding terms.

So, take for f_1 an element of $F(\mu, \nu)$ such that $\int_\Omega f_1 d\mu \geqslant \frac{1}{2} m(\mu, \nu)$. Since $f_1 \in F(\mu, \nu)$, we have $\nu - \mu_{f_1} \geqslant 0$. Now choose a function $f_2 \in F(\mu, \nu - \mu_{f_1})$ such that $\int_\Omega f_2 d\mu \geqslant \frac{1}{2} m(\mu, \nu - \mu_{f_1})$. Since $f_2 \in F(\mu, \nu - \mu_{f_1})$, we have $\nu - \mu_{f_1 + f_2} \geqslant 0$. Continuing this process, at the $(n + 1)$-th step we choose a function $f_{n+1} \in F(\mu, \nu - \mu_{f_1 + f_2 + \cdots + f_n})$ such that

$$\int_\Omega f_{n+1} d\mu \geqslant \frac{1}{2} m(\mu, \nu - \mu_{f_1 + f_2 + \cdots + f_n}).$$

Proceeding in this way, at each step we have the inequality $\nu - \mu_{f_1+f_2+\cdots+f_n} \geqslant 0$ or, equivalently, $\mu_{f_1+f_2+\cdots+f_n} \leqslant \nu$. Since all the functions f_n are non-negative, the partial sums of the series $\sum_{n=1}^{\infty} f_n$ form an increasing sequence of functions. By the version of Levi's theorem given in Theorem 3 of Subsection 7.1.4, the series $\sum_{n=1}^{\infty} f_n$ converges μ-almost everywhere to a function $f \in L_1(\Omega, \Sigma, \mu)$, and $\mu_f \leqslant \nu$. By construction, $\mu_{f_1+f_2+\cdots+f_n} \leqslant \mu_f$ for all n. Consequently,

$$
m(\mu, \nu - \mu_f) \leqslant m(\mu, \nu - \mu_{f_1+f_2+\cdots+f_n}) \leqslant 2 \int_\Omega f_{n+1} d\mu \to 0 \quad \text{as } n \to \infty.
$$

That is, $m(\mu, \nu - \mu_f) = 0$ and, by the lemma in Subsection 7.1.5, $\nu - \mu_f \perp \mu$. To complete the proof, it remains to put $\eta_1 = \mu_f$ and $\eta_2 = \nu - \mu_f$. $\qquad \square$

Let us apply this last theorem to the particular case when $\nu \ll \mu$. In this case, thanks to the uniqueness of the representation $\nu = \eta_1 + \eta_2$, we have $\nu = \eta_1$ and $\eta_2 = 0$. Accordingly, the charge ν itself has the form $\nu = \mu_f$. Decoding the definition of the charge μ_f, we obtain the following deep result due to Radon and Nikodým.

Theorem 2 (Radon–Nikodým theorem). *Suppose the charge ν is absolutely continuous with respect to the measure μ. Then there exists a function $f \in L_1(\Omega, \Sigma, \mu)$ such that $\int_A f \, d\mu = \nu(A)$ for all $A \in \Sigma$.* $\qquad \square$

One can readily verify (for instance, by referring to Theorem 2 of Subsection 7.1.4) that the function f in the Radon–Nikodým theorem is uniquely determined by the measure μ and the charge ν up to equality μ-a.e. This function f is called the *Radon–Nikodým derivative* of the charge ν with respect to the measure μ and is denoted by $\dfrac{d\nu}{d\mu}$.

Exercises

1 Justify the fact that $\int_\Omega f_{n+1} d\mu \to 0$ as $n \to \infty$ in the proof of Theorem 1.

2 Suppose that $\nu \ll \mu$. Show that a measurable function g on Ω is integrable with respect to the charge ν if and only if the function $g \frac{d\nu}{d\mu}$ is integrable with respect to the measure μ. In this case, $\int_\Omega g \frac{d\nu}{d\mu} d\mu = \int_\Omega g \, d\nu$.

7.2 Derivative and Integral on an Interval

In calculus the *Newton–Leibniz formula*

$$
\int_a^b f'(t) dt = f(b) - f(a)
$$

is usually proved only for continuously differentiable functions. Below we provide a complete description of all the functions f for which the Newton–Leibniz formula holds if the integral is understood in the sense of Lebesgue. We will see that such a description is closely related to the subject discussed above, namely, charges and the Radon–Nikodým theorem. Throughout this section we will consider real-valued functions of a real variable and λ will be the Lebesgue measure on the interval considered or on the real line.

7.2.1 The Integral of a Derivative

Theorem 1. *Let* $f : [a, b] \to \mathbb{R}$ *be an increasing function. Then* $f' \in L_1[a, b]$, *and* $\int_a^b f'(t)dt \leqslant f(b) - f(a)$.

Proof. Fix a sequence $\varepsilon_n > 0$ that converges to zero and consider the divided differences $f_n(t) = (f(t + \varepsilon_n) - f(t))/\varepsilon_n$. To ensure that this expression has a meaning on the whole interval $[a, b]$, we put $f(x) = f(b)$ for $x > b$. The functions f_n are integrable, and

$$\int_a^b f_n(t)dt = \frac{1}{\varepsilon_n} \int_a^b (f(t + \varepsilon_n) - f(t))dt = \frac{1}{\varepsilon_n} \left(\int_{a+\varepsilon_n}^{b+\varepsilon_n} f(t)dt - \int_a^b f(t)dt \right)$$

$$= \frac{1}{\varepsilon_n} \left(\int_b^{b+\varepsilon_n} f(t)dt - \int_a^{a+\varepsilon_n} f(t)dt \right) \leqslant f(b) - f(a).$$

Further, $f_n \xrightarrow{\text{a.e.}} f'$ on $[a, b]$. Hence, by Fatou's lemma, the function f' is integrable on $[a, b]$, and

$$\int_a^b f'(t)dt \leqslant \varliminf_{n \to \infty} \int_a^b f_n(t)dt \leqslant f(b) - f(a). \qquad \square$$

Exercises

1. Under the assumptions of the preceding theorem, prove the inequality $\int_a^b f'(t)dt \leqslant f(b - 0) - f(a + 0)$.

2. An example of an increasing function for which $\int_a^b f'(t)dt \neq f(b) - f(a)$ is given by $f = \mathbb{1}_{[1,2]}$, $[a, b] = [0, 2]$.

3. The Newton–Leibniz formula can fail even for continuous increasing functions f: an example is provided by the Cantor staircase (Subsection 2.3.6).

7.2.2　The Derivative of an Integral as a Function of the Upper Integration Limit

Let $f \in L_1[a, b]$ and let F be the "primitive function" of f, i.e., $F(x) = \int_a^x f(t)dt$.

Theorem 1. *For any $f \in L_1[a, b]$ the function F is continuous on $[a, b]$.*

Proof. By Theorem 1 in Subsection 7.1.4, for every $\varepsilon > 0$ there exists a $\delta > 0$ such that for any set $A \in \Sigma$, if $\mu(A) < \delta$, then $\int_A |f|d\mu < \varepsilon$. Applying this assertion to $A = [x, x + \delta]$, we obtain the inequality $|F(x) - F(x + \delta)| < \varepsilon$, which proves the claimed continuity.　　□

Lemma 1. *For any $f \in L_1[a, b]$ the function F is differentiable almost everywhere, $F' \in L_1[a, b]$, and $\int_a^b |F'(t)|dt \leqslant \int_a^b |f(t)|dt$.*

Proof. Write F as the difference $F = F_1 - F_2$, where $F_1(x) = \int_a^x f^+(t)dt$ and $F_2(x) = \int_a^x f^-(t)dt$. The functions F_1 and F_2 are monotone, hence they are differentiable almost everywhere, and so is F; moreover, the derivatives of these three functions are integrable. Further,

$$\int_a^b |F'(t)|dt = \int_a^b |F_1'(t) - F_2'(t)|dt \leqslant \int_a^b F_1'(t)dt + \int_a^b F_2'(t)dt$$

$$\leqslant F_1(b) - F_1(a) + F_2(b) - F_2(a) = \int_a^b f^+(t)dt + \int_a^b f^-(t)dt = \int_a^b |f(t)|dt.$$

□

Theorem 2. $F' \overset{\text{a.e.}}{=} f$ *for any function $f \in L_1[a, b]$. In other words, the derivative of a Lebesgue integral, regarded as a function of the upper integration limit, is equal a.e. to the integrand.*

Proof. Consider the linear operator T acting from $L_1[a, b]$ to $L_1[a, b]$ by the rule $Tf = F' - f$. By the preceding lemma,

$$\|Tf\| = \int_a^b |F'(t) - f(t)|dt \leqslant 2 \int_a^b |f(t)|dt = 2\|f\|,$$

i.e., the operator T is continuous. By a known theorem of calculus, if $f \in C[a, b]$, then $Tf = 0$. Since the set $C[a, b]$ of continuous functions is dense in $L_1[a, b]$, it follows that $T = 0$ on the whole space $L_1[a, b]$, as we needed to prove.　　□

Exercises

1. Let A be a measurable subset of an interval. Applying the last theorem to the function $f = \mathbb{1}_A$, solve Exercise 4 of Subsection 2.3.3 on density points of a set.

2. Show that the Cantor staircase cannot serve as the primitive F for an integrable function f.

7.2.3 *Functions of Bounded Variation and the General Form of a Borel Charge on the Interval*

Since the basic properties of functions of bounded variation are usually presented in calculus courses (as a foundation for the theory of the Stieltjes integral), we recall here these properties with no proof. Readers who for some reason or another have not encountered the notions discussed below are advised to treat the material in the current section as a series of exercises, solving them on their own, or read the detailed presentation in, say, the textbook of A. Kolmogorov and S. Fomin [24].

Definition 1. The *variation of the function f on the interval* $[a, b]$ is the quantity

$$V_a^b(f) = \sup\left\{\sum_{k=1}^n |f(b_k) - f(a_k)|\right\},$$

where the supremum is taken over all finite disjoint collections of open subintervals $(a_k, b_k) \subset [a, b]$. If $V_a^b(f) < \infty$, then f is called a *function of bounded variation* on $[a, b]$.

By its definition, the variation of a function is a nonnegative quantity.

1. In the definition of the variation $V_a^b(f)$ it suffices to take the supremum over the collections of subintervals $(a_k, b_k) \subset [a, b]$ such that $a = a_1 < b_1 = a_2 < \cdots < b_n = b$: in other words, over disjoint collections whose union is the whole interval $[a, b]$, except for a finite number of points. This is precisely the way in which the variation is most often defined in the literature.

2. The set of functions of bounded variation on a fixed interval $[a, b]$ constitutes a linear space, and V_a^b is a convex functional on that space.

3. Monotone functions have bounded variation: $V_a^b(f) = |f(b) - f(a)|$. Consequently, linear combinations of monotone functions also have bounded variation.

4. $V_a^b(f) = V_a^c(f) + V_c^b(f)$ for any $a < c < b$.

5. Every function f of bounded variation is representable as the difference of two increasing functions: $f = f_1 - f_2$, where $f_1(t) = V_a^t(f)$ and $f_2(t) = V_a^t(f) - f(t)$.

Lemma 1. *Every right-continuous function f of bounded variation on the interval*
[a, b] is representable as the difference of two increasing right-continuous functions.

Proof. First write f in some way as a difference $f_1 - f_2$ of two increasing functions.
Now define the sought-for functions g_1 and g_2 by the respective formulas $g_1(t) = \lim_{x \to t+0} f_1(x)$ and $g_2(t) = \lim_{x \to t+0} f_2(x)$ for $t \in [a, b)$, and put $g_1(b) = f_1(b)$
and $g_2(b) = f_2(b)$. Then the functions g_1 and g_2 are right continuous, and $f(t) = \lim_{x \to t+0} f(x) = \lim_{x \to t+0} (f_1(x) - f_2(x)) = g_1(t) - g_2(t)$.   □

Definition 2. A *Borel charge* on the interval $[a, b]$ is a charge given on the family of
all Borel subsets of $[a, b]$. The *distribution function* of the Borel charge ν on $[a, b]$
is the function $F_\nu(t)$ given by $F_\nu(t) = \nu([a, t])$.

In Subsection 2.3.5 we proved that by associating to each Borel measure μ on
the interval $[a, b]$ its distribution function $F_\mu(t) = \mu([a, t])$ one establishes a one-
to-one correspondence between the Borel measures on $[a, b]$ and the increasing
right-continuous functions on $[a, b]$. This fact in conjunction with the preceding
lemma yields the following result.

Theorem 1. *The mapping $\nu \mapsto F_\nu$ establishes a bijective correspondence between*
the family of all Borel charges on the segment $[a, b]$ and the set of all right-continuous
functions of bounded variation on $[a, b]$.

Proof. The linearity relations $F_{\nu_1 + \nu_2} = F_{\nu_1} + F_{\nu_2}$ and $F_{c\nu} = cF_\nu$ follow directly
from the definition of the distribution function of a Borel charge. Further, $F_\nu = F_{\nu^+} - F_{\nu^-}$, i.e., the function F_ν is representable as the difference of two distribution functions of measures — increasing right-continuous functions. Hence, F_ν is
a right-continuous function of bounded variation. Conversely, any right-continuous
function f of bounded variation can be written as the difference $f = f_1 - f_2$ of two
increasing right-continuous functions, each of which, in its turn, can be regarded as
the distribution function of a corresponding Borel measure: $f_1 = F_{\mu_1}$ and $f_2 = F_{\mu_2}$.
Consequently, $f = F_{\mu_1 - \mu_2}$, which proves the surjectivity of the map under consideration. Finally, let us establish its injectivity. Suppose that $F_\nu = 0$. Then $F_{\nu^+} = F_{\nu^-}$.
By Theorem 2 of Subsection 2.3.5, if the distribution functions of two Borel measures on an interval coincide, then so do the measures themselves. Hence, $\nu^+ = \nu^-$,
and so $\nu = \nu^+ - \nu^- = 0$.   □

Exercises

1. Adding a constant to a function does not modify its variation.

2. The notions of variation of a charge (Definition 1 in Subsection 7.1.1) and variation of a function are compatible: if ν is a Borel charge on $[a, b]$ and the point a is
not an atom of ν, then $V_a^b(F_\nu) = |\nu|([a, b])$.

3. In the general case, $V_a^t(F_\nu) = F_{|\nu|}(t) - |\nu|\,(\{a\})$.

4. Suppose the function f of bounded variation is continuous on $[a, b]$. Then the function $f_1(t) = V_a^t(f)$ is also continuous $[a, b]$.

7.2.4 Absolutely Continuous Functions

A function f on the interval $[a, b]$ is said to be *absolutely continuous* if for every $\varepsilon > 0$ there exists a $\delta = \delta(\varepsilon) > 0$ such that for any choice of disjoint open subintervals $(a_k, b_k) \subset [a, b]$, $k = 1, \ldots, n$, satisfying $\sum_{k=1}^n |b_k - a_k| \leqslant \delta$ (i.e., the total length of which does not exceed δ), the total oscillation of the function does not exceed ε: $\sum_{k=1}^n |f(b_k) - f(a_k)| \leqslant \varepsilon$.

Note that every absolutely continuous function is continuous (just apply the definition, taking a single interval of length smaller than δ), and any linear combination of absolutely continuous functions is absolutely continuous.

Theorem 1. *Let f be an absolutely continuous continuous function on $[a, b]$. Then f is a function of bounded variation. Moreover, the functions $f_1(t) = V_a^t(f)$ and $f_2(t) = V_a^t(f) - f(t)$ are also absolutely continuous, i.e., f is representable as the difference $f_1 - f_2$ of two increasing absolutely continuous functions.*

Proof. Take $\delta = \delta(\varepsilon) > 0$ from the definition of absolute continuity. Fix a collection of pairwise disjoint subintervals $(c_n, d_n) \subset [a, b]$, $n = 1, 2, \ldots, N$, such that $\sum_{n=1}^N |d_n - c_n| \leqslant \delta$. In each of the intervals (c_n, d_n) pick a finite collection of disjoint subintervals $(a_{k,n}, b_{k,n}) \subset (c_n, d_n)$, $k = 1, 2, \ldots, m_n$. Then $\sum_{n=1}^N \sum_{k=1}^{m_n} |b_{k,n} - a_{k,n}| \leqslant \delta$, and so $\sum_{n=1}^N \sum_{k=1}^{m_n} |f(b_{k,n}) - f(a_{k,n})| \leqslant \varepsilon$. Taking the supremum over all collections $(a_{k,n}, b_{k,n})$, we obtain the estimate

$$\sum_{n=1}^N V_{c_n}^{d_n}(f) \leqslant \varepsilon \qquad (*)$$

for any collection of disjoint subintervals $(c_n, d_n) \subset [a, b]$ with $\sum_{n=1}^N |d_n - c_n| \leqslant \delta$.

All the assertions of the theorem follow from this estimate. Indeed, divide $[a, b)$ into a finite number of disjoint subintervals $[c_k, d_k) \subset [a, b]$, $k = 1, 2, \ldots, m$ with $|d_k - c_k| < \delta$. Then $V_a^b(f) = \sum_{k=1}^m V_{c_k}^{d_k}(f) \leqslant m\varepsilon < \infty$, i.e., f is a function of bounded variation. Further, condition $(*)$ says that the function f_1 is absolutely continuous, and together with it so is the function $f_2 = f_1 - f$. $\qquad \square$

Exercises

1. Suppose the sequence f_n of functions on $[a, b]$ converges pointwise to the function f. Then $V_a^b(f) \leqslant \sup_{n \in \mathbb{N}} V_a^b(f_n)$.

2. Denote by bv$[a, b]$ the linear space of functions of bounded variation on the interval $[a, b]$. Verify that V_a^b is a seminorm on bv$[a, b]$ (the notion of seminorm is defined just before Exercise 10 of Subsection 6.1.3).

3. Show that the kernel of the seminorm V_a^b consists of constants.

4. Consider the quotient space of bv$[a, b]$ by the subspace X of constants. Define the norm of an equivalence class $[f] \in$ bv$[a, b]/X$ by $\|[f]\| = V_a^b(f)$. Verify that this definition is correct. We denote the resulting normed space bv$[a, b]/X$ by $V[a, b]$.

5. Define the operator $T: V[a, b] \to L_1[a, b]$ by the rule $Tf = f'$ (differentiation, or derivative operator). Verify that T is a continuous linear operator.

6. Based on the preceding exercise, prove Fubini's theorem on the differentiation of series (Exercise 3 in Subsection 2.3.3).

7. Show that $V[a, b]$ is a Banach space.

We let AC$[a, b]$ denote the subset of the space $V[a, b]$ consisting of the equivalence classes of absolutely continuous functions.

8. Show that AC$[a, b]$ is a closed subspace of $V[a, b]$.

7.2.5 Absolutely Continuous Functions and Absolutely Continuous Borel Charges

Theorem 1. *For the distribution function F of a Borel measure μ on the interval $[a, b]$ the following conditions are equivalent:*

A. *F is absolutely continuous and $F(a) = 0$;*

B. *The measure μ is absolutely continuous with respect to the Lebesgue measure λ.*

Proof. B $\Longrightarrow$ A. First, recall that the measure of a semi-open subinterval can be calculated by the formula $\mu((t_1, t_2]) = F(t_2) - F(t_1)$. If $\mu \ll \lambda$, then by the theorem in Subsection 7.1.3, for any $\varepsilon > 0$ there exists a $\delta > 0$ such that for any Borel set $A \subset [a, b]$ satisfying $\lambda(A) < \delta$ we have $\mu(A) < \varepsilon$. Consider a collection of disjoint open subintervals $(a_k, b_k) \subset [a, b]$, $k = 1, 2, \ldots, n$, such that $\sum_{k=1}^n |b_k - a_k| \leqslant \delta$, and put $A = \bigcup_{k=1}^n (a_k, b_k]$. Then $\lambda(A) < \delta$, and hence $\mu(A) = \sum_{k=1}^n (F(b_k) - F(a_k)) < \varepsilon$. This establishes the absolute continuity of the function F. The equality $F(a) = 0$ follows from the fact that the singleton $\{a\}$ has Lebesgue measure zero, and so $\mu(\{a\}) = 0$.

A $\Longrightarrow$ B. Let F be absolutely continuous, $\varepsilon > 0$ be an arbitrary number, and take $\delta = \delta(\varepsilon) > 0$ as in the definition of an absolutely continuous function. First, let us show that $\mu(A) \leqslant \varepsilon$ for any open subset $A \subset [a, b]$ satisfying $\lambda(A) < \delta$. Indeed, such an open set can be written as $A = \bigsqcup_{k=1}^{\infty} (a_k, b_k)$ with $\sum_{k=1}^{\infty} |b_k - a_k| \leqslant \delta$. (To be completely rigorous, two of these intervals, the ones lying at the beginning and the end of the interval $[a, b]$, can be semi-open.) Then $\sum_{k=1}^{n} (F(b_k) - F(a_k)) < \varepsilon$ for all $n \in \mathbb{N}$. Correspondingly, $\mu(A) \leqslant \sum_{k=1}^{\infty} (F(b_k) - F(a_k)) \leqslant \varepsilon$. Now let us prove the requisite absolute continuity of the measure μ with respect to the Lebesgue measure. Let $D \subset [a, b]$ be a set of Lebesgue measure zero. Then for any $\varepsilon > 0$ there exists an open set $A \subset [a, b]$ such that $A \supset D$ and $\lambda(A) < \delta(\varepsilon)$. Therefore, $\mu(D) \leqslant \mu(A) \leqslant \varepsilon$. Since ε is arbitrary, $\mu(D) = 0$. $\qquad\square$

Corollary 1. *Let f be an absolutely continuous function on $[a, b]$ with $f(a) = 0$. Then there exists a Borel charge ν, absolutely continuous with respect to the Lebesgue measure and connected on $[a, b]$ with f by the equality $f(t) = \nu([a, t])$.*

Proof. Write f as the difference $f_1 - f_2$ of two increasing absolutely continuous functions (according to the main theorem of the preceding subsection) that vanish at the point a. For each of these functions f_j, construct a Borel measure μ_j that has f_j as distribution function. Now define the sought-for charge as $\nu = \mu_1 - \mu_2$. $\qquad\square$

Exercises

1. Why in the last proof above can the functions f_1 and f_2 be chosen so that they vanish at the point a?

2. The Cantor staircase (see Subsection 2.3.6) is an example of a continuous, but not absolutely continuous function.

7.2.6 Recovering a Function From its Derivative

Having done much preparatory work, we are now finally ready to deal with the Newton–Leibniz formula for the Lebesgue integral. We call the reader's attention to the fact that the proof, at the surface simple and short, makes essential use of deep results revolving around the Radon–Nikodým theorem.

Theorem 1. *For a function F on the interval $[a, b]$ the following conditions are equivalent:*

(1) *F is absolutely continuous;*
(2) *F is representable as $F(x) = F(a) + \int_a^x f(t)dt$ for some $f \in L_1[a, b]$ (that is, in the terminology of Subsection 7.2.2, $F(x) - F(a)$ is the "primitive" of some function from $L_1[a, b]$);*

(3) *F is differentiable almost everywhere on $[a, b]$, $F' \in L_1[a, b]$, and for any $x \in$ $[a, b]$ there holds the Newton–Leibniz formula $F(x) - F(a) = \int_a^x F'(t)dt$.*

Proof. (1) $\Longrightarrow$ (2). By the corollary in the preceding subsection, there exists a Borel charge μ on $[a, b]$ which has $F(x) - F(a)$ as its distribution function. Since μ is absolutely continuous with respect to the Lebesgue measure λ (by a theorem in the preceding subsection), we can use the Radon–Nikodým theorem (Theorem 2 in Subsection 7.1.6). By this theorem, there exists a function $f \in L_1[a, b]$ such that $\int_A f d\lambda = \mu(A)$ for all Borel subsets of the interval $[a, b]$. Taking for A subintervals $[a, x]$, we conclude that $F(x) - F(a) = \mu([a, x]) = \int_a^x f(t)dt$.

(2) $\Longrightarrow$ (1). The condition $F(x) - F(a) = \int_a^x f(t)dt$ means that $F(x) - F(a)$ is the distribution function of the charge λ_f (see Subsection 7.1.4), given by the formula $\lambda_f(A) = \int_A f \, d\lambda$. Such a charge is absolutely continuous with respect to the Lebesgue measure, hence its distribution function is also absolutely continuous.

(2) $\Longrightarrow$ (3). It suffices to use Theorem 2 of Subsection 7.2.2 on the derivative of an integral as function of the upper integration limit.

(3) $\Longrightarrow$ (2). For f one needs to take F'. □

7.2.7 Exercises: Change of Variables in the Lebesgue Integral

Let Ω and Ω_1 be sets, Σ a σ-algebra on Ω, and $F: \Omega \to \Omega_1$ an arbitrary mapping. We introduce a family Σ_1 of subsets of Ω_1 by the rule: $A \in \Sigma_1$ if and only if $F^{-1}(A) \in \Sigma$.

1. The family Σ_1 is a σ-algebra on Ω_1.

Further, let μ be a charge on Σ. The *image of the charge μ under the mapping F* is the set function $F(\mu): \Sigma_1 \to \mathbb{R}$ defined by the rule $(F(\mu))(A) = \mu(F^{-1}(A))$.

2. $F(\mu)$ is a countably-additive charge; moreover, if μ is a measure, then so is $F(\mu)$.

3. For fixed F, the mapping $\mu \mapsto F(\mu)$ is linear.

4. Give an example where $F(\mu)$ is a measure, but the charge μ is not.

5. If (Ω, Σ, μ) is a complete measure space, the measure space $(\Omega_1, \Sigma_1, F(\mu))$ is also complete.

6. Can the measure space $(\Omega_1, \Sigma_1, F(\mu))$ be complete when the measure space (Ω, Σ, μ) is not?

7. Suppose the mapping F is injective and the measure space $(\Omega_1, \Sigma_1, F(\mu))$ is complete. Then the measure space (Ω, Σ, μ) is also complete.

8. Let μ be a measure for which $F(\mu)$ is non-atomic. Then μ is also non-atomic.

9. Let μ be a measure. The function $f : \Omega_1 \to \mathbb{R}$ is integrable with respect to the measure $F(\mu)$ if and only if the composition $f \circ F$ is integrable with respect to the measure μ. Moreover, in this case, $\int_\Omega f \circ F \, d\mu = \int_{F(\Omega)} f \, dF(\mu)$.

Let us consider now a concrete case of the construction described above, taking functions of a real variable and the Lebesgue measure λ.

Let F be an increasing continuous function on the interval $[a, b]$. With no loss of generality, we can assume that F is the distribution function of some non-atomic Borel measure μ on the interval $[a, b]$.

10. For any interval $[c, d] \subset [F(a), F(b)]$ it holds that $(F(\mu))([c, d]) = d - c$. In other words, the restriction of the measure $F(\mu)$ to the σ-algebra of Borel subsets of the interval $[F(a), F(b)]$ coincides with the Lebesgue measure λ on $[F(a), F(b)]$.

11. Under the conditions of the preceding exercise, $\int_a^b f \circ F \, d\mu = \int_{F(a)}^{F(b)} f \, d\lambda$ for all $f \in L_1[F(a), F(b)]$.

Recall that a monotone function F on an interval is absolutely continuous if and only if the measure μ it generates is absolutely continuous with respect to the Lebesgue measure and the Radon–Nikodým derivative $\frac{d\mu}{d\lambda}$ coincides with F'. Therefore, when the function F is absolutely continuous, the formula of the preceding exercise takes on the well-known form:

12. $\int_a^b f(F(x))F'(x) \, dx = \int_{F(a)}^{F(b)} f(t) \, dt$, where the integrals on both sides of this equality are understood in the sense of Lebesgue.

13. Let us make the convention that, as in the case of the Riemann integral, for the Lebesgue integral it holds that $\int_c^d g(x)dx = -\int_d^c g(x)dx$. Show that the formula

$$\int_a^b f(F(x))F'(x) \, dx = \int_{F(a)}^{F(b)} f(t) \, dt$$

remains valid for any, not necessarily monotone, absolutely continuous function F on the interval $[a, b]$ and any function $f \in L_1[F([a, b])]$.

Comments on the Exercises

Subsection 7.1.1

Exercise 5. Let $\Delta_1, \Delta_2, \ldots$ be a sequence of pairwise disjoint measurable sets, and $\Delta = \bigcup_{k=1}^\infty \Delta_k$. Then the sets $A_n = \bigcup_{k=n+1}^\infty \Delta_k$ form a decreasing chain with empty intersection. Consequently, $\lim_{n\to\infty} \left(\nu(\Delta) - \sum_{k=1}^n \nu(\Delta_k)\right) = \lim_{n\to\infty} \nu(A_n) = 0$.

Subsection 7.1.2

Exercises 7–11. See Subsection 8.4.2.

Subsection 7.1.4

Exercise 4. It suffices to prove the assertion for positive functions f: the general case is then deduced by considering separately the sets $\Omega^+ = \{t : f(t) > 0\}$ and $\Omega^- = \{t : f(t) < 0\}$. For the same reason, we can assume that $g \geqslant 0$, too. By the definition of μ_f, for $g = \mathbb{1}_A$ with $A \in \Sigma$, the two integrals exist and coincide; by Levi's theorem on series, the assertion holds for countably-valued functions $g \geqslant 0$. Now let g be an arbitrary non-negative measurable function. By the corollary to Theorem 3 in Subsection 3.1.4, there exists an increasing sequence of countably-valued functions $g_n \geqslant 0$ such that $g_n - g \leqslant 1/n$. Suppose that the integral $\int_\Omega gf \, d\mu$ exists. Then, since the functions $g_n + 1/n$ admit the μ-integrable majorants $g + 1/n$, the integrals $\int_\Omega g_n f \, d\mu$ exist. For countably-valued functions the theorem is already proved, so the integrals $\int_\Omega g_n d\mu_f$ exist; moreover, $\int_\Omega g_n f \, d\mu = \int_\Omega g_n d\mu_f$. Passing to the limit by applying the Levi theorem for sequences, we conclude that the integral $\int_\Omega g \, d\mu_f$ exists and $\int_\Omega gf \, d\mu = \int_\Omega g \, d\mu_f$. The converse implication: (if $\int_\Omega g \, d\mu_f$ exists, then so does $\int_\Omega gf \, d\mu$) is established in the same way.

Subsection 7.1.6

Exercise 1. The inequality $\mu_{f_1+f_2+\cdots+f_n} \leqslant \nu$ implies that $\sum_{n=1}^\infty \int_\Omega f_n d\mu \leqslant \nu(\Omega)$. Hence, the general term of the series converges to zero.

Exercise 2. See Subsection 7.1.4, Exercise 4, and the comment to it.

Subsection 7.2.3

Exercise 4. Let ν be a Borel charge with the property that $f - f(a) = F_\nu$. The charge ν has no atoms (if the point t_0 was an atom, then f would have a discontinuity at t_0), hence the measure $|\nu|$ also has no atoms as well (Exercise 3 in Subsection 7.1.2). It follows that the function $f_1 = F_{|\nu|}$ we are interested in is continuous.

However, here a straightforward argument yields more: if f is right-continuous in some point $t_0 \in [a, b)$, then $f_1(t) = V_a^t(f)$ is also right-continuous at t_0. By symmetry (replacing the $f(x)$ by $f(-x)$), a similar assertion holds true for left-continuous functions.

Indeed, fix an $\varepsilon > 0$ and pick a finite collection of open subintervals (a_k, b_k), $t_0 = a_1 < b_1 = a_2 < \cdots < b_n = b$, of the interval $[t_0, b]$, such that $V_{t_0}^b(f) \leqslant \sum_{k=1}^n |f(b_k) - f(a_k)| + \varepsilon$. Now take $\delta \in (t_0, b_1]$ such that $|f(\delta) - f(t_0)| < \varepsilon$. Then

$$V_\delta^b(f) \geqslant |f(b_1) - f(\delta)| + \sum_{k=2}^{n} |f(b_k) - f(a_k)|$$

$$\geqslant \sum_{k=1}^{n} |f(b_k) - f(a_k)| - \varepsilon \geqslant V_{t_0}^b(f) - 2\varepsilon.$$

Consequently,

$$|f_1(t_0) - f_1(\delta)| = V_{t_0}^\delta(f) = V_{t_0}^b(f) - V_\delta^b(f) \leqslant 2\varepsilon.$$

Chapter 8
The Integral on $C(K)$

In the first three sections of this chapter we treat in detail the integration theory for functions on a compact topological space K. In the last section the results obtained will be applied to the proof of the Riesz–Markov–Kakutani theorem on the general form of linear functionals on the space $C(K)$. In the last two subsections of this chapter we consider the integration of complex-valued functions with respect to complex charges and also functionals on the complex space $C(K)$. Up to that moment all the functions, charges and functionals will be assumed to be real-valued.

8.1 Regular Borel Measures on a Compact Space

8.1.1 Inner Measure and Regularity

Recall that, by definition, a Borel measure on a topological space X is a finite countably-additive measure given on the family of all Borel subsets of X.

Definition 1. Let μ be a Borel measure on the topological space X. For any subset $A \subset X$, we define the *inner measure* $\mu_*(A)$ as the supremum of the measures of all closed sets contained in A. The measure μ is called *regular* if $\mu_*(A) = \mu(A)$ for all open subsets $A \subset X$. In other words, the measure μ is regular if for any open $A \subset X$ and any $\varepsilon > 0$ there exists a closed subset $B \subset A$ such that $\mu(B) \geqslant \mu(A) - \varepsilon$.

Lemma 1. *Let X be a metric space. Then every open subset $A \subset X$ can be represented as the union of an increasing sequence of closed sets.*

Proof. Consider on X the function $f(x) = \rho(x, X \setminus A)$. This function is continuous (Subsection 1.3.2), and consequently the sets $A_n = f^{-1}([1/n, +\infty))$ are closed. The sets A_n form an increasing sequence, and $\bigcup_{n=1}^{\infty} A_n = f^{-1}((0, +\infty)) = A$. $\qquad\square$

© Springer International Publishing AG, part of Springer Nature 2018
V. Kadets, *A Course in Functional Analysis and Measure Theory*,
Universitext, https://doi.org/10.1007/978-3-319-92004-7_8

Theorem 1. *On a metric space every Borel measure is regular.*

Proof. Let μ be a Borel measure on the metric space X, and $A \subset X$ an open subset. By the preceding lemma, there exists an increasing sequence of closed sets A_n, the union of which is the whole set A. By the countable additivity, $\lim_{k\to\infty} \mu(A_k) = \mu(A)$, i.e., the measure of A can be arbitrarily well approximated by measures of its closed subsets. $\qquad\square$

On general topological spaces one also encounters non-regular Borel measures. As an example, consider the interval $[0, 1]$ equipped with the following topology τ: a set is τ-open if its complement is finite or countable. In this special topology the open and closed sets together already form a σ-algebra. Accordingly, the only τ-Borel sets are the τ-open and τ-closed sets. Define the measure μ by the rule $\mu(A) = 0$ if A is τ-closed (i.e., if A is finite or countable), and $\mu(A) = 1$ if A is τ-open. Then $\mu([0, 1]) = 1$, and $\mu_*([0, 1]) = 0.$[1]

In Sect. 8.3 we will prove, as a corollary of general results, that for a regular Borel measure on a compact space the equality $\mu_*(A) = \mu(A)$ holds not only for open, but also for arbitrary Borel sets (see Exercise 1 in Subsection 8.3.2).

Exercises

1. If the set $A \subset K$ is closed, then $\mu_*(A) = \mu(A)$.

2. $\mu_*(A) \leqslant \mu(A)$ for any Borel subset $A \subset K$.

3. If $A \subset B \subset K$, then $\mu_*(A) \leqslant \mu_*(B)$.

4. Let $A_1 \subset A_2 \subset \cdots$ be an increasing chain of subsets of the compact space K, and $A = \bigcup_{k=1}^{\infty} A_k$. Then $\mu_*(A) \geqslant \lim_{n\to\infty} \mu_*(A_k)$.

The *Baire σ-algebra* on the topological space K is defined to be the smallest σ-algebra Σ_0 with respect to which all continuous functions on K are measurable. Show that

5. If K is compact, then the σ-algebra Σ_0 is generated by the family of all open F_σ-sets.

6. If K is a metric compact space, then the σ-algebra Σ_0 coincides with the σ-algebra of Borel sets.

---

[1] For the readers familiar with the theory of ordinal numbers (ordinals): Denote by ω_1 the first uncountable ordinal. Consider the set X of all ordinals that are not larger than ω_1. We call a neighborhood of the ordinal α any subset $U \subset X$ that contains an interval of the form $(\beta, \alpha]$ with $\beta < \alpha$. With the topology thus defined, the space X will be compact. Further, define a Borel measure μ on X as follows: if the Borel set A contains a subset of the form $B \setminus \{\omega_1\}$, where B is a closed set that has ω_1 as a limit point, then put $\mu(A) = 1$; otherwise, put $\mu(A) = 0$. Then we have $\mu((1, \omega_1)) = 1$, while at the same time the measure of any closed subset of the interval $(1, \omega_1)$ is equal to zero. Hence, μ provides an example of a non-regular Borel measure on a compact topological space.

8.1.2 The Support of a Measure

Let μ be a regular Borel measure on the compact space K. A point $x \in K$ is called an *essential point* of the measure μ if any neighborhood of x has nonzero measure. The *support* of the measure μ is the set supp μ of all essential points of this measure.

Theorem 1. *In the above notation, supp μ is a closed set, $\mu(K \setminus \text{supp}\,\mu) = 0$, and no open set of measure zero intersects supp μ.*

Proof. We need to show that the set $V = K \setminus \text{supp}\,\mu$ for all inessential points of the measure μ is open, has measure zero, and every open set of measure zero is contained in V.

First note that if U is an open subset of the compact space K and $\mu(U) = 0$, then $U \subset V$. Indeed, every point of U has a neighborhood of measure zero (namely, U itself), and so no point of U is essential.

Now let $x \in V$. This means that there exists an open neighborhood U of the point x for which $\mu(U) = 0$, and hence $U \subset V$. Thus, we have shown that together with any of its points, V contains a neighborhood of that point, i.e., V is open.

In view of the regularity of the measure μ, to prove that $\mu(V) = 0$ it suffices to verify that all closed subsets of the set V have measure zero. So, let $W \subset V$ be closed. For each point $x \in W$ pick an open neighborhood U_x such that $\mu(U_x) = 0$. The neighborhoods U_x constitute an open cover of the compact set W. Extract a finite subcover $U_{x_1}, U_{x_2}, \ldots, U_{x_n}$. We have $\mu(W) \leqslant \sum_{k=1}^{n} \mu(U_{x_k}) = 0$. $\qquad\square$

Theorem 2. *Let μ be a regular Borel measure on the compact space K. Suppose that two continuous functions, f_1 and f_2, coincide μ-almost everywhere on K. Then f_1 and f_2 coincide at all points of supp μ.*

Proof. The set U of all points x at which $f_1(x) \neq f_2(x)$ is an open set of measure zero. Hence, $U \cap \text{supp}\,\mu = \emptyset$. $\qquad\square$

8.2 Extension of Elementary Integrals

8.2.1 Elementary Integrals

Definition 1. An *elementary integral* on the compact space K is a linear functional $\mathcal{I}$ on $C(K)$ which satisfies the following *positivity condition*: if the function f is non-negative, then $\mathcal{I}(f) \geqslant 0$.

Properties of elementary integrals

(1) For any $f, g \in C(K)$, if $f \geqslant g$, then $\mathcal{I}(f) \geqslant \mathcal{I}(g)$.

(2) If $|f| \leqslant g$, then $|\mathcal{I}(f)| \leqslant \mathcal{I}(g)$.

(3) $\mathcal{I}$ is a continuous functional on $C(K)$ and $\|\mathcal{I}\| = \mathcal{I}(\mathbb{1})$.[2]

Proof. (1) $f \geqslant g \implies f - g \geqslant 0 \implies \mathcal{I}(f - g) \geqslant 0 \implies \mathcal{I}(f) \geqslant \mathcal{I}(g)$.

(2) $|f| \leqslant g \implies -g \leqslant f \leqslant g \implies -\mathcal{I}(g) \leqslant \mathcal{I}(f) \leqslant \mathcal{I}(g) \implies |\mathcal{I}(f)| \leqslant \mathcal{I}(g)$.

(3) Let f be an arbitrary function belonging to the unit sphere of the space $C(K)$. Then $|f| \leqslant 1$ everywhere, and, by property (2), $|\mathcal{I}(f)| \leqslant \mathcal{I}(\mathbb{1})$. $\qquad\square$

Since the continuous functions on K are Borel measurable and bounded, they are integrable with respect to any Borel measure on K. Accordingly, as an example of an elementary integral one can take the integral with respect to an arbitrary fixed Borel measure on K.

Definition 2. The *elementary integral generated by the Borel measure* μ on the compact space K is the linear functional $\mathbf{F}_\mu$ on $C(K)$ given by the formula $\mathbf{F}_\mu(f) = \int_K f \, d\nu$.

We will show below that apart from the functionals of the form $\mathbf{F}_\mu$ there are no other elementary integrals. Moreover, we will prove that for any functional $\mathcal{I}$ there exists a **regular** Borel measure μ which generates this functional. The idea underlying the construction of such a measure μ is simple: one needs to put $\mu(A) = \mathcal{I}(\mathbb{1}_A)$. However, the implementation of this idea runs into an essential obstacle: the functional $\mathcal{I}$ is defined only for continuous functions, whereas the characteristic functions of sets are, as a rule, discontinuous. For this reason our immediate aim is to extend the functional $\mathcal{I}$ to a sufficiently wide class of functions that includes, at least, the characteristic functions of all Borel sets. This will be carried out in the next three subsections.

8.2.2 The Upper Integral of Lower Semi-continuous Functions

Recall that the symbols $\ell_\infty(K)$ and $LSC(K)$ denote the classes of all bounded, respectively all lower semi-continuous functions on K (for the definition and properties of lower semi-continuous functions refer to Subsection 1.2.4). For functions $g \in LSC(K) \cap \ell_\infty(K)$, i.e., lower semi-continuous bounded functions on K, we introduce the quantity

$$\mathcal{I}^*(g) = \sup\{\mathcal{I}(h) : h \in C(K) \text{ and } h < g\}.$$

Theorem 1. *The quantity $\mathcal{I}^*$ has the following properties:*

(1) *if $g \in C(K)$, then $\mathcal{I}^*(g) = \mathcal{I}(g)$;*

---

[2]The symbol $\mathbb{1}$ stands for the function identically equal to 1.

(2) *if* $g_1, g_2 \in LSC(K) \cap \ell_\infty(K)$ *and* $g_1 \leqslant g_2$, *then* $\mathcal{I}^*(g_1) \leqslant \mathcal{I}^*(g_2)$;

(3) $\mathcal{I}^*(\lambda g) = \lambda \mathcal{I}^*(g)$ *for positive scalars* λ;

(4) $\mathcal{I}^*(g_1 + g_2) = \mathcal{I}^*(g_1) + \mathcal{I}^*(g_2)$ *for all* $g_1, g_2 \in LSC(K) \cap \ell_\infty(K)$.

Proof. The first three properties are obvious. Let us prove the fourth one. Fix $\varepsilon > 0$ and choose functions $h_1, h_2 \in C(K), h_1 < g_1, h_2 < g_2$, for which $\mathcal{I}(h_1) \geqslant \mathcal{I}^*(g_1) - \varepsilon$ and $\mathcal{I}(h_2) \geqslant \mathcal{I}^*(g_2) - \varepsilon$. Then we have

$$\mathcal{I}^*(g_1 + g_2) \geqslant \mathcal{I}(h_1 + h_2) \geqslant \mathcal{I}^*(g_1) + \mathcal{I}^*(g_2) - 2\varepsilon.$$

Since ε is arbitrary, the inequality $\mathcal{I}^*(g_1 + g_2) \geqslant \mathcal{I}^*(g_1) + \mathcal{I}^*(g_2)$ is established.

To prove the opposite inequality, pick a function $h \in C(K)$ such that $h < g_1 + g_2$ and $\mathcal{I}(h) \geqslant \mathcal{I}^*(g_1 + g_2) - \varepsilon$. By Theorem 3 in Subsection 1.2.4, applied to the functions g_1, g_2, for any point $x \in K$ there exists functions $h_{1,x}, h_{2,x} \in C(K)$ such that $h_{1,x} < g_1, h_{2,x} < g_2$ at all points of the compact space K, and $h_{1,x}(x) + h_{2,x}(x) > h(x)$. In view of the continuity of all functions figuring in the last inequality, for every point $x \in K$ there exists an open neighborhood U_x such that, as above, $h_{1,x}(t) + h_{2,x}(t) > h(t)$ for all $t \in U_x$. The neighborhoods $U_x, x \in K$, form a cover of the compact space K; hence, one can extract a finite subcover $U_{x_1}, U_{x_2}, \ldots, U_{x_n}$. Set $h_1 = \max_{k \in \{1,2,\ldots,n\}} h_{1,x}$, $h_2 = \max_{k \in \{1,2,\ldots,n\}} h_{2,x}$. These continuous functions obey the inequalities $h_1 < g_1$ and $h_2 < g_2$, and by construction $h_1 + h_2 > h$ already at all points. Consequently,

$$\mathcal{I}^*(g_1 + g_2) \leqslant \mathcal{I}(h) + \varepsilon \leqslant \mathcal{I}(h_1) + \mathcal{I}(h_2) + \varepsilon \leqslant \mathcal{I}^*(g_1) + \mathcal{I}^*(g_2) + \varepsilon.$$

It remains to let $\varepsilon \to 0$. $\qquad\square$

Remark 1. The quantity $\mathcal{I}^*$ cannot be referred to as a linear functional, since its domain of definition is not a linear space. Lower semi-continuity is preserved under addition and multiplication by positive scalars, but can be lost under multiplication by negative scalars or under subtraction. However, if $h \in C(K)$, then also $-h \in C(K) \subset LSC(K) \cap \ell_\infty(K)$. Then for any $g \in LSC(K) \cap \ell_\infty(K)$ the difference $g - h$ lies in the domain of definition of $\mathcal{I}^*$, so by the previous theorem $\mathcal{I}^*(g - h) + \mathcal{I}^*(h) = \mathcal{I}^*(g)$, and consequently $\mathcal{I}^*(g - h) = \mathcal{I}^*(g) - \mathcal{I}^*(h)$.

Theorem 2. *Let* (g_n), *with* $g_n \in LSC(K) \cap \ell_\infty(K)$, *be an increasing and uniformly bounded sequence of functions that converges pointwise to some function* g. *Then* $g \in LSC(K) \cap \ell_\infty(K)$ *and* $\mathcal{I}^*(g) = \lim_{n \to \infty} \mathcal{I}^*(g_n)$.

Proof. The function g is bounded by the same constant as all the functions g_n, so $g \in \ell_\infty(K)$. Further, $g(x) = \sup\{g_n(x) : n \in \mathbb{N}\}$. By Theorem 1 item (4) of Subsection 1.2.4, $g \in LSC(K)$. That is, $g \in LSC(K) \cap \ell_\infty(K)$.

The numbers $\mathcal{I}^*(g_n)$ form a non-decreasing sequence and are bounded from above by the number $\mathcal{I}^*(g)$. Consequently, $\lim_{n \to \infty} \mathcal{I}^*(g_n)$ exists and is not larger than $\mathcal{I}^*(g)$. To establish the opposite inequality is suffices to derive the estimate

$\lim_{n\to\infty} \mathcal{I}^*(g_n) \geqslant \mathcal{I}(h)$ for all continuous functions h that are smaller than g. For fixed h, let $A_n = \{t \in K : g_n(t) > h(t)\}$. The sets A_n are open, form an increasing chain, and together cover the entire compact space K. Hence, there exists an index $n = n_0$ for which $A_{n_0} = K$. This means that $g_{n_0}(t) > h(t)$ everywhere on K. Therefore, $\lim_{n\to\infty} \mathcal{I}^*(g_n) \geqslant \mathcal{I}^*(g_{n_0}) \geqslant \mathcal{I}(h)$. $\qquad\square$

8.2.3 The Upper Integral on $\ell_\infty(K)$

The *upper integral* of a function $f \in \ell_\infty(K)$ is defined as

$$\overline{\mathcal{I}}(f) = \inf\{\mathcal{I}^*(g) : g \in LSC(K) \cap \ell_\infty(K), \text{ and } g \geqslant f\}.$$

Theorem 1. *The quantity $\overline{\mathcal{I}}$ has the following properties:*

(1) *if $f \in LSC(K) \cap \ell_\infty(K)$, then $\overline{\mathcal{I}}(f) = \mathcal{I}^*(f)$;*[3]

(2) *if $f_1, f_2 \in \ell_\infty(K)$ and $f_1 \leqslant f_2$, then $\overline{\mathcal{I}}(f_1) \leqslant \overline{\mathcal{I}}(f_2)$;*

(3) *$\overline{\mathcal{I}}(\lambda f) = \lambda\overline{\mathcal{I}}(f)$ for all positive scalars λ;*

(4) *$\overline{\mathcal{I}}(f_1 + f_2) \leqslant \overline{\mathcal{I}}(f_1) + \overline{\mathcal{I}}(f_2)$ for all $f_1, f_2 \in \ell_\infty(K)$.*

Proof. As in Theorem 1 of the preceding subsection, only property (4) needs to be verified. Fix $\varepsilon > 0$ and pick functions $g_1, g_2 \in LSC(K) \cap \ell_\infty(K)$, $g_i \geqslant f_i$, such that $\mathcal{I}^*(g_i) \leqslant \overline{\mathcal{I}}(f_i) + \varepsilon$, $i = 1, 2$. We have $\overline{\mathcal{I}}(f_1 + f_2) \leqslant \mathcal{I}^*(g_1 + g_2) \leqslant \overline{\mathcal{I}}(f_1) + \overline{\mathcal{I}}(f_2) + 2\varepsilon$. It remains to let $\varepsilon \to 0$. $\qquad\square$

Remark 1. The last theorem says, in particular, that $\overline{\mathcal{I}}$ is a convex functional on $\ell_\infty(K)$; moreover, on $C(K)$ this functional majorizes our elementary integral $\mathcal{I}$. It follows that in order to provide the desired extension of the elementary integral and construct the Borel measure that generates this integral one can resort to the Hahn–Banach theorem. Such an aproach is perfectly feasible, but we will take a different path, which will lead to an explicit construction of the extension. This will complicate slightly the very definition of the sought-for measure, but simplify the proof of its countable additivity and regularity.

Theorem 2. *Let $f, f_n \in \ell_\infty(K)$, $f_n \geqslant 0$, $f \leqslant \sum_{n=1}^\infty f_n$ at all points. Then $\overline{\mathcal{I}}(f) \leqslant \sum_{n=1}^\infty \overline{\mathcal{I}}(f_n)$.*

Proof. Denote $\sup\{f(x) : x \in K\}$ by a. Fix $\varepsilon > 0$, and for each $n \in \mathbb{N}$ pick a function $g_n \in LSC(K) \cap \ell_\infty(K)$ such that $g_n \geqslant f_n$ and $\mathcal{I}^*(g_n) \leqslant \overline{\mathcal{I}}(f_n) + \varepsilon/2^n$. Consider the auxiliary functions $s_n(x) = \min\left\{a, \sum_{j=1}^n g_j(x)\right\}$. Since $g_n \geqslant 0$, the sequence (s_n) is increasing. Further, $s_1 \leqslant s_n \leqslant a$, i.e., the sequence (s_n) is uniformly bounded. Denote the pointwise limit of the sequence (s_n) by s. By Theorem 2 of the preceding subsection, $s \in LSC(K) \cap \ell_\infty(K)$ and $\mathcal{I}^*(s) = \lim_{n\to\infty} \mathcal{I}^*(s_n)$. We have

---

[3]This property justifies using the term "upper integral" also for the quantity $\mathcal{I}^*$.

$$s(x) = \lim_{n \to \infty} s_n(x) = \min\left\{a, \sum_{j=1}^{\infty} g_j(x)\right\}$$

$$\geqslant \min\left\{a, \sum_{j=1}^{\infty} f_j(x)\right\} \geqslant \min\{a, f(x)\} = f(x),$$

whence

$$\overline{\mathcal{I}}(f) \leqslant \mathcal{I}^*(s) = \lim_{n \to \infty} \mathcal{I}^*(s_n) \leqslant \lim_{n \to \infty} \mathcal{I}^*\left(\sum_{j=1}^{n} g_j(x)\right)$$

$$= \sum_{j=1}^{\infty} \mathcal{I}^*(g_j) \leqslant \sum_{j=1}^{\infty} \overline{\mathcal{I}}(f_j) + \varepsilon.$$

Since ε is arbitrary, the theorem is proved.   □

8.2.4 The Space $L(K, \mathcal{I})$

Let us equip $\ell_\infty(K)$ with the seminorm $p(f) = \overline{\mathcal{I}}(|f|)$. Then equipped with the pseudometric $\rho(f_1, f_2) = \overline{\mathcal{I}}(|f_1 - f_2|)$ generated by this norm, $\ell_\infty(K)$ is a pseudometric space. We let $L(K, \mathcal{I})$ denote the closure in $\ell_\infty(K)$, with respect to this pseudometric, of the set $C(K)$ of all continuous functions.

Theorem 1. $L(K, \mathcal{I})$ is a closed linear subspace in $(\ell_\infty(K), p)$, and $LSC(K) \cap \ell_\infty(K) \subset L(K, \mathcal{I})$.

Proof. Since $C(K)$ is a linear subspace, so is its closure. We only need to prove the inclusion $LSC(K) \cap \ell_\infty(K) \subset L(K, \mathcal{I})$. Let $f \in LSC(K) \cap \ell_\infty(K)$. By the definition of the functional $\mathcal{I}^*$, for every n there is a continuous function $f_n < f$ such that $\mathcal{I}^*(f - f_n) = \mathcal{I}^*(f) - \mathcal{I}(f_n) < 1/n$. But then

$$\rho(f, f_n) = \overline{\mathcal{I}}(|f - f_n|) = \mathcal{I}^*(f - f_n) \to 0 \quad \text{as } n \to \infty,$$

i.e., we succeeded in representing the function f as the limit of a sequence of continuous functions.   □

Theorem 2. $L(K, \mathcal{I})$ enjoys the following properties:

(1) if $f \in L(K, \mathcal{I})$, then $|f| \in L(K, \mathcal{I})$;

(2) if $f \in L(K, \mathcal{I})$, then $f^+ \in L(K, \mathcal{I})$ and $f^- \in L(K, \mathcal{I})$;

(3) if $f, g \in L(K, \mathcal{I})$, then $\max\{f, g\} \in L(K, \mathcal{I})$ and $\min\{f, g\} \in L(K, \mathcal{I})$.

Proof. (1) Let $f_n \in C(K)$ be such that $\overline{\mathcal{I}}(|f - f_n|) \to 0$ as $n \to \infty$. Then $|f_n| \in C(K)$ and

$$\rho(|f|, |f_n|) = \overline{\mathcal{I}}(||f| - |f_n||) \leqslant \overline{\mathcal{I}}(|f - f_n|) \to 0 \quad \text{as } n \to \infty,$$

i.e., the function $|f|$ is the limit of the sequence $|f_n| \in C(K)$, and consequently $|f| \in L(K, \mathcal{I})$.

(2) follows from the already established item (1) and the formulas $f^+ = \frac{f + |f|}{2}$, $f^- = (-f)^+$.

(3) follows from (2) and the formulas $\max\{f, g\} = f + (g - f)^+$, $\min\{f, g\} = -\max\{-f, -g\}$. $\qquad\qquad\qquad\qquad\qquad\qquad\qquad\qquad\qquad\qquad\qquad\qquad\quad\square$

The next theorem shows that the elementary integral $\mathcal{I}$ extends to a linear functional on $L(K, \mathcal{I})$, and that as this extension one can take the upper integral.

Theorem 3. *On $L(K, \mathcal{I})$ the upper integral $\overline{\mathcal{I}}$ is a p-continuous linear functional.*

Proof. For any $f, g \in L(K, \mathcal{I})$ one has that $\overline{\mathcal{I}}(f) \leqslant \overline{\mathcal{I}}(|f - g|) + \overline{\mathcal{I}}(g)$ and $\overline{\mathcal{I}}(g) \leqslant \overline{\mathcal{I}}(|f - g|) + \overline{\mathcal{I}}(f)$ (property (4) in Theorem 1 of Subsection 8.2.3). Consequently,

$$|\overline{\mathcal{I}}(f) - \overline{\mathcal{I}}(g)| \leqslant \overline{\mathcal{I}}(|f - g|) = \rho(f, g),$$

i.e., $\overline{\mathcal{I}}$ is continuous (it satisfies the Lipschitz condition). Linearity follows from the continuity and the fact that on $C(K)$ (that is, on a dense subset) $\overline{\mathcal{I}}$ coincides with the linear functional $\mathcal{I}$, and so on the subspace $C(K)$ the upper integral itself is linear. Indeed, let $f, g \in L(K, \mathcal{I})$, a, b be scalars, and (f_n) and (g_n) be sequences of continuous functions which in the pseudometric ρ tend to f and g, respectively. Then

$$\overline{\mathcal{I}}(af + bg) = \lim_{n \to \infty} \overline{\mathcal{I}}(af_n + bg_n) = \lim_{n \to \infty} \mathcal{I}(af_n + bg_n)$$

$$= a \lim_{n \to \infty} \mathcal{I}(f_n) + b \lim_{n \to \infty} \mathcal{I}(g_n) = a\overline{\mathcal{I}}(f) + b\overline{\mathcal{I}}(g). \qquad\qquad\square$$

The next result is an analogue of Levi's theorem on series for the upper integral on $L(K, \mathcal{I})$.

Theorem 4. *Let $f_n \in L(K, \mathcal{I})$, $f \in \ell_\infty(K)$, $f_n \geqslant 0$, and suppose the series $\sum_{n=1}^{\infty} f_n$ converges to f at all points. Then $f \in L(K, \mathcal{I})$ and $\overline{\mathcal{I}}(f) = \sum_{n=1}^{\infty} \overline{\mathcal{I}}(f_n)$.*

Proof. First let us show that $f \in L(K, \mathcal{I})$. Let $a = \sup\{f(x) : x \in K\}$. Fix $\varepsilon > 0$ and approximate each of the functions f_n by a continuous function g_n to within $\varepsilon/2^n$: $\overline{\mathcal{I}}(|f_n - g_n|) \leqslant \varepsilon/2^n$. Replacing, if necessary, g_n by g_n^+, one can assume that all the functions g_n are non-negative. Now put $s(x) = \min\{a, \sum_{j=1}^{\infty} g_j(x)\}$. The function s is lower semi-continuous and bounded, so by Theorem 2, $s \in L(K, \mathcal{I})$. Further, $|s - f| \leqslant \sum_{n=1}^{\infty} |f_n - g_n|$ at all points, and by Theorem 1 of the preceding subsection,

$$\overline{\mathcal{I}}(|s - f|) \leqslant \sum_{n=1}^{\infty} \overline{\mathcal{I}}(|f_n - g_n|) \leqslant \varepsilon.$$

Thus, the function f can be arbitrarily well approximated in the pseudometric ρ by elements of the space $L(K, \mathcal{I})$, which in view of the fact that $L(K, \mathcal{I})$ is closed in $\ell_\infty(K)$ means that $f \in L(K, \mathcal{I})$.

Now let us establish the equality $\overline{\mathcal{I}}(f) = \sum_{n=1}^\infty \overline{\mathcal{I}}(f_n)$. One direction, namely $\overline{\mathcal{I}}(f) \leqslant \sum_{n=1}^\infty \overline{\mathcal{I}}(f_n)$, follows from Theorem 2 of the preceding subsection. The opposite inequality is an easy consequence of the linearity of the upper integral on $L(K, \mathcal{I})$. Indeed, thanks to the positivity of the terms, we have $\sum_{k=1}^n f_k \leqslant f$ for any natural number n. Consequently,

$$\sum_{k=1}^n \overline{\mathcal{I}}(f_k) = \overline{\mathcal{I}}\left(\sum_{k=1}^n f_k\right) \leqslant \overline{\mathcal{I}}(f).$$

It remains to let $n \to \infty$. □

Corollary 1 (analogue of Levi's theorem on sequences). *Let $g_n \in L(K, \mathcal{I})$, $g \in \ell_\infty(K)$, and suppose that pointwise the sequence (g_n) is non-decreasing and converges to g. Then $g \in L(K, \mathcal{I})$ and $\overline{\mathcal{I}}(g) = \lim_{n\to\infty} \overline{\mathcal{I}}(g_n)$.*

Proof. It suffices to apply the preceding theorem to the series $\sum_{n=1}^\infty (g_{n+1} - g_n)$. The partial sums of this series are equal to $g_{n+1} - g_1$, so it converges to $g - g_1$. Therefore, $g - g_1 \in L(K, \mathcal{I})$, hence $g = (g - g_1) + g_1 \in L(K, \mathcal{I})$ and

$$\overline{\mathcal{I}}(g) - \overline{\mathcal{I}}(g_1) = \lim_{n\to\infty} \sum_{k=1}^{n-1} \overline{\mathcal{I}}(g_{k+1} - g_k) = \lim_{n\to\infty} \overline{\mathcal{I}}(g_n) - \overline{\mathcal{I}}(g_1).$$ □

8.3 Regular Borel Measures and the Integral

8.3.1 $\mathcal{I}$-measurable Sets. The Measure Generated by an Integral

A subset $A \subset K$ is said to be $\mathcal{I}$-*measurable* if $\mathbb{1}_A \in L(K, \mathcal{I})$. The family of all $\mathcal{I}$-measurable subsets of the compact space K will be denoted by $\Sigma_\mathcal{I}$. For each set $A \in \Sigma_\mathcal{I}$ we put $\mu_\mathcal{I}(A) = \overline{\mathcal{I}}(\mathbb{1}_A)$.

Theorem 1. *The family $\Sigma_\mathcal{I}$ is a σ-algebra, and $\mu_\mathcal{I}$ is a countably-additive measure on $\Sigma_\mathcal{I}$.*

Proof. We rely on Theorems 1–4 of the preceding subsection. Let $A \in \Sigma_\mathcal{I}$. Then $\mathbb{1}_{K\setminus A} = \mathbb{1} - \mathbb{1}_A \in L(K, \mathcal{I})$, and so $K \setminus A \in \Sigma_\mathcal{I}$. Next, let $A, B \in \Sigma_\mathcal{I}$. Then $\mathbb{1}_{A\cap B} = \min\{\mathbb{1}_A, \mathbb{1}_B\} \in L(K, \mathcal{I})$, hence $A \cap B \in \Sigma_\mathcal{I}$. Finally, let $A_n \in \Sigma_\mathcal{I}$ be a disjoint sequence of sets, and $A = \bigcup_{n=1}^\infty A_n$. Then the series $\sum_{n=1}^\infty \mathbb{1}_{A_n}$ converges pointwise to $\mathbb{1}_A$. By Theorem 4 of the preceding subsection, this means that $\mathbb{1}_A \in L(K, \mathcal{I})$ and $\mu_\mathcal{I}(A) = \overline{\mathcal{I}}(\mathbb{1}_A) = \sum_{n=1}^\infty \overline{\mathcal{I}}(\mathbb{1}_{A_n}) = \sum_{n=1}^\infty \mu_\mathcal{I}(A_n)$. □

Theorem 2. *$L(K, \mathcal{I})$ coincides with the family of all bounded $\Sigma_{\mathcal{I}}$-measurable functions, and $\int_K f \, d\mu_{\mathcal{I}} = \overline{\mathcal{I}}(f)$ for all $f \in L(K, \mathcal{I})$.*

Proof. First let us show that $L(K, \mathcal{I})$ consists of $\Sigma_{\mathcal{I}}$-measurable functions. Let $f \in L(K, \mathcal{I})$ and a be an arbitrary real number. Consider the sequence of functions $g_n = \min\{n(f - a)^+, 1\}$. This sequence is pointwise non-decreasing. If $f(t) \leqslant a$, then $g_n(t) = 0$, while if $f(t) > a$, then $n(f - a)^+(t) \to \infty$ as $n \to \infty$, so that starting with some index n, we have $g_n(t) = 1$. In other words, the sequence, being non-decreasing, converges pointwise to the characteristic function of the set $f_{>a}$. By the analogue of Levi's theorem (the corollary at the end of the preceding subsection), this characteristic function belongs to $L(K, \mathcal{I})$, i.e., $f_{>a} \in \Sigma_{\mathcal{I}}$. This establishes the measurability of the function f.

To prove the last assertion of the theorem, denote by X the set of all $f \in L(K, \mathcal{I})$ for which $\int_K f \, d\mu_{\mathcal{I}} = \overline{\mathcal{I}}(f)$. By the definition of the measure $\mu_{\mathcal{I}}$, the characteristic functions of all sets from $\Sigma_{\mathcal{I}}$ lie in X. Since X is a linear space, this implies that all finitely-valued measurable functions (i.e., linear combinations of characteristic functions) also lie in X. By the approximation theorem (the corollary to Theorem 3 in Subsection 3.1.4), every bounded $\Sigma_{\mathcal{I}}$-measurable function can be expressed as the pointwise (and even uniform) limit of an increasing sequence of finitely-valued $\Sigma_{\mathcal{I}}$-measurable functions. By Levi's theorem and its analogue for $L(K, \mathcal{I})$, this implies that all bounded $\Sigma_{\mathcal{I}}$-measurable functions belong to our set X. □

Theorem 3. *The family $\Sigma_{\mathcal{I}}$ contains all Borel subsets of the compact space K, and for any $A \in \Sigma_{\mathcal{I}}$ and any $\varepsilon > 0$, there exists an open set U such that $U \supset A$ and $\mu_{\mathcal{I}}(U) < \mu_{\mathcal{I}}(A) + \varepsilon$.*

Proof. Since the characteristic functions of open subsets of the compact space K are lower semi-continuous, such functions lie in $L(K, \mathcal{I})$; consequently, all open sets lie in $\Sigma_{\mathcal{I}}$. Since $\Sigma_{\mathcal{I}}$ is a σ-algebra, then all Borel subsets lie in $\Sigma_{\mathcal{I}}$.

Now let A be an arbitrary element of the σ-algebra $\Sigma_{\mathcal{I}}$. Using the definition of the upper integral for the function $\mathbb{1}_A$ and the fact that the upper integral on $L(K, \mathcal{I})$ coincides with the integral with respect to the measure $\mu_{\mathcal{I}}$, pick a function $f \in LSC(K) \cap \ell_\infty(K)$ such that $f > \mathbb{1}_A$ and $\int_K f \, d\mu_{\mathcal{I}} < \int_K \mathbb{1}_A \, d\mu_{\mathcal{I}} + \varepsilon = \mu_{\mathcal{I}}(A) + \varepsilon$. Now for the requisite set U take the set $f_{>1}$ of all points $t \in K$ where $f(t) > 1$. Thanks to the semicontinuity of f, the set U is open. For any point $t \in A$ it holds that $f(t) > \mathbb{1}_A(t) = 1$, i.e., $t \in U$. This completes the proof of the inclusion $A \subset U$. Finally, it is readily seen that $\mathbb{1}_U \leqslant f$ at all points, and so $\mu_{\mathcal{I}}(U) = \int_K \mathbb{1}_U \, d\mu_{\mathcal{I}} \leqslant \int_K f \, d\mu_{\mathcal{I}} < \mu_{\mathcal{I}}(A) + \varepsilon$. □

Passing to complements of sets, we obtain the following

Corollary 1. *For any $A \in \Sigma_{\mathcal{I}}$ and any $\varepsilon > 0$ there exists a closed set V such that $V \subset A$ and $\mu_{\mathcal{I}}(V) > \mu_{\mathcal{I}}(A) - \varepsilon$. In other words, the inner measure generated by the measure $\mu_{\mathcal{I}}$ coincides on $\Sigma_{\mathcal{I}}$ precisely with the measure $\mu_{\mathcal{I}}$.*

Exercises

1. $(K, \Sigma_\mathcal{I}, \mu_\mathcal{I})$ is a complete measure space.

2. The completion of the measure space $(K, \mathfrak{B}, \mu_\mathcal{I})$, where $\mathfrak{B}$ is the σ-algebra of Borel subsets of the compact space K, coincides with $(K, \Sigma_\mathcal{I}, \mu_\mathcal{I})$.

3. Let $K = [0, 1]$. As the elementary integral $\mathcal{I}$ take the Riemann integral on the interval K. Verify that in this case $\Sigma_\mathcal{I}$ coincides with the σ-algebra of Lebesgue-measurable subsets of the interval, and $\mu_\mathcal{I}$ coincides with the Lebesgue measure.

8.3.2 The General Form of Elementary Integrals

We say that the Borel measure μ generates the elementary integral $\mathcal{I}$ on $C(K)$ if $\int_K f \, d\mu = \mathcal{I}(f)$ for all functions $f \in C(K)$. (In the notation of Definition 2 in Subsection 8.2.1, $\mathcal{I} = \mathbf{F}_\mu$.)

Theorem 1. *For any elementary integral $\mathcal{I}$ on $C(K)$ there exists a unique regular Borel measure μ that generates $\mathcal{I}$.*

Proof. The existence of the requisite measure has actually been already established: for μ one can take the restriction of the measure $\mu_\mathcal{I}$, constructed in the preceding subsection, to the σ-algebra $\mathfrak{B}$ of Borel subsets. Indeed, by Theorem 3 of Subsection 8.3.1 and its corollary, the measure $\mu_\mathcal{I}$ is defined on the Borel subsets and is regular, while Theorem 2 of the same subsection asserts that the equality $\int_K f \, d\mu_\mathcal{I} = \overline{\mathcal{I}}(f)$ holds not only for continuous functions, but for all $f \in L(K, \mathcal{I})$. Hence, our main task is to establish the uniqueness, i.e., show that if μ is a regular Borel measure obeying the conditions of the theorem, then μ coincides with $\mu_\mathcal{I}$ on $\mathfrak{B}$.

To begin with, note that $\mu_\mathcal{I}(U) \geqslant \mu(U)$ for any open set U. To verify this, we use the regularity of the measure μ. For every $\varepsilon > 0$ there exists a closed set V such that $V \subset U$ and $\mu(V) > \mu(U) - \varepsilon$. Now the Urysohn lemma yields a continuous function f with the following properties: $0 \leqslant f \leqslant 1$ everywhere on K, $f(t) = 1$ on V, and $f(t) = 0$ on $K \setminus U$. We have

$$\mu_\mathcal{I}(U) \geqslant \int_K f \, d\mu_\mathcal{I} = \int_K f \, d\mu \geqslant \mu(V) \geqslant \mu(U) - \varepsilon.$$

It remains to let $\varepsilon \to 0$.

Now let us prove that the inequality $\mu_\mathcal{I}(A) \geqslant \mu(A)$ actually holds for all Borel sets. To this end we again fix an $\varepsilon > 0$ and pick an open set $U \supset A$ such that $\mu_\mathcal{I}(U) < \mu_\mathcal{I}(A) + \varepsilon$ (this is possible according to Theorem 3 of Subsection 8.3.1). We have

$$\mu_{\mathcal{I}}(A) \geqslant \mu_{\mathcal{I}}(U) - \varepsilon \geqslant \mu(U) - \varepsilon \geqslant \mu(A) - \varepsilon,$$

so in view of the arbitrariness of ε, $\mu_{\mathcal{I}}(A) \geqslant \mu(A)$.

Applying the established inequality to the complement of the set A and using the obvious equality $\mu(K) = \int_K d\mu = \int_K d\mu_{\mathcal{I}} = \mu_{\mathcal{I}}(K)$, we obtain the opposite inequality

$$\mu_{\mathcal{I}}(A) = \mu_{\mathcal{I}}(K) - \mu_{\mathcal{I}}(K \setminus A) \leqslant \mu(K) - \mu(K \setminus A) = \mu(A).$$

Thus, we have shown that the measures μ and $\mu_{\mathcal{I}}$ coincide on the Borel sets. $\qquad\square$

Exercises

1. Based on the corollary stated as the end of Subsection 8.3.1 and the uniqueness theorem, show that if μ is a regular Borel measure on a compact space, then for any $A \in \mathfrak{B}$ and any $\varepsilon > 0$ there exists a closed set such that $V \subset A$ and $\mu(V) > \mu(A) - \varepsilon$.

2. Based on the theorem on the general form of a Borel measure on the interval (Subsection 2.3.5), prove the following theorem: *For every elementary integral $\mathcal{I}$ on* $[0, 1]$ *there exists a monotone function F on $[0, 1]$ such that $\mathcal{I}$ is expressed as the Stieltjes integral with respect to dF: $\mathcal{I}(f) = \int_0^1 f \, dF$ for all $f \in C[0, 1]$.*

The *power moment problem* on the interval $[0, 1]$ is the problem of finding, given a sequence of numbers $a_0, a_1, \ldots$, a function F on $[0, 1]$ such that $\int_0^1 t^n dF(t) = a_n$ for all $n = 0, 1, 2, \ldots$. The reader is invited to derive from the preceding exercise the following result.

3. In order for the power moment problem for the sequence $a_0, a_1, \ldots$ to have a solution that is a nondecreasing function, it is necessary and sufficient that $\sum_{k=0}^{n} a_k b_k \geqslant 0$ for any non-negative polynomial $b_0 + b_1 t + \cdots + b_n t^n$ on $[0, 1]$.

8.3.3 Approximation of Measurable Functions by Continuous Functions. Luzin's Theorem

Lemma 1. *Let (Ω, Σ, μ) be a measure space. Then the subset of all bounded functions is dense in $L_1(\Omega, \Sigma, \mu)$.*

Proof. Let $f \in L_1(\Omega, \Sigma, \mu)$ be an arbitrary function. Introduce the sets $A_n = \{t \in \Omega : |f(t)| < n\}$ and consider the functions $f_n = f \cdot \mathbb{1}_{A_n}$. The functions $|f_n - f| = |f| \cdot \mathbb{1}_{\Omega \setminus A_n}$ admit $|f|$ as a common integrable majorant and converge pointwise to zero. Consequently, $\|f_n - f\| = \int_K |f_n - f| d\mu \to 0$ as $n \to \infty$. Since the functions f_n are bounded ($|f_n| < n$ at all points), this establishes the claimed denseness of the set of bounded functions in L_1. $\qquad\square$

Let μ be a regular Borel measure on the compact space K. Denote by $\mathfrak{B}$ the σ-algebra of Borel subsets of K, and by L_1 the Banach space $L_1(K, \mathfrak{B}, \mu)$.

Theorem 1. *The subset $C(K)$ of continuous functions is dense in L_1.*

Proof. Let $\mathcal{I}$ be the elementary integral on $C(K)$ given by the formula $\mathcal{I}(f) = \int_K f \, d\mu$. Then, by the theorem on the general form of elementary integrals proved in the preceding subsection (the uniqueness part), the measure $\mu_\mathcal{I}$ generated by the integral $\mathcal{I}$ coincides on $\mathfrak{B}$ with the original measure μ. Hence, for bounded Borel functions we have $\overline{\mathcal{I}}(f) = \int_K f \, d\mu$ (see Theorem 2 in Subsection 8.3.1). Accordingly, for functions from L_1 the distance ρ, defined in Subsection 8.2.4, coincides with distance in L_1. Since $L(K, \mathcal{I})$ is the ρ-closure of the set $C(K)$, Theorem 2 of Subsection 8.3.1 shows that the closure of the set $C(K)$ in L_1 contains all bounded functions from L_1. It remains to apply Lemma 1 above. $\qquad\square$

Theorem 2. *The subset $C(K)$ of continuous functions is dense in the space L_0 of all Borel-measurable functions on K in the sense of convergence almost everywhere.*

Proof. Using the functions $f_n = f \cdot \mathbb{1}_{A_n}$ introduced in the proof of Lemma 1 it is readily verified, as in the preceding theorem, that L_1 is dense in L_0 is the sense of convergence almost everywhere. Further, by the preceding theorem, $C(K)$ is dense in the space L_1 in the L_1-norm. Since convergence in the L_1-norm implies convergence in measure, and since from any sequence that converges in measure one can extract a subsequence that converges almost everywhere, it follows that $C(K)$ is indeed dense in L_1 in the sense of convergence almost everywhere. To complete the proof, it remains to use that in the present case the property of "being dense in" is transitive (Corollary 2 in Subsection 3.2.3). $\qquad\square$

Theorem 3. (Luzin's theorem). *Let f be a Borel-measurable function on K. Then for any $\varepsilon > 0$ there exists a Borel subset $A \subset K$ with $\mu(K \setminus A) < \varepsilon$ such that the restriction of the function f to A is continuous.*

Proof. By the preceding theorem, there exists a sequence (f_n) of continuous functions that converges to f almost everywhere. Using Egorov's theorem, choose a set $A \subset K$ with $\mu(K \setminus A) < \varepsilon$, on which the sequence (f_n) converges uniformly. Now the restriction of the function f to A is the limit of an already uniformly convergent sequence of continuous functions, and as such it is a continuous function. $\qquad\square$

Exercises

1. Show that the set A in the statement of Luzin's theorem can always be taken to be closed.

2. Give an example (on an interval) in which the set A in the statement of Luzin's theorem cannot be taken to be open.

8.4 The General Form of Linear Functionals on $C(K)$

8.4.1 Regular Borel Charges

A charge (signed measure) ν given on the σ-algebra $\mathfrak{B}$ of Borel subsets of the
compact space K is called a *regular Borel charge* on K if for any $A \in \mathfrak{B}$ and any
$\varepsilon > 0$ there exists a closed subset $C \subset A$ such that $|\nu(A) - \nu(C)| < \varepsilon$.

We note that, according to Exercise 1 in Subsection 8.3.2, every regular Borel measure is simultaneously a regular Borel charge. Further, it follows from the definition
that a linear combination of regular Borel charges is again a regular Borel charge.
In particular, the difference of two regular Borel measures is a regular Borel charge.
The next theorem shows that such differences exhaust all regular Borel charges.

Theorem 1. *For a Borel charge ν on a compact space K the following conditions
are equivalent:*

(1) *ν is a regular Borel charge;*

(2) *ν^+ and ν^- are regular Borel measures;*

(3) *the charge ν can be written as the difference of two regular Borel measures;*

(4) *$|\nu|$ is a regular Borel measure.*

Proof. (1) $\Longrightarrow$ (2). Recall that for every $A \in \mathfrak{B}$ the quantity $\nu^+(A)$ is defined by the
formula $\nu^+(A) = \sup\{\nu(\Delta) : \Delta \in \mathfrak{B}, \Delta \subset A\}$. That is, for every $\varepsilon > 0$ there exists
a Borel subset $\Delta \subset A$ such that $\nu(\Delta) > \nu^+(A) - \varepsilon$. Since the charge ν is regular,
one can find a closed subset $C \subset \Delta$ such that $\nu(C) > \nu^+(A) - \varepsilon$. Using the fact that
$\nu^+(C) \geqslant \nu(C)$, we finally conclude that there exists a closed subset $C \subset A$ such
that $\nu^+(C) > \nu^+(A) - \varepsilon$, i.e., that the measure ν^+ is regular. The regularity of the
measure ν^- now follows from the relation $\nu^- = \nu^+ - \nu$.

(2) $\Longrightarrow$ (3). Indeed, $\nu = \nu^+ - \nu^-$.

(3) $\Longrightarrow$ (1). Let $\nu = \mu_1 - \mu_2$, where μ_1 and μ_2 are regular Borel measures. Then
for every $A \in \mathfrak{B}$ and every $\varepsilon > 0$ there exist closed subsets $C_1, C_2 \subset A$ such that
$\mu_1(C_1) > \mu_1(A) - \varepsilon$ and $\mu_2(C_2) \Theta a > \mu_2(A) - \varepsilon$. Set $C = C_1 \cup C_2$. Then C is a
closed subset of A and the inequalities $\mu_1(A) - \varepsilon < \mu_1(C) < \mu_1(A)$ and $\mu_2(A) -
\varepsilon < \mu_2(C) < \mu_2(A)$ hold. Therefore,

$$|\nu(A) - \nu(C)| \leqslant |\mu_1(A) - \mu_1(C)| + |\mu_2(A) - \mu_2(C)| < 2\varepsilon.$$

Since ε is arbitrary, this establishes the regularity of the charge ν.

(2) $\Longrightarrow$ (4). It suffices to use the equality $|\nu| = \nu^+ + \nu^-$.

(4) $\Longrightarrow$ (1). Let $A \in \mathfrak{B}$, and let $C \subset A$ be a closed set such that $|\nu|(A) -
|\nu|(C) < \varepsilon$. Then $|\nu(A) - \nu(C)| = |\nu(A \backslash C)| \leqslant |\nu|(A \backslash C) = |\nu|(A) - |\nu|(C) < \varepsilon$. $\qquad\square$

Exercises

1. Suppose the Borel charge ν is absolutely continuous with respect to a regular Borel measure μ. Then ν is a regular Borel charge.

2. On a metric compact space every Borel charge is regular.

3. Let $M(K, \mathfrak{B})$ be the space of all Borel charges on the compact space K, endowed with the norm $\|\nu\| = |\nu|(K)$. Denote by $M_r(K)$ the set of all regular Borel charges on K. Prove that $M_r(K)$ is a closed linear subspace of the space $M(K, \mathfrak{B})$. From this and Exercise 6 in Subsection 7.1.1 it follows that in the norm $\|\nu\| = |\nu|(K)$ the space $M_r(K)$ is Banach.

4. Suppose the compact space K is not countable. Then the space $M_r(K)$ is not separable.

8.4.2 Formulation of the Riesz–Markov–Kakutani Theorem. Uniqueness Theorem. Examples

Recall that in the exercises of Subsection 7.1.2, by analogy with the integral with respect to a measure, we defined the integral with respect to a charge as the limit of the corresponding integral sums. To avoid relying on the results of the aforementioned exercises and actually solving them for the reader, we will define here the integral with respect to a charge by reduction to the integral with respect to a measure.

Theorem 1. *Let (Ω, Σ) be a set endowed with a σ-algebra of subsets, $\Delta \in \Sigma$, and $\mu_1, \mu_2 : \Sigma \to [0, +\infty)$ countably-additive measures satisfying $\mu_1 \leqslant \mu_2$. If the function $f : \Delta \to \mathbb{R}$ is μ_2-integrable, then it is also μ_1-integrable.*

Proof. By Theorem 2 in Subsection 4.2.2, the function f is integrable on the set Δ with respect to the measure μ if and only if for every $\varepsilon > 0$ there exists a partition D_ε of Δ such that the corresponding upper and lower integral sums, $\overline{S}_\Delta(f, D_\varepsilon, \mu)$ and $\underline{S}_\Delta(f, D_\varepsilon, \mu)$, of f with respect to the measure μ are defined and differ by less than ε. The assertion of the theorem follows from the inequality

$$|\overline{S}_\Delta(f, D_\varepsilon, \mu_1) - \underline{S}_\Delta(f, D_\varepsilon, \mu_1)| \leqslant |\overline{S}_\Delta(f, D_\varepsilon, \mu_2) - \underline{S}_\Delta(f, D_\varepsilon, \mu_2)|. \qquad \square$$

Definition 1. Let (Ω, Σ) be a set with a σ-algebra on it, ν a charge on Σ, and f a measurable function on Ω. Then f is said to be *integrable on the set $\Delta \in \Sigma$ with respect to the charge ν* (or simply *ν-integrable*) if f is integrable with respect to the variation of the charge ν.

It follows from Theorem 1 and the inequalities $\nu^+ \leqslant |\nu|$ and $\nu^- \leqslant |\nu|$ that if f is integrable with respect to ν, then f is integrable with respect to ν^+, as well as with respect to ν^-.

Definition 2. The *integral of the function f with respect to the charge v on the set* Δ is the quantity

$$\int\limits_{\Delta} f dv = \int\limits_{\Delta} f dv^+ - \int\limits_{\Delta} f dv^-. \tag{1}$$

Using formula (1) one can easily transfer to the integral with respect to a charge basic properties of the integral with respect to a measure such as linearity with respect to the integrand, countable additivity with respect to the integration set, and even the Lebesgue dominated convergence theorem. It is slightly harder to deal with estimates of the integral: the integral with respect to a charge of the bigger of two functions may turn to be smaller than the integral of the smaller function. For this reason, in the case of integration with respect to a charge one uses the inequality

$$\left| \int\limits_{A} f \, dv \right| \leqslant \int\limits_{A} |f| d|v|, \tag{2}$$

which is also an easy consequence of the definition:

$$\left| \int\limits_{A} f dv \right| = \left| \int\limits_{A} f dv^+ - \int\limits_{A} f dv^- \right| \leqslant \int\limits_{A} |f| dv^+ + \int\limits_{A} |f| dv^- = \int\limits_{A} |f| d|v|.$$

Let us show that the definition of the integral with respect to a charge just given agrees with the definition provided in the exercises of Subsection 7.1.2.

Theorem 2. *If the function f is integrable on the set $\Delta \in \Sigma$ with respect to the charge v, then the integral sums of f tend to $\int_{\Delta} f dv$ along the directed set of partitions.*

Proof. Let us apply the Hahn theorem to the charge v on Δ. Let Δ^+ and Δ^- be the corresponding positivity and negativity sets. Then the measure v^+ is concentrated on Δ^+, while v^- is concentrated on Δ^-. Use the definition of the integral with respect to a measure and for any $\varepsilon > 0$ choose partitions D_1 and D_2 of the sets Δ^+ and Δ^-, respectively, such that any integral sums of the function f on Δ^+ with respect to partitions finer than D_1, and on Δ^- with respect to partitions finer than D_2, approximate $\int_{\Delta^+} f dv^+$ and $\int_{\Delta^-} f dv^-$, respectively, to within $\varepsilon/2$. Now construct a partition D of the set Δ, taking as its elements all elements of the two partitions D_1 and D_2. Then every integral sum s of f corresponding to partitions finer than D splits into parts s_1 and s_2, corresponding to the pieces of that partition that are finer than D_1 and D_2, respectively. Consequently,

$$\left| s - \int\limits_{\Delta} f dv \right| \leqslant \left| s_1 - \int\limits_{\Delta} f dv^+ \right| + \left| s_2 - \int\limits_{\Delta} f dv^- \right|$$

$$= \left| s_1 - \int\limits_{\Delta^+} f dv^+ \right| + \left| s_2 - \int\limits_{\Delta^-} f dv^- \right| \leqslant \frac{\varepsilon}{2} + \frac{\varepsilon}{2} = \varepsilon,$$

which establishes the required convergence of integral sums to the integral. $\square$

Theorem 3. *The integral with respect to a charge is linear as a function of the charge: if $v_1, v_2 \colon \Sigma \to \mathbb{R}$ are charges, $a_1, a_2 \in \mathbb{R}$ are scalars, and the function $f \colon \Delta \to \mathbb{R}$ is integrable with respect to the charges v_1 and v_2, then f is also integrable with respect to the charge $a_1 v_1 + a_2 v_2$, and*

$$\int_\Delta f d(a_1 v_1 + a_2 v_2) = a_1 \int_\Delta f dv_1 + a_2 \int_\Delta f dv_2. \tag{3}$$

Proof. The integrability follows from the inequality $|a_1 v_1 + a_2 v_2| \leqslant |a_1| \cdot |v_1| + |a_2| \cdot |v_2|$ and Theorem 1. Equality (3) is readily derived from the analogous equality for integral sums. $\square$

In this section we will devote our attention to the particular case of Borel charges and continuous functions on a compact space K.

Definition 3. Let v be a Borel charge on the compact space K. The *functional generated by the charge v* is the mapping $\mathbf{F}_v \colon C(K) \to \mathbb{R}$ given by the rule $\mathbf{F}_v(f) = \int_K f dv$.

Here the existence of the integral is guaranteed by formula (1) and the fact that every bounded measurable function is integrable with respect to any finite measure (Subsection 4.3.3).

Proposition 1. $\mathbf{F}_v$ *is a continuous linear functional on $C(K)$ and $\|\mathbf{F}_v\| \leqslant \|v\|$ (here and in the sequel we will use the notation $\|v\| = |v|(K)$ introduced in Exercise 3 of Subsection 8.4.1).*

Proof. The linearity of the functional $\mathbf{F}_v$ is obvious. Let us estimate the norm of $\mathbf{F}_v$. Take $f \in C(K)$, $\|f\| = 1$. Then $|f(t)| \leqslant 1$ at all points of the compact space K. Now apply inequality (2):

$$|\mathbf{F}_v(f)| = \left| \int_K f \, dv \right| \leqslant \int_K |f| \, d|v| \leqslant \int_K d|v| = |v|(K) = \|v\|. \qquad \square$$

The following obvious property is stated without proof.

Proposition 2. *Let v_1, v_2 be Borel charges on the compact space K, and let $a_1, a_2 \in \mathbb{R}$. Then $\mathbf{F}_{a_1 v_1 + a_2 v_2} = a_1 \mathbf{F}_{v_1} + a_2 \mathbf{F}_{v_2}$. In other words, the mapping $v \mapsto F_v$ is linear.* $\square$

Proposition 3. *Let v be a regular Borel charge on the compact space K such that $\mathbf{F}_v = 0$. Then v is the null charge. If for two regular Borel charges v_1 and v_2 it holds that $\mathbf{F}_{v_1} = \mathbf{F}_{v_2}$, then $v_1 = v_2$.*

Proof. Suppose $\mathbf{F}_v = 0$. Then

$$\int\limits_K f \, dv^+ - \int\limits_K f \, dv^- = \int\limits_K f \, dv = 0,$$

for all functions $f \in C(K)$. That is, the elementary integrals on $C(K)$ generated by the measures v^+ and v^- coincide. Since v^+ and v^- are regular Borel measures, the uniqueness of the representation in the theorem on the general form of elementary integrals (Subsection 8.3.2) implies that the measures v^+ and v^- coincide. Therefore, $v = v^+ - v^- = 0$.

The second part of the assertion reduces to the first thanks to the linearity of the mapping $v \mapsto \mathbf{F}_v$. Indeed, if $\mathbf{F}_{v_1} = \mathbf{F}_{v_2}$, then for the auxiliary charge $v = v_1 - v_2$ we have $\mathbf{F}_v = \mathbf{F}_{v_1} - \mathbf{F}_{v_2} = 0$. That is, $v = v_1 - v_2 = 0$. $\qquad\square$

Together the propositions given above mean that the mapping $U \colon v \mapsto \mathbf{F}_v$, regarded as an operator acting from the space $M_r(K)$ of all regular Borel charges on K into the space $C(K)^*$, is a continuous injective linear operator satisfying $\|U\| \leqslant 1$. The next theorem can be interpreted as the assertion that U is a bijective isometry of the spaces $M_r(K)$ and $C(K)^*$.

Theorem 4. (general form of linear functionals on $C(K)$). *For any continuous linear functional F on $C(K)$ there exists a unique regular Borel measure v on K which generates F (i.e., such that $F = \mathbf{F}_v$). Moreover, $\|F\| = \|v\|$.*

Parts of the Riesz–Markov–Kakutani theorem just stated, namely the uniqueness of the charge v and the inequality $\|\mathbf{F}_v\| \leqslant \|v\|$, have been already established above. The existence of the sought-for charge will be proved below in Subsection 8.4.3, based on the theorem on the general form of elementary integrals. The idea of the proof is to express the functional F as the difference of two elementary integrals. The formula for the norm will be established in Subsection 8.4.4.

Exercises

1. Prove the following analogue of the Lebesgue dominated convergence theorem: *Suppose the functions f_n are integrable on the set Δ with respect to the charge v, $f_n \to f$ $|v|$-almost everywhere as $n \to \infty$, and there exists a v-integrable function g that dominates all f_n (i.e., $|f_n| \leqslant g$ $|v|$-almost everywhere for all $n \in \mathbb{N}$). Then the function f is also v-integrable, and $\int_\Delta f \, dv = \lim\limits_{n\to\infty} \int_\Delta f_n dv$.*

2. Let v_1 and v_2 be regular Borel charges and f_1 and f_2 be bounded functions on the compact space K such that $\int_K g f_1 \, dv_1 = \int_K g f_2 \, dv_2$ for all $g \in C(K)$. Then $\int_K g f_1 \, dv_1 = \int_K g f_2 \, dv_2$ for all bounded Borel-measurable functions g on K.

For each of the functionals G_j on $C[0, 1]$ listed below: (a) verify its linearity and continuity; (b) calculate its norm; (c) find a representation as $\mathbf{F}_\nu$, where ν is a regular Borel charge on the interval $[0, 1]$; (d) calculate the variation of the obtained charge and verify the formula $\|\mathbf{F}_\nu\| = \|\nu\|$.

3. $G_1(f) = f(0)$.

4. $G_2(f) = f(0) - f(1)$.

5. $G_3(f) = \displaystyle\int_0^{1/2} f(t)dt$.

6. $G_4(f) = \displaystyle\int_0^1 f(t)\left(t - \frac{1}{2}\right)dt$.

8.4.3 Positive and Negative Parts of a Functional $F \in C(K)^*$

Consider now the set $C^+(K) = \{f \in C(K) : f \geqslant 0\}$, the *positive cone* of the space $C(K)$. For each $f \in C^+(K)$ we denote by $[0, f]_c$ the following set of functions:

$$[0, f]_c = \{g \in C(K) : 0 \leqslant g \leqslant f\}.$$

Let F be a continuous linear functional on $C(K)$. Define the *positive part* F^+ of F as follows: for $f \in C^+(K)$, put

$$F^+(f) = \sup\{F(g) : g \in [0, f]_c\}, \tag{I}$$

while for an arbitrary function $f \in C(K)$ put

$$F^+(f) = F^+(f^+) - F^+(f^-). \tag{II}$$

Further, we defined the *negative part* F^- of the functional F as $F^- = F^+ - F$.

The aim of the following chain of assertions is to prove that F^+ and F^- are elementary integrals.

Lemma 1. *Let* $f_1, f_2 \in C^+(K)$. *Then* $[0, f_1 + f_2]_c = [0, f_1]_c + [0, f_2]_c$.

Proof. Let $g_1 \in [0, f_1]_c$ and $g_2 \in [0, f_2]_c$, that is, $0 \leqslant g_1 \leqslant f_1$ and $0 \leqslant g_2 \leqslant f_2$. Then $0 \leqslant g_1 + g_2 \leqslant f_1 + f_2$, that is, $g_1 + g_2 \in [0, f_1 + f_2]_c$. This establishes the inclusion $[0, f_1 + f_2]_c \supset [0, f_1]_c + [0, f_2]_c$.

Now let $g \in [0, f_1 + f_2]_c$. We introduce auxiliary functions g_1 and g_2 by the equalities $g_1 = \min\{g, f_1\}$ and $g_2 = g - g_1$. Since $g_1 \in [0, f_1]_c$ and $g_2 \in [0, f_2]_c$, we have that $g = g_1 + g_2 \in [0, f_1]_c + [0, f_2]_c$. This establishes the inclusion $[0, f_1 + f_2]_c \subset [0, f_1]_c + [0, f_2]_c$. $\qquad\square$

Lemma 2. *Let $f_1, f_2 \in C^+(K)$. Then $F^+(f_1 + f_2) = F^+(f_1) + F^+(f_2)$.*

Proof. We have

$$
\begin{aligned}
F^+(f_1) + F^+(f_2) &= \sup\{F(g_1) : g_1 \in [0, f_1]_c\} + \sup\{F(g_2) : g_2 \in [0, f_2]_c\} \\
&= \sup\{F(g_1 + g_2) : g_1 \in [0, f_1]_c, \ g_2 \in [0, f_2]_c\} \\
&= \sup\{F(g) : g \in [0, f_1]_c + [0, f_2]_c\} \\
&= \sup\{F(g) : g \in [0, f_1 + f_2]_c\} = F^+(f_1 + f_2). \qquad \square
\end{aligned}
$$

Lemma 3. *Let $f \in C^+(K)$ and $a \in \mathbb{R}^+$. Then $F^+(af) = aF^+(f)$.*

Proof. We have
$$
F^+(af) = \sup\{F(g) : g \in [0, af]_c\}
$$

$$
= \sup\{F(ah) : h \in [0, f]_c\} = a \sup\{F(h) : h \in [0, f]_c\} = aF^+(f). \qquad \square
$$

Lemma 4. *Let $f_1, f_2 \in C^+(K)$. Then $F^+(f_1 - f_2) = F^+(f_1) - F^+(f_2)$.*

Proof. Since the function $f = f_1 - f_2$ is not necessarily positive, to calculate $F^+(f)$ we need to use formula (II). Further, since

$$
f^+ - f^- = f = f_1 - f_2, \quad f^+ + f_2 = f_1 + f^-,
$$

Lemma 2 shows that $F^+(f^+) + F^+(f_2) = F^+(f_1) + F^+(f^-)$. Consequently,

$$
F^+(f_1 - f_2) = F^+(f) = F^+(f^+) - F^+(f^-) = F^+(f_1) - F^+(f_2). \qquad \square
$$

Theorem 1. *F^+ is an elementary integral on $C(K)$.*

Proof. First let us verify that F^+ is a linear functional. Let $h_1, h_2 \in C(K)$ be arbitrary functions. Write the function $h_1 + h_2$ in the form $(h_1^+ + h_2^+) - (h_1^- + h_2^-)$ and apply Lemmas 4 and 2 above. Then

$$
\begin{aligned}
F^+(h_1 + h_2) &= F^+(h_1^+ + h_2^+) - F^+(h_1^- + h_2^-) = F^+(h_1^+) - F^+(h_1^-) \\
&\quad + F^+(h_2^+) - F^+(h_2^-) = F^+(h_1) + F^+(h_2).
\end{aligned}
$$

This shows that our functional is additive. The positive homogeneity follows from Lemma 3 and formula (II): if $f \in C(K)$ and $a \in \mathbb{R}^+$, then

$$
F^+(af) = F^+(af^+) - F^+(af^-) = aF^+(f^+) - aF^+(f^-) = aF^+(f).
$$

The fact that the minus sign passes from the argument to the value is a consequence of Lemma 4:

$$
F^+(-f) = F^+(f^- - f^+) = F^+(f^-) - F^+(f^+) = -F^+(f).
$$

To complete the proof it remains to verify that on $C^+(K)$ the functional F^+ takes only non-negative values. Indeed, let $f \in C^+(K)$. Now use formula (I): $F^+(f) = \sup\{F(g) : g \in [0, f]_c\} \geqslant F(0) = 0$. $\qquad\square$

Theorem 2. $F^- = (-F)^+$. *In particular, F^- is an elementary integral on $C(K)$.*

Proof. From the preceding theorem and the relation $F^- = F^+ - F$ it follows that F^- is a linear functional. Hence, it suffices to verify the equality $F^-(f) = (-F)^+(f)$ for $f \in C^+(K)$, because it extends to $C(K)$ by linearity. Thus,

$$F^-(f) = F^+(f) - F(f)$$

$$= \sup\{F(g) - F(f) : g \in [0, f]_c\} = \sup\{-F(f - g) : g \in [0, f]_c\}.$$

Now denoting $f - g$ by h and observing that the conditions $g \in [0, f]_c$ and $h \in [0, f]_c$ are equivalent, we obtain the desired equality $F^-(f) = \sup\{-F(h) : h \in [0, f]_c\} = (-F)^+(f)$. $\qquad\square$

As a corollary of the theorems proved above and the equality $F = F^+ - F^-$, we obtain the existence of the sought-for charge in the theorem on the general form of linear functionals on $C(K)$ (Theorem 4 of Subsection 8.4.2).

Corollary 1. *For every continuous linear functional F on $C(K)$ there exists a regular Borel charge ν on K which generates this functional by the rule $F = F_\nu$.*

Proof. Since F^+ and F^- are elementary integrals (theorem in Subsection 8.3.2), there exist regular Borel measures μ_1 and μ_2 which generate these elementary integrals: $F^+ = \mathbf{F}_{\mu_1}$ and $F^- = \mathbf{F}_{\mu_2}$. Then $F = F^+ - F^- = \mathbf{F}_{\mu_1} - \mathbf{F}_{\mu_2} = \mathbf{F}_{\mu_1 - \mu_2}$, i.e., as the desired charge ν one can take $\mu_1 - \mu_2$. $\qquad\square$

Remark 1. As the reader has undoubtedly noticed, the notions of positive and negative parts can be defined for many quite different objects: numbers, functions, charges — and now also for functionals. The general approach to such objects is part of the theory of semi-ordered spaces, vector and normed lattices. A first impression of this useful and far-developed direction of functional analysis can be gained by consulting the textbook of L. V. Kantorovich (one of the founders of this subject) and G. P. Akilov [22, Chapter 10].

Exercises

1. Show that the formula (II) does not contradict formula (I), i.e., that if for positive functions f one defines the quantity $F^+(f)$ by formula (I), then formula (II) gives the same result.

2. For the functions $g_1 = \min\{g, f_1\}$ and $g_2 = g - g_1$ figuring in the second part of the proof of Lemma 1, verify the conditions $g_1 \in [0, f_1]_c$ and $g_2 \in [0, f_2]_c$.

8.4.4 The Norm of a Functional on $C(K)$

As before, we denote by $\mathbb{1}$ the function identically equal to 1 on K. Let F be a continuous linear functional on $C(K)$, F^+ and F^- its positive and negative parts, and μ_1 and μ_2 the regular Borel measures that generate the elementary integrals F^+ and F^-, respectively: $F^+(f) = \int_K f \, d\mu_1$ and $F^-(f) = \int_K f \, d\mu_2$ for any function $f \in C(K)$.

Lemma 1. $\|F\| = \mu_1(K) + \mu_2(K)$.

Proof. First observe that

$$\mu_1(K) + \mu_2(K) = \int_K d\mu_1 + \int_K d\mu_2 = F^+(\mathbb{1}) + F^-(\mathbb{1}),$$

i.e., what we need to prove is the equality $\|F\| = F^+(\mathbb{1}) + F^-(\mathbb{1})$. We have

$$\|F\| = \sup\{F(f) : f \in S_{C(K)}\} = \sup\{F(f^+ - f^-) : f \in S_{C(K)}\}.$$

Since in the present case f^+ and f^- belong to the set $[0, \mathbb{1}]_c$, we can continue our estimate as follows:

$$\|F\| \leqslant \sup\{F(f_1) - F(f_2) : f_1, f_2 \in [0, \mathbb{1}]_c\}$$
$$= \sup\{F(f_1) : f_1 \in [0, \mathbb{1}]_c\} + \sup\{-F(f_2) : f_2 \in [0, \mathbb{1}]_c\} = F^+(\mathbb{1}) + F^-(\mathbb{1}).$$

Now let us prove the opposite inequality. We have already established in the preceding calculation that $F^+(\mathbb{1}) + F^-(\mathbb{1}) = \sup\{F(f_1 - f_2) : f_1, f_2 \in [0, \mathbb{1}]_c\}$. Since in the last expression we have $-1 \leqslant f_1 - f_2 \leqslant 1$, it follows that $\|f_1 - f_2\| \leqslant 1$ and $F(f_1 - f_2) \leqslant \|F\|$, i.e., $F^+(\mathbb{1}) + F^-(\mathbb{1}) \leqslant \|F\|$. $\square$

Now for the functional F under study we introduce the charge $\nu = \mu_1 - \mu_2$. Then $F = \mathbf{F}_\nu$.

Theorem 1. $\|F\| = \|\nu\|$.

Proof. The estimate $\|F\| \leqslant \|\nu\|$ was established in Proposition 1 of Subsection 8.4.2. Let us prove the opposite inequality. Since the measures μ_1 and μ_2 assume only positive values, $\nu(\Delta) = \mu_1(\Delta) - \mu_2(\Delta) \leqslant \mu_1(\Delta) \leqslant \mu_1(K)$ for all Borel sets $\Delta \subset K$. Consequently,

$$\nu^+(K) = \sup\{\nu(\Delta) : \Delta \in \mathfrak{B}\} \leqslant \mu_1(K).$$

Similarly, $\nu^-(K) \leqslant \mu_2(K)$. It remains to apply the preceding lemma: $\|F\| = \mu_1(K) + \mu_2(K) \geqslant \nu^+(K) + \nu^-(K) = \|\nu\|$. $\square$

This last assertion completes the proof of the theorem on the general form of linear functionals on $C(K)$, formulated in Subsection 8.4.2. Since this theorem establishes an isomorphism (called the canonical isomorphism) between the spaces $C(K)^*$ and $M_r(K)$, it is often stated as the equality $C(K)^* = M_r(K)$.

Exercises

1. Verify that $\overline{B}_{C(K)} = [0, \mathbb{1}]_c - [0, \mathbb{1}]_c$. Then derive from this the equality $\|F\| = F^+(\mathbb{1}) + F^-(\mathbb{1})$ established above.

2. Show that $\|F\| = \|F^+\| + \|F^-\|$.

3. For every (not necessarily regular) Borel charge ν on K consider the elementary integral F_ν generated by ν. Now for the elementary integral $\mathbf{F}_\nu$ construct a regular Borel charge $P(\nu)$ such that $\mathbf{F}_\nu = \mathbf{F}_{P(\nu)}$. Show that the mapping P defined in this manner is a projector of the space $M(K, \mathfrak{B})$ onto the subspace $M_r(K)$ (see Exercise 3 in Subsection 8.4.1). Show that $\|P\| = 1$.

4. On the example of the space $C[0, 1]$, convince yourself that the dual space to a separable Banach space can be non-separable.

5. Let σ be a regular Borel charge on K and g a function on K, integrable with respect to σ. Define the functional $F \in C(K)^*$ by the formula $F(f) = \int_K fg \, d\sigma$. How does one write this functional in the form indicated in the theorem on the general form of linear functionals on $C(K)$? Prove the equality $\|F\| = \int_K |g|d|\sigma|$.

8.4.5 Complex Charges and Integrals

In this subsection we address the integration of complex-valued functions with respect to complex-valued charges. Those assertions for which, in our opinion, the differences from the real-valued case are not essential will be only stated, the proof being left to the reader.

Definition 1. Let (Ω, Σ) be a set endowed with a σ-algebra. A complex-valued set function $\eta \colon \Sigma \to \mathbb{C}$ is called a *complex charge* on Ω if it satisfies the countable additivity condition. For a complex charge η we define in the natural manner the charges $\operatorname{Re} \eta$ and $\operatorname{Im} \eta$: $(\operatorname{Re} \eta)(\Delta) = \operatorname{Re}(\eta(\Delta))$ and $(\operatorname{Im} \eta)(\Delta) = \operatorname{Im}(\eta(\Delta))$ for all $\Delta \in \Sigma$. Then

$$\eta = \operatorname{Re} \eta + i \operatorname{Im} \eta. \qquad (*)$$

Definition 2. The *variation of the complex charge* η on a set $\Delta \in \Sigma$ is the quantity $|\eta|(\Delta)$, defined as the supremum of sums of the form $\sum_{k=1}^{n} |\eta(\Delta_k)|$, taken over all finite collections $\{\Delta_k\}_1^n$ of pairwise disjoint subsets of Δ.

Theorem 1. *For a real-valued charge the value of the variation, calculated by Definition 2 above, coincides with the value calculated by the rule*

$$|\eta|(\Delta) = \eta^+(\Delta) + \eta^-(\Delta).$$

In other words, the new definition of the variation agrees with the already known one. $\qquad \square$

Theorem 2. *The variation of a complex charge η has the following properties:*

(i) $|\eta(\Delta)| \leqslant |\eta|(\Delta)$;

(ii) $\max\{|\mathrm{Re}\,\eta|(\Delta), |\mathrm{Im}\,\eta|(\Delta)\} \leqslant |\eta|(\Delta) \leqslant |\mathrm{Re}\,\eta|(\Delta) + |\mathrm{Im}\,\eta|(\Delta)$;

(iii) $|\eta|$ *is a finite countably-additive measure on* Ω. $\square$

Like in the case of measures, we introduce admissible partitions (those for which $\sum_{k=1}^{\infty} \sup_{t \in \Delta_k} [|f(t)| \cdot |\nu|(\Delta_k)] < \infty$) and the integral sums of a complex-valued function f with respect to a complex charge ν: $S_A(f, D, T, \nu) = \sum_{k=1}^{\infty} f(t_k)\nu(\Delta_k)$. Integrability and the integral of a function with respect to a charge are also defined as the limit of integral sums for increasingly finer partitions.

Theorem 3. *The integral of a complex-valued function with respect to a complex charge depends linearly on the charge for a fixed function:*

$$\int_\Delta f\, d(a_1\eta_1 + a_2\eta_2) = a_1 \int_\Delta f\, d\eta_1 + a_2 \int_\Delta f\, d\eta_2,$$

and depends linearly on the function for a fixed charge:

$$\int_\Delta (a_1 f_1 + a_2 f_2)\, d\eta = a_1 \int_\Delta f_1\, d\eta + a_2 \int_\Delta f_2\, d\eta;$$

here, the existence of the integrals in the right-hand side implies the existence of the integrals in the left-hand side. $\square$

Theorem 4. *A measurable function f is integrable on the set $\Delta \in \Sigma$ with respect to the complex charge η if and only if the integral $\int_\Delta |f|\, d|\eta|$ exists. In this case,*

$$\left| \int_\Delta f\, d\eta \right| \leqslant \int_\Delta |f|\, d|\eta|.$$

Proof. If for a measurable function the integral $\int_\Delta |f|\, d|\eta|$ exists, then in view of the inequalities $|\mathrm{Im}\,\eta|(\Delta) \leqslant |\eta|(\Delta)$ and $|\mathrm{Re}\,\eta|(\Delta) \leqslant |\eta|(\Delta)$, so do the following four integrals: $\int_\Delta \mathrm{Re}\, f\, d\,\mathrm{Re}\,\eta$, $\int_\Delta \mathrm{Re}\, f\, d\,\mathrm{Im}\,\eta$, $\int_\Delta \mathrm{Im}\, f\, d\,\mathrm{Re}\,\eta$, and $\int_\Delta \mathrm{Im}\, f\, d\,\mathrm{Im}\,\eta$. Using the linearity (the preceding theorem), the appropriate combination of these integrals yields $\int_\Delta f\, d\eta$.

Conversely, the existence of the integral $\int_\Delta f\, d\eta$ implies the existence of an admissible partition D for f with respect to ν. This partition will also be admissible for $|f|$ with respect to $|\nu|$, which in view of the measurability of the function $|f|$ and Exercise 2 in Subsection 4.3.3 means that the integral $\int_\Delta |f|\, d|\eta|$ exists. Finally, the inequality $|\int_\Delta f\, d\eta| \leqslant \int_\Delta |f|\, d|\eta|$ is obtained by passing to the limit in the corresponding integral sums. $\square$

Using formula (∗), the main notions and results concerning real charges (for instance, the absolute continuity criterion and the Radon–Nikodým theorem in Sect. 7.1) can be easily transferred to the case of complex charges. Similarly, by separating the real and and imaginary parts it is easy to extend to the complex case properties of the integrals, such as countable additivity with respect to the set and the Lebesgue dominated convergence theorem.

8.4.6 Regular Complex Charges and Functionals on the Complex Space $C(K)$

Definition 1. A *complex Borel charge* on the compact space K is a complex charge defined on the σ-algebra $\mathfrak{B}$ of Borel subsets of K. A complex Borel charge η on the compact space K is called *regular* if $|\eta|$ is a regular Borel measure.

From property (ii) in Theorem 2 of Subsection 8.4.5 and Exercise 1 in Subsection 8.4.1 it follows that a complex Borel charge η on the compact space K is regular if and only if the real charges $\mathrm{Re}\,\eta$ and $\mathrm{Im}\,\eta$ are regular.

Next we consider linear functionals on the complex space $C(K)$, i.e., the space of all complex-valued continuous functions on K.

Definition 2. Let η be a Borel charge on the compact space K. The *functional generated by the charge* η is the mapping $\mathbf{F}_\eta \colon C(K) \to \mathbb{C}$ acting by the rule $\mathbf{F}_\eta(f) = \int_K f \, d\eta$.

Here the existence of the integral is guaranteed by Theorem 4 in the preceding subsection and the integrability criterion for real-valued measurable functions (Subsection 4.3.3).

As in the real case, for complex Borel charges on the compact space K, we will use the notation $\|\eta\| = |\eta|(K)$.

Theorem 1. $\mathbf{F}_\eta$ *is a continuous linear functional on* $C(K)$ *and* $\|\mathbf{F}_\eta\| \leqslant \|\eta\|$. *Moreover, if the charge* η *is regular, then* $\|\mathbf{F}_\eta\| = \|\eta\|$.

Proof. The linearity of the functional $\mathbf{F}_\nu$ is obvious. The upper bound for the norm follows, like in the real version, from the inequality $\left|\int_K f \, d\eta\right| \leqslant \int_K |f| \, d|\eta|$. It remains to prove in the case of a regular charge the inequality $\|\mathbf{F}_\eta\| \geqslant \|\eta\|$.

Fix $\varepsilon > 0$. By the definition of the variation, there exist a finite collection $\{\Delta_k\}_1^n$ of pairwise disjoint subsets of the compact space K such that $\sum_{k=1}^n |\eta(\Delta_k)| > |\eta|(K) - \varepsilon$. Since the charge is regular, we can, with no loss of generality, assume that all sets Δ_k are closed: otherwise, one can replace them by smaller closed sets that have approximately the same charge $\eta(\Delta_k)$. Now consider $K_1 = \bigsqcup_{k=1}^n \Delta_k$. The set K_1 is closed, and

$$|\eta|(K \setminus K_1) = |\eta|(K) - |\eta|(K_1) \leqslant |\eta|(K) - \sum_{k=1}^n |\eta(\Delta_k)| < \varepsilon.$$

Now denote by f the function on K_1 that on each set Δ_k takes the constant value $\alpha_k = e^{-i \arg \eta(\Delta_k)}$. Since the sets Δ_k are pairwise disjoint, the piecewise-constant function f is continuous on K_1. Extend f to a continuous function on the whole space K with preservation of the condition $|f| \leqslant 1$. Then $f \in S_{C(K)}$ and

$$\|\mathbf{F}_\eta\| \geqslant |\mathbf{F}_\eta(f)| = \left| \int\limits_K f \, d\eta \right| \geqslant \left| \int\limits_{K_1} f \, d\eta \right| - \varepsilon$$

$$= \left| \sum_{k=1}^n \alpha_k \eta(\Delta_k) \right| - \varepsilon = \sum_{k=1}^n |\eta(\Delta_k)| - \varepsilon \geqslant \|\eta\| - 2\varepsilon.$$

Letting $\varepsilon \to 0$, we obtain the requisite estimate.   □

The formulation of the theorem on the general form of linear functionals on the complex space $C(K)$ is a verbatim repetition of the real version, and its proof can be done by reduction to the real case.

Theorem 2. *For every continuous linear functional F on the complex space $C(K)$ there exists a unique regular complex Borel charge η on K which generates this functional (i.e., for which $F = \mathbf{F}_\eta$). Moreover, $\|F\| = \|\eta\|$.*

Proof. The equality $\|\mathbf{F}_\eta\| = \|\eta\|$ was established above. It implies the injectivity of the mapping $\eta \mapsto F_\eta$ on the space of regular complex Borel charges, i.e., the uniqueness of the charge η in the theorem. It remains to prove the existence of the requisite charge.

Regard the space $C_{\mathbb{R}}(K)$ of real-valued continuous functions on K as a subset of the complex space $C(K)$. Now define on $C_{\mathbb{R}}(K)$ two linear functionals, F_1 and F_2, by the formulas $F_1(f) = \operatorname{Re} F(f)$ and $F_2(f) = \operatorname{Im} F(f)$. By the real version of the theorem on the general form of linear functionals on $C_{\mathbb{R}}(K)$, applied to F_1 and F_2, there exist regular real Borel charges ν_1 and ν_2, such that for every function $f \in C_{\mathbb{R}}(K)$ it holds that $F_1(f) = \int_K f \, d\nu_1$ and $F_2(f) = \int_K f \, d\nu_2$. Substituting these equalities into the formula $F(f) = F_1(f) + i F_2(f)$ and denoting $\nu_1 + i\nu_2$ by η, we obtain the equality $F(f) = \int_K f \, d\eta$, which holds for all $f \in C_{\mathbb{R}}(K)$. Since both sides of this last equality depend linearly on f, it easily extends to complex functions of the form $f = f_1 + if_2$, with $f_1, f_2 \in C_{\mathbb{R}}(K)$. Therefore, the equality $F(f) = \int_K f \, d\eta$ holds on the entire complex space $C(K)$, and consequently η is the sought-for charge.   □

Remark 1. The reader should note that the argument in the proof of Theorem 1 yields a direct proof of the equality $\|\mathbf{F}_\eta\| = \|\eta\|$ also in the case of real charges.

Exercises

1. The proof of Theorem 1 contains the phrase: "Extend f to a continuous function on the whole space K with preservation of the condition $|f| \leqslant 1$." Why is such an extension possible?

2. Prove the following — historically the first version — of the theorem on the general form of linear functionals on $C[0, 1]$. **F. Riesz's theorem**: *For every linear functional $F \in C[0, 1]^*$ there exists a function $\widetilde{F}$ of bounded variation on $[0, 1]$, with $V_0^1(\widetilde{F}) = \|F\|$, such that the functional F is expressed as a Stieltjes integral with respect to $d\widetilde{F}$: $F(f) = \int_0^1 f\, d\widetilde{F}$ for all $f \in C[0, 1]$.*

3. Is the function $\widetilde{F}$ in F. Riesz's theorem stated above uniquely determined by the functional F?

4. Sharpen F. Riesz's theorem as follows: the function $\widetilde{F}$ can be chosen in the class of functions that are right-continuous on $(0, 1]$ and take the value 0 at the point zero, and in this class $\widetilde{F}$ is uniquely determined by F.

5. Solve the complex version of Exercise 2 in Subsection 8.4.2.

6. Solve the complex version of Exercise 5 in Subsection 8.4.4.

Comments on the Exercises

Subsection 8.4.2

Exercise 2. For the proof one has to take the measure $\mu = |\nu_1| + |\nu_2|$, which majorizes both charges appearing in the formula; represent the function g as a limit of a μ-almost everywhere convergent uniformly bounded sequence (g_n) of continuous functions, and apply the dominated convergence theorem.

Subsection 8.4.4

Exercise 5. See Subsection 7.1.4, Exercise 4, and the comment on it. The requisite charge ν, for which $F = \mathbf{F}_\nu$, is defined as $\nu(\Delta) = \int_\Delta g\, d\sigma$. To prove the formula $\|\nu\| = \int_K |g|\, d|\sigma|$, one can express the positivity and negativity sets of the charge ν, K_ν^+ and K_ν^- in terms of the positivity and negativity sets of the charge σ, K_σ^+ and K_σ^-: $K_\nu^+ = (K_\sigma^+ \cap g_{>0}) \cup (K_\sigma^- \cap g_{\leqslant 0})$, $K_\nu^- = (K_\sigma^+ \cap g_{\leqslant 0}) \cup (K_\sigma^- \cap g_{>0})$, and then use the equality $|\nu|(K) = \nu(K_\nu^+) - \nu(K_\nu^-)$.

Subsection 8.4.6

Exercises 2–4. Use the theorem on the general form of a Borel charge on the interval (Subsection 7.2.3): take $\widetilde{F}$ equal to 0 at the point zero, and in the remaining points of the interval take $\widetilde{F}$ equal to the distribution function of the Borel charge that generates the functional F. A direct proof of F. Riesz's theorem can be found in A. Kolmogorov and S. Fomin's textbook [24, Chapter IV]. Therein one can also find the definition and main properties of the Stieltjes integral, as well as a discussion of the uniqueness of the function $\widetilde{F}$. True, the terminology in [K-F] is not completely identical with the one adopted in the present book (for instance, here the distribution function is defined somewhat differently).

Chapter 9
Continuous Linear Functionals

9.1 The Hahn–Banach Theorem in Normed Spaces

The Hahn–Banach theorem on the extension of a linear functional proved in Subsection 5.4.3 is extremely general in nature and is applicable in any real linear space. In particular, it can also be used in normed spaces. However, this last application requires some refinements. First of all, in normed spaces, among all linear functionals the most interest is attached to continuous functionals. Accordingly, it is desirable that the extension will preserve continuity. Second, we would like to have a version of the theorem that is equally appropriate for real as well as complex spaces.

9.1.1 The Connection Between Real and Complex Functionals

Let X be a complex linear space, i.e., in X multiplication by complex scalars is defined. Then, in particular, multiplication by real scalars is also defined in X, i.e., X can be also regarded as a real linear space. Hence, on X one can talk about two types of linear functionals. Specifically, a functional f on X is called a *real linear functional* if f takes real values, is additive (i.e., $f(x + y) = f(x) + f(y)$ for all $x, y \in X$), and is *real-homogeneous* (i.e., $f(\lambda x) = \lambda f(x)$ for all $x \in X$ and all $\lambda \in \mathbb{R}$), and is called a *complex linear functional* if f takes complex values, is additive, and is *complex-homogeneous* (i.e., $f(\lambda x) = \lambda f(x)$ for all $x \in X$ and all $\lambda \in \mathbb{C}$).

For each complex functional f we define in a natural manner its real and imaginary parts: $(\operatorname{Re} f)(x) = \operatorname{Re}(f(x))$ and $(\operatorname{Im} f)(x) = \operatorname{Im}(f(x))$. The so-defined $\operatorname{Re} f$ and $\operatorname{Im} f$ are real linear functionals, and $f = \operatorname{Re} f + i \operatorname{Im} f$. The following two theorems describe completely the connection between the real and imaginary parts of a complex linear functional.

Theorem 1. *Let f be a complex linear functional on X. Then for any $x \in X$ it holds that* $\operatorname{Im} f(x) = -\operatorname{Re} f(ix)$.

© Springer International Publishing AG, part of Springer Nature 2018
V. Kadets, *A Course in Functional Analysis and Measure Theory*,
Universitext, https://doi.org/10.1007/978-3-319-92004-7_9

Proof. In the equality $f(\lambda x) = \lambda f(x)$ put $\lambda = i$ and calculate the real parts. We obtain $\operatorname{Re} f(ix) = \operatorname{Re}(if(x)) = -\operatorname{Im} f(x)$.   □

Theorem 2. *Let g be a real linear functional on X. Then the functional f defined by the formula $f(x) = g(x) - ig(ix)$ is a complex linear functional.*

Proof. The additivity of f and its real homogeneity are obvious. Let us verify the complex homogeneity. First, note that $f(ix) = g(ix) - ig(-x) = i(g(x) - ig(ix)) = if(x)$. Now let $\lambda = a + ib$ be an arbitrary complex number. Then

$$f((a+ib)x) = f(ax) + f(ibx) = af(x) + ibf(x) = (a+ib)f(x),$$

as needed.   □

Theorems 1 and 2 show that the correspondence $f \mapsto \operatorname{Re} f$ between complex and real linear functionals is bijective. In the case of a normed space X one can say more.

Theorem 3. *Let X be a complex normed space, and f a continuous complex linear functional on X. Then $\|f\| = \|\operatorname{Re} f\|$.*

Proof. Let $\mathbb{T} = \{\lambda \in \mathbb{C} : |\lambda| = 1\}$. Using the definition of the norm of a functional and the fact that the set of products $\{\lambda x : x \in S_X, \lambda \in \mathbb{T}\}$ coincides with S_X, we have

$$\|\operatorname{Re} f\| = \sup_{x \in S_X} |\operatorname{Re} f(x)| = \sup_{x \in S_X, \lambda \in \mathbb{T}} |\operatorname{Re} f(\lambda x)|$$

$$= \sup_{x \in S_X} \left(\sup_{\lambda \in \mathbb{T}} |\operatorname{Re} \lambda f(x)| \right). \tag{$*$}$$

For fixed x, the set of numbers $\{\lambda f(x) : \lambda \in \mathbb{T}\}$ is a circle of radius $|f(x)|$ centered at zero. The real parts of these numbers fill the segment from $-|f(x)|$ to $|f(x)|$. Therefore, $\sup_{\lambda \in \mathbb{T}} |\operatorname{Re} \lambda f(x)| = |f(x)|$. Substituting this relation into $(*)$, we obtain the claimed relation $\|\operatorname{Re} f\| = \sup_{x \in S_X} |f(x)| = \|f\|$.   □

9.1.2 The Hahn–Banach Extension Theorem

Theorem 1. *Let Y be a subspace of the normed linear space X, and let $f \in Y^*$. Then there exists a functional $\widetilde{f} \in X^*$ such that $\widetilde{f}(y) = f(y)$ for all $y \in Y$ and $\|\widetilde{f}\| = \|f\|$. In other words, every continuous linear functional given on a subspace of a normed space extends to the entire space with preservation of its norm.*

Proof. The theorem as stated covers both the real and the complex case, but we need separate proofs. We begin with the real case. Define on X a convex functional p by the formula $p(x) = \|f\| \cdot \|x\|$. Then the functional f obeys the majorization condition $f(y) \leqslant p(y)$ for all $y \in Y$. Let us apply the analytic form of the Hahn–Banach theorem given in Subsection 5.4.3. So, let $\widehat{f}$ be an extension of the functional

f to the entire space X with preservation of the majorization condition. Let $x \in X$ be an arbitary element. Writing the majorization condition for x and $-x$, we obtain two inequalities: $\widetilde{f}(x) \leqslant p(x)$ and $-\widetilde{f}(x) \leqslant p(x)$. This means that $|\widetilde{f}(x)| \leqslant p(x) = \|f\| \cdot \|x\|$ for all $x \in X$. Hence, $\|\widetilde{f}\| \leqslant \|f\|$, and consequently $\widetilde{f}$ is continuous. The opposite inequality $\|\widetilde{f}\| \geqslant \|f\|$ follows from the fact that the functional $\widetilde{f}$ is an extension of the functional f:

$$\|\widetilde{f}\| = \sup_{x \in S_X} \|\widetilde{f}(x)\| \geqslant \sup_{x \in S_Y} \|\widetilde{f}(x)\| = \sup_{x \in S_Y} \|f(x)\| = \|f\|.$$

Now let us address the complex case. Let X be a complex normed space and f a continuous complex linear functional on Y. Then $g = \operatorname{Re} f$ is already a real functional, and by what we just proved, there exists a real functional $\widetilde{g}$ on X such that $\widetilde{g}(y) = g(y)$ for all $y \in Y$ and $\|\widetilde{g}\| = \|g\|$. Now define the sought-for functional $\widetilde{f}$ by the formula $\widetilde{f}(x) = \widetilde{g}(x) - i\widetilde{g}(ix)$. By Theorem 2 of the preceding Subsection 9.1.1, $\widetilde{f}$ is a complex linear functional. Further, for any $y \in Y$, Theorem 1 of the preceding subsection shows that $\operatorname{Im} f(y) = -\operatorname{Re} f(iy) = -g(iy)$. Therefore,

$$\widetilde{f}(y) = \widetilde{g}(y) - i\widetilde{g}(iy) = g(y) - ig(iy) = f(y).$$

Finally, by Theorem 3 of Subsection 9.1.1, $\|\widetilde{f}\| = \|\widetilde{g}\| = \|g\| = \|f\|$. $\square$

Exercises

1. Let X be a normed space, let f be a linear functional on X, and let $A = \{x \in X : f(x) = 1\}$. Show that $\rho(0, A) = \|f\|^{-1}$. In particular, if $\rho(0, A) \neq 0$, then the functional f is continuous.

2. Let X be a normed space, $x \in X$, $f \in X^*$, a an arbitrary scalar, and $A = \{x \in X : f(x) = a\}$. Then $\rho(x, A) = |f(x) - a|/\|f\|$.

3. For a non-zero linear functional f on a normed space X, the following conditions are equivalent:

— f is continuous;

— the kernel of f is closed;

— the kernel of f is not dense in X.

4. Let $1 \leqslant p < \infty$. Denote by $C_p[a, b]$ the subspace of the normed space $L_p[a, b]$ consisting of all the functions continuous on $[a, b]$. Let $t_0 \in [a, b]$ be a given point. Prove that the linear functional δ_{t_0} on $C_p[a, b]$, acting by the rule $\delta_{t_0}(f) = f(t_0)$, is discontinuous. From this fact and the preceding exercise deduce that the subset of functions satisfying the condition $f(0) = 0$ is dense in $C_p[a, b]$.

5. Solve Exercise 10 of Subsection 6.2.2: for $1 \leqslant p < \infty$, the set of continuous functions satisfying the condition $f(0) = 0$ is dense in $L_p[0, 1]$.

6. Call a set $\Delta \in [a, b]$ "very small" if the subspace V_Δ of the space $C_p[a, b]$ consisting of the functions that vanish identically on Δ is dense in $C_p[a, b]$. Prove that the set Δ is very small if and only if its closure has measure 0.

9.2   Applications

9.2.1   Supporting Functionals

Let X be a normed space, and $x_0 \in X \setminus \{0\}$. A functional $f_0 \in X^*$ is called a *supporting functional at the point* x_0 if $\|f_0\| = 1$ and $f_0(x_0) = \|x_0\|$.

Theorem 1. *For any point $x_0 \in X \setminus \{0\}$ there exists a supporting functional at x_0.*

Proof. Consider the subspace $Y = \text{Lin}\{x_0\}$. Then Y is one-dimensional and x_0 is a basis in Y. Define a linear functional f on Y by setting $f(x_0) = \|x_0\|$. In other words, for each $y = \lambda x_0$ put $f(y) = \lambda \|x_0\|$. Let us compute the norm of f. If $y = \lambda x_0 \in S_Y$, then $|\lambda| \|x_0\| = 1$. Therefore,

$$\|f\| = \sup_{\lambda x_0 \in S_Y} |f(\lambda x_0)| = \sup_{\lambda x_0 \in S_Y} |\lambda| \|x_0\| = 1.$$

Now use the Hahn–Banach theorem from the preceding subsection and extend the functional f to a functional $f_0 \in X^*$ preserving its norm. The resulting extension is the sought-for supporting functional, because $\|f_0\| = \|f\| = 1$ and $f_0(x_0) = f(x_0) = \|x_0\|$. $\square$

Corollary 1. *If X is a normed space and $X \neq \{0\}$, then also $X^* \neq \{0\}$.* $\square$

At this point, let us recall some facts from linear algebra. If X is a finite-dimensional linear space and X' is the space of all linear functionals on X, then $\dim X' = \dim X$. If $E \subset X'$ is a subspace and $E \neq X'$, then there exists an element $x_0 \in X \setminus \{0\}$ that is annihilated by all functionals from E: $f(x_0) = 0$ for all $f \in E$ (this last fact is equivalently restated as follows: if in a system of linear homogeneous equations the number of unknowns is larger than the number of equations, then the system has a non-zero solution).

Corollary 2. *On a finite-dimensional normed space every linear functional is continuous.*

Proof. Let X be a finite-dimensional normed space. Consider the space X^* of continuous linear functionals on X as a linear subspace of the space X' of *all* linear functionals on X. Now suppose that $X^* \neq X'$. Then there exists an $x_0 \in X \setminus \{0\}$ such that $f(x_0) = 0$ for all $f \in X^*$. Hence, for this element there is no supporting functional. This contradicts Theorem 1. $\square$

Theorem 2. *Let X and E be normed spaces, with X finite-dimensional. Then any linear operator T acting from X into E is continuous.*

Proof. Pick a basis $\{x_k\}_{k=1}^n$ in X. For each $x \in X$ denote by $\{x_k^*(x)\}_{k=1}^n$ the coefficients of the decomposition of the element x in the basis $\{x_k\}_{k=1}^n$: $x = \sum_{k=1}^n x_k^*(x)x_k$. We claim that x_k^* are linear functionals. Indeed, for any $x, y \in X$ and any scalars a, b we have that

$$\sum_{k=1}^n (ax_k^*(x) + bx_k^*(y))x_k = a \sum_{k=1}^n x_k^*(x)x_k + b \sum_{k=1}^n x_k^*(y)x_k$$

$$= ax + by = \sum_{k=1}^n x_k^*(ax + by)x_k.$$

In view of the uniqueness of the decomposition with respect to a basis, $ax_k^*(x) + bx_k^*(y) = x_k^*(ax + by)$, i.e., linearity is established. By the preceding corollary, $x_k^* \in X^*, k = 1, 2, \ldots, n$. Using the linearity of the operator T and the triangle inequality, we get the estimate

$$\|Tx\| = \left\| \sum_{k=1}^n x_k^*(x)Tx_k \right\| \leqslant \sum_{k=1}^n |x_k^*(x)| \cdot \|Tx_k\| \leqslant \sum_{k=1}^n \|x_k^*\| \cdot \|Tx_k\| \cdot \|x\|,$$

for all $x \in X$, i.e., $\|T\| \leqslant \sum_{k=1}^n \|x_k^*\| \cdot \|Tx_k\| < \infty$. $\qquad\square$

Exercises

1. On any infinite-dimensional normed space there exists a discontinuous linear functional.

2. Consider $\ell_1^{(2)}$, the two-dimensional analogue of the space ℓ_1. That is, $\ell_1^{(2)}$ is the space of vectors $\bar{x} = (x_1, x_2)$, equipped with the norm $\|\bar{x}\| = |x_1| + |x_2|$. Show that the supporting functional at the point $\bar{x}_0 = (1, 0)$ is not unique. Describe all supporting functionals at this point.

3. Take for the normed space X the space $\mathbb{R}^2$, endowed with some norm. The unit sphere with respect to this norm is a convex closed curve γ in $\mathbb{R}^2$. Prove the equivalence of the following conditions: (1) at each nonzero point of the space X there exists a unique supporting functional; (2) the curve γ has at each of its points a unique tangent line.

9.2.2   The Annihilator of a Subspace

Let A be a subset of the normed space X. The *annihilator of the subset* A is the set of functionals

$$A^{\perp} = \{f \in X^* : f(y) = 0 \text{ for all } y \in A\}.$$

Theorem 1. *$A^{\perp}$ is a closed subspace of the normed space X^*.*

Proof. Let $f_1, f_2 \in A^{\perp}$. Then for any $y \in A$ and any λ_1, λ_2 we have $(\lambda_1 f_1 + \lambda_2 f_2)(y) = \lambda_1 f_1(y) + \lambda_2 f_2(y) = 0$, so $\lambda_1 f_1 + \lambda_2 f_2 \in A^{\perp}$. Linearity is established. Let us prove that $A^{\perp}$ is closed. Let $f_1, f_2, f_3, \ldots \in A^{\perp}$, $f = \lim_{n\to\infty} f_n$, and $y \in A$. Then

$$|f(y)| = |f(y) - f_n(y)| = |(f - f_n)(y)| \leqslant \|f - f_n\| \cdot \|y\| \to 0 \quad \text{as } n \to \infty,$$

and so $f \in A^{\perp}$. Therefore, a limit of functionals from $A^{\perp}$ itself lies in $A^{\perp}$.   □

Let us list the simplest properties of the annihilator.

Theorem 2. (1) *If $A \subset B$, then $A^{\perp} \supset B^{\perp}$.*

(2) $A^{\perp} = (\text{Lin } A)^{\perp}$.

(3) *Let $\overline{B}$ denote the closure of the set B. Then $(\overline{B})^{\perp} = B^{\perp}$.*

(4) $A^{\perp} = (\overline{\text{Lin }} A)^{\perp}$.

Proof. (1) If $f \in B^{\perp}$, then f annihilates all elements of the set B, and hence all elements of the set A.

(2) The inclusion $A^{\perp} \supset (\text{Lin } A)^{\perp}$ follows from the first property. To prove the opposite inclusion, let $f \in A^{\perp}$, and let $x = \sum_{k=1}^{n} \lambda_k x_k$ be an arbitrary linear combination of elements of the set A. Then $f(x) = \sum_{k=1}^{n} \lambda_k f(x_k) = 0$. Hence, $f \in (\text{Lin } A)^{\perp}$.

(3) If the functional f vanishes on the whole set B and is continuous, then f also vanishes on $\overline{B}$. Therefore, $(\overline{B})^{\perp} \supset B^{\perp}$. The opposite inclusion follows from the first property.

(4) This assertion follows from properties (2) and (3).   □

Theorem 3. *For a closed subset Y of the normed space X, the following conditions are equivalent:*

1. $Y = X$.

2. $Y^{\perp} = \{0\}$.

Proof. We only need to prove that $2. \Longrightarrow 1.$ So suppose that condition 1. is not satisfied, i.e., Y is strictly included in X. Then the quotient space X/Y does not reduce to zero, and by Corollary 1 in Subsection 9.2.1, on X/Y there exists a non-zero continuous linear functional g. Let $q: X \to X/Y$ be the quotient mapping, $q(x) = [x]$. Define the functional f as the composition $f(x) = g(q(x))$. Since q is surjective and g is not identically zero, f is also not identically zero. At the same time, $f \in Y^{\perp}$, so we have reached a contradiction. □

Exercises

1. Let A and B be subsets of the normed space X. Then $(A \cup B)^{\perp} = A^{\perp} \cap B^{\perp}$.

2. Extend the result of the preceding exercise to the annihilator of the union of an arbitrary (even infinite) number of sets.

3. Show that $A^{\perp}$ is closed in X^* not only in the sense of convergence in norm, but also in the sense of pointwise convergence.

4. Give an example of two subsets A and B for which $(A \cap B)^{\perp} \neq A^{\perp} \cup B^{\perp}$.

5. Prove that $(A \cap B)^{\perp} \supset A^{\perp} \cup B^{\perp}$. Deduce from this the inclusion $(A \cap B)^{\perp} \supset \overline{\mathrm{Lin}}\,(A^{\perp} \cup B^{\perp})$.

6. Give an example of two subsets A and B for which $(A \cap B)^{\perp} \neq \overline{\mathrm{Lin}}\,(A^{\perp} \cup B^{\perp})$.

7. Give an example of two closed subspaces $A, B \subset X$ for which $(A \cap B)^{\perp} \neq \overline{\mathrm{Lin}}\,(A^{\perp} \cup B^{\perp})$.

9.2.3 Complete Systems of Elements

A subset A of the normed space X is called a *complete system of elements of the normed space X* if the closure of the linear span of the set A coincides with the whole space X.[1]

Complete systems arise in various problems of mathematical analysis, when certain functions are to be approximated by simpler ones. For example, Weierstrass' theorem asserting that the set of polynomials is dense in the space of continuous functions on an interval can be stated as follows: the sequence of power functions $\{1, t, t^2, \ldots\}$ is complete in $C[a, b]$. In the theory of trigonometric series one establishes the completeness in the (complex) space $C[0, 2\pi]$ of the systems $\{e^{ikt}\}_{k=-\infty}^{\infty}$ and $\{1, \cos t, \sin t, \cos 2t, \ldots\}$. Important examples of complete system are encountered in mathematical physics as systems of eigenfunctions of various differential operators.

---

[1]Not to be confused with the notion of a complete system of elements of a linear space (see Subsection 5.1.1), the definition of which does not include the word "closure".

Theorem 3 of the preceding subsection can be recast as the following *completeness criterion* for systems of elements.

Theorem 1. *Let X be a normed space. A set $A \subset X$ is a complete system of elements if and only if $A^\perp = \{0\}$.*

Proof. By item (4) in Theorem 2 of Subsection 9.2.2, the condition $A^\perp = \{0\}$ is equivalent to the condition $(\overline{\mathrm{Lin}}\, A)^\perp = \{0\}$, which in turn, by Theorem 3 of Subsection 9.2.2, is equivalent to the equality $\overline{\mathrm{Lin}}\, A = X$, i.e., the completeness of the system of elements A.   □

This criterion often allows one to reduce the question of the completeness of a system of elements in a complex normed space to problems in the theory of functions of a complex variable. This is precisely the way in which one proves Müntz's theorem asserting the completeness of the system of power functions (let $b > a > 0$, $\lambda_k > 0$; then the system $\{f_k(t) = t^{\lambda_k}\}_{k=1}^\infty$ is complete in $C[a, b]$ if and only if $\sum_{k=1}^\infty \lambda_k^{-1} = \infty$); Levinson's theorem [27] asserting the completeness of the system of exponential functions (see the monograph of B. Ya. Levin [26], Appendix 3); and many other results.

Example 1. Let $b > a > 0$, $\lambda_k \in \mathbb{C}$, $k = 1, 2, \ldots$, suppose the sequence (λ_k) has a limit point and define $f_k(t) = t^{\lambda_k}$, $k = 1, 2, \ldots$. Then the system $\{f_k\}_{k=1}^\infty$ is complete in $C[a, b]$.

Proof. Consider a functional $x^* \in (C[a, b])^*$ that annihilates all f_k. Next, let the function $g_z \in C[a, b]$ be given by $g_z(t) = t^z$ and consider the function of a complex variable $F(z) = x^*(g_z)$. We claim that F is holomorphic for all $z \in \mathbb{C}$. Indeed,

$$\frac{F(z + \Delta z) - F(z)}{\Delta z} = x^*\left(\frac{g_{z+\Delta z} - g_z}{\Delta z}\right).$$

Since as $\Delta z \to 0$ the function $(g_{z+\Delta z}(t) - g_z(t))/\Delta z = (t^{z+\Delta z} - t^z)/\Delta z$ converges uniformly on $[a, b]$ to $f(t) := \partial(t^z)/\partial z = t^z \ln t$, and since the functional x^* is continuous precisely with respect to uniform convergence, we see that

$$\frac{F(z + \Delta z) - F(z)}{\Delta z} \to x^*(f) \quad \text{as } \Delta z \to 0.$$

Hence, holomophicity is established. Further, by construction, $F(\lambda_k) = x^*(g_{\lambda_k}) = x^*(f_k) = 0$. That is, the holomorphic function F vanishes on a sequence that has a limit in its domain of holomorphy. By the uniqueness theorem, $F(z) \equiv 0$. In particular, $F(n) = 0$, $n = 0, 1, 2, \ldots$. This means that the functional x^* annihilates all the elements of the complete system of functions $\{1, t, t^2, \ldots\}$. Hence, $x^* = 0$. We have thus proved that the annihilator of the system $\{f_k\}_{k=1}^\infty$ reduces to the null functional. Therefore, by the proved criterion, the system $\{f_k\}_{k=1}^\infty$ is complete.   □

Exercises

1. A normed space is separable if and only if it contains a countable complete system of elements.

2. The system $\{f_k\}_{k=1}^{\infty} \subset C[a, b]$ from the example given above has the following unusual property called *overcompleteness*: every infinite subsystem of it is again complete.

9.3 Convex Sets and the Hahn–Banach Theorem in Geometric Form

Throughout this section X will be a real normed space, and A, B will be nonempty subsets of X. Accordingly, all linear functionals will be assumed to be real.

9.3.1 Some Lemmas

Lemma 1. 1. *If A is an open set, then for any $b \in X$ the set $A + b$ is also open; if A is a neighborhood of a point $x \in X$, then $A + b$ is a neighborhood of the point $x + b$.*

2. *If $\lambda \neq 0$ and A is a neighborhood of the point $x \in X$, then λA is a neighborhood of the point λx.*

Proof. 1. The (parallel) translation map $x \mapsto x + b$ is bijective and preserves the distance between elements. Hence, translation takes balls into balls (of the same radius), and open sets into open sets.

2. The homothety $x \mapsto \lambda x$ is bijective and multiplies the distance between elements by the factor $|\lambda|$. Hence, it takes balls again into balls, even though the radius changes. $\square$

Lemma 2. *If A is an open set and $B \subset X$, then the set $A + B$ is also open.*

Proof. $A + B = \bigcup_{b \in B} (A + b)$, so $A + B$ is represented as a union of open sets. $\square$

Lemma 3. *If A and B are convex, then so is $A + B$.*

Proof. Let $x_1, x_2 \in A + B$ and $\lambda \in [0, 1]$. By the definition of the sum of two sets, there exist elements $a_1, a_2 \in A$ and $b_1, b_2 \in B$ such that $x_1 = a_1 + b_1$ and $x_2 = a_2 + b_2$. Consequently,

$$\lambda x_1 + (1 - \lambda)x_2 = \lambda(a_1 + b_1) + (1 - \lambda)(a_2 + b_2)$$

$$= (\lambda a_1 + (1 - \lambda)a_2) + (\lambda b_1 + (1 - \lambda)b_2) \in A + B. \qquad \square$$

Lemma 4. *Let A be convex, $a_1, a_2 \in A$, $\lambda \in (0, 1)$, and suppose a_1 is an interior point of A. Then the convex combination $\lambda a_1 + (1 - \lambda)a_2$ is also an interior point of A.*

Proof. The set A is a neighborhood of the point a_1. By Lemma 1, λA is a neighborhood of the point λa_1, and $\lambda A + (1 - \lambda)a_2$ is a neighborhood of the point $\lambda a_1 + (1 - \lambda)a_2$. At the same time, in view of the convexity of the set A, $\lambda A + (1 - \lambda)a_2 \subset \lambda A + (1 - \lambda)A \subset A$. Hence, we have found a neighborhood of the point $\lambda a_1 + (1 - \lambda)a_2$ that is entirely contained in A. $\qquad \square$

Corollary 1. *The interior of a convex set A is convex.*

Proof. It suffices to apply Lemma 4 to the case when both a_1 and a_2 are interior points. $\qquad \square$

Corollary 2. *If the interior $\overset{\circ}{A}$ of the convex set A is not empty, then $\overset{\circ}{A}$ is dense in A.*

Proof. Let a be an arbitrary point of A. Fix $x \in \overset{\circ}{A}$. By Lemma 4, the whole segment $\lambda a + (1 - \lambda)x$ with $\lambda \in (0, 1)$ consists of interior points. Since $\lambda a + (1 - \lambda)x \to a$ as $\lambda \to 1$, we conclude that the point a belongs to the closure of the set $\overset{\circ}{A}$. $\qquad \square$

Lemma 5. *Suppose the linear functional f on X does not vanish identically. Let $\theta \in \mathbb{R}$, let $A \subset X$ be a set with nonempty interior, and suppose that $f(a) \leqslant \theta$ for all $a \in A$. Then for all $x \in \overset{\circ}{A}$ the strict inequality $f(x) < \theta$ holds. In particular, if A is open, then $f(x) < \theta$ for all $x \in A$.*

Proof. By hypothesis, there exists a vector $e \in X$ such that $f(e) > 0$. Let $x \in A$ be an interior point. Choose an $\varepsilon > 0$ small enough for the point $x + \varepsilon e$ to lie in A. Then $f(x) < f(x) + \varepsilon f(e) = f(x + \varepsilon e) \leqslant \theta$. $\qquad \square$

We note that in Lemma 5 the functional may be discontinuous, and the assumption that the set A is open can be replaced by the more general algebraic assumption that for every point $x \in A$ the set $A - x$ is absorbing.

9.3.2 The Separation Theorem for Convex Sets

All the statements collected in this subsection can be regarded as generalizations of the following statement from ordinary geometry: Let A and B be disjoint convex subsets in the plane. Then one can draw a line l such that A and B lie on different sides with respect to l. To separate bodies in three-dimensional space we already need a plane instead of a line. In higher dimensions (in particular, in an infinite-dimensional space), the role of the separating line or plane is played by a hyperplane — a constant level set of a linear functional. The reader can find more about hyperplanes in Subsection 5.3.3 and the exercises therein.

Lemma 1. *Let A be an open convex subset of the normed space X and $x_0 \in X \setminus A$. Then there exists a functional $f \in X^* \setminus \{0\}$, such that $f(a) \leqslant f(x_0)$ for all $a \in A$.*

Proof. First we prove the lemma under the additional assumption that A contains the zero element of the space X. In this case A will also contain some ball $r B_X$, and so is an absorbing set. Accordingly, the Minkowski functional φ_A of the set A is a convex functional (see Subsection 5.4.2). The inclusion $r B_X \subset A$ means that $\varphi_A(x) \leqslant \|x\|/r$. As in the proof of Theorem 1 of Subsection 9.2.1, we consider the space $Y = \mathrm{Lin}\{x_0\}$. We define on Y a linear functional f such that $f(x_0) = \varphi_A(x_0)$. We claim that on Y the linear functional f is majorized by the convex functional φ_A. Indeed, for $\lambda \geqslant 0$ the positive homogeneity of the Minkowski functional implies that $f(\lambda x_0) = \varphi_A(\lambda x_0)$. At the same time, $f(-\lambda x_0) = -f(\lambda x_0) = -\varphi_A(\lambda x_0) \leqslant 0 \leqslant \varphi_A(-\lambda x_0)$. Hence, the majorization condition $f(t x_0) \leqslant \varphi_A(t x_0)$ is proved for positive as well as negative t, i.e., it is satisfied for all elements of the space Y.

Now using the analytic form of the Hahn–Banach theorem, we extend f to the entire space X with preservation of the linearity and the majorization condition. The majorization condition means, in particular, that $f(x) \leqslant \varphi_A(x) \leqslant \|x\|/r$, i.e., $f \in X^*$. Recall that, by the definition of the Minkowski functional, $\varphi_A(a) \leqslant 1$ for all $a \in A$, and since $x_0 \notin A$, we have $\varphi_A(x_0) \geqslant 1$. Combining these conditions, we see that $f(a) \leqslant \varphi_A(a) \leqslant 1 \leqslant \varphi_A(x_0) = f(x_0)$ for all $a \in A$. Moreover, f does not vanish identically since, as we just verified, $f(x_0) \geqslant 1$.

Thus, the lemma is established under the additional assumption that $0 \in A$. The general case reduces to the one already treated by translation. Namely, suppose $a_0 \in A$. Consider the auxiliary set $B = A - a_0$. Then B is a convex open set which contains 0, and $x_0 - a_0 \notin B$. As we already proved, there exists a functional $f \in X^* \setminus \{0\}$ such that $f(b) \leqslant f(x_0 - a_0)$ for all $b \in B$. Putting $b = a - a_0$, where $a \in A$, in the last inequality, we obtain $f(a) - f(a_0) = f(a - a_0) \leqslant f(x_0 - a_0) = f(x_0) - f(a_0)$, i.e., $f(a) \leqslant f(x_0)$. □

Theorem 1 (Hahn–Banach theorem — geometric form). *Let A and B be disjoint convex sets of the normed space X and let A be open. Then there exist a functional $f \in X^* \setminus \{0\}$ and a scalar $\theta \in \mathbb{R}$ such that $f(a) < \theta$ for all $a \in A$ and $f(b) \geqslant \theta$ for all $b \in B$.*

Proof. Consider the auxiliary set $C = A - B$. By Lemmas 2 and 3 of Subsection 9.3.1, C is open and convex. Since A and B are disjoint, $0 \notin C$. Applying the last lemma to the set C and the point $x_0 = 0$, we conclude that there exists a functional $f \in X^* \setminus \{0\}$ such that $f(a - b) \leqslant 0$ for all $a \in A$ and $b \in B$. Put $\theta = \sup_{a \in A} f(a)$. Since the inequality $f(a) \leqslant f(b)$ holds for all $a \in A$ and $b \in B$, we also have $\theta = \sup_{a \in A} f(a) \leqslant f(b)$ for all $b \in B$. The fact that $f(a) < \theta$ for all $a \in A$ follows from the obvious inequality $f(a) \leqslant \theta$ and Lemma 5 of Subsection 9.3.1. □

Corollary 1. *Suppose A and B are disjoint convex subsets of the normed space X and A has nonempty interior. Then there exist a functional $f \in X^* \setminus \{0\}$ and a scalar $\theta \in \mathbb{R}$ such that $f(a) \leqslant \theta$ for all $a \in A$ and $f(b) \geqslant \theta$ for all $b \in B$. Moreover, at the interior points of the set A there holds the strict inequality $f(x) < \theta$.*

Proof. The interior $\overset{\circ}{A}$ of the set A is a convex open set. It remains to use the preceding theorem and the fact that $\overset{\circ}{A}$ is dense in A.   □

By a direct application of Lemma 5 of Subsection 9.3.1, we deduce from the main theorem the following

Corollary 2. *Let A and B be a disjoint convex open subsets of the normed space X. Then there exist a functional $f \in X^* \setminus \{0\}$ and a scalar $\theta \in \mathbb{R}$ such that $f(a) < \theta$ for all $a \in A$ and $f(b) > \theta$ for all $b \in B$.*   □

Finally, we have

Corollary 3. *Let A and B be a disjoint convex closed subsets of the normed space X and suppose that one of these sets is compact. Then there exist a functional $f \in X^* \setminus \{0\}$ and a scalar $\theta \in \mathbb{R}$ such that $f(a) < \theta$ for all $a \in A$ and $f(b) > \theta$ for all $b \in B$.*

Proof. Denote by $r = \inf_{a \in A, b \in B} \|a - b\|$ the *distance between the sets A and B*. From the hypotheses it follows that $r > 0$. Consider the auxiliary sets $A + \frac{r}{3} B_X$ and $B + \frac{r}{3} B_X$, i.e., the $\frac{r}{3}$-neighborhoods of the sets A and B, respectively. These auxiliary sets are open, convex, and disjoint, so the preceding corollary applies.   □

9.3.3 Examples

The fact that some of the conditions imposed on the sets A and B in the formulation of the geometric form of the Hahn–Banach theorem are essential is obvious. For instance, the sets cannot be separated if they intersect. Further, if for one of the sets one takes a circle, and for the other the center of this circle, then it becomes clear why the theorem is not valid for non-convex sets. At the same time, the importance of the assumptions of topological character, such as, say, some of the sets being open, closed, or compact, is not that obvious. Below we provide examples that demonstrate the role of such conditions.

Example 1. Consider in the plane $\mathbb{R}^2$ the sets $A = \{(x, y) : y \leqslant 0\}$ (the lower half-plane) and $B = \{(x, y) : x > 0, y \geqslant \frac{1}{x}\}$ (the part of the first quadrant lying over the graph $y = 1/x$). Then A and B are closed and disjoint, but to strictly separate them by a line so that neither A nor B will intersect that line is impossible. That is, one cannot drop the assumption that one of the sets in Corollary 3 of Subsection 9.3.2 is compact. True, in this example the abscissa axis separates the sets in the sense of the geometric form of the Hahn–Banach theorem: the sets lie on different sides of the line, and only one of them intersects the line. In the next example even this type of separation fails.

Example 2. Consider in the plane $\mathbb{R}^2$ the sets $A = \{(x, 0) : x \geqslant 0\}$ (the closed positive half-line of the abscissa axis) and $B = \{(x, y) : y < 0\} \cup \{(x, 0) : x < 0\}$ (the union of the open lower half-plane and the strictly negative half-line of the abscissa axis). These are disjoint convex sets, one of which is closed. As the same time, they cannot be strictly separated by a line: the only line with respect to which A and B lie (non-strictly) on different sides is the abscissa axis, but both sets intersect this axis.

Finally, let us give an example of disjoint convex sets for which no hyperplane that separates them even non-strictly exists. Such an example can only live in an infinite-dimensional space.

Example 3. Consider the linear space $\mathcal{P}$ of all polynomials with real coefficients. Take for A the set of all polynomials with strictly negative leading coefficient, and for B the set of all polynomials with non-negative ($\geqslant 0$). These sets are convex and disjoint. We will show that for any non-zero linear functional f on $\mathcal{P}$ there exists no scalar $\theta \in \mathbb{R}$ such that $f(a) \leqslant \theta$ for all $a \in A$ and $f(b) \geqslant \theta$ for all $b \in B$. First we note that the monomials $p_n(t) = t^n$, $n = 0, 1, 2, \ldots$, form a Hamel basis in $\mathcal{P}$. Hence, a functional f is uniquely determined by its values on p_n; we denote $f(p_n)$ by f_n.

Suppose that $f(a) \leqslant \theta$ for all $a \in A$ and $f(b) \geqslant \theta$ for all $b \in B$. Then, in particular, since $0 \in B$, it follows that $0 = f(0) \geqslant \theta$, i.e., $\theta \leqslant 0$. Next, for any $\varepsilon > 0$ we have $\varepsilon\, p_0 \in A$ and $\varepsilon\, f_0 = f(\varepsilon\, p_0) \leqslant \theta$. Letting $\varepsilon \to 0$, we conclude that $\theta \geqslant 0$, i.e., $\theta = 0$. Further, each of the monomials p_n lies in B, and so $f_n = f(p_n) \geqslant 0$. On the other hand, for any $\varepsilon > 0$ we have $p_n - \varepsilon p_{n+1} \in A$. Consequently, $f_n - \varepsilon\, f_{n+1} = f(p_n - \varepsilon\, p_{n+1}) \leqslant 0$, whence, upon letting $\varepsilon \to 0$, we obtain that $f_n \leqslant 0$. Therefore, all f_n are equal to 0, and so $f = 0$.

Exercises

1. Verify that in Examples 1–3 of Subsection 9.3.3 the sets A and B are convex.

A family of sets is said to be *linked* if any two members of the family intersect. We say that a Banach space X has the *linked balls property* if every linked family of non-empty closed balls (with arbitrary centers and arbitrary radii) has non-empty intersection. Prove that

2. The real line $\mathbb{R}$ has the linked balls property, but $\mathbb{C}$ does not.

3. The real space ℓ_∞ has the linked balls property.

A Banach space E is said to be 1-*injective* if for operators acting in E, the following analogue of the Hahn–Banach extension theorem holds: for any subspace Y of an arbitrary normed space X and any operator $T \in L(Y, E)$, there exists an operator $\widetilde{T} \in L(X, E)$ such that $\widetilde{T}(y) = T(y)$ for all $y \in Y$ and $\|\widetilde{T}\| = \|T\|$.

If in this last definition we remove the condition $\|\widetilde{T}\| = \|T\|$, then we arrive at the more general concept of *injective space*.

4. If the Banach space E has the linked balls property, then E is 1-injective (I. Nachbin, 1950). In particular, the space ℓ_∞ is 1-injective.

5. Prove the injectivity of the space ℓ_∞ without using the linked balls property, based only on the Hahn–Banach extension theorem.

6. Suppose that in the plane there are given N convex closed bounded sets, any three of which intersect. Then all N sets have a common point (E. Helly, 1936).

7. Show that in the preceding exercise the assumption that the sets are closed can be discarded.

8. Suppose that in the plane there is given an infinite family of convex closed sets, one of which is bounded, and with the property that any three sets of the family intersect. Then the whole family has a non-empty intersection.

9. Give an example showing that in the formulation of the preceding statement the boundedness assumption cannot be discarded.

10. Formulate and prove a version of Helly's theorem for sets in n-dimensional space.

11. Using Helly's theorem, solve Exercise 10 of Subsection 1.3.3: in the Euclidean plane any set A of unit diameter can be included in a disc U of radius $1/\sqrt{3}$ (Jung's Theorem).

9.4   Adjoint Operators

9.4.1   The Connection Between Properties of an Operator and Those of Its Adjoint

Let X and E be normed spaces, and $T \in L(X, E)$. The *adjoint* (also referred to in the literature as the *dual*, or *conjugate*) of the operator T is the operator $T^* : E^* \to X^*$ that sends each functional $f \in E^*$ into the functional $T^* f = f \circ T$. In other words, the functional $T^* f \in X^*$ acts according to the rule $(T^* f)(x) = f(Tx)$.

Lemma 1. *For any element $e \in E$ we have $\|e\| = \sup_{f \in S_{E^*}} |f(e)|$.*

Proof. For every $f \in S_{E^*}$, it holds that $|f(e)| \leqslant \|f\| \cdot \|e\| = \|e\|$, and so $\sup_{f \in S_{E^*}} |f(e)| \leqslant \|e\|$. To obtain the opposite inequality, we use the existence of a supporting functional f_0 at the point e. By the definition of a supporting functional (Subsection 9.2.1), $f_0 \in S_{E^*}$ and $f_0(e) = \|e\|$. We have $\sup_{f \in S_{E^*}} |f(e)| \geqslant |f_0(e)| = \|e\|$. □

Theorem 1. *The operator T^* is continuous and $\|T^*\| = \|T\|$.*

Proof. Indeed,

$$\|T^*\| = \sup_{f \in S_{E^*}} \|T^* f\| = \sup_{f \in S_{E^*}} \sup_{x \in S_X} |(T^* f)(x)|$$

$$= \sup_{x \in S_X} \sup_{f \in S_{E^*}} |f(Tx)| = \sup_{x \in S_X} \|Tx\| = \|T\|. \qquad □$$

Theorem 2. *The image and kernel of the operators T and T^* are connected by the following relations:*

(1) $\operatorname{Ker} T^* = (T(X))^\perp$;
(2) $T^*(E^*) \subset (\operatorname{Ker} T)^\perp$.

Proof. (1)

$$(f \in \operatorname{Ker} T^*) \Longleftrightarrow (T^* f = 0) \Longleftrightarrow ((T^* f)x = 0 \ \forall x \in X)$$
$$\Longleftrightarrow (f(Tx) = 0 \ \forall x \in X) \Longleftrightarrow (f \in (T(X))^\perp).$$

(2) Let $g \in T^*(E^*)$, i.e., $g = T^* f$ for some $f \in E^*$. Then for every $x \in \operatorname{Ker} T$ we have $g(x) = (T^* f)(x) = f(Tx) = 0$, i.e., $g \in (\operatorname{Ker} T)^\perp$. □

Corollary 1. *For the operator T^* to be injective it is necessary and sufficient that T have dense image. In particular, if T is surjective, then T^* is injective.* □

Corollary 2. *If the operator T^* is surjective, then T injective.* □

To prove Corollary 1, it suffices to apply the first part of the preceding Theorems 2 and 3 of Subsection 9.2.2. Corollary 2 follows from the second part of Theorem 2.

Exercises

1. Let X, Y and Z be normed spaces, $T_1 \in L(X, Y)$ and $T_2 \in L(Y, Z)$. Then $(T_2 T_1)^* = T_1^* T_2^*$.

2. Give an example of an operator $T \in L(X, E)$ for which $T^*(E^*) \neq (\operatorname{Ker} T)^{\perp}$.

3. If the operator T^* has dense image, then T is injective. The converse assertion is not true. (In Chap. 17 we will see that the injectivity of the operator T is equivalent to the image of T^* being dense, not in the norm topology, but instead in the weak* topology $\sigma(X^*, X)$ introduced in that chapter.)

4. Let $T \in L(X, Y)$ be a bijective operator, $T^{-1} \in L(Y, X)$. Then $(T^{-1})^* = (T^*)^{-1}$.

9.4.2   The Duality Between Subspaces and Quotient Spaces

Let X be a normed space and Y be a subspace of X. Consider the operator $R: X^* \to Y^*$ (the restriction operator) which sends each functional $f \in X^*$ to its restriction to the subspace Y. Since every functional given on Y can be extended to a functional on X, the operator R is surjective. The kernel of R coincides with $Y^{\perp}$. Denote by U the injectivization of the operator R. By the definition of the injectivization, $U \in L(X^*/Y^{\perp}, Y^*)$, and if $f \in X^*$ and $[f]$ is the corresponding element in the quotient space $X^*/Y^{\perp}$, the functional $U[f]$ acts on the element $y \in Y$ by the rule $(U[f])(y) = f(y)$. The operator U is bijective (being the injectivization of a surjective operator) and is called the *canonical isomorphism of the spaces $X^*/Y^{\perp}$ and Y^**.

Theorem 1. *The canonical isomorphism of the spaces $X^*/Y^{\perp}$ and Y^* is an isometry, i.e., $\|U[f]\| = \|[f]\|$ for all $[f] \in X^*/Y^{\perp}$.*

Proof. $U[f]$ is a continuous linear functional given on the space Y. By the Hahn–Banach theorem, $U[f]$ admits an extension g to the whole space X with $\|g\| = \|U[f]\|$. Since the functionals g and f coincide on Y, $[f] = [g]$. We have $\|U[f]\| = \|g\| \geqslant \|[g]\| = \|[f]\|$. Conversely, the restriction operator R does not increase the norm of functionals, i.e., $\|R\| \leqslant 1$. Since U is the injectivization of the operator R, we also have $\|U\| \leqslant 1$. Therefore, $\|U[f]\| \leqslant \|[f]\|$.   □

The equivalence class $[f]$ is often identified with the functional $U[f]$, and one says that $[f]$ acts on the element $y \in Y$ by the rule $[f](y) = f(y)$. In the framework of this convention one can say that $X^*/Y^{\perp}$ and Y^* are one and the same space: $X^*/Y^{\perp} = Y^*$.

An analogous description holds for the dual of a quotient space. The reader will obtain this description, expressed by the equality $(X/Y)^* = Y^{\perp}$, by solving the exercises below.

Exercises

Let $q: X \to X/Y$ be the quotient mapping (i.e., $q(x) = [x]$ for all $x \in X$), and let $q^*: (X/Y)^* \to X^*$ be the adjoint operator. Show that:

1. The operator q^* acts by the rule $(q^* f)(x) = f([x])$.

2. The image of the operator q^* coincides with $Y^\perp$.

3. The operator q^* effects a bijective isometry between the spaces $(X/Y)^*$ and $Y^\perp$.

Let j be the natural operator which embeds the subspace Y in the ambient space X ($j(y) = y$ for all $y \in Y$), and $j^*: X^* \to Y^*$ be its adjoint.

4. Verify that j^* coincides with the restriction operator R.

Comments on the Exercises

Subsection 9.1.2

Exercise 3. Theorem 4 in Subsection 16.2.3 gives this result in the general setting of topological vector spaces.

Exercise 6. First, it is sufficient to consider $[a, b] = [0, 1]$. If a continuous function vanishes on a set, then it also vanishes on the closure of that set. Hence, with no loss of generality we can assume that the set Δ is closed. Suppose $\lambda(\Delta) \neq 0$, where λ is the Lebesgue measure on $[0, 1]$. Then the formula $F_\Delta(f) = \int_\Delta f \, d\lambda$ defines a non-zero continuous functional on the space $C_p[0, 1]$; Ker F_Δ is not a dense set in $C_p[0, 1]$, and consequently $V_\Delta \subset$ Ker F_Δ is not dense. For the converse assertion, suppose $\lambda(\Delta) = 0$. Fix $\varepsilon > 0$ and an element $f \in C_p[0, 1]$. Let $C = \max\{|f(t)|: t \in [0, 1]\}$. The regularity of the Lebesgue measure gives us a closed subset $K \subset [0, 1] \setminus \Delta$ with $\lambda(K) > 1 - \varepsilon$. Consider on the closed set $\Delta \sqcup K$ the continuous function g that equals 0 on Δ and equals f on K. Extend g to a continuous function h on the whole segment $[0, 1]$ with preservation of the condition $|h| \leqslant C$. Then $h \in V_\Delta$ and $\|f - h\|^p \leqslant (2C)^p \varepsilon$, which by the arbitrariness of ε establishes the desired density of V_Δ in $C_p[0, 1]$.

Subsection 9.3.3

Exercise 2. Let $[a_\gamma, b_\gamma]$, $\gamma \in \Gamma$, be a family of pairwise intersecting intervals (the balls in $\mathbb{R}$ are intervals). Then for any $\gamma_1, \gamma_2 \in \Gamma$ one has $a_{\gamma_1} \leqslant b_{\gamma_2}$ (otherwise the corresponding intervals would not intersect). This means that the number $a = \sup_{\gamma \in \Gamma} a_\gamma$ lies to the right of all left, and to the left of all right endpoints of the segments $[a_\gamma, b_\gamma]$, i.e., it lies in the intersection of all these intervals.

Exercise 4. For the proof of this theorem of Nachbin and the main facts about 1-injective subspaces, the reader is referred to the textbook by L. Kantorovich and G. Akilov [22, Chapter 5, Section 8.3]. For injective spaces one can consult the book by J. Lindenstrauss and L. Tzafriri [28, Vol. 1, §2.f].

Exercise 6. One can proceed by induction on N. Let $A_1, \ldots, A_N$ be convex sets satisfying the assumption of the exercise. By the induction hypothesis, $B = \bigcap_{k=1}^{N-1} A_k$ is a nonempty closed convex set. Suppose the assertion is not true, i.e., B does not intersect A_N. Then there exists a line ℓ such that B and A_N lie strictly on different sides with respect to ℓ. Consider the sets $C_k = A_k \cap \ell$, $k = 1, 2, \ldots, N - 1$. Since each of the sets $A_k \cap A_j$, $1 \leqslant k \leqslant j \leqslant N - 1$, intersects both B and A_N, convexity implies that all the sets $A_k \cap A_j$, $1 \leqslant k \leqslant j \leqslant N - 1$, intersect the line ℓ. That is to say, the sets C_k are nonempty and any two of them intersect. Since C_k are closed segments on ℓ, it follows (see Exercise 2 and the comments to it) that the intersection of all sets C_k, $k = 1, 2, \ldots, N - 1$, is not empty. Then the set $B \cap \ell = \bigcap_{k=1}^{N-1} C_k$ is also nonempty, which contradicts our choice of the line ℓ.

Exercise 10. Statement: let $A_1, \ldots, A_N$ be convex, closed, bounded subsets of $\mathbb{R}^n$, each $n + 1$ of which intersect. Then all the N sets have a common point. The proof is completely analogous to the proof, given above, of the planar version of Helly's theorem.

For other versions of Helly's theorem and its applications one can consult the short book by Hadwiger and Debrunner [17].

Exercise 11. Consider the collection of all discs of radius $1/\sqrt{3}$ centered at points of A. Any common point of this collection can serve as the center of the required disc U.

Chapter 10
Classical Theorems on Continuous Operators

10.1 Open Mappings

Let X, Y be Banach spaces, and $T \in L(X, Y)$. We say that T is an *open operator*, or an *open mapping*, if the image $T(A)$ of any open set $A \subset X$ is an open set in Y.

Let us list a few elementary properties of open operators.

— Any open operator is surjective. Indeed, the image $T(X)$ of the operator is open in Y and forms a linear subspace. Hence, $T(X)$ contains the linear span of some ball in Y, i.e., $T(X) = Y$.

— If an open operator T is injective, then it is also bijective, and T^{-1} is a continuous operator. (This is an immediate consequence of the definition of continuity in terms of preimages of open sets.)

10.1.1 An Openness Criterion

Theorem 1. *The operator $T \in L(X, Y)$ is open if and only if the image $T(B_X)$ of the unit ball contains some ball $r B_Y$ with $r > 0$.*

Proof. The necessity of the condition in the theorem follows from the fact that the image $T(B_X)$ of the unit ball under an open mapping is an open set that contains the zero element of the space Y. Now let us prove the sufficiency. Let $A \subset X$ be an arbitrary open subset, and $x_0 \in A$. Pick $\alpha > 0$ such that the ball $B_X(x_0, \alpha) = x_0 + \alpha B_X$ is also contained in A. Then

$$T(A) \supset T x_0 + \alpha T(B_X) \supset T x_0 + \alpha r B_Y,$$

that is, any point $T x_0$ of the set $T(A)$ has an entire neighborhood contained in $T(A)$. Hence, the set $T(A)$ is open, and since A was arbitrary, this shows that the operator T is open. $\qquad \square$

© Springer International Publishing AG, part of Springer Nature 2018

V. Kadets, *A Course in Functional Analysis and Measure Theory*, Universitext, https://doi.org/10.1007/978-3-319-92004-7_10

Exercises

1. Let X be a normed space, X_1 a closed subspace of X. Then the quotient mapping $q\colon X \to X/X_1$ is an open operator.

2. Is the set of open operators acting from the normed space X into the normed space Y a linear subspace of $L(X, Y)$? Is this set closed in $L(X, Y)$? Open?

10.1.2   Ball-Like Sets

A subset A of a Banach space X is called *ball-like* if for any sequence of points $x_n \in A$ and any sequence of scalars λ_n that satisfy the condition $\sum_{n=1}^\infty |\lambda_n| \leqslant 1$, the series $\sum_{n=1}^\infty \lambda_n x_n$ converges to an element of A.

The verification of the properties of ball-like sets listed below is left to the reader.

1. Ball-like sets are bounded.
2. Any closed, convex, bounded, and balanced set in a Banach space is ball-like.
3. The open unit ball of a Banach space is ball-like. Hence, ball-like sets are not necessarily closed.
4. The image of a ball-like set under a continuous linear operator is again a ball-like set.

Theorem 2. *Suppose the closure $\overline{A}$ of the ball-like set A in the Banach space X contains the ball $r B_X$, where r is some positive number. Then the set A itself contains the ball $r B_X$.*

Proof. With no loss of generality, we can assume that $r = 1$ (one is reduced to this case by replacing the set A by $\frac{1}{r}A$). Fix $x \in B_X$. We claim that $x \in A$. Indeed, take an $\varepsilon > 0$ such that $\frac{1}{1-\varepsilon}x \in B_X$, and put $x_0 = \frac{1}{1-\varepsilon}x$. By hypothesis, $x_0 \in \overline{A}$. Now take a point $y_0 \in A$ that approximates x_0 to within ε, i.e., $\|x_0 - y_0\| < \varepsilon$. The vector $x_1 = x_0 - y_0$ lies in εB_X, which in turn is contained in $\varepsilon \overline{A}$. Pick $y_1 \in A$ such that $\|x_1 - \varepsilon y_1\| < \varepsilon^2$. Then the vector $x_2 = x_1 - \varepsilon y_1 = x_0 - y_0 - \varepsilon y_1$ lies in $\varepsilon^2 \overline{A}$. Continuing this process, we obtain vectors $y_n \in A$, such that $\|x_0 - y_0 - \varepsilon y_1 - \cdots - \varepsilon^n y_n\| < \varepsilon^{n+1}$. Then the series $\sum_{n=0}^\infty \varepsilon^n y_n$ converges to x_0. Hence, since the set A is ball-like, the vector $x = (1 - \varepsilon)x_0 = \sum_0^\infty (1 - \varepsilon)\varepsilon^n y_n$ lies in A, as we needed to prove. $\square$

Exercise

Let X be a separable Banach space. Using the properties of ball-like sets and Exercise 4 in Subsection 6.4.2, show that there exists a continuous linear operator $T\colon \ell_1 \to X$ for which $T\left(B_{\ell_1}\right) = B_X$. This implies the *quotient universality* of the space ℓ_1: for any separable Banach space X there exists a subspace $Y \subset \ell_1$ such that the quotient space ℓ_1/Y is isometric to the space X (see Exercise 7 in Subsection 6.4.2).

Remark 1. The idea of considering ball-like sets and using them in solving the exercise above, as well as the proof of the Banach open mapping theorem given below, are taken from a paper by T. Banakh, W.E. Lyantse, and Ya.V. Mykytyuk [46].

10.1.3 The Banach Open Mapping Theorem

Theorem 1. *Let* X, Y *be Banach spaces and* $T \in L(X, Y)$ *be a surjective operator. Then* T *is an open mapping.*

Proof. Consider the set $A = T(B_X)$. By the openness criterion for mappings (see Subsection 10.1.1), it suffices to show that A contains some open ball $r B_Y$ with $r > 0$. Since A is a ball-like set (see Subsection 10.1.2, properties 3 and 4), Theorem 2 of Subsection 10.1.2 shows that it suffices to prove that the closure $\overline{A}$ of the set A contains such a ball $r B_Y$. To do this, we resort to Baire's theorem.

We use the surjectivity of the operator T and write the space Y as

$$Y = T(X) = \bigcup_{n=1}^{\infty} T(n B_X) = \bigcup_{n=1}^{\infty} n A.$$

By Baire's theorem, A cannot be nowhere dense in Y, that is, $\overline{A}$ must contain some ball of the form $y_0 + r B_Y$. Since $\overline{A}$ is convex and symmetric, this implies that

$$\overline{A} \supset \frac{1}{2}\left(\overline{A} - \overline{A}\right) \supset \frac{1}{2}\left((y_0 + r B_Y) - (y_0 + r B_Y)\right) \supset r B_Y.$$

The theorem is proved. $\qquad\square$

In the literature the Open Mapping Theorem just proved is sometimes called the "Banach–Schauder[1] theorem".

---

[1]To Juliusz Schauder, a prominent member of the Lviv school of mathematics, we owe many fruitful ideas. For example, Schauder was the first to use Baire's theorem to prove the open mapping theorem; to him also belongs the theorem on the compactness of the adjoint operator (Theorem 3 in Subsection 11.3.2), which lies at the foundations of the theory of compact operators. Also, it is clearly hard to overestimate the importance of the fixed-point principle (Subsection 15.1.4) and of the concept of a Schauder basis (Subsection 10.5). Thus, whenever hearing Schauder's name, the reader should be ready to grasp something valuable for his mathematical culture.

10.2   Invertibility of Operators and Isomorphisms

10.2.1   Isomorphisms. Equivalent Norms

Definition 1. Let X and Y be normed spaces. A linear operator $T : X \to Y$ is called an *isomorphism* if it is continuous, bijective, and the inverse operator $T^{-1} : Y \to X$ is also continuous. The normed spaces X and Y are said to be *isomorphic* (denoted in this book as $X \approx Y$), if there exists an isomorphism $T : X \twoheadrightarrow Y$. A particular case of isomorphism, isometry, was considered earlier in Subsection 6.1.2.

As follows from the definition, an isomorphism preserves all topological structures: it maps opens sets into open sets, closed sets into closed sets, convergent sequences into convergent sequences. The next theorem exhibits yet another example of structure preserved by isomorphisms.

Theorem 1. *Let X and Y be isomorphic normed spaces and let X be complete. Then the space Y is also complete.*

Proof. By hypothesis, there exists an isomorphism $T : X \to Y$. To show that the space Y is complete, take an arbitrary Cauchy sequence $y_n \in Y$. Set $x_n = T^{-1} y_n$. Thanks to the linearity and continuity of the operator T^{-1}, the vectors x_n also form a Cauchy sequence:

$$\|x_n - x_m\| = \|T^{-1}(y_n - y_m)\| \leqslant \|T^{-1}\| \cdot \|y_n - y_m\| \to 0 \quad \text{as } n, m \to \infty.$$

Since the space X is complete, the sequence (x_n) has a limit, which we denote by x. In view of the continuity of the operator T, $Tx = \lim_{n \to \infty} Tx_n = \lim_{n \to \infty} y_n$. Hence, the sequence y_n has a limit. $\qquad\square$

Theorem 2. *Let X and Y be finite-dimensional normed spaces with $\dim X = \dim Y$. Then $X \approx Y$.*

Proof. Since X and Y have the same dimension, there exists a bijective linear operator $T : X \to Y$. By Theorem 2 of Subsection 9.2.1, every operator on a finite-dimensional space is continuous. In particular, the operators T and T^{-1} are continuous, i.e., T is an isomorphism. $\qquad\square$

Corollary 1. *Every finite-dimensional normed space is complete. Every finite-dimensional subspace of an arbitrary normed space is closed.*

Proof. Let X be a finite-dimensional normed space, $\dim X = n$. By the preceding theorem, the space $\mathbb{R}^n$ (and in the complex case, $\mathbb{C}^n$), equipped with the standard norm $\|x\| = \left(\sum_{k=1}^{n} |x_k|^2\right)^{1/2}$, is isomorphic to the space X. Since $\mathbb{R}^n$ is complete, Theorem 1 shows that so is X. The fact that every finite-dimensional subspace is closed is a particular case of the assertion that every complete subspace of a metric space is closed. $\qquad\square$

In contrast to the finite-dimensional spaces, many of the infinite-dimensional spaces introduced above in the text are pairwise not isomorphic. Thus, for instance, among the spaces $L_p[0, 1]$ and ℓ_q with $1 \leqslant p, q \leqslant \infty$ there are only two pairs of isomorphic spaces: $L_2[0, 1] \approx \ell_2$ (this fact will be established in Chapter 12), and $L_\infty[0, 1] \approx \ell_\infty$ (the proof of this by far not obvious theorem of A. Pełczyński can be found, e.g., in the first volume of the book by Lindenstrauss and Tzafriri [28, p. 111]).

Definition 2. Two norms, $\| \cdot \|_1$ and $\| \cdot \|_2$, on the linear space X are said to be *equivalent* (and then one writes $\| \cdot \|_1 \sim \| \cdot \|_2$) if there exist constants $C_1, C_2 > 0$ such that $C_1 \|x\|_1 \leqslant \|x\|_2 \leqslant C_2 \|x\|_1$ for all $x \in X$.

Theorem 3. *For two norms on a linear space X, $\| \cdot \|_1$ and $\| \cdot \|_2$, the following conditions are equivalent:*

(1) $\| \cdot \|_1 \sim \| \cdot \|_2$;

(2) *the identity operator I on X, regarded as an operator acting from the normed space $(X, \| \cdot \|_1)$ into the normed space $(X, \| \cdot \|_2)$, is an isomorphism;*

(3) *the norms $\| \cdot \|_1$ and $\| \cdot \|_2$ give the same topology on X.*

Proof. In order for the operator I, regarded as acting from $(X, \| \cdot \|_1)$ to $(X, \| \cdot \|_2)$, to be continuous, it is necessary and sufficient that there exist a constant $C_2 > 0$, such that $\|x\|_2 \leqslant C_2 \|x\|_1$ for all $x \in X$. Similarly, for the operator I^{-1} to be continuous, it is necessary and sufficient that there exist a constant $C_1 > 0$ such that $C_1 \|x\|_1 \leqslant \|x\|_2$ for all $x \in X$. This establishes the equivalence of conditions (1) and (2).

Condition (3) means that the spaces $(X, \| \cdot \|_1)$ and $(X, \| \cdot \|_2)$ have the same families of open sets. It can be reformulated as follows: a set A is open in $(X, \| \cdot \|_1)$ if and only if the set $I(A)$ is open in $(X, \| \cdot \|_2)$. But since the operator I is bijective, this is equivalent to the simultaneous continuity of the operators I and I^{-1}. $\square$

Theorem 4. *Let X be a finite-dimensional linear space. Then all the norms on X are equivalent.*

Proof. Since every operator on a finite-dimensional space is continuous, it follows that for any two norms $\| \cdot \|_1$ and $\| \cdot \|_2$ on X, the operator I figuring in item (2) of Theorem 3 above is an isomorphism. Hence, $\| \cdot \|_1 \sim \| \cdot \|_2$.

Exercises

1. On any collection of normed spaces the isomorphism relation $\approx$ is an equivalence relation.

2. Let X, Y be normed spaces. If the operator $T: X \to Y$ is an isomorphism, then the adjoint operator $T^*: Y^* \to X^*$ is also an isomorphism. For non-complete spaces the converse statement is not true (for the case of Banach spaces, see Exercises 4 and 5 in Subsection 10.2.3 below).

3. Show that on the set of all norms given on a fixed linear space, the equivalence of norms is a genuine equivalence relation.

4. On every infinite-dimensional linear space there exist non-equivalent norms.

5. For each pair among the three norms on $\mathbb{R}^n$ listed below, prove their equivalence and calculate the best possible constants C_1, C_2 appearing in the equivalence of those norms:

$$\|x\|_1 = \sum_{k=1}^n |x_k|, \quad \|x\|_2 = \left(\sum_{k=1}^n |x_k|^2 \right)^{1/2}, \quad \text{and} \quad \|x\|_\infty = \max_{1 \leqslant k \leqslant n} |x_k|.$$

6. For each pair among the three norms on $C[0, 1]$ listed below prove that they are not equivalent:

$$\|f\|_1 = \int_0^1 |f(t)| dt, \quad \|f\|_2 = \left(\int_0^1 |f(t)|^2 dt \right)^{1/2}, \quad \text{and} \quad \|f\|_\infty = \max_{t \in [0,1]} |f(t)|.$$

10.2.2 The Banach Inverse Operator Theorem

Theorem 1. *Let X, Y be Banach spaces and $T \in L(X, Y)$ be a bijective operator. Then the operator T^{-1} is continuous, i.e., T is an isomorphism.*

Proof. Since T, in particular, is surjective, it is an open operator. And as we already remarked, if an open operator is bijective, then T^{-1} is a continuous operator. □

This theorem is also referred to as the *Banach inverse mapping theorem*.

The inverse operator theorem admits the following useful reformulation: Let X, Y be Banach spaces, $T \in L(X, Y)$. Suppose that for any $b \in Y$ the equation $Tx = b$ has a solution, and this solution is unique. Then the solution depends continuously on the right-hand side. In other words, the solution is *stable* under small perturbations of the right-hand side.

Exercises

1. Using the injectivization of the operator (Subsection 5.2.3 and Exercises 5–6 in Subsection 6.4.2), deduce the open mapping theorem from the inverse operator theorem.

2. Suppose that on the linear space X there are given two norms, $\|\cdot\|_1$ and $\|\cdot\|_2$ such that $\|\cdot\|_1 \leqslant \|\cdot\|_2$, and the space is complete with respect to both these norms. Then $\|\cdot\|_1 \sim \|\cdot\|_2$.

3. In the infinite-dimensional Banach space X choose a linearly independent sequence of vectors $\{e_n\}_{n=1}^\infty$, and choose a set $A \subset X$, such that $A \cup \{e_n\}_{n=1}^\infty$ is a Hamel basis of X. Now define the operator $T : X \to X$ by the following rule: for $x \in A$ put $Tx = x$, and for the vectors $\{e_n\}_{n=1}^\infty$ put $Te_n = \frac{1}{n}e_n$, and then extend T to the whole space X by continuity. Prove that T is a bijective linear operator, but not an isomorphism. Which of the conditions of the inverse operator theorem does our operator T fail to satisfy?

4. On every infinite-dimensional Banach space there exists a norm that is not equivalent to the original norm, yet in which the space nevertheless remains complete.

The exercises given below show, in particular, that in the inverse operator theorem — and hence in the open mapping theorem as well — the completeness of the space is an essential condition.

5. Let $\mathcal{P}$ be the space of all polynomials (of arbitrarily large degree) with real coefficients, equipped with the norm $\|a_0 + a_1 t + \cdots + a_n t^n\| = |a_0| + |a_1| + \cdots + |a_n|$. Define the operator $T : \mathcal{P} \to \mathcal{P}$ by the formula

$$T\left(a_0 + a_1 t + \cdots + a_n t^n\right) = a_0 + \frac{a_1}{2}t + \cdots + \frac{a_n}{n+1}t^n.$$

Show that T is continuous, but T^{-1} is discontinuous.

6. In the setting of Exercise 6 in Subsection 10.2.1, verify that $\|\cdot\|_1 \leqslant \|\cdot\|_\infty$. Using the fact that these norms are not equivalent, show that in the inverse operator theorem one cannot just require that the space X be complete without imposing further constraints on the space Y.

7. In the infinite-dimensional real Banach space X, choose a Hamel basis A. Multiplying, if necessary, the elements of this basis by positive scalars, one can ensure that $A \subset B_X$. Now as the unit ball of a new norm $\|\cdot\|_1$ take the set $\mathrm{conv}(A \cup (-A))$. Find the explicit expression of the norm $\|\cdot\|_1$ in terms of the coefficients of the expansion in the Hamel basis A. Prove that $\|\cdot\|_1$ majorizes the original norm, but that with respect to $\|\cdot\|_1$ the space X is not complete. Based on this example, show that in the inverse operator theorem one cannot merely require that the space Y be complete without imposing further constraints on the space X.

10.2.3 Bounded Below Operators. Closedness of the Image Criterion

Let X, Y be normed spaces. The operator $T \in L(X, Y)$ is said to be *bounded below* if there exists a constant $c > 0$ such that $\|Tx\| \geqslant c\|x\|$ for all $x \in X$.

We immediately note that every bounded below operator is injective. Indeed, if $Tx = 0$ for some $x \in X$, then the inequality $0 = \|Tx\| \geqslant c\|x\|$ means that $x = 0$. The reader will find examples of injective, but not bounded below operators in the exercises.

Theorem 1. *The operator T is not bounded below if and only if there exists a sequence $x_n \in S_X$ such that $T x_n \to 0$ as $n \to \infty$.*

Proof. If the operator is bounded below with some constant $c > 0$ and $x_n \in S_X$, then since $\|T x_n\| \geqslant c\|x_n\| = c$, the images $T x_n$ of the elements x_n cannot converge to zero. Conversely, if the operator is not bounded below, then, in particular, it is not bounded below with constant $c = 1/n$. That is, for each n there exists a y_n such that $\|T y_n\| < \frac{1}{n}\|y_n\|$. Put $x_n = \frac{y_n}{\|y_n\|}$. This is the sought-for sequence: $x_n \in S_X$ and $\|T x_n\| < \frac{1}{n} \to 0$ as $n \to \infty$. $\square$

Theorem 2. *The operator T is bounded below if and only if T establishes an isomorphism between the normed spaces X and $T(X)$.*

Proof. If T is bounded below, then it is injective. Hence, regarded as an operator acting from X to $T(X)$, T is bijective. Let $c > 0$ be the constant in the definition of boundedness below. Then for any $y \in T(X)$ we have

$$\|T^{-1}y\| \leqslant \frac{1}{c}\|T(T^{-1}y)\| = \frac{1}{c}\|y\|,$$

i.e., the operator $T^{-1}\colon T(X) \to X$ is continuous. Conversely, if T is an isomorphism of the spaces X and $T(X)$, then there exists the operator $T^{-1}\colon T(X) \to X$ and $\|T^{-1}\| < +\infty$. Then for every $x \in X$ we have that $\|x\| = \|T^{-1}(Tx)\| \leqslant \|T^{-1}\|\|Tx\|$, i.e., T is bounded below with the constant $c = 1/\|T^{-1}\|$. $\square$

In view of the last theorem, bounded below operators are also called *isomorphic embeddings*.

Theorem 3. *Let X be a Banach space and let the operator $T \in L(X, Y)$ be bounded below. Then the image of T is closed in Y.*

Proof. By the preceding theorem, the subspace $T(X)$ is isomorphic to the space X. Hence, $T(X)$ is complete, and we know that complete subspaces are closed. $\square$

In Banach spaces, for injective operators the converse also holds.

Theorem 4. *If X and Y are Banach spaces and the operator $T \in L(X, Y)$ has closed image, then T is bounded below.*

Proof. Since a closed subspace of a complete space is itself complete, $T(X)$ is a Banach space. By the Banach inverse operator theorem, the operator T is an isomorphism of the spaces X and $T(X)$. ☐

Recall (see Subsection 5.2.3 and also Exercises 5 and 6 in Subsection 6.4.2) that the *injectivization* of the operator $T \in L(X, Y)$ is the operator $\widetilde{T} : X/\operatorname{Ker} T \to Y$ that acts on the elements of the quotient space by the rule $\widetilde{T}[x] = Tx$. The operator $\widetilde{T}$ is continuous and $\|\widetilde{T}\| = \|T\|$. Since the image of the injectivization coincides with the image of the original operator, we obtain the following assertion.

Corollary 1. *Let X and Y be Banach spaces. The operator $T \in L(X, Y)$ has closed image if and only if its injectivization $\widetilde{T}$ is bounded below. In other words, the image of the operator T is not closed if and only if there exists a sequence $[x_n] \in S_{X/\operatorname{Ker} T}$, such that $T x_n \to 0$ as $n \to \infty$.*

Using the fact that $\|[x_n]\| = \operatorname{dist}(x_n, \operatorname{Ker} T)$, we restate the last assertion without using the term "quotient space".

Theorem 5. *Let X and Y be Banach spaces. An operator $T \in L(X, Y)$ has a non-closed image if and only if there exists a sequence $x_n \in X$ with the following properties:*

(1) $\operatorname{dist}(x_n, \operatorname{Ker} T) = 1$;

(2) $\|T x_n\| \to 0$ *as $n \to \infty$.*

If the norm of an equivalence class is equal to 1, then in that class there are representatives with norm arbitrarily close to 1. Hence, one can add another condition:

(3) $\|x_n\| \to 1$ *as $n \to \infty$.* ☐

Exercises

Let X, E be Banach spaces and $T \in L(X, E)$.

1. For the operator T^* to be bounded below with a constant c it is necessary and sufficient that $T(B_X) \supset c B_E$.

2. For the operator T to be bounded below with constant c it is necessary and sufficient that $T^*(B_{E^*}) \supset c B_{X^*}$.

3. For the operator T^* to be surjective it is necessary and sufficient that the operator T be bounded below.

4. For the operator T to be surjective it is necessary and sufficient that the operator T^* be bounded below.

5. For the operator T to be an isomorphism it is necessary and sufficient that the operator T^* be an isomorphism.

6. For the operator $T \in L(X, E)$ to be non-invertible it is necessary and sufficient that one of the following mutually exclusive cases holds:

— T is not injective;

— T is injective, but not bounded below;

— T is bounded below, but not surjective.

7. The last of the cases listed in the preceding exercise means, in particular, that the operator T^* is not injective.

8. On the example of the integration operator $T : C[0, 1] \to C[0, 1]$, $(Tf)(t) = \int_0^t f(\tau)d\tau$, convince yourself that a continuous linear operator can be simultaneously injective and not bounded below. Compare this example with the result of Exercise 2 in Subsection 6.4.1.

Let $g \in C[0, 1]$ be a fixed function. Define the operator $T_g : C[0, 1] \to C[0, 1]$ of multiplication by the function g by the rule $T_g(f) = f \cdot g$. Verify that:

9. T_g is continuous and $\|T_g\| = \|g\|$.

10. T_g is injective if and only if the set $g^{-1}(0)$ has no interior points.

11. T_g is bounded below if and only if the function g is everywhere different from zero.

12. Now consider the multiplication operator T_g by the function $g \in C[0, 1]$ as an operator acting from $L_1[0, 1]$ to $L_1[0, 1]$. What is the norm of this operator equal to? What do the injectivity and boundedness below criteria look like in the present case? And the closedness of the image criterion? Do the answers change if $g \in L_\infty[0, 1]$?

10.3   The Graph of an Operator

10.3.1 · The Closed Graph Theorem

Let X and Y be normed spaces. Then their Cartesian product $X \times Y$ is a linear space with respect to componentwise addition and multiplication by scalars. We define a norm on $X \times Y$ by the formula $\|(x, y)\| = \|x\| + \|y\|$. It is readily verified that this expression satisfies the norm axioms; that convergence in this norm coincides with the component-wise convergence: $(x_n, y_n) \to (x, y)$ in $X \times Y$ if and only if $x_n \to x$ in X and $y_n \to y$ in Y; and that if X and Y are Banach spaces, then their Cartesian product $X \times Y$ is also a Banach space.

The *graph* of the linear operator $T : X \to Y$ is the set $\Gamma(T) = \{(x, Tx) : x \in X\}$. We leave it to the reader to verify that the graph of a linear operator is a linear subspace of the space $X \times Y$.

Theorem 1 (Closed graph theorem). *Let X and Y be Banach spaces. The linear operator $T : X \to Y$ is continuous if and only if its graph is closed in $X \times Y$.*

This theorem is usually applied in the following, more detailed formulation: T is continuous if and only if, for any sequence $x_n \in X$, if $x_n \to x$ in X and $Tx_n \to y$ in Y as $n \to \infty$, then $y = Tx$ (in other words, if the sequence (x_n, Tx_n) of points of the graph $\Gamma(T)$ converges to the point $(x, y) \in X \times Y$, then $(x, y) \in \Gamma(T)$).

Proof. Suppose T is continuous and $x_n \to x$. Then $Tx_n \to Tx$. If we also have $Tx_n \to y$, then $y = Tx$.

Conversely, suppose $\Gamma(T)$ is a closed subspace of the space $X \times Y$. Then $\Gamma(T)$ is a Banach space. Consider the auxiliary operator $U : \Gamma(T) \to X$ acting by the rule $U(x, Tx) = x$. Since $\|U(x, Tx)\| = \|x\| \leqslant \|x\| + \|Tx\| = \|(x, Tx)\|$, the operator U is continuous (and $\|U\| \leqslant 1$). Moreover, since U is bijective, the inverse operator theorem shows that the norm $\|U^{-1}\|$ is finite. Therefore, $\|Tx\| \leqslant \|(x, Tx)\| = \|U^{-1}x\| \leqslant \|U^{-1}\| \cdot \|x\|$, which proves the requisite continuity of the operator T. $\square$

Exercises

1. Verify that the definitions of the Cartesian product of normed spaces and of the graph of an operator agree with the definitions considered in Exercises 4–7 of Subsection 1.3.1.

2. According to Exercise 6 of Subsection 1.3.1, for non-linear mappings the closed graph theorem fails. Where in the proof of the closed graph theorem is the linearity of the operator T used?

3. Give an example showing that in the closed graph theorem the completeness assumption cannot be discarded.

4. The space $\ell_1 \times \ell_1$ is isometric to ℓ_1.

5. The space $c_0 \times c_0$ is isomorphic, but not isometric, to c_0.

6. Show that the following norms on $X \times Y$ are equivalent to the original norm: $\|(x, y)\|_\infty = \max\{\|x\|, \|y\|\}$, and $\|(x, y)\|_2 = (\|x\|^2 + \|y\|^2)^{1/2}$.

10.3.2   Complemented Subspaces

Let X be a linear space and X_1, X_2 be subspaces of X. We say that the space X *decomposes into the direct sum of the subspaces X_1 and X_2* (brief notation: $X = X_1 \oplus X_2$), if any element $x \in X$ admits a **unique** representation as a sum $x = x_1 + x_2$ with $x_1 \in X_1$ and $x_2 \in X_2$.

Theorem 1. *For the decomposition $X = X_1 \oplus X_2$ to hold it is necessary and sufficient that the following two conditions be simultaneously satisfied:*

(1) $X = X_1 + X_2$, *i.e., every $x \in X$ can be written in the form $x = x_1 + x_2$, with $x_1 \in X_1$ and $x_2 \in X_2$;*

(2) $X_1 \cap X_2 = \{0\}$, *i.e., the subspaces have trivial intersection.*

Proof. Let us show that condition (2) is equivalent to the uniqueness of the representation $x = x_1 + x_2$. Suppose first uniqueness holds. Consider an arbitrary element $x \in X_1 \cap X_2$. Then we can write the two equalities $0 = 0 + 0$ and $0 = x + (-x)$. In both equalities the first term lies in X_1, and the second in X_2. Then the uniqueness yields $x = 0$.

Conversely, let $X_1 \cap X_2 = \{0\}$. Suppose that some $x \in X$ admits two decompositions, $x = x_1 + x_2$ and $x = \widetilde{x}_1 + \widetilde{x}_2$, where $x_1, \widetilde{x}_1 \in X_1$ and $x_2, \widetilde{x}_2 \in X_2$. Then $x_1 - \widetilde{x}_1 \in X_1$, and at the same time $x_1 - \widetilde{x}_1 = \widetilde{x}_2 - x_2 \in X_2$. Consequently, $x_1 - \widetilde{x}_1 \in X_1 \cap X_2 = \{0\}$, and so $x_1 = \widetilde{x}_1$. Similarly, $x_2 = \widetilde{x}_2$, and the uniqueness is established. $\square$

Recall (Subsection 6.5.2) that a linear operator $P \colon X \to X$ is called a *projector onto the subspace* X_1 if $P(X) \subset X_1$ and $Px = x$ for all $x \in X_1$. Clearly, if P is a projector, then $P(Px) = Px$ for all $x \in X$, i.e., $P^2 = P$.

Theorem 2. *Suppose the operator $P \colon X \to X$ satisfies the equality $P^2 = P$. Then P is a projector onto the subspace $P(X)$; moreover, $Q = I - P$ is a projector onto the subspace $\operatorname{Ker} P$, and $X = P(X) \oplus \operatorname{Ker} P$.*

Proof. Let $y \in P(X)$ be an arbitrary element of the image of P. Then $y = Px$ with $x \in X$, and $Py = P(Px) = P^2 x = Px = y$. This shows that P is a projector onto $P(X)$. Since the operator Q also satisfies the relation $Q^2 = Q$, because $(I - P)^2 = I - 2P + P^2 = I - P$, Q is a projector onto $Q(X)$. We claim that $Q(X) = \operatorname{Ker} P$. Indeed, $x \in \operatorname{Ker} P \iff Px = 0 \iff Qx = x \iff x \in Q(X)$.

It remains to verify that $X = P(X) \oplus \operatorname{Ker} P$. First, every element $x \in X$ has the representation $x = Px + Qx$. Since $Px \in P(X)$ and $Qx \in \operatorname{Ker} P$, we conclude that $X = P(X) + \operatorname{Ker} P$. By Theorem 1, to complete the proof it remains to show that $P(X) \cap \operatorname{Ker} P = \{0\}$. Suppose that $x \in P(X) \cap \operatorname{Ker} P$. Then, on the one hand, $x \in P(X)$, so that $x = Px$, and on the other hand $x \in \operatorname{Ker} P$, so that $Px = 0$. Hence, $x = 0$. $\square$

Let X be a linear space, and let X_1 and X_2 be subspaces of X such that $X = X_1 \oplus X_2$. Define the operator $P \colon X \to X$ by the following rule. Given $x \in X$, decompose it as $x = x_1 + x_2$, where $x_1 \in X_1, x_2 \in X_2$ (by assumption, this decomposition exists and is unique), and then put $P(x) = x_1$. Next, if $x = x_1 + x_2$ and $y = y_1 + y_2$ are the decompositions of the vectors x and y, respectively, then $ax + by = (ax_1 + by_1) + (ax_2 + by_2)$ is the decomposition of the vector $ax + by$. Therefore, $P(ax + by) = ax_1 + by_1 = aPx + bPy$, so P is a linear operator. Now, by construction, $P(X) \subset X_1$, and for every $x_1 \in X_1$ the decomposition $x_1 = x_1 + 0$ means that $Px_1 = x_1$. Hence, P is a projector onto the subspace X_1; it is called the *projector onto the subspace X_1 parallel to the subspace X_2*.

Theorem 3. *Let X be a Banach space, X_1 and X_2 be closed[2] subspaces of X, and $X = X_1 \oplus X_2$. Then the projector P onto X_1 parallel to X_2 is a continuous operator.*

Proof. This statement — the central result of the present subsection — can be regarded as a typical example of application of the closed graph theorem.

We need to show that for every sequence $x_n \in X$, if $x_n \to x$ and $Px_n \to y$ as $n \to \infty$, then $y = Px$. Write $x_n = x_{n,1} + x_{n,2}$, with $x_{n,1} \in X_1$ and $x_{n,2} \in X_2$. By definition, $Px_n = x_{n,1}$. It follows that $x_{n,1} \to y$ as $n \to \infty$, and since the subspace X_1 is closed, $y \in X_1$. Next, $x_{n,2} = x_n - x_{n,1} \to x - y$ as $n \to \infty$. Again because of closedness, but now of the space X_2, we have $x - y \in X_2$. Thus, the obvious equality $x = y + (x - y)$ yields the decomposition of x as a sum of vectors from X_1 and from X_2. The first of these terms will be the projection of the vector x to X_1 parallel to X_2: $y = Px$.   □

A closed subspace X_1 of the Banach space X is said to be *complemented* if there exists a closed subspace $X_2 \subset X$ (called a *complement of X_1*) such that $X = X_1 \oplus X_2$. By the preceding theorem, the subspace $X_1 \subset X$ is complemented if and only if there exists a projector $P \in L(X, X)$ with $P(X) = X_1$. Complemented subspaces play an important role in the extension of operators (Subsection 6.5.2). It is easy to provide examples of complemented subspaces (see the exercises below), but the justification of each particular example of non-complemented subspace requires substantial efforts. A classical example of a non-complemented subspace is c_0 as a subspace of ℓ_∞. Another example, which arises naturally in the theory of Fourier series, is provided in the exercises of Subsection 10.4.3.

Exercises

1. Using the operator $U: X_1 \times X_2 \to X$, $U(x_1, x_2) = x_1 + x_2$, reduce Theorem 1 to the following assertion: a linear operator is injective if and only if its kernel reduces to zero.

---

[2]By definition, a subspace of a Banach space is a closed linear subspace, so the word "closed" in the statement of this theorem is superfluous. We emphasize the closedness here because in the previous chain of results we were speaking about subspaces of linear spaces, which were just linear subspaces. Also, for a continuous projector P in a normed space X its image $P(X)$ is automatically closed, because it is equal to $\mathrm{Ker}(I - P)$.

2. Suppose the Banach space X decomposes into a direct sum of two closed subspaces X_1 and X_2. Then $X_1 \times X_2 \approx X$.

3. Show that the adjoint of a projector is also a projector. Describe the kernel and image of the adjoint projector in terms of the kernel and image of the original projector.

4. Give an example showing that in the formulation of Theorem 3 the completeness assumption on the space X cannot be discarded.

5. Any one-dimensional subspace of a Banach space is complemented, and the corresponding projector can be chosen so that $\|P\| = 1$.

6. Any finite-dimensional subspace of a Banach space is complemented.

7. Any closed subspace of finite codimension is complemented.

8. Let X_1 be a subspace of the Banach space X, and X_2 be a subspace of X_1 such that X_2 is complemented in X. Then X_2 is complemented in X_1.

9. The space ℓ_∞ is complemented in any ambient Banach space.

10. A more general result: Suppose the subspace X_1 of the Banach space X is injective (see exercises in Subsection 9.3.3). Then X_1 is complemented in X.

11. Show that in $C[-1, 1]$ the subspace of all even functions is complemented.

12. Show that in $C[-1, 1]$ the subspace of all odd functions is complemented.

13. Show that in $C[-1, 1]$ the subspace of all functions that vanish on the interval $[-1, 0]$ is complemented.

14. Let X_1 be a complemented subspace of the Banach space X. Then any complement of X_1 is isomorphic to the quotient space X/X_1.

10.4   The Uniform Boundedness Principle and Applications

10.4.1   The Banach–Steinhaus Theorem on Pointwise Bounded Families of Operators

Definition 1. Let X and Y be normed spaces. A family $G \subset L(X, Y)$ of continuous linear operators is said to be *pointwise bounded* if for every $x \in X$ it holds that $\sup_{T \in G} \|Tx\| < \infty$.

The family G is said to be *uniformly bounded* if $\sup_{T \in G} \|T\| < \infty$.

Theorem 1 (Uniform boundedness principle). *Any pointwise bounded family of continuous linear operators from a Banach space X into a Banach space Y is uniformly bounded.*

Proof. Let $G \subset L(X, Y)$ be a pointwise bounded family. For each $x \in X$, define $M_x = \sup_{T \in G} \|Tx\|$ and $A_n = \{x \in X : M_x \leqslant n\}$. The sets A_n are closed and their union is the whole space X. Therefore, by Baire's theorem, at least one of the sets A_n is not nowhere dense, and consequently contains some ball. That is, there exist a number $n \in \mathbb{N}$ and a ball of the form $B_X(x_0, r) = x_0 + r B_X$ such that for any $x \in x_0 + r B_X$ the inequality $\|Tx\| \leqslant n$ holds for all $T \in G$. Then for any $x \in B_X$ and any $T \in G$,

$$\|Tx\| = \left\| \frac{1}{r} T(x_0 + rx) - Tx_0 \right\| \leqslant \frac{2n}{r}.$$

Taking here the supremum over all $x \in B_X$, we obtain that $\|T\| \leqslant 2n/r$ for all $T \in G$, which establishes the claimed uniform boundedness of the family G. $\square$

Exercises

1. Verify that the sets A_n introduced above are indeed closed.

2. Where in the proof have we used the fact that the space X is Banach?

3. Let $\mathcal{P}_1$ be the space of polynomials considered in Exercise 9 of Subsection 6.4.2, and $D_n : \mathcal{P}_1 \to \mathcal{P}_1$ be the n-th derivative operator. Show that the operators $n D_n$ form a pointwise bounded, but not uniformly bounded family.

4. Give an example of a pointwise bounded, but not uniformly bounded family of functions on the interval $[0,1]$.

5. Deduce the uniform boundedness principle established above from the closed graph theorem by using the following recipe. Let $G \subset L(X, Y)$ be a pointwise bounded family. Consider the auxiliary space $\ell_\infty(G \times B_{Y^*})$ of all bounded functions on $G \times B_{Y^*}$, endowed with the sup-norm. Define the operator $U : X \to \ell_\infty(G \times B_{Y^*})$ by the formula $(Ux)(T, y^*) = y^*(Tx)$. This operator has a closed graph, and so it is continuous. We have

$$\sup_{T \in G} \|T\| = \sup_{x \in B_X} \sup_{T \in G} \|Tx\| = \sup_{x \in B_X} \sup_{T \in G} \sup_{y^* \in B_{Y^*}} |y^*(Tx)|$$

$$= \sup_{x \in B_X} \|Ux\| = \|U\| < \infty.$$

10.4.2 Pointwise Convergence of Operators

Recall (Subsection 6.4.3) that a sequence of operators $T_n \in L(X, Y)$ is said to *converge pointwise* to the operator $T \in L(X, Y)$, if for any $x \in X$ the sequence $T_n x$ converges to Tx as $n \to \infty$. Since a pointwise converging sequence is also pointwise bounded, the next theorem is a direct consequence of the uniform boundedness principle.

Theorem 1 (Banach–Steinhaus theorem). *Any pointwise convergent sequence of operators $T_n \in L(X, Y)$, acting from a Banach space X to a normed space Y, is uniformly bounded.* □

Let X and Y be normed spaces and A a subset of X. Naturally, the sequence of operators $T_n \in L(X, Y)$ is said to *converge pointwise on A* to the operator $T \in L(X, Y)$ if $\lim_{n \to \infty} T_n x = Tx$ for all $x \in A$.

Theorem 2 (pointwise convergence criterion). *If the uniformly bounded sequence of $T_n \in L(X, Y)$, acting from the normed space X into the normed space Y, converges pointwise on a dense subset $A \subset X$ to an operator $T \in L(X, Y)$, then $T_n \to T$ pointwise as $n \to \infty$ on the whole space X.*

Proof. Let $M = \sup_n \|T_n - T\|$. Fix $x \in X$ and $\varepsilon > 0$. Suppose $a \in A$ approximates x to within ε / M, i.e., $\|x - a\| \leqslant \varepsilon / M$. Then

$$\varlimsup_{n \to \infty} \|(T_n - T)x\| \leqslant \varlimsup_{n \to \infty} \|(T_n - T)(x - a)\| + \varlimsup_{n \to \infty} \|(T_n - T)a\|$$

$$= \varlimsup_{n \to \infty} \|(T_n - T)(x - a)\| \leqslant M \frac{\varepsilon}{M} = \varepsilon.$$

Since ε is arbitrary, this means that

$$\varlimsup_{n \to \infty} \|(T_n - T)x\| = 0,$$

i.e., $T_n x \to Tx$ as $n \to \infty$.

Exercises

1. Let $T_n, T \in L(X, Y)$ and $\|T_n - T\| \to 0$ as $n \to \infty$. Then the sequence T_n convergence pointwise to T.

2. Suppose that X, Y are Banach spaces, $T_n, T \in L(X, Y)$, $U_n, U \in L(Y, Z)$, the sequence T_n converges pointwise to T and the sequence U_n converges pointwise to U. Then the sequence $U_n T_n$ converges pointwise to UT.

3. Deduce the pointwise convergence criterion based on Exercise 9 in Subsection 1.2.1, using the following recipe. Consider the space $\ell_\infty(\mathbb{N}, Y)$ of all bounded sequences of elements of the space Y, equipped with the norm $\left\| (y_n)_{n=1}^\infty \right\| = \sup_n \|y_n\|$. Then consider the subspace $c_0(\mathbb{N}, Y) \subset \ell_\infty(\mathbb{N}, Y)$ consisting of all sequences that converge to 0 and show that it is closed. Define the operator $U : X \to \ell_\infty(\mathbb{N}, Y)$ by the rule $Ux = (T_1 x, T_2 x, \ldots)$. If the sequence (T_n) converges pointwise to 0 on a dense subset A, then $U(A) \subset c_0(\mathbb{N}, Y)$. If, in addition, (T_n) is a bounded sequence of operators, then the operator U is continuous. Therefore, $U(X) \subset c_0(\mathbb{N}, Y)$, i.e., the sequence (T_n) converges pointwise to 0 on the whole space X.

4. Suppose X, Y are metric spaces, A is a dense subset of X, and $f_n, f : X \to Y$ are functions satisfying the Lipschitz condition with the same constant C. Then the pointwise convergence of the sequence (f_n) to f on A implies its pointwise convergence on the whole space X.

5. Suppose X, Y are Banach spaces and A is a dense subset of X. Then for the sequence of operators $T_n \in L(X, Y)$ to converge pointwise on X and its limit to be a continuous operator, it is necessary and sufficient that the following conditions be satisfied simultaneously:

(1) the sequence (T_n) is bounded;

(2) for any $x \in A$, the sequence of values $(T_n x)$ is Cauchy.

6. Give an example of a pointwise bounded sequence of continuous functions on the interval $[0, 1]$ that converges to 0 on a dense subset, but does not converge to 0 on the whole interval.

7. Suppose the sequence of operators $T_n \in L(X, Y)$ converges pointwise on the subset $A \subset X$ to an operator $T \in L(X, Y)$. Then T_n converges to T on Lin A.

Remark 1. In view of this last exercise, in the pointwise convergence criterion established above, the requirement that set is dense can be relaxed: it suffices to require that the set Lin A is dense, i.e., A forms a complete system of elements.

8. For the sequence of operators $T_n \in L(\ell_1, Y)$ to converge pointwise to the operator $T \in L(\ell_1, Y)$, it is necessary and sufficient that the following two conditions be satisfied simultaneously: $\sup_{n,m} \|T_n e_m\| < \infty$, and for every $m \in \mathbb{N}$, $T_n e_m \to T e_m$ as $n \to \infty$ (for the terminology/notation, see Exercise 6 in Subsection 6.3.3; see also Exercise 4 in Subsection 6.4.2).

9. Consider the functionals $f_n \in \ell_1^*$, acting as $f_n(a_1, a_2, \ldots) = a_n$. Show that the sequence (f_n) converges pointwise to 0, but $\|f_n\| = 1$, and so this sequence does not converge to 0 in norm.

10.4.3   Two Theorems on Fourier Series on an Interval

Theorem 1. *Let (g_n) be a uniformly bounded sequence of measurable functions on the interval $[a, b]$ which obeys the following condition: $\int_\Delta g_n(t)dt \to 0$ as $n \to \infty$ for any interval $\Delta \subset [a, b]$. Then $\int_{[a,b]} f(t)g_n(t)dt \to 0$ as $n \to \infty$ for any function $f \in L_1[a, b]$.*

Proof. Suppose $M := \sup_{n,t} |g_n(t)| < \infty$. Define the linear functionals F_n on $L_1[a, b]$ by the rule $F_n(f) = \int_{[a,b]} f(t)g_n(t)dt$. Since $|F_n(f)| \leqslant M \int_{[a,b]} |f(t)| \, dt = M \|f\|$, the functionals F_n are continuous and $\|F_n\| \leqslant M$. By hypothesis, the functionals F_n converge to 0 on any function of the form $\mathbb{1}_\Delta$. Consequently, the sequence (F_n) converges to 0 on all piecewise-constant functions. Since the set of piecewise-constant functions is dense in $L_1[a, b]$ (see Exercise 1 below), to complete the proof it remains to apply the pointwise convergence criterion established in Subsection 10.4.2. $\square$

For a function $f \in L_1[-\pi, \pi]$ one defines its *Fourier coefficients* $\widehat{f}_n$, $n \in \mathbb{Z}$, by the formula

$$\widehat{f}_n = \frac{1}{2\pi} \int_{-\pi}^\pi f(t)e^{-int} dt.$$

Corollary 1. *For any integrable function f on $[a, b]$ the integrals of the form $\int_a^b f(t)e^{i\alpha t} dt$, $\int_a^b f(t) \sin \alpha t \, dt$, or $\int_a^b f(t) \cos \alpha t \, dt$ tend to 0 as $\alpha \to \pm\infty$. In particular, the Fourier coefficients of any function $f \in L_1[-\pi, \pi]$ tend to zero as $n \to \infty$.*

Proof. We apply the preceding theorem on the interval $[a, b]$ to the functions $g_n(t) = e^{i\alpha_n t}$, $g_n(t) = \sin \alpha_n t$, or $g_n(t) = \cos \alpha_n t$, with real α_n such that $\alpha_n \to \infty$ as $n \to \infty$. The fact that the condition $\int_\Delta g_n(t)dt \to 0$ as $n \to \infty$ is satisfied for any interval $\Delta = [c, d]$ is verified by direct calculation of the corresponding integral. $\square$

Denote by $C(\mathbb{T}) \subset C[-\pi, \pi]$ the subspace consisting of the functions g that satisfy $g(-\pi) = g(\pi)$.[3] Every function $f \in C(\mathbb{T})$ can be extended from the interval $[-\pi, \pi]$ to the whole real line as a continuous 2π-periodic function. Accordingly, the elements of the space $C(\mathbb{T})$ can be regarded as 2π-periodic functions defined on the whole real line. We let $S_n f$ denote the partial sums of the Fourier series of f: $(S_n f)(t) = \sum_{k=-n}^n \widehat{f}_k e^{ikt}$. For completeness of the exposition, we recall a formula the reader is undoubtedly familiar with from calculus:

$$(S_n f)(t) = \frac{1}{2\pi} \int_{-\pi}^\pi f(t + \tau) \frac{\sin((n + 1/2)\tau)}{\sin(\tau/2)} d\tau. \tag{1}$$

To prove this relation, we insert in the definition of the partial sums $S_n g$ the expressions of the Fourier coefficients:

$$(S_n f)(t) = \sum_{k=-n}^n \widehat{f}_k e^{ikt} = \frac{1}{2\pi} \int_{-\pi}^\pi f(x) \sum_{k=-n}^n e^{ik(t-x)} dx.$$

---

[3]The usage of the notation $C(\mathbb{T})$ is explained in Exercise 2 below.

Now making the change of variables $\tau = x - t$ and using the fact that the integral of a 2π-periodic function over any interval of length 2π coincides with the integral over $[-\pi, \pi]$, we obtain

$$(S_n f)(t) = \frac{1}{2\pi} \int_{-\pi}^{\pi} f(t + \tau) \sum_{k=-n}^{n} e^{ik\tau} d\tau = \frac{1}{2\pi} \int_{-\pi}^{\pi} f(t + \tau) \frac{e^{-in\tau} - e^{i(n+1)\tau}}{1 - e^{i\tau}} d\tau.$$

To obtain formula (1) it remains to divide the numerator and denominator of the integrand by $e^{it/2}$ and use the formula $e^{ix} - e^{-ix} = 2i \sin x$.

As one knows from calculus, the Fourier series $\sum_{n=-\infty}^{+\infty} \hat{f}_n e^{int}$ of any continuously differentiable function f converges uniformly to f. On the other hand, if the differentiability assumption is discarded this assertion is no longer true. Moreover, there exist continuous functions whose Fourier series do not converge pointwise. The next theorem shows how to establish the existence of such examples without constructing them explicitly. A similar reasoning proves particularly useful in situations where the explicit construction and justification of an example is associated with considerable difficulties.

Theorem 2. *There exists a function $g \in C(\mathbb{T})$ for which the sequence of values $(S_n g)(0)$ of the partial sums of its Fourier series is not bounded.*

Proof. Introduce the linear functionals G_n on $C(\mathbb{T})$ by the rule $G_n(g) = (S_n g)(0)$. We need to show that the sequence of functionals (G_n) is not pointwise bounded. In view of the uniform boundedness principle, to this end it suffices to verify that the functionals G_n are continuous, but their norms are not jointly bounded. By formula (1),

$$G_n(f) = \frac{1}{2\pi} \int_{-\pi}^{\pi} f(t) \frac{\sin((n + 1/2)t)}{\sin(t/2)} dt.$$

Therefore (see Exercise 3 below),

$$\|G_n\| = \frac{1}{2\pi} \int_{-\pi}^{\pi} \left| \frac{\sin((n + 1/2)t)}{\sin(t/2)} \right| dt = \frac{1}{\pi} \int_{0}^{\pi} \left| \frac{\sin((n + 1/2)t)}{\sin(t/2)} \right| dt < \infty,$$

i.e., the functionals G_n are continuous. Let us estimate their norms from below:

$$\|G_n\| \geqslant \frac{2}{\pi} \int_{0}^{\pi} \left| \frac{\sin((n + 1/2)t)}{t} \right| dt \geqslant \frac{2}{\pi} \sum_{k=1}^{n-1} \int_{\pi k/(n+1/2)}^{\pi(k+1)/(n+1/2)} \left| \frac{\sin((n + 1/2)t)}{t} \right| dt$$

$$\geqslant \frac{2}{\pi} \sum_{k=1}^{n-1} \frac{n + 1/2}{\pi(k + 1)} \int_{\pi k/(n+1/2)}^{\pi(k+1)/(n+1/2)} |\sin((n + 1/2)t)| dt$$

$$= \frac{2}{\pi} \sum_{k=1}^{n-1} \frac{1}{\pi(k + 1)} \int_{\pi k}^{\pi(k+1)} |\sin \tau| d\tau = \frac{4}{\pi^2} \sum_{k=1}^{n-1} \frac{1}{k + 1} \geqslant \frac{4}{\pi^2} \int_{2}^{n+1} \frac{dx}{x} = \frac{4}{\pi^2} \ln \frac{n + 1}{2}.$$

Hence, $\|G_n\| \geqslant \dfrac{4}{\pi^2} \ln \dfrac{n+1}{2} \to \infty$ as $n \to \infty$. $\qquad\qquad\qquad\qquad\qquad\qquad$ □

Exercises

1. Prove that the set of piecewise-continuous functions is dense in $L_1[-\pi, \pi]$ by using the following recipe. First, $C[-\pi, \pi]$ is dense in $L_1[-\pi, \pi]$ (this we already know, and even in a more general situation — see Theorem 1 is Subsection 8.3.3). Further, every continuous function can be approximated in the metric of $L_1[-\pi, \pi]$ (and even uniformly) by piecewise-constant functions.

2. Every function $g \in C[-\pi, \pi]$ that satisfies $g(-\pi) = g(\pi)$ corresponds to a continuous function f on the unit circle $\mathbb{T} = \{e^{it} : t \in [-\pi, \pi]\}$ via the formula $f(e^{it}) = g(t)$. Verify that the map $g \mapsto f$ is a bijective isometry between the subspace of functions $g \in C[-\pi, \pi]$ that satisfy $g(-\pi) = g(\pi)$ and the space of all continuous functions on $\mathbb{T}$.

3. (A particular case of Exercise 5 in Subsection 8.4.4.)   Given a function $v \in L_1[-\pi, \pi]$, define a linear functional V on $C(\mathbb{T})$ by the formula $V(g) = \int_{-\pi}^{\pi} g(t)v(t)dt$. Based on the theorem on the general form of linear functionals on $C(K)$, show that $\|V\| = \int_{-\pi}^{\pi} |v(t)|dt$.

4. Suppose the function f is Lebesgue integrable on the interval $[-\pi, \pi]$ and satisfies the *Dini condition* in the point $x_0 \in [-\pi, \pi]$: $(f(x_0 + t) - f(x_0))/t \in L_1[-\pi, \pi]$. Then the Fourier series of the function f converges at the point x_0 to $f(x_0)$.

By solving the next chain of exercises the reader will be able, in particular, to justify the fact that the subspace $A(\mathbb{T})$ of $C(\mathbb{T})$, defined as the closure in $C(\mathbb{T})$ of the span of the sequence of functions $\{e^{ikt}\}_{k=0}^{\infty}$, is not complemented in $C(\mathbb{T})$.

5. Interpret the partial sum S_n of the Fourier series as an operator acting from $C(\mathbb{T})$ into $C(\mathbb{T})$: $(S_n g)(t) = \sum_{k=-n}^{n} \widehat{f}_k e^{ikt}$. Prove that S_n is a projector onto the subspace $E_n = \mathrm{Lin}\{e^{ikt}\}_{k=-n}^{n}$. Prove that $\|S_n\|$ coincides with the norm of the functional G_n figuring in the proof of Theorem 2, and consequently $\|S_n\| \geqslant \frac{4}{\pi^2} \ln \left(\frac{n+1}{2}\right)$.

6. Let U_τ be the shift operator by τ in $C(\mathbb{T})$: $(U_\tau f)(t) = f(t + \tau)$. Verify that U_τ maps $C(\mathbb{T})$ bijectively into $C(\mathbb{T})$ and that $\|U_\tau f\| = \|f\|$ for all $f \in C(\mathbb{T})$.

Let $P \in L(C(\mathbb{T}), C(\mathbb{T}))$ be a projector onto $A(\mathbb{T})$. For each function $f \in C(\mathbb{T})$ consider the function $(\widetilde{P} f)(t) = \frac{1}{2\pi} \int_{-\pi}^{\pi} (U_\tau P U_{-\tau} f)(t)d\tau$. Show that this "shift-averaged" operator P enjoys the following properties

7. $\widetilde{P} f \in C(\mathbb{T})$ for all $f \in C(\mathbb{T})$.

8. $\widetilde{P} \in L(C(\mathbb{T}), C(\mathbb{T}))$ and $\|\widetilde{P}\| \leqslant \|P\|$.

9. $\widetilde{P}$ is a projector onto $A(\mathbb{T})$.

10. The operator $\widetilde{P}$ commutes with shifts: $\widetilde{P}U_\tau = U_\tau\widetilde{P}$, for all τ. In particular, if we define $g_k = \widetilde{P}(e^{ikt})$, then $g_k(t + \tau) = g_k(t)e^{ik\tau}$. For $t = 0$ we have $g_k(\tau) = g_k(0)e^{ik\tau}$.

11. From Exercise 9 above and the last equality it follows that $\widetilde{P}(e^{ikt}) = e^{ikt}$ for $k \geqslant 0$, while $\widetilde{P}(e^{ikt}) = 0$ for $k < 0$.

12. Let $f \in C(\mathbb{T})$ and $g = \widetilde{P}f$. Then $\widehat{g}_k = \widehat{f}_k$ for $k \geqslant 0$, while $\widehat{g}_k = 0$ for $k < 0$.

13. Denote by U_n the multiplication operator by e^{int}: $(U_n f)(t) = f(t)e^{int}$. The operators $U_n \in L(C(\mathbb{T}))$ are bijective isometries. The operator $\widetilde{P}$ is related to the partial sum of Fourier series operator S_n by the identity

$$S_n = U_{n+1}(I - \widetilde{P})U_{-(2n+1)}\widetilde{P}U_n$$

Consequently, $\|I - \widetilde{P}\| \cdot \|\widetilde{P}\| \geqslant \|S_n\|$.

14. In view of the arbitrariness of n in the preceding exercise and Exercise 5, the operator $\widetilde{P}$ is discontinuous. This contradicts Exercise 8 above. Thus, we proved that there exists no projector of $C(\mathbb{T})$ onto $A(\mathbb{T})$, i.e., that the subspace $A(\mathbb{T})$ of $C(\mathbb{T})$ is not complemented.

15. Let $P_n \in L(C(\mathbb{T}), C(\mathbb{T}))$ be a projector onto $E_n = \mathrm{Lin}\{e^{ikt}\}_{k=-n}^n$. Consider again the shift-averaged operator, i.e., the projector

$$(\widetilde{P}_n f)(t) = \frac{1}{2\pi}\int_{-\pi}^{\pi}(U_\tau P_n U_{-\tau} f)(t)d\tau.$$

Prove that

$$\|P_n\| \geqslant \frac{4}{\pi^2}\ln\left(\frac{n+1}{2}\right).$$

That is, this estimate is satisfied by any projector onto E_n, and not only by the partial sum of Fourier series operator.

16. For each $n \in \mathbb{N}$, fix some collection $K_n \subset [-\pi, \pi)$ of $2n + 1$ points of the interval. For any function $f \in C(\mathbb{T})$, denote by $T_n f$ the trigonometric interpolation polynomial of f: $T_n f \in \mathrm{Lin}\{e^{ikt}\}_{k=-n}^n$, and $(T_n f)(t) = f(t)$ in each point $t \in K_n$. Based on the preceding exercise and the Banach–Steinhaus theorem, show that there exists a function $f \in C(\mathbb{T})$ for which the sequence $T_n f$ of interpolants does not converge to f.

10.5   The Concept of a Schauder Basis

10.5.1   Definition and Simplest Properties

The reader is already familiar with one generalization of the concept of basis to the infinite-dimensional case, namely, the Hamel basis, the existence of which was established in Subsection 5.1.3. Although any linear space contains a Hamel basis, for Banach spaces such Hamel bases turn out to be rather inconvenient in applications. First of all, Hamel bases in infinite-dimensional Banach spaces are not countable (Exercise 4 in Subsection 6.3.3). Further, despite the existence theorem, there is no concrete infinite-dimensional Banach space in which an example of a Hamel basis is known. Finally, Hamel bases are not at all connected with the topological structure of the ambient space. For instance, if a sequence of elements of a Banach space converges to a limit, the coefficients of the decomposition of its terms in a Hamel basis do not necessarily converge to the coefficients of the limit element. For these reasons, in the theory of Banach spaces the fundamental notion of basis is not that of a Hamel basis, but that of a Schauder basis, the study of which we now begin.

Definition 1. The sequence of elements $\{e_n\}_1^\infty$ of a Banach space X is called a *basis* (or, equivalently, a *Schauder basis*) of X, if for any element $x \in X$ there exists a unique sequence of coefficients $\{a_n\}_1^\infty$ such that the series $\sum_{n=1}^\infty a_n e_n$ converges to x. The series $\sum_{n=1}^\infty a_n e_n$ is called the *decomposition of the element x in the* (or *with respect to*) *the basis* $\{e_n\}_1^\infty$, and the numbers $\{a_n\}_1^\infty$ are called the *coefficients* of the decomposition.

An example of a basis is provided by any orthonormal basis in a Hilbert space. This example will be treated in detail in Subsection 12.3.3. Another example is the standard basis in the space ℓ_1 that we already encountered in Exercise 6 of Subsection 6.3.3. In Theorem 2 of Subsection 14.3.3 we will show that the trigonometric system $\{1, e^{it}, e^{-it}, e^{2it}, e^{-2it}, \ldots\}$ forms a basis in $L_p[0, 2\pi]$ for $1 < p < \infty$. Numerous examples of bases in all classical separable Banach spaces, various classes of bases, and their generalizations, can be found in the fundamental two-volume treatise by I. Singer [39, 40]. A modern survey on various classes of bases in function spaces is provided by T. Figiel and T. Wojtaszczyk in Chap. 14 of the collection [20].

Exercises

1. The elements of a basis are linearly independent. In particular, no element of a basis can be equal to 0.

2. Each basis $\{e_n\}_1^\infty$ of a Banach space X is a complete system of elements in X: $\overline{\text{Lin}}\,\{e_n\}_1^\infty = X$.

3. Using the properties of Taylor series, show that the sequence $1, t, t^2, t^3, \ldots$ does not form a basis in the space $C[0, 1]$. This will prove that, in contrast to the finite-dimensional case, completeness and linear independence together are not sufficient for the basis property to hold.

4. If the Banach space X has a Schauder basis, then X is necessarily infinite-dimensional and separable.

Let X be one of the sequence spaces ℓ_p ($1 \leqslant p \leqslant \infty$) or c_0. The *standard basis* (also referred to as the *canonical*, or sometimes as the *natural* basis) of the space X is the system of vectors $\{e_n\}_1^\infty$, where $e_1 = (1, 0, 0, \ldots)$, $e_2 = (0, 1, 0, \ldots),\ldots.$ Prove that:

5. The standard basis of the space c_0 is a basis.

6. The standard basis of the space ℓ_∞ is not a basis of ℓ_∞.

7. For $1 \leqslant p < \infty$, the standard basis of the space ℓ_p is a basis.

8. The space ℓ_∞ is not separable, hence it contains no Schauder basis.

The following question, formulated already by S. Banach, turned out to be far from simple: does every separable infinite-dimensional Banach space have a basis? A negative answer was provided in 1973 by P. Enflo [53] (see also [28, Sec. 2.d]).

10.5.2 Coordinate Functionals and Partial Sum Operators

Definition 1. Let $\{e_n\}_1^\infty$ be a basis of the Banach space X, and let $x \in X$. Denote by $e_n^*(x)$ the coefficients of the decomposition of x in the basis $\{e_n\}_1^\infty$, and by $S_n(x)$ the n-th partial sum of the decomposition, i.e., $S_n(x) = \sum_{k=1}^{n} e_k^*(x) e_k$.

Proposition 1. e_n^* *are linear functionals on X, and S_n are linear operators acting from X into X.*

Proof. Let $x, y \in X$, and let a, b be arbitrary scalars. Then we have the following decompositions:

$$x = \sum_{k=1}^{\infty} e_k^*(x) e_k, \qquad y = \sum_{k=1}^{\infty} e_k^*(y) e_k, \qquad ax + by = \sum_{k=1}^{\infty} e_k^*(ax + by) e_k.$$

Hence,

$$\sum_{k=1}^{\infty} (a e_k^*(x) + b e_k^*(y)) e_k = \sum_{k=1}^{\infty} e_k^*(ax + by) e_k.$$

In view of the uniqueness of the decomposition of an element in a basis, we conclude that $a e_k^*(x) + b e_k^*(y) = e_k^*(ax + by)$. $\qquad\square$

Henceforth the functionals e_n^* will be referred to as the *coordinate functionals*, and the operators S_n as the *partial sum operators* with respect to the basis $\{e_n\}_1^\infty$.

Theorem 1 (Banach's theorem). *Let* $\{e_n\}_1^\infty$ *be a basis of the Banach space* X. *Then the partial sum operators* S_n *are continuous and* $\sup_n \|S_n\| = C < \infty$.

Proof. Let us introduce the auxiliary space E of all numerical sequences $a = (a_n)_1^\infty$ for which the series $\sum_{n=1}^\infty a_n e_n$ converges. We equip the space E with the norm

$$\|a\| = \sup \left\{ \left\| \sum_{n=1}^N a_n e_n \right\| : N = 1, 2, \ldots \right\}.$$

As we observed in Exercise 1 of Subsection 6.3.3, E is a Banach space. Define the operator $T : E \to X$ by the rule

$$Ta = \sum_{n=1}^\infty a_n e_n.$$

Then T is bijective, because $\{e_n\}_1^\infty$ is a basis. In view of the obvious inequality $\|Ta\| \leqslant \|a\|$, T is continuous. Hence, by the Banach inverse operator theorem, the operator T^{-1} is also continuous. This means that there exists a constant C such that $\|a\| \leqslant C\|Ta\|$ for all $a \in E$. In other words, for any $a \in E$ it holds that

$$\sup \left\{ \left\| \sum_{n=1}^N a_n e_n \right\| : N = 1, 2, \ldots \right\} \leqslant C \left\| \sum_{n=1}^\infty a_n e_n \right\|.$$

This last inequality establishes the requisite continuity and joint boundedness of the partial sum operators.   □

Corollary 1. *Let* $\{e_n\}_1^\infty$ *be a basis of the Banach space* X. *Then the coordinate functionals* e_n^* *are continuous and* $\sup_n \left\{ \|e_n\| \cdot \|e_n^*\| \right\} < \infty$.

Proof. It suffices to use the estimate

$$\left\| e_n^*(x)e_n \right\| = \left\| (S_n - S_{n-1})(x) \right\| \leqslant (\|S_n\| + \|S_{n-1}\|) \|x\| \leqslant 2C \|x\|,$$

where C is the constant from the preceding theorem.   □

10.5.3 Linear Functionals on a Space with a Basis

Let X be a Banach space with a basis $\{e_n\}_1^\infty$. Then any linear functional $f \in X^*$ is uniquely determined by its values on the elements of the basis. In other words, each functional f can be identified with the numerical sequence $(f(e_1), f(e_2), \ldots)$, and accordingly the space X^* can be identified with the set of all such sequences. This observation is more rigorously formulated in the following theorem, the easy proof of which is left to the reader.

Theorem 1. *For every $f \in X^*$, put $Uf = (f(e_1), f(e_2), \ldots)$. Denote by $\widetilde{X}$ the set of all numerical sequences of the form Uf, $f \in X^*$. Then $\widetilde{X}$ is a linear space with respect to the coordinatewise operations, and U is a bijective linear mapping of the space X^* onto the space $\widetilde{X}$. Further, let $f \in X^*$, $Uf = (f_1, f_2, \ldots)$. Then for any $x = \sum_{n=1}^{\infty} x_n e_n \in X$ the action of the functional f on the element x can be calculated by the rule $f(x) = \sum_{n=1}^{\infty} x_n f_n$.* $\qquad\square$

Theorem 2. *Let $(f_1, f_2, \ldots)$ be a sequence of numbers. Set*

$$\|(f_1, f_2, \ldots)\|_{\widetilde{X}} = \sup \left\{ \left| \sum_{n=1}^{N} a_n f_n \right| : N \in \mathbb{N}, \; \left\| \sum_{n=1}^{N} a_n e_n \right\| \leqslant 1 \right\}.$$

Then for the sequence of numbers $(f_1, f_2, \ldots)$ to belong to the space $\widetilde{X}$ it is necessary and sufficient that $\|(f_1, f_2, \ldots)\|_{\widetilde{X}} < \infty$. Moreover, if $f \in X^*$ is a functional that generates the sequence $(f_1, f_2, \ldots)$, then $\|f\| = \|(f_1, f_2, \ldots)\|_{\widetilde{X}}$.

Proof. The (non-closed) linear subspace $Y = \mathrm{Lin}\{e_n\}_{n=1}^{\infty}$ is dense in X. It is readily seen that the condition $\|(f_1, f_2, \ldots)\|_{\widetilde{X}} < \infty$ is equivalent to the following: the linear functional f given on Y by the rule

$$f\left(\sum_{n=1}^{N} a_n e_n \right) = \sum_{n=1}^{N} a_n f_n$$

is continuous. The norm of this functional is equal to $\|(f_1, f_2, \ldots)\|_{\widetilde{X}}$. Since every linear functional given on Y uniquely extends by continuity (see Subsection 6.5.1) to the whole space X, this is equivalent with the existence of a functional $f \in X^*$, acting on linear combinations of vectors of the basis by the rule $f\left(\sum_{n=1}^{N} a_n e_n \right) = \sum_{n=1}^{N} a_n f_n$. Since $Uf = (f_1, f_2, \ldots)$, the last condition is equivalent to $(f_1, f_2, \ldots) \in \widetilde{X}$. Finally, the equality $\|f\| = \|(f_1, f_2, \ldots)\|_{\widetilde{X}}$ simply means that the norm of the restriction of the functional $f \in X^*$ to the dense subspace Y, that is, the number $\|(f_1, f_2, \ldots)\|_{\widetilde{X}}$, coincides with $\|f\|$. $\qquad\square$

In the examples given below $\widetilde{X}$ will be regarded as a normed space equipped with the norm from Theorem 2. Theorem 2 means, in particular, that the normed spaces X^* and $\widetilde{X}$ are isomorphic, and the operator U effects this isomorphism (and is even isometric).

Example 1. Let $X = c_0$ with the standard basis $\{e_n\}_1^{\infty}$: $e_1 = (1, 0, 0, \ldots)$, $e_2 = (0, 1, 0, \ldots), \ldots$. Then $\|(f_1, f_2, \ldots)\|_{\widetilde{X}} = \sum_{n=1}^{\infty} |f_n|$, and so the space $\widetilde{X}$ coincides with ℓ_1. Indeed, in the present case the condition $\left\| \sum_{n=1}^{N} a_n e_n \right\| \leqslant 1$ simply means that all a_n are smaller than or equal in modulus to 1. Under this condition, the largest possible value of the quantity $\left|\sum_{n=1}^{N} a_n f_n\right|$ is $\sum_{n=1}^{N} |f_n|$ (this value is attained for $a_n = \operatorname{sign} f_n$). We have

$$\|(f_1, f_2, \ldots,)\|_{\widetilde{X}} = \sup\left\{ \left|\sum_{n=1}^{N} a_n f_n\right| : N \in \mathbb{N}, \; \left\|\sum_{n=1}^{N} a_n e_n\right\| \leqslant 1 \right\}$$

$$= \sup\left\{ \sum_{n=1}^{N} |f_n| : N \in \mathbb{N} \right\} = \sum_{n=1}^{\infty} |f_n|.$$

Since, by Theorem 2, the space $\widetilde{X}$ can be identified with the dual space, the result of the last example can be briefly expressed by the equality $(c_0)^* = \ell_1$. In detail, this is stated as the theorem on the general form of linear functionals on the space c_0: every element $(f_1, f_2, \ldots)$ of the space ℓ_1 generates a continuous linear functional $f(x) = \sum_{n=1}^{\infty} x_n f_n$ on c_0, and the norm of the functional f coincides with the norm of the element $(f_1, f_2, \ldots)$ in ℓ_1. Conversely, every functional $f \in (c_0)^*$ is generated by an element $(f_1, f_2, \ldots) \in \ell_1$ by the rule described above; moreover, the element $(f_1, f_2, \ldots)$ is uniquely determined by the functional f.

Example 2. Let $X = \ell_1$ with the standard basis $\{e_n\}_1^{\infty}$. Then

$$\|(f_1, f_2, \ldots)\|_{\widetilde{X}} = \sup_{n \in \mathbb{N}} |f_n|,$$

hence the space $\widetilde{X}$ coincides with ℓ_{∞}. In other words, $(\ell_1)^* = \ell_{\infty}$.

Indeed, in the present case, if $\left\|\sum_{n=1}^{N} a_n e_n\right\| \leqslant 1$, then $\sum_{n=1}^{N} |a_n| \leqslant 1$, and $\left|\sum_{n=1}^{N} a_n f_n\right| \leqslant \sup_{n \in \mathbb{N}} |f_n|$. Accordingly, $\|(f_1, f_2, \ldots,)\|_{\widetilde{X}} \leqslant \sup_{n \in \mathbb{N}} |f_n|$. On the other hand, if in the definition of the norm on the space $\widetilde{X}$ we replace the linear combination $\sum_{n=1}^{N} a_n e_n$ by a single basis vector e_n, then we obtain the estimate $\|(f_1, f_2, \ldots)\|_{\widetilde{X}} \geqslant |f_n|$. Taking the supremum over all n, we obtain the inequality $\|(f_1, f_2, \ldots,)\|_{\widetilde{X}} \geqslant \sup_{n \in \mathbb{N}} |f_n|$.

Exercises

1. Recast the equality $(\ell_1)^* = \ell_{\infty}$ from Example 2 as a theorem on the general form of linear functionals on the space ℓ_1.

2. Based on the completeness of the dual space (Subsection 6.4.4) and Theorem 2, establish the completeness of the spaces ℓ_1 and ℓ_{∞}.

3. Let $1 < p < \infty$, and let p' be the *conjugate* (or *dual*) *index* of p: $\frac{1}{p} + \frac{1}{p'} = 1$. Using *Hölder's inequality* for finite sums,

$$\left|\sum_{n=1}^{N} a_n f_n\right| \leqslant \left(\sum_{n=1}^{N} |a_n|^p\right)^{1/p} \left(\sum_{n=1}^{N} |f_n|^{p'}\right)^{1/p'},$$

derive the theorem on the general form of linear functionals on the space ℓ_p, $1 < p < \infty$: $(\ell_p)^* = \ell_{p'}$. Prove the completeness of the space ℓ_p for $1 < p < \infty$.

4. As we already observed in Exercise 6 of Subsection 10.5.1, the standard basis of the space ℓ_∞ is not a basis of this space. Accordingly, for ℓ_∞ we cannot use the description of functionals on a space with a basis. Show that there exists a functional $f \in (\ell_\infty)^*$ that cannot be represented in the form $f(x) = \sum_{n=1}^{\infty} x_n f_n$ of a "scalar product" with a fixed numerical sequence.

Comments on the Exercises

Subsection 10.4.2

Exercise 2. By the Banach–Steinhaus theorem, $\sup_n \|U_n\| = M < \infty$. For every $x \in X$ it holds that

$$\|(U_n T_n - UT)x\| \leqslant \|U_n(T_n - T)x\| + \|(U_n - U)Tx\|$$
$$\leqslant M\|(T_n - T)x\| + \|(U_n - U)Tx\| \to 0 \quad \text{as } n \to \infty.$$

Subsection 10.4.3

Exercise 4. Denote $f(x_0)$ by a. We have

$$(S_n f)(x_0) - a = (S_n(f - a))(x_0) = \frac{1}{2\pi} \int\limits_{-\pi}^{\pi} (f(x_0 + t) - a)\frac{\sin(n + 1/2)t}{\sin(1/2)t}\,dt.$$

Applying Theorem 1 of Subsection 10.4.3 to the functions $g_n(t) = \sin(n + (1/2))t$, we obtain the requisite condition $(S_n f)(x_0) \to a$. For an alternative formulation of this argument, see Subsection 14.2.1.

Exercises 7–12. The shift-averaging construction described here is treated in a more general form in the textbook by W. Rudin [38, Chapter 5, Sections 5.15–5.19]. The idea of using this construction to prove that the subspace $A(\mathbb{T})$ is not complemented in $C(\mathbb{T})$ is also due to Rudin.

Subsection 10.5.3

Exercise 4. From Theorem 3 of Subsection 9.2.2 applied to $X = \ell_\infty$ and $Y = c_0$ it follows that there exists a nonzero functional $f \in (\ell_\infty)^*$ which annihilates the whole space c_0. This provides the sought-for example.

Chapter 11
Elements of Spectral Theory
of Operators. Compact Operators

11.1 Algebra of Operators

Let X and Y be Banach spaces. The space $L(X, Y)$ of continuous operators is itself a Banach space when equipped with the operator norm. In this space the operations of addition of operators and multiplication of operators by numbers (scalars) are defined in a natural manner. The reader is familiar with yet another operation, namely, the composition (multiplication) of operators. We note that, generally speaking, the composition is not defined for all pairs of elements of the space $L(X, Y)$. The composition $A \circ B$ is not defined if the operator A is not defined on the image of the operator B. The situation changes radically in the case when $X = Y$. The multiplication turns out to be well-defined, and the space $L(X, X)$ (henceforth denoted for simplicity by $L(X)$) is an algebra with respect to the operations of addition and multiplication of operators. That is to say, with operators acting in one space one can in some sense work as we would with numbers: add, subtract, multiply, pass to a limit. This analogy with numbers proves rather fruitful. In many cases it allows us to find simple reasoning methods that lead to important and useful results. To understand how to use this analogy, it is convenient to forget for a while that we are dealing with operators, and make acquaintance with the general properties of Banach algebras.

11.1.1 Banach Algebras: Axiomatics and Examples

A complex Banach space **A** endowed with a supplementary operation of multiplication of its elements is called a *Banach algebra* if the multiplication obeys the following axioms:

— $a(bc) = (ab)c$ for any $a, b, c \in \mathbf{A}$ (associativity);
— $a(\lambda b) = (\lambda a)b = \lambda(ab)$ for any $a, b \in \mathbf{A}$ and any scalar λ;
— $a(b + c) = ab + ac$ and $(a + b)c = ac + bc$ (distributivity);

© Springer International Publishing AG, part of Springer Nature 2018
V. Kadets, *A Course in Functional Analysis and Measure Theory*,
Universitext, https://doi.org/10.1007/978-3-319-92004-7_11

— $\|ab\| \leqslant \|a\| \cdot \|b\|$ for all $a, b \in \mathbf{A}$ (multiplicative triangle inequality);
— there exists an element $e \in \mathbf{A}$ (e is called the *unit element* of the algebra $\mathbf{A}$), such that multiplication by e leaves elements unchanged: $ea = ae = a$ for all $a \in \mathbf{A}$;
— $\|e\| = 1$.

One introduces in a natural manner the notions of *subalgebra* (a closed linear subspace $X \subset \mathbf{A}$ that is stable under multiplication and contains the element e) and of *isomorphism of Banach algebras* (i.e., an isomorphism T of Banach spaces which additionally satisfies the condition $T(ab) = T(a)T(b)$).

One example of a Banach algebra has already been mentioned above — the algebra $L(X)$ of continuous linear operators in the Banach space X. The role of the multiplication in $L(X)$ is played by the composition of operators, and the identity operator serves as the identity element of the algebra $L(X)$. Let us list a few more examples, leaving the verification of the Banach algebra axioms in these examples to the reader.

Examples

1. The space $C(K)$ endowed with the usual multiplication of functions. The unit element here is the function identically equal to 1.

2. The space $L_\infty(\Omega, \Sigma, \mu)$ with the usual mutiplication.

3. The space ℓ_∞ with coordinatewise multiplication.

4. The space ℓ_1, where for the multiplication one takes the convolution of sequences: recall that for $x, y \in \ell_1$, $x = (x_0, x_1, \ldots)$, and $y = (y_0, y_1, \ldots)$, the *convolution* $x * y$ is defined as the vector $((x * y)_0, (x * y)_1, \ldots, (x * y)_n, \ldots)$ whose coordinates are calculated by the rule $(x * y)_n = \sum_{k=0}^{n} x_k y_{n-k}$. The unit element is the vector $e = (1, 0, 0, \ldots)$.

5. The space W consisting of the functions $g \in C(\mathbb{T})$ whose Fourier coefficients obey the condition $\sum_{n=-\infty}^{+\infty} |\widehat{f_n}| < \infty$. On this space one considers the usual multiplication, and the norm is given by the formula $\|f\| = \sum_{n=-\infty}^{+\infty} |\widehat{f_n}|$.

In examples 1–5, the multiplication is commutative: $ab = ba$ for any elements $a, b \in \mathbf{A}$. Note, however, that commutativity of multiplication is not included among the axioms of an algebra. Moreover, commutativity does not hold in the most important example for us here: the algebra of operators $L(X)$.

As in the case of numbers, for the elements of a Banach algebra the following theorem on the limit of a product holds true.

Theorem 1. *The multiplication in a Banach algebra is continuous as a function of two variables. In other words, if $a_n \to a$ and $b_n \to b$ as $n \to \infty$, then $a_n b_n \to ab$ as $n \to \infty$.*

Proof. We have

$$\|a_n b_n - ab\| \leqslant \|a_n b_n - a_n b\| + \|a_n b - ab\|$$

$$\leq \|a_n\| \cdot \|b_n - b\| + \|a_n - a\| \cdot \|b\| \to 0 \quad \text{as } n \to \infty. \qquad \square$$

Exercises

1. A Banach algebra has only one unit element.

2. The spaces $L_2[0, 1]$ and $L_1[0, 1]$ are not Banach algebras with respect to the multiplication of functions.

3. The spaces ℓ_2 and ℓ_1 are not Banach algebras with respect to the coordinatewise multiplication of vectors.

4. The space W described in Example 5 is isomorphic, as a Banach space, to the space ℓ_1. How do we need to define a multiplication operation on the space ℓ_1, so that ℓ_1 with this operation will be isomorphic to W also as a Banach algebra?

5. For each $a \in \mathbf{A}$, define the operator $T_a \colon \mathbf{A} \to \mathbf{A}$ by the formula $T_a(b) = ab$. Verify that T_a is a continuous linear operator and $\|T_a\| = \|a\|$.

6. Every Banach algebra $\mathbf{A}$ is isomorphic to a subalgebra of the algebra $L(\mathbf{A})$. Hence, in a certain sense that algebra of operators is a universal example of a Banach algebra.

7. Give an example of two operators in the two-dimensional space $\mathbb{C}^2$ that do not commute.

11.1.2 Invertibility in Banach Algebras

An element a of the Banach algebra $\mathbf{A}$ is said to be *invertible* if there exists an element $a^{-1} \in \mathbf{A}$, called the *inverse* of a, such that $a^{-1}a = aa^{-1} = e$. If the inverse element exists, it is unique. Indeed, if in addition to a^{-1} there is another inverse $b \in \mathbf{A}$ of a, then $b = be = baa^{-1} = ea^{-1} = a^{-1}$.

Note that if two elements $a, b \in \mathbf{A}$ are invertible, then their product ab is also invertible, and $(ab)^{-1} = b^{-1}a^{-1}$.

Lemma 1. *If for two elements $a, b \in \mathbf{A}$ both products ab and ba are invertible, then the elements a and b themselves are invertible. In particular, if $a, b \in \mathbf{A}$ commute and their product ab is invertible, then a and b are invertible.*

Proof. In view of the symmetry of the condition, it suffices to verify that a is invertible. Let us show that the element $g = b(ab)^{-1}$ is the inverse of a. First, $ag = ab(ab)^{-1} = e$. On the other hand, $ga = b(ab)^{-1}a = b(ab)^{-1}aba(ba)^{-1} = ba(ba)^{-1} = e$. $\qquad\square$

Lemma 2 (on small perturbations of the unit element). *Suppose the element a of the Banach algebra $\mathbf{A}$ satisfies the condition $\|a\| < 1$. Then the element $e - a$ is invertible, and the following inversion formula holds:*

$$(e - a)^{-1} = e + a + a^2 + \cdots + a^n + \cdots.$$

Proof. Let us prove the inversion formula. Since $\|a^n\| \leqslant \|a\|^n$, the series $e + a + a^2 + \cdots + a^n + \cdots$ is majorized by a convergent geometric progression, and hence converges to some element $b \in \mathbf{A}$. It remains to verify that $b(e - a) = (e - a)b = e$. The two required equalities are derived by simply opening the parentheses:

$$(e + a + a^2 + \cdots)(e - a) = (e + a + a^2 + \cdots) - (a + a^2 + \cdots) = e,$$

$$(e - a)(e + a + a^2 + \cdots) = (e + a + a^2 + \cdots) - (a + a^2 + \cdots) = e. \qquad \square$$

The inversion formula is the natural analogue of the formula for the sum of a geometric progression. At the same time, care should be exercised when similar analogies are used: unlike numbers, elements of an algebra do not necessarily commute. Other complications may also arise, connected with non-invertibility, the impossibility of expressing the norm of a product in terms of the norms of its factors, etc.

Theorem 1   (on small perturbations of an invertible element). *Suppose $a, b \in \mathbf{A}$, a is invertible, and $\|b\| < 1/\|a^{-1}\|$. Then the element $a - b$ is also invertible. In other words, if a is invertible, then the whole ball of radius $r = 1/\|a^{-1}\|$ centered at a consists of invertible elements.*

Proof. Write the element $a - b$ as the product $a - b = a(e - a^{-1}b)$. The first factor is invertible by assumption, and the second satisfies the conditions of Lemma 2: $\|a^{-1}b\| \leqslant \|a^{-1}\| \cdot \|b\| < 1$. Hence, the second factor is also invertible, which establishes the invertibility of the element $a - b$. $\qquad \square$

Corollary 1. *The set of all invertible (respectively, non-invertible) elements of a Banach algebra $\mathbf{A}$ is open (respectively, closed) in $\mathbf{A}$.* $\qquad \square$

Theorem 2. *Let $a \in \mathbf{A}$ be an invertible element and $\{a_n\}$ be a sequence such that $a_n \to a$ as $n \to \infty$. Then for n large enough all the elements a_n are also invertible and $a_n^{-1} \to a^{-1}$ as $n \to \infty$. In other words, the operation of passing to the inverse is continuous in its domain of definition.*

Proof. Via multiplication by a^{-1}, the problem reduces to the case $a = e$. Thus, let $a_n \to e$ as $n \to \infty$. Denote $e - a_n$ by b_n. Fix a number N such that $\|b_n\| < 1/2$ for $n > N$. Then, thanks to Lemma 2, for $n > N$ all the a_n are invertible and

$$\|a_n^{-1} - e\| = \|(e - b_n)^{-1} - e\| = \|(e + b_n + b_n^2 + \cdots) - e\| = \|b_n + b_n^2 + \cdots\|$$

$$\leqslant \|b_n\| \cdot \|e + b_n + b_n^2 + \cdots\| \leqslant \|b_n\| \left(1 + \frac{1}{2} + \frac{1}{4} + \cdots \right) = 2\|b_n\| \to 0$$

as $n \to \infty$. $\qquad \square$

Exercises

1. If the series $e + a + a^2 + \cdots + a^n + \cdots$ converges, then the inversion formula $(e - a)^{-1} = e + a + a^2 + \cdots + a^n + \cdots$ holds.

2. On the example of the operator on $\mathbb{C}^2$ with a matrix of the form $\begin{pmatrix} 0 & 0 \\ M & 0 \end{pmatrix}$ show that in the two-dimensional Euclidean space $\mathbb{C}^2$ there exist operators T with arbitrarily large norm for which the series $I + T + T^2 + \cdots$ nevertheless converges. This demonstrates that, in contrast to the scalar case, the inversion formula is applicable to some elements with large norm.

3. In $L(\ell_2)$ the set of all non-invertible operators has nonempty interior.

4. In $L(\ell_2)$ the set of all non-invertible operators is not closed in the sense of pointwise convergence.

5. If the space X is finite-dimensional, then the set of all non-invertible operators has empty interior in $L(X)$.

6. $A(\mathbb{T})$ is a subalgebra of the algebra $C(\mathbb{T})$ (for the definition, see the exercises in Subsection 10.4.3).

7. Give an example of a function $f \in A(\mathbb{T})$ for which $\frac{1}{f} \in C(\mathbb{T}) \setminus A(\mathbb{T})$.
This shows that an element can be non-invertible in a subalgebra, yet be invertible in a wider algebra.

8. Give an example of a function $f \in C[0, 1]$ that is non-invertible not only in $C[0, 1]$, but also in any other wider algebra.

9. Let $a \in \mathbf{A}$ and suppose the operator T_a from Exercise 5 of Subsection 11.1.1 is not bounded below. Then the element a is non-invertible not only in $\mathbf{A}$, but also in any other wider Banach algebra.

10. Based on the preceding exercise and Exercise 11 in Subsection 10.2.3, show that if the element $f \in C[0, 1]$ is non-invertible in $C[0, 1]$, then it also is non-invertible in any other wider Banach algebra.

11. Based on the theorem on small perturbations of an invertible element, prove that if a_n are invertible elements, $a_n \to a$ as $n \to \infty$, and $\sup_n \|a_n^{-1}\| < \infty$, then a is also invertible.

An element $a \in \mathbf{A}$ is said to be *right invertible* if there exists an element $b \in \mathbf{A}$ (called a *right inverse*) such that $ab = e$. Similarly, an element $a \in \mathbf{A}$ is said to be *left invertible* if there exists an element $d \in \mathbf{A}$ (called a *left inverse*) such that $da = e$.

12. If the element $a \in \mathbf{A}$ is both right and left invertible, then it is invertible, and both its right inverse and its left inverse coincide with a^{-1}.

13. On the example of the right-shift operator $S_r \in L(\ell_2)$, acting as $S_r(x_1, x_2, \ldots) = (0, x_1, x_2, \ldots)$, show that an operator can be left invertible, but not right invertible, and that the left inverse is not necessarily unique.

14. On the example of the left-shift operator $S_l \in L(\ell_2)$, acting as $S_l(x_1, x_2, \ldots) = (x_2, x_3, \ldots)$, show that an operator can be right invertible, but not left invertible, and that the right inverse is not necessarily unique.

11.1.3   The Spectrum

Definition 1. A complex number λ is said to *belong to the spectrum* of the element $a \in \mathbf{A}$ if the element $a - \lambda e$ is not invertible. The set of all such points is called the *spectrum* of the element a, and is denoted by $\sigma(a)$. A complex number that does not belong to the spectrum of a is called a *regular point* of the element a.

Theorem 1. *The spectrum of any element $a \in \mathbf{A}$ has the following properties:*

— $\sigma(a)$ *is closed in* $\mathbb{C}$;

— $\sigma(a)$ *is bounded and lies in the closed disc of radius* $\|a\|$ *centered at* 0.

Proof. Closedness. Let $\lambda_n \in \sigma(a)$ be such that $\lambda_n \to 0$ as $n \to \infty$. Then the elements $a - \lambda_n e$ are not invertible, and neither is their limit $a - \lambda e$, since the set of non-invertible elements of a Banach algebra is closed (Corollary 1, Subsection 11.1.2). Hence, $\lambda \in \sigma(a)$, so the spectrum is closed, as claimed.

Boundedness. Let $|\lambda| > \|a\|$. Then $a - \lambda e = -\lambda(e - \lambda^{-1}a)$, and the element $e - \lambda^{-1}a$ is invertible by the lemma on small perturbations of the unit element. Therefore, all complex numbers that are larger in modulus than the norm $\|a\|$ of a are regular points, hence all the points of the spectrum are not larger in modulus than $\|a\|$.

Exercises

1. Prove that the spectrum of any element $f \in C[0, 1]$ coincides with the set of values (range) of the function f.

2. Describe the spectrum of an element of the algebra $L_\infty[0, 1]$.

The *spectral radius* of the element a is the number

$$r(a) = \varlimsup_{n \to \infty} \|a^n\|^{1/n}.$$

3. Show that $\sigma(a) \subset r(a)D \subset \|a\| D$, where D is the closed unit disc in the complex plane (to do this, sharpen the lemma on small perturbations of the unit element).

4. Show that in the formula for the spectral radius the upper limit can be replaced by the ordinary limit.

5. Show that $r(a)$ is the minimal radius of a disc that contains the spectrum of the element a.

6. Show that $\sigma(a + te) = \sigma(a) + t$, and $\sigma(tA) = t\sigma(A)$.

11.1.4 The Resolvent and Non-emptyness of the Spectrum

The *resolvent* of the element $a \in \mathbf{A}$ is the function $R_a \colon \mathbb{C} \setminus \sigma(a) \to \mathbf{A}$ defined by the formula

$$R_a(\lambda) = (a - \lambda e)^{-1}.$$

Properties of the resolvent: 1. *Main* (or *first*) *resolvent identity*:

$$R_a(\lambda) - R_a(\mu) = (\lambda - \mu) R_a(\lambda) R_a(\mu).$$

Proof. We have

$$(\lambda - \mu)(a - \lambda e)^{-1}(a - \mu e)^{-1} = (a - \lambda e)^{-1} \left((a - \mu e) - (a - \lambda e) \right) (a - \mu e)^{-1}$$

$$= (a - \lambda e)^{-1} - (a - \mu e)^{-1}. \qquad \square$$

2. *Commutativity*: $R_a(\lambda) R_a(\mu) = R_a(\mu) R_a(\lambda)$. This is an obvious consequence of the main identity.

3. *Continuity* at each point λ of the domain of definition. This follows from the continuity of the operations of addition, multiplication, and passage to the inverse element (concerning the latter, see Theorem 2 in Subsection 11.1.2).

4. *The resolvent converges to 0 at infinity.*

Proof. Letting $\lambda \to \infty$ is allowed, since the spectrum is a bounded set; in the proof we will take $|\lambda| > 2\|a\|$. Let us rewrite the resolvent as

$$R_a(\lambda) = -\lambda^{-1}(e - \lambda^{-1}a)^{-1}.$$

Since $|\lambda| > 2\|a\|$, we have $\|\lambda^{-1}a\| < 1/2$, and so we can apply to the element $e - \lambda^{-1}a$ the inversion formula:

$$R_a(\lambda) = -\lambda^{-1}\left(e + \lambda^{-1}a + \lambda^{-2}a^2 + \cdots\right).$$

Passing to norms and using the triangle inequality, we obtain

$$\|R_a(\lambda)\| < \frac{1}{|\lambda|}\left(1 + \frac{1}{2} + \frac{1}{4} + \cdots\right) = \frac{2}{|\lambda|},$$

which obviously converges to 0 as $\lambda \to \infty$. $\square$

Definition 1. Let $D \subset \mathbb{C}$ be an open set and E be a complex Banach space. A function $F: D \to E$ is said to be *differentiable at the point* $\lambda_0 \in D$ if the limit

$$\lim_{\lambda \to \lambda_0} \frac{F(\lambda) - F(\lambda_0)}{\lambda - \lambda_0}$$

exists. As in the scalar case, this limit is called the *derivative* of the function F at the point λ_0 and is denoted by $F'(\lambda_0)$. The function $F(\lambda)$ is said to be *analytic* in the domain D if it is differentiable at all points of D.

Proposition 1. *The resolvent is analytic in its domain of definition.*

Proof. We use the main resolvent identity to calculate the needed limit:

$$\lim_{\lambda \to \lambda_0} \frac{R_a(\lambda) - R_a(\lambda_0)}{\lambda - \lambda_0} = \lim_{\lambda \to \lambda_0} R_a(\lambda) R_a(\lambda_0) = (R_a(\lambda_0))^2.$$ $\square$

Theorem 1 (Liouville's theorem for Banach space-valued functions). *If the function* $F: \mathbb{C} \to E$ *is analytic and bounded, then it is constant.*

Proof. Suppose $F(z_1) \neq F(z_2)$ for some points $z_1, z_2 \in \mathbb{C}$. By the Hahn–Banach theorem, there exists a functional $f \in E^*$ such that $f(F(z_1)) \neq f(F(z_2))$. Consider the auxiliary function $g: \mathbb{C} \to \mathbb{C}$ given by $g(z) = f(F(z))$.

In view of the continuity, the functional f can be switched with the limit symbol, so the function g is analytic. Moreover, $\sup_{z \in \mathbb{C}} |g(z)| \leqslant \|f\| \cdot \sup_{z \in \mathbb{C}} \|F(z)\| < \infty$. Hence, by Liouville's theorem for scalar-valued functions, g is a constant, so $g(z_1) = g(z_2)$. The contradiction we reached completes the proof of the theorem. $\square$

We are now ready to prove the theorem for the sake of which we introduced the notion of resolvent.

Theorem 2 (Non-emptyness of the spectrum). *The spectrum of any element of a Banach algebra is not empty.*

Proof. We argue by reduction ad absurdum. Suppose the spectrum of the element $a \in \mathbf{A}$ is empty. Then the domain of definition of the resolvent R_a is the entire complex plane. Since the resolvent is continuous and tends to 0 at ∞, it follows that R_a is bounded in the entire plane $\mathbb{C}$. And since R_a is analytic in $\mathbb{C}$, the conditions of Liouville's theorem are satisfied. Hence, $R_a = $ const. But $\lim_{\lambda \to \infty} R_a(\lambda) = 0$, so necessarily $R_a(\lambda) \equiv 0$, which contradicts the definition of the resolvent: the values of R_a are necessarily invertible elements of the algebra.   □

Remark 1. The application of Liouville's theorem here is not accidental. The analogous theorem establishing that the spectrum of a square matrix is not empty is based on the existence of a root of the equation $\det(A - \lambda I) = 0$, which follows from the fundamental theorem of algebra, in its turn most frequently proved by using Liouville's theorem.

Exercises

1. Calculate the spectrum and resolvent of the unit element.

2. Suppose the element $a \in \mathbf{A}$ satisfies the equation $a^2 = a$ (such elements are called *idempotent elements*, or simply *idempotents*). Using the inversion formula, calculate the resolvent of a. What kind of spectrum can an idempotent have?

3. Suppose the element $a \in \mathbf{A}$ satisfies for some $n \in \mathbb{N}$ the equation $a^n = 0$ (such elements are called *nilpotent elements*, or simply *nilpotents*). Using the inversion formula, calculate the resolvent and the spectrum of a.

4. Let $\lambda \in \sigma(a)$, and let λ_n be regular points of the element a such that $\lambda_n \to \lambda$ as $n \to \infty$. Then $\|R_a(\lambda_n)\| \to \infty$ as $n \to \infty$.

5. Let $\sigma(a) = \{0\}$ and suppose $\|R_a(\lambda)\| \leqslant C|\lambda|^{-1}$, for all $\lambda \in \mathbb{C} \setminus \{0\}$, where $C > 0$ is some constant. Show that $a = 0$.

6. Suppose the Banach algebra $\mathbf{A}$ has the property that all its nonzero elements are invertible. Based on the theorem on the non-emptiness of the spectrum, show that in this case every element $a \in \mathbf{A}$ has the form $a = \lambda e$ for some $\lambda \in \mathbb{C}$. In other words, up to isomorphism, the only Banach algebra that is a field is the field $\mathbb{C}$ of complex numbers.

11.1.5 The Spectrum and Eigenvalues of an Operator

Henceforth, up to the moment we will pass to the theme of "operators in Hilbert space", the letter X will stand only for a complex Banach space. In the present subsection we will consider continuous linear operators in a space X, i.e., elements of the algebra $L(X)$. The *spectrum* of an operator is a particular case of the spectrum of an element in an algebra: the number λ belongs to the spectrum of the operator $T \in L(X)$ if the operator $T - \lambda I$ is not invertible.

The number $\lambda \in \mathbb{C}$ is called an *eigenvalue* of the operator $T \in L(X)$ if there exists a non-zero element $x \in X$, called an *eigenvector corresponding* (or *associated*, or *belonging) to the eigenvalue* λ, if $Tx = \lambda x$. In this case $(T - \lambda I)x = 0$, i.e., the operator $T - \lambda I$ is not injective, and hence is not invertible. Therefore, any eigenvalue of the operator T belongs to the spectrum of T. However, besides non-injectivity, a reason for the operator $T - \lambda I$ to be non-invertible is lack of surjectivity. Therefore, generally speaking, the spectrum of an operator is not exhausted by its eigenvalues. Moreover, even rather simple operators may have no eigenvalues, while as we already know, the spectrum is always non-empty.

Example 1. Consider the operator $T \in L(C[0, 1])$, acting by the rule $(Tf)(x) = \int_0^x f(t)dt$. Suppose that T has an eigenvalue λ with eigenvector f. Then $\int_0^x f(t)dt = \lambda f(x)$, and so $f(x) = \lambda f'(x)$, $f(0) = 0$. This Cauchy problem has the unique solution $f \equiv 0$, so the operator T has no eigenvalues.

It goes without saying that similar examples are possible only in infinite-dimensional spaces. As we know from linear algebra, every operator in $\mathbb{C}^n$ is given by a square matrix A, and the search for its eigenvalues reduces to solving the equation $\det(A - \lambda I) = 0$, which in turn is always solvable.

Let λ be an eigenvalue of the operator $T \in L(X)$. The *eigenspace* (or *eigensubspace*) corresponding (or associated, or belonging) to the eigenvalue λ is defined to be the set $\text{Ker}(T - \lambda I)$. In other words, the eigenspace corresponding to the eigenvalue λ consists of the eigenvectors corresponding to λ, and zero.

Definition 1. A subspace $Y \subset X$ is called an *invariant subspace* of the operator $T \in L(X)$, if $T(Y) \subset Y$.

Eigenspaces are obvious examples of invariant subspaces. Conversely, knowing invariant subspaces of an operator can help in the search for eigenvectors and eigenvalues. For example, if the operator $T \in L(X)$ has a finite-dimensional invariant subspace Y, then the restriction of T to Y is already an operator in a finite-dimensional space, so it has eigenvectors in this space.

Theorem 1. *Suppose the operators A and T commute. Then any eigenspace of one of these two operators is an invariant subspace of the other operator.*

Proof. Let λ be an eigenvalue of A and $Y = \text{Ker}(A - \lambda I)$ be the corresponding eigensubspace. Take an arbitrary eigenvector $x \in Y$. We claim that $Tx \in Y$, i.e., Tx is also an eigenvalue of A corresponding to the eigenvalue λ. Indeed, $A(Tx) = T(Ax) = T(\lambda x) = \lambda(Tx)$. $\square$

Exercises

1. Verify that for the operator T in Example 1, $0 \in \sigma(T)$.

2. In a finite-dimensional space X the injectivity and surjectivity of an operator $T \in L(X)$ are equivalent.

3. In a finite-dimensional space the spectrum of an operator and the set of its eigenvalues coincide.

4. In a finite-dimensional space the invertibility of an operator is equivalent to its right invertibility (do not confuse with the case of operators acting from one space to another!).

5. The assertions of Exercises 2 and 4 above fail in the space ℓ_2 for the left-shift operator V acting as $V(x_1, x_2, \ldots) = (x_2, x_3, \ldots)$.

Let us introduce two more notions. A point λ is called an *approximate eigenvalue* of the operator T if there exists a sequence of elements $x_n \in S_X$ such that $Tx_n - \lambda x_n \to 0$ as $n \to \infty$. For large n the elements x_n are "almost" eigenvalues of the operator T, though a genuine eigenvector corresponding to the number λ does not necessarily exist.

6. The number λ belongs to the spectrum of the operator T if and only if one of the following cases occur:

— λ is an eigenvalue of T, i.e., the operator $T - \lambda I$ is not invertible.
— λ is an approximate eigenvalue of T, i.e., the operator $T - \lambda I$ is not bounded below.
— λ is an eigenvalue of the adjoint operator T^*, i.e., the operator $(T - \lambda I)^*$ is not injective.

7. Calculate the spectrum and resolvent of the unit operator.

8. Let T be a projector in the space X. Describe the spectrum and resolvent of T.

9. Determine the spectrum for the following operators:

— the right-shift operator $U(a_1, a_2, \ldots) = (0, a_1, a_2, \ldots)$ and the left-shift operator $V(a_1, a_2, \ldots) = (a_2, a_3, \ldots)$ in sequence spaces for the particular spaces $X = \ell_2$ and $X = \ell_\infty$;
— the shift operators $(T_\tau f)(t) = f(t + \tau)$ in spaces of functions for the following particular spaces: X is the space $C_b(\mathbb{R})$ of all bounded continuous functions on the real line equipped with the sup-norm, and $X = L_2(\mathbb{R})$.

10. Prove that the left-shift operator $U \in L(\ell_2)$ defined in Exercise 9 above is an interior point in the set of all non-invertible operators in $L(\ell_2)$. This enables one to solve Exercise 3 in Subsection 11.1.2. A more general fact: every bounded below non-invertible operator $T \in L(X, Y)$ is surrounded by a neighborhood consisting of non-invertible operators.

11. Prove that the boundary points of the spectrum of an operator are either eigenvalues, or approximate eigenvalues.

12. Prove that in the space $L(X, Y)$ (although this space is not an algebra) the set of invertible operators is open.

11.1.6   The Matrix of an Operator

In linear algebra operators are often given by their matrices. The notion of the matrix of an operator is also meaningful in infinite-dimensional spaces (needless to say, with finite matrices replaced by infinite matrices).

Let X, Y be Banach spaces, $A \in L(X, Y)$ an operator, and $\{e_m\}_{m \in \mathbb{N}}$ and $\{g_n\}_{n \in \mathbb{N}}$ bases in the spaces X and Y, respectively. The operator A is determined if we know the images Ae_m of all elements of the basis, since the linear span of the basis is a dense subset in X, and the operator A is continuous. Denote by g_n^* the coordinate functionals of the basis $\{g_n\}$ in Y. Each element of the space Y can be expressed as a series $y = \sum_{n=1}^{\infty} g_n^*(y) g_n$; in particular, $Ae_m = \sum_{n=1}^{\infty} g_n^*(Ae_m) g_n$.

Therefore, the numbers $a_{n,m} = g_m^*(Ae_n)$ completely determine the operator A.

Definition 1. The collection of numbers $a_{n,m} = g_n^*(Ae_m)$, $n, m \in \mathbb{N}$, is called the *matrix of the operator* A in the bases $\{e_m\}$, $\{g_n\}$. With this notation, $Ae_m = \sum_{n=1}^{\infty} a_{n,m} g_n$, hence, the n-th column of the matrix A consists of the coefficients of the expansion of the element Ae_n in the basis $\{g_m\}$.

Up to this point we have studied spectral theory having in our possession a rather restricted supply of examples. The next exercises show how, by using the notion of the matrix of an operator, one can construct wide classes of operators in classical spaces.

Exercises

1. Let the numbers $a_{n,m}$ form the matrix of the operator A in the bases $\{e_m\}$, $\{g_n\}$. Let $x = \sum_{m=1}^{\infty} x_m e_m$. Then

$$Ax = \sum_{m=1}^{\infty} \left(\sum_{n=1}^{\infty} a_{nm} x_m g_n \right) = \sum_{n=1}^{\infty} \left(\sum_{m=1}^{\infty} a_{nm} x_m g_n \right).$$

2. Suppose that in the standard basis of the space ℓ_2 the matrix of the operator $T \in L(\ell_2)$ has diagonal form. Show that the spectrum of such an operator T coincides with the closure of the set of diagonal elements of its matrix.

3. Using the preceding exercise, show that every bounded nonempty closed set of complex numbers can serve as the spectrum of some operator.

4. Under the conditions of Exercise 2 above, let all diagonal elements of the matrix of the operator T be distinct. Describe all operators that commute with T.

5. Suppose the matrix A has the two-diagonal form

$$
\begin{pmatrix}
1 & 0 & 0 & 0 & \cdots \\
1 & 1 & 0 & 0 & \cdots \\
0 & 1 & 1 & 0 & \\
0 & 0 & 1 & 1 & \\
\vdots & \vdots & & \ddots & \ddots
\end{pmatrix}
$$

and so on to infinity. Show that in the space ℓ_2 there exists a continuous operator whose matrix in the standard basis of ℓ_2 is A. Calculate the norm of this operator and its spectrum. Are the points of the spectrum eigenvalues?

6. Prove that the numbers $a_{n,m}$ form the matrix of a continuous operator in the standard basis of the space ℓ_1 if and only if the quantity $\sup_m \sum_n |a_{n,m}|$ is finite. Express the norm of such an operator through the elements of this matrix.

7. By analogy with the preceding exercise, describe the operators in the space c_0. To this end consider the matrix of the adjoint operator and use the result of the preceding exercise.

8. The matrix $A = \{a_{n,m}\}$ is called a *Hilbert–Schmidt* matrix, if $\sum_{n,m \in \mathbb{N}} |a_{nm}|^2 < \infty$. Show that any Hilbert–Schmidt matrix A is the matrix of a some continuous operator in the standard basis of the space ℓ_2.

9. Give an example of a continuous operator A in ℓ_2 whose matrix is not Hilbert–Schmidt.

Remark 1. Unfortunately, specifying an operator by a matrix becomes less convenient when dealing with infinite-dimensional spaces. Verifying that a matrix does indeed give a continuous operator is often not an easy task, and sometimes is practically impossible. It is for this reason that in infinite-dimensional spaces matrices are applied far less frequently that in the finite-dimensional setting.

11.2 Compact Sets in Banach Spaces

An important task of functional analysis is to reveal new effects that appear when one passes from finite-dimensional spaces to infinite-dimensional spaces. Knowledge of these effects allows one to avoid mistakes that result when one superficially applies analogies with the finite-dimensional case or, say, through the unsubstantiated "approximation" of an infinite-dimensional object by finite-dimensional objects when looking for numerical solutions of equations or optimization problems in Banach spaces. Even more important (especially in applications) is to learn to single-out classes of objects for which the analogy with the finite-dimensional case is, though not to a full extent, nevertheless applicable. Below, in Section 11.3, we study one such object — the class of compact (completely continuous) operators. Compactness will play an important role in fixed-point theorems (Chap. 16), in the study of the duality between a space and its conjugate (dual) space, as well as in many other sections of our book. Although the main properties of compact sets in topological and metric spaces were already recalled in Chap. 1, it is not redundant to pause and discuss special features emerging in the study of compact sets in Banach spaces.

11.2.1 Precompactness: General Results

As in the case of an arbitrary complete metric space (Theorem 2 of Subsection 1.4.1), for a closed subset A of a Banach space X the following conditions are equivalent:

— A is compact;
— A is precompact;
— from any sequence of elements of A one can extract a convergent subsequence.

However, in a Banach space there are structures that are not present in an arbitrary complete metric space. These are the operations of addition, multiplication by scalars, and notions they generate, such as dimension, linear subspace, linear operator, etc. The connections between precompactness and these linear structures are discussed in this subsection.

Theorem 1. *Precompactness is stable under the linear operations:*

(a) *if A, B are precompact in a normed space, then $A + B$ is precompact;*
(b) *if $A \subset X$ is precompact and $T \in L(X, Y)$, then $T(A)$ is precompact in Y;*
(c) *in particular, precompactness is preserved under multiplication by a scalar.*

Proof. (a) Let A_1 and B_1 be finite $\varepsilon/2$-nets of the sets A and B, respectively. Then $A_1 + B_1$ is a finite ε-net of the set $A + B$.

(b) If A_1 is a finite $\varepsilon/\|T\|$-net of the set A, then $T(A_1)$ is a finite ε-net of the set $T(A)$.

(c) Multiplication by a fixed scalar is a continuous linear operator.   □

Theorem 2. *For a bounded subset A of a normed space X the following conditions are equivalent:*

(1) *A is precompact;*
(2) *for any $\varepsilon > 0$ there exists a finite-dimensional subspace $Y \subset X$ that constitutes an ε-net for A.*

Proof. (1) $\Longrightarrow$ (2). Let A_1 be a finite ε-net of the set A. Then $\operatorname{Lin} A_1$ is a finite-dimensional subspace that provides the requisite ε-net.

(2) $\Longrightarrow$ (1). Let the subspace $Y \subset X$ be finite-dimensional and an ε-net for A. Denote by r a positive number such that $A \subset r B_X$. The set $(r + \varepsilon) B_Y$ is a bounded subset of the finite-dimensional space Y, and hence is precompact. Since Y is an ε-net for A, for any $x \in A$ there exists a $y \in Y$ such that $\|x - y\| < \varepsilon$. But this element

y lies in $(r + \varepsilon)B_Y$, because $\|y\| \leqslant \|x\| + \|y - x\| < r + \varepsilon$. Therefore, $(r + \varepsilon)B_Y$ is an ε-net for A. That is, for every $\varepsilon > 0$ we found a precompact ε-net for A; this means (Lemma 2 of Subsection 1.4.1) that A is precompact. $\qquad\square$

Theorem 3. *The convex hull of any precompact set is precompact.*

Proof. We use the preceding theorem. Let A be a precompact set, and let the subspace $Y \subset X$ be finite-dimensional and be an ε-net for A. We claim that Y is also an ε-net for conv A. Indeed, let $x = \sum_{k=1}^{n} \lambda_k x_k$ be an arbitrary convex combination of elements $x_k \in A$. Since Y is an ε-net for A, one can choose elements $y_k \in Y$ such that $\|x_k - y_k\| < \varepsilon$. Put $y = \sum_{k=1}^{n} \lambda_k y_k$. Then we have

$$\|x - y\| = \left\| \sum_{k=1}^{n} \lambda_k(x_k - y_k) \right\| \leqslant \sum_{k=1}^{n} \lambda_k \|x_k - y_k\| < \varepsilon \sum_{k=1}^{n} \lambda_k = \varepsilon.$$

But $y \in Y$ (because Y is a linear subspace!), so Y is an ε-net for the set conv A, and conv A is a precompact. $\qquad\square$

Theorem 4. *Let X, Y be Banach spaces, and let $T_n, T \in L(X, Y)$. Suppose the sequence of operators (T_n) converges to T at every point of a precompact set $A \subset X$ and the norms of all the operators T_n and T are bounded from above by some constant $C < \infty$. Then the sequence (T_n) converges to T uniformly on A. In particular, by the Banach–Steinhaus theorem, if T_n converge to T pointwise on the whole X, then the boundedness condition is satisfied, so the pointwise convergence on X implies uniform convergence on every precompact set.*

Proof. Fix $\varepsilon > 0$, and choose in the precompact set A a finite $\varepsilon/(4C)$-net B. The set B is finite, and at each of its points y, $T_n y \to Ty$ as $n \to \infty$. Hence, one can choose a number m such that for any $n > m$ the inequality $\|(T_n - T)y\| < \varepsilon/2$ holds for all $y \in B$. Since B is an $\varepsilon/(4C)$-net for A, for any $x \in A$ there exists a $y \in Y$ such that $\|x - y\| < \varepsilon/(4C)$. It follows that for any $n > m$ and any $x \in A$,

$$\|(T_n - T)x\| \leqslant \|(T_n - T)y\| + \|(T_n - T)(x - y)\|$$
$$< \frac{\varepsilon}{2} + \|T_n - T\| \cdot \|x - y\| < \frac{\varepsilon}{2} + 2C\frac{\varepsilon}{4C} = \varepsilon.$$

This establishes the uniform convergence on A. $\qquad\square$

The next theorem should be regarded as an important call for caution: **in an infinite-dimensional space bounded sets are not necessarily precompact!**

Theorem 5 (F. Riesz's theorem). *The unit ball in an infinite-dimensional normed space cannot be precompact.*

Proof. Suppose, by contradiction, that B_X is precompact. Fix $\varepsilon \in (0, 1)$. Let $Y \subset X$ be an arbitrary finite-dimensional subspace. Since $Y \neq X$, the quotient space X/Y is nontrivial. Pick an element $[x] \in X/Y$ such that $\varepsilon < \|[x]\| < 1$. By the definition of the norm in the quotient space, there exists a representative $z \in [x]$ such that $\|z\| < 1$. Then $z \in B_X$, but at the same time $\inf_{y \in Y} \|z - y\| = \|[z]\| = \|[x]\| > \varepsilon$, that is, Y is not an ε-net for B_X. Thus, we have shown that no finite-dimensional space can be an ε-net for the precompact set B_X, which contradicts Theorem 2. $\square$

Exercises

1. Reduce assertion (b) of Theorem 1 to the theorem asserting that the image of a compact set under a continuous map is compact.

2. Using the operator $U \colon X_1 \times X_2 \to X$ given by $U(x_1, x_2) = x_1 + x_2$, reduce assertion (a) of Theorem 1 to assertion (b) and the fact that a Cartesian product of compact sets is compact.

3. Derive the following generalization of Theorem 4: Let X, Y be metric spaces, $T_n \colon X \to Y$, and suppose the sequence of mappings $\{T_n\}_{n=1}^{\infty}$ is uniformly continuous. If the sequence (T_n) converges pointwise to a mapping T, then on any precompact set $A \subset X$ the sequence (T_n) converges to T uniformly.

Definition 1. A metric space X is said to have the *small ball property* (and one writes $X \in$ SBP), if for any $\varepsilon > 0$ the space X can be covered by a sequence of balls $B(x_n, r_n)$ whose radii satisfy the conditions $\sup_n r_n < \varepsilon$ and $r_n \to 0$ as $n \to \infty$.

4. If X is precompact, then $X \in$ SBP.

5. If $X = \bigcup_{n=1}^{\infty} X_n$ and all $X_n \in$ SBP, then $X \in$ SBP.

6. If X is a Banach space and $X \in$ SBP, then X is finite-dimensional.

7. The closure of the convex hull of a compact set is compact.

8. Give an example showing that the convex hull of a compact set is not necessarily compact.

9. For any precompact set K in a Banach space X there exists a sequence (x_n) of elements of X, with $x_n \to 0$, such that K is included in the closure of the convex hull of the sequence (x_n).

11.2.2 Finite-Rank Operators and the Approximation Property

A continuous operator is called a *finite-rank operator* (or *finite-dimensional operator*) if its image is finite-dimensional. We have already encoutered various examples of finite-rank operators: linear functionals, partial-sum of Fourier series operators (Subsection 10.4.3), partial-sum operators with respect to a Schauder basis (Subsection 10.5.2).

A sequence of operators $T_n \in L(X)$ is called an *approximate identity* if all the operators T_n are of finite rank and $T_n \to I$ pointwise. The space X is said to have the *pointwise approximation property*[1] if X admits an approximate identity.

Examples

1. Let $\{e_n\}_1^\infty$ be a basis of the Banach space X. Then the partial-sum operators S_n form an approximate identity. Indeed, let $x \in X$ be an arbitrary element and $x = \sum_{k=1}^\infty a_k e_k$ be its decomposition in the basis $\{e_n\}_1^\infty$. Then $S_n(x) = \sum_{k=1}^n a_k e_k$ and $S_n(x) \to x$ as $n \to \infty$. Therefore, any space with a basis has the pointwise approximation property.

2. For any function $f \in C[0, 1]$ denote by $L_n(f)$ the piecewise-linear continuous function that coincides with f at the points 0, $\frac{1}{n}$, $\frac{2}{n}$, $\frac{3}{n}$, ..., 1, and is linear on the intervals $\left[\frac{k}{n}, \frac{k+1}{n}\right]$. The operators L_n thus defined form an approximate identity in the space $C[0, 1]$.

3. Let $\Delta_{n,k} = \left[\frac{k-1}{n}, \frac{k}{n}\right]$, $k = 1, 2, \ldots, n$. For any function $f \in L_1[0, 1]$, put

$$E_n(f) = \sum_{k=1}^n \left(n \int_{\Delta_{n,k}} f(t)dt \, \mathbb{1}_{\Delta_{n,k}} \right).$$

In other words, $E_n(f)$ is a piecewise-constant function whose value on the interval $\Delta_{n,k}$ is the mean value of the function f on $\Delta_{n,k}$, $k = 1, 2, \ldots, n$. The operators E_n so defined (called *averaging operators*) form an approximate identity in $L_1[0, 1]$.

Theorem 1. *Let X be a Banach space with the pointwise approximation property. Let the sequence $T_n \in L(X)$ be some fixed approximate identity in the space X. Then for any set $D \subset X$ the following two conditions are equivalent:*

(1) *D is precompact;*

(2) *D is bounded, and the sequence (T_n) converges uniformly on D to the identity operator.*

---

[1] Actually, there is an entire group of properties of "approximation property" type (see [28]). What we call here the pointwise approximation property should be more accurately referred to as the *bounded approximation property for separable spaces*.

Proof. The implication (1) $\Longrightarrow$ (2) is a direct consequence of Theorem 4 of Subsection 11.2.1. Let us prove the converse implication. Suppose that $T_n \to I$ on D as $n \to \infty$. Then for any $\varepsilon > 0$ there exists a number $m = m(\varepsilon)$ such that $\|T_m x - x\| < \varepsilon$ for all $x \in D$. This means, in particular, that the subspace $T_m(X)$ is an ε-net for D. Since by the definition of an approximate identity all the operators T_n are of finite rank, the subspace $T_m(X)$ is finite-dimensional. To complete the proof it remains to apply Theorem 2 of Subsection 11.2.1. $\square$

Exercises

1. Prove that if an operator is given by a matrix with only finitely many non-zero elements, then the operator is of finite rank. Give an example of a matrix with infinitely many non-zero elements which defines a finite-rank operator in the space ℓ_2.

2. Let X, Y be Banach spaces, and let $\{y_k\}_{k=1}^n \subset Y$ and $\{f_k\}_{k=1}^n \subset X^*$ be finite collections of vectors and functionals, respectively. Define the operator $T \in L(X, Y)$, written $T = \sum_{k=1}^n f_k \otimes y_k$, by the rule $Tx = \sum_{k=1}^n f_k(x) y_k$.[2] Prove that T is a finite-rank operator, and that every finite-rank operator can be represented in the above form, where moreover the vectors $\{y_k\}_{k=1}^n$ and the functionals $\{f_k\}_{k=1}^n$ can be chosen to be linearly independent.

3. Verify that the mappings L_n constructed in the second example above are indeed finite-rank continuous operators that form an approximate identity in $C[0, 1]$.

4. Justify the third example above by using the following recipe: Verify that the averaging operators E_n are linear and dim $E_n(L_1[0, 1]) = n$; prove that $\|E_n\| = 1$. Prove that for any continuous f, $E_n f \to f$ as $n \to \infty$ uniformly on $[0, 1]$ (and hence, also in the metric of $L_1[0, 1]$). To prove the pointwise convergence of E_n to I on the whole space $L_1[0, 1]$, use the pointwise convergence criterion (Subsection 10.4.2).

5. Prove that the operator $T \in L(C[0, 1])$ acting as $(Tf)(t) = \int_0^1 (x + t) f(x) dx$ has finite rank.

6. Let $T \in L(X)$ be a finite-rank operator. Then the image of T is a finite-dimensional invariant subspace. Since every eigenvector with non-zero eigenvalue lies in $T(X)$, the search for non-zero eigenvalues reduces to the study of the action of the operator T in the subspace $T(X)$. Based on these considerations, find all non-zero eigenvalues and the corresponding eigenvectors of the operator of the preceding exercise.

---

[2] Here the symbol $\otimes$ is read as *tensor product*.

11.2.3 Compactness Criteria for Sets in Specific Spaces

Since the compactness criterion (closedness + boundedness) known from calculus fails in infinite-dimensional spaces, verifying compactenss often becomes a non-trivial problem. In the attempt to solve this problem, knowing the specifics of the ambient space in which the set tested for compactness lies is of help. Each concrete space comes with its own compactness criterion. The reader is already familiar with one such compactness criterion, Arzelà's theorem. As we remarked in Subsection 1.4.2, in the space $C(K)$ of continuous functions on a compact metric space, precompactness of a set is equivalent to two conditions being satisfied simultaneously: uniform boundedness and equicontinuity. All the other classical compactness criteria are based on the theorem given in the preceding subsection. The recipe for producing such criteria is sufficiently simple: one needs to choose an approximate identity and write out in detail what uniform convergence on the set in question means.

Theorem 1 (Compactness criterion in c_0). *For a set $D \subset c_0$ to be precompact it is necessary and sufficient that there exists an element $z \in c_0$ which majorizes coordinatewise all elements of D, i.e., $z = (z_1, z_2, \ldots) \in c_0$ and $z_n \geqslant |x_n|$ for all vectors $x = (x_1, x_2, \ldots) \in D$ and all $n \in \mathbb{N}$.*

Proof. Since both precompactness and the existence of a majorant imply boundedness, it suffice to carry out the proof for bounded sets. So, consider the sequence of operators $P_n \in L(c_0)$, acting by the rule $P_n((x_j)_{j=1}^\infty) = (x_1, \ldots, x_n, 0, 0, 0, \ldots)$. Clearly, the operators P_n form an approximate identity in c_0. According to the theorem in the preceding subsection, we need to prove that for a bounded set D the existence of a majorant $z \in c_0$ is equivalent to the uniform convergence of the operators P_n to I on D.

By definition, the uniform convergence $P_n \to I$ on D means that for any $\varepsilon > 0$ there exists a number $n(\varepsilon)$ such that for every vector $x = (x_1, x_2, \ldots) \in D$ and any $n \geqslant n(\varepsilon)$, we have $\|x - P_n x\| \leqslant \varepsilon$. Decoding what the norm in c_0 and the definition of the operators P_n mean, we obtain an equivalent formulation: for any $\varepsilon > 0$ there exists a number $n(\varepsilon)$ such that for any $n \geqslant n(\varepsilon)$,

$$\sup\{|x_j| : x = (x_1, x_2, \ldots) \in D, \ j \geqslant n+1\} \leqslant \varepsilon.$$

If we define $y_n = \sup\{|x_j| : x = (x_1, x_2, \ldots) \in D, \ j \geqslant n\}$, the last condition means that $y_n \to 0$ as $n \to \infty$, i.e., the vector $y = (y_1, y_2, \ldots)$ is an element of the space c_0. It remains to observe that convergence of y_n to zero is equivalent to the existence of the requisite majorant. Indeed, the vector $y = (y_1, y_2, \ldots)$ is an element of c_0 and $y_n \geqslant |x_n|$ for all vectors $x = (x_1, x_2, \ldots) \in D$, i.e., y is a majorant of all elements of the set D. Conversely, if D admits a majorant $z = (z_1, z_2, \ldots) \in c_0$, then

$$y_n \leqslant \sup\{|z_j| : j \geqslant n\} \to 0 \quad \text{as } n \to \infty. \qquad \square$$

Applying the same operators $P_n((x_j)_{j=1}^\infty) = (x_1, \ldots, x_n, 0, 0, 0, \ldots)$ in the space ℓ_p, we obtain the following result.

Theorem 2 (Compactness criterion in ℓ_p). *Let $1 \leqslant p < \infty$. Then for a bounded set $D \subset \ell_p$ to be precompact, it is necessary and sufficient that for any $\varepsilon > 0$ there exists a number $n(\varepsilon)$ such that $\sum_{k=n(\varepsilon)}^\infty |x_k|^p \leqslant \varepsilon$ for all $x = (x_1, x_2, \ldots) \in D$.*

Upon considering partial-sum operators one arrives at the following compactness criterion in a space with a basis:

Theorem 3. *Let X be a Banach space with a basis $\{e_n\}_1^\infty$. Then for a set $D \subset X$ to be precompact, it is necessary and sufficient that for any $\varepsilon > 0$ there exists a number $n(\varepsilon)$ such that $\left\| \sum_{k=n(\varepsilon)}^\infty x_k e_k \right\| \leqslant \varepsilon$ for all vectors $x = \sum_{k=1}^\infty x_k e_k \in D$.*

In exactly the same way we can obtain a compactness criterion in the space $L_1[0, 1]$, based on the averaging operators E_n from the third example of an approximate identity given in the preceding subsection. This criterion is indeed valid, and in many cases proves sufficiently convenient. However, with no major effort, we can come up with a more elegant formulation.

All functions $f \in L_1[0, 1]$ will be considered to be defined not only on the interval $[0, 1]$, but also on the whole real line. To this end we extend them by periodicity with period 1. Now for each $\tau \in \mathbb{R}$ define the *shift* (or *translation*) operator $L_\tau \colon L_1[0, 1] \to L_1[0, 1]$ by the rule $(L_\tau f)(t) = f(t + \tau)$. As one can readily see, L_τ is a bijective isometry; in particular, $\|L_\tau\| = 1$.

Lemma 1. $L_\tau f \to f$ *as $\tau \to 0$, for all $f \in L_1[0, 1]$.*

Proof. Since the operators L_τ are jointly bounded, it suffices to establish the pointwise convergence to the unit operator not on all of $L_1[0, 1]$, but only on a dense subset (Subsection 10.4.2). As such a subset we take the set E of all 1-periodic continuous functions. For $f \in E$, denote $\|f\|_\infty = \sup_{t \in [0,1]} |f(t)|$. Then for all $t, \tau \in [0, 1]$ we have $|f(t + \tau) - f(t)| \leq 2\|f\|_\infty$, and by continuity $|f(t + \tau) - f(t)| \to 0$ as $\tau \to 0$. Now the Lebesgue dominated convergence theorem gives us the desired property

$$\|L_\tau f - f\| = \int_0^1 |f(t + \tau) - f(t)| dt \to 0$$

as $\tau \to 0$. $\qquad\qquad\square$

The lemma just proved can be restated as follows:

For any function $f \in L_1[0, 1]$ and any $\varepsilon > 0$, there exists a $\delta > 0$ such that for every $\tau \in [-\delta, \delta]$ it holds that $\int_0^1 |f(t + \tau) - f(t)| dt < \varepsilon$.
This property of a function f is called *continuity in the mean*.

Definition 1. A family of functions $D \subset L_1[0, 1]$ is said to be *equicontinuous in the mean* if for any $\varepsilon > 0$ there exists a $\delta > 0$ such that for every function $f \in D$ and every $\tau \in [-\delta, \delta]$ it holds that $\int_0^1 |f(t + \tau) - f(t)| dt < \varepsilon$.

Theorem 4 (Compactness criterion in $L_1[0, 1]$). *For a bounded subset $D \subset L_1[0, 1]$ to be precompact it is necessary and sufficient that it is equicontinuous in the mean.*

Proof. Suppose D is precompact. Since, by the lemma, $L_\tau \to I$ pointwise, one has that $L_\tau \to I$ uniformly on D. This uniform convergence is just another formulation of the requisite equicontinuity in the mean.

Now the converse. Suppose the set D is equicontinuous in the mean. Let us show that in this case the averaging operators E_n introduced in Example 3 of Subsection 11.2.2 converge uniformly on D to the unit operator. Since the sequence (E_n) is an approximate identity in $L_1[0, 1]$, this will establish the precompactness of D. Thus, given an arbitrary $\varepsilon > 0$, take a $\delta > 0$ as in the definition of the equicontinuity in the mean: $\int_0^1 |f(t + \tau) - f(t)| dt < \varepsilon$ for all $f \in D$ and all $\tau \in [-\delta, \delta]$. Then for any $n > 1/\delta$ and any $f \in D$ we have

$$\|E_n(f) - f\| = \int_0^1 \left| \sum_{k=1}^n n \int_{\Delta_{n,k}} f(x) dx \mathbb{1}_{\Delta_{n,k}}(t) - f(t) \right| dt$$

$$= \int_0^1 \left| \sum_{k=1}^n n \left(\int_{\Delta_{n,k}} [f(x) - f(t)] dx \right) \mathbb{1}_{\Delta_{n,k}}(t) \right| dt$$

$$\leqslant \int_0^1 \sum_{k=1}^n n \int_{\Delta_{n,k}} |f(x) - f(t)| dx \mathbb{1}_{\Delta_{n,k}}(t) dt = n \sum_{k=1}^n \int_{\Delta_{n,k}} \int_{\Delta_{n,k}} |f(x) - f(t)| dx dt.$$

Using the fact that all pairs $(x, t) \in \bigcup_{k=1}^n \Delta_{n,k} \times \Delta_{n,k}$ obey the conditions $0 \leqslant t \leqslant 1$ and $t - \frac{1}{n} \leqslant x \leqslant t + \frac{1}{n}$, and making the change of variables $x \to t + \tau$, we complete the estimate to

$$\|E_n(f) - f\| \leqslant \int_{[-1/n, 1/n]} \int_0^1 |f(t + \tau) - f(t)| dt d\tau < 2\varepsilon. \qquad \square$$

Exercises

1. In the definitions of continuity and equicontinuity in the mean, instead of $\tau \in [-\delta, \delta]$ one can write $\tau \in [0, \delta]$.

2. In the space ℓ_∞, the operators P_n, acting as $P_n((x_j)_{j=1}^\infty) = (x_1, \dots, x_n, 0, 0, \dots)$, do not form an approximate identity.

3. The space ℓ_∞ admits no approximate identity (indeed, we note that in ℓ_∞ no convenient compactness criterion is known).

4. Give an example of a precompact set $D \subset \ell_p$ that admits no joint majorant $z \in \ell_p$.

5. For the subsets of $C[0, 1]$ listed below, determine whether or not they are (a) bounded, (b) convex, (c) closed, (d) precompact. Also, find (e) the interior and (f) the boundary of these sets.

(i) The set of all non-decreasing functions $f \in C[0, 1]$ that satisfy the condition $0 \leqslant f \leqslant 1$.

(ii) The set of all non-decreasing functions $f \in C[0, 1]$ that satisfy the conditions $f \geqslant 0$ and $\int_{[0,1]} f(t)dt \leqslant 1$.

(iii) The set of all continuously differentiable functions f that satisfy the conditions $f(0) = 0$ and $\int_0^1 |f'(t)|dt \leqslant 1$.

(iv) The set of all continuously differentiable functions f that satisfy the conditions $f(0) = 0$ and $\int_0^1 |f'(t)|^2 dt \leqslant 1$.

6. For the sets in $L_1[0, 1]$ listed below, determine whether or not they are (a) bounded, (b) convex, (c) closed, (d) precompact. Also, find (e) the interior and (f) the boundary of these sets.

(i) The set of all non-decreasing functions $f \in L_1[0, 1]$ that satisfy the condition $0 \leqslant f \leqslant 1$.

(ii) The set of all non-decreasing functions $f \in L_1[0, 1]$ that satisfy the conditions $f \geqslant 0$ and $\int_{[0,1]} f(t)dt \leqslant 1$.

(iii) The set of all continuously differentiable functions f that satisfy the conditions $f(0) = 0$ and $\int_0^1 |f'(t)|dt \leqslant 1$.

(iv) The set of all continuously differentiable functions f that satisfy the conditions $f(0) = 0$ and $\int_0^1 |f'(t)|^2 dt \leqslant 1$.

11.3 Compact (Completely Continuous) Operators

In this section the letters X, Y, and Z will be used exclusively to denote Banach spaces.

11.3.1 Definition and Examples

Definition 1. The operator $T: X \to Y$ is called *compact*, or *completely continuous*, if the image $T(B_X)$ of the unit ball of the space X under T is precompact in the space Y. The set of all compact operators acting from the space X into the space Y will be denoted by $K(X, Y)$.

Since every bounded set is contained in some ball, the image of any bounded set under a compact operator is contained in a set of the form $rT(B_X)$, and hence is precompact.

Example 1. Consider the kernel operator (or integration/integral operator with kernel) $T : C[0, 1] \to C[0, 1]$, acting as

$$(Tf(t))(x) = \int_0^1 K(t, x) f(t) dt,$$

where the kernel $K : [0, 1] \times [0, 1] \to \mathbb{C}$ is jointly continuous in its variables.[3]

To verify that a kernel operator is compact in $C[0, 1]$, i.e, that the set $T(B_{C[0,1]})$ is precompact, we resort to Arzelà's theorem. First, we have

$$\|Tf\| \leqslant \max_{x \in [0,1]} \int_0^1 |K(t, x)| \cdot |f(t)| \, dt \leqslant \|f\| \max_{t, x \in [0,1]} |K(t, x)|,$$

which proves the boundedness of the set $T(B_{C[0,1]})$. To prove the equicontinuity of $T(B_{C[0,1]})$, we first, given an arbitrary $\varepsilon > 0$, choose a $\delta(\varepsilon) > 0$ such that for any $x_1, x_2 \in [0, 1]$ satisfying $|x_1 - x_2| < \delta(\varepsilon)$ the estimate $|K(t, x_1) - K(t, x_2)| < \varepsilon$ holds. Now let $g \in T(B_{C[0,1]})$ be an arbitrary element. Then $g = Tf$, with $f \in B_{C[0,1]}$. Accordingly, for any $x_1, x_2 \in [0, 1]$ satisfying $|x_1 - x_2| < \delta(\varepsilon)$, we have

$$|g(x_1) - g(x_2)| = |(Tf)(x_1) - (Tf)(x_2)|$$
$$\leqslant \int_0^1 |K(t, x_1) - K(t, x_2)| \cdot |f(t)| \, dt < \varepsilon \int_0^1 |f(t)| \, dt \leqslant \varepsilon \|f\| \leqslant \varepsilon,$$

so the set $T(B_{C[0,1]})$ is indeed equicontinuous.

Example 2. If the space X is infinite-dimensional, then the unit operator X is not compact. Indeed, $I(B_X) = B_X$, and the unit ball of an infinite-dimensional space is not precompact.

Exercises

1. Show that every finite-rank operator is compact.

2. A diagonal operator $T \in L(\ell_p)$ (i.e., an operator whose matrix in the standard basis is diagonal) is compact if and only if the diagonal elements of its matrix form a sequence that converges to zero.

3. Calculate the norm of the operator from Exercise 2.

---

[3] Clearly, it is unfortunate that we use the same letter K to denote the class of compact operators as well as the kernel of a kernel operator, not to speak of using it to denote compact spaces when we are dealing with the space $C(K)$. But what can we do: these notations are widely accepted. To make the reader even happier, we could, as customary, also denote a compact operator by K! But enough is enough.

4. Show that for an integration operator T with kernel K in $C[0, 1]$ it holds that $\|T\| = \max_{x \in [0,1]} \int_0^1 |K(t, x)| \, dt$ (see Example 1 above).

5. Show that the integration operator with kernel considered in Example 1 above is compact as an operator acting from $L_1[0, 1]$ to $C[0, 1]$.

6. Show that the integration operator appearing in the example in Subsection 11.1.5 is compact.

7. Let $K: [0, 1] \times [0, 1] \to \mathbb{C}$ be a function of two variables which satisfies the following condition: for each fixed $x \in [0, 1]$, the function $K_x(t) = K(t, x)$ of the variable t is integrable. Suppose further that the mapping $x \mapsto K_x$ is continuous from $[0, 1]$ to $L_1[0, 1]$. Then the integration operator with kernel $K(t, x)$ is compact in $C[0, 1]$.

11.3.2 Properties of Compact Operators

Theorem 1. *The set $K(X, Y)$ of compact operators has the following properties:*

(1) *$K(X, Y)$ is a linear subspace of $L(X, Y)$.*

(2) *$K(X, Y)$ is an operator ideal, i.e., if $T \in K(X, Y)$, then the products AT and TA are compact for every continuous operator A for which the composition is defined.*

(3) *The set $K(X, Y)$ is closed in $L(X, Y)$ in the sense of convergence in norm.*[4]

Proof. (1) Stability under multiplication by a scalar is obvious, while stability under addition follows from the relation $(T_1 + T_2)(B_X) \subset T_1(B_X) + T_2(B_X)$ and the fact that a sum of precompact sets is precompact.

(2) Let $A \in L(Z, X)$. Then every bounded subset of the space Z is mapped by the operator A into a bounded set, which, in its turn, is mapped by the operator T into a precompact set. Therefore, the operator TA is compact. Now let $A \in L(Y, Z)$. Then every bounded subset of the space X is mapped by T into a precompact set, which in turn is mapped by A also into a precompact set.

(3) Suppose the sequence of compact operators T_n converges to an operator T. We need to show that the limit T is also compact. To this end, we fix an $\varepsilon > 0$ and construct a precompact ε-net for $T(B_X)$. Choose a number n such that $\|T - T_n\| < \varepsilon$. Consider the precompact set $K = T_n(B_X)$. Then for any $x \in B_X$ it holds that $\|Tx - T_n x\| \leqslant \|T - T_n\| \cdot \|x\| < \varepsilon$. Hence, K provides the sought-for ε-net. □

Corollary 1. *If the compact operator $A \in L(X, Y)$ is invertible, then the spaces X and Y are finite-dimensional.*

---

[4]The reader should note that here closedness in the sense of pointwise convergence does not hold.

Proof. In the present case $I_X = A^{-1}A$ and $I_Y = AA^{-1}$, where I_X and I_Y are the identity operators in the respective spaces X and Y. Hence, by property (2), I_X and I_Y are compact operators, which for infinite-dimensional spaces X and Y is not possible, as we remarked in Example 2 of the preceding subsection. □

The next theorem shows that in a wide class of spaces (which practically includes all spaces encountered in applications) the compact operators can be approximated in norm by finite-rank operators. The theory of compact operators emerged from Fredholm's investigations on the theory of integral equations. In Fredholm's work the approach to integral operators relied on the approximation of integral operators by finite-rank operators. The contemporary treatment, more general and technically less complicated, is based on other ideas, due in the first place to F. Riesz. Nevertheless, the approximation by finite-rank operators remains a useful tool for the investigation of specific operators, in particular, in problems of numerical resolution of equations involving compact operators.

Theorem 2. *Let Y be a Banach space with the pointwise approximation property (for example, a space with a basis) and $T \in K(X, Y)$. Suppose that the operators S_n form an approximate identity in Y. Then the operators $T_n = S_n T$ form a sequence of finite-rank operators that converges in norm to the operator T.*

Proof. The operators T_n are of finite rank, because so are the operators S_n. Moreover, the sequence (S_n) converges pointwise to the identity operator I, therefore (Theorem 4 in Subsection 11.2.1), one also has uniform convergence on the precompact set $T(B_X)$. Consequently,

$$\|T_n - T\| = \sup_{x \in B_X} \|T_n x - Tx\| = \sup_{x \in B_X} \|(S_n - I)Tx\| = \sup_{y \in T(B_X)} \|(S_n - I)y\| \to 0$$

as $n \to \infty$. □

Theorem 3 (Schauder's theorem on compactness of the adjoint operator). *Let X, Y be Banach spaces, and $T \in K(X, Y)$. Then $T^* \in K(Y^*, X^*)$.*

Proof. According to the definition, we need to show that the sets $G = \overline{T^*(B_{Y^*})}$ are precompact in X^*. Let $K = \overline{T(B_X)}$. By assumption, K is compact in Y. Consider the space $C(K)$ of continuous functions on this compact set and the mapping $U: G \to C(K)$ which acts by the rule $U(T^* f) = f|_K$, i.e., sends the functional $T^* f \in X^*$ into the restriction of the functional $f \in Y^*$ to K.

We claim that U is an isometry. Indeed, let $T^* f_1$ and $T^* f_2$ be two arbitrary elements of the set G. We have

$$\|T^* f_1 - T^* f_2\| = \|T^*(f_1 - f_2)\| = \sup_{x \in B_X} |T^*(f_1 - f_2)x|$$

$$= \sup_{x \in B_X} |(f_1 - f_2)(Tx)| = \sup_{y \in T(B_X)} |(f_1 - f_2)y|$$

$$= \sup_{y \in K} |(f_1 - f_2)y| = \|f_1|_K - f_2|_K\|_{C(K)}.$$

This estimate also shows that the mapping U is well defined: if $T^* f_1 = T^* f_2$, then

$$\|U(T^* f_1) - U(T^* f_2)\| = \|f_1|_K - f_2|_K\| = 0,$$

that is, $U(T^* f_1) = U(T^* f_2)$.

Since the mapping U is an isometry, the precompactness of the set G is equivalent to the precompactness in $C(K)$ of its image $U(G) = \{f|_K : f \in B(Y^*)\}$. Now the set $U(G)$ is uniformly bounded, since for any $f \in B(Y^*)$ we have

$$\|f|_K\|_{C(K)} = \sup_{x \in B_X} |f|_K (Tx)| \leqslant \|f\| \cdot \sup_{x \in B_X} \|Tx\| \leqslant \|T\|.$$

On the other hand, the set $U(G)$ is equicontinuous, since its elements are functions that satisfy the Lipschitz condition with constant 1:

$$|f|_K(y_1) - f|_K(y_2)| = |f(y_1) - f(y_2)| \leqslant \|f\| \cdot \|y_1 - y_2\| \leqslant \|y_1 - y_2\|.$$

To complete the proof, it remains to apply Arzelà's theorem. □

Exercises

1. Let T be the kernel operator (see Example 1 and Exercise 4 in Subsection 11.3.1), with a kernel of the form $K(t, \tau) = \sum_{j=1}^{n} f_j(t) g_j(\tau)$. Show that this operator has finite rank.

2. Show that the kernel operator from Example 1 of Subsection 11.3.1 can be arbitrarily well approximated by finite-rank kernel operators.

3. Represent the operator from Exercise 1 above in the form $T = \sum_{k=1}^{n} f_k \otimes y_k$, as in Exercise 2 of Subsection 11.2.2.

4. The adjoint of a finite-rank operator is itself a finite-rank operator.

5. For any finite-rank operator $A \in L(X, Y)$ it holds that

$$\text{codim}_X \text{Ker } A = \text{codim}_{Y^*} \text{Ker } A^* = \dim A(X) = \dim A^*(X^*).$$

If the operator A has infinite rank, then all these four numerical characteristics are infinite (recall that for a subspace E of a linear space Z its codimension $\text{codim}_Z E$ is defined as $\dim(Z/F)$, see Definition 1 and Exercises 8–15 in Subsection 5.3.3).

6. The image of a compact operator $A \in L(X, Y)$ is closed if and only if A is of finite rank (i.e., as a rule, the image of a compact operator is not closed).

7. Give an example of a noncompact operator whose image is not closed.

8. Prove the following theorem, due to I. K. Daugavet: Let $T : C[0, 1] \to C[0, 1]$ be a compact operator. Then $\|I + T\| = 1 + \|T\|$. For details about Daugavet's theorem and its role in the theory of Banach spaces, see [57, 62].

9. Prove that the *Daugavet equality* $\|I + T\| = 1 + \|T\|$ for an operator T in the space ℓ_2 holds if and only if $\|T\| \in \sigma(T)$.

11.3.3 Operators of the Form $I - T$ with T a Compact Operator

In the study of the spectrum of a compact operator a central role is played by the following assertion.

Theorem 1. *Let $T \in K(X, X)$. Then*

(1) *The image of the operator $I - T$ is a closed subspace.*

(2) *If the operator $I - T$ is injective, then it is surjective.*

(3) *If the operator $I - T$ is surjective, then it is injective.*

We note that properties (2) and (3) together mean that the operator $I - T$ is either invertible, or simultaneously not injective and not surjective.

The present subsection is devoted to the proof of properties (1)–(3), devoting to each of them a separate proposition.

Proposition 1. *Let $T \in K(X, X)$. Then the image of the operator $A = I - T$ is a closed subspace.*

Proof. We argue by contradiction. Let $Y = \operatorname{Ker} A$. By the criterion established in Theorem 5 of Subsection 10.2.3, the image not being closed means that there exists a sequence (x_n) in X with the following properties

1. $\operatorname{dist}(x_n, Y) = 1$;

2. $\|Ax_n\| \to 0$ as $n \to \infty$;

3. $\|x_n\| \to 1$.

By the third property, the sequence (x_n) is bounded, i.e., it lies in $r B_X$ for some $r > 0$. In view of the compactness of the operator T, the sequence (Tx_n) lies in the precompact set $r T(B_X)$. Passing, if necessary, to a subsequence, one can ensure that the sequence (Tx_n) has a limit, which we denote by s. Since $x_n = Ax_n + Tx_n \to s$ as $n \to \infty$, we have $\operatorname{dist}(s, Y) = 1$. On the other hand, $As = \lim_{n \to \infty} Ax_n = 0$, so that s lies in the kernel of the operator A, i.e., in the subspace Y. The contradiction we reached completes the proof. $\qquad \square$

Proposition 2. *If the operator $A = I - T$, where $T \in K(X, X)$, is injective, then it is surjective.*

Proof. We again argue by contradiction. So, suppose $A(X) \neq X$. Consider the subspaces $Y_n = A^n(X), n = 0, 1, 2, \ldots$. First, let us prove that the Y_n's form a sequence $Y_0 \supset Y_1 \supset Y_2 \supset \cdots$ of strictly included subspaces. To this end we proceed by induction.

The strict inclusion $Y_0 \supset Y_1$ is obvious, since $Y_0 = X$ and $Y_1 = A(X)$. Now assume that the inclusion $Y_{n-1} \supset Y_n$ is strict. The injective operator A preserves strict inclusions, hence $Y_n = A(Y_{n-1}) \supset A(Y_n) = Y_{n+1}$, as needed.

Let us continue our reasoning. In view of the strict inclusions established above, there exist functionals $\widetilde{f}_n \in S(Y_n^*)$ such that $\widetilde{f}_n(Y_{n+1}) = 0$ for $n = 0, 1, 2, \ldots$. Extend these functionals to the whole space X, preserving their norms. Let $f_n \in X^*$ be the extended functionals. By the theorem on the compactness of the adjoint operator, the operator T^* maps the set $\{f_n\}_{n=1}^{\infty}$ into a precompact set. Now let us prove that the sequence $(T^* f_n)$ is separated, i.e., $\inf_{m \neq n \in \mathbb{N}} \|T^* f_m - T^* f_n\| > 0$. This way we will contradict the precompactness and complete the proof of the proposition.

Let $n > m$. Then

$$
\begin{aligned}
\|T^* f_m - T^* f_n\| &\geqslant \sup_{x \in B_{Y_n}} |(T^* f_m - T^* f_n)(x)| \\
&= \sup_{x \in B_{Y_n}} |((T^* - I^*)f_n + f_n + (I^* - T^*)f_m - f_m)(x)| \\
&= \sup_{x \in B_{Y_n}} |f_n((T - I)x) + f_m((I - T)x) + f_n(x) - f_m(x)| \\
&= \sup_{x \in B_{Y_n}} |-f_n(A(x)) + f_m(A(x)) + f_n(x) - f_m(x)| \\
&= \sup_{x \in B_{Y_n}} |f_n(x)| = \|\widetilde{f}_n\| = 1.
\end{aligned}
$$

Here we used the fact that $Ax \in Y_{n+1}$ for $x \in Y_n$ and the functionals f_j are equal to 0 on the subspaces with bigger index. $\qquad\square$

Let us record an important consequence of the last proposition.

Corollary 1. *Suppose the operator* $\overset{\cdot}{A} = I - T$, *where T is compact, is not invertible. Then A is not injective.*

Proof. Suppose, by contradiction, that A is injective. By the preceding proposition, A is also surjective, and injectivity plus surjectivity means invertibility. $\qquad\square$

Proposition 3. *If the operator* $A = I - T$, *where $T \in K(X, X)$ is surjective, then it is also injective.*

Proof. Suppose the operator A is surjective. Then A^* is injective (Corollary 1 in Subsection 9.4.1). Since A^* is again of the form "identity minus compact", the preceding theorem shows that the operator A^* is also surjective. Therefore, the operator A is injective (Corollary 2 in Subsection 9.4.1), as we needed to prove. $\qquad\square$

Remark 1. The main results of the present section are particular cases of the more general Fredholm theorem, formulated below in Exercise 2. Fredholm studied corresponding phenomena for integral operators. Fredholm's results were transferred to the more general case of compact operators by F. Riesz.

Exercises

1. Show that for an operator of the form $A = I - T$, with T compact, the following conditions are equivalent:

— the equation $Ax = b$ is solvable for any right-hand side;
— the homogeneous equation $Ax = 0$ does not have non-zero solutions;
— the equation $Ax = b$ is solvable for any right-hand side, and moreover the solution is unique.

2. Fredholm's theorem. Let $A = I - T$, where the operator $T \in L(X)$ is compact. Then dim Ker $A =$ dim Ker $A^* =$ codim $A(X) =$ codim $A^*(X^*)$.

Operators of the form "scalar + compact". An operator $A \in L(X)$ is said to be of *scalar + compact form* if it can be represented as $A = \lambda I + T$, with $T \in K(X, X)$ and $\lambda \in \mathbb{C}$.

3. In an infinite-dimensional space, an operator cannot have two distinct scalar + compact representations $A = \lambda I + T$.

4. The operators of scalar + compact form form a subalgebra in $L(X)$.

5. Let T be a diagonal operator in the space ℓ_2 (see Exercise 2 in Subsection 11.1.6), and let λ_n be the diagonal elements of its matrix in the standard basis $\{e_n\}_{n=1}^\infty$ of ℓ_2 (in other words, $Te_n = \lambda_n e_n, n = 1, 2, \ldots$). Prove that T is of scalar + compact form if and only if the sequence (λ_n) has a limit.

6. Let T be a projector. Then T is of scalar + compact form if and only if either its image, or its kernel, has finite dimension.

A linear functional f on a Banach algebra **A** is called a *multiplicative functional* (another term used is that of *complex homomorphism*), if $f(e) = 1$ and $f(xy) = f(x)f(y)$ for all $x, y \in$ **A**.

7. A multiplicative functional can vanish only on non-invertible elements of the algebra.

8. Every multiplicative functional on a Banach algebra is continuous and its norm is equal to 1.

9. Construct a multiplicative functional on the algebra of all scalar + compact operators acting on a Banach space X.

10. Show that on $L(\ell_2)$ there exist no multiplicative functionals.

11. For multiplicative functionals the analogue of the Hahn–Banach theorem fails: not every multiplicative functional, given on a subalgebra, can be extended to the entire Banach algebra with preservation of linearity and multiplicativity.

12. Let X be an infinite-dimensional Banach space. We say that X has the *scalar + compact property* if every operator $A \in L(X)$ is of scalar + compact form. Prove that none of the spaces $C(K)$, ℓ_p, and L_p has this property.

11.3.4   The Structure of the Spectrum of a Compact Operator

Theorem 1. *Let $T \in L(X)$ be a compact operator acting on an infinite-dimensional Banach space. Then:*

(1) *The spectrum of T is either finite, or consists of a sequence of complex numbers that converges to 0.*

(2) *0 belongs to the spectrum.*

(3) *If $\lambda \neq 0$ belongs to the spectrum, then λ is an eigenvalue of T; the eigenspaces corresponding to the nonzero eigenvalues are finite-dimensional.*

Proof. We begin by establishing the last property. Suppose $\lambda \neq 0$ does not belong to the spectrum of T. Then the operator $(T - \lambda I) = -\lambda(I - \lambda^{-1}T)$ is not invertible. Hence, by the corollary to Proposition 2 of Subsection 11.3.3, this operator is not injective, i.e., there exists a non-zero element $x \in X$ such that $(T - \lambda I)x = 0$. This means that x is an eigenvector, and λ an eigenvalue, of the operator T. Further, let Y be the eigenspace corresponding to the eigenvalue λ. Since the restriction of T to Y is a bijective compact operator, Corollary 1 in Subsection 11.3.2 yields the required finite-dimensionality.

The fact that 0 belongs to the spectrum, i.e., that the operator T is not invertible, follows from the same Corollary 1 in Subsection 11.3.2.

Finally, let us prove property (1). We proceed by reductio ad absurdum. So, suppose the spectrum of the operator T is infinite and has a non-zero limit point $a \in \mathbb{C}$. Further, let $\lambda_k \neq a$ be a sequence of eigenvalues that converges to a, and x_k be corresponding eigenvectors. Consider the closed linear span $Y = \overline{\mathrm{Lin}}\{x_k\}_{k=1}^{\infty}$ of this sequence of eigenvectors and denote by $A \in L(Y)$ the restriction of the operator $I - a^{-1}T$ to Y. Notice that $Ax_k = (1 - a^{-1}\lambda_k)x_k$. Therefore, the image of the operator A contains all the vectors x_k, so this image is dense in Y. By the main theorem of Subsection 11.3.3, the image is closed, that is, the operator A is surjective, and hence bijective. But this is impossible, because A is not bounded below:

$$\frac{\|Ax_k\|}{\|x_k\|} = \left|1 - \frac{\lambda_k}{a}\right| \to 0 \quad \text{as } k \to \infty. \qquad \square$$

Exercises

1. Show that a compact operator cannot be an interior point of the set of invertible operators in $L(X)$. Use this fact to solve Exercise 5 in Subsection 11.1.2.

Find the spectra of the following operators:

2. The integration operator from the example in Subsection 11.1.5.

3. $T \in L(C[0, 1])$, $Tf(t) = \int_0^1 (x + t) f(x) dx$.

4. $T \in L(C[0, 1])$, $Tf(t) = f(t) + \int_0^1 (x + t) f(x) dx$.

5. $T \in L(C[0, 2\pi])$, $Tf(t) = \int_0^{2\pi} \cos(x + t) f(x) dx$.

6. $T \in L(C[0, 1])$, $Tf(t) = \int_0^1 (x + t) f(t) dx$.

Comments on the Exercises

References for the theme "Banach algebras": the textbooks [44, Chapter 9] and [38, Ch. 10].

Subsection 11.1.1

Exercise 2. The product may not lie in the same space.

Exercise 3. There is no identity element.

Subsection 11.1.4

Exercise 4. Use Exercise 11 in Subsection 11.1.2.

Subsection 11.1.5

Exercise 6. See Exercises 6 and 7 in Subsection 10.2.3.

Exercise 11. Use Exercise 4 in Subsection 11.1.4.

Subsection 11.1.6

Exercise 5. See the indications at the end of Subsection 12.3.5 (just below the isomorphism theorem).

Subsection 11.2.1

Exercises 4–6. Details on this property are found in the paper [48]. As communicated to the author by Zbigniew Lipecki, this property was studied earlier by a series of authors: E. Szpilrajn-Marczewski, Fund. Math. **15** (1930), 126–127 and Fund. Math. **22** (1934), 303–311; R. Duda and R. Telgársky, Czechoslovak Math. J. **18(93)** (1968), 66–82; Ch. Bandt, Mathematika **28** (1981), 206–210.

Exercise 9. See [28, V.1, Proposition 1.e.2].

Subsection 11.2.2

Exercise 2. We need to prove only the second part of the assertion. Let T be a finite-rank operator and $\{y_k\}_{k=1}^n$ be a basis of the finite-dimensional space $T(X)$. Further, let $g_k \in T(X)^*$ be the coordinate functionals corresponding to the basis $\{y_k\}_{k=1}^n$. Then we have $Tx = \sum_1^n g_k(Tx)y_k$, i.e., as the sought-for $\{f_k\}_{k=1}^n$ one can take $f_k(x) = g_k(Tx)$.

Exercise 5. In view of the equality $(Tf)(t) = \int_0^1 xf(x)dx + t\int_0^1 f(x)dx$, the image of the operator is contained in the two-dimensional space of functions of the form $g(t) = a + bt$.

Subsection 11.3.2

Exercise 5. Using the representation $Ax = \sum_{k=1}^n f_k(x)y_k$ with linearly independent collections $\{y_k\}_{k=1}^n$ and $\{f_k\}_{k=1}^n$ (Exercise 2 of Subsection 11.2.2 and the comments to it), find the expression of A^*. Deduce from this that $\dim A(X) = \dim A^*(X^*) = n$. To convince yourself that the codimension of the kernel is equal to the dimension of the image, use the injectivization of a linear operator (Subsection 5.2.3).

Exercise 6. If the image is finite-dimensional, then it is closed, since a finite-dimensional space is closed in any ambient space. Conversely, suppose the image is closed. Then $A(X)$ is a Banach space. By the open mapping theorem, the image of the unit ball is an open set in $A(X)$; hence, the precompact set $A(B(X)))$ contains some ball U of the space $A(X)$. Then this ball U is precompact, and so the unit ball of the space $A(X)$, which is obtained from U by parallel translation and multiplication by a scalar, is also precompact. Therefore, $A(X)$ is a finite-dimensional space.

Subsection 11.3.3

Exercise 2. See [29, Chapter 6, pp. 281–284]. Hint: arguing as in Proposition 2 of Subsection 11.3.3, deduce that among the subspaces $Y_n = A^n(X)$, $n = 0, 1, 2, \ldots$, only finitely many of them can be pairwise distinct. Convince yourself that $Y = \bigcap_n Y_n$ is a finite-codimensional subspace that is mapped by the operator A bijectively into itself. Consider the subspace $Z = \bigcup_n \operatorname{Ker} A^n$. Prove that $X = Y \oplus Z$. Determine how A acts on each of these components.

Exercise 12. In 2006, when the Russian version of this book was published, the long-standing open problem of the existence of such spaces with the scalar + compact property was still open. It was solved in the affirmative in 2011 by S. Argyros and R. Haydon, Acta Math. **206**, No. 1, 1–54 (2011), Zbl 1223.46007.

Subsection 11.3.4

Exercise 2. $\sigma(T) = \{0\}$. Indeed, the operator is compact, and so $0 \in \sigma(T)$. There are no non-zero points in the spectrum, since by the theorem in Subsection 11.3.4, such points would be eigenvalues, and as was shown in Subsection 11.1.5, our operator has no eigenvalues.

Chapter 12
Hilbert Spaces

Among the infinite-dimensional Banach spaces, Hilbert spaces are distinguished by their relative simplicity. In Hilbert spaces we are able to use our geometric intuition to its fullest potential: measuring angles between vectors, applying Pythagoras' theorem, and using orthogonal projections. Here we do not run into anomalous phenomena[1] such as non-complemented subspaces or, say, linear functionals that do not attain their upper bound on the unit sphere. All separable infinite-dimensional Hilbert spaces are isomorphic to one another. Thanks to this relative simplicity, Hilbert spaces are often used in applications. In fact, whenever possible (true, this is not always the case), one seeks to use the language of Hilbert spaces rather than that of general Banach or topological vector spaces. The theory of operators in Hilbert spaces is developed in much more depth than that in the general case, which is yet another reason why this technique is frequently employed in applications.

12.1 The Norm Generated by a Scalar Product

12.1.1 Scalar Product

Let X be a complex linear space. A function $\langle \cdot, \cdot \rangle : X \times X \to \mathbb{C}$, which associates to each pair of elements x, y of the space X a complex number, is called a *scalar product* (or, frequently, *inner product*) if it obeys the following axioms:

(1) $\langle x, x \rangle \geqslant 0$ for all $x \in X$ (positivity);

(2) if $\langle x, x \rangle = 0$, then $x = 0$ (non-degeneracy);

---

[1] Although as far as we know, flying saucers, poltergeists, telepathy, and Big Foot are not encountered even in general Banach spaces!

© Springer International Publishing AG, part of Springer Nature 2018
V. Kadets, *A Course in Functional Analysis and Measure Theory*,
Universitext, https://doi.org/10.1007/978-3-319-92004-7_12

(3) $\langle a_1 x_1 + a_2 x_2, y \rangle = a_1 \langle x_1, y \rangle + a_2 \langle x_2, y \rangle$ for all $x_1, x_2, y \in X$ and all $a_1, a_2 \in \mathbb{C}$ (linearity in the first variable);

(4) $\langle x, y \rangle = \overline{\langle y, x \rangle}$ for all $x, y \in X$ (Hermitian, or complex conjugate, symmetry) (the overline in the last formula stands for complex conjugation).

Note that the third and fourth axioms imply the rules of opening the brackets with respect to the second variable:

(5) $\langle x, y_1 + y_2 \rangle = \langle x, y_1 \rangle + \langle x, y_2 \rangle$;

(6) $\langle x, ay \rangle = \overline{a} \langle x, y \rangle$.

Two elements x, y for which $\langle x, y \rangle = 0$ are said to be *orthogonal* (to one another; this is abbreviated as $x \perp y$).

Examples

I. Scalar product in $\mathbb{C}^n$: let $x, y \in \mathbb{C}^n$, $x = (x_1, \ldots, x_n)$, $y = (y_1, \ldots, y_n)$. Put $\langle x, y \rangle = \sum_{k=1}^{n} x_k \overline{y}_k$.

II. Scalar product in ℓ_2: let $x, y \in \ell_2$, $x = (x_1, x_2, \ldots)$, $y = (y_1, y_2, \ldots)$. Put $\langle x, y \rangle = \sum_{k=1}^{\infty} x_k \overline{y}_k$.

III. Scalar product in $L_2(\Omega, \Sigma, \mu)$: let $f, g \in L_2$. Put $\langle f, g \rangle = \int_{\Omega} f \overline{g} \, d\mu$.

Although, needless to say, in all the spaces listed above there exist other scalar products, from now, unless otherwise stipulated, by the scalar products in the spaces $\mathbb{C}^n$, ℓ_2 and L_2 we will mean precisely the examples given above.

In this book we will focus on scalar products in complex spaces. This is motivated mainly by applications to operator theory, where the non-emptiness of spectrum theorem requires the field of complex numbers. Nevertheless, the scalar product axioms and the above examples also make sense in real spaces. The only changes are in fact simplifications: in the real case $\langle x, y \rangle$ is defined to take real values, so all the complex conjugations in the axioms and examples just disappear. All the results of this chapter remain valid for real spaces with the real version of the scalar product. The significant differences appear in the next chapter when we study unitary operators (Subsection 13.2.2) and the polar representation (Subsection 13.2.3). In that part of the theory complex numbers are unavoidable.

Exercises

1. Based on the numerical inequality $|ab| \leqslant |a|^2 + |b|^2$ ($a, b \in \mathbb{C}$), prove the convergence of the series and the existence of the integral in the definition of the scalar product in ℓ_2 and $L_2[0, 1]$, respectively.

2. Verify that all the examples of scalar products given above do indeed satisfy the necessary axioms.

3. What must the interval $[a, b]$ be in order for the functions $f(t) = e^{it}$ and $g(t) = e^{2it}$ to be orthogonal in $L_2[a, b]$?

4. Can two polynomials of degree one be orthogonal in $L_2[0, 1]$?

5. Can two positive functions be orthogonal in $L_2[0, 1]$? Two negative functions? Two sign-changing functions?

6. Is the orthogonality relation $\perp$ an equivalence relation? An order relation?

7. Derive the following (*square of a sum*) formula:

$$\langle x + y, x + y \rangle = \langle x, x \rangle + 2\,\mathrm{Re}\,\langle x, y \rangle + \langle y, y \rangle.$$

Note! This formula will be repeatedly used in the sequel.

12.1.2 The Cauchy–Schwarz Inequality

Theorem 1. *Let X be a space with a scalar product. Then for any $x, y \in X$ the following* Cauchy–Schwarz inequality (*also known as the* Cauchy–Bunyakovsky *or* Cauchy–Bunyakovsky–Schwarz inequality) *holds:*

$$|\langle x, y \rangle| \leqslant \langle x, x \rangle^{1/2} \langle y, y \rangle^{1/2}.$$

Proof. By the positivity axiom,

$$\langle x + ty, x + ty \rangle \geqslant 0$$

for all $t \in \mathbb{R}$. Opening the brackets, we see that for every $t \in \mathbb{R}$,

$$\langle x, x \rangle + 2t\,\mathrm{Re}\,\langle x, y \rangle + t^2 \langle y, y \rangle \geqslant 0.$$

A quadratic polynomial with real coefficients can be non-negative on the whole real line only if its discriminant is non-positive. Hence, we have shown that, for any elements $x, y \in X$,

$$\mathrm{Re}\,\langle x, y \rangle \leqslant \langle x, x \rangle^{1/2} \langle y, y \rangle^{1/2}.$$

To derive from this the required Cauchy–Schwarz inequality, we observe that

$$|\langle x, y \rangle| = \mathrm{Re}\,\langle x, e^{i\,\arg\langle x, y \rangle} y \rangle$$

$$\leqslant \langle x, x \rangle^{1/2} \langle e^{i\,\arg\langle x, y \rangle} y, e^{i\,\arg\langle x, y \rangle} y \rangle^{1/2} = \langle x, x \rangle^{1/2} \langle y, y \rangle^{1/2}. \qquad \square$$

Exercises

1. For the functions $f(t) = t$ and $g(t) = t^2$ in $L_2[0, 1]$ calculate in $L_2[0, 1]$ the scalar products $\langle f, g \rangle$, $\langle f, f \rangle$, and $\langle g, g \rangle$. Verify in this example the Cauchy–Schwarz inequality.

2. Suppose that for two elements x, y of a space with scalar product the Cauchy–Schwarz inequality becomes an equality: $|\langle x, y \rangle| = \langle x, x \rangle^{1/2} \langle y, y \rangle^{1/2}$. Then x and y are linearly dependent.

3. Based on the examples II and III in Subsection 12.1.1, derive the following versions of the Cauchy–Schwarz inequality:[2]

$$\left| \sum_{k=1}^{\infty} x_k y_k \right| \leqslant \left(\sum_{k=1}^{\infty} |x_k|^2 \right)^{1/2} \left(\sum_{k=1}^{\infty} |y_k|^2 \right)^{1/2}$$

and

$$\left| \int_{\Omega} fg \, d\mu \right| \leqslant \left(\int_{\Omega} |f|^2 d\mu \right)^{1/2} \left(\int_{\Omega} |g|^2 d\mu \right)^{1/2}.$$

4. Suppose the function F of two variables on the linear space X obeys all the scalar product axioms, except for non-degeneracy. Verify that in this case, too, the Cauchy–Schwarz inequality $|F(x, y)| \leqslant F(x, x)^{1/2} F(y, y)^{1/2}$ holds. (**Note!** This fact will be used later in the study of self-adjoint operators.)

12.1.3 The Concept of Hilbert Space

Definition 1. Let H be a space equipped with a scalar product. The quantity $\|x\| = \sqrt{\langle x, x \rangle}$ is called the *norm generated by the scalar product* $\langle \cdot, \cdot \rangle$.

With the notation just introduced, the Cauchy–Schwarz inequality can be recast as $|\langle x, y \rangle| \leqslant \|x\| \cdot \|y\|$, and the square of a sum formula takes on the form $\|x + y\|^2 = \|x\|^2 + 2\mathrm{Re}\,\langle x, y \rangle + \|y\|^2$.

Let us verify that the norm generated by the scalar product satisfies the triangle inequality. To this end we square the required inequality $\|x + y\| \leqslant \|x\| + \|y\|$ and open the parentheses:

$$\|x\|^2 + 2\mathrm{Re}\,\langle x, y \rangle + \|y\|^2 \leqslant \|x\|^2 + 2\|x\| \cdot \|y\| + \|y\|^2.$$

Reducing here the like terms, this last inequality becomes $\mathrm{Re}\,\langle x, y \rangle \leqslant \|x\| \cdot \|y\|$, a simple consequence of the Cauchy–Schwarz inequality. We leave to the reader the verification of the remaining norm axioms for the norm generated by the scalar product.

---

[2]Citation from the Wikipedia article titled "Cauchy–Schwarz inequality" that explains the different names used for this inequality: "The inequality for sums was published by Augustin-Louis Cauchy (1821), while the corresponding inequality for integrals was first proved by Viktor Bunyakovsky (1859). The modern proof of the integral inequality was given by Hermann Amandus Schwarz (1888)".

The following *parallelogram identity*, also known as the *parallelogram law*,[3] is an important additional property of the norm generated by a scalar product.

Proposition 1 (parallelogram identity). *Let H be a space equipped with a scalar product, then for any two elements $x, y \in H$*

$$\|x + y\|^2 + \|x - y\|^2 = 2\|x\|^2 + 2\|y\|^2.$$

Proof. The square of a sum formula gives us $\|x + y\|^2 = \|x\|^2 + 2\mathrm{Re}\,\langle x, y \rangle + \|y\|^2$. Taking $-y$ instead of y we also obtain the square of a difference formula $\|x - y\|^2 = \|x\|^2 - 2\mathrm{Re}\,\langle x, y \rangle + \|y\|^2$. It remains to put these two formulas together. □

Theorem 1. *Equip the space H with the norm generated by the scalar product. Let $h \in H$ and define the mapping $F: H \to \mathbb{C}$ by the formula $F(x) = \langle x, h \rangle$. Then F is a continuous linear functional and $\|F\| = \|h\|$.*

Proof. The linearity of the functional F is just axiom (3) of the scalar product. The continuity of F and the inequality $\|F\| \leqslant \|h\|$ follow from the Cauchy–Schwarz inequality, rewriting the latter as $|F(x)| = |\langle x, h \rangle| \leqslant \|x\| \cdot \|h\|$. To estimate the norm of the functional F from below, evaluate F on the element $h/\|h\| \in S_H$:

$$\|F\| \geqslant |F(h/\|h\|)| = |\langle h, h \rangle / \|h\|\,| = \|h\|^2/\|h\| = \|h\|.$$

The theorem is proved. □

Definition 2. A space H with scalar product is called a *Hilbert space* if it is complete in the norm generated by the scalar product.[4]

As in the case of general Banach spaces, in the theory of Hilbert spaces subspaces are understood to be closed linear subspaces. On a subspace of a Hilbert space there is defined the same scalar product as on the whole space. With respect to this scalar product any subspace of a Hilbert space is itself a Hilbert space.

Exercises

1. Suppose $x \perp y$. Then $\|x + y\|^2 = \|x\|^2 + \|y\|^2$ (the analogue of Pythagoras' theorem). Is the converse true, i.e., if $\|x + y\|^2 = \|x\|^2 + \|y\|^2$, does it follow that $x \perp y$?

---

[3]The "Parallelogram Law" was approved by the Athens Popular Assembly around 345 B.C. and stated that in a parallelogram the sum of the squared lengths of the diagonals should be equal to the sum of the squares of the lengths of all four sides.☺

[4]Motivated by this terminology, many authors call general spaces with a scalar product "pre-Hilbert spaces".

2. Verify that the norms on the spaces ℓ_2 and L_2 introduced earlier in Subsection 6.2.2 are generated by the corresponding scalar products (see examples II and III in Subsection 12.1.1).

3. Let x_n, x, y_n, and y be elements of H, and let $x_n \to x$ and $y_n \to y$ as $n \to \infty$. Then $\langle x_n, y_n \rangle \to \langle x, y \rangle$ as $n \to \infty$.

4. Verify that in the Banach spaces $C[0, 1]$, $L_1[0, 1]$, c_0, and ℓ_1 the parallelogram identity does not hold.

5. Show that if the parallelogram identity is satisfied for any two elements x, y of the normed space X, then the norm in X is generated by a scalar product.

6. Verify that the following useful formula holds for every element h of the Hilbert space H: $\|h\| = \sup_{y \in S_H} |\mathrm{Re}\,\langle h, y \rangle| = \sup_{y \in S_H} \mathrm{Re}\,\langle h, y \rangle$. This can be done either directly, using the Cauchy–Schwarz inequality, or deduced from Theorem 1 and the formula $\|F\| = \|\mathrm{Re}\,F\|$ (Theorem 3 in Subsection 9.1.1).

12.2 Hilbert Space Geometry

12.2.1 The Best Approximation Theorem

Theorem 1. *Let H be a Hilbert space and $A \subset H$ a convex closed set. Then for any element $h \in H$ there exists in A a unique element closest to h. In other words, there exists a unique $a_0 \in A$ such that $\|h - a_0\| = \rho(h, A)$.*

Proof. Since $\rho(h, A) = \rho(0, A - h)$, it suffices to prove the theorem for the case when $h = 0$. Denote $\rho(0, A)$ by r and consider the set

$$A_n = \left\{ a \in A : \|a\| \leqslant r + \frac{1}{n} \right\} = A \cap \left(r + \frac{1}{n} \right) \overline{B}_H.$$

The intersection of all the sets A_n is the set of all elements lying at distance r from zero. Hence, we have to prove that $\bigcap_{n=1}^{\infty} A_n$ consists of a single point. To this end we use the nested sets theorem (Subsection 1.3.3). The A_n's form a decreasing chain of convex closed sets. It remains to show that $\mathrm{diam}(A_n) \to 0$ as $n \to \infty$. To estimate the diameter, we take two arbitrary points $x, y \in A_n$ and apply the parallelogram identity and the inequality $r \leqslant \|e\| \leqslant r + 1/n$, which holds for all $e \in A_n$. This yields

$$\|x - y\|^2 = 2\|x\|^2 + 2\|y\|^2 - \|x + y\|^2 = 2 \left(\|x\|^2 + \|y\|^2 - 2 \left\| \frac{x + y}{2} \right\|^2 \right)$$

$$\leqslant 2\big((r + 1/n)^2 + (r + 1/n)^2 - 2r^2 \big) = \frac{8r}{n} + \frac{4}{n^2}.$$

Therefore,

$$\operatorname{diam}(A_n) \leqslant \left(\frac{8r}{n} + \frac{4}{n^2}\right)^{1/2} \to 0 \quad \text{as } n \to \infty. \qquad \square$$

Exercises

1. Verify that the sets A_n figuring in the proof of the best approximation theorem are convex and closed.

2. Where in the proof was the convexity of the sets A_n used?

3. Where in the proof was the fact that the norm of the Hilbert space is generated by the scalar product used?

4. On the space $C[0, 1]$ consider the functional F given by the rule $F(x) = \int_0^{1/2} x(t)dt - \int_{1/2}^1 x(t)dt$. Then $A = \{x \in C[0, 1] : F(x) = 1\}$ is a convex closed set which contains no element closest to zero (see also Subsection 6.4.4, Exercise 4).

5. In the space $C[0, 1]$ consider the set A of all functions that take the value 1 at 0. Verify that the distance from A to 0 is equal to 1, that this distance is attained, but there is more than one point in A closest to 0.

6. For a normed space X the following properties are equivalent:

— in any convex subset $A \subset X$, for any element $x \in X$, if A contains an element that is closest to x, then this element is unique;

— for any two non-collinear vectors $x, y \in X$ there holds the *strict triangle inequality* $\|x + y\| < \|x\| + \|y\|$;

— $\|x + y\| < 2$ for any two distinct vectors $x, y \in S_X$;

— the unit sphere of the space does not contain rectilinear segments of non-zero length (this property of a space is called *strict convexity*).

7. Suppose that for a subset A of the normed space X the assertion of the best approximation theorem holds true: For any element $h \in X$, A contains an element that is closest to h. Then A is closed.

8. Suppose that for a subset A of a finite-dimensional Hilbert space H the assertion of the best approximation theorem holds true. Then A is convex.

9. Generalize the assertion of the last exercise to the case of an infinite-dimensional Hilbert space.[5]

---

[5]If you succeed in doing so, publish the result! At least at the time these lecture notes were written, the problem was still open. The subsets $A \subset H$ enjoying the property that for every $h \in H$ there exists in A a unique element closest to h are called *Chebyshev sets*. A Google search with the keywords "convexity of Chebyshev sets" will give the reader dozens of references to papers about this challenging open problem.

12.2.2   *Orthogonal Complements and Orthogonal Projectors*

Definition 1. Let H be a Hilbert space. An element $h \in H$ is said to be *orthogonal* to the subset $X \subset H$ (in which case one writes $h \perp X$) if h is orthogonal to all the elements of X. The set of all elements orthogonal to the subset X is called the *orthogonal complement* to X and is denoted by $X^\perp$.

Proposition 1. $X^\perp$ *is a subspace of H (recall that in Banach spaces, and hence also in Hilbert spaces, we agreed to use the term* subspace *to mean, without special mention,* closed linear subspace).

Proof. We have $X^\perp = \bigcap_{x \in X} x^\perp$. Hence, it is sufficient to prove that for any element x the set $x^\perp$ is a (closed!) subspace of H. But $x^\perp$ coincides with the kernel $F^{-1}(0)$ of the continuous linear functional $F \colon y \mapsto \langle y, x \rangle$ (see Theorem 1 of Subsection 12.1.3).   □

The next assertion is a direct generalization of a fact that is already well-known from school geometry: to find in a subspace X the closest element to a point h, we need to drop a perpendicular onto the subspace.

Proposition 2. *Let X be a subspace of the Hilbert space H, $h \in H$, and $h_0 \in X$. Then the following conditions are equivalent:*

(a) h_0 *is the element of X closest to h;*

(b) $h - h_0 \in X^\perp$.

Proof. (a) $\Longrightarrow$ (b). Suppose that condition (b) is not satisfied. Then there exists an element $x \in X$ for which $\langle x, h - h_0 \rangle \neq 0$. Multiplying x, if necessary, by a constant, we obtain a vector for which $\langle x, h - h_0 \rangle = 1$. By condition (a), for every $t > 0$ it holds that

$$\|h - h_0\|^2 \leqslant \|h - h_0 - tx\|^2 = \|h - h_0\|^2 - 2t + t^2 \|x\|^2.$$

That is, for every $t > 0$ one has $t^2 \|x\|^2 - 2t \geqslant 0$, which obviously does not hold for small t.

(b) $\Longrightarrow$ (a). Let $h_1 \in X$ be an arbitrary element. Then

$$\|h - h_1\|^2 = \|(h - h_0) - (h_0 - h_1)\|^2 = \|h - h_0\|^2 + \|h_0 - h_1\|^2 \geqslant \|h - h_0\|^2.$$

This means that h_0 is the closest element to h in X, as we needed to prove.   □

Theorem 1. *Let X be a subspace of the Hilbert space H. The H decomposes into the direct sum of the subspaces X and $X^\perp$: $H = X \oplus X^\perp$.*

Proof. We have to show that 1) $X \cap X^\perp = \{0\}$, and 2) for any $h \in H$ there exist elements $x \in X$ and $y \in X^\perp$ such that $h = x + y$.

1) Suppose some element x belongs simultaneously to the subspaces X and $X^\perp$. Then $x \perp x$, i.e., $\langle x, x \rangle = 0$, and so $x = 0$.

2) Let $h \in H$ be an arbitrary element. Denote by x the element of the subspace X closest to h and put $y = h - x$. Then $x \in X$, $y \in X^\perp$ (according to Proposition 2 above), and $h = x + y$. $\qquad\square$

Since $H = X \oplus X^\perp$, there exists (see Subsection 10.3.2) a bounded projector P onto the subspace X with Ker $P = X^\perp$. The action of this projector can be described directly as follows: for each $h \in H$, write the decomposition $h = x + y$, with $x \in X$ and $y \in X^\perp$. Then $Ph = x$. Such a projector P is called the *orthogonal projector* (or *orthoprojector*) onto the subspace X. Since, as was shown in the second step of the proof of Theorem 1, the element x in the above decomposition $h = x + y$ is the closest in X to h, another equivalent way to define P is to say that it sends each element $h \in H$ into the element $Ph \in X$ closest to h. The reader should familiarize himself with the properties of orthogonal projectors by solving the exercises given below.

Remark 1. As we just proved, in a Hilbert space for any subspace there exists a bounded projector onto that subspace. The converse statement, called the Lindenstrauss–Tzafriri theorem [67], is also true: if in the Banach space X any subspace is complemented, then X is isomorphic to a Hilbert space.

Exercises

1. The norm of the orthogonal projector onto a nonzero subspace is equal to 1.

2. If a projector P onto the subspace X is such that $\|P\| = 1$, then P is an orthogonal projector.

3. Based directly on the definition, show that the orthogonal complement to a subset is a closed linear subspace.

4. Let X be a subspace of the Hilbert space H. Then $(X^\perp)^\perp = X$.

5. For any subset $X \subset H$, $(X^\perp)^\perp$ coincides with the closure of the linear span of the set X.

6. Let $e_k \in H$, $k = 1, 2, \ldots$, and put $X = \overline{\text{Lin}}\{e_k\}_1^\infty$. Then for an element $y \in H$ to belong to $X^\perp$, it is necessary and sufficient that y is orthogonal to all e_k. In other words, $X^\perp = \bigcap_{k \in \mathbb{N}} e_k^\perp$.

12.2.3 The General Form of Linear Functionals on a Hilbert Space

Theorem 1. *For any linear functional F on a Hilbert space H there exists an element $h \in H$ such that*

$$F(x) = \langle x, h \rangle \tag{1}$$

for all $x \in H$. The element h is uniquely determined, and $\|F\| = \|h\|$.

Proof. In the case of the trivial (identically equal to 0) functional F the assertion is obvious. Now suppose $F \neq 0$. Denote the kernel of F by X. There exists a norm-1 element e that is orthogonal to X. As the sought-for element take $h = \overline{F(e)}e$. For this choice, if we take $x \in X$, then both $F(x) = 0$ and $\langle x, h \rangle = 0$, so (1) holds for all $x \in X$. The equality (1) also holds for $x = e$:

$$\langle e, h \rangle = F(e)\langle e, e \rangle = F(e).$$

Consequently, by linearity, $F(x) = \langle x, h \rangle$ for any $x \in \mathrm{Lin}\{e, X\}$. But X is a hyper-subspace in H (see Subsection 5.3.3), therefore $\mathrm{Lin}\{e, X\} = H$ and (1) holds on the whole space H. The equality $\|F\| = \|h\|$ was established above, in Theorem 1 of Subsection 12.1.3. It remains to verify the uniqueness of the element h. Let $h_1 \in H$ be another element such that $F(x) = \langle x, h_1 \rangle$ for all $x \in H$. Then $\langle x, h - h_1 \rangle = 0$ for all $x \in H$; in particular, $\langle h - h_1, h - h_1 \rangle = 0$. That is, $h - h_1 = 0$ and $h = h_1$. The theorem is proved. $\qquad\square$

Remark 1. The final argument of the proof can be formulated as an individual propo-sition: If $\langle x, h_1 \rangle = \langle x, h_2 \rangle$ for all $x \in H$, then $h_1 = h_2$. This assertion will be used repeatedly in the sequel.

Exercises

1. In the proof above we dealt only with the case $F \neq 0$ (by the way, where was this condition implicitly used?). Treat the case $F = 0$ yourself.

2. In what way was the continuity of the functional F used?

3. Why does the fact that the equality $F(x) = \langle x, h \rangle$ holds for all $x \in X$ and for $x = e$ imply that it holds for all $x \in \mathrm{Lin}\{e, X\}$?

4. Is the completeness of the space H important in the last theorem?

5. Define the map $U : H \to H^*$ by the rule: for $h \in H$, $(Uh)(x) = \langle x, h \rangle$. Show that U is bijective, additive (i.e., $U(h_1 + h_2) = Uh_1 + Uh_2$), but not homogeneous. Instead of homogeneity, one has that $U(\lambda h) = \overline{\lambda} U(h)$.

6. Let H be the subspace of $L_2[0, 1]$ consisting of the polynomials of degree at most 2. Represent the functional F on H given by $F(f) = f(0)$, as $F(f) = \langle f, h \rangle$, as indicated in the theorem on the general form of linear functionals on H. Do the same for the functional F_1 on H given by the formula $F_1(f) = f'(0)$.

7. Based on the theorem in Subsection 12.1.3 and Exercise 2 in Subsection 9.1.2, prove the following n-dimensional analogue of the well-known formula for computing the distance of a point to a plane. Suppose the hyperplane A in the n-dimensional coordinate space is given by the equation $a_1x_1 + \cdots + a_nx_n = b$. Then the distance of any vector $x = (x_1, \ldots, x_n)$ to the hyperplane A is given by the formula

$$\rho(x, A) = \frac{|a_1x_1 + \cdots + a_nx_n - b|}{\sqrt{a_1^2 + \cdots + a_n^2}}.$$

8. Suppose the numerical sequence $a = (a_1, a_2, \ldots)$ has the property that the series $\sum_{m=1}^{\infty} a_m x_m$ converges for any $x = (x_1, x_2, \ldots) \in \ell_2$. Then $a \in \ell_2$ and the formula $f(x) = \sum_{m=1}^{\infty} a_m x_m$ gives a linear functional on ℓ_2.

By using the preceding exercise, the following theorem is reduced to the closed graph theorem.

9. Suppose the infinite matrix $A = (a_{n,m})_{n,m=1}^{\infty}$ has the following property: for any $x = (x_1, x_2, \ldots) \in \ell_2$ and any $n \in \mathbb{N}$, the series $\sum_{m=1}^{\infty} a_{n,m} x_m$ converges, and the numerical sequence $Ax = \left(\sum_{m=1}^{\infty} a_{n,m} x_m \right)_{n=1}^{\infty}$ belongs to ℓ_2 (in other words, the operator of multiplication by the matrix A maps ℓ_2 into ℓ_2). Then the operator of multiplication by the matrix A is continuous as an operator mapping ℓ_2 into ℓ_2.

12.3   Orthogonal Series

We next turn to the study of a subject that to some degree is discussed in other university courses: calculus — in the treatment of trigonometric series, and linear algebra — in the construction of orthonormal bases in finite-dimensional Euclidean spaces.

12.3.1   A Convergence Criterion for Orthogonal Series

Lemma 1 (**n-dimensional Pythagoras' theorem**). *Suppose the elements $(x_k)_1^n$ of the Hilbert space H are pairwise orthogonal: $\langle x_k, x_j \rangle = 0$ for all $k \neq j$. Then $\left\| \sum_{k=1}^{n} x_k \right\|^2 = \sum_{k=1}^{n} \|x_k\|^2$.*

Proof. We need to use the definition of the norm, open the brackets, and discard the null terms:

$$\left\| \sum_{k=1}^{n} x_k \right\|^2 = \left\langle \sum_{k=1}^{n} x_k, \sum_{k=1}^{n} x_k \right\rangle = \sum_{k=1}^{n} \langle x_k, x_k \rangle + \sum_{k,j=1,\, k \neq j}^{n} \langle x_k, x_j \rangle = \sum_{k=1}^{n} \|x_k\|^2. \quad \square$$

Theorem 1 (convergence criterion for orthogonal series). *Let* $(x_k)_1^\infty$ *be a sequence of pairwise-orthogonal elements of the Hilbert space H. Then for the series $\sum_{k=1}^\infty x_k$ to converge it is necessary and sufficient that the numerical series $\sum_{k=1}^\infty \|x_k\|^2$ converges.*

Proof. By the Cauchy criterion, for the series $\sum_{k=1}^\infty x_k$ to converge it is necessary and sufficient that $\left\|\sum_{k=n}^m x_k\right\|^2 \to 0$ as $n, m \to \infty$. In view of the lemma proved above, this is equivalent to the condition $\sum_{k=n}^m \|x_k\|^2 \to 0$ as $n, m \to \infty$, which again by the Cauchy criterion is equivalent to the convergence of the series $\sum_{k=1}^\infty \|x_k\|^2$. $\square$

Remark 1. If the series of pairwise-orthogonal terms $\sum_{k=1}^\infty x_k$ converges, then by simply letting $n \to \infty$ in the relation $\left\|\sum_{k=1}^n x_k\right\|^2 = \sum_{k=1}^n \|x_k\|^2$ we obtain an infinite-dimensional version of Pythagoras' theorem: $\left\|\sum_{k=1}^\infty x_k\right\|^2 = \sum_{k=1}^\infty \|x_k\|^2$.

Exercises

Let $\sum_{k=-\infty}^{+\infty} a_k e^{ikt}$ be a series in $L_2[0, 2\pi]$. Under what conditions on the coefficients a_k does this series converge in $L_2[0, 2\pi]$? What is the norm of the sum of this series equal to?

12.3.2 Orthonormal Systems. Bessel's Inequality

For the sake of simplicity, throughout this subsection Γ will stand for a finite or countable set of indices. However, the exposition will proceed in such a manner that the reader should be able, with no major effort, to extend the main assertions to the case of uncountable sets Γ.

Definition 1. A system $\{x_k\}_{k\in\Gamma}$ of elements of the Hilbert space H is said to be *orthonormal* if the elements x_k are pairwise orthogonal and the norms of all x_k are equal to 1. These two conditions can be written as a single relation $\langle x_k, x_j \rangle = \delta_{k,j}$, where $\delta_{k,j}$ is the Kronecker symbol. The *Fourier coefficients* of the element $h \in H$ in (with respect to) the orthonormal system $\{x_k\}_{k\in\Gamma}$ are the numbers $\widehat{h}_k = \langle h, x_k \rangle$, $k \in \Gamma$.

The next statement clarifies the role of the Fourier coefficients.

Proposition 1. *Let $\{x_k\}_{k\in\Gamma}$ be an orthonormal system, $\{a_k\}_{k\in\Gamma}$ be scalars, and $h = \sum_{k\in\Gamma} a_k x_k$ (if in this last sum there are infinitely many non-zero terms, then the sum is understood as a series, written in some fixed order). Then $a_k = \widehat{h}_k$ for all $k \in \Gamma$.*

Proof. Taking the scalar product of both sides of the equality $h = \sum_{k\in\Gamma} a_k x_k$ with x_j, we obtain $\langle h, x_j \rangle = \sum_{k\in\Gamma} a_k \langle x_k, x_j \rangle$. Since $\langle x_k, x_j \rangle = \delta_{k,j}$, the sum in the right-hand side of the last equality reduces to a single term, a_j. This yields the needed equality $a_j = \langle h, x_j \rangle$. $\square$

Proposition 2. *Let $\{x_k\}_{k\in\Gamma}$ be an orthonormal system of elements of the Hilbert space H. Let the subspace X be the closed linear span of the set $\{x_k\}_{k\in\Gamma}$, $h \in H$. Then for $h_0 \in X$ to be the closest element to h in X it is necessary and sufficient that the Fourier coefficients of h_0 with respect to the system $\{x_k\}_{k\in\Gamma}$ coincide with the Fourier coefficients of h.*

Proof. By Proposition 2 of Subsection 12.2.2, h_0 is the closest element to h in X if and only if $h - h_0 \in X^\perp$. Since $X = \overline{\mathrm{Lin}}\{x_k\}_{k\in\Gamma}$, the condition $h - h_0 \in X^\perp$ is equivalent to the equalities $\langle h - h_0, x_k \rangle = 0$ holding simultaneously for all $k \in \Gamma$ (Exercise 6 in Subsection 12.2.2), which in turn means that $\langle h_0, x_k \rangle = \langle h, x_k \rangle$ for all $k \in \Gamma$, as needed. $\qquad\square$

Proposition 3. *Let in the above proposition Γ be finite. Denote by P the orthogonal projector onto the subspace $X = \mathrm{Lin}\{e_k\}_{k\in\Gamma}$. Then $Ph = \sum_{k\in\Gamma} \widehat{h}_k x_k$ for every $h \in H$.*

Proof. Consider the element $x = \sum_{k\in\Gamma} \widehat{h}_k x_k$. By Proposition 1, $\widehat{x}_k = \widehat{h}_k$ for all $k \in \Gamma$. By Proposition 2, x is the closest element to h in the subspace X, i.e., $Ph = x$. $\qquad\square$

Theorem 1 (Bessel's inequality). *Let $\{x_k\}_{k\in\Gamma}$ be an orthonormal system in the Hilbert space H, and let $h \in H$. Then*

$$\sum_{k\in\Gamma} |\widehat{h}_k|^2 \leqslant \|h\|^2.$$

Proof. It suffices to prove Bessel's inequality for finite orthonormal systems: indeed, if $\sum_{k\in\Delta} |\widehat{h}_k|^2 \leqslant \|h\|^2$ for every finite subset $\Delta \subset \Gamma$, then the sum over the whole Γ is bounded by the same number.

So, suppose the set Γ is finite. In the notation of Proposition 3, $Ph = \sum_{k\in\Gamma} \widehat{h}_k x_k$. The definition of the orthoprojector implies that $(h - Ph) \perp Ph$, and consequently $\|Ph\|^2 + \|h - Ph\|^2 = \|h\|^2$. Therefore, $\sum_{k\in\Gamma} |\widehat{h}_k|^2 = \|Ph\|^2 \leqslant \|h\|^2$. $\qquad\square$

Exercises

1. Why in the proof of Proposition 1 were we allowed to move the scalar product inside the sum?

2. Why in the formulation of Bessel's inequality does the sum $\sum_{k\in\Gamma} |\widehat{h}_k|^2$ not depend on the order in which the terms are written?

Let Γ be some, possibly uncountable, index set. By definition, a series $\sum_{k\in\Gamma} x_k$ of elements of a Banach space is said to *converge unconditionally* to an element x if for any $\varepsilon > 0$ there exists a finite subset $\Delta \subset \Gamma$ such that for any finite subset $\Delta_1 \subset \Gamma$, if $\Delta_1 \supset \Delta$, then $\|x - \sum_{k\in\Delta_1} x_k\| < \varepsilon$. (Essentially, we are dealing here with convergence along a certain directed set. Compare with Exercise 5 in Subsection 4.1.2.)

Suppose the series $\sum_{k \in \Gamma} x_k$ of elements of a Banach space converges unconditionally to the element x. Show that:

3. For every $\varepsilon > 0$, the number of elements of norm larger than ε is finite.

4. The number of non-zero terms of the series is at most countable.

5. If we write in an arbitrary way all non-zero terms of the series as a sequence $x_{k_1}, x_{k_2}, \ldots$, then the resulting series $\sum_{n=1}^{\infty} x_{k_n}$ converges to x in the ordinary sense.

6. Verify that if the convergence of a series with uncountable many terms is understood as unconditional convergence, then the assertions about orthogonal series and orthonormal systems proved in the preceding two subsections remain valid for uncountable systems and series.

12.3.3 Fourier Series, Orthonormal Bases, and the Parseval Identity

The series $\sum_{n=1}^{\infty} \widehat{h}_n e_n$, where $\widehat{h}_n$ are the corresponding Fourier coefficients, is called the *Fourier series* of the element $h \in H$ in (with respect to) the system $\{e_n\}_1^{\infty} \subset H$.

Theorem 1. *Let $\{e_n\}_1^{\infty}$ be an orthonormal system in the Hilbert space H, $X = \overline{\mathrm{Lin}}\{e_k\}_1^{\infty}$, and P be the orthogonal projector on the subspace X. Then the Fourier series of any element $h \in H$ converges and its sum coincides with Ph.*

Proof. The convergence of the series $\sum_{n=1}^{\infty} \widehat{h}_n e_n$ follows from the Bessel inequality (Subsection 12.3.2) and the convergence criterion for orthogonal series (Subsection 12.3.1). Further, for the element $x := \sum_{n=1}^{\infty} \widehat{h}_n e_n \in X$ its Fourier coefficients $\widehat{x}_k$ are equal to the corresponding $\widehat{h}_k$ (Proposition 1 of Subsection 12.3.2). By Proposition 2 of Subsection 12.3.2, this means that x is the closest element to h in X, that is, $Ph = x$. $\square$

Definition 1. A complete orthonormal system $\{e_n\}_{n \in \Gamma}$ in the Hilbert space H is called an *orthonormal basis*. In other words, the elements $\{e_n\}_{n \in \Gamma}$ constitute an orthonormal basis if $\langle e_k, e_j \rangle = \delta_{k,j}$ for all $k, j \in \Gamma$ and $\overline{\mathrm{Lin}}\{e_n\}_{n \in \Gamma} = H$.

If under the assumptions of Theorem 1 $\{e_n\}_1^{\infty}$ is an orthonormal basis, then $X = H$ and the orthogonal projector P is simply the identity operator. This implies the following result.

Theorem 2. *Let $\{e_n\}_1^{\infty}$ be an orthonormal basis in the Hilbert space H. Then the Fourier series of any element $h \in H$ converges and $\sum_{n=1}^{\infty} \widehat{h}_n e_n = h$.* $\square$

Exercises

1. Let H be a Hilbert space, and $\sum_{n=1}^{\infty} \widehat{h}_n e_n$ be the Fourier series of the element $h \in H$. The following conditions are equivalent:

— $\sum_{n=1}^{\infty} \widehat{h}_n e_n = h$;

— the *Parseval identity* (also referred to as the *Parseval equality*) holds for h:

$$\sum_{n=1}^{\infty} |\widehat{h}_k|^2 = \|h\|^2.$$

2. If $\{e_n\}_1^{\infty}$ is an orthonormal basis, then the Parseval identity holds for all elements $h \in H$.

3. Choose the coefficients a_k so that the functions $f_k = a_k e^{ikt}$, $k \in \mathbb{Z}$, form an orthonormal system in $L_2[0, 2\pi]$.

4. Suppose the sequence (f_k) of continuously differentiable functions constitutes an orthonormal system in $L_2[0, 2\pi]$. Prove that the derivatives of the functions (f_k) cannot be jointly bounded.

5. Give an example of a sequence (f_k) of continuously differentiable functions with jointly bounded derivatives such that (f_k) constitutes an orthonormal system in $L_2(-\infty, +\infty)$.

12.3.4 Gram–Schmidt Orthogonalization and the Existence of Orthonormal Bases

Theorem 1. *Let H be a Hilbert space, $\{x_n\}_1^{\infty} \subset H$ a linearly independent sequence, and $X_n = \mathrm{Lin}\{x_k\}_1^n$. Then there exists an orthonormal system $\{e_n\}_1^{\infty}$ with the property that $\mathrm{Lin}\{e_k\}_1^n = X_n$ for all n. The orthonormal system $\{e_n\}_1^{\infty}$ is called the Gram–Schmidt orthogonalization of the system $\{x_n\}_1^{\infty}$.*

Proof. Denote by P_n the orthogonal projector onto the subspace X_n. Set

$$e_1 = \frac{x_1}{\|x_1\|}, \quad e_2 = \frac{x_2 - P_1 x_2}{\|x_2 - P_1 x_2\|}, \quad \ldots, e_{n+1} = \frac{x_{n+1} - P_n x_{n+1}}{\|x_{n+1} - P_n x_{n+1}\|}, \ldots.$$

All elements e_k with $k \leqslant n$ thus defined lie in X_n, and $e_{n+1} \perp X_n$. Hence, for every n the vector e_{n+1} is orthogonal to all the preceding e_k. Moreover, $\|e_k\| = 1$ for all k. Therefore, $\{e_n\}_1^{\infty}$ is an orthonormal system. By construction, $\mathrm{Lin}\{e_k\}_1^n \subset X_n$ and the dimensions of these spaces coincide. Consequently, $\mathrm{Lin}\{e_k\}_1^n = X_n$. $\qquad\square$

Remark 1. Since the elements $\{e_k\}_1^n$ form an orthonormal basis in X_n, Proposition 3 of Subsection 12.3.2 shows that $P_n h = \sum_{k=1}^n \widehat{h}_k e_k$ for all $h \in H$. That is, e_k can be constructed from the elements $\{x_n\}_1^\infty$ in explicit form, by means of the recursion formula

$$e_{n+1} = \frac{x_{n+1} - \sum_{k=1}^n \langle x_{n+1}, e_k \rangle e_k}{\left\| x_{n+1} - \sum_{k=1}^n \langle x_{n+1}, e_k \rangle e_k \right\|}.$$

Theorem 2. *In every infinite-dimensional separable Hilbert space there exists an orthonormal basis.*

Proof. Since the Hilbert space H is separable, it contains a countable dense sequence. Discarding from this sequence the elements that are linearly dependent on the preceding ones, we obtain a linearly independent sequence $\{x_n\}_1^\infty \subset H$ that is complete in H. Let $\{e_n\}_1^\infty$ be the Gram–Schmidt orthogonalization of the sequence $\{x_n\}_1^\infty$. Then $\{e_n\}_1^\infty$ is an orthonormal system, and $\mathrm{Lin}\{e_k\}_1^\infty = \mathrm{Lin}\{x_k\}_1^\infty$. Therefore, the constructed system $\{e_n\}_1^\infty$ is complete in H, which by definition means that $\{e_n\}_1^\infty$ is an orthonormal basis of the space H. $\qquad\square$

Exercises

1. In the proof of Theorem 1 we stated that the vector e_{n+1} is orthogonal to all the preceding vectors e_k. Why is e_{n+1} also orthogonal to all the succeeding vectors e_k?

2. Why in the formula $e_{n+1} = \dfrac{x_{n+1} - P_n x_{n+1}}{\| x_{n+1} - P_n x_{n+1} \|}$ can the denominator be never equal to zero?

3. In Theorem 1 we proved the existence of a Gram–Schmidt orthogonalization of the sequence $(x_n)_1^\infty$. Is this orthogonalization unique?

4. What property of the projector P_n was used when we asserted that e_{n+1} is orthogonal to X_n?

5. Justify the equality $\mathrm{Lin}\{e_k\}_1^\infty = \mathrm{Lin}\{x_k\}_1^\infty$ in the proof of Theorem 2.

6. Prove that the family of all orthonormal systems in the Hilbert space H, ordered by inclusion, satisfies the conditions of Zorn's lemma. From this one can derive both the theorem on the existence of an orthonormal basis in a separable Hilbert space and its analogue for the non-separable case. The drawback of this argument is that it does not provide an explicit construction of a basis.

7. Let H be a Hilbert space. Then any two orthonormal bases in H have the same cardinality ("number of elements").

12.3.5 The Isomorphism Theorem

Definition 1. Let H_1 and H_2 be Hilbert spaces. The operator $T : H_1 \to H_2$ is called a *Hilbert space isomorphism* if T is bijective and $\langle Tx, Ty \rangle = \langle x, y \rangle$ for all $x, y \in H_1$. The Hilbert spaces H_1, H_2 are said to be *isomorphic* if there exists a Hilbert space isomorphism $T : H_1 \to H_2$.

Theorem 1. *Every separable infinite-dimensional Hilbert space is isomorphic to the space ℓ_2.*

Proof. Let $\{e_n\}_1^\infty$ be an orthonormal basis in H. Define the operator $T : H \to \ell_2$ by the rule $Th = (\langle h, e_1 \rangle, \langle h, e_2 \rangle, \ldots, \langle h, e_n \rangle, \ldots)$. That is, to each element $h \in H$ we associate the sequence of its Fourier coefficients in the basis $\{e_n\}_1^\infty$. By the Bessel inequality, $Th \in \ell_2$ and $\|Th\| \leqslant \|h\|$. The operator T has an inverse T^{-1}, which acts from ℓ_2 to H as $T^{-1}(a_n)_1^\infty = \sum_{n=1}^\infty a_n e_n$. Therefore, the operator T is invertible, and hence bijective. It remains to verify that $\langle Tx, Ty \rangle = \langle x, y \rangle$.

So, let $x = \sum_{k=1}^\infty a_k e_k$ and $y = \sum_{j=1}^\infty b_j e_j$ be arbitrary elements of H. We have

$$\langle x, y \rangle = \left\langle \sum_{k=1}^\infty a_k e_k, \sum_{j=1}^\infty b_j e_j \right\rangle$$

$$= \sum_{k=1}^\infty \left(\sum_{j=1}^\infty a_k \overline{b}_j \langle e_k, e_j \rangle \right) = \sum_{k=1}^\infty a_k \overline{b}_k = \langle Tx, Ty \rangle,$$

as needed. $\qquad\square$

Since all separable infinite-dimensional Hilbert spaces are isomorphic to each other, whenever one needs to solve a concrete problem one can choose a space that is more convenient for that problem. Let us give an example. In Exercise 5 of Subsection 11.1.6, the reader was asked to establish the existence and calculate the spectrum of an operator A which in the standard basis of the Hilbert space ℓ_2 has a two-diagonal matrix, with all elements on the two diagonals equal to 1. Let H_2 be the subspace of $L_2[0, 1]$ obtained by taking the closure of the linear span of the orthonormal system $e_n = e^{2\pi i n t}$, $n = 0, 1, 2, \ldots$. Now in H_2 consider the operator T of multiplication by the function $g(t) = 1 + e^{2\pi i t}$, acting as $Tf = g \cdot f$. In the considered basis e_n the operator T has the matrix

$$\begin{pmatrix} 1 & 0 & 0 & 0 & \cdots \\ 1 & 1 & 0 & 0 & \cdots \\ 0 & 1 & 1 & 0 & \cdots \\ 0 & 0 & 1 & 1 & \cdots \\ \vdots & \vdots & \vdots & \ddots & \ddots \end{pmatrix},$$

i.e., the same matrix as A. Hence, instead of considering the operator A in ℓ_2, one can consider the operator T in H_2, which enjoys the same properties. However, the operator T is already given by a simple explicit expression, and its properties are much easier to investigate. We invite the reader to solve Exercise 5 in Subsection 11.1.6 using the idea described above of replacing ℓ_2 by the space H_2.

Exercises

8. The proof of the isomorphism theorem started with the words: "Let $\{e_n\}_1^\infty$ be an orthonormal basis in H". Why does such a basis exist in H?

9. Show that the operator T figuring in the proof of the isomorphism theorem is linear.

10. Why does the series $\sum_{n=1}^\infty a_n e_n$ figuring in the definition of the operator T^{-1} converge?

11. Verify that the operator T^{-1} acting as $T^{-1}(a_n)_1^\infty = \sum_{n=1}^\infty a_n e_n$ is indeed the inverse of T.

12. Why does every element of the space H admit the representation $x = \sum_{n=1}^\infty a_n e_n$ as the sum of a convergent series with respect to the system $\{e_n\}_1^\infty$?

13. Justify the convergence of the series $\sum_{k=1}^\infty \left(\sum_{j=1}^\infty a_k \overline{b}_j \langle e_k, e_j \rangle \right)$ and the equality $\langle \sum_{k=1}^\infty a_k e_k, \sum_{j=1}^\infty b_j e_j \rangle = \sum_{k=1}^\infty \left(\sum_{j=1}^\infty a_k \overline{b}_j \langle e_k, e_j \rangle \right)$ in the proof of the isomorphism theorem.

14. Construct a concrete Hilbert space isomorphism between the spaces $L_2[0, 1]$ and ℓ_2.

15. Any Hilbert space isomorphism T of two Hilbert spaces H_1 and H_2 is an isometry: $\|Th\| = \|h\|$ for all $h \in H_1$.

16. Every bijective isometry between two Hilbert spaces H_1 and H_2 is an isomorphism of Hilbert spaces.

17. Can a finite-dimensional Hilbert space be isomorphic to an infinite-dimensional one?

18. Can a separable Hilbert space be isomorphic to a non-separable one?

19. Suppose the Hilbert spaces H_1 and H_2 have complete orthonormal systems of the same cardinality. Then H_1 and H_2 are isomorphic.

12.4 Self-adjoint Operators

12.4.1 Bilinear Forms on a Hilbert Space

Definition 1. Let H be a Hilbert space. A mapping $F \colon H \times H \to \mathbb{C}$ is called a *bilinear form* if for any elements $x, y, x_1, x_2, y_1, y_2 \in H$ and any complex numbers λ_1, λ_2 the following relations hold:
— $F(\lambda_1 x_1 + \lambda_2 x_2, y) = \lambda_1 F(x_1, y) + \lambda_2 F(x_2, y)$;
— $F(x, \lambda_1 y_1 + \lambda_2 y_2) = \overline{\lambda}_1 F(x, y_1) + \overline{\lambda}_2 F(x, y_2)$.

Using the term "bilinear form" here is not entirely correct: linearity in the second variable is modified. In many textbooks the so-modified condition is called *sesquilinearity*, and the form is called *sesquilinear* instead of bilinear.

Examples

1. In the finite-dimensional coordinate space $\mathbb{C}^n$ every bilinear form can be expressed as $F(x, y) = \sum_{k,j=1}^{n} a_{k,j} x_k \overline{y}_j$, where x_k and y_j are the coordinates of the vectors x and y, respectively, and $a_{k,j}$ are elements of a complex $n \times n$ matrix A.

2. Let A be a linear operator in H. The expressions $F(x, y) = \langle x, Ay \rangle$ and $G(x, y) = \langle Ax, y \rangle$ give bilinear forms.

Definition 2. A bilinear form is said to be *continuous* if it is continuous in each of its variables.

Theorem 1 (General form of continuous bilinear forms[6] on a Hilbert space). *Let F be a continuous bilinear form on a Hilbert space H. Then there exists a continuous linear operator $A \in L(H)$ such that*

$$F(x, y) = \langle x, Ay \rangle$$

for all $x, y \in H$. The operator A is uniquely determined by the form F.

Proof Fix an element $y \in H$. Then the mapping $x \mapsto F(x, y)$ is a continuous linear functional. By the theorem on the general form of continuous linear functionals on a Hilbert space, there exists an element $A(y) \in H$ such that $F(x, y) = \langle x, A(y) \rangle$ for all $x \in H$. Moreover, the element $A(y)$ is uniquely determined by y. It remains to verify that the mapping $y \mapsto A(y)$ is linear and continuous.

Linearity. We have $\langle x, A(\lambda_1 y_1 + \lambda_2 y_2) \rangle = F(x, \lambda_1 y_1 + \lambda_2 y_2) = \lambda_1 F(x, y_1) + \lambda_2 F(x, y_2) = \langle x, \lambda_1 A(y_1) + \lambda_2 A(y_2) \rangle$. Since the element x is arbitrary, this means that $A(\lambda_1 y_1 + \lambda_2 y_2) = \lambda_1 A(y_1) + \lambda_2 A(y_2)$.

Continuity. For each $x \in H$ the expression $F(x, y) = \langle x, Ay \rangle$ is continuous in y. That is, for any sequence (y_n) in H that converges to zero, we have $\langle x, Ay_n \rangle \to 0$ as $n \to \infty$. This means that the sequence of functionals $f_n(x) = \langle x, Ay_n \rangle$ converges pointwise to zero. By the Banach–Steinhaus theorem, the sequence (f_n) is bounded. Since $\|f_n\| = \|Ay_n\|$, we deduce that the operator A takes any sequence (y_n) that converges to zero into a bounded sequence. But by Theorem 1 of Subsection 6.4.1, this property is equivalent to the continuity of the operator A. The theorem is proved.

$\square$

---

[6]It's funny to speak about the "form of forms". Mathematical language has many such pearls. One of my favorite ones is the inequality $n > N$ in the standard definition of the limit: $\lim_{n \to \infty} a_n = a$ if $\forall \varepsilon > 0 \exists N \in \mathbb{N} \forall n > N \ |a_n - a| < \varepsilon$.

12.4.2 The Adjoint of a Hilbert Space Operator

The reader is already familiar with the notion of an adjoint (conjugate, dual) operator, defined as an operator acting between the corresponding dual spaces; but since in a Hilbert space the continuous linear functionals are identified with elements of the space itself, here it is natural to introduce the notion of adjoint operator in a somewhat different way.

Definition 1. Let $A \in L(H)$. The *adjoint* of the operator A is the operator A^* such that $\langle Ax, y \rangle = \langle x, A^*y \rangle$ for all elements $x, y \in H$.

The correctness of this definition, i.e., the existence and uniqueness of the operator A^*, is guaranteed by the theorem on the general form of bilinear forms.

Let us list the basic properties of the operation of passing to the adjoint operator:

1. $(A_1 + A_2)^* = A_1^* + A_2^*$;

2. $I^* = I$;

3. $(\lambda A)^* = \bar{\lambda} A^*$;

4. $(A_1 A_2)^* = A_2^* A_1^*$;

5. $(A^*)^* = A$.

All the listed properties are verified by following the same scheme. For example, let us establish property 1. We have to verify that $(A_1 + A_2)^* y = A_1^* y + A_2^* y$ for all $y \in H$. To this end we need, in turn, to verify that $\langle x, (A_1 + A_2)^* y \rangle = \langle x, A_1^* y + A_2^* y \rangle$ for all $x \in H$. We have

$$\langle x, (A_1 + A_2)^* y \rangle = \langle (A_1 + A_2)x, y \rangle = \langle A_1 x, y \rangle + \langle A_2 x, y \rangle$$
$$= \langle x, A_1^* y \rangle + \langle x, A_2^* y \rangle = \langle x, A_1^* y + A_2^* y \rangle.$$

Remark 1. Although in the setting of Hilbert space theory the adjoint operator is defined differently than for general Banach spaces (Subsection 9.4.1), we are in fact dealing with a particular case of that general definition. Indeed, if one identifies each element $y \in H$ with the linear functional on H that it generates, i.e., $y(x) = \langle x, y \rangle$, the definition $\langle Ax, y \rangle = \langle x, A^*y \rangle$ takes on the familiar form $(A^*y)(x) = y(Ax)$. Accordingly, the general theorems proved in Subsection 9.4.1 concerning the connections between images, kernels, injectivity and surjectivity of an operator and of its adjoint, as well as the formula $\|A^*\| = \|A\|$, remain in force. (Prove this!) In particular, if the operator A^* is injective, then the image of the operator A is dense in H.

Lemma 1. *Suppose the operators A and A^* are bounded below. Then A is invertible.*

Proof. Since A is bounded below, it is injective, and its image is closed. Next, since A^* is bounded below, it also is injective, and so the image of the operator A is dense in H. Since the image of A is closed and dense in H, we have that $A(H) = H$, i.e., A is surjective. Injectivity + surjectivity = invertibility. $\qquad\square$

Exercises

1. Verify properties 2–5 of the operation of passing to the adjoint operator.

2. If the definition of an adjoint operator in a Hilbert space is a particular case of that of an adjoint operator defined for general Banach spaces, then why does property 3 of the operation of passage to the adjoint operator in a Hilbert space differ from the analogous property in Banach spaces? Where does the complex conjugation bar over λ come from?

3. Does the analogue of property 5 hold for general Banach spaces?

4. Prove the relations $\sigma(U^{-1}) = \{1/\lambda : \lambda \in \sigma(U)\}$ and $\sigma(U^*) = \{\bar{\lambda} : \lambda \in \sigma(U)\}$, the first for invertible, and the second for arbitrary operators $U \in L(H)$.

5. Verify that a continuous bilinear form is jointly continuous in its variables.

6. Will the above definition of the adjoint operator be correct if H is a non-complete space with scalar product?

7. Let $P \in L(H)$ be a projector. Then P^* is also a projector. Onto what subspace?

The *Hilbert–Schmidt norm* of the operator A is defined as

$$\|A\|_{HS} = \left(\sum_{n,m \in \mathbb{N}} |\langle Ae_n, g_m \rangle|^2 \right)^{1/2},$$

where $\{e_n\}_1^\infty$, $\{g_n\}_1^\infty$ is a fixed pair of orthonormal bases of the Hilbert space. The operator A is called a *Hilbert–Schmidt operator* if $\|A\|_{HS} < \infty$, i.e., if its matrix in this pair of bases is Hilbert–Schmidt. Show that:

8. $\|A\| \leqslant \|A\|_{HS}$.

9. $\left(\sum_{n \in \mathbb{N}} \|A^* g_n\|^2 \right)^{1/2} = \|A\|_{HS} = \left(\sum_{n \in \mathbb{N}} \|A e_n\|^2 \right)^{1/2}$.

10. $\|A\|_{HS}$ does not depend on the choice of the pair of orthonormal bases $\{e_n\}_1^\infty$, $\{g_n\}_1^\infty$, and $\|A\|_{HS} = \|A^*\|_{HS}$.

11. Use the preceding exercise to establish the following fact: let U be a fixed ellipsoid in a finite-dimensional Euclidean space. Then all rectangular parallelepipeds circumscribing U have the same diameter. Note that even in the three-dimensional case proving this fact by methods of analytic geometry is far from simple.

12. Any Hilbert–Schmidt operator is compact.

12.4.3 Self-adjoint Operators and Their Quadratic Forms

Definition 1. The operator A in a Hilbert space H is called *self-adjoint* if $A^* = A$ or, equivalently, if $\langle Ax, y \rangle = \langle x, Ay \rangle$ for all $x, y \in H$.

We mentioned earlier the analogy between operators and complex numbers. For Hilbert space operators, thanks to the notion of an adjoint operator, this analogy is far more effective than in the general case. In particular, self-adjoint operators ($A^* = A$) are the analogues of real numbers ($z = \bar{z}$). At the same time, one has to be careful with this analogy, since operators, in contrast to numbers, may not commute!

The next theorem provides nontrivial examples of self-adjoint operators.

Theorem 1. *For a projector $P \in L(H)$ the following conditions are equivalent:*

(1) *P is a self-adjoint operator;*

(2) *P is an orthogonal projector.*

Proof. Since P is a projector, H decomposes into the direct sum $H = H_1 \oplus H_2$, where H_1 and H_2 are respectively the kernel and the image of P. First, let us prove that (1) $\implies$ (2), i.e., that if P is a self-adjoint operator, then $H_1 \perp H_2$. Indeed, let $h_1 \in H_1$ and $h_2 \in H_2$. Then $\langle h_1, h_2 \rangle = \langle h_1, Ph_2 \rangle = \langle Ph_1, h_2 \rangle = \langle 0, h_2 \rangle = 0$.

Now let us show that (2) $\implies$ (1), i.e., that if $H_1 \perp H_2$, then P is self-adjoint. To this end, take elements $x, y \in H$ and decompose them as $x = x_1 + x_2$ and $y = y_1 + y_2$, where $x_1, y_1 \in H_1$ and $x_2, y_2 \in H_2$. Then we have $\langle Px, y \rangle = \langle x_2, y \rangle = \langle x_2, y_2 \rangle = \langle x, y_2 \rangle = \langle x, Py \rangle$, as needed. $\square$

Definition 2. Let A be a self-adjoint operator. The *bilinear form of* (or *associated to*) the operator A is the function $F(x, y) = \langle Ax, y \rangle$; the function $g(x) = \langle Ax, x \rangle$ is called the *quadratic form of* (or *associated to*) the operator A.

Note that $g(x) = \langle Ax, x \rangle = \langle x, Ax \rangle = \overline{\langle Ax, x \rangle} = \overline{g(x)}$, hence the quadratic form of a self-adjoint operator takes **only** real values. It is not hard to verify that

$$\operatorname{Re} \langle Ax, y \rangle = \frac{1}{4} \left(\langle A(x + y), (x + y) \rangle - \langle A(x - y), (x - y) \rangle \right). \tag{1}$$

In a similar way one can find also the imaginary part of $\langle Ax, y \rangle$. Hence, the bilinear form is uniquely determined by the quadratic form. By the theorem on the general form of bilinear forms, from the bilinear form one can in turn recover the operator itself. Therefore, all the information about a self-adjoint operator can be obtained if one knows the properties of its quadratic form. As an example, we give the following very useful *formula for the norm of a self-adjoint operator*.

Theorem 2. *Let A be a self-adjoint operator. Then*

$$\|A\| = \sup_{x \in S_H} |\langle Ax, x \rangle|. \tag{2}$$

Proof. Let $q = \sup_{x \in S_H} |\langle Ax, x \rangle|$. Then for every $z \in S_H$ one has the estimate $|\langle Az, z \rangle| \leqslant q\|z\|^2$. By homogeneity, this estimate also holds for all $z \in H$. We need to show that $\|A\| = q$. Using formula $\|h\| = \sup_{y \in S_H} \mathrm{Re}\,\langle h, y \rangle$ (Subsection 12.1.3, Exercise 6) for $h = Ax$ and formula (2) we deduce that

$$\|A\| = \sup_{x \in S_H} \|Ax\| = \sup_{x,y \in S_H} \mathrm{Re}\,\langle Ax, y \rangle$$

$$= \sup_{x,y \in S_H} \frac{1}{4}\big(\langle A(x+y), (x+y)\rangle - \langle A(x-y), (x-y)\rangle\big)$$

$$\leqslant \sup_{x,y \in S_H} \frac{q}{4}\big(\|x+y\|^2 + \|x-y\|^2\big) = \sup_{x,y \in S_H} \frac{2q}{4}\big(\|x\|^2 + \|y\|^2\big) = q,$$

i.e., $\|A\| \leqslant q$. The converse inequality is an immediate consequence of the Cauchy–Schwarz inequality:

$$q = \sup_{x \in S_H} |\langle Ax, x \rangle| \leqslant \sup_{x \in S_H} \|Ax\| = \|A\|. \qquad \square$$

Exercises

1. For which scalars λ is the operator λI self-adjoint?

2. Calculate the norm of the operator A given by the matrix $\begin{pmatrix} 0 & 0 \\ 1 & 0 \end{pmatrix}$ in the two-dimensional coordinate space (equipped with the standard Euclidean norm). Is formula (2) valid for this operator?

3. Where in the proof of formula (2) was the self-adjointness of the operator A — without which (2) is not valid — used?

4. If A is different from zero, can the function $g(x) = \langle Ax, x \rangle$ be identically equal to zero?

5. Verify that the self-adjoint operators form a closed linear subspace in $L(H)$. Is this subspace closed in the sense of pointwise convergence?

6. Verify that a product of self-adjoint operators is self-adjoint if and only if the operators commute. (**N.B. In the sequel we will use the results of the last two exercises!**)

7. How must the subspaces H_1, $H_2 \subset H$ be related in order for the orthogonal projectors P_1, P_2 onto these subspace to commute?

8. Calculate the adjoint of the kernel integral operator T on $L_2[0, 1]$, given by $(Tf)(x) = \int_0^1 K(t, x) f(t)\,dt$, where $K \in L_2([0, 1] \times [0, 1])$. Such operators are called *Hilbert–Schmidt integral operators*. Under what conditions on K will the operator T be self-adjoint? Using the expansion of the kernel K in a double Fourier series, prove that a Hilbert–Schmidt integral operator is a Hilbert–Schmidt operator in the sense described in Exercises 8–12 of Subsection 12.4.2.

9. Let the operator T be given by its matrix in an orthonormal basis. How are the matrices of the operators T and T^* connected?

10. Given an operator $T \in L(H)$, define its *real* and *imaginary parts* by the formulas $\operatorname{Re} T = \frac{1}{2}(T + T^*)$ and $\operatorname{Im} T = \frac{1}{2i}(T - T^*)$. Verify that $\operatorname{Re} T$ and $\operatorname{Im} T$ are self-adjoint operators and $T = \operatorname{Re} T + i \operatorname{Im} T$.

11. Suppose that T and T^* commute (operators T with this property are called *normal*). Then $\|T\| = \sqrt{\|(\operatorname{Re} T)^2 + (\operatorname{Im} T)^2\|}$.

12.4.4 Operator Inequalities

The following definition helps us take further the analogy between operators and numbers.

Definition 1. The operator $A \in L(H)$ is said to be *positive* (and one writes $A \geqslant 0$) if it is self-adjoint and its quadratic form is non-negative (i.e., $\langle Ax, x \rangle \geqslant 0$ for all $x \in H$). Let $A, B \in L(H)$. We say that $A \geqslant B$, if $A - B \geqslant 0$.

Theorem 1. *Let $A \in L(H)$ be a positive operator. Then for all $x \in H$*

$$\|Ax\| \leqslant \sqrt{\|A\|} \sqrt{\langle Ax, x \rangle}.$$

Proof. The bilinear form of the operator A satisfies all the scalar product axioms, except for non-degeneracy. Hence, as noted in Exercise 4 of Subsection 12.1.2, the Cauchy–Schwarz inequality holds for it: $|\langle Ax, y \rangle| \leqslant \sqrt{\langle Ax, x \rangle} \sqrt{\langle Ay, y \rangle}$. Taking here the supremum over $y \in S_H$, we obtain the required estimate. $\square$

The main objective of this subsection is to prove for operators a result analogous to the theorem asserting the existence of a limit for a bounded monotone sequence of numbers.

Theorem 2. *Suppose the operators $A_n \in L(H)$ form a bounded increasing sequence, i.e., $A_1 \leqslant A_2 \leqslant \cdots$ and $\sup_n \|A_n\| < \infty$. Then this sequence has a pointwise limit.*

Proof. Fix an arbitrary vector $x \in H$. The sequence of non-negative reals $a_n := \langle A_n x - A_1 x, x \rangle$ is non-decreasing and bounded. Therefore, it has a limit, and so

$$a_n - a_m = \langle (A_n - A_m)x, x \rangle \to 0 \quad \text{as } n, m \to \infty.$$

Applying Theorem 1, we obtain the claimed pointwise convergence:

$$\|A_n x - A_m x\| \leqslant \sqrt{\|A_n - A_m\|}\,\sqrt{\langle (A_n - A_m)x, x\rangle} \to 0 \quad \text{as } n, m \to \infty. \qquad \Box$$

Exercises

1. For which scalars λ is λI a positive operator?

2. Any orthogonal projector is a positive operator.

3. Let $\{e_n\}_{n=1}^{\infty}$ be an orthonormal basis in H. Verify that the partial sum operators S_n form a non-decreasing bounded sequence. Describe its pointwise limit.

4. Under the conditions of Theorem 2, can the sequence A_n fail to converge in the norm of the space $L(H)$?

5. Let A, B be positive operators and $A + B = 0$. Then $A = B = 0$.

6. Prove that the product of two commuting positive operators is a positive operator.[7]

12.4.5 The Spectrum of a Self-adjoint Operator

Since the quadratic form of a self-adjoint operator takes only real values, the eigenvalues of the operator are necessarily real (this reasoning is well known from linear algebra). Although the spectrum of an operator in an infinite-dimensional space is not exhausted by its eigenvalues, the assertion about the positivity of the spectrum remains in force.

Theorem 1 (Structure of the spectrum of a self-adjoint operator). *Let* $A \in L(H)$ *be a self-adjoint operator. Define*

$$\alpha_- = \alpha_-(A) = \inf\{\langle Ax, x\rangle : x \in S_H\},$$
$$\alpha_+ = \alpha_+(A) = \sup\{\langle Ax, x\rangle : x \in S_H\}.$$

Then

(i) *the spectrum of the operator A consists only of real numbers;*
(ii) $\sigma(A) \subset [\alpha_-, \alpha_+]$;
(iii) *the endpoints of the interval $[\alpha_-, \alpha_+]$ belong to the spectrum.*

---

[7] Despite the simplicity of its formulation, this exercise is not simple.

Proof. (i) Let $\lambda = a + ib$ be a complex number with imaginary part $b \neq 0$. We have to show that the operator $A - \lambda I$ is invertible. First we show that $A - \lambda I$ is bounded below. Indeed,

$$\|(A - \lambda I)x\|^2 = \|(Ax - ax) - ibx\|^2$$
$$= \|Ax - ax\|^2 - 2\operatorname{Re}\langle Ax - ax, ibx \rangle + b^2 \|x\|^2.$$

The quantities $\langle Ax, x \rangle$ and $\langle x, x \rangle$ are real, and so $2\operatorname{Re}\langle Ax - ax, ibx \rangle = 0$ and $\|(A - \lambda I)x\|^2 \geqslant b^2 \|x\|^2$, i.e., $A - \lambda I$ is bounded below. For the same reason the operator $(A - \lambda I)^* = A - \bar{\lambda} I$ is also bounded below (it has the same form, only with a different coefficient λ). Hence, by Lemma 1 of Subsection 12.4.2, the operator $A - \lambda I$ is invertible.

(ii) Notice that the values of the quadratic form of the operator $A - \alpha_+ I$ are non-positive. Indeed, for any $x \in S_H$ we have $\langle (A - \alpha_+ I)x, x \rangle = \langle Ax, x \rangle - \alpha_+ \leqslant 0$. Now let $\lambda > \alpha_+$ and $\varepsilon = \lambda - \alpha_+$. Let us show that the operator $A - \lambda I = (A - \alpha_+ I) - \varepsilon I$ is invertible. To this end, since this operator is self-adjoint, it suffices to show that it is bounded below (here we use again Lemma 1 of Subsection 12.4.2). We have

$$\|(A - \lambda I)x\|^2 = \|(A - \alpha_+ I)x\|^2 - 2\operatorname{Re}\langle (A - \alpha_+ I)x, \varepsilon x \rangle + \varepsilon^2 \|x\|^2 \geqslant \varepsilon^2 \|x\|^2.$$

Hence, the operator $A - \lambda I$ is invertible, so λ does not belong to the spectrum of A. This proves that $\sigma(A) \subset (-\infty, \alpha_+]$ for any self-adjoint operator A. Replacing A by $-A$, we get that $-\sigma(A) = \sigma(-A) \subset (-\infty, \alpha_+(-A)] = (-\infty, -\alpha_-(A)]$, i.e., $\sigma(A) \subset [\alpha_-(A), +\infty)$.

(iii) Let us show that $\alpha_- \in \sigma(A)$, that is, that the operator $B = A - \alpha_- I$ is not invertible. Here comes to help the following obvious property of the quadratic form of the operator B: $\inf_{x \in S_H} \langle Bx, x \rangle = \inf_{x \in S_H} \langle Ax, x \rangle - \alpha_- = 0$; in particular, B is positive. Now Theorem 1 of Subsection 12.4.4 yields

$$\inf_{x \in S_H} \|Bx\| \leqslant \|B\|^{1/2} \inf_{x \in S_H} \langle Bx, x \rangle^{1/2} = 0.$$

Thus, our operator is not bounded below, and hence is not invertible. The fact that α_+ also belongs to the spectrum is readily obtained upon replacing the operator A by $-A$, as we did above. The theorem is proved. $\qquad\square$

Let us list several corollaries of the theorem just proved.

Corollary 1. *A self-adjoint operator is positive if and only if its spectrum consists only of non-negative numbers.*

Proof. The operator A is positive if and and only if $\langle Ax, x \rangle \geqslant 0$ for all $x \in S_H$. This is equivalent to the condition $\alpha_-(A) \geqslant 0$. $\qquad\square$

Corollary 2. *Let A be a self-adjoint operator. Then*

$$\|A\| = \sup\{|t| : t \in \sigma(A)\}.$$

Proof. Indeed, $\|A\| = \sup\{|\langle Ax, x\rangle| : x \in S_H\} = \max\{|\alpha_-(A)|, |\alpha_+(A)|\} = \sup\{|t| : t \in \sigma(A)\}$. □

Corollary 3. *Let A be a self-adjoint operator such that $\sigma(A) = \{0\}$. Then $A = 0$.*

Proof. By Corollary 2, if $\sigma(A) = \{0\}$, then $\|A\| = 0$. □

We remark that for non-self-adjoint operators the corollaries given above may fail.

Example. In the two-dimensional space, consider the operator given by the matrix $\begin{pmatrix} 0 & 0 \\ 1 & 0 \end{pmatrix}$. Its spectrum is $\{0\}$, but the operator is not equal to zero.

Exercises

1. Let $A \in L(H)$ be such that $\sigma(A) = \{-2, 1\}$. Can the norm of A be equal to 3? Does the answer change if A is self-adjoint?

2. Let A be a self-adjoint operator such that $\sigma(A) = \{\lambda_0\}$. Then $A = \lambda_0 I$.

3. Let K be an arbitrary closed bounded set of real numbers. Construct a self-adjoint operator A for which $\sigma(A) = K$.

12.4.6 Compact Self-adjoint Operators

In this subsection we will prove that the matrix of any compact self-adjoint operator can be reduced, by means of an appropriate choice of orthonormal basis, to diagonal form. In other words, for any compact self-adjoint operator there exists an orthonormal basis consisting of eigenvectors.

Recall that a subspace $X \subset H$ is said to be *invariant* under the operator A if $A(X) \subset X$.

Theorem 1. *If the subspace X is invariant under the self-adjoint operator A, then its orthogonal complement $X^\perp$ is also invariant under A.*

Proof. We need to show that for every element $y \in X^\perp$ its image Ay again lies in $X^\perp$. To this end we need to verify that $\langle x, Ay\rangle = 0$ for all $x \in X$. But $\langle x, Ay\rangle = \langle Ax, y\rangle$, and $\langle Ax, y\rangle = 0$, because $Ax \in X$ (by the invariance of the subspace X), and $y \in X^\perp$. □

Theorem 2. *Let $X_1 = \text{Ker}\,(A - \lambda_1 I)$ and $X_2 = \text{Ker}\,(A - \lambda_2 I)$ be eigenspaces of the self-adjoint operator A corresponding to two distinct eigenvalues $\lambda_1 \neq \lambda_2$. Then $X_1 \perp X_2$.*

Proof. For any $x_1 \in X_1$ and $x_2 \in X_2$, we have

$$\lambda_1 \langle x_1, x_2 \rangle = \langle Ax_1, x_2 \rangle = \langle x_1, Ax_2 \rangle = \lambda_2 \langle x_1, x_2 \rangle,$$

which is possible only if $\langle x_1, x_2 \rangle = 0$. $\qquad\qquad\qquad\qquad\qquad\qquad\square$

Lemma 1. *Suppose the compact self-adjoint operator A in the Hilbert space Y has no eigenvectors. Then Y consists only of the element 0.*

Proof. Suppose, by contradiction, that Y is not zero-dimensional. By the theorem on the structure of the spectrum of a compact operator (Subsection 11.3.4), the spectrum of the operator A consists only of 0 (otherwise A would have eigenvectors). By Corollary 3 in Subsection 12.4.7, $A = 0$. But then the whole space Y consists of eigenvectors with eigenvalue 0. $\qquad\qquad\qquad\qquad\qquad\qquad\square$

Theorem 3. *Let H be a separable Hilbert space. Then for any compact self-adjoint operator $A \in L(H)$ there exists an orthonormal basis in H consisting of eigenvectors of A.*

Proof. Choose in each of the eigenspaces of the operator A an orthonormal basis and write all these bases as one sequence $\{e_n\}_{n=1}^{\infty}$. All vectors e_n are eigenvectors of A; moreover, by Theorem 2 above, they form an orthonormal system. It remains to show that the system $\{e_n\}_{n=1}^{\infty}$ is complete. Denote $\overline{\text{Lin}}\{e_n\}_{n=1}^{\infty}$ by X. The subspace X is invariant under the operator A and contains all its eigenvectors. Hence, the subspace $X^{\perp}$ is also invariant under A, and contains no eigenvectors of A. By the preceding lemma, this means that the $X^{\perp} = \{0\}$, that is, $X = H$. $\qquad\square$

Remark 1. In the theorem above both the compactness and the self-adjointness of the operator are essential. Indeed, for a non-self-adjoint operator the system of its eigenvectors may fail to be complete even in the two-dimensional case (Example at the end of Subsection 12.4.5). Here is an example of a self-adjoint operator that has no eigenvalues.

Consider the operator $A \in L(L_2[0, 1])$ acting by the rule $(Af)(t) = tf(t)$. Suppose f is an eigenvector[8] of the operator A with eigenvalue λ, i.e., $\lambda f(t) = tf(t)$ almost everywhere. This equality can hold only if $f \overset{\text{a.e.}}{=} 0$. Hence, the operator A has no eigenvectors.

---

[8]Since here the elements of the space are functions, "eigenvectors" are mostly referred to as "eigenfunctions".

Finally, let us collect in one theorem all essential properties of the spectrum of a compact self-adjoint operator that were established in Theorem 1 of Subsection 11.3.4, Theorem 1 of Subsection 12.4.5 and in Theorems 2 and 3 of the present subsection.

Theorem 4 (Spectral theorem for compact self-adjoint operators). *Let H be a separable infinite-dimensional Hilbert space, and $A \in L(H)$ be a compact self-adjoint operator. Then*

(1) *The spectrum of A is either a finite set of reals, or is a sequence of real numbers that converge to 0.*

(2) *0 belongs to the spectrum.*

(3) *If $\lambda \neq 0$ belongs to the spectrum, then λ is an eigenvalue of A; the eigenspaces corresponding to the nonzero eigenvalues are finite-dimensional.*

(4) *Eigenspaces corresponding to distinct eigenvalues are mutually orthogonal.*

(5) *There exists an orthonormal basis of H consisting of eigenvectors of A.* □

Exercises

Fix a function $g \in L_1[0, 1]$ and for every $f \in L_2[0, 1]$ define $(A_g f)(t) = g(t) f(t)$.

1. Prove that if the image of the operator A_g again lies in $L_2[0, 1]$, then $A_g \in L(L_2[0, 1])$. (Hint: the most economical way of proving continuity in similar cases is to use the closed graph theorem.)

2. In the setting of the preceding exercise, calculate the norm of the operator A_g. Use the result to prove that $A_g \in L(L_2[0, 1])$ if and only if $g \in L_\infty[0, 1]$.

3. Calculate A_g^*. Under what conditions is the operator A_g self-adjoint? Positive?

4. Calculate the spectrum of A_g.

5. Characterize the functions g for which the operator A_g has eigenvalues and eigenvectors.

6. Suppose g is not identically a constant. Can the operator A_g have a complete system of eigenvectors? Consider the same question for a continuous function g.

7. In the setting of Exercise 8 of Subsection 12.4.3, suppose that the kernel K is jointly continuous. Then the eigenfunctions of the corresponding Hilbert–Schmidt integral operator that belong to non-zero eigenvalues are continuous.

Comments on the Exercises

Subsection 12.1.3

Exercise 5. P. Jordan, J. von Neumann, 1935. For a reference to the corresponding work and a survey of various characterizations of the norm generated by a scalar product, see [11, Chapter 7, §3].

Subsection 12.3.4

Exercise 7. Let H be an infinite-dimensional Hilbert space. Denote by $\alpha(H)$ the smallest possible cardinality of a dense subset of H. Fix a dense subset G of H with card$(G) = \alpha(H)$. Let us prove that the cardinality of any orthonormal basis E in H is equal to $\alpha(H)$. For each element $e \in E$, consider an open ball U_e of radius $\sqrt{2}/2$ and centered at e. Since the distance between any two distinct elements of E is equal to $\sqrt{2}$, the considered balls are pairwise disjoint. At the same time, every ball U_e must intersect the dense set G. Pick for each $e \in E$ a point $f(e) \in U_e \cap G$. Then $f : E \to G$ is an injective mapping, and consequently card$(E) \leqslant$ card$(G) = \alpha(H)$.

Conversely, consider $\mathrm{Lin}_{\mathbb{Q}} E$, the set of finite linear combinations with rational coefficients of elements of the basis E. Since $\mathrm{Lin}_{\mathbb{Q}} E$ is dense in H, card$(\mathrm{Lin}_{\mathbb{Q}} E) \geqslant \alpha(H)$. At the same time, card$(\mathrm{Lin}_{\mathbb{Q}} E) =$ card(E).

Subsection 12.4.2

Exercise 2. The solution of the riddle is hidden in Exercise 5 of Subsection 12.2.3.

Exercise 3. If the operator A acts from X to Y, then the operator $(A^*)^*$ acts from $(X^*)^*$ to $(Y^*)^*$, i.e., generally speaking, $A \neq (A^*)^*$. Nevertheless, some analogy with the equality $(A^*)^* = A$ is preserved. To understand what this analogy amounts to, one needs to work out the connection between a space and its second dual, which will be done in Subsection 17.2.2.

Exercise 11. This elegant lemma was proved by W. Banaszczyk in 1990, J. Reine Angew. Math. **403**, 187–200. See also the proof in [21, Lemma 18.3.1].

Subsection 12.4.3

Exercise 4. In complex Hilbert spaces, which are the ones we are dealing with, this is not possible. Moreover, every operator A satisfies the inequality

$$\sup_{x \in S_H} |\langle Ax, x \rangle| \geqslant \frac{1}{2} \|A\|.$$

In real Hilbert spaces, however, such examples do indeed exist: it suffices to consider the operator of rotation by 90 degrees in $\mathbb{R}^2$. Similar phenomena served as the starting point of an interesting direction in the study of operators and Banach algebras — the theory of the numerical range (see the monographs [8, 9], and also the survey [64]).

Exercise 11. See Lemma 1 in Subsection 13.1.2.

Subsection 12.4.4

Exercise 5. By our assumptions, $\langle Ax, x \rangle + \langle Bx, x \rangle = \langle (A + B)x, x \rangle = 0$ for all $x \in H$ and the quantities $\langle Ax, x \rangle$ and $\langle Bx, x \rangle$ are nonnegative. Therefore, $\langle Ax, x \rangle = \langle Bx, x \rangle = 0$ for all $x \in H$, and to complete the proof it remains to use the formula for the norm of a self-adjoint operator (Theorem 2 in Subsection 12.4.3).

Chapter 13
Functions of an Operator

One of the most fruitful applications of the aforementioned analogy between operators and numbers is encountered in the study of differential equations. As it turns out, the solution of the equation $y' = Ay$ can be written in the form $y = e^{At}y_0$ not only for scalar-valued functions and a numerical parameter A, but also for vector-valued functions and an operator A, respectively. The apparatus of functions of an operator was created precisely to enable the free use of such analogies.

13.1   Continuous Functions of an Operator

13.1.1   Polynomials in an Operator

In this subsection we consider operators in an arbitrary complex Banach space X.

Definition 1. Given a polynomial $p = a_0 + a_1 t + \cdots + a_n t^n$ and an operator $T \in L(X)$, an operator of the form $p(T) = a_0 I + a_1 T + \cdots + a_n T^n$ is called a *polynomial in the operator T*.

Let us list some readily verifiable properties of polynomials in operators.

Theorem 1. *Let p_1, p_2 be polynomials, $T \in L(X)$, and $\lambda_1, \lambda_2 \in \mathbb{C}$. Then*

(i) $(\lambda_1 p_1 + \lambda_2 p_2)(T) = \lambda_1 p_1(T) + \lambda_2 p_2(T)$;

(ii) $(p_1 p_2)(T) = p_1(T) p_2(T)$.

Further,
(iii) *suppose the operators $T_1, T_2 \in L(X)$ commute, and p_1, p_2 are polynomials. Then the operators $p_1(T_1)$ and $p_2(T_2)$ also commute.* □

© Springer International Publishing AG, part of Springer Nature 2018
V. Kadets, *A Course in Functional Analysis and Measure Theory*,
Universitext, https://doi.org/10.1007/978-3-319-92004-7_13

Theorem 2. *The operator $p(T)$ is invertible if and only if the polynomial p does not vanish at any of the points of the spectrum of the operator T.*

Proof. Let $t_1, \ldots, t_n$ be the roots of the polynomial p, i.e., $p(t) = a_n(t - t_1) \cdots (t - t_n)$. Then $p(T) = a_n(T - t_1 I) \cdots (T - t_n I)$. By Lemma 1 of Subsection 11.1.2, the invertibility of a product of commuting operators is equivalent to the simultaneous invertibility of its factors. Therefore, the invertibility of the operator $p(T)$ is equivalent to the simultaneous invertibility of the factors $T - t_i I$, i.e., to the fact that none of the roots t_i of the polynomial p lie in the spectrum of the operator T. $\square$

Theorem 3 (Spectral mapping theorem for polynomials in an operator). *The spectrum of the polynomial $p(T)$ consists of the values of the polynomial in the points of the spectrum of the operator T, i.e., $\sigma(p(T)) = p(\sigma(T))$.*

Proof. Let us show that $\lambda \in \sigma(p(T))$ if and only $\lambda \in p(\sigma(T))$. Indeed, the condition $\lambda \in \sigma(p(T))$ means that the operator $p(T) - \lambda I = (p - \lambda)(T)$ is not invertible. By the preceding theorem, this is equivalent to the polynomial $p - \lambda$ vanishing at some point of the spectrum: there exists a $t \in \sigma(T)$ such that $p(t) = \lambda$. This in turn is equivalent to the requisite condition $\lambda \in p(\sigma(T))$. $\square$

Exercises

1. Let p_1, p_2 be a pair of coprime polynomials and assume that $p_1 p_2(T) = 0$. Prove that the whole space X decomposes into the direct sum of its subspaces $X_1 = \operatorname{Ker} p_1(T)$ and $X_2 = \operatorname{Ker} p_2(T)$.

2. By analogy with calculus, introduce the concepts of derivative and differentiability for functions $f: [0, 1] \to E$, where E is a Banach space. Verify for differentiable functions $f, g: [0, 1] \to E$, that $(f + g)' = f' + g'$.

3. Let $f: [0, 1] \to L(X)$ be a differentiable function. Prove that $\dfrac{d}{dt}\left[f^2(t)\right] = f'(t)f(t) + f(t)f'(t)$.

4. Prove that if all the values of a function $f: [0, 1] \to L(X)$ pairwise commute, then the values of f and f' also commute.

5. For any operator $A \in L(X)$, define e^A by the formula

$$e^A = 1 + A + \frac{1}{2!}A^2 + \frac{1}{3!}A^3 + \cdots .$$

Is it true that if $f: [0, 1] \to L(X)$ is a differentiable function, then the function $y = e^{f(t)}$ is a solution of the differential equation $y' = f'(t)y$? Why is the particular case $y = e^{tA}$ of this formula successfully used for equations $y' = Ay$ with constant coefficients $A \in L(X)$?

6. Suppose that in some basis the matrix of the operator $A \in L(X)$ is diagonal. What will the matrix of the operator $p(A)$, where p is a polynomial, look like in this basis? How about the matrix of the operator e^A?

7. By analogy with the above, define polynomials in the elements of a Banach algebra **A**. Prove that all properties of polynomials in an operator considered above carry over to polynomials in elements of a Banach algebra.

The reader interested in the theory of functions of elements of a Banach algebra in the general case is referred to W. Rudin's textbook [38].

13.1.2 Polynomials in a Self-adjoint Operator

From here on till the end of the chapter we will consider only operators in a Hilbert space.

Lemma 1. Let $A, B \in L(H)$ be commuting self-adjoint operators. Then $\|A + iB\| = \sqrt{\|A^2 + B^2\|}$.

Proof. Since A and B commute, their product is a self-adjoint operator. Hence, $\langle Ax, Bx \rangle = \langle BAx, x \rangle$ is a real number for all $x \in H$. Therefore,

$$\|(A + iB)x\|^2 = \|Ax\|^2 + 2\operatorname{Re}(-i)\langle Ax, Bx \rangle + \|Bx\|^2 = \|Ax\|^2 + \|Bx\|^2,$$

and so

$$\|A + iB\|^2 = \sup_{x \in S_H} \|(A + iB)x\|^2 = \sup_{x \in S_H} (\|Ax\|^2 + \|Bx\|^2)$$

$$= \sup_{x \in S_H} (\langle Ax, Ax \rangle + \langle Bx, Bx \rangle) = \sup_{x \in S_H} \langle (A^2 + B^2)x, x \rangle = \|A^2 + B^2\|. \qquad \square$$

Theorem 1. Let $A \in L(H)$ be a self-adjoint operator and $p = a_0 + a_1 t + \cdots + a_n t^n$ be a polynomial. Then the operator $p(A)$ has the following properties:

(i) $(p(A))^* = \overline{p}(A)$, where $\overline{p} = \overline{a}_0 + \overline{a}_1 t + \cdots + \overline{a}_n t^n$. In particular, if all the coefficients of p are real, then $p(A)$ is a self-adjoint operator.

(ii) $\|p(A)\| = \sup_{t \in \sigma(A)} |p(t)|$.

Proof.
(i) $(p(A))^* = \overline{a}_0 (I)^* + \overline{a}_1 (A)^* + \cdots + \overline{a}_n (A^n)^* = \overline{p}(A)$.

(ii) Consider first the case of a polynomial with real coefficients. By Corollary 2 in Subsection 12.4.5 and the spectral mapping theorem for polynomials in an operator (Theorem 3 of Subsection 13.1.1),

$$\|p(A)\| = \sup_{\tau \in \sigma(p(A))} |\tau| = \sup_{\tau \in p(\sigma(A))} |\tau|.$$

To obtain the required formula, it remains to define $\tau = p(t)$ and observe that as t runs through $\sigma(A)$, τ runs through $p(\sigma(A))$.

Now suppose that the coefficients of the polynomial p have the form $a_j = u_j + iv_j$, $u_j, v_j \in \mathbb{R}$. Put $p_1 = u_0 + u_1 t + \cdots + u_n t^n$ and $p_2 = v_0 + v_1 t + \cdots + v_n t^n$. Using the lemma and the case of real polynomials treated above, we have

$$\|p(A)\| = \|p_1(A) + ip_2(A)\| = \sqrt{\|p_1(A)^2 + p_2(A)^2\|}$$

$$= \sqrt{\|(p_1^2 + p_1^2)(A)\|} = \sqrt{\sup_{t \in \sigma(A)} |(p_1^2 + p_1^2)(t)|} = \sup_{t \in \sigma(A)} |p(t)|. \qquad \square$$

Exercises

1. Give an example of a pair of self-adjoint operators $A, B \in L(H)$, for which $\|A + iB\| \neq \sqrt{\|A^2 + B^2\|}$.

2. Give an example of a pair of commuting self-adjoint operators $A, B \in L(H)$, for which $\|A + iB\| \neq \sqrt{\|A\|^2 + \|B\|^2}$.

3. Let $A \in L(H)$ be a self-adjoint operator and p_1, p_2 be polynomials such that $p_1(t) = p_2(t)$ for all $t \in \sigma(A)$. Then $p_1(A) = p_2(A)$.

13.1.3 Definition of a Continuous Function of a Self-adjoint Operator

Lemma 1. *Let $K \subset \mathbb{R}$ be a compact subset, and let $[a, b]$ be the smallest interval containing K. Then every function $f \in C(K)$ can be extended to a continuous function on $[a, b]$.*

Proof. The set $[a, b] \setminus K$ can be written as a union of open intervals with endpoints in K. Now redefine the function f on each such interval $(c, d) \subset [a, b] \setminus K$ by linear interpolation: $f(t) = f(c) + (t - c)\dfrac{f(d) - f(c)}{d - c}$. $\qquad \square$

Lemma 2. *Let $K \subset \mathbb{R}$ be a compact subset. Then for any function $f \in C(K)$ there exists a sequence of polynomials (p_n) which converges to f uniformly on K.*

Proof. Let $[a, b]$ be the smallest interval containing K. Then by the preceding lemma, we may assume that f is defined on the whole interval $[a, b]$. By the Weierstrass theorem, there exists a sequence of polynomials (p_n) which converges uniformly to f on $[a, b]$. This sequence (p_n) will also converge to f on K, a subset of $[a, b]$. $\qquad \square$

Lemma 3. (a) *Let A be a self-adjoint operator, and let (p_n) be a sequence of polynomials which converges uniformly on $\sigma(A)$. Then the sequence of operators $p_n(A)$ converges in norm.*

(b) *If the sequences of polynomials (p_n) and (q_n) converge uniformly on $\sigma(A)$ to one and the same limit, then $p_n(A)$ and $q_n(A)$ also converge to one and the same limit.*

Proof. We use assertion (ii) of the theorem proved in the preceding subsection:

$$\|p_n(A) - p_m(A)\| = \sup_{t \in \sigma(A)} |(p_n - p_m)(t)| \to 0 \quad \text{as } n, m \to \infty.$$

Since the space of operators is complete, this proves assertion (a). Assertion (b) is proved in exactly the same way:

$$\|p_n(A) - q_n(A)\| = \sup_{t \in \sigma(A)} |(p_n - q_n)(t)| \to 0 \quad \text{as } n \to \infty. \qquad \square$$

Definition 1. Let A be a self-adjoint operator, and $f \in C(\sigma(A))$ be a continuous function given on the spectrum of the operator A. The *function f of the operator A* is defined as

$$f(A) = \lim_{n \to \infty} p_n(A),$$

where (p_n) is an arbitrary sequence of polynomials that converges uniformly on $\sigma(A)$ to f.

The relevance of this definition is justified by Lemmas 2 and 3 proved above.

Exercises

1. Deduce Lemma 1 from Tietze's extension theorem (Theorem 3 in Subsection 1.2.3).

2. Consider in $C(\sigma(A))$ the subspace $\mathcal{P}$ consisting of all polynomials. Define the operator $U : \mathcal{P} \to L(H)$ by the formula $U(p) = p(A)$. Verify that U is a continuous linear operator. What is the norm of U equal to?

3. Applying the theorem of extension by continuity (Subsection 6.5.1) to the operator U, extend it to the whole space $C(\sigma(A))$. Verify that the equality $U(p) = p(A)$ holds not only for polynomials, but also for arbitrary continuous functions.[1]

---

[1] We could have used the extension of the operator U to $C(\sigma(A))$ and *defined* continuous functions of the operator A by the equality $f(A) = U(f)$. However, such a definition would be unnecessarily abstract and require additional interpretation.

13.1.4 Properties of Continuous Functions of a Self-adjoint Operator

First we will present properties that are obtained by direct passage to the limit from polynomials to continuous functions of a self-adjoint operator.

Theorem 1. *Let A be a self-adjoint operator, $f_1, f_2 \in C(\sigma(A))$, and $\lambda_1, \lambda_2 \in \mathbb{C}$. Then*

(1) $(\lambda_1 f_1 + \lambda_2 f_2)(A) = \lambda_1 f_1(A) + \lambda_2 f_2(A)$, *and*

(2) $(f_1 f_2)(A) = f_1(A) f_2(A)$.

   Further, let $f \in C(\sigma(A))$. Then

(3) $(f(A))^* = \overline{f}(A)$. *In particular, if the function f takes only real values on $\sigma(A)$, then $f(A)$ is a self-adjoint operator.*

(4) $\|f(A)\| = \sup_{t \in \sigma(A)} |f(t)|$.

   Finally,

(5) *suppose the operators A and B commute, and let f and g be continuous functions on the spectra of the operators A and B, respectively. Then $f(A)$ and $g(B)$ also commute.*   □

The following property already needs justification.

Theorem 2 (Invertibility criterion). *Let f be a continuous function defined on the spectrum of the self-adjoint operator A. Then for the operator $f(A)$ to be invertible it is necessary and sufficient that the function f has no zeros on the spectrum of A.*

Proof. Suppose first that the function f has no zeros on $\sigma(A)$. Then $g := 1/f$ is also a continuous function, $gf = 1$ and by assertion (2) of the preceding theorem, the operator $g(A)$ is the inverse of the operator $f(A)$. Now suppose that f vanishes at some point $t_0 \in \sigma(A)$. Pick a sequence of polynomials (p_n) which converges to f uniformly on $\sigma(A)$. With no loss of generality, we can assume that $p_n(t_0) = 0$ (otherwise, we replace $p_n(t)$ by $\widetilde{p}_n(t) = p_n(t) - p_n(t_0)$). By Theorem 2 of Subsection 13.1.1, the operators $p_n(A)$ are not invertible. Hence, since the set of non-invertible operators is closed (see the corollary to Theorem 1 of Subsection 11.1.2), the operator $f(A) = \lim_{n \to \infty} p_n(A)$ is also non-invertible.

Theorem 3 (Spectral mapping theorem for continuous functions). *Let A be a self-adjoint operator and $f \in C(\sigma(A))$. Then $\sigma(f(A)) = f(\sigma(A))$.*

Proof. We repeat the argument used earlier for polynomials (Theorem 3 of Subsection 13.1.1). The condition $\lambda \in \sigma(f(A))$ means that the operator $f(A) - \lambda I = (f - \lambda)(A)$ is not invertible. By the preceding assertion, this is equivalent to the existence of a point $t \in \sigma(A)$ such that $f(t) - \lambda = 0$. In turn, this is equivalent to the required condition $\lambda \in f(\sigma(A))$. $\qquad\square$

Theorem 4. *Under the conditions of the preceding theorem, if $f \geqslant 0$ on the spectrum of the operator A, then $f(A) \geqslant 0$.*

Proof. By the spectral mapping theorem, $\sigma(p(A)) \subset [0, +\infty)$. It remains to use Corollary 1 in Subsection 12.4.5. $\qquad\square$

Exercises

1. Suppose that in some basis the matrix of the operator A is diagonal. What will the matrix of the operator $f(A)$, where f is a continuous function, look like in that basis?

2. Does the definition of the operator e^A, given above in Exercise 5 of Subsection 13.1.1, agree with the definition as a continuous function of a self-adjoint operator?

3. Suppose $A \in L(H)$ is self-adjoint, $f \in C(\sigma(A))$, $f(A) = (f(A))^*$, and the function g is continuous on the spectrum of the operator $f(A)$. Prove that $g(f(A)) = (g \circ f)(A)$.

13.1.5 Applications of Continuous Functions of an Operator

Theorem 1. *The product of two commuting positive operators is a positive operator.*

Proof. Let $A, B \in L(H)$ be a pair of commuting positive operators. Since the spectrum of any positive operator lies on the positive half-line, the function $\sqrt{t}$ is continuous on the spectra of both operators A and B and takes positive values there. Consequently, the operators $\sqrt{A}$ and $\sqrt{B}$ are self-adjoint, and by property (5) in Theorem 1 of Subsection 13.1.4, $\sqrt{A}$ and $\sqrt{B}$ commute. We have

$$\langle ABx, x \rangle = \langle (\sqrt{A}\sqrt{A})(\sqrt{B}\sqrt{B})x, x \rangle = \langle (\sqrt{A}\sqrt{B})(\sqrt{A}\sqrt{B})x, x \rangle$$

$$= \langle (\sqrt{A}\sqrt{B})x, (\sqrt{A}\sqrt{B})x \rangle = \left\| (\sqrt{A}\sqrt{B})x \right\|^2 \geqslant 0. \qquad\square$$

Lemma 1. *Let $A \in L(H)$ be a self-adjoint operator, and let $f \in C(\sigma(A))$ be such that $f(\sigma(A)) = \{0, 1\}$. Then $f(A)$ is an orthogonal projector onto a non-trivial (i.e., different from $\{0\}$ and the whole H) subspace.*

Proof. Since the function f satisfies the condition $f^2 = f$, we have $f^2(A) = f(A)$, and so the operator $f(A)$ is a projector. Since A is self-adjoint, $f(A)$ is an orthogonal

projector. Finally, since $\sigma(f(A)) = f(\sigma(A)) = \{0, 1\}$, $f(A)$ cannot coincide with
the zero operator or with the identity operator. That is, the image $f(A)$ is a nontrivial
subspace.   □

Definition 1. Let $H = H_1 \oplus H_2$, $A_1 \in L(H_1)$, $A_2 \in L(H_2)$. The operator $A \in L(H)$
that coincides with A_j on the space H_j, $j = 1, 2$, is called the *direct sum of the operators A_1 and A_2 with respect to the decomposition $H = H_1 \oplus H_2$*, and is denoted by
$A = A_1 \oplus A_2$. In other words, if $h_1 \in H_1$ and $h_2 \in H_2$, then $(A_1 \oplus A_2)(h_1 + h_2) = A_1 h_1 + A_2 h_2$.

The reader is encouraged to verify on his own that the operator $A = A_1 \oplus A_2$ is
invertible if and only if both operators A_1 and A_2 are invertible. This readily implies
that $\sigma(A_1) \cup \sigma(A_2) = \sigma(A_1 \oplus A_2)$.

Theorem 2. *Let $A \in L(H)$ be a self-adjoint operator whose spectrum is the union
of two disjoint closed sets: $\sigma(A) = K_1 \cup K_2$. Then the space H admits an orthogonal
direct sum decomposition $H = H_1 \oplus H_2$ into two nontrivial A-invariant subspaces,
and the operator A decomposes into the direct sum $A = A_1 \oplus A_2$ of two operators
$A_1 \in L(H_1)$ and $A_2 \in L(H_2)$, such that $\sigma(A_1) = K_1$ and $\sigma(A_2) = K_2$.*

Proof. The functions $f_1 = \mathbb{1}_{K_1}$ and $f_2 = \mathbb{1}_{K_2}$ are continuous on $\sigma(A)$. Consider
the operators $P_1 = f_1(A)$ and $P_2 = f_2(A)$. By Lemma 1, P_1 and P_2 are othogonal
projectors. Since $f_1 + f_2 \equiv 1$ on $\sigma(A)$, we have $P_1 + P_2 = I$. Put $H_1 = P_1(H)$
and $H_2 = \text{Ker } P_1$. Then one has the direct sum decomposition $H = H_1 \oplus H_2$, with
$H_2 = P_2(H)$ (Theorem 2 of Subsection 10.3.2); moreover, $H_1 \perp H_2$, because P_1 is
an orthogonal projector. H_j is the eigensubspace of the operator P_j corresponding to
the eigenvalue 1. Since a function of an operator commutes with the operator itself
this implies (Theorem 1 of Subsection 11.1.5) that the subspaces H_j are invariant
under A.

We define the sought-for operators $A_j \in L(H_j)$, $j = 1, 2$, as the restrictions of
the operator A to the subspaces H_j. With this definition, we obviously have that
$A = A_1 \oplus A_2$ and $\sigma(A_1) \cup \sigma(A_2) = \sigma(A) = K_1 \cup K_2$. To complete the proof, it
remains to verify the inclusions $\sigma(A_j) \subset K_j$, $j = 1, 2$. By symmetry, it suffices to
consider the case $j = 1$. Let $\lambda \notin K_1$. Consider the function $g(t)$ equal to $\frac{1}{t - \lambda}$ for
$t \in K_1$ and to 0 on K_2. Then, for every $x \in H_1$,

$$g(A)(A_1 - \lambda I)x = g(A)(A - \lambda I)x = f_1(A)x = P_1 x = x.$$

(Here and below I denotes the identity operator in the whole space, as well as in the
subspaces H_1 and H_2.)
The subspace H_1 is invariant under $g(A)$ (again by Theorem 1 of Subsection 11.1.5);
hence, thanks to commutativity, the last equality means that the restriction of the
operator $g(A)$ to the subspace H_1 is the inverse of $A_1 - \lambda I$. Thus, we have shown
that $\lambda \notin K_1$ implies that $\lambda \notin \sigma(A_1)$, which is equivalent to the inclusion $\sigma(A_1) \subset K_1$.
                                                                                               □

Corollary 1. *Let λ_0 be an isolated point of the spectrum of the self-adjoint operator A. Then λ_0 is an eigenvalue of A.*

Proof. Apply the preceding result, taking $K_1 = \{\lambda_0\}$ and $K_2 = \sigma(A) \setminus \{\lambda_0\}$. In this case $\sigma(A_1) = \{\lambda_0\}$, that is (Corollary 3 in Subsection 12.4.5), $A_1 - \lambda_0 I = 0$, and any element of the subspace H_1 provides the required eigenvector. □

Exercises

1. In the last corollary, why can't the subspace H_1 be equal to $\{0\}$?

2. The self-adjoint operator $A \in L(H)$ is positive if and only if there exists a self-adjoint operator $B \in L(H)$ such that $B^2 = A$.

3. Suppose $B \geqslant 0$ and $B^2 = A$. Then $B = \sqrt{A}$.

4. Suppose dim $H \geqslant 2$. Then there exist infinitely many self-adjoint operators $B \in L(H)$ such that $B^2 = I$.

13.2 Unitary Operators and the Polar Representation

13.2.1 The Absolute Value of an Operator

Let $T \in L(H)$ be an arbitrary operator. Following the analogy with numbers, one can conjecture that T^*T will be a positive self-adjoint operator. Let us verify that this is the case. Since $(T^*T)^* = T^*(T^*)^* = T^*T$, self-adjointness holds. Positivity is a consequence of the scalar product axioms: $\langle T^*Tx, x \rangle = \langle Tx, Tx \rangle \geqslant 0$. Now since the operator T^*T is positive, the function $\sqrt{t}$ is continuous on its spectrum, which enables us to define the *absolute value of the operator* T as $|T| = \sqrt{T^*T}$. The absolute value of an operator is a positive operator.

Theorem 1. *For any element $x \in H$, $\||T|x\| = \|Tx\|$. In particular, $|T|x = 0$ if and only if $Tx = 0$.*

Proof. Indeed,

$$\||T|x\|^2 = \langle |T|x, |T|x \rangle = \langle |T|^2x, x \rangle = \langle T^*Tx, x \rangle = \langle Tx, Tx \rangle = \|Tx\|^2. \qquad □$$

For the ensuing material we need the following reformulation.

Theorem 2 (Weak polar representation). *Let X be the image of the operator $|T|$, and Y the image of the operator T. Then there exists an isometric bijective operator $V \in L(X, Y)$ such that $T = V \circ |T|$. Moreover, the operator V is unique.*

Proof. First, the uniqueness. Let $x = |T|(h)$ be an arbitrary element of the space X. For the equality $T = V \circ |T|$ to hold when both terms are evaluated on h, it is necessary and sufficient that the operator V satisfies the condition $Vx = Th$. Hence, V is uniquely determined. By the preceding theorem, if the element x admits two representations $x = |T|(h_1) = |T|(h_2)$, then $\|Th_1 - Th_2\| = 0$, i.e., the condition $Vx = Th$ can be taken as the definition of the operator V. As h runs through[2] the entire Hilbert space H, the elements $x = |T|(h)$ and $Vx = Th$ run through the entire spaces X and Y, respectively. Therefore, the operator V is bijective. Finally, V is an isometry, because $\|Vx\| = \|Th\| = \| |T|(h) \| = \|x\|$. $\qquad\qquad\qquad\square$

Exercises

Calculate the absolute values of the following operators:

1. The multiplication operator $A_g \in L(H)$ by a bounded function g: $(A_g f)(t) = g(t)f(t)$.

2. The right-shift operator $S_r \in L(\ell_2)$, acting as $S_r(x_1, x_2, \ldots) = (0, x_1, x_2, \ldots)$;

3. The left-shift operator $S_l \in L(\ell_2)$, acting as $S_l(x_1, x_2, \ldots) = (x_2, x_3, \ldots)$.

13.2.2 Definition and Simplest Properties of Unitary Operators

A complex number lying on the unit circle satisfies the equation $z \cdot \bar{z} = 1$. Developing further the analogy between operators and numbers, it is natural to introduce the corresponding class of operators.

Definition 1. The operator $U \in L(H)$ is called *unitary* if $UU^* = U^*U = I$. In other words, the operator U is unitary if it is invertible and $U^{-1} = U^*$.

Theorem 1. *Unitary operators preserve the scalar product: $\langle Ux, Uy \rangle = \langle x, y \rangle$ for all $x, y \in H$. Consequently, unitary operators preserve orthogonality: if $x \perp y$, then $Ux \perp Uy$.*

Proof. Indeed, $\langle Ux, Uy \rangle = \langle x, U^*Uy \rangle = \langle x, y \rangle$. $\qquad\qquad\qquad\square$

---

[2]If one ponders over this generally accepted expression, then one is struck by the disparity with the picture arising here. Indeed, to "run through" even a domain in the plane, an element requires considerable effort. It is true that in this case "he" could actually perform this task by moving along the Peano curve (though in his place I would look for a more interesting activity). As for the infinite-dimensional case, "running through" the entire space is in fact impossible. Indeed, prove that a continuous mapping $f : [0, +\infty) \to H$ cannot be surjective.

Theorem 2 (Unitarity criterion). *An operator is unitary if and only if it is a bijective isometry.*

Proof. Let U be unitary. Then U is invertible, and hence bijective. Further, $\|Ux\|^2 = \langle Ux, Ux \rangle = \langle x, x \rangle = \|x\|^2$, i.e., U is an isometry. Conversely, suppose that U is an isometry. Then

$$\langle x, x \rangle = \|x\|^2 = \|Ux\|^2 = \langle Ux, Ux \rangle = \langle x, U^*Ux \rangle.$$

Thus, the quadratic forms of the operators U^*U and I coincide, hence so do the operators themselves: $I = U^*U$. And since for a bijective operator the notions of right inverse and left inverse coincide, we conclude that $UU^* = I$, as needed. $\square$

Theorem 3. *The spectrum of any unitary operator U lies on the unit circle.*

Proof. Since U is isometric, $\|U\| = \|U^{-1}\| = 1$. Therefore, if $|\lambda| < 1$, then, by the theorem on small perturbations of an invertible element (Theorem 1 in Subsection 11.1.2), the operator $U - \lambda I$ is invertible, while if $|\lambda| > 1$, the invertibility of the operator $U - \lambda I = \lambda(I - \lambda^{-1}U)$ is guaranteed by the lemma on the invertibility of small perturbations of the identity element (Lemma 2 in Subsection 11.1.2). Therefore, $U - \lambda I$ can be non-invertible only if $|\lambda| = 1$. $\square$

In Sect. 13.3 we will pursue further the analogy between unitary operators and numbers of absolute magnitude 1: a chain of exercises culminating in Exercise 23 of that section will show that every unitary operator U can be represented as $U = e^{iA}$ with A self-adjoint.

Exercises

1. Under what conditions on the function g will the multiplication operator $A_g \in L(L_2[0, 1])$, $(A_g f)(t) = g(t) f(t)$, be unitary?

2. Show that for every closed subset K of the unit circle there exists a unitary operator U such that $\sigma(U) = K$.

3. Suppose the operator $U \in L(H)$ is an isometric embedding (i.e., $\|Ux\| = \|x\|$ for all $x \in H$) with dense image. Then U is unitary.

13.2.3 Polar Decomposition

A *polar decomposition* of the operator T is a representation of the operator as $T = UA$, where U is a unitary operator and A is a positive self-adjoint operator. That is, the polar decomposition of an operator is the analogue of the polar decomposition for complex numbers, $z = e^{i \arg z} \cdot |z|$. In contrast to the scalar case, for operators such a decomposition is not always possible. To find conditions for the existence of the polar decomposition, we regard it as an equation in the unknown operators U and A.

So, suppose that U and A are solutions of the equation $T = UA$ with the required properties. Then $T^* = AU^*$ (we used the self-adjointness of A), and thanks to the unitarity of U, we have $T^*T = A^2$. Extracting the square root, we obtain the value of one of the unknowns:

$$A = |T|.$$

To determine the second unknown, we have the equation $T = U \circ |T|$. How does this condition differ from the analogous condition on the operator V in Theorem 2 of Subsection 13.2.1? Only by the fact that the operator U needs to be defined not merely on the subspace $X = |T|(H)$, but on the whole space H, with preservation of the isometry and bijectivity properties that the operator V enjoyed. Let us examine when such an extension is possible. To formulate the result, we need to define more precisely what the dimension of a Hilbert space means. For a finite-dimensional space, the dimension was defined as the number of elements in a basis of the space. Generalizing this definition to the infinite-dimensional case, the *dimension of a Hilbert space* is the cardinality of an orthonormal basis of the space (see Exercise 7 in Subsection 12.3.4). Two Hilbert spaces have the same dimension if and only if they are isomorphic (Exercise 12 in Subsection 12.3.5).

Lemma 1. *Let X, Y be linear subspaces of the space H, and $V \in L(X, Y)$ be a bijective isometry. Then the following conditions are equivalent:*

(1) *the mapping V can be extended to a unitary operator $U \in L(H)$.*

(2) $\dim X^{\perp} = \dim Y^{\perp}$, *where the dimensions may be finite or infinite.*

Proof. (1) $\Longrightarrow$ (2). Since on X the operators U and V coincide, $U(X) = Y$. By Theorem 1 of Subsection 13.2.2, a unitary operator preserves orthogonality, and so $U(X^{\perp}) = Y^{\perp}$. In view of the injectivity of the operator, this yields the required equality of the dimensions.

(2) $\Longrightarrow$ (1). With no loss of generality, we can assume that the subspaces X and Y are closed (otherwise we extend the operator V by continuity to the closure of the subspace X). Thanks to the equality of dimensions, there exists a bijective isometry $W: X^{\perp} \to Y^{\perp}$. Given an arbitrary $x \in H$, we decompose it as $x = x_1 + x_2$, with $x_1 \in X$ and $x_2 \in X^{\perp}$. Now define the requisite operator U by the rule $Ux = Vx_1 + Wx_2$. $\square$

Theorem 1. *For the existence of a polar decomposition of the operator T it is necessary and sufficient that the equality of dimensions*

$$\dim \operatorname{Ker} T = \dim \operatorname{Ker} T^*$$

holds.

Proof. By the last lemma above and the arguments that precede it, the required necessary and sufficient condition is the equality $\dim((|T|(H))^{\perp} = \dim(T(H))^{\perp}$. To reduce this condition to the one in the lemma, we observe that $(T(H))^{\perp} = \mathrm{Ker}\,T^*$. On the other hand, self-adjointness implies that $(|T|(H))^{\perp} = \mathrm{Ker}|T|$. In turn, by Theorem 1 of Subsection 13.2.1, $\mathrm{Ker}|T| = \mathrm{Ker}\,T$. $\qquad\qquad\square$

Let us mention several useful sufficient conditions for the existence of the polar decomposition.

Corollary 1.

1. *For the operator T to admit a polar decomposition it is sufficient that T be invertible.*

2. *Let T be a normal operator, i.e., T commutes with T^*. Then T admits a polar decomposition.*

3. *Let T be a scalar $+$ compact operator. Then T admits a polar decomposition.*

Proof.

1. If T in invertible, then so is T^*. Consequently, $\dim \mathrm{Ker}\,T = \dim \mathrm{Ker}\,T^* = 0$.

2. $\mathrm{Ker}\,T = \mathrm{Ker}\,|T| = \mathrm{Ker}\,\sqrt{T^*T} = \mathrm{Ker}\,\sqrt{TT^*} = \mathrm{Ker}\,|T^*| = \mathrm{Ker}\,(T^*)$.

3. This follows from the Fredholm theorem (see Exercise 2 in Subsection 11.3.3).

Exercises

1. Prove that any operator $T \in L(H)$ is representable, and in fact in a unique way, as $T = A + iB$, where A and B are self-adjoint operators. Moreover, the operator T will be normal if and only if A and B commute. The stated representation serves as the starting point of one of the ways of constructing functions of a normal operator (see [4]).

2. Show that the operator T is normal if and only if it admits a polar decomposition $T = UA$ with commuting operators A and U.

3. Show that if an operator T has a non-commuting polar decomposition, then T is not normal.

4. Describe the operators for which the polar decomposition is unique.

5. Justify the following fact that was already used, without drawing attention to it, in the present subsection: if two positive operators A and B satisfy the equality $A^2 = B^2$, then $A = B$. Does this assertion remain true if we discard the positivity assumption? Where specifically did we use this fact?

13.3   Borel Functions of an Operator

Using the fact that any continuous function can be approximated by polynomials, we were able to construct continuous functions of a self-adjoint operator. Below we show that functions of a self-adjoint operator can also be defined in a considerably more general situation, namely, for any bounded Borel-measurable function. The construction is based on the possibility of unique extension of linear functionals from the space of continuous functions to the wider space of bounded Borel-measurable functions.

Let $A \in L(H)$ be a fixed self-adjoint operator and K be its spectrum.

Given two arbitrary elements $x, y \in H$, define the linear functional $F_{x,y} \in C(K)^*$ by the formula $F_{x,y}(f) = \langle f(A)x, y \rangle$. Clearly, in addition to the linearity in f, the following relations, characteristic of bilinear forms, hold: $F_{a_1 x_1 + a_2 x_2, y} = a_1 F_{x_1, y} + a_2 F_{x_2, y}$ and $F_{y,x} = \overline{F}_{x,y}$. It is also readily verified that $\|F_{x,y}\| \leqslant \|x\| \cdot \|y\|$; indeed,

$$|F_{x,y}(f)| = |\langle f(A)x, y \rangle| \leqslant \|f(A)\| \cdot \|x\| \cdot \|y\| = \|f\| \cdot \|x\| \cdot \|y\|.$$

By the theorem on the general form of continuous linear functionals on the space $C(K)$, there exists a regular Borel charge $\sigma_{x,y}$ on K such that

$$F_{x,y}(f) = \int_K f \, d\sigma_{x,y}.$$

Since the indicated correspondence between functionals on $C(K)$ and charges is a bijective isometry, the relations for functionals written above remain valid for charges:

$$\|\sigma_{x,y}\| \leqslant \|x\| \cdot \|y\|, \quad \sigma_{a_1 x_1 + a_2 x_2, y} = a_1 \sigma_{x_1, y} + a_2 \sigma_{x_2, y}, \quad \text{and } \sigma_{y,x} = \overline{\sigma}_{x,y}.$$

Definition 1. Let f be a bounded Borel function on K, and $\sigma_{x,y}$ be the Borel charges defined above. Define the operator $f(A)$ by the equality

$$\langle f(A)x, y \rangle = \int_K f \, d\sigma_{x,y}.$$

In view of the theorem in Subsection 12.4.1 (with a change in the order of factors) the above definition is correct: the expression on the right-hand side of the equality is a continuous bilinear form.

Note that the last definition is consistent with the definition of a continuous function of an operator (i.e., the two definitions give the same result), and many of the properties of continuous functions of an operator listed in Subsection 13.1.4 remain valid in the more general situation.

Theorem 1. *For bounded Borel functions on K the following relations hold:*

1. $(\lambda_1 f_1 + \lambda_2 f_2)(A) = \lambda_1 f_1(A) + \lambda_2 f_2(A)$;

2. $(f_1 f_2)(A) = f_1(A) f_2(A)$;

3. $(f(A))^* = \overline{f}(A)$. *In particular, if on $\sigma(A)$ the function f takes only real values, then $f(A)$ is a self-adjoint operator.*

Proof. Property 1. follows from the linearity of the integral.

To verify property 3, we use the relation $\sigma_{y,x} = \overline{\sigma}_{x,y}$:

$$\langle (f(A))^* x, y \rangle = \langle x, f(A) y \rangle = \overline{\langle f(A) y, x \rangle}$$
$$= \int_K \overline{f} \, d\overline{\sigma}_{y,x} = \int_K \overline{f} \, d\sigma_{x,y} = \langle \overline{f}(A) x, y \rangle.$$

It remains to verify property 2. The equality $(f_1 f_2)(A) = f_1(A) f_2(A)$ is already known to hold for continuous functions. Threfore, for $f_1, f_2 \in C(K)$ we have the equality of bilinear forms

$$\langle (f_1 f_2)(A) x, y \rangle = \langle f_1(A)(f_2(A) x), y \rangle. \tag{1}$$

Using the definition of the charges $\sigma_{x,y}$, we recast (1) as

$$\int_K f_1 f_2 \, d\sigma_{x,y} = \int_K f_1 \, d\sigma_{f_2(A) x, y}.$$

Since the integrals above are equal for any continuous function f_1, they will also be equal for any bounded Borel-measurable function.[3] Going in the opposite direction, we deduce that the equality (1) again holds not only for a continuous, but also for an arbitrary bounded Borel function f_1. Rewriting (1) in the form

$$\langle (f_1 f_2)(A) x, y \rangle = \langle f_2(A) x, \overline{f}_1(A) y \rangle$$

and using the definition, we conclude that for any bounded Borel function f_1 the equality of integrals

$$\int_K f_1 f_2 d\sigma_{x,y} = \int_K f_2 d\sigma_{x, \overline{f}_1(A) y} \tag{2}$$

is valid for any continuous function f_2. Extending equality (2) to the more general class of bounded Borel-measurable functions and passing again to bilinear forms, we see that relation (1) is valid for all bounded Borel-measurable functions f_1 and f_2. The coincidence of the bilinear forms implies the coincidence of the corresponding operators. Thus, the required mutiplicativity relation is proved. $\square$

The remaining properties of continuous functions of operators discussed in Subsection 13.1.4 are not fully valid for Borel functions. The main reason for this is that two different functions, f_1, f_2, that coincide almost everywhere with respect to all charges $\sigma_{x,y}$, generate the same operator: $f_1(A) = f_2(A)$.

---

[3]To prove this, take a measure μ that dominates the variations of both charges figuring in the equality; represent the Borel function as the limit of a μ-almost everywhere convergent sequence of uniformly bounded continuous functions and apply the dominated convergence theorem.

Theorem 2 (Sufficient conditions for invertibility). *If the bounded Borel function f is separated away from 0 on $\sigma(A)$ (i.e., there exists an $\varepsilon > 0$ such that $|f(t)| \geqslant \varepsilon$ for all $t \in \sigma(A)$), then the operator $f(A)$ is invertible.*

Proof. The operator $\dfrac{1}{f}(A)$ is the inverse to $f(A)$.   □

Theorem 3 (Spectral mapping theorem for bounded Borel functions of an operator). *The spectrum of a function of an operator is contained in the closure of the image of the spectrum of the original operator: $\sigma(f(A)) \subset \overline{f(\sigma(A))}$.*

Proof. Let $\lambda \notin \overline{f(\sigma(A))}$. Then the function $f - \lambda$ satisfies the conditions of the preceding theorem. Hence, the operator $(f - \lambda)(A) = f(A) - \lambda I$ is invertible, i.e., $\lambda \notin \sigma(f(A))$. The theorem is proved.   □

Theorem 4 (Estimate of the norm of a function of an operator). *Let f be a bounded Borel function on $K = \sigma(A)$. Then*

$$\|f(A)\| \leqslant \sup_{t \in \sigma(A)} |f(t)|.$$

Proof. We use the condition $|\sigma_{x,y}|(K) = \|\sigma_{x,y}\| \leqslant \|x\| \cdot \|y\|$ and the estimate of the integral through the variation of the charge (Theorem 4 in Subsection 8.4.5) to obtain

$$\|f(A)\| = \sup_{x,y \in S_H} |\langle f(A)x, y \rangle| \leqslant \sup_{x,y \in S_H} \int_{\sigma(A)} |f| \, d \, |\sigma_{x,y}| \leqslant \sup_{t \in \sigma(A)} |f(t)|. \quad □$$

Theorem 5. *Let (f_n) be a monotonically increasing uniformly bounded sequence of real-valued Borel functions on $K = \sigma(A)$ that converges at each point to the function f. Then the sequence of operators $(f_n(A))$ converges pointwise to the operator $f(A)$.*

Proof. The operators $f_n(A)$ form a monotone bounded sequence. By Theorem 2 of Subsection 12.4.4, there exists the pointwise limit of the sequence $f_n(A)$, which we denote by T. To establish the claimed equality $f(A) = T$, we compare the bilinear forms of the operators:

$$\langle Tx, y \rangle = \lim_{n \to \infty} \langle f_n(A)x, y \rangle = \lim_{n \to \infty} \int_{\sigma(A)} f_n \, d\sigma_{x,y} = \int_{\sigma(A)} f \, d\sigma_{x,y} = \langle f(A)x, y \rangle.$$

Here we used the Lebesgue dominated convergence theorem.   □

Discontinuous Borel functions of an operator can be calculated, rather than using the (quite abstract) definition, by using approximation (in one sense or another) by continuous functions. For example, if the bounded Borel function f on $\sigma(A)$ is representable as the pointwise limit of an increasing sequence (f_n) of continuous functions,[4] the last theorem enables us to calculate $f(A)$ as the pointwise limit of the sequence $(f_n(A))$. For more details on such an approach to functions of an operator, we refer to the exercises below.

---

[4] Such a representation is possible only for lower-semicontinuous functions.

Exercises

1. Suppose the operators A and B commute. Then each of them commutes with the bounded Borel functions of the other.

2. Suppose the operators A and B commute, and let f and g be Borel functions on the spectra of the operators A and B, respectively. Then the operators $f(A)$ and $g(B)$ also commute.

3. Suppose that in some basis the matrix of the self-adjoint operator A is diagonal. If f is a bounded Borel function, what will the matrix of the operator $f(A)$ look like in that basis?

4. Describe the functions of the multiplication operator A_g by a bounded function g: $A_g \in L(L_2[0, 1])$, $(A_g f)(t) = g(t) f(t)$.

Definition. We say that the sequence of operators $A_n \in L(H)$ *form converges*[5] to the operator $A \in L(H)$ if the corresponding bilinear forms converge:

$$\langle A_n x, y \rangle \to \langle A x, y \rangle \quad \text{as } n \to \infty$$

for all $x, y \in H$. Notation: $A_n \xrightarrow{\text{form}} A$.

5. Pointwise convergence of operators implies form convergence.

6. Let $\{e_n\}_1^\infty$ be an orthormal system in H. Then the operators A_n, acting by the formula $A_n x = \langle x, e_1 \rangle e_n$, form converge to 0, but do not converge pointwise.

7. If $A_n \xrightarrow{\text{form}} A$, then $\|A\| \leqslant \sup_n \|A_n\| < \infty$.

8. If $A_n \xrightarrow{\text{form}} A$, $B_n \xrightarrow{\text{form}} B$, and $a, b \in \mathbb{C}$, then $a A_n + b B_n \xrightarrow{\text{form}} a A + b B$.

9. If $A_n \xrightarrow{\text{form}} A$ and $B \in L(H)$, then $A_n B \xrightarrow{\text{form}} AB$ and $B A_n \xrightarrow{\text{form}} BA$.

10. If $A_n \xrightarrow{\text{form}} A$, then $A_n^* \xrightarrow{\text{form}} A^*$. Does the analogous property hold for pointwise convergence?

11. Provide an example in which $A_n \xrightarrow{\text{form}} A$ and $B_n \xrightarrow{\text{form}} B$, but $A_n B_n$ does not form converge to AB.

Let $A \in L(H)$ be a self-adjoint operator and K be its spectrum. The Borel measure μ on K is said to be a *control measure* of the operator A if all the charges $\sigma_{x,y}$ generated by the operator A are absolutely continuous with respect to μ.

---

[5]The generally accepted name for this type of convergence of operators is "weak pointwise convergence", because in this case $A_n x$ weakly converge to Ax for all $x \in H$. The meaning of word "weakly" here is "when evaluated by every continuous linear functional". We will speak a lot about weak convergence and weak topology in Chap. 17.

12. For any self-adjoint operator in a separable Hilbert space H there exists a control measure. Hint: choose a sequence of pairs (x_n, y_n) that is dense in $S_H \times S_H$. Define the control measure by the formula $\mu = \sum_{n=1}^{\infty} \frac{1}{2^n} |\sigma_{x_n, y_n}|$.

13. Let $A \in L(H)$ be a self-adjoint operator and K be its spectrum. Let μ be a control measure for A, and let (f_n) be a uniformly bounded sequence of Borel functions on K which converges μ-almost everywhere to f. Then $f_n(A) \xrightarrow{\text{form}} f(A)$.

14. Let $A \in L(H)$ be a self-adjoint operator and K be its spectrum. Then for any bounded Borel function f on K there exists a sequence (f_n) of continuous functions on K such that $f_n(A) \xrightarrow{\text{form}} f(A)$.

The next chain of exercises will enable the reader to construct on her/his own a theory of functions of a unitary operator. Throughout this part U will denote a fixed unitary operator and S the spectrum of U.

The main difference compared to the case of self-adjoint operators is that the polynomials are not dense in the space of continuous functions on the unit circle: the closure of the set of polynomials in the uniform metric contains only the boundary values of functions analytic in the unit disc. For instance, the function $1/z$ does not belong to this closure.

15. Find the distance of the function $1/z$ to the set of polynomials in the space of continuous functions on the unit circle.

To circumvent this difficulty, we introduce the set of generalized polynomials, which also include negative powers of the indeterminate:

$$\mathcal{P}_* = \{p \in C(S) : p(z) = \sum_{k=-n}^{n_1} a_k z^k\}.$$

By analogy with the case of ordinary polynomials, we put

$$p(U) = \sum_{k=-n}^{n} a_k U^k.$$

16. Verify that the mapping $p \mapsto p(U)$ enjoys the linearity and multiplicativity properties.

17. Establish an invertibility criterion for the operator $p(U)$. Prove the spectral mapping theorem.

A generalized polynomial $p(z) = \sum_{k=-n}^{n} a_k z^k$ is said to be symmetric,[6] if $a_{-k} = \overline{a}_k$ for all indices k.

---

[6]This term is used by convention and has nothing to do with the symmetric polynomials of several variables.

18. Prove that a symmetric polynomial takes only real values on the unit circle. Conversely, if S is an infinite subset of the unit circle and the polynomial p takes only real values on S, then p is symmetric.

19. For any generalized polynomial p on the unit circle, the functions Re p and Im p are symmetric polynomials.

20. Prove that every real-valued continuous function on the unit circle can be arbitrarily well approximated in the uniform metric by symmetric polynomials.

21. Prove that a symmetric polynomial of a unitary operator is a self-adjoint operator. Deduce for this case the formula $\|p(U)\| = \sup_{t \in \sigma(U)} |p(t)|$.

22. From this point on, the entire scheme for constructing functions of a self-adjoint operator carries over with no modifications to functions of a unitary operator. Verify this!

23. Prove that every unitary operator U can be represented as $U = e^{iA}$, where A is a self-adjoint operator. (Hint: Pick a branch f of the argument on the unit circle. Take $f(U)$ for A.) Is this representation unique?

24. Suppose the generalized polynomial p takes on the unit circle $\mathbb{T}$ only positive values. Then there is another generalized polynomial g such that $p(z) = |g(z)|^2$ for all $z \in \mathbb{T}$.

13.4 Functions of a Self-adjoint Operator and the Spectral Measure

13.4.1 The Integral with Respect to a Vector Measure

Suppose given a set Ω, an algebra Σ of subsets of Ω, and a Banach space X. A mapping $\mu \colon \Sigma \to X$ is called an *X-valued measure* if it has the finite additivity property: $\mu(D_1 \cup D_2) = \mu(D_1) + \mu(D_2)$ for all disjoint subsets $D_1, D_2 \in \Sigma$. Measures with values in Banach spaces are also called *vector measures*.

The basic case we will be dealing with is that of complex scalars. The real case is practically identical.

Example. Let $X = \mathbb{C}^n$ be the space of rows. Then every X-valued measure μ can be written as $\mu(D) = (\mu_1(D), \mu_2(D), \dots, \mu_n(D))$, where μ_j are finitely-additive complex charges.

We define the integral of a scalar function with respect to a vector measure by analogy with how we proceeded in Sect. 4.2 for the ordinary integral. The difference here is not only that we are dealing with a vector measure, but also that the measure is only finitely — and not countably — additive. For this reason, in all definitions we will work only with finite partitions of sets into subsets.

So, let $\Delta \in \Sigma$ and $f : \Delta \to \mathbb{C}$ be a function. Let $D = \{\Delta_k\}_{k=1}^n$ be a finite partition of the set Δ into subsets $\Delta_k \in \Sigma$, and $T = \{t_k\}_1^n$ be a collection of marked points. The *integral sum* of the function f on the set Δ with respect to the pair (D, T) is the vector

$$S_\Delta(f, D, T) = \sum_{k=1}^n f(t_k)\mu(\Delta_k) \in X.$$

The element $x \in X$ is called the *integral* of the function f on the set Δ with respect to the vector measure μ (notation: $x = \int_\Delta f \, d\mu$) if for any $\varepsilon > 0$ there exists a finite partition D_ε of the set Δ such that for any finite partition D refining D_ε and any choice of marked points T for D, it holds that $\|x - S_\Delta(f, D, T)\| \leqslant \varepsilon$. The function $f : \Delta \to \mathbb{C}$ is said to be *integrable* on the set Δ with respect to the measure μ, or μ-*integrable*, if the corresponding integral exists.

In other words, the function f is integrable on Δ if its integral sums have a limit along the directed set of finite partitions with marked points, analogous to that described in Subsection 4.1.3.

Let us list, with no proofs, a number of simple properties of the integral.

(1) Linearity in the function: if the functions f and g are integrable on Δ and a, b are scalars, then the function $af + bg$ is also integrable, and $\int_\Delta (af + bg)d\mu = a \int_\Delta f \, d\mu + b \int_\Delta g \, d\mu$.

(2) Set additivity: if $\Delta_1 \cap \Delta_2 = \emptyset$ and f is integrable on both sets Δ_1 and Δ_2, then f is integrable on their union, and $\int_{\Delta_1} f \, d\mu + \int_{\Delta_2} f \, d\mu = \int_{\Delta_1 \sqcup \Delta_2} f \, d\mu$.

(3) The characteristic function of any set $\Delta \in \Sigma$ is integrable, and $\int_\Omega \mathbb{1}_\Delta \, d\mu = \mu(\Delta)$.

(4) For any collection $\{\Delta_k\}_1^n$ of measurable subsets and any collection of scalars $\{a_k\}_1^n$ the step function $f = \sum_{k=1}^n a_k \mathbb{1}_{\Delta_k}$ is integrable, and $\int_\Omega f \, d\mu = \sum_{k=1}^n a_k\mu(\Delta_k)$.

(5) Let $G \in L(X, Y)$ be a continuous linear operator and $\mu : \Sigma \to X$ be an X-valued measure. Then the composition $G \circ \mu$ is a Y-valued measure. Every μ-integrable function f is also $(G \circ \mu)$-integrable, and $G\left(\int_\Delta f \, d\mu\right) = \int_\Delta f \, d(G \circ \mu)$.

13.4.2 Semivariation and Existence of the Integral

Definition 1. Let $\mu : \Sigma \to X$ be a vector measure. For each $\Delta \in \Sigma$ we define the *semivariation of the measure* μ on the set Δ, denoted by $\|\mu\|(\Delta)$, as the supremum of the quantity $\left\|\sum_{k=1}^n a_k\mu(\Delta_k)\right\|$ over all finite partitions $\{\Delta_k\}_{k=1}^n$ of the set Δ into measurable subsets and all finite collections of scalars $\{a_k\}_{k=1}^n$ that satisfy the condition $|a_k| \leqslant 1$. We define $\|\mu\| = \|\mu\|(\Omega)$. The measure μ is said to be *bounded* if $\|\mu\| < \infty$. Throughout the remaining part of this subsection the measure μ will be assumed to be bounded.

Lemma 1. *For any bounded function f on the set $\Delta \in \Sigma$, any finite partition $D = \{\Delta_k\}_{k=1}^n$ of Δ into subsets $\Delta_k \in \Sigma$, and any collection $T = \{t_k\}_1^n$ of marked points $t_k \in \Delta_k$, one has the estimate*

$$\|S_\Delta(f, D, T)\| \leqslant \|\mu\|(\Delta) \cdot \sup_{t \in \Delta} |f(t)|.$$

Proof. Denote $\sup_{t \in \Delta} |f(t)|$ by M and $a_k = f(t_k)/M$. Since $|a_k| \leqslant 1, k = 1, 2, \ldots, n$, the required estimate follows:

$$\left\| \sum_{k=1}^n f(t_k)\mu(\Delta_k) \right\| = M \left\| \sum_{k=1}^n a_k \mu(\Delta_k) \right\| \leqslant M\|\mu\|(\Delta) = \|\mu\|(\Delta) \cdot \sup_{t \in \Delta} |f(t)|. \quad \square$$

Letting the integral sums converge to the integral, we obtain the following assertion.

Theorem 1. *The inequality*

$$\left\| \int_\Delta f \, d\mu \right\| \leqslant \|\mu\|(\Delta) \cdot \sup_{t \in \Delta} |f(t)|$$

holds for all bounded integrable functions f on Δ. $\qquad \square$

By analogy with Subsection 4.3.2 we prove the following uniform limit theorem.

Theorem 2. *Let f and f_n be scalar-valued functions on Δ and μ be a bounded vector measure. Suppose that the functions f_n are μ-integrable on Δ and the sequence (f_n) converges uniformly on Δ to f. Then f is integrable and $\int_\Delta f \, d\mu = \lim_{n \to \infty} \int_\Delta f_n \, d\mu$.*

Proof. Let $x_n = \int_\Delta f_n d\mu$. The sequence (x_n) is Cauchy: indeed,

$$\|x_n - x_m\| = \left\| \int_\Delta (f_n - f_m) d\mu \right\| \leqslant \sup_{t \in \Delta} \|f_n(t) - f_m(t)\| \cdot \|\mu\|(\Delta) \to 0 \quad \text{as } n, m \to \infty.$$

Denote the limit of the sequence x_n by x. Fix $\varepsilon > 0$ and choose an $n \in \mathbb{N}$ such that $\sup_{t \in \Delta} \|f_n(t) - f(t)\| < \varepsilon/(3\|\mu\|(\Delta))$ and $\|x - x_n\| < \varepsilon/3$. Further, let D_ε be a partition such that, starting with D_ε, we have $\|x_n - S_\Delta(f_n, D, T)\| \leqslant \varepsilon$. Then for any partition $D \succ D_\varepsilon$ and any collection of marked points T corresponding to D,

$$\|x - S_\Delta(f, D, T)\| \leqslant \|x - x_n\| + \|x_n - S_\Delta(f_n, D, T)\|$$
$$+ \|S_\Delta(f_n - f, D, T)\| \leqslant \frac{\varepsilon}{3} + \frac{\varepsilon}{3} + \frac{\varepsilon}{3} = \varepsilon.$$

Therefore, f is integrable and $\int_\Delta f \, d\mu = x$. It remains to recall that, by construction,

$$x = \lim_{n \to \infty} \int_\Delta f_n \, d\mu. \qquad \square$$

Theorem 3. *Let* $\mu\colon \Sigma \to X$ *be a bounded vector measure and* $\Delta \in \Sigma$. *Then every bounded measurable function* f *is integrable on* Δ.

Proof. The function f can be represented as the limit of a uniformly convergent sequence f_n of finitely-valued functions (Corollary 1 in Subsection 3.1.4). Since any finitely-valued measurable function is integrable, it remains to apply Theorem 2 on uniform limit. $\square$

We have provided the basic definitions and simplest properties of vector measures that are required for the theory of operators. To simplify the exposition, we did not aim at maximal generality in definitions and statements. The theory of vector measures is itself an extensive domain of functional analysis, rich in deep results and applications. For an introduction to the theory of vector measures we refer to the monograph of J. Diestel and J.J. Uhl [13].

Exercises

1. Prove that for real-valued charges the semivariation coincides with the variation of the charge familiar from Definition 1 in Subsection 7.1.1.

2. Verify that the expression $\|\mu\| = \|\mu\|(\Omega)$ gives a norm on the space $M(\Omega, \Sigma, X)$ of all bounded X-valued measures on Σ. Prove that the normed space $M(\Omega, \Sigma, X)$ is complete.

3. A vector measure is bounded if and only if its range (set of all its values) is bounded.

4. Prove that if a vector measure is given on a σ-algebra and is countably-additive, then it is bounded.

5. Let Σ be a σ-algebra, $\mu\colon \Sigma \to X$ be a vector measure, and let μ be weakly countably-additive (i.e., $x^* \circ \mu$ be a countably-additive charge for every $x^* \in X^*$), then μ is also countably-additive in the ordinary sense.

Let (Ω, Σ, ν) be a space with (ordinary finite scalar-valued positive) measure. We say that the vector measure $\mu\colon \Sigma \to X$ is *absolutely continuous with respect to* ν if $\mu(\Delta) = 0$ for all sets $\Delta \in \Sigma$ such that $\nu(\Delta) = 0$.

6. Prove that if the vector measure $\mu\colon \Sigma \to X$ is countably-additive on the σ-algebra Σ and absolutely continuous with respect to ν, then for any $\varepsilon > 0$ there exists a $\delta > 0$ such that $\|\mu\|(\Delta) < \varepsilon$ for all $\Delta \in \Sigma$ with $\nu(\Delta) < \delta$.

7. Under the conditions of the preceding exercise, the following analogue of the dominated convergence theorem holds true: If the uniformly bounded sequence (f_n) of measurable functions converges ν-almost everywhere to the function f, then $\int_\Omega f \, d\mu = \lim_{n \to \infty} \int_\Omega f_n \, d\mu$.

8. Suppose the vector measure $\mu : \Sigma \to X$ is countably-additive on the σ-algebra Σ. Then on Σ there exists a scalar-valued measure ν with respect to which μ is absolutely continuous.

13.4.3 The Spectral Measure and Spectral Projectors

Let $A \in L(H)$ be a fixed self-adjoint operator, and let $\mathfrak{B}$ be the σ-algebra of Borel sets on its spectrum $\sigma(A)$.

Definition 1. The *spectral measure* of the operator A is the vector measure $\mu_A : \mathfrak{B} \to L(H)$ defined by the rule $\mu_A(\Delta) = \mathbb{1}_\Delta(A)$.

We note that using the term "measure" here is correct, since $\mathbb{1}_{\Delta_1} + \mathbb{1}_{\Delta_2} = \mathbb{1}_{\Delta_1 \cup \Delta_2}$ for any pair of disjoint sets Δ_1 and Δ_2.

Lemma 1. Let $f = \sum_{k=1}^{n} \alpha_k \mathbb{1}_{\Delta_k}$ be a finitely-valued Borel function on $\sigma(A)$. Then $f(A) = \int_{\sigma(A)} f \, d\mu_A$.

Proof. Indeed,

$$\int_{\sigma(A)} f \, d\mu_A = \sum_{k=1}^{n} \alpha_k \mu_A(\Delta_k) = \sum_{k=1}^{n} \alpha_k \mathbb{1}_{\Delta_k}(A) = f(A). \qquad \square$$

Theorem 1. *The spectral measure of any self-adjoint operator A is bounded, and $\|\mu_A\|(\Delta) \leqslant 1$ for all Borel subsets $\Delta \subset \sigma(A)$.*

Proof. By definition, $\|\mu_A\|(\Delta) = \sup \left\| \sum_{k=1}^{n} a_k \mu_A(\Delta_k) \right\|$, where the supremum is taken over all finite partitions $\{\Delta_k\}_{k=1}^{n}$ of the set Δ into Borel subsets and all finite collections of scalars $\{a_k\}_{k=1}^{n}$ that satisfy the condition $|a_k| \leqslant 1$. Consider the function $f = \sum_{k=1}^{n} a_k \mathbb{1}_{\Delta_k}$. By Theorem 4 of Sect. 13.3,

$$\left\| \sum_{k=1}^{n} a_k \mu_A(\Delta_k) \right\| = \|f(A)\| \leqslant \sup_{t \in \sigma(A)} |f(t)| = \sup_{1 \leqslant k \leqslant n} |a_k| \leqslant 1. \qquad \square$$

Theorem 2 (Main identity for the spectral measure). *For every bounded Borel function f on $\sigma(A)$,*

$$f(A) = \int_{\sigma(A)} f \, d\mu_A.$$

Proof. For finitely-valued functions the required relation was proved in Lemma 1 above. Now let f be a bounded Borel function and let the sequence (f_n) of finitely-valued functions converge uniformly to f on $\sigma(A)$. Then

$$\|f_n(A) - f(A)\| = \|(f_n - f)(A)\| \leqslant \sup_{t \in \sigma(A)} |(f_n - f)(t)| \to 0 \quad \text{as } n \to \infty,$$

i.e., $f_n(A) \to f(A)$. On the other hand, by the uniform limit theorem (Theorem 2 of Subsection 13.4.2),

$$f_n(A) = \int_{\sigma(A)} f_n \, d\mu_A \to \int_{\sigma(A)} f \, d\mu_A \quad \text{as } n \to \infty.$$

The theorem is proved.   □

Corollary 1. $A = \int_{\sigma(A)} t \, d\mu_A(t).$ □

If an operator has a complete system of eigenvectors (i.e., the matrix of the operator is diagonalizable), then the structure of the operator becomes clear once its eigenvectors are calculated. For operators that are often encountered in various problems, it proves quite reasonable to carry out this work of calculating the eigenvalues and eigenvectors, even if not simple, so that subsequently the results of this investigation could be applied whenever needed. The spectral measure and the integral decompositions with respect to this measure play for self-adjoint operators the same role as the eigenvector expansions do for diagonalizable operators.

Exercises: Properties of the Spectral Measure

1. All values of the spectral measure are orthogonal projectors (called *spectral projectors*).

2. $\mu_A(\sigma(A)) = I.$

3. If the operator T commutes with A, then the spectral projectors of the operator A commute with T.

4. The image of every spectral projector is an invariant subspace of A.

5. $\int_{\sigma(A)} f_1 \, d\mu_A \cdot \int_{\sigma(A)} f_2 \, d\mu_A = f_1(A) f_2(A) = (f_1 f_2)(A) = \int_{\sigma(A)} (f_1 f_2) \, d\mu_A;$ in particular, $\mu_A(D_1)\mu_A(D_2) = \mu_A(D_1 \cap D_2).$

6. We note that the last property looks rather unusual: say, for a regular scalar-valued Borel charge μ on a compact space it can be satisfied (prove this as an exercise) only if μ is a probability measure concentrated at a single point.

7. Denote by X the image of the operator $\mu_A(D)$. Prove that $\sigma(A|_X) \subset \overline{D}.$

8. The point $\lambda \in \sigma(A)$ is an eigenvalue of the operator A if and only if $\mu_A(\{\lambda\}) \neq 0.$

9. The spectral measure of an operator with infinite spectrum does not possess the countable additivity property in the sense of norm convergence. At the same time, pointwise countable additivity does hold.

10. Let A be a diagonalizable operator. Prove that for any set $D \subset \sigma(A)$ the image of the projector $\mu_A(D)$ is the closure of the linear span of the set of eigenvectors of A associated to the eigenvalues that lie in D.

11. Let A be the multiplication operator in $L_2[0, 1]$ by the function $g(t) = t$: $(Af)(t) = tf(t)$. Show that for any set $D \subset \sigma(A)$ the image of the projector $\mu_A(D)$ is the set of all functions from $L_2[0, 1]$ that vanish identically in the complement of D.

12. Prove that the operator $\mu_A(\{0\})$ is the orthogonal projector onto the kernel of the operator A.

13. Prove that the set of invertible operators in $L(H)$ is connected.

13.4.4 Linear Equations

If $T \in L(H)$ is an invertible operator and $y \in H$, then the problem of solving the equation $Tx = y$ is equivalent to the same problem for the equation $T^*Tx = T^*y$, where T^*T, as we know, is a positive self-adjoint operator. For non-invertible T possessing a polar decomposition $T = UA$ with $A \geqslant 0$, the equation $Tx = y$ is equivalent to $Ax = U^*b$. These are some of the reasons why the most important linear equations in Hilbert space are of the form $Ax = b$, where $A \in L(H)$ is a given positive operator, $b \in H$ is a given element, and $x \in H$ is the unknown. So, in this subsection we consider linear equations with a positive self-adjoint operator A. As in the previous subsection, μ_A denotes the spectral measure of A.

Lemma 1. *Denote by P the orthogonal projector $\mu_A(\sigma(A) \setminus \{0\})$. Then $PA = A$.*

Proof. The equality $\mathbb{1}_{\sigma(A)\setminus\{0\}}(t) \cdot t = t$ holds everywhere on $\sigma(A)$. It remains to plug the operator A in it. □

Corollary 1. *For the equation $Ax = b$ to be solvable it is necessary that the element b satisfies the condition $Pb = b$.*

Proof. Indeed, if $Ax = b$ for some $x \in H$, then $Pb = PAx = Ax = b$. □

If one observes that the operator $Q = I - P = \mu_A(\{0\})$ is the orthogonal projector onto the kernel of the operator A, the condition $Pb = b$ can be written in the more familiar form $b \perp \mathrm{Ker}\, A$.

Lemma 2. *Let (f_n) be a non-decreasing sequence of bounded Borel functions which converge pointwise on the set $\sigma(A) \setminus \{0\}$ to the function $1/t$. Then the sequence of operators $(Af_n(A))$ converges pointwise to the operator P from the preceding lemma.*

Proof. Apply Theorem 5 of Sect. 13.3. □

Theorem 1. *Let* (f_n) *be a non-decreasing sequence of bounded Borel functions that converges pointwise on* $\sigma(A) \setminus \{0\}$ *to the function* $1/t$, *and let* $b \in H$ *be an element which satisfies the condition* $Pb = b$. *Then for the solvability of the equation* $Ax = b$ *it is necessary and sufficient that the sequence of elements* $(f_n(A)(b))$ *converges. In this case, the limit of the sequence will be one of the solutions of the equation.*

Proof. Suppose the equation $Ax = b$ is solvable and let x_0 be a solution. Then, by Lemma 2, $f_n(A)(b) = f_n(A)(Ax_0) = Af_n(A)x_0 \to Px_0$, so the convergence is proved. Conversely, suppose the sequence $f_n(A)(b)$ converges to some element x_0. Then by the same lemma, $Ax_0 = \lim_{n\to\infty} Af_n(A)(b) = Pb = b$. $\qquad \square$

We note that if the operator A is injective, then $P = I$ and the condition $Pb = b$ is automatically satisfied. Further, if the operator A is invertible, then 0 does not lie in the spectrum of A and the function $1/t$ is continuous on $\sigma(A)$. Accordingly, one can take for (f_n) a uniformly convergent sequence of polynomials, and the rate of convergence of the elements $f_n(A)(b)$ to a solution will be estimated by the rate of convergence of the polynomials f_n (in this case no monotonicity of the sequence (f_n) is required). If the operator is given by an explicit expression, then the polynomials in this operator can also be written explicitly. Therefore, in the case of an invertible operator A, Theorem 1 provides a completely feasible method for solving the equation $Ax = b$ approximately. Needless to say, the lower the degree of the polynomial, the easier is to compute its value on an operator. Hence, here the most appropriate approach is to take as f_n the best approximation polynomials of the function $1/t$ on $\sigma(A)$.

In the case of an non-invertible operator the problem of finding approximate solutions to the equation $Ax = b$ is considerably more difficult. This problem belongs to the class of so-called *ill-posed problems*: arbitrarily small perturbations of the right-hand side can make the problem unsolvable, or strongly modify its solution. Since in approximate calculations all the initial data are usually also known only approximately, this issue is quite crucial. Help in solving this problem can come from exploiting a priori information about the solution which is not contained in the equation. In so doing, in any case the accuracy of the solution depends on the magnitude of the error in the right-hand side of the equation, and letting $n \to \infty$ in the sequence $(f_n(A)(b))$ does not lead to convergence to the solution. Moreover, as a rule, the approximating sequence approaches the solution only up to some moment, after which its behavior is in no way related to the true solution. Using the a priori information in order to find a reasonable step in the approximation is one of the available ideas for *regularizing* an ill-posed problem. For details on this subject one can consult the monograph by A. Tikhonov and V. Arsenin [41].

Exercises

1. What property of the spectrum of the operator A ensures the existence of a sequence (f_n) satisfying the conditions of Lemma 2 and Theorem 1?

2. Using Exercises 7 and 8 of Subsection 13.4.2 and the pointwise countable additivity of the spectral measure (Exercise 9 in Subsection 13.4.3), replace the monotonicity condition in Theorem 1 by the condition that the sequence $t f_n(t)$ is uniformly bounded on the spectrum of the operator A.

3. Let $\mu: \Sigma \to X$ be a vector measure. The set $D \in \Sigma$ is said to be *negligible* with respect to the measure μ if $\|\mu\|(D) = 0$. Is the set D being μ-negligible equivalent to the equality $\mu(D) = 0$? Does the answer change if μ is the spectral measure of a self-adjoint operator?

4. Prove Theorem 1 when pointwise convergence of the functions f_n is replaced by convergence almost everywhere with respect to the measure μ_A.

5. Is the set of noninvertible operators in $L(H)$ connected?

Comments on the Exercises

Subsection 13.1.1

Exercise 1. Use the following algebraic result: if p_1, p_2 is a pair of coprime polynomials, then there are polynomials q_1, q_2 such that $p_1 q_1 + p_2 q_2 = 1$. Substituting T one obtains $p_1(T)q_1(T) + p_2(T)q_2(T) = I$. For every $x \in X$ one gets the decomposition $x = x_1 + x_2$, where $x_1 = p_1(T)q_1(T)x$ and $x_2 = p_2(T)q_2(T)x$. It remains to show that $x_1 \in \mathrm{Ker}\, p_2(T)$, $x_2 \in \mathrm{Ker}\, p_1(T)$ and that $\mathrm{Ker}\, p_2(T) \cap \mathrm{Ker}\, p_1(T) = \{0\}$.

Exercise 5. For the differentiation formula $(e^f(t))' = f'(t)e^{f(t)}$ to hold, it is necessary that all values of the function f pairwise commute. In particular, this condition is satisfied when $f(t) = tA$, where $A \in L(X)$ is a fixed operator. In this case the differential equation $y' = f'(t)y$ becomes the equation with the constant operator coefficient $y' = Ay$.

Subsection 13.1.5

Exercise 3. By assumption, $B^2 = A$, and so B commutes with A. Then B also commutes with $\sqrt{A}$. We have $(B - \sqrt{A})(B + \sqrt{A}) = B^2 - A = 0$. Define the subspace $X \subset H$ as the closure of the image of the operator $B + \sqrt{A}$. The equality $(B - \sqrt{A})(B + \sqrt{A}) = 0$ means that on X the operator $B - \sqrt{A}$ is equal to zero. It remains to show that $B - \sqrt{A}$ is equal to zero on $X^\perp$. Since $B + \sqrt{A}$ is self-adjoint, the orthogonal complement of its image is its kernel: $X^\perp = \mathrm{Ker}(B + \sqrt{A})$. Again due to the commutativity, $\mathrm{Ker}(B + \sqrt{A})$ is an invariant subspace for the operators B and $\sqrt{A}$. Since B and $\sqrt{A}$ are positive operators and since on $X^\perp$ the operator $B + \sqrt{A}$ vanishes, it follows that $B = \sqrt{A} = 0$ on $X^\perp$ (Exercise 5 in Subsection 12.4.4). Therefore, $B - \sqrt{A}$ is indeed the zero operator on $X^\perp$.

Another way to solve this exercise is to apply Exercise 3 of Subsection 13.1.4 to the operator B and the functions $f(t) = t^2$ and $g(t) = \sqrt{t}$.

Subsection 13.2.3

Exercise 1. See the related Exercise 10 in Subsection 12.4.3.

Exercise 5. See Exercise 3 in Subsection 13.1.5 and its solution. This fact was used at the very beginning of the subsection to pass from the equality $T^*T = A^2$ to the equality $A = |T|$.

Section 13.3

Exercise 1. Use the fact that this property has already been established for continuous functions of an operator.

Exercise 2. By Exercise 1 in this subsection, A also commutes with $g(B)$. Applying again Exercise 1, but now to the operators $g(B)$ and A, we obtain the required assertion.

Subsection 13.4.2

Exercise 5. This classical Orlicz–Pettis theorem can be found, for example, in [13, p. 22].

Exercise 8. See [13, p. 14].

Subsection 13.4.3

Exercise 1. The equality $(\mathbb{1}_\Delta)^2 = \mathbb{1}_\Delta$ means that $\mu_A(\Delta)$ is a projector; thanks to self-adjointness, $\mu_A(\Delta)$ is an orthogonal projector.

Exercise 3. Use Exercise 1 of Sect. 13.3.

Exercise 4. Apply Exercise 3 and Theorem 1 of Subsection 11.1.5.

Exercises 7 and 8. Argue as in Theorem 2 of Subsection 13.1.5.

Exercises 9. The pointwise countable additivity follows from Theorem 5 of Sect. 13.3.

Exercise 13. Let $T \in L(H)$ be an invertible operator and $T = e^{iA}|T|$ be its polar decomposition (see Subsection 13.2.3 and Exercise 24 in Sect. 13.3). Define the continuous curve $F: [0, 1] \to L(H)$ by the formula $F(t) = e^{itA}|T|$. This curve passes only through invertible operators and connects the operator $|T|$ to the operator T. Further, the curve $G: [0, 1] \to L(H)$ given by the formula $G(t) = (1 - t)|T| + tI$ connects $|T|$ with the identity operator. Hence, in the set of invertible operators every operator can be joined to the identity operator by a continuous curve.

Chapter 14
Operators in L_p

In this chapter (Ω, Σ, μ) will be a measure space (with μ finite or σ-finite), and the parameter p will be assumed to satisfy the condition $1 \leqslant p \leqslant \infty$. For the sake of convenience, in applications to Fourier series and Fourier integrals, $L_p(\Omega, \Sigma, \mu)$ will be regarded as a space of complex-valued functions. We note that in the majority of problems in the theory of L_p-spaces the differences between the real-valued and complex-valued cases are insignificant.

14.1 Linear Functionals on L_p

The main problem addressed in this section is the proof of the theorem on the general form of linear functionals on L_p.

14.1.1 The Hölder Inequality

Definition 1. Given any $1 < p < \infty$, its *conjugate exponent* is the number $p' = \frac{p}{p-1}$. For $p = 1$ we set $p' = +\infty$, while for $p = +\infty$ we set $p' = 1$.

The exponents p and p' are connected by the relation $\frac{1}{p} + \frac{1}{p'} = 1$. Note that $(p')' = p$, and if $1 \leqslant p \leqslant 2$, then $2 \leqslant p' \leqslant \infty$; finally, $2' = 2$.

Lemma 1. *For any scalars $a, b \geqslant 0$ and $1 < p < \infty$ there holds the inequality*

$$ab \leqslant \frac{a^p}{p} + \frac{b^{p'}}{p'}. \tag{1}$$

Proof. The left-hand side of inequality (1) is equal to the area of the parallelogram $[0, a] \times [0, b]$, while the terms in the right-hand side are equal to the areas of the figures $S_1 = \{(x, y) : 0 \leqslant x \leqslant a, \ 0 \leqslant y \leqslant x^{p-1}\}$ and $S_2 = \{(x, y) : 0 \leqslant y \leqslant b,$

$0 \leqslant x \leqslant y^{p'-1}\}$, respectively. Since $p' - 1 = \frac{1}{p-1}$, the boundaries $y = x^{p-1}$ and $x = y^{p'-1}$ of these figures are given by the same curve. It remains to observe that $[0, a] \times [0, b] \subset S_1 \cup S_2$ (drawing the corresponding picture is left to the reader). $\square$

Theorem 1. *Let $f \in L_p(\Omega, \Sigma, \mu)$ and $g \in L_{p'}(\Omega, \Sigma, \mu)$. Then $fg \in L_1(\Omega, \Sigma, \mu)$ and the Hölder inequality $\|fg\|_1 \leqslant \|f\|_p \|g\|_{p'}$ holds. In detailed form:*

$$\int_\Omega |fg| d\mu \leqslant \left(\int_\Omega |f|^p d\mu \right)^{1/p} \left(\int_\Omega |g|^{p'} d\mu \right)^{1/p'}.$$

Proof. We consider the case $1 < p < \infty$, leaving the simple cases $p = 1$ and $p = \infty$ to the reader. If now in inequality (1) we replace a by $|f(t)|$ and b by $|g(t)|$, we obtain the estimate

$$|f(t)g(t)| \leqslant \frac{|f(t)|^p}{p} + \frac{|g(t)|^{p'}}{p'}. \tag{2}$$

That is, the measurable function fg has an integrable majorant, and hence is itself integrable. Further, the Hölder inequality is stable under the multiplication of the functions f and g by scalars. Hence, it suffices to prove it in the case where $\|f\|_p = \|g\|_{p'} = 1$. Now we use inequality (2) and get

$$\|fg\|_1 = \int_\Omega |fg| d\mu \leqslant \int_\Omega \left(\frac{|f|^p}{p} + \frac{|g|^{p'}}{p'} \right) d\mu = \frac{1}{p} + \frac{1}{p'} = 1 = \|f\|_p \|g\|_{p'}. \quad \square$$

Exercises

1. Deduce from the Hölder inequality the Cauchy–Schwarz inequality in L_2.

2. Derive as a particular case of the Hölder inequality the following Hölder inequality for series: $\sum_{k=1}^\infty |a_k b_k| \leqslant \left(\sum_{k=1}^\infty |a_k|^p \right)^{1/p} \left(\sum_{k=1}^\infty |b_k|^{p'} \right)^{1/p'}$.

3. Describe the pairs of functions (f, g) for which the Hölder inequality turns into an equality.

4. In the proof of the Hölder inequality, the assumption that the measure is σ-finite was not used, and hence is not essential.

14.1.2 Connections Between the Spaces L_p for Different Values of p

Theorem 1. *Let (Ω, Σ, μ) be a finite measure space, and let $1 \leqslant p_1 < p_2 \leqslant \infty$. Then $L_{p_1}(\Omega, \Sigma, \mu) \supset L_{p_2}(\Omega, \Sigma, \mu)$ and*

$$\|f\|_{p_1} \leqslant \|f\|_{p_2} \mu(\Omega)^{1/p_1 - 1/p_2} \tag{3}$$

for all $f \in L_{p_2}(\Omega, \Sigma, \mu)$.

Proof. Let $p_2 < \infty$ and $f \in L_{p_2}(\Omega, \Sigma, \mu)$. To estimate the integral $\int_\Omega |f|^{p_1} d\mu$ we use the Hölder inequality with the exponent $p = p_2/p_1$ and, correspondingly, $p' = p_2/(p_2 - p_1)$:

$$\int_\Omega |f|^{p_1} d\mu = \int_\Omega |f|^{p_1} \cdot 1 \, d\mu \leqslant \left(\int_\Omega |f|^{p_1 p} d\mu \right)^{1/p} \cdot \left(\int_\Omega 1 \, d\mu \right)^{1/p'}$$

$$= \left(\int_\Omega |f|^{p_2} d\mu \right)^{p_1/p_2} \cdot (\mu(\Omega))^{(p_2 - p_1)/p_2}.$$

It remains to raise both sides of the inequality to the power $1/p_1$.

The case $p_2 = \infty$ is easily dealt. Let $f \in L_\infty(\Omega, \Sigma, \mu)$. Then $|f(t)| \leqslant \|f\|_\infty$ for almost all $t \in \Omega$. Consequently,

$$\left(\int_\Omega |f|^{p_1} d\mu \right)^{1/p_1} \leqslant \|f\|_\infty \left(\int_\Omega d\mu \right)^{1/p_1} = \|f\|_\infty \mu(\Omega)^{1/p_1}. \qquad \square$$

We note that the inequality (3) takes its simplest form in the case of probability spaces (Ω, Σ, μ), when $\mu(\Omega) = 1$, and accordingly $\|f\|_{p_1} \leqslant \|f\|_{p_2}$.

Corollary 1. *Let (Ω, Σ, μ) be a finite measure space, $1 \leqslant p_1 \leqslant p_2 \leqslant \infty$, and $f_n, f \in L_{p_2}(\Omega, \Sigma, \mu)$ be such that $\|f_n - f\|_{p_2} \to 0$ as $n \to \infty$.*

Then $\|f_n - f\|_{p_1} \to 0$ as $n \to \infty$. In other words, convergence in L_p for some value of p implies convergence in L_p with a smaller value of p. $\qquad \square$

Corollary 2. *Let (Ω, Σ, μ) be a finite measure space, $1 \leqslant p_1 \leqslant p_2 \leqslant \infty$, $A \subset L_{p_2}(\Omega, \Sigma, \mu)$, and suppose A is closed in the metric of the space $L_{p_1}(\Omega, \Sigma, \mu)$. Then A is also closed in the metric of the space $L_{p_2}(\Omega, \Sigma, \mu)$.* $\qquad \square$

Theorem 2. *Let $1 \leqslant p_1 \leqslant p_2 \leqslant \infty$. Then $\ell_{p_1} \subset \ell_{p_2}$ and*

$$\|x\|_{p_2} \leqslant \|x\|_{p_1} \tag{4}$$

for all $x \in \ell_{p_1}$.

Proof. In view of the connection between the unit ball and the norm of a space, we need to show the inclusion $B_{\ell_{p_1}} \subset B_{\ell_{p_2}}$. Let $x = (x_k)_{k \in \mathbb{N}} \in B_{\ell_{p_1}}$. Then $\sum_{k=1}^\infty |x_k|^{p_1} < 1$. In particular, $|x_k| < 1$ and $|x_k|^{p_2} \leqslant |x_k|^{p_1}$ for all k. Consequently,

$$\sum_{k=1}^\infty |x_k|^{p_2} \leqslant \sum_{k=1}^\infty |x_k|^{p_1} < 1,$$

i.e, $x \in B_{\ell_{p_2}}$. This takes care of the case $p_2 < \infty$. For $p_2 = \infty$ inequality (4) takes the form $\sup_k |x_k| \leqslant \left(\sum_{k=1}^{\infty} |x_k|^{p_1}\right)^{1/p_1}$ and follows from the fact that a sum of positive terms is larger than any of the terms. $\square$

Despite the fact that convergence in L_p cannot be described in terms of the notions of convergence used in the theory of measure and integral,[1] a connection with those types of convergence, already familiar to us, nevertheless exists. The next theorem helps one understand better the structure of the convergence in L_p.

Theorem 3 (Vitali's convergence theorem). *Let* (Ω, Σ, μ) *be a measure space (with μ finite or σ-finite), $p \in [1, \infty)$, and $f_n, f \in L_p(\Omega, \Sigma, \mu)$. Then*

(i) *If $f_n \to f$ in the L_p-metric, it holds that:*

  (a) *$f_n \to f$ in measure;*

  (b) *there exists a subsequence of the sequence (f_n) that converges to f almost everywhere on Ω;*

  (c) *for every $\varepsilon > 0$ there exists a subset $A \in \Sigma$ of finite measure such that*

$$\int_{\Omega \setminus A} |f_n|^p d\mu < \varepsilon^p \tag{$*$}$$

  for all $n \in \mathbb{N}$;

  (d) *for every $\varepsilon > 0$ there exists a $\delta > 0$ such that*

$$\int_D |f_n|^p d\mu < \varepsilon^p \tag{$**$}$$

  for every subset $D \in \Sigma$ of $\mu(D) < \delta$ and all $n \in \mathbb{N}$.

(ii) *The conditions (a), (c) and (d) together imply that $f_n \to f$ in the L_p-metric.*

(iii) *If $f_n \to f$ almost everywhere on Ω and all f_n have a common majorant $g \in L_p(\Omega, \Sigma, \mu)$, then $f_n \to f$ in the L_p-metric.*

Proof Assertion (a) of item (i) follows from the Chebyshev inequality (Lemma 1 in Subsection 4.3.1). Indeed, for any $\varepsilon > 0$, let $B_{n,\varepsilon} = \{t \in \Omega : |f_n(t) - f(t)| > \varepsilon\}$. Applying the Chebyshev inequality to the function $g = |f_n - f|^p$, which is larger than the number ε^p on the set $B_{n,\varepsilon}$, we obtain the estimate

$$\mu(B_{n,\varepsilon}) \leqslant \frac{1}{\varepsilon^p} \int_{\Omega} |f_n - f|^p d\mu = \frac{1}{\varepsilon^p} \|f_n - f\|^p \to 0 \quad \text{as } n \to \infty,$$

which means precisely convergence in measure.

---

[1] Except for the space L_∞, in which a sequence (f_n) converges if and only if there exists a set of measure 0 in the complement of which (f_n) converges uniformly.

Assertion (b) follows from (a) because a sequence converging in measure has a subsequence which converges almost everywhere (Subsection 3.2.3, Theorem 2 and Exercise 12).

Let us demonstrate (c). In the case of $\mu(\Omega) < \infty$ take $A = \Omega$ and the job is done. Now consider the case of (Ω, Σ, μ) being a σ-finite measure space. For the given $\varepsilon > 0$ find such an $N \in \mathbb{N}$ that $\|f - f_n\| < \varepsilon/2$ for all $n > N$. Write Ω as a union of sets $\Omega_1 \subset \Omega_2 \subset \dots, m = 1, 2, \dots$, of finite measure. Since by the Lebesgue dominated convergence theorem

$$\int_\Omega |f_n|^p \mathbb{1}_{\Omega \setminus \Omega_k} d\mu \xrightarrow[k \to \infty]{} 0, \quad \text{and} \quad \int_\Omega |f|^p \mathbb{1}_{\Omega \setminus \Omega_k} d\mu \xrightarrow[k \to \infty]{} 0,$$

there is an n_0 such that

$$\int_{\Omega \setminus \Omega_{n_0}} |f_n|^p d\mu < \varepsilon^p, \ n = 1, \dots, N, \quad \text{and} \quad \int_{\Omega \setminus \Omega_{n_0}} |f|^p d\mu < \left(\frac{\varepsilon}{2}\right)^p.$$

Put $A = \Omega_{n_0}$. Then for $n = 1, \dots, N$ the condition (*) is evident, and for $n > N$,

$$\left(\int_{\Omega \setminus A} |f_n|^p d\mu\right)^{1/p} = \|f_n \mathbb{1}_{\Omega \setminus A}\| \leqslant \|f_n \mathbb{1}_{\Omega \setminus A} - f \mathbb{1}_{\Omega \setminus A}\| + \|f \mathbb{1}_{\Omega \setminus A}\|$$

$$\leqslant \|f_n - f\| + \left(\int_{\Omega \setminus A} |f|^p d\mu\right)^{1/p} < \frac{\varepsilon}{2} + \frac{\varepsilon}{2} = \varepsilon,$$

which implies (*).

Assertion (d) of item (i) follows from the absolute continuity of integral (Subsection 7.1.4, Theorem 1 and Exercise 5). Namely, again fix an $N \in \mathbb{N}$ such that $\|f - f_n\| < \varepsilon/2$ for all $n > N$, and pick a $\delta > 0$ in such a way that

$$\int_D |f_n|^p d\mu < \varepsilon^p, \ n = 1, \dots, N, \quad \text{and} \quad \int_D |f|^p d\mu < \left(\frac{\varepsilon}{2}\right)^p$$

for every $D \in \Sigma$, which has $\mu(D) < \delta$. Then for such a D the condition (**) is evident when $n \leqslant N$. On the other hand,

$$\left(\int_D |f_n|^p d\mu\right)^{1/p} = \|f_n \mathbb{1}_D\| \leqslant \|f_n \mathbb{1}_D - f \mathbb{1}_D\| + \|f \mathbb{1}_D\|$$

$$\leqslant \|f_n - f\| + \left(\int_D |f|^p d\mu\right)^{1/p} < \frac{\varepsilon}{2} + \frac{\varepsilon}{2} = \varepsilon,$$

which implies (**) for $n > N$.

Let us prove (ii). Assume contrary that $\|f_n - f\|$ does not tend to 0. Then there is an $a > 0$ and a subsequence $(g_n) \subset (f_n)$ such that $\|g_n - f\| > a$ for all n. Passing to a subsequence, we may assume that $g_n \xrightarrow{a.e.} f$. Applying (c) and (d) with $\varepsilon = a/8$ we obtain the corresponding subset $A \in \Sigma$ of finite measure and $\delta > 0$ such that

$$\int_{\Omega\setminus A}|g_n|^p d\mu < \left(\frac{a}{8}\right)^p, \quad \text{and} \quad \int_D |g_n|^p d\mu < \left(\frac{a}{8}\right)^p$$

for every subset $D \in \Sigma$ of $\mu(D) < \delta$ and all $n \in \mathbb{N}$. The Fatou lemma implies the same conditions for the limiting function f. According to Egorov's theorem there is a subset $B \in \Sigma_A$ with $\mu(B) < \delta$ such that (g_n) converges uniformly to f on $A \setminus B$. Then

$$a < \|g_n - f\| \leqslant \|(g_n - f)\mathbb{1}_{A\setminus B}\| + \|g_n\mathbb{1}_B\| + \|f\mathbb{1}_B\| + \|g_n\mathbb{1}_{\Omega\setminus A}\| + \|f\mathbb{1}_{\Omega\setminus A}\|$$

$$\leqslant \|(g_n - f)\mathbb{1}_{A\setminus B}\| + \frac{a}{2} \longrightarrow \frac{a}{2} \quad \text{as } n \to \infty,$$

which gives us the desired contradiction.

Finally, (iii) is an obvious consequence of the Lebesgue dominated convergence theorem. $\qquad\square$

Exercises

1. Let $p_0 \in [1, \infty)$ be a fixed number, and let $x \in \ell_{p_0}$. Then $\|x\|_p \to \|x\|_\infty$ as $p \to \infty$. In some sense, this justifies the notations $\|x\|_\infty$ and ℓ_∞.

2. Let $p_0 \in [1, \infty)$ be a fixed number, $x \in \ell_{p_0}$. Then for $p \in [p_0, \infty)$ the quantity $\|x\|_p$ depends continuously on p.

3. Formulate and prove an analogous assertion for the space $L_p(\Omega, \Sigma, \mu)$ in the case of a finite measure μ.

4. In Theorems 1 and 2 we proved that in the case of a finite measure, the space $L_p(\Omega, \Sigma, \mu)$ decreases as a set as p increases, while the space ℓ_p (a particular case of $L_p(\Omega, \Sigma, \mu)$ with an infinite measure μ) grows as p increases. Show that for $L_p[0, \infty)$ no increase and no decrease occurs when p grows.

5. Show that $\ell_{p_1} \neq \ell_{p_2}$ for $p_1 \neq p_2$, so that the growth of ℓ_p as a set when p grows is strict.

6. The analogous statement for $L_p[0, 1]$.

7. Suppose $p_0 \in (1, \infty)$. Show that $L_{p_0}[0, 1] \neq \bigcap_{p<p_0} L_p[0, 1]$ and that $L_{p_0}[0, 1] \neq \bigcup_{p>p_0} L_p[0, 1]$.

8. Show that $\ell_{p_0} \neq \bigcap_{p>p_0} \ell_p$ and $\ell_{p_0} \neq \bigcup_{p<p_0} \ell_p$.

9. Let (Ω, Σ, μ) be a finite measure space. Then $L_\infty(\Omega, \Sigma, \mu)$ is a dense subset in any of the spaces $L_p(\Omega, \Sigma, \mu)$.

10. Let (Ω, Σ, μ) be a finite measure space. Then the set of finitely-valued measurable functions is dense in any of the spaces $L_p(\Omega, \Sigma, \mu)$.

11. Let (Ω, Σ, μ) be a σ-finite measure space. Then the set of finitely-valued measurable functions with support of finite measure is dense in any of the spaces $L_p(\Omega, \Sigma, \mu)$ with $p \in [1, \infty)$. If $\mu(\Omega) = \infty$, then in $L_\infty(\Omega, \Sigma, \mu)$ this set is already not dense.

12. The set of piecewise-constant functions $f = \sum_{k=1}^{n} c_k \mathbb{1}_{(a_k, b_k)}$ is dense for $p \in [1, \infty)$ in any of the spaces L_p on the interval, as well as in the spaces L_p on the real line.

13. Let (Ω, Σ, μ) be a σ-finite measure space, Σ a countably-generated σ-algebra (i.e., there exists a countable family of sets that generates Σ in the sense of Definition 1, Subsection 2.1.2), and $p \in [1, \infty)$. Then $L_p(\Omega, \Sigma, \mu)$ is a separable normed space.

14. The space $L_\infty[0, 1]$ is not separable.

14.1.3 Weighted Integration Functionals

Let $1 \leqslant p \leqslant \infty, g \in L_{p'}(\Omega, \Sigma, \mu)$. Define on $L_p(\Omega, \Sigma, \mu)$ the *weighted integration functional* W_g by the rule $W_g(f) = \int_\Omega fg\, d\mu$. By Theorem 1 of Subsection 14.1.1, for any $f \in L_p(\Omega, \Sigma, \mu)$ the product fg is integrable, i.e., the functional W_g is well defined. The linearity of this functional is obvious.

Theorem 1. *For $g \in L_{p'}$ the functional W_g is continuous on L_p, and $\|W_g\| = \|g\|_{p'}$.*

Proof. The inequality $|W_g(f)| \leqslant \|f\|_p \|g\|_{p'}$, and together with it the estimate $\|W_g\| \leqslant \|g\|_{p'}$, follow from the Hölder inequality. Let us prove the opposite estimate. Thanks to homogeneity, is suffices to consider the case where $\|g\|_{p'} = 1$.

Let $1 < p < \infty$. Consider the function $f = |g|^{p'/p} e^{-i \arg g}$. This function lies in $L_p(\Omega, \Sigma, \mu)$, and $\|f\|_p = 1$. Therefore,

$$\|W_g\| \geqslant |W_g(f)| = \int_\Omega |g|^{p'/p+1} d\mu = \int_\Omega |g|^{p'} d\mu = 1 = \|g\|_{p'}.$$

If $p = \infty$, then in the argument above one needs to take for f the function $e^{-i \arg g}$. The case $p = 1$ is somewhat more complicated. In this case the functional W_g, generally speaking, does not attain its supremum on the unit sphere of the space $L_p = L_1$, and to estimate its norm from below it is not enough to use one successfully chosen function. So let us deal with this last remaining case. Since $p = 1$, we have $p' = +\infty$, and, as we assumed, $\|g\|_\infty = 1$. Fix $\varepsilon > 0$. The set $|g|_{>1-\varepsilon} = \{t \in \Omega : |g(t)| > 1 - \varepsilon\}$ has positive measure (otherwise, $\|g\|_\infty$ would not exceed $1 - \varepsilon$). Now pick in the set $|g|_{>1-\varepsilon}$ a measurable subset Δ of finite non-zero measure. Consider the function $f = \frac{1}{\mu(\Delta)} \mathbb{1}_\Delta e^{-i \arg g} \in L_1(\Omega, \Sigma, \mu)$. Since $\|f\|_1 = 1$, we have

$$\|W_g\| \geqslant |W_g(f)| = \frac{1}{\mu(\Delta)} \int_\Omega \mathbb{1}_\Delta |g| d\mu \geqslant \frac{1-\varepsilon}{\mu(\Delta)} \int_\Delta d\mu = 1 - \varepsilon = \|g\|_\infty - \varepsilon.$$

It remains to let here $\varepsilon \to 0$. $\qquad\qquad\qquad\qquad\qquad\qquad\qquad\qquad\qquad$ $\square$

Exercises

1. Give an example of a function $g \in L_\infty[0, 1]$ such that the corresponding functional W_g does not attain its norm on $S_{L_1[0,1]}$.

2. Which property of the measure μ, following from σ-finiteness, was used in the proof of the above theorem? And where exactly? Which part of the assertion of the theorem is valid for any countably-additive (including also non-σ-finite) measure?

As we will show below, for $1 \leqslant p < \infty$ there are no other continuous linear functionals on $L_p(\Omega, \Sigma, \mu)$ apart from the weighted integration functionals W_g with weights $g \in L_{p'}(\Omega, \Sigma, \mu)$. As the exercises proposed below will show, for $p = \infty$ the situation changes drastically: a large proportion of the functionals on L_∞ are not weighted integration functionals.

3. Let ν be a Borel charge (finite, since according to the axioms we adopted, charges take only finite values) on $[0, 1]$, and let F_ν be the functional on $C[0, 1]$ given by the formula $F_\nu(f) = \int_K f \, d\nu$. Using the Hahn–Banach theorem, we extend F_ν to a functional $\widetilde{F}_\nu$ on the entire space $L_\infty[0, 1]$. Prove that the functional $\widetilde{F}_\nu$ can have the form W_g only if the charge ν is absolutely continuous with respect to the Lebesgue measure.

On the interval $[0, 1]$ consider the σ-algebra $2^{[0,1]}$ of all subsets and define the measure μ as follows: the measure of any finite set is equal to the number of elements (cardinality) of the set, while the measure of any infinite set is taken to be $+\infty$. The space $L_1([0, 1], 2^{[0,1]}, \mu)$ is usually denoted by $\ell_1[0, 1]$. In other words, the elements of the space $\ell_1[0, 1]$ are those functions with countable support for which $\|f\| = \sum_{t \in [0,1]} |f(t)| < \infty$. Prove that:

4. The cardinality of the dual space $\ell_1[0, 1]^*$ is larger than the cardinality of the continuum.

5. The space ℓ_∞ has a subspace isometric to $\ell_1[0, 1]$.

6. The cardinality of the dual space $(\ell_\infty)^*$ is larger than the cardinality of the continuum.

7. The set of functionals on ℓ_∞ of the "weighted integral" (in the present case — "weighted sum") form has the cardinality of the continuum.

As M. I. Kadets has shown in 1967, any two separable infinite-dimensional Banach spaces are homeomorphic as topological spaces[2] (do not confuse with isomorphism!). The next exercise provides an example of how nonlinear homeomorphisms of Banach spaces can be constructed.

8. Let $1 \leqslant p_1 < p_2 < \infty$. Consider the mapping $M \colon L_{p_1}[0, 1] \to L_{p_2}[0, 1]$ introduced by S. Mazur by the formula $M(g) = |g|^{p_1/p_2} e^{i \arg g}$. Prove that this mapping effects a (nonlinear) homeomorphism of the spaces $L_{p_1}[0, 1]$ and $L_{p_2}[0, 1]$, i.e., M is bijective, and both M and M^{-1} are continuous.

14.1.4 The General Form of Linear Functionals on L_p

Theorem 1. *Let $1 \leqslant p < \infty$. Then every linear functional $G \in L_p(\Omega, \Sigma, \mu)^*$ is uniquely representable as a weighted integration functional W_g, where the function $g \in L_{p'}(\Omega, \Sigma, \mu)$. Moreover, $\|G\| = \|g\|_{p'}$.*

Proof. The formula for the norm of the functional W_g has already been proved in the preceding subsection. It remains to establish the existence and uniqueness of the sought-for function $g \in L_{p'}(\Omega, \Sigma, \mu)$. We begin with the uniqueness. Suppose $G = W_{g_1} = W_{g_2}$. Then $W_{g_1-g_2} = W_{g_1} - W_{g_2} = 0$ and $\|g_1 - g_2\|_{p'} = \|W_{g_1-g_2}\| = 0$, i.e., the elements g_1 and g_2 of the space $L_{p'}(\Omega, \Sigma, \mu)$ coincide.

The proof of the existence requires some effort so we break it into several lemmas. We first treat the particular case of a finite measure. The general case will be reduced to it by means of a device that we already encountered in Subsection 4.6.2.

Lemma 1. *Let $1 \leqslant p < \infty$, $G \in L_p(\Omega, \Sigma, \mu)^*$, and $\mu(\Omega) < \infty$. Then there exists a function $g \in L_1(\Omega, \Sigma, \mu)$ such that*

$$G(\mathbb{1}_\Delta) = \int_\Omega \mathbb{1}_\Delta g \, d\mu$$

for all $\Delta \in \Sigma$.

Proof. As the reader has probably guessed already, the argument relies on the Radon–Nikodým theorem.

Define a set function ν on the σ-algebra Σ by the rule $\nu(\Delta) = G(\mathbb{1}_\Delta)$. Thanks to the linearity of the functional G and the equality $\mathbb{1}_{\Delta_1} + \mathbb{1}_{\Delta_2} = \mathbb{1}_{\Delta_1 \sqcup \Delta_2}$, which holds for any disjoint pair $\Delta_1, \Delta_2 \in \Sigma$, the set function ν is finitely additive. Let us show that in fact ν is countably additive. To this end, according to Exercise 5 in Subsection 7.1.1, we need to show that for any collection of sets $A_k \in \Sigma$, $k \geqslant 1$, forming a decreasing chain with empty intersection, $\lim_{k\to\infty} \nu(A_k) = 0$. Indeed, in this case

---

[2]Making this reference, I am glad to use the occasion to pay tribute to my father Mikhail Kadets (1923–2011, the name is often spelled as Kadec). He was a prominent mathematician and excellent teacher, who had a great influence not only on me (as a person and as a mathematician), but also on the development of Banach space theory in general.

$\|\mathbb{1}_{A_k}\|_p = (\mu(A_k))^{1/p} \to 0$ as $k \to \infty$, and so, thanks to the continuity of the functional G, also $\nu(A_k) = G(\mathbb{1}_{A_k}) \to 0$ as $k \to \infty$. Therefore, ν is a charge. Further, the inequality $|\nu(\Delta)| \leqslant \|G\| \cdot \|\mathbb{1}_\Delta\|_p = \|G\| \, (\mu(\Delta))^{1/p}$ means that the charge ν is absolutely continuous with respect to the measure μ. Now define the sought-for function $g \in L_1(\Omega, \Sigma, \mu)$ as the Radon–Nikodým derivative of the charge ν with respect to the measure μ. Then

$$G(\mathbb{1}_\Delta) = \nu(\Delta) = \int_\Delta g \, d\mu = \int_\Omega \mathbb{1}_\Delta \, g \, d\mu. \qquad \Box$$

Lemma 2. *Under the assumptions of the preceding lemma,*

$$G(f) = \int_\Omega f g \, d\mu \tag{5}$$

for all functions $f \in L_\infty(\Omega, \Sigma, \mu)$.

Proof. Denote by F the linear functional on $L_\infty(\Omega, \Sigma, \mu)$ acting by the rule $F(f) = G(f) - \int_\Omega f g \, d\mu$. Since all functions of the form $\mathbb{1}_\Delta$ with $\Delta \in \Sigma$ lie in Ker F, it follows that Ker F also contains all functions $\sum_{k=1}^n a_k \mathbb{1}_{\Delta_k}$ with $\Delta_k \in \Sigma$ (all finitely-valued functions). Therefore (see Exercise 10 in Subsection 14.1.2), the kernel of the functional F is dense in $L_\infty(\Omega, \Sigma, \mu)$. Further,

$$|F(f)| \leqslant \|G\| \, \|f\|_p + \|f\|_\infty \|g\|_1 \leqslant \|G\| \, \|f\|_\infty (\mu(\Omega))^{1/p} + \|f\|_\infty \|g\|_1$$
$$= \big(\|G\| \, (\mu(\Omega))^{1/p} + \|g\|_1 \big) \, \|f\|_\infty,$$

that is, $\|F\| \leqslant \|G\| \cdot (\mu(\Omega))^{1/p} + \|g\|_1 < \infty$ and the functional F is continuous. A continuous functional that vanishes on a dense set vanishes on the whole space. $\Box$

Lemma 3. *The constructed function g lies in $L_{p'}(\Omega, \Sigma, \mu)$.*

Proof. Using the preceding lemma and the continuity of the functional G in the norm $\| \cdot \|_p$, we deduce that every function $f \in L_\infty(\Omega, \Sigma, \mu)$ obeys the estimate

$$\left| \int_\Omega f g \, d\mu \right| = |G(f)| \leqslant \|G\| \cdot \|f\|_p. \tag{6}$$

We consider first the case $p > 1$, and so $p' \neq \infty$. Fix $N > 0$ and substitute in (6) the function $f = |g|^{p'-1} \mathbb{1}_{|g|<N} \cdot e^{-i \arg g}$. Then we have

$$\int_{|g|<N} |g|^{p'} d\mu = \left| \int_\Omega f g \, d\mu \right| \leqslant \|G\| \cdot \|f\|_p$$

$$= \|G\| \cdot \left(\int_{|g|<N} |g|^{(p'-1)p} d\mu \right)^{1/p} = \|G\| \cdot \left(\int_{|g|<N} |g|^{p'} d\mu \right)^{1/p}.$$

Dividing both sides of this inequality by $\left(\int_{|g|<N} |g|^{p'} d\mu \right)^{1/p}$ and raising the result to the power p', we obtain the inequality $\int_{|g|<N} |g|^{p'} d\mu \leqslant \|G\|^{p'}$. Upon letting the parameter N go to infinity, this last estimate becomes $\int_{\Omega} |g|^{p'} d\mu \leqslant \|G\|^{p'}$, which says, in particular, that $g \in L_{p'}(\Omega, \Sigma, \mu)$.

Let us pass to the case $p = 1$, $p' = \infty$. Suppose that $g \notin L_{\infty}(\Omega, \Sigma, \mu)$. Then for any $N > 0$ the set $|g|_{>N}$ has non-zero measure. Let us substitute in (6) the function $f = \mathbb{1}_{|g|_{>N}} e^{-i \arg g}$. We obtain

$$N\mu(|g|_{>N}) \leqslant \int_{|g|>N} |g| d\mu = \left| \int_{\Omega} fg \, d\mu \right| \leqslant \|G\| \cdot \|f\|_1 = \|G\| \cdot \mu(|g|_{>N}).$$

Therefore, $N \leqslant \|G\|$ for all $N > 0$, so we have reached a contradiction. $\qquad\square$

Completion of the Proof of the Theorem. Thus, in the case of a finite measure μ, we have established the existence of a function $g \in L_{p'}(\Omega, \Sigma, \mu)$ such that for all functions $f \in L_{\infty}(\Omega, \Sigma, \mu)$ relation (5), which can be written as $G(f) = W_g(f)$, holds. Hence, G and W_g are continuous linear functionals on the space $L_p(\Omega, \Sigma, \mu)$ that coincide on the dense (Exercise 9 in Subsection 14.1.2) subset $L_{\infty}(\Omega, \Sigma, \mu) \subset L_p(\Omega, \Sigma, \mu)$. Consequently, $G(f) = W_g(f)$ for all $f \in L_p(\Omega, \Sigma, \mu)$.

Let us address now the case of a σ-finite measure μ. Fix some decomposition $\Omega = \bigsqcup_{n=1}^{\infty} \Omega_n$, where $0 < \mu(\Omega_n) < \infty$ and $\Omega_n \in \Sigma$. Introduce the numbers $a_n = 2^n \mu(\Omega_n)$. Further, introduce on (Ω, Σ) a new measure μ_1 by the formula $\mu_1(B) = \sum_{n=1}^{\infty} \mu(B \cap \Omega_n)/a_n$. With this definition, the triple (Ω, Σ, μ_1) is a finite measure space. Now consider on Ω the function $h = \sum_{n=1}^{\infty} a_n \mathbb{1}_{\Omega_n}$. Recall (Subsection 4.6.2) that a function f is μ-integrable on Ω if and only if the function $f \cdot h$ is μ_1-integrable on Ω, and then $\int_{\Omega} f \, d\mu = \int_{\Omega} fh \, d\mu_1$.

Define the linear operator $T: L_p(\Omega, \Sigma, \mu_1) \to L_p(\Omega, \Sigma, \mu)$ by the formula $Tf = f \cdot h^{-1/p}$. Then T effects a bijective isometry of the spaces $L_p(\Omega, \Sigma, \mu_1)$ and $L_p(\Omega, \Sigma, \mu)$:

$$\|Tf\|^p = \int_{\Omega} |f|^p h^{-1} d\mu = \int_{\Omega} |f|^p d\mu_1 = \|f\|^p.$$

Next, to the functional $G \in L_p(\Omega, \Sigma, \mu)^*$ we associate the functional $T^*G \in L_p(\Omega, \Sigma, \mu_1)^*$ by the recipe $(T^*G)(f) = G(Tf)$. Since μ_1 is a finite measure, the functional T^*G falls under the conditions of the already established particular case of the theorem: there exists a function $g_1 \in L_{p'}(\Omega, \Sigma, \mu_1)$ such that $(T^*G)(f) = \int_{\Omega} fg_1 \, d\mu_1$ for all functions $f \in L_p(\Omega, \Sigma, \mu_1)$. Decoded, this condition says that that for any $f \in L_p(\Omega, \Sigma, \mu_1)$ we have

$$G(Tf) = \int_{\Omega} fg_1 h \, d\mu = \int_{\Omega} (Tf) \cdot g_1 h^{1/p'} d\mu.$$

Re-denote Tf by $\widetilde{f}$ and $g_1 h^{1/p'}$ by g. Then, as required, $g \in L_{p'}(\Omega, \Sigma, \mu)$, and the equality $G(\widetilde{f}) = \int_\Omega \widetilde{f} g \, d\mu$, as desired, holds for all functions $\widetilde{f} \in L_p(\Omega, \Sigma, \mu)$. $\square$

Remark 1. Since the correspondence $g \mapsto W_g$ effects a bijective isometry of the spaces $L_{p'}(\Omega, \Sigma, \mu)$ and $L_p(\Omega, \Sigma, \mu)^*$, the function g is usually identified with the functional W_g it generates. Under this identification, the theorem on the general form of linear functionals on L_p, $1 \leqslant p < \infty$, is expressed by the equality $L_p(\Omega, \Sigma, \mu)^* = L_{p'}(\Omega, \Sigma, \mu)$.

Remark 2. Since the exponents p and p' play symmetric roles (we have $(p')' = p$), the equality $L_p(\Omega, \Sigma, \mu)^* = L_{p'}(\Omega, \Sigma, \mu)$, $1 \leqslant p < \infty$, can be recast as $L_{p'}(\Omega, \Sigma, \mu)^* = L_p(\Omega, \Sigma, \mu)$, $1 < p \leqslant \infty$. Since the dual space of any normed space is complete, this implies, in particular, that $L_p(\Omega, \Sigma, \mu)$ is complete for all $1 < p \leqslant \infty$.

Remark 3. For $1 < p < \infty$ both relations, $L_p(\Omega, \Sigma, \mu)^* = L_{p'}(\Omega, \Sigma, \mu)$ and $L_{p'}(\Omega, \Sigma, \mu)^* = L_p(\Omega, \Sigma, \mu)$, hold. Combining them, we conclude that for $1 < p < \infty$ we have $(L_p(\Omega, \Sigma, \mu)^*)^* = L_p(\Omega, \Sigma, \mu)$. This property of spaces, called *reflexivity*, plays an important role in the theory of Banach spaces, and will be discussed in the subsequent chapters.

Remark 4. Since the spaces ℓ_p constitute particular cases of the spaces $L_p(\Omega, \Sigma, \mu)$, in which the role of the integral is played by the sum of the terms of a sequence, we obtain the following theorem on the general form of linear functionals on ℓ_p.

Theorem 1. *Let* $1 \leqslant p < \infty$. *Then for any* $f = (f_1, f_2, \ldots) \in \ell_p$ *and* $g = (g_1, g_2, \ldots) \in \ell_{p'}$, *the expression* $W_g(f) = \sum_{n=1}^\infty f_n g_n$ *is well defined. For fixed* $g \in \ell_{p'}$, W_g *is a continuous linear functional on* ℓ_p, *and* $\|W_g\| = \|g\|_{p'}$. *Further, for any linear functional* $G \in \ell_p^*$ *there exists a unique element* $g \in \ell_{p'}$, *such that* $G = W_g$.

Under the identification $g \to W_g$, the theorem on the general form of linear functionals on ℓ_p for $1 \leqslant p < \infty$ can be stated as the equality $\ell_p^* = \ell_{p'}$.

Remark 5. Although the theorem on the general form of linear functionals on the space L_p is no longer true for $p = \infty$, the first step in the proof, namely the introduction of the set function ν on the σ-algebra Σ by the rule $\nu(\Delta) = G(\mathbb{1}_\Delta)$, also makes sense for a functional $G \in L_\infty(\Omega, \Sigma, \mu)^*$. The main difference compared with the case $p < \infty$ is that here ν is only a finitely-additive charge, rather than a countably-additive one. For this reason, despite the fact that the absolute continuity condition $\mu(\Delta) = 0 \implies \nu(\Delta) = 0$ is satisfied, the Radon–Nikodým theorem is no longer applicable. Nevertheless, it is not difficult to show that the set function ν has a finite semivariation and, by Theorem 3 of Subsection newexlinkssec:13semivariat13.4.213 (where the more general case of a vector measure ν was treated), any function $f \in L_\infty(\Omega, \Sigma, \mu)$ will be ν-integrable. Hence, we have the following result.

Theorem 2 (General form of linear functionals on L_∞). *For any functional $G \in L_\infty(\Omega, \Sigma, \mu)^*$ there exists a unique finitely-additive bounded charge $v \colon \Sigma \to \mathbb{C}$ which vanishes on μ-negligible sets, and generates the functional G by the rule $G(f) = \int_\Omega f\, dv$. Conversely, every such charge defines, via the indicated rule, a continuous linear functional on $L_\infty(\Omega, \Sigma, \mu)$. The norm of the functional G coincides with the semivariation on Ω of the charge that generates it.*

Exercises

1. The theorem on the general form of linear functionals on the space L_2 can be derived as a particular case of the theorem on the general form of linear functionals on L_p. At the same time, L_2 is a Hilbert space, so for it the theorem on the general form of linear functionals on a Hilbert space is valid. Don't these theorems contradict one another?

2. For $1 < p < \infty$ the condition that the measure is σ-finite in Theorem 1 is redundant.

3. For $p = 1$, although the condition that the measure is σ-finite in Theorem 1, can in principle be weakened, it cannot be eliminated completely. Show that in the case where $\Omega = [0, 1]$, $\Sigma = \mathfrak{B}$, and μ is the counting measure (i.e., the measure of any finite set is the number of its elements, and the measure of any infinite set is equal to $+\infty$), $L_1(\Omega, \Sigma, \mu)^* \neq L_\infty(\Omega, \Sigma, \mu)$. Specifically, $L_1(\Omega, \Sigma, \mu)^*$ consists of all functionals W_g, where g are arbitrary bounded functions, and not only bounded Borel-measurable functions on $[0, 1]$.

4. Fill in the details of the theorem on the general form of linear functionals on L_∞.

5. Prove the following theorem on the general form of a linear operator on L_∞.

Let X be a Banach space. For every continuous linear operator $T \colon L_\infty(\Omega, \Sigma, \mu) \to X$ there exists a unique finitely-additive bounded vector measure $v \colon \Sigma \to X$ which vanishes on μ-negligible sets, and which generates the operator T by the rule $Tf = \int_\Omega f\, dv$. Conversely, every such vector measure generates via the indicated rule a continuous operator $T \colon L_\infty(\Omega, \Sigma, \mu) \to X$. The norm of the operator T is equal to the semivariation on Ω of the measure that generates it.

6. Let $K \colon [0, 1] \times [0, 1] \to \mathbb{C}$ be a Borel-measurable function of two variables. Denote by $\widetilde{K}$ the function $\widetilde{K}(t) = \int_0^1 |K(t, x)|\, dx$. Show that if $\|\widetilde{K}\|_\infty < \infty$, then the expression $(Tf(t))(x) = \int_0^1 K(t, x) f(t)\, dt$ defines a continuous linear operator $T \colon L_1[0, 1] \to L_1[0, 1]$, and $\|T\| \leqslant \|\widetilde{K}\|_\infty$. This operator on $L_1[0, 1]$ is called the integration operator with kernel K.

7. Under the assumptions of the preceding exercise, prove that $\|T\| = \|\widetilde{K}\|_\infty$.

8. Prove that in the space $L_1[0, 1]$ every finite-rank operator is representable as an integration operator with a kernel.

9. Using the approximation of a compact operator by finite-rank operators, prove that any compact operator $T \in L(L_1[0, 1])$ can be represented as an integration operator with a kernel.

10. Let $[a_n, b_n] \subset [0, 1]$, $n = 1, 2, \ldots$, be a sequence of pairwise disjoint intervals of nonzero length. Take as kernel the function $K = \sum_{k=1}^{\infty} \frac{1}{b_k - a_k} \mathbb{1}_{[a_k, b_k] \times [a_k, b_k]}$. Then the integration operator with the kernel K in $L_1[0, 1]$ is not a compact operator. Compare with the Exercises 8 in Subsection 12.4.3 and 12 in Subsection 12.4.2.

11. Suppose that the kernel in Exercise 6 above obeys the condition $\|\widetilde{K}\|_\infty < \infty$. Consider the function $K_t \colon [0, 1] \to \mathbb{C}$, $K_t(x) = K(t, x)$. If the family of functions $\{K_t\}_{t \in [0,1]}$ forms a precompact set in $L_1[0, 1]$, then the integration operator with the kernel K in $L_1[0, 1]$ is a compact operator.

12. Provide an example of a finite-rank operator in $C[0, 1]$ that is not representable as an integration operator with kernel.

In the light of the theorem on the general form of linear functionals on L_p, if the operator T acts from L_p to L_r and $p, r \in [1, \infty)$, then its adjoint (dual) should be considered as an operator acting from $L_{r'}$ to $L_{p'}$. For the operators listed below, prove that they are linear and continuous, and calculate the adjoint operator.

13. The identity embedding operator $j_{p,r}$ of the space $L_p[0, 1]$ into $L_r[0, 1]$, where $r \leqslant p$.

14. Any integration operator with kernel acting in $L_p[0, 1]$.

15. The multiplication operator $T_g \colon L_p[0, 1] \to L_1[0, 1]$ by a function $g \in L_{p'}[0, 1]$: $T_g(f) = f \cdot g$.

16. The operator $S_g \colon L_1[0, 1] \to L_1[0, 1]$ of composition with a monotone continuous function $g \colon [0, 1] \to [0, 1]$, i.e., $S_g(f) = f \circ g$. Under what conditions on g will the operator indeed act from $L_1[0, 1]$ to $L_1[0, 1]$?

14.2 The Fourier Transform on the Real Line

Formally, to define an operator $T \colon X \to Y$ we first need to give the spaces X and Y and only then specify how the operator T act on the elements of the space X. In real life[3] everything goes in the opposite order: first some analytical expression that one naturally should treat as a linear operator arises, and only afterwards does one fit it

---

[3]One wonders to what degree solving mathematical problems can be considered as "real life"?.

into the formal scheme. In this process, of course, the specification of the operator by the given analytical expression is not unambiguous, and depending on the choice of the spaces one obtains operators with different properties.

In this section we treat one such example, where for the analytical expression one takes the Fourier integral on the real line, well known from calculus.

14.2.1 δ-Sequences and the Dini Theorem

In this subsection we discuss a useful device for proving limit theorems for integral expressions. In less rigorous terms, this device can be described as follows: if a sequence of functions (g_n) converges in some sense to the δ-measure supported at the point t_0, then for a large number of functions f it holds that $\int_\Omega f g_n \, d\mu \to f(t_0)$ as $n \to \infty$.

Definition 1. Let (Ω, Σ, μ) be a space with a nonatomic measure (finite or σ-finite). The sequence of functions $g_n \in L_\infty(\Omega, \Sigma, \mu)$ is called a δ-*sequence* if there exists an increasing sequence of sets $\Omega_m \in \Sigma$ of finite measure such that

(i) $\mu(\Omega \setminus \bigcup_{m=1}^\infty \Omega_m) = 0$;

(ii) $\int_{\Omega_m} g_n \, d\mu \to 1$ as $m \to \infty$, for every $n \in \mathbb{N}$;

(iii) $\int_{\Omega_m} h g_n \, d\mu \to 0$ for any function h integrable on Ω_m.

Remark 1. If the functions g_n are integrable on the whole space Ω, then condition (ii) can be replaced by the simpler condition $\int_\Omega g_n d\mu = 1$.

Definition 2. Let (g_n) be a δ-sequence. We call the measurable function g an *appropriate multiplier* if it is almost everywhere different from zero and $\sup_{n \in \mathbb{N}} \|g_n g\|_\infty = M < \infty$.

Theorem 1. *Let (g_n) be a δ-sequence, f be a function integrable on Ω, and a be an arbitrary scalar. Further, suppose there exists an appropriate multiplier g such that $\frac{f-a}{g} \in L_1(\Omega, \Sigma, \mu)$ Then $\int_\Omega f g_n \, d\mu \to a$ as $n \to \infty$.*

Proof. Since the function $(f - a)/g$ is integrable on Ω, and $\|g_n g\|_\infty < \infty$, the product $(f - a)g_n$ is also integrable on Ω. By conditions (i) and (ii) of Definition 1,

$$\int_\Omega f g_n d\mu - a = \lim_{m \to \infty} \left(\int_{\Omega_m} f g_n d\mu - a \right)$$

$$= \lim_{m \to \infty} \int_{\Omega_m} (f - a) g_n \, d\mu = \int_\Omega \frac{f - a}{g} g g_n \, d\mu. \tag{1}$$

Next, by the Lebesgue dominated convergence theorem,

$$\int_{\Omega\setminus\Omega_m}\left|\frac{f-a}{g}\right|d\mu = \int_\Omega \left|\frac{f-a}{g}\right|\mathbb{1}_{\Omega\setminus\Omega_m}d\mu \to 0 \quad \text{as } m \to \infty.$$

Fix $\varepsilon > 0$ and choose a number m such that the estimate $\int_{\Omega\setminus\Omega_m}\left|\frac{f-a}{g}\right|d\mu < \varepsilon$ holds. Using formula (1) and the fact that, by condition (iii) of Definition 1, $\lim_{n\to\infty}\left|\int_{\Omega_m}(f-a)g_n\,d\mu\right| = 0$, we have

$$\overline{\lim_{n\to\infty}}\left|\int_\Omega fg_n\,d\mu - a\right| = \overline{\lim_{n\to\infty}}\left|\int_\Omega \frac{f-a}{g}gg_n\,d\mu\right| = \overline{\lim_{n\to\infty}}\left|\int_{\Omega\setminus\Omega_m}\frac{f-a}{g}gg_n\,d\mu\right| \leqslant M\varepsilon,$$

which in view of the arbitrariness of ε yields the required limit relation. □

Recall that the function f on the (finite or infinite) interval (a, b) satisfies the *Dini condition* at the point $x_0 \in (a, b)$ if the function $(f(x_0 + t) - f(x_0))/t$ is integrable on some interval of the form $[-\varepsilon, +\varepsilon]$. Clearly, if $f \in L_1(-\infty, \infty)$ satisfies the Dini condition at x_0, then $(f(x_0 + t) - f(x_0))/t \in L_1(-\infty, \infty)$. We note that the Dini condition is not too restrictive. Indeed, say, differentiability at the point x_0, the Lipschitz condition, or the Hölder condition with exponent $p > 0$ impose more serious constraints on the behavior of a function.

Theorem 2. *Suppose $u \in L_1(-\infty, \infty)$ and satisfies the Dini condition at the point x. Then*

$$\lim_{N\to\infty}\frac{1}{\pi}\int_{-\infty}^{\infty}u(x+t)\frac{\sin Nt}{t}dt = u(x).$$

Proof. We apply Theorem 1 to $\Omega = \mathbb{R}$ with the Lebesgue measure, $f(t) = u(x + t)$, $g_N(t) = \frac{1}{\pi}\frac{\sin Nt}{t}$, $\Omega_m = \{t \in \mathbb{R}: \frac{1}{m} \leqslant |t| \leqslant m\}$, and the multiplier $g(t) = t$. For this choice, condition (i) in the definition of a δ-sequence is obviously satisfied, while condition (ii) follows from the known formula $\lim_{m\to\infty}\frac{1}{\pi}\int_{-m}^{m}\frac{\sin t}{t}dt = 1$, usually given in complex analysis courses as a nice example of application of residue theory.[4] Let us verify that property (iii) holds for g_N. Let h be a function integrable on Ω_m. Since the function $1/t$ is bounded on Ω_m, the function $h(t)/t$ is integrable on Ω_m. To obtain the relation $\lim_{N\to\infty}\int_{\Omega_m}h(t)\frac{\sin Nt}{t}dt = 0$, it remains to apply Corollary 1 of Subsection 10.4.3. The remaining conditions of Theorem 1 in our case are the boundedness of the sine function and the Dini condition. □

Applying Theorem 1 with $\Omega_m = \Omega = [-\pi, \pi]$ to the partial sums of the Fourier series,

---

[4] See, e.g., [42, Subsection 3.1.2.2].

$$(S_n f)(t) = \frac{1}{2\pi} \int\limits_{-\pi}^{\pi} f(t+\tau) \frac{\sin((n+1/2)\tau)}{\sin(\tau/2)} d\tau,$$

one can readily obtain the following result, formulated earlier in Subsection 10.4.3, Exercise 4.

Theorem 3. *Suppose the function f is Lebesgue integrable on the interval $[-\pi, \pi]$ and satisfies the Dini condition at the point $x_0 \in [-\pi, \pi]$. Then the Fourier series of f converges at the point x_0 to $f(x_0)$.* $\square$

Exercises

1. Where in the statement or proof of Theorem 1 does the condition $g_n \in L_\infty(\Omega, \Sigma, \mu)$ of Definition 1 play a role?

2. For each of the conditions (i)–(iii) of Definition 1, find its role in Theorem 1.

14.2.2 The Fourier Transform in L_1 on the Real Line

Let $f \in L_1(-\infty, \infty)$. The *Fourier transform* of the function f is the function $\widehat{f}(t)$ defined as

$$\widehat{f}(t) = \int\limits_{-\infty}^{\infty} f(\tau) e^{-it\tau} d\tau. \tag{2}$$

Since $|e^{-it\tau}| = 1$ for $t, \tau \in \mathbb{R}$, the integrand in formula (2) is an integrable function, so the function $\widehat{f}$ is defined for all $t \in \mathbb{R}$. Moreover,

$$|\widehat{f}(t)| \leqslant \int\limits_{-\infty}^{\infty} |f(\tau) e^{-it\tau}| d\tau = \int\limits_{-\infty}^{\infty} |f(\tau)| d\tau = \|f\|_1.$$

Taking the supremum over all $t \in \mathbb{R}$, we see that $\widehat{f}$ is a bounded function and

$$\sup_{-\infty < t < +\infty} |\widehat{f}(t)| \leqslant \|f\|_1 \tag{3}$$

for all functions $f \in L_1(-\infty, \infty)$.

Let us list the Fourier transforms for several concrete functions. The required calculations are left to the reader as exercises on the subject "methods for computing integrals".[5]

----

[5] See [24, Ch. 8, §4].

Example 1. Let $f = \mathbb{1}_{[a,b]}$. Then $\widehat{f}(t) = (e^{-ita} - e^{-itb})/(it)$. In particular, if $f = \mathbb{1}_{[-a,a]}$, then $\widehat{f}(t) = 2t^{-1}\sin at$.

Example 2. Let $f(t) = e^{-a|t|}$, where $a > 0$. Then $\widehat{f}(t) = \dfrac{2a}{a^2 + t^2}$.

Example 3. Let $f(t) = e^{-at^2}$, where $a > 0$. Then $\widehat{f}(t) = \sqrt{\dfrac{\pi}{a}} \cdot e^{-\frac{t^2}{4a}}$.

We denote by $\ell_\infty(-\infty, \infty)$ the space of all bounded complex-valued functions on the line, equipped with the norm $\|f\| = \sup_{-\infty < t < +\infty} |f(t)|$. Convergence in this space coincides with the uniform convergence on the line. We denote by $C_0(-\infty, \infty)$ the subspace of $\ell_\infty(-\infty, \infty)$ consisting of all continuous functions that tend to zero at infinity. Clearly, the subspace $C_0(-\infty, \infty)$ is closed in $\ell_\infty(-\infty, \infty)$.

Definition 1. The *Fourier transform operator* in $L_1(-\infty, \infty)$ is the mapping $F: L_1(-\infty, \infty) \to \ell_\infty(-\infty, \infty)$ acting by the rule $F(f) = \widehat{f}$. When it is clear from the context that one is dealing with the operator and not just with an individual value $\widehat{f}$, one uses the short name "Fourier transform" also for the Fourier transform operator.

Theorem 1. *The Fourier transform F in $L_1(-\infty, \infty)$ is a linear operator, $\|F\| = 1$, and for any function $f \in L_1(-\infty, \infty)$ its image $\widehat{f}$ is a continuous function that tends to zero at infinity.*

Proof. The inequality $\|F\| \leqslant 1$, and together with it the continuity of the operator F, follow from (3). To obtain the opposite inequality, we apply the transformation to an arbitrary positive function $f \in S_{L_1(-\infty,\infty)}$ and obtain

$$\|F\| \geqslant \|F(f)\| = \|\widehat{f}\| \geqslant |\widehat{f}(0)| = \int\limits_{-\infty}^{\infty} f(\tau)d\tau = \|f\|_1 = 1.$$

It remains to show that $F(L_1(-\infty, \infty)) \subset C_0(-\infty, \infty)$. By Example 1 above, $F(f) \in C_0(-\infty, \infty)$ for any function of the form $f = \mathbb{1}_{[a,b]}$. Hence, by linearity, $F_1(f) \in C_0(-\infty, \infty)$ for any piecewise-constant function f on the line (i.e., any function of the form $f = \sum_{k=1}^n c_k \mathbb{1}_{[a_k, b_k]})$. Since the set of piecewise-constant functions is dense in $L_1(-\infty, \infty)$ and the mapping F is continuous, this implies (see Exercise 9 in Subsection 1.2.1) that the whole image of the transformation F lies in the subspace $C_0(-\infty, \infty)$. $\qquad\square$

Remark 1. Since $C_0(-\infty, \infty)$ is a subspace not only of $\ell_\infty(-\infty, \infty)$, but also of $L_\infty(-\infty, \infty)$, the mapping F can be regarded, whenever convenient, as acting from $L_1(-\infty, \infty)$ into $L_\infty(-\infty, \infty)$.

Recall that in Subsection 4.6.3 we have proved that for any two functions $f, g \in L_1(-\infty, \infty)$ their convolution $(f * g)(t) = \int_{-\infty}^{\infty} f(\tau)g(t - \tau)d\tau$ is defined almost everywhere and is integrable on the real line.

Theorem 2. *Let $f, g \in L_1(-\infty, \infty)$. Then $F(f * g) = F(f) \cdot F(g)$. In other words, the Fourier transform maps the convolution into the ordinary product.*

Proof. We have

$$[F(f * g)](t) = \int_{-\infty}^{\infty} (f * g)(\tau) \cdot e^{-it\tau} d\tau$$

$$= \int_{-\infty}^{\infty} \left(\int_{-\infty}^{\infty} f(x)g(\tau - x)dx \right) e^{-it\tau} d\tau.$$

Switching the order of integration in the double integral and making the change of variables $\tau - x = y$, we obtain

$$[F(f * g)](t) = \int_{-\infty}^{\infty} \left(\int_{-\infty}^{\infty} g(\tau - x)e^{-it\tau} d\tau \right) f(x)dx$$

$$= \int_{-\infty}^{\infty} \left(\int_{-\infty}^{\infty} g(y)e^{-it(x+y)} dy \right) f(x) dx$$

$$= \int_{-\infty}^{\infty} \left(\int_{-\infty}^{\infty} g(y)e^{-ity} dy \right) e^{-itx} f(x)dx = \widehat{f}(t)\widehat{g}(t). \qquad \square$$

The property just established explains why the Fourier transform is often used in probability theory, in particular in the study of sums of independent random variables, to prove limit theorems for such sums. The point is that the density of the distribution of a sum of independent random variables is the convolution of the corresponding individual densities of distributions, and after taking the Fourier transform the convolution becomes the much more convenient to handle usual multiplication operation.

Exercises

1. Carry out the necessary calculations in Examples 1–3.

2. Why is the set of piecewise-constant functions dense in $L_1(-\infty, \infty)$?

3. Taking Subsection 4.6.3 as a model, justify the correctness of the switch of order of integration in the double integral used in the proof of Theorem 2.

4. Let $g \in L_1(-\infty, \infty)$, $a \in \mathbb{R}$, and $f(t) = g(t + a)$. Then $\widehat{f}(t) = e^{ita}\widehat{g}(t)$.

5. Let $g \in L_1(-\infty, \infty)$, $a \in \mathbb{R}$, and $f(t) = e^{ita}g(t)$. Then $\widehat{f}(t) = \widehat{g}(t - a)$.

6. Let $g \in L_1(-\infty, \infty)$ and $f(t) = \overline{g(-t)}$. Then $\widehat{f}(t) = \overline{\widehat{g}(t)}$.

7. Let $f \in L_1(-\infty, \infty)$ be an even real-valued function. Then $\widehat{f}$ is real-valued.

8. The operator F is not bounded below.

14.2.3 Inversion Formulas

In this subsection we study the problem of recovering a function from its Fourier transform.

Recall that the *integral in the sense of principal value* of a function f on the line is defined as

$$\text{p.v.-}\int_{-\infty}^{\infty} f(t)\,dt = \lim_{m \to \infty} \int_{-m}^{m} f(t)\,dt.$$

Clearly, if $f \in L_1(-\infty, \infty)$, then p.v.-$\int_{-\infty}^{\infty} f(t)\,dt = \int_{-\infty}^{\infty} f(t)\,dt$, but the p.v.-integral can also exist for functions that are not Lebesgue integrable on the line. For instance, p.v.-$\int_{-\infty}^{\infty} f(t)\,dt = 0$ for any odd locally-integrable (i.e., integrable on any finite interval) function, but among such functions there are also some that do not lie in $L_1(-\infty, \infty)$ (for instance, $f(t) = t$).

Theorem 1. *Suppose the function f is integrable on the line and satisfies the Dini condition at some point $x \in \mathbb{R}$. Then the value of f at x is recovered by the following Fourier formula:*

$$2\pi f(x) = \text{p.v.-}\int_{-\infty}^{\infty} \widehat{f}(t) e^{itx}\,dt.$$

Proof. We transform the quantity $a_m = \int_{-m}^{m} \widehat{f}(t) e^{itx}\,dt$ as follows:

$$a_m = \int_{-m}^{m} \left(\int_{-\infty}^{\infty} f(\tau) e^{-it\tau}\,d\tau \right) e^{itx}\,dt = \int_{-\infty}^{\infty} f(\tau) \int_{-m}^{m} e^{it(x-\tau)}\,dx\,d\tau$$

$$= 2\int_{-\infty}^{\infty} f(\tau) \frac{\sin m(\tau - x)}{\tau - x}\,d\tau = 2\int_{-\infty}^{\infty} f(y + \tau) \frac{\sin my}{y}\,d\tau.$$

To derive the desired relation $2\pi f(x) = \lim_{m \to \infty} a_m$ it remains to apply Theorem 2 of Subsection 14.2.1. $\qquad\square$

If under the assumptions of the theorem not only the function f is integrable, but also its Fourier transform $\widehat{f}$, then the *Fourier formula* takes on the form

$$f(x) = \frac{1}{2\pi} \int_{-\infty}^{\infty} \widehat{f}(t) e^{itx}\,dt. \qquad (4)$$

Unfortunately, the Fourier formula, despite its simplicity and elegance, is applicable only to relatively "nice" functions. To obtain a formula that recovers a function from its Fourier transform and works for all functions $f \in L_1(-\infty, \infty)$, one resorts to the following procedure. First, one modifies the function by convolution with a "nice" function so as to achieve a sufficient degree of smoothness, then to the "so-corrected" function one applies the Fourier formula, and then one already recovers the initial function f. Based on this procedure, one can obtain various recovery formulas. Here we give only one of them.

Theorem 2. *For any function $f \in L_1(-\infty, \infty)$, the equality*

$$f(x) = \lim_{n \to \infty} \text{p.v.-} \int_{-\infty}^{\infty} \widehat{f}(t) \frac{n \sin(t/n)}{2\pi t} e^{itx} dt \tag{5}$$

holds for almost all $x \in \mathbb{R}$.

Proof. Fix $\varepsilon > 0$ and consider the auxiliary function $g_\varepsilon(x) = \frac{1}{2\varepsilon} \int_{x-\varepsilon}^{x+\varepsilon} f(t) dt$. Since $g_\varepsilon = \frac{1}{2\varepsilon} \mathbb{1}_{[-\varepsilon,\varepsilon]} * f$, we have $g_\varepsilon \in L_1(-\infty, \infty)$. Further, the Fourier transform maps convolution into multiplication, so $\widehat{g_\varepsilon} = \widehat{f}(t) \sin(\varepsilon t)/(\varepsilon t)$ (see also Example 1 in Subsection 14.2.2). The function $g_\varepsilon(x)$ satisfies the Dini condition at almost every point $x \in \mathbb{R}$ (an integral is almost everywhere differentiable with respect to the upper integration limit), hence Theorem 1 applies to it:

$$g_\varepsilon(x) = \text{p.v.-} \int_{-\infty}^{\infty} \widehat{f}(t) \frac{\sin \varepsilon t}{2\pi \varepsilon t} e^{itx} dt \quad \text{a.e.} \tag{6}$$

Further, by the same theorem on the differentiability of an integral with respect to the upper integration limit, $f(x) = \lim_{\varepsilon \to 0} g_\varepsilon(x)$ for almost all $x \in \mathbb{R}$. Putting $\varepsilon = 1/n$ in (6), we obtain the desired inversion formula (5). $\quad\square$

Corollary 1 (uniqueness theorem for the Fourier transform). *If $f, g \in L_1(-\infty, \infty)$ and $\widehat{f} = \widehat{g}$, then $f = g$ a.e. In other words, the Fourier transform in L_1 is an injective operator.* $\quad\square$

Exercises

1. The set $F(L_1(-\infty, \infty))$ is not closed in $C_0(-\infty, \infty)$.

2. Endow the set $\mathbb{Z}$ of integers with the counting measure μ: the measure of a set is equal to the number of its elements. In this case the space $L_1(\mathbb{Z}, \mu)$ can be identified with $\ell_1(\mathbb{Z})$. For functions $f \in L_1(\mathbb{Z}, \mu)$ we define the Fourier transform $\widehat{f}: [0, 2\pi] \to \mathbb{C}$ by the formula $\widehat{f}(t) = \int_{\mathbb{Z}} f(x) e^{itx} d\mu(x) = \sum_{n \in \mathbb{Z}} f(n) e^{int}$. Prove that $\widehat{f} \in C[0, 2\pi]$.

3. Define the operator $F_{\mathbb{Z}} \colon \ell_1(\mathbb{Z}) \to C[0, 2\pi]$ by the rule $F_{\mathbb{Z}}(f) = \widehat{f}$. Verify that $F_{\mathbb{Z}}$ is linear, and $\|F_{\mathbb{Z}}\| = 1$. What does the analogue of the Fourier formula look like in this case?

4. Prove that the operator $F_{\mathbb{Z}}$ is injective. Is it bounded below?

5. The image of the operator $F_{\mathbb{Z}}$ coincides with the set W appearing in Example 5 of Subsection 11.11. Show that the image of $F_{\mathbb{Z}}$ is not closed and is not dense in $C[0, 2\pi]$.

6. The closure in $C[0, 2\pi]$ of the image of the operator $F_{\mathbb{Z}}$ coincides with $C(\mathbb{T})$.

7. Define the convolution for functions from $\ell_1(\mathbb{Z})$ and prove that the operator $F_{\mathbb{Z}}$ takes convolution into multiplication.

14.2.4 The Fourier Transform and Differentiation

Lemma 1. *Let $f \in L_1(-\infty, \infty)$ be such that f is absolutely continuous on every finite interval $[a, b] \subset \mathbb{R}$ and $f' \in L_1(-\infty, \infty)$. Then $f(x) \to 0$ as $x \to \infty$.*

Proof. By the absolute continuity assumption, $f(x) = f(0) + \int_0^x f'(t)\,dt$. Hence, f has limits both as $x \to +\infty$ and $x \to -\infty$, namely, $f(0) + \int_0^{+\infty} f'(t)\,dt$ and $f(0) + \int_0^{-\infty} f'(t)\,dt$, respectively. These limits cannot be different from 0, since otherwise the function f would not be integrable on the line. $\qquad\square$

Theorem 1. *Let $f \in L_1(-\infty, \infty)$ be such that f is absolutely integrable on every finite interval $[a, b] \subset \mathbb{R}$ and $f' \in L_1(-\infty, \infty)$. Then $F(f') = itF(f)$.*

Proof. One applies the integration by parts formula and the preceding lemma to obtain

$$\int_{-\infty}^{\infty} f'(\tau)e^{-it\tau}\,d\tau = f(\tau)e^{-it\tau}\Big|_{-\infty}^{+\infty} + it \int_{-\infty}^{\infty} f(\tau)e^{-it\tau}\,d\tau = it\widehat{f}(t). \qquad\square$$

As already noted, the Fourier transform of an integrable function tends to 0 at infinity. From the preceding theorem one can readily obtain an estimate of the rate of convergence of the function $\widehat{f}$ to 0 in terms of the degree of smoothness of the function f:

Corollary 1. *Suppose the function f is n-times continuously differentiable, and f as well as all its derivatives of order up to and including n are integrable on the real line. Then $\widehat{f}(t) = o(t^{-n})$ as $n \to \infty$.*

Proof. Applying Theorem 1 n times we deduce that $F(f^{(n)}) = (it)^n F(f)$. Accordingly, $|\widehat{f}(t) \cdot t^n| = |F(f^{(n)})|(t) \to 0$ as $t \to \infty$. $\qquad\square$

Denote by $\mathcal{D}^2(\mathbb{R})$ the set of all twice continuously differentiable functions on the real line with bounded support (i.e, the support of each function $f \in \mathcal{D}^2(\mathbb{R})$ lies in a finite interval, which depends on f).

Corollary 2. *Let $f \in \mathcal{D}^2(\mathbb{R})$. Then $\widehat{f} \in L_p(-\infty, \infty)$ for all $p \geqslant 1$.* □

Proof. The function $\widehat{f}$ is continuous, and by Corollary 1 above, $\widehat{f}(t) = o(t^{-2})$. □

Exercises

1. Suppose the functions $f(t)$ and $g(t) = tf(t)$ lie in $L_1(-\infty, \infty)$. Then the function $\widehat{f}$ is differentiable everywhere and $\dfrac{d}{dt}\widehat{f} = -i\widehat{g}$.

2. Suppose the functions $f(t), tf(t), \ldots, t^n f(t)$ lie in $L_1(-\infty, \infty)$. Then the function $\widehat{f}$ possesses an n-th derivative and $\dfrac{d^n}{dt^n}\widehat{f}(t) = (-i)^n F(t^n f(t))$.

3. Denote by $s_{(-\delta,\delta)}$ the horizontal strip in $\mathbb{C}$ consisting of all points z for which $-\delta < \operatorname{Im} z < \delta$. Suppose $f \in L_1(-\infty, \infty)$ and for some $\delta > 0$ the function $f(t)e^{\delta|t|}$ is also integrable. Then $\int_{-\infty}^{\infty} f(\tau)e^{-iz\tau}d\tau$ exists for all $z \in s_{(-\delta,\delta)}$ and is an analytic function of the variable z in $s_{(-\delta,\delta)}$. In other words, under our assumptions, $\widehat{f}$ can be regarded as an analytic function in $s_{(-\delta,\delta)}$.

4. Under the assumptions of the preceding exercise, suppose the function f takes only non-negative real values. Then $\widehat{f}$ obeys the so-called *ridge condition*: $|\widehat{f}(z)| \leqslant \widehat{f}(i \operatorname{Im} z)$ for all $z \in s_{(-\delta,\delta)}$.

5. Fill in the omitted details in the proof of the following result:

Theorem 2. *Let (a, b) be a finite or infinite interval of the real line, $p \in [1, \infty)$, and f a function that is different from zero almost everywhere on (a, b), and satisfies for some $C, \delta > 0$ the inequality $|f(t)| \leqslant Ce^{-\delta|t|}$. Then the functions $f_n(t) = t^n f(t)$, $n = 0, 1, 2, \ldots$, form a complete system in $L_p(a, b)$.*

Proof. By the completeness criterion given in Subsection 9.2.3, we need to show that any element $g \in L_{p'}(a, b)$ that annihilates all functions f_n is equal to zero. Here annihilation means that

$$\int_{(a,b)} f_n g \, d\lambda = \int_{-\infty}^{\infty} t^n f(t)g(t)\mathbb{1}_{(a,b)}(t)dt = 0 \tag{7}$$

for all $n = 0, 1, 2, \ldots$ Introducing the auxiliary function $h(t) = f(t)g(t)\mathbb{1}_{(a,b)}(t)$, we rewrite (7) in the form $(d^n/dt^n)\widehat{h}(0) = 0$, $n = 0, 1, 2, \ldots$, which in view of the analyticity of the function $\widehat{h}$ means that $\widehat{h}(t) \equiv 0$. By the uniqueness theorem, this implies that $h = 0$, and so $g = 0$, too. □

6. Using the preceding theorem, establish the completeness of:

— the set of polynomials in $L_p[0, 1]$, $1 \leqslant p < \infty$;

— the system $f_n(t) = t^n e^{-t^2/2}$, $n = 0, 1, 2, \ldots$, in $L_p(-\infty, \infty)$, $1 \leqslant p < \infty$;

— the system $f_n(t) = t^n e^{-t}$, $n = 0, 1, 2, \ldots$, in $L_p(0, \infty)$, $1 \leqslant p < \infty$.

7. None of the systems of functions listed in the preceding exercise is complete in the corresponding space L_∞.

14.2.5   The Fourier Transform in L_2 on the Real Line

The aim of this subsection is to define the Fourier transforms on functions belonging to the space $L_2(-\infty, \infty)$. Although the definition itself, proposed by Plancherel in 1910, is more complicated than its analogue in $L_1(-\infty, \infty)$, the Fourier transform (operator) on $L_2(-\infty, \infty)$ turns out to be considerably simpler and more convenient as regards its properties than the Fourier transform on $L_1(-\infty, \infty)$. Specifically, up to a constant factor, it is a unitary operator and has a complete system of eigenvectors.

Definition 1. Let X, Y be normed spaces, the linear operations in which agree on the intersection $X \cap Y$. In this case the set $X \cap Y$ will be regarded as a normed space endowed with the norm $\|u\| = \|u\|_X + \|u\|_Y$. It is readily seen that $u_n \to u$ as $n \to \infty$ in $X \cap Y$ if and only if $u_n \to u$ as $n \to \infty$ in the metric of the space X, as well as in that of the space Y.

Lemma 1. *Let X_1, X_2 be normed spaces and (Ω, Σ, μ) be a σ-finite measure space, $1 \leqslant p_1 \leqslant p_2 \leqslant \infty$. Further, let $T_j : X_j \to L_{p_j}(\Omega, \Sigma, \mu)$, $j = 1, 2$, be continuous linear operators which coincide on a dense subset E of the space $X_1 \cap X_2$. Then $T_1 x = T_2 x$ for all $x \in X_1 \cap X_2$.*

Proof. Let $x \in X_1 \cap X_2$, $x_n \in E$, such that $x_n \to x$ as $n \to \infty$ in the metric of the space $X_1 \cap X_2$. Using the assumption that $T_1 = T_2$ on E, we define $f_n = T_1 x_n = T_2 x_n$. Since $x_n \to x$ in the metric of each of the spaces X_1, X_2, it follows that $f_n = T_j x_n \to T_j x$, as $n \to \infty$, $j = 1, 2$, in the metric of the corresponding space L_{p_j}. By assertion (i) of Theorem 3 in Subsection 14.1.2, this means that $T_1 x$ and $T_2 x$ coincide almost everywhere on Ω.   □

The particular case of intersection of spaces we are interested in here is the space $L_1(-\infty, \infty) \cap L_2(-\infty, \infty)$, equipped with the norm $\|f\| = \|f\|_1 + \|f\|_2$.

Theorem 1. *The following sets are dense in $L_1(-\infty, \infty) \cap L_2(-\infty, \infty)$:*

(i)   *the set E_1 of functions with supports contained in finite intervals;*

(ii) *the set E_2 of bounded functions with supports contained in finite intervals;*

(iii) *the set E_3 of finite-valued measurable functions with supports contained in finite intervals;*

(iv) *the set E_4 of piecewise-constant functions;*

(v) *the set $\mathcal{D}^2(\mathbb{R})$ (see Subsection 14.2.4).*

Proof. (i) Let $f \in L_1(-\infty, \infty) \cap L_2(-\infty, \infty)$. Then $f_n = \mathbb{1}_{[-n,n]} f$ lies in E_1 and converges to f.

(ii) Let $f \in E_1$. Then put $f_n = f \cdot \mathbb{1}_{A_n}$, where $A_n = \{t \in \mathbb{R} : |f|(t) < n\}$.

(iii) Any bounded function $f \in E_2$ can be approximated on an interval by finitely-valued functions, and in fact even uniformly (Theorem 3 of Subsection 3.1.4).

(iv) By (iii), it suffices to show that the closure of the set E_4 includes all the functions of the form $\mathbb{1}_A$, where A is a measurable subset of some finite interval $[a, b]$. Since the Lebesgue measure was constructed via extension from the algebra of finite unions of subintervals, for any $\varepsilon > 0$ there exists a set B, representable as the union of a finite disjoint collection of subintervals, such that $\lambda(A \triangle B) < \varepsilon$. Accordingly, $\|\mathbb{1}_A - \mathbb{1}_B\| < \varepsilon + \varepsilon^{1/2}$. It remains to observe that $\mathbb{1}_B$ is a piecewise-constant function.

(v) By (iv), is suffices to show that any characteristic function of the kind $\mathbb{1}_{[a,b]}$ can be approximated by functions from $\mathcal{D}^2(\mathbb{R})$. To this end, it in turn suffices to "smooth" the function $\mathbb{1}_{[a,b]}$ so that the area under its graph will remain almost unchanged. The unconvinced reader should produce the requisite approximation by himself. $\square$

Lemma 2. *The following* Plancherel formula *holds for all functions $f, g \in \mathcal{D}^2(\mathbb{R})$:*

$$\langle f, g \rangle = \frac{1}{2\pi} \langle \widehat{f}, \widehat{g} \rangle.$$

In particular (taking $g = f$),

$$\|f\|_2 = \frac{1}{\sqrt{2\pi}} \|\widehat{f}\|_2.$$

Proof. First, since $f, g \in \mathcal{D}^2(\mathbb{R}) \subset L_2(-\infty, \infty)$, also $\widehat{f}, \widehat{g} \in L_2(-\infty, \infty)$ (Corollary 2 in Subsection 14.2.4). Therefore, both scalar products $\langle f, g \rangle$ and $\langle \widehat{f}, \widehat{g} \rangle$ are well defined. Since the functions considered lie in $L_1(-\infty, \infty)$ and are differentiable, and since, by the same Corollary 2 in Subsection 14.2.4, their Fourier transforms are integrable, one can apply to them the Fourier formula (4), Subsection 14.2.3, and get

$$2\pi \langle f, g \rangle = 2\pi \int_{-\infty}^{\infty} f(x) \overline{g(x)} \, dx = \int_{-\infty}^{\infty} \left(\int_{-\infty}^{\infty} \widehat{f}(t) e^{itx} dt \right) \overline{g(x)} \, dx$$

$$= \int_{-\infty}^{\infty} \widehat{f}(t) \left(\overline{\int_{-\infty}^{\infty} g(x) e^{-itx} dx} \right) dt = \langle \widehat{f}, \widehat{g} \rangle. \qquad \square$$

From this lemma it follows that the mapping $f \mapsto \widehat{f}$, regarded as an operator acting from the subspace $\mathcal{D}^2(\mathbb{R}) \subset L_2(-\infty, \infty)$ to $L_2(-\infty, \infty)$, is continuous.

Further, since $\mathcal{D}^2(\mathbb{R})$ is dense in $L_2(-\infty, \infty)$, Theorem 1 of Subsection 6.5.1 shows that this mapping extends uniquely to a continuous operator acting from $L_2(-\infty, \infty)$ to $L_2(-\infty, \infty)$. This reasoning justifies the correctness of the following definition.

Definition 2. The *Fourier transform (operator)* in $L_2(-\infty, \infty)$ is the continuous linear mapping $F_2 \colon L_2(-\infty, \infty) \to L_2(-\infty, \infty)$ which acts on any function $f \in \mathcal{D}^2(\mathbb{R})$ by the rule $F_2(f) = \widehat{f}$.

Theorem 2. *The Fourier transform in $L_2(-\infty, \infty)$ has the following properties:*

A. $F_2(f) = \widehat{f}$ *for any function* $f \in L_1(-\infty, \infty) \cap L_2(-\infty, \infty)$.

B. *The Plancherel formula* $\langle f, g \rangle = \frac{1}{2\pi}\langle F_2(f), F_2(g) \rangle$ *holds for all functions* $f, g \in L_2(-\infty, \infty)$. *In particular,* $\|f\|_2 = \frac{1}{\sqrt{2\pi}}\|F_2(f)\|_2$ *for all* $f \in L_2(-\infty, \infty)$.

Proof. Assertion A follow from Lemma 1 and the fact that the set $\mathcal{D}^2(\mathbb{R})$ is dense in $L_1(-\infty, \infty) \cap L_2(-\infty, \infty)$. Assertion B is the extension by continuity of Lemma 2 to $L_2(-\infty, \infty)$. $\qquad\square$

The next theorem provides a more constructive definition of the Fourier transform in L_2.

Theorem 3. *Let* $f \in L_2(-\infty, \infty)$. *Define* $(F_{2,n}f)(t) = \int_{-n}^{n} f(\tau)e^{-it\tau}d\tau$. *Then* $F_2 f = \lim_{n\to\infty} F_{2,n} f$, *where the limit is understood in the metric of the space* $L_2(-\infty, \infty)$.

Proof. Consider the functions $f_n = f \cdot \mathbb{1}_{[-n,n]}$. Clearly, the sequence f_n converges to f in the L_2-metric. Thanks to the continuity of the operator F_2, the sequence $F_{2,n}(f) = F_2(f_n)$ converges to $F_2(f)$. $\qquad\square$

Remark 1. It is natural to denote the limit of the integrals $\int_{-n}^{n} f(\tau)e^{-it\tau}d\tau$ as $n \to \infty$ by $\int_{-\infty}^{\infty} f(\tau)e^{-it\tau}d\tau$. Accordingly, for the Fourier transform in $L_2(-\infty, \infty)$ one uses the same notation $\widehat{f}(t) = \int_{-\infty}^{\infty} f(\tau)e^{-it\tau}d\tau$ as in $L_1(-\infty, \infty)$, except that here the integral is understood not in the sense of Lebesgue, but in the sense of Theorem 3, as the limit in $L_2(-\infty, \infty)$ of the functions $\int_{-n}^{n} f(\tau)e^{-it\tau}d\tau$.

To study further the Fourier operator F_2, it is reasonable to look at polynomials in this operator, and the simplest of them is the operator $(F_2)^2$.

Lemma 3. *The operator $(F_2)^2$ is self-adjoint.*

Proof. By Theorem 3 (see also Exercise 2 of Subsection 10.4.3), the operator $(F_2)^2$ is the pointwise limit of the operators $(F_{2,n})^2$. The latter can be easily written as integration operators with symmetric kernels:

$$(F_{2,n} \circ F_{2,n}f)(t) = \int\limits_{-n}^{n}\int\limits_{-n}^{n} f(\tau)e^{-ix(t+\tau)}dxd\tau = \int\limits_{-n}^{n} f(\tau)\frac{2\sin n(t+\tau)}{t+\tau}d\tau.$$

Therefore, $(F_{2,n})^2$ are self-adjoint operators, hence so is their pointwise limit. $\qquad\square$

Theorem 4. *The operator $\widetilde{F}_2 = \frac{1}{\sqrt{2\pi}} F_2$ is unitary.*

Proof. By assertion B of Theorem 2, the operator $\widetilde{F}_2$ effects an isometric embedding, and hence, in particular, is bounded below. Therefore (by the preceding lemma), $(\widetilde{F}_2)^2$ is a self-adjoint bounded below operator. It follows (see Lemma 1 in Subsection 12.4.2) that $(\widetilde{F}_2)^2$ is bijective. Then the operator $\widetilde{F}_2$ itself is also bijective, and any bijective isometry of a Hilbert space is a unitary operator (Theorem 2 of Subsection 13.2.2). $\qquad\square$

Now let us discuss the spectral properties of the Fourier transform F_2.

Theorem 5. *The operator $\widetilde{F}_2 = \frac{1}{\sqrt{2\pi}} F_2$ satisfies the relation $(\widetilde{F}_2)^4 = I$. Consequently, the spectrum of the operator F_2 lies in the set $\{1, i, -1, -i\}$, and the space $L_2(-\infty, \infty)$ decomposes into the orthogonal direct sum of the eigenspaces of the operator F_2.*

Proof. By Lemma 3 and Theorem 4, the operator $(\widetilde{F}_2)^2$ is both self-adjoint and unitary. Hence, its spectrum lies simultaneously on the real line and on the unit circle. That is, $\sigma((\widetilde{F}_2)^2) \subset \{-1, 1\}$. By the spectral mapping theorem, $\sigma((\widetilde{F}_2)^4) = \{1\}$. In view of the self-adjointness, this means that $(\widetilde{F}_2)^4 = I$. By the same spectral mapping theorem, $\sigma(\widetilde{F}_2) \subset \{1, i, -1, -i\}$. The last assertion of the theorem can be derived from Exercise 1 of Subsection 13.1.1, or alternatively by solving the Exercises 1–5 below, which provide an explicit construction of a complete system of eigenvectors of the operator F_2. $\qquad\square$

Exercises

Consider the sequence of functions $x_n(t) = t^n e^{-t^2/2}$ and the subspaces $X_n = \mathrm{Lin}\{x_k\}_{k=0}^n$, $n = 0, 1, 2, \ldots$. In other words, the subspace X_n consists of the functions $P_n(t) e^{t^2/2}$, where P_n is a polynomial of degree at most n.

1. Show that $\bigcup_{n=1}^{\infty} X_n$ is dense in $L_2(-\infty, \infty)$.

2. The subspaces X_n, $n = 0, 1, 2, \ldots$ are invariant under the operator F_2.

3. $X_0, X_0^{\perp} \cap X_1, X_1^{\perp} \cap X_2, X_2^{\perp} \cap X_3, \ldots$ is a sequence of one-dimensional invariant subspaces of the operator F_2.

4. If in each of the one-dimensional subspaces from the preceding exercise we choose one non-zero element h_n; then the resulting sequence will be an orthogonal basis consisting of eigenvectors of the operator F_2. (Actually, the contruction of the system involves the Gram–Schmidt orthogonalization of the system $\{x_n\}_0^{\infty}$.)

5. A concrete example of choice of the required elements h_n is provided by the sequence of *Hermite functions* $h_n(x) = \frac{d^n(e^{-x^2})}{dx^n} \cdot e^{x^2/2}$. Verify the inclusion $h_n \in X_{n-1}^{\perp} \cap X_n$.

The Fourier transform in $\mathbb{R}^n$ is constructed by analogy with the one-dimensional case: for $f \in L_1(\mathbb{R}^n)$ we put $\widehat{f}(\mathbf{x}) = \int_{\mathbb{R}^n} f(\mathbf{y})e^{-i\langle \mathbf{x}, \mathbf{y}\rangle}d\mathbf{y}$.

6. Construct for the Fourier transform in $\mathbb{R}^n$ the analogue of the theory developed above for the one-dimensional case.

7. If the measure μ is σ-finite, then with respect to the norm introduced in Definition 1, $L_{p_0}(\Omega, \Sigma, \mu) \cap L_{p_1}(\Omega, \Sigma, \mu)$ is a Banach space for all $p_0, p_1 \in [1, +\infty]$.

8. Show that there exist Banach spaces X and Y for which $X \cap Y$ is not a Banach space.

9. Think of an additional condition that ensures the completeness of the space $X \cap Y$, and which will be satisfied by the subspaces in Example 7 above.

14.3   The Riesz–Thorin Interpolation Theorem and its Consequences

14.3.1   The Hadamard Three-Lines Theorem

In this subsection we prove a variant of the Phragmén–Lindelöf principle, which estimates a function analytic in the strip $\Pi = \{z \in \mathbb{C} : 0 \leqslant \operatorname{Re} z \leqslant 1\}$ in terms of its values on the boundary of Π.

For a function $f : \Pi \to \mathbb{C}$, let $M_\theta = M_\theta(f) = \sup\{|f(\theta + iy)| : y \in \mathbb{R}\}$, where $\theta \in [0, 1]$.

Theorem 1 (Hadamard's theorem). *Let* $f : \Pi \to \mathbb{C}$ *be a bounded function analytic in* Π *and continuous up to the boundary. Then for any* $\theta \in [0, 1]$ *one has the estimate* $M_\theta \leqslant M_0^{1-\theta} M_1^\theta$. *In other words,* $\ln M_\theta$ *is a convex function.*

We break the proof into several auxiliary assertions.

Lemma 1. *Under the assumptions of the theorem, suppose that*

(i) *the function* $f(\theta + iy)$ *tends to zero as* $y \to \infty$, *uniformly in* $\theta \in [0, 1]$, *and*

(ii) $|f(z)| \leqslant 1$ *on the boundary of the strip* Π.

Then $|f(z)| \leqslant 1$ *in the whole strip* Π.

Proof. Using (i), choose $r > 0$ large enough such that the estimate $|f(z)| \leqslant 1$ holds on the horizontal segments $\{z = \theta \pm ir : \theta \in [0, 1]\}$. Then the estimate $|f(z)| \leqslant 1$ holds on the entire boundary of the rectangular domain $\Pi_r = \{z \in \Pi : |\operatorname{Im} z| \leqslant r\}$. Hence, by the maximum principle, $|f(z)| \leqslant 1$ for all $z \in \Pi_r$. Since as r grows the domain Π_r captures all the points of the strip Π, the inequality $|f(z)| \leqslant 1$ holds for all $z \in \Pi$. $\qquad\square$

Lemma 2. *Under the assumptions of the theorem, suppose the function f is bounded in modulus by some number C in the entire strip Π and $|f(z)| \leqslant 1$ on the boundary of Π. Then $|f(z)| \leqslant 1$ in the entire strip Π.*

Proof. Consider the function $g_\varepsilon(z) = 1 + \varepsilon z$, with $\varepsilon > 0$. This function is everywhere in Π larger in modulus than 1 and $|g_\varepsilon(\theta + iy)| \geqslant \varepsilon|y|$. Therefore, the auxiliary function f/g_ε satisfies all the conditions of Lemma 1. Applying Lemma 1, we conclude that the estimate $\left|\frac{f(z)}{1+\varepsilon z}\right| \leqslant 1$ holds for all $z \in \Pi$. It remains to let ε go to zero.   □

Proof of the Theorem. Consider the function $g(z) = f(z)M_0^{z-1}M_1^{-z}$. Since

$$|g(z)| = |f(z)|M_0^{\operatorname{Re}z - 1}M_1^{-\operatorname{Re}z},$$

the function g satisfies the conditions of Lemma 2. Hence, $|f(\theta+iy)|M_0^{\theta-1}M_1^{-\theta} \leqslant 1$. As we can assume that M_0 and M_1 are different from zero (otherwise $f \equiv 0$), this yields the desired inequality $|f(\theta + iy)| \leqslant M_0^{1-\theta}M_1^\theta$.   □

Exercise. Prove the following variant of the Phragmén–Lindelöf principle for an analytic function f, given in the half-plane $D = \{z : \operatorname{Re}z \geqslant 0\}$: suppose $|f(z)| \leqslant 1$ on the imaginary axis and for some function g such that $g(r)/r \to 0$ as $r \to \infty$, the estimate $|f(z)| \leqslant e^{g(|z|)}$ holds for all $z \in D$. Then $|f(z)| \leqslant 1$ for all $z \in D$.

14.3.2 The Riesz–Thorin Theorem

Throughout this subsection, $(\Omega_1, \Sigma_1, \mu_1)$ and $(\Omega_2, \Sigma_2, \mu_2)$ are finite or σ-finite measure spaces, $p_0, p_1, q_0, q_1 \in [1, +\infty]$, $\theta \in [0, 1]$; the exponents p_θ, q_θ are defined by the relations $\frac{1}{p_\theta} = \frac{1-\theta}{p_0} + \frac{\theta}{p_1}$ and $\frac{1}{q_\theta} = \frac{1-\theta}{q_0} + \frac{\theta}{q_1}$. Let $E \subset L_p(\Omega_1, \Sigma_1, \mu_1)$ be a linear subspace. The norm of a linear operator $T \colon (E, \|\cdot\|_p) \to L_q(\Omega_2, \Sigma_2, \mu_2)$ will be denoted by $\|T\|_{p,q}$.

Lemma 1. *Let E be the space of all finitely-valued measurable functions on Ω_1 with supports of finite measure, and let $T \colon E \to \bigcap_{\theta \in [0,1]} L_{q_\theta}(\Omega_2, \Sigma_2, \mu_2)$ be a linear operator. Then for any $\theta \in [0, 1]$ one has the estimate*

$$\|T\|_{p_\theta, q_\theta} \leqslant \left(\|T\|_{p_0, q_0}\right)^{1-\theta}\left(\|T\|_{p_1, q_1}\right)^\theta. \tag{1}$$

Proof. The idea of the proof, proposed by Thorin (1939), is to reduce inequality (1) to Hadamard's three-lines theorem. Let us define $M_0 = \|T\|_{p_0, q_0}$, $M_1 = \|T\|_{p_1, q_1}$. We need to show that for any finitely-valued function $u = \sum_{k=1}^n u_k \mathbb{1}_{U_k} \in E$ with $\|u\|_{p_\theta} = 1$ (u_k are scalars, and $U_k \in \Sigma_1$, $k = 1, 2, \ldots, n$, is a collection of sets of finite measure) it holds that $\|Tu\|_{q_\theta} \leqslant M_0^{1-\theta}M_1^\theta$. Since the norm of the element Tu in $L_{q_\theta}(\Omega_2, \Sigma_2, \mu_2)$ coincides with the norm of the "weighted integral functional" that it generates on the space $L_{q_\theta'}(\Omega_2, \Sigma_2, \mu_2)$, the lemma reduces to proving that the inequality

$$\left| \int_{\Omega_2} w \cdot T u \, d\mu_2 \right| \leqslant M_0^{1-\theta} M_1^{\theta} \tag{2}$$

holds for all $w \in L_{q'_\theta}(\Omega_2, \Sigma_2, \mu_2)$ with $\|w\|_{q'_\theta} = 1$. Finally, it suffices to verify (2) for finitely-valued functions $w = \sum_{k=1}^m w_k \mathbb{1}_{W_k}$.

Let us redefine the numbers p_θ, q'_θ for complex values of the parameter θ:

$$\frac{1}{q'_z} = \frac{1-z}{q'_0} + \frac{z}{q'_1} \quad \text{and} \quad \frac{1}{p_z} = \frac{1-z}{p_0} + \frac{z}{p_1}.$$

For any complex number ξ, put $\operatorname{sign} \xi = \xi/|\xi|$, if $\xi \neq 0$, and $\operatorname{sign} 0 = 0$. For fixed $\theta \in [0, 1]$, set

$$u_z = \sum_{k=1}^n |u_k|^{p_\theta/p_z} \cdot \operatorname{sign} u_k \cdot \mathbb{1}_{U_k}, \quad w_z = \sum_{k=1}^n |w_k|^{q'_\theta/q'_z} \cdot \operatorname{sign} w_k \cdot \mathbb{1}_{W_k}.$$

By direct substitution in the definition of the L_p-norm using the fact that $|a^z| = a^{\operatorname{Re} z}$ for $a > 0$, the conditions $\|u\|_{p_\theta} = \|w\|_{q'_\theta} = 1$ yield the equalities

$$\|u_{iy}\|_{p_0} = \|u_{1+iy}\|_{p_1} = \|w_{iy}\|_{q'_0} = \|w_{1+iy}\|_{q'_1} = 1. \tag{3}$$

Now define the function f of a complex variable by the formula

$$f(z) = \int_{\Omega_2} w_z \cdot T u_z \, d\mu_2. \tag{4}$$

Substituting in (4) the expressions for u_z and w_z, one can easily verify that $f(z)$ is a linear combination of functions of the form a^z with $a > 0$:

$$f(z) = \sum_{k=1}^n \sum_{j=1}^m |u_k|^{p_\theta/p_z} \cdot |w_j|^{q'_\theta/q'_z} \operatorname{sign} u_k w_j \int_{\Omega_2} T(\mathbb{1}_{U_k})_{W_j} d\mu_2.$$

Since each of these power functions is analytic and bounded in the strip $\Pi = \{z \in \mathbb{C} : 0 \leqslant \operatorname{Re} z \leqslant 1\}$, the same properties are enjoyed by the function $f(z)$. Further, the Hölder inequality and conditions (3) imply that

$$|f(iy)| = \left| \int_{\Omega_2} w_{iy} \cdot T u_{iy} \, d\mu_2 \right| \leqslant \|T u_{iy}\|_{q_0} \|w_{iy}\|_{q'_0} \leqslant M_0 \|u_{iy}\|_{p_0} \|w_{iy}\|_{q'_0} = M_0.$$

In much the same way, $|f(1 + iy)| \leqslant M_1$. Applying the Hadamard three-lines theorem we obtain, in particular, the estimate $|f(\theta)| \leqslant M_0^{1-\theta} M_1^{\theta}$. To obtain the required inequality (2) it remains to observe that $u_\theta = u$, $w_\theta = w$, and hence $f(\theta) = \int_{\Omega_2} w \cdot Tu \, d\mu_2$. □

Corollary 1 (Lyapunov's inequality). *Let* $f \in L_{p_0}(\Omega, \Sigma, \mu) \cap L_{p_1}(\Omega, \Sigma, \mu)$. *Then for any* $\theta \in [0, 1]$,

$$\|f\|_{p_\theta} \leqslant \left(\|f\|_{p_0}\right)^{1-\theta} \left(\|f\|_{p_1}\right)^{\theta}. \tag{5}$$

In particular, $L_{p_0}(\Omega, \Sigma, \mu) \cap L_{p_1}(\Omega, \Sigma, \mu) \subset L_{p_\theta}(\Omega, \Sigma, \mu)$.

Proof. A linear functional is also an operator, except it takes values in the field of scalars $\mathbb{C}$. Since $\mathbb{C}$ can be identified with $L_2(\Omega_2, \Sigma_2, \mu_2)$, where for Ω_2 one takes a single-point set and $\mu_2(\Omega_2) = 1$, the preceding lemma can be applied with $q_0 = q_1 = 2$ to the linear functional W_f of integration with the weight f, acting on finitely-valued functions. As $(\Omega_1, \Sigma_1, \mu_1)$ one can take (Ω, Σ, μ), and as the exponents to which one applies the lemma one needs to take not p_0, p_1, and p_θ themselves, but their conjugates. □

Corollary 2. *Let X be a dense subset of the space* $L_{p_0}(\Omega_1, \Sigma_1, \mu_1) \cap L_{p_1}(\Omega_1, \Sigma_1, \mu_1)$. *Then for any* $\theta \in [0, 1]$ *the set X is dense in* $L_{p_\theta}(\Omega_1, \Sigma_1, \mu_1)$.

Proof. From (5) it follows that the identity embedding operator $J: L_{p_0}(\Omega_1, \Sigma_1, \mu_1) \cap L_{p_1}(\Omega_1, \Sigma_1, \mu_1) \to L_{p_\theta}(\Omega_1, \Sigma_1, \mu_1)$, $Jf = f$, is continuous (for example, it takes sequences that converge to zero into sequences that converge to zero). $L_{p_0}(\Omega_1, \Sigma_1, \mu_1) \cap L_{p_1}(\Omega_1, \Sigma_1, \mu_1)$, which is the image of the operator J, is a dense subset of the space $L_{p_\theta}(\Omega_1, \Sigma_1, \mu_1)$, and hence (Exercise 8 of Subsection 1.2.1) the operator J maps the dense subset $X \subset L_{p_0}(\Omega_1, \Sigma_1, \mu_1) \cap L_{p_1}(\Omega_1, \Sigma_1, \mu_1)$ into the dense subset $X \subset L_{p_\theta}(\Omega_1, \Sigma_1, \mu_1)$. □

Theorem 1 (Riesz–Thorin theorem). *Let X be a dense linear subspace in* $L_{p_0}(\Omega_1, \Sigma_1, \mu_1) \cap L_{p_1}(\Omega_1, \Sigma_1, \mu_1)$ *and let* $T: X \to L_{q_0}(\Omega_2, \Sigma_2, \mu_2) \cap L_{q_1}(\Omega_2, \Sigma_2, \mu_2)$ *be a linear operator. Then for every* $\theta \in [0, 1]$ *the following estimate holds:*

$$\|T\|_{p_\theta, q_\theta} \leqslant \left(\|T\|_{p_0, q_0}\right)^{1-\theta} \left(\|T\|_{p_1, q_1}\right)^{\theta}.$$

Proof. We need to prove the inequality only when the quantities $\|T\|_{p_0, q_0}$ and $\|T\|_{p_1, q_1}$ are finite. In this case the operator $T: X \to L_{q_0}(\Omega_2, \Sigma_2, \mu_2) \cap L_{q_1}(\Omega_2, \Sigma_2, \mu_2)$ is continuous, and so it uniquely extends to a continuous operator

$$T: L_{p_0}(\Omega_1, \Sigma_1, \mu_1) \cap L_{p_1}(\Omega_1, \Sigma_1, \mu_1) \to L_{q_0}(\Omega_2, \Sigma_2, \mu_2) \cap L_{q_1}(\Omega_2, \Sigma_2, \mu_2).$$

That is, it suffices to consider the case when $X = L_{p_0}(\Omega_1, \Sigma_1, \mu_1) \cap L_{p_1}(\Omega_1, \Sigma_1, \mu_1)$. Further, if $p_\theta = +\infty$, then also $p_0 = p_1 = +\infty$, and the required estimate follows from the Lyapunov inequality:

$$\|Tf\|_{q_\theta} \leqslant \left(\|Tf\|_{q_0}\right)^{1-\theta} \left(\|Tf\|_{q_1}\right)^{\theta} \leqslant \left(\|T\|_{\infty, q_0}\right)^{1-\theta} \left(\|T\|_{\infty, q_1}\right)^{\theta} \|f\|_{\infty}.$$

Thus, we can assume that $p_\theta \neq +\infty$. Denote by $\widetilde{T}$ the restriction of the operator T to the subspace E from Lemma 1. By Corollary 1,

$$T(E) \subset L_{q_0}(\Omega_2, \Sigma_2, \mu_2) \cap L_{q_1}(\Omega_2, \Sigma_2, \mu_2) \subset \bigcap_{\theta \in [0,1]} L_{q_\theta}(\Omega_2, \Sigma_2, \mu_2),$$

so Lemma 1 applies to the operator $\widetilde{T}$:

$$\|\widetilde{T}\|_{p_\theta, q_\theta} \leqslant \left(\|\widetilde{T}\|_{p_0, q_0}\right)^{1-\theta} \left(\|\widetilde{T}\|_{p_1, q_1}\right)^\theta \leqslant \left(\|T\|_{p_0, q_0}\right)^{1-\theta} \left(\|T\|_{p_1, q_1}\right)^\theta.$$

It remains to use the fact that the subspace E is dense in $L_{p_\theta}(\Omega_1, \Sigma_1, \mu_1)$, in view of which $\|\widetilde{T}\|_{p_\theta, q_\theta} = \|T\|_{p_\theta, q_\theta}$. $\qquad\square$

The assertion just proved can be interpreted as follows. Let $D \subset [0, 1] \times [0, 1]$ be the set of all pairs (x, y) for which $X \subset L_{1/x}(\Omega_1, \Sigma_1, \mu_1)$, $T(X) \subset L_{1/y}(\Omega_2, \Sigma_2, \mu_2)$, and $\|T\|_{1/x, 1/y} < \infty$. Define the function $F: D \to \mathbb{C}$ by the formula $F(x, y) = \ln \|T\|_{1/x, 1/y}$. Then D is a convex set, and F is a convex function.

Exercises

1. In Lemma 1 and below we work with parameters p, q that are allowed to take finite, as well as infinite values. Provide a meaning for the relations $\frac{1}{p_\theta} = \frac{1-\theta}{p_0} + \frac{\theta}{p_1}$ and $\frac{1}{q_\theta} = \frac{1-\theta}{q_0} + \frac{\theta}{q_1}$ in the case where one of the parameters takes the value $+\infty$.

2. Using the "multiplication convention" $\infty \cdot 0 = 1$, provide a meaning for the expressions u_z and w_z in the case when one of the parameters figuring in their definition takes the value $+\infty$. Verify the correctness of the proof of Lemma 1 for such values of the parameters.

3. Why in the proof of Theorem 1 must the case $p_\theta = +\infty$ be treated separately?

4. Justify the last phrase in the proof of Theorem 1: "It remains to use the fact that the subspace...".

14.3.3 Applications to Fourier Series and the Fourier Transform

In this subsection we will prove, in particular, that for $1 < p < \infty$ the trigonometric system $\{1, e^{it}, e^{-it}, e^{2it}, e^{-2it}, \ldots\}$ constitutes a Schauder basis in $L_p[0, 2\pi]$ (Theorem 2). The proof relies on a very elegant argument proposed by V. Yudin [74].

Definition 1. Define

$$E[0, 2\pi] = \text{Lin}\{1, e^{it}, e^{-it}, e^{2it}, e^{-2it}, \ldots\},$$

$$E_+[0, 2\pi] = \mathrm{Lin}\{1, e^{it}, e^{2it}, e^{3it}, \ldots\},$$
$$E_-[0, 2\pi] = \mathrm{Lin}\{e^{-it}, e^{-2it}, e^{-3it}, \ldots\}.$$

The *Riesz operator* is the mapping $R\colon E[0, 2\pi] \to E[0, 2\pi]$ that acts by the rule

$$R\left(\sum_{k=-m}^{n} a_k e^{ikt}\right) = \sum_{k=0}^{n} a_k e^{ikt}.$$

In other words, the Riesz operator is the projector of the space $E[0, 2\pi]$ onto $E_+[0, 2\pi]$ parallel to the subspace $E_-[0, 2\pi]$.

Lemma 1 (Yudin). *Let $p = 2k$, where $k \in \mathbb{N}$. Let $f, g \in L_p(\Omega, \Sigma, \mu)$ be such that $\int_\Omega f^k \bar{g}^k d\mu = 0$. Then the L_p-norms of the functions f and g satisfy the inequality*

$$\|f\|^p + \|g\|^p \leqslant k^p \|f - g\|^p.$$

Proof. From the assumption that $\int_\Omega f^k \bar{g}^k d\mu = 0$ it follows that

$$\|f\|^p + \|g\|^p = \int_\Omega (|f|^{2k} + |g|^{2k}) d\mu = \int_\Omega |f^k - g^k|^2 d\mu$$

$$= \int_\Omega |f - g|^2 |f^{k-1} + f^{k-2}g + \cdots + g^{k-1}|^2 d\mu$$

$$\leqslant k^2 \int_\Omega |f - g|^2 \max\{|f|^{k-1}, |g|^{k-1}\}^2 d\mu.$$

Applying the Hölder inequality with the exponents k and $k' = \frac{k}{k-1}$, we obtain

$$\|f\|^p + \|g\|^p \leqslant k^2 \left(\int_\Omega |f - g|^{2k} d\mu\right)^{1/k} \left(\int_\Omega \max\{|f|^{2k}|g|^{2k}\} d\mu\right)^{(k-1)/k}$$

$$\leqslant k^2 \left(\int_\Omega |f - g|^{2k} d\mu\right)^{1/k} \left(\int_\Omega (|f|^{2k} + |g|^{2k}) d\mu\right)^{(k-1)/k}$$

$$= k^2 \left(\int_\Omega |f - g|^{2k} d\mu\right)^{1/k} \left(\|f\|^p + \|g\|^p\right)^{(k-1)/k}.$$

It remains to divide both sides of the last inequality by $(\|f\|^p + \|g\|^p)^{(k-1)/k}$ and raise the result to the power k.   □

Lemma 2. *Let $f \in E_+[0, 2\pi]$, $g \in E_-[0, 2\pi]$, and $k \in \mathbb{N}$. Then $\int_0^{2\pi} f^k(t) \bar{g}^k(t) dt = 0$.*

Proof. Since the system of exponentials is orthogonal, the functions $E_+[0, 2\pi]$ and $E_-[0, 2\pi]$ are orthogonal to one another in $L_2[0, 2\pi]$. At the same time, $f^k \in E_+[0, 2\pi]$, while $g^k \in E_-[0, 2\pi]$. $\square$

Theorem 1 (M. Riesz). *Let* $1 < p < \infty$. *Then the Riesz operator R extends to a continuous operator acting from* $L_p[0, 2\pi]$ *into* $L_p[0, 2\pi]$.

Proof. Since $E[0, 2\pi]$ is a subspace dense in $L_p[0, 2\pi]$, for the existence of the claimed extension it is necessary and sufficient that $\|R\|_{p,p} < \infty$.

We consider first the case $p = 2k$, where $k \in \mathbb{N}$. Let $h \in E[0, 2\pi]$ be an arbitrary element. Consider the functions $f = Rh \in E_+[0, 2\pi]$ and $g = Rh - h \in E_-[0, 2\pi]$. Then, by Lemma 1,

$$\|Rh\|^p = \|f\|^p \leqslant k^p \|f - g\|^p = k^p \|h\|^p,$$

i.e., $\|R\|_{p,p} \leqslant k < \infty$.

Applying the Riesz–Thorin interpolation theorem to intervals of the form $[2k, 2k + 2]$, we deduce that $\|R\|_{p,p} < \infty$ for all $p \geqslant 2$. It remains to examine the case $1 < p < 2$. In this case $p' \in (2, +\infty)$ and, what we already proved, $\|R\|_{p',p'} < \infty$. We remark that for any $h \in E[0, 2\pi]$ we have

$$\|Rh\|_p = \sup \left\{ \left| \int\limits_0^{2\pi} (Rh)(t)\overline{u}(t)dt \right| : \overline{u} \in E[0, 2\pi], \|\overline{u}\|_{p'} \leqslant 1 \right\}. \qquad (6)$$

Defining $f = Rh, g = Rh - h, v = Ru, w = Ru - v$, and taking into account that $f, v \in E_+[0, 2\pi]$, $g, w \in E_-[0, 2\pi]$, we obtain:

$$\int\limits_0^{2\pi} (Rh)(t)\overline{u}(t)dt = \int\limits_0^{2\pi} f(t)(\overline{v}(t) - \overline{w}(t))dt$$

$$= \int\limits_0^{2\pi} (f(t) - g(t))\overline{v}(t)dt = \int\limits_0^{2\pi} h(t)\overline{(Ru)(t)}dt.$$

Substituting into (6) and using the Hölder inequality, we conclude that $\|Rh\|_p \leqslant \|h\|_p \|R\|_{p',p'}$, and so $\|R\|_{p,p} \leqslant \|R\|_{p',p'} < \infty$. $\square$

For $p \in [1, \infty)$ the closure of $E_+[0, 2\pi]$ in $L_p[0, 2\pi]$ is the space $\mathcal{H}_p$, well-known in complex analysis. The closure of $E_+[0, 2\pi]$ in $L_\infty[0, 2\pi]$ is the space $A(\mathbb{T})$, already mentioned in Subsection 10.4.3. M. Riesz's theorem means that $\mathcal{H}_p$ is complemented in $L_p[0, 2\pi]$ for $1 < p < \infty$, with the extension of the Riesz operator R being the corresponding bounded projector. It was shown in the exercises of Subsection 10.4.3 that $A(\mathbb{T})$ is not complemented in $C(\mathbb{T})$ (and consequently is not complemented in $L_\infty[0, 2\pi]$). By a duality argument one can show that $\mathcal{H}_1$ is not complemented in $L_1[0, 2\pi]$.

The next result fulfills our promise made in Subsection 10.5.1: it implies that the trigonometric system with its natural ordering $\{1, e^{it}, e^{-it}, e^{2it}, e^{-2it}, \ldots\}$ constitutes a basis in $L_p[0, 2\pi]$ for $1 < p < \infty$. Incidentally, it is amazing that the property of being a basis depends on the ordering, and except in the case of $p = 2$, there are orderings in which the trigonometric system is no longer a basis in $L_p[0, 2\pi]$.

Theorem 2. *Let* $1 < p < \infty$. *Then for any function* $f \in L_p[0, 2\pi]$ *the partial sums of its Fourier series* $S_n f$ *converge to* f *in the* L_p-*metric.*

Proof. Denote by U_n the multiplication operator by e^{int}: $(U_n f)(t) = f(t)e^{int}$. The operators $U_n \in L(L_p[0, 2\pi])$ are bijective isometries. The operator R is related to the partial sum of Fourier series operator S_n by the identity

$$S_n = U_{n+1}(I - R)U_{-(2n+1)}RU_n$$

(which is readily verified for the exponentials e^{imt}, and then extended by linearity and continuity to the whole space $L_p[0, 2\pi]$). Consequently, $\|S_n\| \leqslant \|I - R\| \cdot \|R\|$, that is, the sequence (S_n) of operators is uniformly bounded. Moreover, (S_n) converges pointwise to the identity operator I on the dense subspace $E[0, 2\pi]$ of $L_p[0, 2\pi]$, which for a bounded sequence of operators means pointwise convergence on the whole space. $\qquad\square$

The reader will find further applications in the exercises below.

Exercises

1. Let $1 \leqslant p \leqslant 2$. Then the Fourier transform F extends from $L_1(-\infty, \infty) \cap L_2(-\infty, \infty)$ to a continuous operator acting from $L_p(-\infty, \infty)$ to $L_{p'}(-\infty, \infty)$.

2. Let $1 < p \leqslant 2$ and $f \in L_p[0, 2\pi]$. Then $\sum_{n=-\infty}^{\infty} |\widehat{f_n}|^{p'} < \infty$.

3. Let $1 \leqslant p \leqslant 2$, $a_n \in \mathbb{C}$, and $\sum_{n=-\infty}^{\infty} |a_n|^p < \infty$. Then the series $\sum_{n=-\infty}^{\infty} a_n e^{int}$ converges in the metric of $L_{p'}[0, 2\pi]$.

4. Consider the Riesz operator R and partial sums of its Fourier series S_n as operators acting from $L_p[0, 2\pi]$ into $L_p[0, 2\pi]$. Calculate their adjoints.

5. In Exercises 11–14 of Subsection 10.4.3 it was proved that for the Riesz operator (there this operator appeared under the pseudonym $\widetilde{P}$) it holds that $\|R\|_{\infty,\infty} = \infty$. Based on the duality between L_∞ and L_1, prove that $\|R\|_{1,1} = \infty$.

6. Analogously to Exercises 6–14 of Subsection 10.4.3, show that the subspace $\mathcal{H}_1$ is not complemented in $L_1[0, 2\pi]$.

7. Describe the pairs (p, q) for which $\|R\|_{p,q} < \infty$.

Comments on the Exercises

Subsection 14.1.2

Exercise 9. Follow the proof of Lemma 1 in Subsection 8.3.3.

Exercise 10. Use the preceding exercise and Theorem 3 of Subsection 3.1.4.

Exercise 12. By the definition of the Lebesgue measure through the outer measure, every measurable set A on the interval can be approximated by an open set B such that $A \subset B$ and $\lambda(B \setminus A) < \varepsilon$. In turn, every open set can be approximated from inside by a finite union of intervals. Moreover, the characteristic functions will approximate the corresponding characteristic functions in the L_p-metric. Passing to linear combinations, we conclude that any finitely-valued function can be approximated in L_p by piecewise-constant functions.

Exercise 14. Consider in $L_\infty[0, 1]$ the set of functions of the form $\mathbb{1}_{[c,d]}$, where $[c, d] \subset [0, 1]$. This is an uncountable set, in which the distance between any two elements is equal to 1. Such "beasts" cannot live in a separable space.

Subsection 14.1.4

Exercise 6. One proceeds by analogy with the theorem on convolution in L_1 (Subsection 4.6.3).

Exercise 7. Consider the space E of kernels K that obey the inequality $\|\widetilde{K}\|_\infty < \infty$, endowed with the norm $\|K\| = \|\widetilde{K}\|_\infty$. Each element $K \in E$ can be identified with the corresponding function $s_K : [0, 1] \to L_1[0, 1]$ given by the rule $[s_K(t)](x) = K(t, x)$. Prove that the function s_K is Borel measurable. From this, by analogy with the theorem on the approximation of a measurable function by simple functions, deduce that kernels of the form $K(t, x) = \sum_{k=1}^{\infty} \mathbb{1}_{A_k}(t) g_k(x)$, where A_k are pairwise disjoint sets and (g_k) is a bounded sequence of elements of the space $L_1[0, 1]$, form a dense set in E. Next, prove the claimed formula $\|T\| = \|\widetilde{K}\|_\infty$ for kernels of the form $K(t, x) = \sum_{k=1}^{\infty} \mathbb{1}_{A_k}(t) g_k(x)$. Then the general result is obtained by passing to the limit.

Subsection 14.2.5

Exercise 1. Use Exercise 6 in Subsection 14.2.4.

Exercise 2. First note that the functions $x_n(t)$ are infinitely differentiable and rapidly decaying at infinity. It follows that these functions and their linear combinations lie in $L_1(-\infty, \infty) \cap L_2(-\infty, \infty)$, and for the functions we are interested in, $F_2 = F$. Moreover, to these functions one can apply the formula $F(f') = it \cdot F(f)$ (Theorem 1 of Subsection 14.2.4).

The proof is carried out by induction on n. For $n = 0$ the assertion follows from the equality $F(x_0) = \sqrt{2\pi}\, x_0$ (Example 3 is Subsection 14.2.2). Now suppose that $F(X_n) \subset X_n$, and let us prove that $F(X_{n+1}) \subset X_{n+1}$. Calculating the derivative $x'_n = n x_{n-1} - x_{n+1}$ and using the Fourier transform of derivative formula, we have

$$F(x_{n+1}) = F(n x_{n-1} - x'_n) = n F(x_{n-1}) - it F(x_n) \subset X_{n-1} + t X_n \subset X_{n+1}.$$

Since by the induction hypothesis the functions $F(x_k)$, $k = 0, 1, \ldots, n$, also lie in X_{n+1}, this establishes the desired inclusion.

Exercise 8. Consider in ℓ_2 some non-closed linear subspace E (for instance, take $E = \text{Lin}\{e_n\}_1^\infty$, where $\{e_n\}_1^\infty$ is the standard basis in ℓ_2). Introduce the natural embedding operators $U_i \colon \ell_2 \to \ell_2 \times \ell_2$, acting as $U_1 x = (x, 0)$ and $U_2 x = (0, x)$. Now in the linear space $\ell_2 \times \ell_2$ consider the subspace $F = \{(x, -x) : x \in E\}$ and the quotient mapping $j \colon \ell_2 \times \ell_2 \to (\ell_2 \times \ell_2)/F$. The required subspaces X, Y will be the following subspaces of the linear space $(\ell_2 \times \ell_2)/F$: $X = j U_1(\ell_2)$ and $Y = j U_2(\ell_2)$, endowed with the norms inherited from ℓ_2: $\|j U_1(s)\|_X = \|j U_2(s)\|_Y = \|s\|_{\ell_2}$. The norms are well defined and the spaces X and Y are subspaces isomorphic to ℓ_2: the isomorphism are effected by the operators $j U_1$ and $j U_2$. Moreover, on E the operators $j U_1$ and $j U_2$ coincide and map E into $X \cap Y$. Therefore, $X \cap Y$ will be isomorphic to the normed space E and will not be a Banach space. This construction can be described as follows: take two copies of the space ℓ_2 and "glue" them along the non-complete subspace E.

Exercise 9. Such a condition reads a follows: X and Y are linear subspaces of some linear space G; and on G there is defined a convergence $\xrightarrow{G}$, with the following properties:

— if $g_n, g \in X$ and $\|g_n - g\|_X \to 0$ as $n \to \infty$, then for some subsequence of indices $g_{n_k} \xrightarrow{G} g$;

— if $g_n, g \in Y$ and $\|g_n - g\|_Y \to 0$ $(n \to \infty)$, then for some subsequence of indices $g_{n_j} \xrightarrow{G} g$;

— if $g_n, g, h \in G$, $g_n \xrightarrow{G} g$, and $g_n \xrightarrow{G} h$ as $n \to \infty$, then $g = h$.

In Exercise 7 the role of such a space G can be played by the space $L_0(\Omega, \Sigma, \mu)$ of (equivalence classes of) measurable functions, equipped with almost everywhere convergence.

Subsection 14.3.2

Exercise 4. Use Lemma 1 of Subsection 14.2.5.

Chapter 15
Fixed Point Theorems and Applications

Suppose that on the set X there is given a mapping $f : X \to X$. An element $x \in X$ is called a *fixed point* of the mapping f if $f(x) = x$. Many problems, looking rather dissimilar at a first glance, from various domains of mathematics, can be reduced to the search for fixed points of appropriate mappings. For this reason each of the theorems on existence of fixed points discussed in the present chapter has numerous and often very elegant applications.

15.1 Some Classical Theorems

15.1.1 Contractive Mappings

Let X be a metric space. A mapping $f : X \to X$ is called *contractive* if there exists a constant $C \in [0, 1)$ such that for any $x_1, x_2 \in X$,

$$\rho(f(x_1), f(x_2)) \leqslant C\rho(x_1, x_2). \tag{1}$$

Theorem 1 (S. Banach). *Let X be a complete metric space and $f : X \to X$ a contractive mapping. Then f has a unique fixed point $x_0 \in X$. Moreover, for any point $y_0 \in X$, the sequence (y_n) of its iterates, defined by the recursion formula $y_n = f(y_{n-1})$, converges to x_0.*

Proof. First, the uniqueness. Let $x_0, x_1 \in X$ be fixed points of the mapping f. Then

$$\rho(x_0, x_1) = \rho(f(x_0), f(x_1)) \leqslant C\rho(x_0, x_1),$$

where $C < 1$ is the constant figuring in the definition of contractivity. But the inequality $\rho(x_0, x_1) \leqslant C\rho(x_0, x_1)$ can hold only if $\rho(x_0, x_1) = 0$.

© Springer International Publishing AG, part of Springer Nature 2018
V. Kadets, *A Course in Functional Analysis and Measure Theory*,
Universitext, https://doi.org/10.1007/978-3-319-92004-7_15

Now let us examine the properties of the sequence y_n. Let $d = \rho(y_0, y_1)$. Taking successively $n = 1, 2, \ldots$ in the estimate $\rho(y_n, y_{n+1}) = \rho\left(f(y_{n-1}), f(y_n)\right) \leqslant C\rho(y_{n-1}, y_n)$, we conclude that $\rho(y_1, y_2) \leqslant Cd, \rho(y_2, y_3) \leqslant C^2 d, \ldots, \rho(y_n, y_{n+1}) \leqslant C^n d$. It follows that for any $n < m$ we have

$$\rho(y_n, y_m) \leqslant \rho(y_n, y_{n+1}) + \rho(y_{n+1}, y_{n+2}) + \cdots + \rho(y_{m-1}, y_m)$$
$$\leqslant \left(C^n + C^{n+1} + \cdots\right) d = \frac{C^n d}{1 - C} \to 0 \quad \text{as } n, m \to \infty.$$

Since the space X is complete, this means that the sequence (y_n) converges. Denote its limit by x_0. It remains to prove that x_0 is a fixed point of f. Indeed,

$$\rho(x_0, f(x_0)) \leqslant \rho(x_0, f(y_n)) + \rho(f(y_n), f(x_0))$$
$$\leqslant \rho(x_0, y_{n+1}) + C\rho(y_n, x_0) \to 0 \quad \text{as } n \to \infty,$$

that is, $\rho(x_0, f(x_0)) = 0$, and $x_0 = f(x_0)$.   □

Exercises

1. Every contractive mapping is continuous.

2. Let X be a normed space. For a linear operator $T : X \to X$ to be a contractive mapping it is necessary and sufficient that $\|T\| < 1$. What is the fixed point in this case?

3. Banach's theorem not only establishes the existence of the fixed point, but also provides a way to compute it approximately. Prove the following estimate of the rate of convergence of the approximations y_n to the fixed point x_0: $\rho(y_n, x_0) \leqslant C^n d/(1 - C)$, where C is the constant figuring in (1), and $d = \rho(y_0, y_1)$. Provide an example of a contractive mapping on the real line for which this estimate is sharp (i.e., cannot be improved).

4. Give an example of a mapping $f : \mathbb{R} \to \mathbb{R}$ that has no fixed points and obeys the following weaker variant of condition (1): for every $x_1, x_2 \in X$ such that $x_1 \neq x_2$, one has $\rho(f(x_1), f(x_2)) < \rho(x_1, x_2)$.

5. The contractivity property of a map in a normed space can be invalidated by changing the original norm to an equivalent one. Provide an example.

6. Describe the mappings $f : \mathbb{R}^2 \to \mathbb{R}^2$ that are contractive in *all* the norms on $\mathbb{R}^2$.

7. Let X be a complete metric space, K a compact space, and $F : K \times X \to X$ a continuous function that is uniformly contractive in the second variable: there exists a constant $C \in [0, 1)$ such that for any $x_1, x_2 \in X$ and any $t \in K$, it holds that $\rho(F(t, x_1), F(t, x_2)) \leqslant C\rho(x_1, x_2)$. Show that for every $t \in K$ there exists a unique solution x of the equation $F(t, x) = x$; moreover, this solution $x(t)$ depends continuously on t.

8. Prove the following *implicit function theorem*: Suppose that the function $\Phi(t, x)$ is defined and continuous in the strip $\Pi = \{(t, x) : t \in [a, b], x \in \mathbb{R}\}$, and also that Φ is continuously differentiable in the second variable and the derivative Φ'_x obeys in the whole strip the inequality $m \leqslant \Phi'_x \leqslant M$, where $m, M \in (0, +\infty)$. Then for every $t \in [a, b]$ there exists a unique solution of the equation $\Phi(t, x) = 0$; moreover, this solution depends continuously on t.

9. Recall that the invertibility of an operator $U \in L(X)$ can be interpreted as the existence and uniqueness of a solution to the equation $Ux = b$ for every right-hand side $b \in X$. Deduce the theorem on small perturbations of the unit operator (asserting that if $T \in L(X)$ and $\|T\| < 1$, then the operator $I - T$ is invertible) from the fixed point theorem.

15.1.2   The Fixed Point Property. Brouwer's Theorem

Definition 1. A topological space X is said to have the *fixed point property* if every continuous mapping $f : X \to X$ has at least one fixed point.

Example 1. The interval $[0, 1]$ has the fixed point property.

Indeed, let $f : [0, 1] \to [0, 1]$ be a continuous function. Consider the two sets $A = \{t \in [0, 1] : f(t) \leqslant t\}$ and $B = \{t \in [0, 1] : f(t) \geqslant t\}$. These sets are closed and $A \cup B = [0, 1]$. Hence, since the interval is connected, A and B must intersect. Any point of the set $A \cap B$ provides the required fixed point.

Example 2. A circle in the plane does not have the fixed point property. As a mapping with no fixed points one can take, say, a central symmetry of the circle.

An important class of examples is provided by the following theorem of Brouwer.

Theorem 1. *Every convex compact set in a finite-dimensional normed space has the fixed point property.*

A relatively elementary proof of this theorem in terms of combinatorial properties of simplices, which does not resort to complicated topological properties, can be found in the textbook by Kuratowski [25, V. 1, §28.1][1]

Let us give an example of application of Brouwer's theorem.

---

[1] The proof mentioned here was proposed by B. Knaster, S. Mazurkiewicz, and K. Kuratowski, so that the treatment in [25] belongs to one of these authors. Generally, we find it very useful to refer beginners to textbooks written not simply by pedagogues who are familiar with and know well how to treat the material, but by people who contributed in an essential manner to the creation and development of the corresponding fields of mathematics. In such texts the reader has the opportunity to make acquaintance not only with the results, but also — even more importantly — with the ways people who demonstrated the fruitfulness of their approach to mathematical research think.

Theorem 2. *Let $A: \mathbb{R}^n \to \mathbb{R}^n$ be a linear operator given by a matrix whose elements $a_{i,j}$ are all positive. Then A has an eigenvector with non-negative components.*

Proof. Let $\mathbb{R}^n_+ = \{x = (x_1, \ldots, x_n) : x_k \geqslant 0, \ k = 1, \ldots, n\}$. If $\mathbb{R}^n_+$ contains a nonzero vector x such that $Ax = 0$, then x will already provide the required eigenvector (with eigenvalue 0). Hence, one can assume that $Ax \neq 0$ for all $x \in \mathbb{R}^n_+ \setminus \{0\}$. By assumption, $A(\mathbb{R}^n_+) \subset \mathbb{R}^n_+$. Consider the sum of coordinates functional $s(x) = \sum_{k=1}^n x_k$ and the compact set $K = \{x \in \mathbb{R}^n_+ : s(x) = 1\}$. By Brouwer's theorem, the mapping $f: K \to K$ given by the formula $f(x) = Ax/s(Ax)$ must have a fixed point $x_0 \in K$. For this point, $Ax_0/s(Ax_0) = x_0$, i.e., x_0 is an eigenvector with eigenvalue $s(Ax_0)$. $\square$

Exercises

1. If the topological space X is not connected (i.e., it can be decomposed into the union of two disjoint closed sets), then X does not have the fixed point property.

2. If X is homeomorphic to a space with the fixed point property, then X itself has this property.

In the theory of topological spaces, an analogue of the notion of a complemented subspace is the notion of a retract. A subspace Y of a topological space X is called a *retract* if there exists a continuous mapping $P: X \to Y$ (called a *retraction*) such that $Py = y$ for all $y \in Y$.

3. Any retract of a topological space with the fixed point property itself has the fixed point property.

4. Based on the preceding exercise, give an example of a compact set in $\mathbb{R}^2$ that has the fixed point property, but is not homeomorphic with a convex compact set.

5. The unit sphere in a finite-dimensional normed space is not a retract of the closed unit ball.

6. Construct a retraction of the closed unit ball of the space ℓ_2 to the unit sphere of ℓ_2.

7. The closed unit ball in the space ℓ_2 does not have the fixed point property.

15.1.3 Partitions of Unity and Approximation of Continuous Mappings by Finite-Dimensional Mappings

Definition 1. Let K be a nonempty subset of the metric space X and U_j, $j = 1, 2, \ldots, n$, be open sets such that $\bigcup_{j=1}^n U_j \supset K$. A family of continuous functions $\varphi_j: K \to \mathbb{R}$, $j = 1, 2, \ldots, n$, is called a *partition of unity on K subordinate to the cover* $\{U_j\}_1^n$ if $\sum_{j=1}^n \varphi_j \equiv 1$, $\varphi_j \geqslant 0$, and $\operatorname{supp} \varphi_j \subset U_j$ for all j.

Theorem 1. *Under the conditions listed above, there exists a partition of unity on K subordinate to the cover $\{U_j\}_1^n$.*

Proof. The sets $U_j^c = X \setminus U_j$ are closed. If at least one of the U_j^c's is empty, then $U_j = X \supset K$ and the problem is trivially solved: for this index j, put $\varphi_j \equiv 1$, and for $k \neq j$ put $\varphi_k \equiv 0$. Hence, we can assume that all sets U_j^c are non-empty. Consider on K the functions $g_j(x) = \rho(x, U_j^c)$. These functions are nonnegative, continuous (Proposition 1 of Subsection 1.3.2), and have the following property: $x \in U_j$ if and only if $g_j(x) \neq 0$. Since each point $x \in K$ lies in at least one of the sets U_j, the function $g = \sum_{j=1}^n g_j$ is different from zero everywhere on K. Now as the sought-for functions φ_j one can take $\varphi_j = g_j/g$. These functions are continuous (the denominators do not vanish), non-negative, and satisfy $\operatorname{supp} \varphi_j = \operatorname{supp} g_j = U_j$ and $\sum_{j=1}^n \varphi_j = (1/g) \sum_{j=1}^n g_j = 1$. $\qquad\square$

The next definition generalizes to the nonlinear case the notion of a finite-rank operator.

Definition 2. A mapping g of the set X into a linear space Y is said to be *finite-dimensional*, if $\dim \operatorname{Lin}(g(X)) < \infty$.

Theorem 2. *Let K be a precompact set in the normed space Y. Then for any $\varepsilon > 0$ there exists a continuous finite-dimensional mapping $g_\varepsilon \colon K \to Y$ with $g_\varepsilon(K) \subset \operatorname{conv} K$ such that g approximates on K the unit operator to within ε: $\sup_{x \in K} \|g_\varepsilon(x) - x\| \leqslant \varepsilon$.*

Proof. Pick a finite ε-net $y_1, y_2, \ldots, y_n$ in K. Then the open balls $U_k = B(y_k, \varepsilon)$ form a cover of the set K. Let $\varphi_j \colon K \to \mathbb{R}$, $j = 1, 2, \ldots, n$, be a partition of unity on K subordinate to the cover $\{U_j\}_1^n$. Put $g_\varepsilon(x) = \sum_{j=1}^n \varphi_j(x) y_j$. The continuity of the mapping $g_\varepsilon(x)$ so defined follows from the continuity of all φ_j. Further, $g_\varepsilon(K) \subset \operatorname{conv} K$, since the sums $\sum_{j=1}^n \varphi_j(x) y_j$ are convex combinations of points $y_j \in K$. It remains to verify that $\sup_{x \in K} \|g_\varepsilon(x) - x\| \leqslant \varepsilon$.

Now let $x \in K$ be an arbitrary element. Denote by N the set of indices $1 \leqslant j \leqslant n$ for which $\varphi_j(x) \neq 0$. By the definition of a partition of unity, for $j \in N$ one has the inclusion $x \in B(y_k, \varepsilon)$, i.e., $\|x - y_j\| < \varepsilon$. Therefore,

$$\|g_\varepsilon(x) - x\| = \left\| \sum_{j=1}^n \varphi_j(x) y_j - \sum_{j=1}^n \varphi_j(x) \cdot x \right\| = \left\| \sum_{j=1}^n \varphi_j(x)(y_j - x) \right\|$$

$$\leqslant \sum_{j \in N} \varphi_j(x) \|y_j - x\| < \varepsilon. \qquad\square$$

Theorem 3. *Let K be a convex compact set in a normed space Y. Then every continuous mapping $f \colon K \to K$ can be arbitrarily well uniformly approximated by finite-dimensional continuous mappings of the set K into itself.*

Proof. Let g_ε be the function provided by Theorem 2. In view of the convexity, $g_\varepsilon(K) \subset K$. It is readily seen that the composition $g_\varepsilon \circ f \colon K \to K$ is a finite-dimensional mapping that approximates f to within ε: $\sup_{x \in K} \|g_\varepsilon(f(x)) - f(x)\| \leqslant \sup_{y \in K} \|g_\varepsilon(y) - y\| \leqslant \varepsilon$. $\qquad\square$

Exercises

Although Theorem 2 talks about the approximation of the unit (hence a linear) operator, the mapping g_ε cannot be always taken to be a linear operator (more precisely, the restriction of a linear operator to K).

1. If under the assumptions of Theorem 2 the space Y has the pointwise approximation property (Subsection 11.2.2), then for the mapping g_ε one can take the restriction to K of a linear operator.

2. If for any precompact set $K \subset Y$ the function g_ε figuring in Theorem 2 can be taken to be a continuous linear operator of finite rank, then for such a Y the following analogue of Theorem 2 of Subsection 11.3.2 holds true: for any normed space X and any compact operator $T \in L(X, Y)$ there exists a sequence of finite-dimensional operators $T_n \in L(X, Y)$ which converges in norm to T.

15.1.4 The Schauder's Principle

In this subsection Brouwer's fixed point theorem will be extended from the finite-dimensional case to the infinite-dimensional one.

Definition 1. Let X be a metric space. An element $x \in X$ is called an *ε-fixed point* of the mapping $f \colon X \to X$ if $\rho(f(x), x) < \varepsilon$.

Lemma 1. *Let X be a compact metric space. Then for a continuous mapping $f \colon X \to X$ to have a fixed point it suffices that for every $\varepsilon > 0$ the mapping f has an ε-fixed point.*

Proof. Using the existence of an ε-fixed point for $\varepsilon = \frac{1}{n}$, $n \in \mathbb{N}$, we obtain a sequence $x_n \in X$ such that $\rho(f(x_n), x_n) \to 0$ as $n \to \infty$. With no loss of generality we may assume that the sequence (x_n) converges (otherwise, we replace (x_n) by a subsequence). Denote $\lim_{n \to \infty} x_n$ by x. Then $f(x) = \lim_{n \to \infty} f(x_n)$ and $\rho(f(x), x) = \lim_{n \to \infty} \rho(f(x_n), x_n) = 0$. That is, $f(x) = x$, and x is the required fixed point. $\qquad\square$

Lemma 2. *Let Y be a normed space, $K \subset Y$ a convex compact set, and $f \colon K \to K$ a continuous finite-dimensional mapping. Then f has a fixed point.*

Proof. Let $X = \mathrm{Lin}(f(K))$, and $\widetilde{K} = X \cap K$. Then X is a finite-dimensional normed space, $\widetilde{K} \subset X$ is a convex compact set, and $f(\widetilde{K}) \subset f(K) \subset X \cap K = \widetilde{K}$. By Brouwer's theorem, the mapping f has a fixed point in $\widetilde{K}$. Of course, this fixed point also lies in K. $\qquad\qquad\square$

Theorem 1 (Schauder's principle). *Every convex compact set in a normed space has the fixed point property.*

Proof. Let Y be a normed space, $K \subset Y$ a convex compact subset, and $f: K \to K$ a continuous mapping. By Lemma 1, it suffices to show that for any $\varepsilon > 0$, f has an ε-fixed point. Applying Theorem 3 of Subsection 15.1.3, we find a finite-dimensional continuous mapping $f_\varepsilon: K \to K$ such that $\rho(f(x), f_\varepsilon(x)) < \varepsilon$ for all $x \in K$. By Lemma 2, f_ε has a fixed point x_ε. This fixed point x_ε is an ε-fixed point for the mapping f; indeed, $\rho(x_\varepsilon, f(x_\varepsilon)) = \rho(f_\varepsilon(x_\varepsilon), f(x_\varepsilon)) < \varepsilon$. $\qquad\square$

Let us give a reformulation of Schauder's principle that proves convenient in applications.

Theorem 2. *Let V be a convex, closed, and bounded subset of a Banach space, and $F: V \to V$ a continuous mapping such that $F(V)$ is precompact. Then F has a fixed point.*

Proof. Let K denote the closure of the convex hull of the set $F(V)$. By assumption, K is a convex compact set and $F(K) \subset F(V) \subset K$, hence F can be regarded as a mapping of the compact set K into itself. It remains to apply Theorem 1. $\qquad\square$

Let us mention here that Schauder's principle (discovered in 1927) is valid for convex compact sets not only in normed spaces, but also in locally convex topological vector spaces (Leray, Schauder, 1934). It was also extended to multi-valued mappings (Kakutani, 1941, see the textbook by L. Kantorovich and G. Akilov [22, Chap. 16, §5], where applications to mathematical economics are also given).

Exercises

A set V in a linear space X is called a *convex cone* if it is stable under addition of elements and multiplication by positive scalars. Let $F: X \to \mathbb{R}$ be a linear functional with the property that $F(v) > 0$ for all $v \in V \setminus \{0\}$. Then the set $V_F = \{v \in V: F(v) = 1\}$ is called the *base of the cone V*.

1. Convex cones and their bases are convex sets.

2. (Abstract version of Theorem 2 of Subsection 15.1.2). Suppose V is a convex cone in the Banach space X which has a compact base, $T \in L(X)$, and $T(V) \subset V$. Then the operator T has an eigenvector that lies in V.

3. Suppose V is a closed convex cone in the Banach space X, $T: X \to X$ is a compact operator, and $T(V) \subset V$. Further, suppose $F \in X^*$ is a functional such that $F(v) > 0$ for all $v \in V \setminus \{0\}$. Then if the base V_F is bounded and $\inf\{x^*(Tv) : v \in V_F\} > 0$, the operator T has an eigenvector that lies in V.

4. Does the cone of all non-negative functions from $L_1[0, 1]$ have a bounded closed base? Does this cone have a compact base?

5. Consider the same questions for the cone of all non-negative functions in $L_2[0, 1]$.

6. Consider the integral operator $T: L_1[0, 1] \to L_1[0, 1]$, $(Tf)(x) = \int_0^x f(t)dt$, which has no eigenfunctions (see Example 1 in Subsection 11.1.5). Doesn't this example contradict Exercise 3 if one takes as V the cone of all nonnegative functions and as F the integration functional over the interval, i.e., $F(f) = \int_0^1 f(t)dt$?

15.2   Applications to Differential Equations and Operator Theory

15.2.1   The Picard and Peano Theorems on the Existence of a Solution to the Cauchy Problem for Differential Equations

Recall that the *Cauchy problem* for the differential equation $y' = f(t, y)$ is the problem of finding a continuously differentiable function $y(t)$ which is defined in a neighborhood of the point t_0, and which satisfies the equation as well as the initial condition $y(t_0) = y_0$. In the case where the function $f(t, y)$ is continuous, the Cauchy problem is equivalent to the integral equation

$$y(t) = y_0 + \int_{t_0}^t f(s, y(s))ds. \tag{1}$$

Theorem 1 (Picard's theorem). *Assume the function $f: [t_0, T] \times [y_0 - \theta, y_0 + \theta] \to [-M, M]$ is measurable and satisfies the Lipschitz condition in the second variable with a constant $\gamma > 0$ that does not depend on the first variable. Then there exists a $\tau > 0$ such that on the interval $t \in [t_0, t_0 + \tau]$ equation (1) has a solution, and this solution is unique. Moreover, for τ one can take any number smaller than $\tau_0 = \min\{\theta/M, 1/\gamma, T - t_0\}$.*

Proof. Consider the Banach space $C[t_0, t_0 + \tau]$ and its subset U consisting of all functions y that satisfy on $[t_0, t_0 + \tau]$ the condition $|y(t) - y_0| \leqslant \theta$. Define the mapping $F: U \to C[t_0, t_0 + \tau]$ by the rule

$$(F(y))(t) = y_0 + \int_{t_0}^t f(s, y(s))ds.$$

Then the solutions of equation (1) are the fixed points of the mapping F.

Let us verify that F is a contractive mapping of the set U into itself. First, for any $y \in U$ we have that

$$|(F(y))(t) - y_0| = \left| \int_{t_0}^{t} f(s, y(s)) ds \right| \leqslant M\tau \leqslant \theta,$$

i.e., $F(y) \in U$. Consequently, $F(U) \subset U$. Further, for any $y_1, y_2 \in U$ we have

$$\|F(y_1) - F(y_2)\| = \max_{t \in [t_0, t_0+\tau]} \left| \int_{t_0}^{t} [f(s, y_1(s)) - f(s, y_2(s))] ds \right| \leqslant \gamma\tau \|y_1 - y_2\|,$$

and, by construction, $\gamma\tau < 1$. Hence, F is a contractive mapping. Finally, the set U is closed in $C[t_0, t_0 + \tau]$, and so U is a complete metric space with respect to the metric under consideration. It follows that Banach's contractive mapping theorem applies, which establishes the existence and uniqueness of a fixed point. $\qquad\square$

Theorem 2 (Peano's theorem). *Suppose the function* $f : [t_0, T] \times [y_0 - \theta, y_0 + \theta] \to [-M, M]$ *is measurable and is continuous in the second variable, uniformly in the first variable. In other words, for every* $\varepsilon > 0$ *there exists a* $\delta = \delta(\varepsilon) > 0$, *such that for any* $t \in [t_0, T]$ *and any points* $y_1, y_2 \in [y_0 - \theta, y_0 + \theta]$, *if* $|y_1 - y_2| \leqslant \delta$, *then* $|f(t, y_1) - f(t, y_2)| \leqslant \varepsilon$. *Then equation (1) has a solution on the segment* $t \in [t_0, t_0 + \tau]$, *where for* τ *one can take* $\min\{\theta/M, T - t_0\}$.

Proof. Consider the same set U and mapping F as in the preceding proof. In contrast to Theorem 1, here the existence of a fixed point is deduced not from the contractive mapping theorem, but from the Schauder's principle in the formulation of Theorem 2 of Subsection 15.1.4. In particular, this is the reason why the theorem asserts the existence of a solution, but not its uniqueness.

Let us verify that the conditions of Theorem 2 of Subsection 15.1.4 are satisfied in the present case. The set U is a closed ball in the space $C[t_0, t_0 + \tau]$, of radius θ and centered at the function identically equal to y_0. Hence, U is a convex, closed, and bounded subset of $C[t_0, t_0 + \tau]$. The proof of the inclusion $F(U) \subset U$ given for Picard's theorem remains in force. Let us verify that the mapping F is continuous. For every $\varepsilon > 0$, take the $\delta(\varepsilon)$ from the condition of uniform continuity in y of the function $f(t, y)$. Then for any functions $y_1, y_2 \in U$ with $\|y_1 - y_2\| < \delta(\varepsilon/\tau)$ it holds that $|f(s, y_1(s)) - f(s, y_2(s))| \leqslant \varepsilon/\tau$, and so

$$\|F(y_1) - F(y_2)\| = \max_{t \in [t_0, t_0+\tau]} \left| \int_{t_0}^{t} [f(s, y_1(s)) - f(s, y_2(s))] ds \right| \leqslant \varepsilon.$$

Finally, let us verify that the set $F(U)$ is precompact. Since $F(U) \subset U$ and U is a ball, $F(U)$ is a bounded set. By Arzelà's theorem, it remains to show that the family of functions $F(U)$ is equicontinuous. For every function $g \in F(U)$ there exists a function $y \in U$ such that $F(y) = g$. Therefore, for any $t_1, t_2 \in [t_0, t_0 + \tau]$ it holds that

$$|g(t_1) - g(t_2)| = |(F(y))(t_1) - (F(y))(t_1)| = \left| \int\limits_{t_1}^{t_2} f(s, y(s))ds \right| \leqslant M|t_1 - t_2|.$$

Hence, the family $F(U)$ is not just uniformly continuous, but obeys the Lipschitz condition with the common constant M.

Therefore, all conditions of Theorem 2 of Subsection 15.1.4 are verified, which establishes the existence of the fixed point.   □

Exercises

1. On the example of the Cauchy problem $y' = 2\sqrt{|y|}$, $y(0) = 0$, convince yourself that, indeed, the assumptions of Peano's theorem don't guarantee the uniqueness of the solution.

2. Based on Exercise 7 of Subsection 15.1.1, prove that in Picard's theorem the solution of the Cauchy problem depends continuously on the initial condition y_0.

3. Provide some solvability conditions in $C[a, b]$ for the integral equation $y(t) = \int_a^b f(s, t, y(s))ds$ by following the recipe: If the kernel f is small and nice, then the equation has a solution in a given ball centered at zero.

15.2.2 The Lomonosov Invariant Subspace Theorem

Recall that a closed subspace Y of the space X is called an *invariant subspace* for the operator $A \in L(X)$ if $A(Y) \subset Y$. The subspace $Y \subset X$ is called *non-trivial* if it is different from zero and from the whole space X. Knowledge of the invariant subspaces helps one understand the structure of the operator. Thus, for example, in linear algebra, to construct the Jordan form one exhibits the root subspaces; the decomposition of the space into a direct sum of invariant subspaces allows one to reduce the problem of solving the equation $Ax = b$ to equations in the corresponding subspaces. Very likely, currently the most important unsolved problem in operator theory is the *invariant subspace problem*: does every bounded operator in a Hilbert space have a nontrivial invariant subspace?

A large number of works are devoted to the invariant subspace problem (see, e.g., the monograph [2] and the surveys [45, 71]). Examples of continuous operators with no nontrivial invariant subspaces are known in various Banach spaces (for example, in ℓ_1). There are also positive results, the first of which was the von Neumann theorem, asserting that every compact operator in a Hilbert space has a non-trivial invariant subspace. Von Neumann's theorem was proved in the 1930s, but was first published 20 years later by Aronszajn and Smith, who extended the result to the case of Banach spaces. Below we prove a theorem on the existence of an invariant subspace due to the former Kharkiv mathematician Victor Lomonosov [68] (Feb 7, 1946 – Mar 29, 2018. From 1990 to his death, V. Lomonosov worked at Kent State University, Ohio). Lomonosov's theorem is distinguished by the generality and elegance of both its formulation and proof.

Remarks

(i) Let G be a subset of X such that $A(G) \subset G$. Then $A(\operatorname{Lin} G) \subset \operatorname{Lin} G$.

(ii) Let E be a linear subspace of the space X such that $A(E) \subset E$. Then the closure of the subspace E is also invariant under A.

(iii) The kernel of an operator and the closure of its image are invariant subspaces.

(iv) For any element $x \in X \setminus \{0\}$, the set $G = \{A^n x\}_{n=1}^{\infty}$ satisfies the condition of Remark (i). Hence, the closure of the linear span of the set G is an invariant subspace of the operator A.

(v) A more general example. Let M be a subalgebra of the algebra $L(X)$ (i.e., a subspace that together with any two of its elements also contains their product), and let $x \in X. \setminus \{0\}$. Define the *orbit of the element* x as the set $M(x) = \{Tx : T \in M\}$. Then the closure of the orbit $M(x)$ is an invariant subspace for every operator from the subalgebra M.

N.B. Verify this! We will use this example.

Theorem 1 (Lomonosov's theorem). *Let* $A \in L(X) \setminus \{0\}$ *be a compact operator in an infinite-dimensional complex Banach space* X. *Then all operators that commute with* A *have a common nontrivial invariant subspace.*

Proof. We will argue by contradiction. Thus, let us assume that there is no such invariant subspace. Fix an open ball U in the space X such that the closure of the set $A(U)$ does not contain zero. Denote the closure of $A(U)$ by K. Then K is a compact set and $0 \notin K$. Let $M \subset L(X)$ be the subalgebra consisting of all operators that commute with A. Note that the orbit $M(x)$ of any non-zero element is dense in X: otherwise, by Remark (v), the closure of the orbit will be a non-trivial invariant subspace for the operators from M. Hence, for every point $s \in K$ one can find an operator $T_s \in M$ such that $T_s(s) \in U$. Then the operator T_s also maps some neighborhood V_s of the element s into U. As s runs through K, the neighborhoods V_s form a cover of the compact set K. Hence, we can extract from it a finite subcover, i.e., there exists a finite subset $J \subset K$ such that $\bigcup_{s \in J} V_s \supset K$.

Let $\varphi_s \in C(K)$, $s \in J$, be a partition of unity subordinate to the cover $\bigcup_{s \in J} V_s$ of K.[2] Now consider the mapping $F \colon K \to X$ defined by the rule

---

[2]That is, $\varphi_s \geqslant 0$, $\sum_{s \in J} \varphi_s \equiv 1$, and the support of the function φ_s lies in the corresponding set V_s, see Subsection 15.1.3.

$$F(x) = A\left(\sum_{s \in J} \varphi_s(x) \cdot T_s(x)\right).$$

For each point $x \in K$ in the last sum only the terms for which $\varphi_s(x) \neq 0$ are different from zero, i.e., only those for which V_s contains the element x. If $x \in V_s$, then, by construction, $T_s(x) \in U$, and so $F(x) \in K$. Thus, $F(K) \subset K$, and we are under the conditions of Schauder's principle. Denote the resulting fixed point of the mapping F by x_0. Then x_0 will also be a fixed point for the compact operator $T \in M$ defined by

$$T = A\left(\sum_{s \in J} \varphi_s(x_0) T_s\right).$$

Now consider the eigenspace $Y = \mathrm{Ker}(T - I)$. By the theorem in Subsection 11.1.5, the subspace Y is invariant for the operator A. Thanks to the compactness of the operator T, the subspace Y is finite-dimensional. Since any operator in a finite-dimensional space has eigenvalues, the restriction of the operator A to the subspace Y also has an eigenvalue μ. Denote by E the eigenspace of A corresponding to the eigenvalue μ. By the same theorem of Subsection 11.1.5, the subspace E is invariant for all operators that commute with A. $\qquad\square$

Corollary 1. *If the operator T commutes with at least one compact operator, then T has a non-trivial invariant subspace.*

Exercises

1. Let $A \in L(X)$ and let Y be an invariant subspace of the operator A. Define the operators $A_1 \in L(Y)$ and $A_2 \in L(X/Y)$ as the restriction and quotient of A, respectively: $A_1(y) = A(y)$ and $A_1([x]) = [Ax]$. Can the operator A be recovered from the operators A_1 and A_2? Are the spectra of these operators connected in some way or another? Does the answer depend on the dimensions of the three subspaces involved?

Verify that:

2. The shift operator $U(a_1, a_2, \dots) = (0, a_1, a_2, \dots)$ in ℓ_2 does not commute with any compact operator, yet it has non-trivial invariant subspaces.

3. Show that the subset M of the algebra $L(X)$ that was introduced in the proof of Lomonosov's theorem is indeed a subalgebra of $L(X)$.

4. The set M is closed in the sense of pointwise convergence of operators. Is it also closed in norm?

Fill in the omitted details in the proof of Lomonosov's theorem:

5. Why is the required choice of the ball U possible?

6. Will F be a linear operator?

7. Why is Schauder's principle applicable?

8. Where was the infinite-dimensionality of the space used? (In finite-dimensional spaces the theorem is not true. Counterexample: $A = I$.)

15.3 Common Fixed Points of a Family of Mappings

15.3.1 Kakutani's Theorem

Recall that the diameter of a set V in a metric space X is defined as the number $\mathrm{diam}(V) = \sup\{\rho(x, y) : x, y \in V\}$.

Definition 1. The *radius of the set V at the point $x \in V$* is the smallest radius $r_x(V)$ of a closed ball centered at x that contains the whole set V. Equivalent definition: $r_x(V) = \sup\{\rho(x, y) : y \in V\}$. Obviously,

$$\mathrm{diam}(V) = \sup_{x \in V} r_x(V). \qquad (*)$$

The point $x \in V$ is called a *diametral point* (or *element*) of the set V if $r_x(V) = \mathrm{diam}(V)$.

Lemma 1. *Let V be a convex compact set in the normed space X such that V consists of more that one point. Then the set V contains a non-diametral point, i.e., there exists an $x \in V$ for which $r_x(V) < \mathrm{diam}(V)$.*

Proof. Fix a positive number $\varepsilon < \mathrm{diam}(V)$ and choose in V an ε-net $x_1, x_2, \ldots, x_n$. For the required non-diametral point x we take the arithmetic mean of the points of the ε-net: $x = \frac{1}{n}(x_1 + x_2 + \cdots + x_n)$. Consider an arbitrary $y \in V$. Then

$$\|x - y\| = \frac{1}{n}\left\|\sum_{k=1}^{n}(x_k - y)\right\| \leqslant \frac{1}{n}\sum_{k=1}^{n}\|x_k - y\|.$$

At least one of the terms in the last sum does not exceed ε, and all the rest are bounded above by $\mathrm{diam}(V)$. Therefore, $\|x - y\| \leqslant \frac{n-1}{n}\mathrm{diam}(V) + \frac{1}{n}\varepsilon$. Taking here the supremum over $y \in V$, we obtain

$$r_x(V) \leqslant \frac{n-1}{n}\mathrm{diam}(V) + \frac{1}{n}\varepsilon < \mathrm{diam}(V). \qquad \square$$

Lemma 2. *Let V be a convex compact set in the normed space X such that V consists of more than one point. Then there exists a non-empty convex compact subset $V_0 \subset V$,*

$V_0 \neq V$, that is invariant under all bijective isometries of the compact set V into itself.

Proof. Using the preceding lemma, pick $x_0 \in V$ with $r_{x_0}(V) < \text{diam}(V)$. Let $r_0 = r_{x_0}(V)$ and take for the required V_0 the set of all points $x \in V$ for which $r_x(V) \leqslant r_0$. By construction, $x_0 \in V_0$, so that V_0 is not empty. Further, by $(*)$, V contains a point for which $r_x(V) > r_0$ (indeed, thanks to the compactness of V, even such that $r_x(V) = \text{diam}(V)$). Therefore, $V_0 \neq V$.

Notice that a point $x \in V$ lies in V_0 if and only if the distance from x to all $y \in V$ is not larger than r_0. That is, V_0 can be written as an intersection $V_0 = V \cap \left(\bigcap_{y \in V} \bar{B}(y, r_0) \right)$ of convex closed sets, hence V_0 is itself convex and closed. It remains to verify the invariance under all bijective isometries $T : V \to V$. Let $x \in V$; we need to prove that $T(x) \in V$, that is, that $\|T(x) - y\| \leqslant r_0$ for all $y \in V$. Indeed, since T is bijective, the point y has the form $y = T(z)$ for some $z \in V$. Therefore, $\|T(x) - y\| = \|T(x) - T(z)\| = \|x - z\| \leqslant r_0$.   □

Theorem 1 (Kakutani's theorem). *Let K be a nonempty convex compact subset of the normed space X. Then all bijective isometries of the compact set K into itself have a common fixed point.*

Proof. Consider the family $\mathcal{W}$ of all nonempty convex closed subsets V of the compact set K with the property that

$$T(V) \subset V \text{ for any bijective isometry } T : K \to K. \qquad (**)$$

Since T^{-1} is a bijective isometry of K whenever T is, property $(**)$ means that $T(V) = V$ for every bijective isometry of K, i.e., that T is also a bijective isometry of V. We equip the family $\mathcal{W}$ with the decreasing order of sets. We leave it to the reader to verify that the intersection of any chain of elements of the family $\mathcal{W}$ is again an element of $\mathcal{W}$, i.e., that $\mathcal{W}$ is inductively ordered (that the intersection of the elements of a chain is not empty is guaranteed by the compactness of K). By Zorn's lemma, $\mathcal{W}$ contains a minimal element with respect to inclusion, V. It follows from Lemma 2 that this minimal set V cannot contain more than one point. Hence, V consists of a single point $x_0 \in K$, and the condition $(**)$ means that x_0 is a fixed point for all bijective isometries $T : K \to K$.   □

Exercises

1. Let K be a compact metric space. Then every isometric embedding $T : K \to K$ is bijective, that is the word "bijective" in the statements of Lemma 2 and Theorem 2 can be omitted.

2. Let V be some ball in a normed space. Then (a) the center of V is a common fixed point for all bijective isometries $T : V \to V$; (b) there are no other common fixed points of all bijective isometries $T : V \to V$.

3. Give an example of a ball in a metric space in which assertion (a) of the preceding exercise is not true. Do the same for (b). Give an example in which both (a) and (b) are not true.

By definition, a Banach space X has *normal structure* if for any convex, closed, bounded subset $V \subset X$ which consists of more than one point there exists a non-diametral element.

4. Every finite-dimensional Banach space has normal structure.

5. Every Hilbert space has normal structure.

6. The space ℓ_1 does not have normal structure. A set all points of which are diametral is

$$V = \{x = (x_1, x_2, \ldots) \in \ell_1 : \sum_{k=1}^{\infty} x_k = 1, \ x_k \geqslant 0 \ \forall k \in \mathbb{N}\}.$$

7. The space c_0 does not have normal structure. Indeed, consider the set $V = \{x = (x_1, x_2, \ldots) \in c_0 : 1 \geqslant x_1 \geqslant x_2 \geqslant \cdots\}$.

8. The spaces $C[0, 1]$, $L_\infty[0, 1]$, and $L_1[0, 1]$ do not have normal structure.

15.3.2 Topological Groups

Definition 1. A group G endowed with a Hausdorff topology τ is called a *topological group* if the topology is compatible with the group structure in the following sense:

1) the multiplication operation $(x, y) \mapsto x \cdot y$, $x, y \in G$, is jointly continuous in its variables;
2) the inversion operation $x \mapsto x^{-1}$ is continuous.

Examples of topological groups are the normed spaces with the addition operation, the unit circle $\mathbb{T}$ in $\mathbb{C}$ with the multiplication operation, the set of all unitary matrices of order n with the multiplication of matrices as operation and equipped with the metric inherited from the space of operators, the group of invertible elements of any Banach algebra, and many other groups that arise naturally in problems of analysis.

For groups one can perform operations on subsets, analogous to those introduced in Subsection 5.1.4 for linear spaces:

$$A_1 A_2 = \{a_1 a_2 : a_1 \in A_1, a_2 \in A_2\} \quad \text{and} \quad A^{-1} = \{a^{-1} : a \in A\}.$$

We denote by e the unit element of the topological group G, and by $\mathfrak{N}_e$ the family of all neighborhoods of e. We leave to the reader the verification of the following properties of topological groups. The proof of quite similar results will be given in Subsection 16.2.1 below under the theme "Topological vector spaces".

(i) For any $x \in G$, the sets xU with $U \in \mathfrak{N}_e$ form a neighborhood basis of the element x.

(ii) The family of sets $U \cdot x$ with $U \in \mathfrak{N}_e$ enjoys the same property.

(iii) For any neighborhood $W \in \mathfrak{N}_e$ there exists a neighborhood $U \in \mathfrak{N}_e$ such that $U \cdot U \subset W$.

(iv) For any neighborhood $W \in \mathfrak{N}_e$ there exists a neighborhood $U \in \mathfrak{N}_e$ such that $U^{-1} \subset W$.

(v) For any neighborhood $W \in \mathfrak{N}_e$ there exists a neighborhood $U \in \mathfrak{N}_e$ such that $U \cdot U^{-1} \subset W$, $U^{-1}U \subset W$, and $U \cdot U \subset W$.

Properties (i) and (ii) mean that the degree of closeness of the elements of the group can be "measured" by means of neighborhoods of the unit element. If $U \in \mathfrak{N}_e$, then the condition $x^{-1}y \in U$ can be treated as saying that "x approximates y to within U".

Definition 2. Let G be a topological group and Z a metric space. A mapping $f: G \to Z$ is said to be *uniformly continuous* if for every $\varepsilon > 0$ there exists a neighborhood $U \in \mathfrak{N}_e$ such that the images of any U-close points $x, y \in G$ are close to within ε: $x^{-1}y \in U \implies \rho(f(x), f(y)) < \varepsilon$.

Theorem 1. *Let G be a compact topological group and Z a metric space. Then any continuous mapping $f: G \to Z$ is uniformly continuous.*

Proof. Fix $\varepsilon > 0$. Using the continuity of the mapping f, pick for each $x \in G$ a neighborhood $W_x \in \mathfrak{N}_e$ such that for every $y \in G$ satisfying $y \in xW_x$ it holds that $\rho(f(x), f(y)) < \varepsilon/2$. Further, by property (iii) of topological groups, for any $x \in G$ one can choose an open neighborhood $U_x \in \mathfrak{N}_e$ such that $U_x U_x \subset W_x$. Since the sets xU_x, $x \in G$, form an open cover of the compact set G, one can extract from it a finite subcover. That is, there exists a finite set $A \subset X$ such that $\bigcup_{x \in A} xU_x \supset G$. Set $U = \bigcap_{x \in A} U_x$. Let us verify that U is the neighborhood required in the definition of uniform continuity. Let $x, y \in G$ and $x^{-1}y \in U$. Pick an $x_0 \in A$ such that $x \in x_0 U_{x_0}$. In particular, $x \in x_0 W_{x_0}$, i.e., $\rho(f(x), f(x_0)) < \varepsilon/2$. Further, $y \in xU \subset x_0 U_{x_0} U_{x_0} \subset x_0 W_{x_0}$, i.e., $\rho(f(y), f(x_0)) < \varepsilon/2$, and by the triangle inequality, $\rho(f(y), f(x)) < \varepsilon$. $\qquad\square$

We leave it to the reader to verify that the following analogue of Arzelà's theorem (Subsection 1.4.2) holds.

Theorem 2. *Let G be a compact topological group. In order for the family $\mathcal{F}$ of continuous scalar-valued functions on G to be precompact in $C(G)$, it is necessary and sufficient that (a) $\mathcal{F}$ is uniformly bounded, and (b) $\mathcal{F}$ obeys the following equicontinuity condition: for any $\varepsilon > 0$ there exists a neighborhood $U \in \mathfrak{N}_e$ such that for any function $f \in \mathcal{F}$ and any two points $x, y \in G$, if $x^{-1}y \in U$, then $\rho(f(x), f(y)) < \varepsilon$.* $\qquad\square$

15.3.3 Haar Measure

Definition 1. A *Haar measure* on the topological group G is a nonzero regular Borel measure μ on G that is invariant under translations and symmetry (inversion), i.e., $\mu(s\Delta) = \mu(\Delta s) = \mu(\Delta^{-1}) = \mu(\Delta)$ for any Borel subset $\Delta \subset G$ and any $s \in G$. A nonzero regular Borel measure on G that is invariant under left (resp., right) translations is called a *left* (resp., *right*) *Haar measure*.

From this point on till the end of this subsection, G will be a compact topological group. Our main goal is to establish the existence of a Haar measure on such a group. The proof relies on Kakutani's theorem.

Let us consider the Banach space $C(G)$ of continuous real-valued functions on G, equipped, as usual with the max-norm. For each $s \in G$ we define the *left* and *right translations operators* $L_s, R_s : C(G) \to C(G)$ by the rules $(L_s f)(x) = f(sx)$ and $(R_s f)(x) = f(xs)$. We also introduce the *symmetry operator* $\Psi : C(G) \to C(G)$ by the formula $(\Psi f)(x) = f(x^{-1})$.

Lemma 1. *The translation operators have the following obvious properties:*

I. $L_e = I$, $L_s L_t = L_{ts}$; *in particular,* $(L_s)^{-1} = L_{s^{-1}}$.

II. $R_e = I$, $R_s R_t = R_{st}$; *in particular,* $(R_s)^{-1} = R_{s^{-1}}$.

III. $L_s R_t = R_t L_s$.

IV. *The operators* L_s, R_s *and* Ψ *are bijective isometries of the space* $C(G)$.

V. $L_{s^{-1}} \Psi = \Psi R_s$.   □

Further, for any function $f \in C(G)$ we introduce the sets $c_L(f)$ and $c_R(f)$ as the closures of the convex hulls of all left, respectively right, translates of f:

$$c_L(f) = \overline{\mathrm{conv}}\{L_s f : s \in G\} \quad \text{and} \quad c_R(f) = \overline{\mathrm{conv}}\{R_s f : s \in G\}.$$

Finally, we use the symbol $\mathbb{1}$ to denote the function $\mathbb{1}_G$ identically equal to 1 on G.

Lemma 2. *Let G be a compact topological group. Then*

A. *For every function $f \in C(G)$ the set $c_L(f)$ is compact in $C(G)$.*

B. *The set $c_L(f)$ is invariant under all left translation operators, and these operators act bijectively on $c_L(f)$.*

C. *If $g \in c_L(f)$, then $c_L(g) \subset c_L(f)$.*

D. *For every function $f \in C(G)$ there exists a scalar a such that $a \cdot \mathbb{1} \in c_L(f)$.*

The set $c_R(f)$ enjoys analogous properties:

A'. *For every function $f \in C(G)$ the set $c_R(f)$ is compact in $C(G)$.*

B'. *The set $c_R(f)$ is invariant under all right translation operators, and these operators act bijectively on $c_R(f)$.*

C'. *If $g \in c_R(f)$, then $c_R(g) \subset c_R(f)$.*

D'. *For every function $f \in C(G)$ there exists a scalar b such that $b \cdot 1 \in c_R(f)$.*

Proof. A. In order to prove compactness of $c_L(f)$ we need to establish the precompactness of the set $\text{conv}\{L_s f : s \in G\}$. Consider the family of functions $H := \{L_s f : s \in G\}$ as a subset of the normed space $C(G)$. Since translations are isometries, all elements $L_s f$ with $s \in G$ have the same norm, equal to $\|f\|$. This establishes the boundedness of the set H.

Using the uniform continuity of the function f (Theorem 1 of the preceding subsection), given any $\varepsilon > 0$ we choose a neighborhood $U \in 1_e$ such that for any $x, y \in G$ with $x^{-1}y \in U$ we have the estimate $|f(x) - f(y)| \leqslant \varepsilon$. Since for any $s \in G$ the elements sx and sy are also close (indeed, $(sx)^{-1}(sy) = x^{-1}s^{-1}sy = x^{-1}y \in U$), the last estimate also holds for the function $L_s f$: $|(L_s f)(x) - (L_s f)(y)| = |f(sx) - f(sy)| \leqslant \varepsilon$. This establishes the equicontinuity of the family H, and by Theorem 2 of the preceding subsection, the precompactness of H. Since the operation of taking the convex hull preserves precompactness (Theorem 3 of Subsection 11.2.1), the precompactness of $\text{conv}\,H$ and, consequently, the compactness of $c_L(f)$, are established.

B. The family $H = \{L_s f : s \in G\}$ is invariant under the left translation operator L_t: $L_t H = \{L_t L_s f : s \in G\} = \{L_{st} f : s \in G\} \subset H$. In view of the linearity and continuity of the operator L_t, taking the convex hull and the closure does not spoil the invariance. The bijectivity of L_t on the set $c_L(f)$ follows from the existence of the inverse operator $L_{t^{-1}}$, which also leaves $c_L(f)$ invariant.

C. If $g \in c_L(f)$, then according to item B, $L_s g \in c_L(f)$ for any $s \in G$. That is, $\{L_s g : s \in G\} \subset c_L(f)$. It remains to use the fact that the set $c_L(f)$ is convex and closed.

D. Applying Kakutani's theorem of Subsection 15.3.1 to the convex compact set $c_L(f)$, we see that there exists an element $g \in c_L(f)$ that is fixed under all isometries of the compact set $c_L(f)$. In particular, g is a fixed point of all left translation operators. Defining $a = g(e)$, we claim that $a \cdot 1 = g$, i.e., that g is the sought-for identically constant function belonging to $c_L(f)$. Indeed, for any $s \in G$ we have $g(s) = (L_s g)(e) = g(e) = a$.

The properties A' – D' of the set $c_R(f)$ can be proved in much the same manner, or by reducing them to the properties A–D by means of the relation $c_R(f) = \Psi(c_L(\Psi f))$. $\qquad\qquad\qquad\qquad\qquad\qquad\qquad\qquad\qquad\qquad\qquad\qquad\qquad\square$

Let us strengthen the assertions D and D' of the last lemma.

Lemma 3. *For any function $f \in C(G)$ there exists only one scalar a for which $a \cdot 1 \in c_L(f)$, and only one scalar b for which $b \cdot 1 \in c_R(f)$; moreover, $a = b$.*

Proof. Let us denote the set of all scalars a (respectively, b) for which $a \cdot 1 \in c_L(f)$ (respectively, $b \cdot 1 \in c_R(f)$) by A_f (respectively, B_f). We will prove that $a = b$ for all $a \in A_f$ and all $b \in B_f$. This will establish that $A_f = B_f$ and both these sets consist of a single point.

To this end we fix an $\varepsilon > 0$ and pick convex combinations of translates of the function f which approximate $a \cdot \mathbb{1}$ and $b \cdot \mathbb{1}$ to within ε:

$$\left\| a \cdot \mathbb{1} - \sum_{k=1}^{n} \lambda_k L_{s_k} f \right\| < \varepsilon \tag{1}$$

and

$$\left\| b \cdot \mathbb{1} - \sum_{j=1}^{m} \mu_j R_{t_j} f \right\| < \varepsilon. \tag{2}$$

Applying the operator $\mu_j R_{t_j}$ to the function $a \cdot \mathbb{1} - \sum_{k=1}^{n} \lambda_k L_{s_k} f$, summing the result over j, and taking into account that $R_{t_j} \mathbb{1} = \mathbb{1}$, $\mu_j \geqslant 0$, and $\sum_{j=1}^{m} \mu_j = 1$, we deduce from (1) that

$$\left\| a \cdot \mathbb{1} - \sum_{k=1}^{n} \sum_{j=1}^{m} \lambda_k \mu_j R_{t_j} L_{s_k} f \right\| < \varepsilon.$$

Similarly, (2) implies that

$$\left\| b \cdot \mathbb{1} - \sum_{k=1}^{n} \sum_{j=1}^{m} \lambda_k \mu_j R_{t_j} L_{s_k} f \right\| < \varepsilon$$

(don't forget that the left and right translations operators commute!). Therefore, $\| a \cdot \mathbb{1} - b \cdot \mathbb{1} \| < 2\varepsilon$, which in view of the arbitrariness of ε means that $a = b$, as we needed to show. $\qquad\square$

Theorem 1 (A. Haar 1933, J. von Neumann, 1934). *On every compact topological group G there exists a unique Borel probability measure μ which is a left Haar measure. This measure is simultaneously a Haar measure on G.*

Proof. By the theorem on the general form of an elementary integral (Subsection 8.3.2.), there exists a bijective correspondence between regular Borel measures on G and elementary integrals. Let us reformulate the problem of searching for a left Haar measure in terms of an elementary integral. Specifically, we need to find a linear functional $\mathcal{I}$ on $C(G)$, called a *left-invariant mean,* such that

(i) if $f \geqslant 0$, then $\mathcal{I}(f) \geqslant 0$.
(ii) $\mathcal{I}(\mathbb{1}) = 1$.
(iii) $\mathcal{I}(L_s f) = \mathcal{I}(f)$ for all $s \in G$ and all functions $f \in C(G)$.

We have already encountered such a functional in Subsection 5.5.1, where we established its existence for a commutative (semi)group G; in fact, the functional was defined not only on $C(G)$, but also on $\ell_\infty(G)$. Here this earlier result does not work: the group can be non-commutative, and moreover, we need not only existence, but also uniqueness.

We begin with the uniqueness. Suppose that a functional with the requisite properties exists. Then for any function $f \in C(G)$ and any $g \in c_L(f)$ we have, in view of (iii), that $\mathcal{I}(g) = \mathcal{I}(f)$. Hence, $\mathcal{I}(f)$ is equal to that constant a for which $a \cdot \mathbb{1} \in c_L(f)$. This argument not only proves the uniqueness of the functional $\mathcal{I}$, but also suggests how to construct it.

So, let us establish the existence of the required functional. For each $f \in C(G)$ we choose the number $\mathcal{I}(f)$ such that $\mathcal{I}(f) \cdot \mathbb{1} \in c_L(f)$. By the preceding two lemmas, such a choice is indeed possible and is unique.

Let $f \geqslant 0$. Then $c_L(f)$ consists only of non-negative functions. In particular, $\mathcal{I}(f) \cdot \mathbb{1} \geqslant 0$, and so $\mathcal{I}(f) \geqslant 0$. This proves condition (i) in the definition of a left-invariant mean. Further, $\mathbb{1} \in c_L(\mathbb{1})$, i.e., $\mathcal{I}(\mathbb{1}) = 1$, which proves condition (ii). Finally, from item C of Lemma 2 and the uniqueness of the choice of $\mathcal{I}(f)$ it follows that $\mathcal{I}(g) = \mathcal{I}(f)$ for all $g \in c_L(f)$. In particular, this ensures that condition (iii) in the definition of a left-invariant mean is satisfied.

Now let us prove the linearity of the functional $\mathcal{I}$. The homogeneity is obvious, so let us prove the additivity. Let $f, g \in C(G)$. By construction, there exists a convex combination of left translates of the function f which approximates $\mathcal{I}(f) \cdot \mathbb{1}$ to within ε:

$$\left\| \mathcal{I}(f) \cdot \mathbb{1} - \sum_{k=1}^{n} \lambda_k L_{s_k} f \right\| < \varepsilon. \tag{3}$$

Consider the auxiliary function $\widetilde{g} = \sum_{k=1}^{n} \lambda_k L_{s_k} g$. Since $\widetilde{g} \in c_L(g)$, we have $\mathcal{I}(\widetilde{g}) = \mathcal{I}(g)$. Therefore, there exists a function of the form $\sum_{j=1}^{m} \mu_j L_{t_j} \widetilde{g}$, i.e., a convex combination of left translates of the function $\widetilde{g}$, which approximates $\mathcal{I}(g) \cdot \mathbb{1}$:

$$\left\| \mathcal{I}(g) \cdot \mathbb{1} - \sum_{k=1}^{n} \sum_{j=1}^{m} \lambda_k \mu_j L_{t_j} L_{s_k} g \right\| < \varepsilon. \tag{4}$$

But from (3) one can readily deduce, as we did above in the proof of Lemma 3, that

$$\left\| \mathcal{I}(f) \cdot \mathbb{1} - \sum_{k=1}^{n} \sum_{j=1}^{m} \lambda_k \mu_j L_{t_j} L_{s_k} f \right\| < \varepsilon. \tag{5}$$

Combining (4) and (5), we get

$$\left\| (\mathcal{I}(f) + \mathcal{I}(g)) \cdot \mathbb{1} - \sum_{k=1}^{n} \sum_{j=1}^{m} \lambda_k \mu_j L_{t_j s_k} (f + g) \right\| < 2\varepsilon.$$

Since $\sum_{k=1}^{n} \sum_{j=1}^{m} \lambda_k \mu_j L_{t_j s_k} (f + g)$ is a convex combination of translates of the function $f + g$, and in view of the arbitrariness of ε, the last condition means that $(\mathcal{I}(f) + \mathcal{I}(g)) \cdot \mathbb{1} \in c_L(f + g)$, i.e., $\mathcal{I}(f + g) = \mathcal{I}(f) + \mathcal{I}(g)$.

Thus, we established the existence and uniqueness of a left-invariant mean, and hence the existence and uniqueness of a left Haar measure. Now we observe that, by Lemma 3, for any $f \in C(G)$ the function $\mathcal{I}(f) \cdot \mathbb{1}$ lies not only in $c_L(f)$, but also in $c_R(f)$. Therefore, $\mathcal{I}(R_t f) \cdot \mathbb{1} \in c_R(R_t f) \subset c_R(f)$. By the same Lemma 3, in $c_R(f)$ there is only one function of the form $a \cdot \mathbb{1}$. Hence, $\mathcal{I}(R_t f) = \mathcal{I}(f)$. This completes the proof of the right-invariance of the mean and of the measure it generates. Finally, the functional $\widetilde{\mathcal{I}}(f) = \mathcal{I}(\Psi f)$ is also a left-invariant mean, so that in view of the uniqueness of the left-invariant mean, $\mathcal{I}(f) = \mathcal{I}(\Psi f)$. This implies the invariance of the Haar measure under the symmetry mapping $s \mapsto s^{-1}$. $\qquad\square$

Remark 1. The left and right Haar measures exist not only on compact, but also on locally compact groups (see [19, Chap. 4]), but in the latter case the left and right Haar measures may not coincide, or even may not be finite measures, and the existence proof is more complicated than in the compact case.

Exercises

1. In the proof of the last theorem we treated as obvious the following fact: let $u\colon G \to G$ be a homeomorphism of the compact space G, and let $U\colon C(G) \to C(G)$ be the composition operator acting by the rule $(Uf)(s) = f(u(s))$. Suppose the elementary integral $\mathcal{I}$ is invariant under U, that is, $\mathcal{I}(Uf) = \mathcal{I}(f)$ for all $f \in C(G)$. Then the measure $\mu_{\mathcal{I}}$ generated by the integral $\mathcal{I}$ is u-invariant: $\mu_{\mathcal{I}}(u(\Delta)) = \mu_{\mathcal{I}}(\Delta)$ for any Borel set $\Delta \subset G$. Prove this fact using the change of variable formula for the Lebesgue integral (Exercise 9 of Subsection 7.2.7) and the bijectivity of the correspondence $\mathcal{I} \mapsto \mu_{\mathcal{I}}$ between the set of all elementary integrals and the set of all regular Borel measures on the compact space G.

2. Let G be a finite group. What is its Haar measure?

3. Let G be the unit circle $\mathbb{T} \subset \mathbb{C}$ with the operation of multiplication of complex numbers. What is the Haar measure in this case?

4. Let K be a compact metric space. Then any isometric mapping $u\colon K \to K$ is bijective.

5. Let K be a compact metric space. Denote by $\Theta(K)$ the set of all isometries $u\colon K \to K$, equipped with the operation of composition. Prove that with respect to the uniform metric $\Theta(K)$ is a compact topological group.

6. Let K be a compact metric space with the following properties: for any two points $x, y \in K$ there exists an isometry $u\colon K \to K$ that maps x into y. Then on K there exists a unique regular Borel measure ν that is invariant under all isometries of K. For any point $x_0 \in K$, this measure is connected with the Haar measure μ on $\Theta(K)$ by the relation $\nu(A) = \mu\{u \in \Theta(K) : u(x_0) \in A\}$.

7. Let S^2 be the unit sphere in the three-dimensional Euclidean space, λ the standard Lebesgue measure on S^2, and A a measurable set with $\lambda(A) < \lambda(S^2)/n$. Then for any collection $x_1, \ldots, x_n \in S^2$ of n points one can find an isometry $u \colon S^2 \to S^2$ with the property that none of the points $x_1, \ldots, x_n$ will lie in $u(A)$.

Comments on the Exercises

Subsection 15.1.1

Exercise 7. The existence and uniqueness of the solution follow from Banach's theorem. The continuity can be deduced from the same theorem by means of the following device. Consider the space $C(K, X)$ of continuous X-valued functions on K with the uniform metric. Define the contractive mapping $G \colon C(K, X) \to C(K, X)$ by the formula $[G(f)](t) = F(t, f(t))$. The fixed point f of this mapping will be a continuous function satisfying the condition $F(t, f(t)) = f(t)$.

Exercise 8. Reduce to the preceding exercise by taking $K = [a, b]$, $X = \mathbb{R}$, and
$$F(t, x) = x - \frac{2}{m + M} \Phi(t, x).$$

Subsection 15.1.2

Exercises 5 and 7. Let $P \colon \overline{B}_X \to S_X$ be a retraction. Then the mapping $Q = -P$ has no fixed points.

Exercise 6. It is more convenient to construct the requested retraction in $L_2[0, +\infty)$ instead of ℓ_2. The problems are equivalent because these two Hilbert spaces are isometric. For every f in the unit ball of $L_2[0, +\infty)$ define $\widetilde{f}(t) = 0$ for $t \in [0, 1 - \|f\|)$ and $\widetilde{f}(t) = f(t - 1 + \|f\|)$ for $t \geqslant 1 - \|f\|$. The mapping $f \mapsto \sqrt{1 + \|f\|} \cdot \mathbb{1}_{[0, 1 - \|f\|)} + \widetilde{f}$ will be the requested retraction.

Subsection 15.1.4

Exercises 4 and 5. The functional F from Exercise 6 generates a bounded closed base in the case $L_1[0, 1]$. In the case $L_2[0, 1]$, this cone has no bounded closed base. The lack of existence of a compact base follows indirectly from Exercises 2 and 6.

Subsection 15.3.1

Exercise 1. Assume $T(K) \neq K$. Take an $x_0 \in K \setminus T(K)$ and consider the sequence (x_k) of iterations: $x_{k+1} = T(x_k)$. Since for $k \geqslant 1$ all x_k belong to $T(K)$, and $T(K)$ is compact, there is an $\varepsilon > 0$ such that all the distances $\rho(x_0, x_k) > \varepsilon$, $k = 1, 2, \ldots$. Then $\rho(x_n, x_{n+k}) = \rho(T^n(x_0), T^n(x_k)) = \rho(x_0, x_k) > \varepsilon$ for all $n, k \in \mathbb{N}$, so the sequence (x_k) does not contain convergent subsequences.

Chapter 16
Topological Vector Spaces

16.1 Supplementary Material from Topology

We have already encountered a very general type of convergence — convergence along a directed set. We now turn to yet another type, convergence along a filter, and apply this new technique to the study of compact topological spaces. Throughout this chapter we will have to frequently deal, within one and the same argument, with sets as well as some families of subsets. To make it easier to distinguish these objects, we will denote sets by upper case Roman italic letters A, B, X, Y, and so on, and use for families Gothic letters $\mathfrak{A}$, $\mathfrak{C}$, $\mathfrak{D}$, $\mathfrak{F}$. Of course, the difference here is rather conventional, since any family of sets is itself a set.

16.1.1 Filters and Filter Bases

Definition 1. A family $\mathfrak{F}$ of subsets of a set X is called a *filter on X* if it satisfies the following axioms:

(i) $\mathfrak{F}$ is not empty;

(ii) $\emptyset \notin \mathfrak{F}$;

(iii) if A, $B \in \mathfrak{F}$, then $A \cap B \in \mathfrak{F}$;

(iv) if $A \in \mathfrak{F}$ and $A \subset B \subset X$, then $B \in \mathfrak{F}$.
   Let us note several consequences of the filter axioms:

(v) $X \in \mathfrak{F}$ (follows from (i) and (iv));

(vi) in view of (iii), the intersection of any finite number of elements of a filter is again an element of that filter; from (ii) we deduce that

(vii) the intersection of any finite number of elements of a filter is not empty.

© Springer International Publishing AG, part of Springer Nature 2018
V. Kadets, *A Course in Functional Analysis and Measure Theory*,
Universitext, https://doi.org/10.1007/978-3-319-92004-7_16

An example of a filter is provided by the family $\mathfrak{N}_x$ of all neighborhoods of a point x in a topological space X.

Definition 2. A non-empty family $\mathfrak{D}$ of subsets of a set X is called a *filter basis* (also *base* in the literature) if

(a) $\emptyset \notin \mathfrak{D}$, and

(b) for any sets $A, B \in \mathfrak{D}$ there exists a set $C \in \mathfrak{D}$ such that $C \subset A \cap B$.

Let $\mathfrak{D}$ be a filter basis. The *filter generated by the basis* $\mathfrak{D}$ is the family $\mathfrak{F}$ of all sets $A \subset X$ such that A contains as a subset at least one element of $\mathfrak{D}$.

We leave to the reader to verify that this definition is correct, i.e., that what we call the filter $\mathfrak{F}$ generated by the basis $\mathfrak{D}$ is indeed a filter.

If X is a topological space and $x_0 \in X$, and as a basis $\mathfrak{D}$ we take the family of all open sets that contain x_0, then the filter generated by the basis $\mathfrak{D}$ is precisely the filter $\mathfrak{N}_{x_0}$ of all neighborhoods of the point x_0.

Let us give one more example. Let $(x_n)_{n=1}^{\infty}$ be a sequence of elements of the set X. Then the family $\mathfrak{D}_{(x_n)}$ of "tails" of the sequence (x_n) (i.e., the family of sets of the form $\{x_n\}_{n=N}^{\infty}$, $N \in \mathbb{N}$) is a filter basis. The filter $\mathfrak{F}_{(x_n)}$ generated by the basis $\mathfrak{D}_{(x_n)}$ is called the *filter associated with the sequence* (x_n).

Theorem 1. *Let X, Y be sets, $f : X \to Y$ a mapping, and $\mathfrak{D}$ a filter basis in X. Then the family $f(\mathfrak{D})$ of all images $f(A)$ with $A \in \mathfrak{D}$ is a filter basis in Y.*

Proof. Axiom (a) in the definition of a filter basis is obvious. Further, let $f(A)$ and $f(B)$ be arbitrary elements of $f(\mathfrak{D})$, $A, B \in \mathfrak{D}$. By axiom (b), there exists a $C \in \mathfrak{D}$, such that $C \subset A \cap B$. Then $f(C) \subset f(A) \cap f(B)$, which proves (b) for $f(\mathfrak{D})$. □

In particular, if $\mathfrak{F}$ is a filter on X, then $f(\mathfrak{F})$ is a filter basis in Y.

Definition 3. The *image of the filter* $\mathfrak{F}$ under the mapping f is the filter $f[\mathfrak{F}]$ generated by the filter basis $f(\mathfrak{F})$. Equivalently, $A \in f[\mathfrak{F}]$ if and only if $f^{-1}(A) \in \mathfrak{F}$.

Recall (see Subsection 1.2.3) that a family of sets $\mathfrak{C}$ is said to be *centered* if the intersection of any finite collection of members of $\mathfrak{C}$ is not empty.

Theorem 2. *Let $\mathfrak{C} \subset 2^X$ be a non-empty family of sets. For the existence of a filter $\mathfrak{F}$ such that $\mathfrak{F} \supset \mathfrak{C}$ (i.e., such that all elements of $\mathfrak{C}$ are also elements of the filter $\mathfrak{F}$) it is necessary and sufficient that $\mathfrak{C}$ be a centered family.*

Proof. If $\mathfrak{F}$ is a filter and $\mathfrak{F} \supset \mathfrak{C}$, then any finite collection $A_1, \ldots, A_n$ of elements of the family $\mathfrak{C}$ will consist of elements of the filter $\mathfrak{F}$. Hence (property (vii) of filters), $\bigcap_{k=1}^{n} A_k \neq \emptyset$. Necessity is thus proved. Conversely, suppose $\mathfrak{C}$ is a centered family. Then the family $\mathfrak{D}$ of all sets of the form $\bigcap_{k=1}^{n} A_k$, where $n \in \mathbb{N}$ and $A_1, \ldots, A_n \in \mathfrak{C}$, is a filter basis. Now for $\mathfrak{F}$ one needs to take the filter generated by the basis $\mathfrak{D}$. □

Definition 4. Let $\mathfrak{F}$ be a filter on X. A family $\mathfrak{D}$ of subsets is said to be a *basis of the filter* $\mathfrak{F}$ if $\mathfrak{D}$ is a filter basis and the filter generated by $\mathfrak{D}$ coincides with $\mathfrak{F}$.

Theorem 3. *For the family $\mathfrak{D}$ to be a basis of the filter $\mathfrak{F}$, it is necessary and sufficient that the following two conditions be satisfied:*

$-\ \mathfrak{D} \subset \mathfrak{F}$;

$-\ $ *for any $A \in \mathfrak{F}$ there exists a $B \in \mathfrak{D}$ such that $B \subset A$.*

Definition 5. Let $\mathfrak{F}$ be a filter on X and $A \subset X$. The *trace of the filter* $\mathfrak{F}$ on A is the family of subsets $\mathfrak{F}_A = \{A \cap B : B \in \mathfrak{F}\}$.

Theorem 4. *For the family $\mathfrak{F}_A$ to be a filter on A, it is necessary and sufficient that all intersections $A \cap B$ with $B \in \mathfrak{F}$ are non-empty. In particular, $\mathfrak{F}_A$ will be a filter whenever $A \in \mathfrak{F}$.*

Exercises

1. Prove Theorems 3 and 4.

Below we give examples of filters and filter bases. Many of these examples will be used in the sequel. The reader is invited to verify the corresponding axioms.

2. The *Fréchet filter* on $\mathbb{N}$: the elements of this filter are the complements of the finite sets of natural numbers. A basis of the Fréchet filter is provided by the sets $A_1 = \{1, 2, 3, \ldots\}$, $A_2 = \{2, 3, 4, \ldots\}, \ldots, A_n = \{n, n + 1, n + 2, \ldots\}, \ldots$.

3. The *neighborhood filter of infinity* in a normed space X: the set $A \subset X$ lies in this filter if the set $X \setminus A$ is bounded.

4. The filter $\mathfrak{N}_x^0$ of *deleted* (or *punctured*) *neighborhoods* of a given point x in a topolgical space X: a basis of this filter consists of the sets of the form $U \setminus \{x\}$, where U is a neighborhood of x. For this definition to be correct, it is necessary that the point x is not isolated.

5. The neighborhood filters of the point $+\infty$ in $\mathbb{R}$: a basis of the filter consists of the intervals $(a, +\infty)$ with $a \in \mathbb{R}$.

6. The filter of deleted neighborhoods of the "point" $a + 0$ in $\mathbb{R}$: a basis of this filter consists of the sets (a, b) with $b \in (a, +\infty)$.

7. The *statistical filter* $\mathfrak{F}_s$ on $\mathbb{N}$: $A \in \mathfrak{F}_s$ if $\lim_{n \to \infty} |A \cap \{1, 2, \ldots, n\}|/n = 1$. Here $|B|$ denotes the cardinality of the set B.

8. Let $(G, \succ)$ be a directed set. The *section filter* on G is the filter $\mathfrak{F}_{\succ}$, a basis of which consists of all sets of the form $\{x \in G : x \succ a\}$ with $a \in G$.

Prove that

9. The set of all filters on $\mathbb{N}$ is not countable. In fact, the cardinality of this set is bigger than the cardinality of the continuum.

10. The filters in Execises 3, 5, and 6 have countable bases.

11. The statistical filter (Exercise 7) does not have a countable basis.

12. Let $(x_n)_{n=1}^{\infty}$ be a sequence in X, and let the function $f : \mathbb{N} \to X$ act by the rule $f(n) = x_n$. Then the image of the Fréchet filter from Exercise 2 under f is the filter $\mathfrak{F}_{(x_n)}$ associated with the sequence (x_n).

16.1.2   Limits, Limit Points, and Comparison of Filters

Definition 1. Suppose given two filters $\mathfrak{F}_1$ and $\mathfrak{F}_2$ on the topological space X. We say that $\mathfrak{F}_1$ *majorizes* $\mathfrak{F}_2$ if $\mathfrak{F}_1 \supset \mathfrak{F}_2$; in other words, if every element of the filter $\mathfrak{F}_2$ is also an element of the filter $\mathfrak{F}_1$.

Example 1. Let $(x_n)_{n \in \mathbb{N}}$ be a sequence in X, and $(x_{n_k})_{k \in \mathbb{N}}$ be a subsequence of $(x_n)_{n \in \mathbb{N}}$. Then the filter $\mathfrak{F}_{(x_{n_k})}$ associated with the subsequence majorizes the filter $\mathfrak{F}_{(x_n)}$ associated with the sequence itself. Indeed, let $A \in \mathfrak{F}_{(x_n)}$. Then there exists an $N \in \mathbb{N}$ such that $\{x_n\}_{n=N}^{\infty} \subset A$. But then also $\{x_{n_k}\}_{k=N}^{\infty} \subset A$, that is, $A \in \mathfrak{F}_{(x_{n_k})}$.

Definition 2. Let X be a topological space, and $\mathfrak{F}$ a filter on X. The point $x \in X$ is called the *limit of the filter* $\mathfrak{F}$ (denoted $x = \lim \mathfrak{F}$) if $\mathfrak{F}$ majorizes the neighborhood filter of the point x. In other words, $x = \lim \mathfrak{F}$ if every neighborhood of the point x belongs to the filter $\mathfrak{F}$.

The point $x \in X$ is said to be a *limit point of the filter* $\mathfrak{F}$ if every neighborhood of x intersects all elements of the filter $\mathfrak{F}$. The set of all limit points of the filter $\mathfrak{F}$ is denoted by $\mathrm{LIM}(\mathfrak{F})$.

Example 2. Let $(x_n)_{n \in \mathbb{N}}$ be a sequence in the topological space X. Then $\lim \mathfrak{F}_{(x_n)} = \lim_{n \to \infty} x_n$, and $\mathrm{LIM}(\mathfrak{F}_{(x_n)})$ coincides with the set of limit points of the sequence $(x_n)_{n \in \mathbb{N}}$.

Theorem 1. *Let $\mathfrak{F}$ be a filter on the topological space X, and $\mathfrak{D}$ be a basis for the filter $\mathfrak{F}$. Then*

(a) *$x = \lim \mathfrak{F}$ if and only if for any neighborhood U of the point x there exists an element $A \in \mathfrak{D}$ such that $A \subset U$.*

(b) *If $x = \lim \mathfrak{F}$, then x is a limit point of the filter $\mathfrak{F}$. If, in addition, X is a Hausdorff space, then the filter $\mathfrak{F}$ has no other limit points. In particular, if a filter in a Hausdorff space has a limit, then this limit is unique.*

(c) *The set $\mathrm{LIM}(\mathfrak{F})$ coincides with the intersection of the closures of all elements of the filter $\mathfrak{F}$.*

Proof. (a) By definition, $x = \lim \mathfrak{F}$ if any neighborhood U of the point x belongs to the filter $\mathfrak{F}$. In its turn, $U \in \mathfrak{F}$ if and only if U contains some set $A \in \mathfrak{D}$.

(b) Let $x = \lim \mathfrak{F}$, and let U be a neighborhood of x. Then $U \in \mathfrak{F}$, hence any set $A \in \mathfrak{F}$ intersects U. That is, $x \in \mathrm{LIM}(\mathfrak{F})$.

Further, let $x = \lim \mathfrak{F}$, $y \in \mathrm{LIM}(\mathfrak{F})$, and let U and V be arbitrary neighborhoods of the points x and y, respectively. Then $U \in \mathfrak{F}$, and since any neighborhood of a limit point intersects all elements of the filter $\mathfrak{F}$, $U \cap V \neq \emptyset$. Since the space is Hausdorff, this is possible only if $x = y$.

(c) By definition, $x \in \mathrm{LIM}(\mathfrak{F})$ if and only if every element $A \in \mathfrak{F}$ intersects all neighborhoods of the point x. This is equivalent to x belonging to the closure of every element $A \in \mathfrak{F}$. □

Theorem 2. *Suppose $\mathfrak{F}_1$, $\mathfrak{F}_2$ are filters in the topological space X, and $\mathfrak{F}_1 \subset \mathfrak{F}_2$. Then:*

(i) *if $x = \lim \mathfrak{F}_1$, then $x = \lim \mathfrak{F}_2$;*

(ii) *if $x \in \mathrm{LIM}(\mathfrak{F}_2)$, then $x \in \mathrm{LIM}(\mathfrak{F}_1)$. In particular,*

(iii) *if $x = \lim \mathfrak{F}_2$, then $x \in \mathrm{LIM}(\mathfrak{F}_1)$.*

Proof. (i) $\mathfrak{F}_1$ majorizes the neighborhood filter $\mathfrak{N}_x$ of the point x and $\mathfrak{F}_1 \subset \mathfrak{F}_2$, therefore $\mathfrak{N}_x \subset \mathfrak{F}_2$.

(ii) Since as the collection of sets increases their intersection decreases, we have $\mathrm{LIM}(\mathfrak{F}_2) = \bigcap_{A \in \mathfrak{F}_2} \overline{A} \subset \bigcap_{A \in \mathfrak{F}_1} \overline{A} = \mathrm{LIM}(\mathfrak{F}_1)$. □

Definition 3. Let X be a set, Y a topological space, and $\mathfrak{F}$ a filter in X. The point $y \in Y$ is called the *limit of the mapping $f : X \to Y$ with respect to the filter $\mathfrak{F}$* (denoted $y = \lim\limits_{\mathfrak{F}} f$), if $x = \lim f[\mathfrak{F}]$. In other words, $y = \lim\limits_{\mathfrak{F}} f$ if for any neighborhood U of the point y there exists an element $A \in \mathfrak{F}$ such that $f(A) \subset U$.

The point $y \in Y$ is called a *limit point of the mapping $f : X \to Y$ with respect to the filter $\mathfrak{F}$* if $y \in \mathrm{LIM}(f[F])$, i.e., if any neighborhood of the point y intersects the images of all elements of the filter $\mathfrak{F}$ under f.

Example 3. Let X be a topological space, $f : \mathbb{N} \to X$, and $\mathfrak{F}$ be the Fréchet filter on $\mathbb{N}$ (see Exercise 2 in Subsection 16.1.1). Then $\lim\limits_{\mathfrak{F}} f = \lim\limits_{n \to \infty} f(n)$.

Theorem 3. *Let X, Y be topological spaces, $\mathfrak{F}$ a filter in X, $x = \lim \mathfrak{F}$, and $f : X \to Y$ a continuous mapping. Then $f(x) = \lim\limits_{\mathfrak{F}} f$.*

Proof. Let U be an arbitrary neighborhood of the point $f(x)$. Then there exists a neighborhood V of the point x such that $f(V) \subset U$. The condition $x = \lim \mathfrak{F}$ means that $V \in \mathfrak{F}$. That is, for any neighborhood U of the point $f(x)$ we have found the required element $V \in \mathfrak{F}$ for which $f(V) \subset U$. $\square$

Exercises

To avoid complicating the formulations connected with the possible non-uniqueness of the limit, in the exercises below all the topological spaces are assumed to be Hausdorff.

1. Let $(G, \succ)$ be a directed set, X a topological space, $f : G \to X$ a mapping, and $\mathfrak{F}_\succ$ the section filter on G (see Exercise 8 in Subsection 16.1.1). Then $\lim\limits_{\mathfrak{F}_\succ} f = \lim\limits_{(G,\succ)} f$. Thus, the limit with respect to a directed set is a particular case of a limit with respect to a filter.

2. Suppose the subspace A of the topological space X intersects all the elements of the filter $\mathfrak{F}$. Let $\mathfrak{F}_A = \{A \cap B : B \in \mathfrak{F}\}$ be the trace of the filter $\mathfrak{F}$ on A. Then $\mathrm{LIM}(\mathfrak{F}_A) \subset \mathrm{LIM}(\mathfrak{F})$.

3. Let $A \in \mathfrak{F}$. Then the existence of $\lim \mathfrak{F}_A$ in the topology induced on A implies the existence of $\lim \mathfrak{F}$ and $\lim \mathfrak{F}_A = \lim \mathfrak{F}$.

4. Let $\lim \mathfrak{F} = a \in A$. Then $\lim \mathfrak{F}_A = a$.

5. Let X and Y be topological spaces, and $\mathfrak{N}_x$ be the neighborhood filter of the point $x \in X$. A mapping $f : X \to Y$ is continuous at the point x if and only if the limit $\lim\limits_{\mathfrak{N}_x} f$ exists. If this limit exists, then it is equal to $f(x)$.

6. Let X and Y be topological spaces, and $\mathfrak{N}_x^0$ be the filter of deleted neighborhoods of the point $x \in X$ (see Exercise 4 in Subsection 16.1.1), and suppose x is not an isolated point. Then the continuity of the mapping $f : X \to Y$ at the point x is equivalent to the condition $\lim\limits_{\mathfrak{N}_x^0} f = f(x)$.

7. In the topological space X, consider the filter $\mathfrak{F}$ consisting of all sets that contain a fixed set $A \subset X$. Then $\mathrm{LIM}(\mathfrak{F})$ coincides with the closure of the set A.

8. Based on Exercises 5 and 6 of Subsection 16.1.1, write for a function f of a real variable the expressions $\lim\limits_{x \to +\infty} f(x)$ and $\lim\limits_{x \to a+0} f(x)$ as the limits of the function with respect to appropriately chosen filters on $\mathbb{R}$.

9. For a function f of a real variable, write the expressions $\lim\limits_{x \to \infty} f(x)$, $\lim\limits_{x \to a} f(x)$, $\lim\limits_{x \to a-0} f(x)$, and $\lim\limits_{x \to -\infty} f(x)$ as limits with respect to a filter.

10. Let $\mathfrak{F}$ be a filter on the set X. A sequence $x_n \in X$ is said to be *cofinal* for the filter $\mathfrak{F}$ if $\mathfrak{F}_{\{x_n\}} \supset \mathfrak{F}$. If the filter $\mathfrak{F}$ has a countable basis, then there exists a cofinal sequence for $\mathfrak{F}$.

11. For the statistical filter $\mathfrak{F}_s$ (Exercise 7 in Subsection 16.1.1) there exists no cofinal sequence.

12. On the interval $[0, 1]$ consider the filter consisting of all the sets with finite complement. This filter does not have a countable basis, yet it possesses a cofinal sequence. (More precisely, any sequence $x_n \in [0, 1]$ of pairwise distinct numbers is cofinal for this filter.)

13. Let X be a set, Y a topological space, $f : X \to Y$ a mapping, and $y = \lim_{\mathfrak{F}} f$. If the sequence $x_n \in X$ is cofinal for the filter $\mathfrak{F}$, then $f(x_n) \to y$ as $n \to \infty$. In particular, if the filter $\mathfrak{F}$ on the set X has a countable basis, then there exists a sequence $x_n \in X$ such that $f(x_n) \to y$ as $n \to \infty$.

14. If the filter $\mathfrak{F}$ on the set X does not possess a cofinal sequence, then there exist a topological space Y and a mapping $f : X \to Y$, which has a limit y with respect to $\mathfrak{F}$, such that no sequence of the form $(f(x_n))$, with $x_n \in X$, converges to y.

16.1.3 Ultrafilters. Compactness Criteria

In the preceding subsection, we introduced the order relation $\supset$ on the family of filters given on a set X. The next lemma justifies the application of Zorn's lemma to the family of filters.

Lemma 1. *Let $\mathfrak{M}$ be a linearly ordered non-empty family of filters given on the set X, i.e., for any $\mathfrak{F}_1, \mathfrak{F}_2 \in \mathfrak{M}$, either $\mathfrak{F}_1 \supset \mathfrak{F}_2$, or $\mathfrak{F}_2 \supset \mathfrak{F}_1$. Then the union $\mathfrak{F}$ of all filters in the family $\mathfrak{M}$ is again a filter on X.*

Proof. We need to verify that the family of sets $\mathfrak{F}$ satisfies the filter axioms. The axioms (i) and (ii) are obvious here, so let us establish that the remaining two are satisfied.

(iii) Let $A, B \in \mathfrak{F}$. Then there exist filters $\mathfrak{F}_1, \mathfrak{F}_2 \in \mathfrak{M}$, such that $A \in \mathfrak{F}_1$ and $B \in \mathfrak{F}_2$. By hypothesis, one of the filters $\mathfrak{F}_1, \mathfrak{F}_2$ majorizes the other. Suppose, for instance, that $\mathfrak{F}_2 \supset \mathfrak{F}_1$. Then both sets A, B lie in $\mathfrak{F}_2$, and since $\mathfrak{F}_2$ is a filter, it follows that $A \cap B \in \mathfrak{F}_2 \subset \mathfrak{F}$.

(iv) Let $A \in \mathfrak{F}$ and $A \subset B \subset X$. Then there exists a filter $\mathfrak{F}_1 \in \mathfrak{M}$ such that $A \in \mathfrak{F}_1$. Since $\mathfrak{F}_1$ is a filter, also $B \in \mathfrak{F}_1 \subset \mathfrak{F}$. $\square$

Definition 1. An *ultrafilter* on X is a filter on X that is maximal with respect to inclusion. In detail, the filter $\mathfrak{A}$ on X is called an ultrafilter if any filter $\mathfrak{F}$ on X that majorizes $\mathfrak{A}$ necessarily coincides with $\mathfrak{A}$.

Zorn's lemma yields the following existence theorem.

Theorem 1. *For any filter $\mathfrak{F}$ on X there exists an ultrafilter that majorizes it.*   □

Lemma 2. *Suppose $\mathfrak{A}$ is an ultrafilter, $A \subset X$, and all elements of $\mathfrak{A}$ intersect A. Then $A \in \mathfrak{A}$.*

Proof. It is readily seen that when one adds to the family of sets $\mathfrak{A}$ the set A as a new element one obtains a centered family of sets. By Theorem 2 in Subsection 16.1.1, there exists a filter $\mathfrak{F}$ which contains all elements of this centered family. We have that $\mathfrak{F} \supset \mathfrak{A}$, and $\mathfrak{A}$ is an ultrafilter, that is, $\mathfrak{F} = \mathfrak{A}$. On the other hand, by construction, $A \in \mathfrak{F}$. Hence, $A \in \mathfrak{A}$.   □

Theorem 2 (ultrafilter criterion). *For the filter $\mathfrak{A}$ on X to be an ultrafilter it is necessary and sufficient that for any set $A \subset X$, either A itself or $X \setminus A$ belongs to $\mathfrak{A}$.*

Proof. Necessity. Suppose $\mathfrak{A}$ is an ultrafilter and $X \setminus A \notin \mathfrak{A}$. Then no set $B \in \mathfrak{A}$ is entirely contained in $X \setminus A$, i.e., every $B \in \mathfrak{A}$ intersects A. Hence, by Lemma 2, $A \in \mathfrak{A}$.

Sufficiency. Suppose that $\mathfrak{A}$ is not an ultrafilter. Then there exist a filter $\mathfrak{F} \supset \mathfrak{A}$ and a set $A \in \mathfrak{F} \setminus \mathfrak{A}$. By construction, $A \notin \mathfrak{A}$. On the other hand, $X \setminus A$ does not intersect A, $A \in \mathfrak{F}$, and consequently $X \setminus A$ cannot belong to the filter $\mathfrak{F}$, and the more so not to the filter $\mathfrak{A}$, which is smaller than $\mathfrak{F}$.   □

Corollary 1. *The image of any ultrafilter is an ultrafilter.*

Proof. Let $f : X \to Y$ and let $\mathfrak{A}$ be an ultrafilter on X. Consider an arbitrary set $A \subset Y$. Then either $f^{-1}(A)$, or $f^{-1}(Y \setminus A) = X \setminus f^{-1}(A)$ belongs to $\mathfrak{A}$. It follows that either A or $Y \setminus A$ belongs to $f[\mathfrak{A}]$.   □

Lemma 3. *Let $\mathfrak{A}$ be an ultrafilter on the Hausdorff topological space X and $x \in \mathrm{LIM}(\mathfrak{A})$. Then $x = \lim \mathfrak{A}$. In particular, an ultrafilter can have at most one limit point.*

Proof. Let U be an arbitrary neighborhood of the point x. Then, by the definition of a limit point, U intersects all elements of $\mathfrak{A}$. By Lemma 2, $U \in \mathfrak{A}$.   □

Theorem 3 (compactness criteria in terms of filters). *For a Hausdorff topological space X, the following conditions are equivalent:*

(1) *X is compact;*

(2) *every filter on X has a limit point;*

(3) *every ultrafilter on X has a limit.*

Proof. We will successively establish the equivalence of the listed conditions.

(1) $\implies$ (2). The filter $\mathfrak{F}$ is a centered family of sets. All the more the family of closures of the elements of the filter is also centered. Consequently (Theorem 1 of Subsection 1.2.3), the intersection $\mathrm{LIM}(\mathfrak{F})$ of these closures is not empty.

(2) $\implies$ (1). Let $\mathfrak{C}$ be an arbitrary centered family of closed subsets of the space X. By Theorem 2 of Subsection 16.1.1, there exists a filter $\mathfrak{F} \supset \mathfrak{C}$. Then $\bigcap_{A \in \mathfrak{C}} \overline{A} \supset \bigcap_{A \in \mathfrak{F}} \overline{A} = \mathrm{LIM}(\mathfrak{F}) \neq \emptyset$.

(2) $\implies$ (3). By condition (2), every ultrafilter has a limit point, and by Lemma 3 this point is the limit of the ultrafilter.

(3) $\implies$ (2). Consider an arbitrary filter $\mathfrak{F}$ on X and choose (Theorem 1) an ultra-filter $\mathfrak{A} \supset \mathfrak{F}$. By (3), the ultrafilter $\mathfrak{A}$ has a limit $x \in X$. By assertion (iii) of Theorem 2 in Subsection 16.1.2, x is a limit point of the filter $\mathfrak{F}$. $\qquad\square$

Corollary 2. *Suppose $\mathfrak{A}$ is an ultrafilter on E, X a topological space, and the image of the mapping $f : E \to X$ lies in a compact subset $K \subset X$. Then there exists the limit $\lim_{\mathfrak{A}} f$.*

Proof. Consider f as a mapping acting from E into K. Since (Corollary 1) $f[\mathfrak{A}]$ is an ultrafilter on the compact space K, there exists the limit $\lim f[\mathfrak{A}]$. But, by definition, $\lim_{\mathfrak{A}} f = \lim f[\mathfrak{A}]$. $\qquad\square$

Exercises

1. Let E be a set and e be an element of E. Verify that the family $\mathfrak{A}_e \subset 2^E$ of all sets containing e constitutes an ultrafilter on E. Ultrafilters of this form are called *trivial ultrafilters*.

2. Let E be a set, X a topological space, and $e \in E$. Then $f(e) = \lim_{\mathfrak{A}_e} f$ for any mapping $f : E \to X$.

3. Prove that on any infinite set there exist non-trivial ultrafilters. It is interesting that to construct an explicit example of a non-trivial ultrafilter is in principle impossible: such a construction necessarily relies on the Axiom of Choice or on Zorn's lemma.

4. Let $\mathfrak{A}$ be an ultrafilter on E. Use induction on n to show that if an element $A \in \mathfrak{A}$ is covered a finite number of sets: $A \subset \bigcup_{k=1}^{n} A_k$, then at least one of the sets A_k belongs to $\mathfrak{A}$.

5. Every ultrafilter on a finite set E is trivial.

6. Let $\mathfrak{A}$ be an ultrafilter on $\mathbb{N}$. Then either $\mathfrak{A}$ is trivial, or $\mathfrak{A}$ majorizes the Fréchet filter.

7. Let $\mathfrak{A}$ be an ultrafilter on $\mathbb{N}$ which majorizes the Fréchet filter. Then $f \mapsto \lim_{\mathfrak{A}} f$ is a continuous linear functional on ℓ_∞ (recall that sequences $f = (f_1, f_2, \ldots)$, i.e., elements of the space ℓ_∞, can be regarded as bounded functions on $\mathbb{N}$ with values in the corresponding field of scalars $\mathbb{R}$ or $\mathbb{C}$). Based on this example, show that $(\ell_\infty)^* \neq \ell_1$.

8. Let $\mathfrak{A}_1, \mathfrak{A}_2$ be ultrafilters on $\mathbb{N}$, $\mathfrak{A}_1 \neq \mathfrak{A}_2$. Then there exists an $f \in \ell_\infty$ for which $\lim_{\mathfrak{A}_1} f \neq \lim_{\mathfrak{A}_2} f$.

9. For each set $A \subset \mathbb{N}$ we denote by $\mathcal{U}_A$ the family of all ultrafilters on $\mathbb{N}$ that have A as an element. We equip the set $\beta\mathbb{N}$ of all ultrafilters on $\mathbb{N}$ with the following topology: the neighborhoods of the ultrafilter $\mathfrak{A}$ are all sets $\mathcal{U}_A$ with $A \in \mathfrak{A}$, as well as all the larger sets. More formally: the topology on $\beta\mathbb{N}$ is specified by neighborhood bases (see Subsection 1.2.1); as a neighborhood basis of the element $\mathfrak{A} \in \beta\mathbb{N}$ one takes the family $\mathcal{U}_{\mathfrak{A}} = \{\mathcal{U}_A : A \in \mathfrak{A}\}$.[1] Verify the axioms given in Subsection 1.2.1, the satisfaction of which is necessary for the specification of a topology by means of neighborhoods.

10. Identify the trivial ultrafilter $\mathfrak{A}_n$, generated by the point $n \in \mathbb{N}$, with the point n itself. Under this identification, $\mathbb{N} \subset \beta\mathbb{N}$. Prove that $\mathbb{N}$ is a dense subset of the topological space $\beta\mathbb{N}$, i.e., $\beta\mathbb{N}$ is separable.

11. Let $\mathfrak{A}$ be an ultrafilter on $\mathbb{N}$ which majorizes the Fréchet filter. For each $x = (x_1, x_2, \ldots) \in \ell_\infty$ define $F(x)$ as the limit with respect to $\mathfrak{A}$ of the function f given by $f(n) = (x_1 + x_2 + \cdots + x_n)/n$. Verify that the functional F is invariant under translations. By this construction you will obtain a proof of the existence of the generalized Banach limit (see the exercises in Subsection 5.5.2) that does not resort to the Hahn–Banach theorem.

16.1.4 The Topology Generated by a Family of Mappings. The Tikhonov Product

Suppose that on the set X there is given a family of mappings $\mathcal{F}$, where the mappings $f \in \mathcal{F}$ act in respective (possibly different) topological spaces $f(X)$. For any point $x \in X$, any finite family of mappings $\{f_k\}_{k=1}^n \subset \mathcal{F}$, and any open neighborhoods V_k of the points $f_k(x)$ in the spaces $f_k(X)$, respectively, we introduce the sets

$$U_{n, \{f_k\}_{k=1}^n, \{V_k\}_{k=1}^n}(x) = \bigcap_{k=1}^n f_k^{-1}(V_k).$$

---

[1] What a splendid thing is the modern system of notations: $\mathcal{U}_{\mathfrak{A}}$ is a familiy of neighborhoods. Each neighborhood is a set of ultrafilters. Each ultrafilter is a family of sets of natural numbers. Thus, with one symbol $\mathcal{U}_{\mathfrak{A}}$ we managed to denote a set of sets of sets of sets of natural numbers!

Recall (Subsection 1.2.1) the following fact: Suppose that for each point $x \in X$ there is given a non-empty family $\mathcal{U}_x$ of subsets with the following properties:

— if $U \in \mathcal{U}_x$, then $x \in U$;

— if $U_1, U_2 \in \mathcal{U}_x$, then there exists a $U_3 \in \mathcal{U}_x$ such that $U_3 \subset U_1 \cap U_2$;

— if $U \in \mathcal{U}_x$ and $y \in U$, then there exists a set $V \in \mathcal{U}_y$ such that $V \subset U$.

Then there exists a topology τ on X for which the families $\mathcal{U}_x$ are neighborhood bases of the corresponding points.

Consequently, on X there exist a topology (possibly not separated) in which the sets $U_{n,\{f_k\}_{k=1}^n,\{V_k\}_{k=1}^n}(x)$ constitute a neighborhood basis of the point x, for any $x \in X$. We denote this topology by $\sigma(X, \mathcal{F})$. In particular, among the neighborhoods of the point $x \in X$ in the topology $\sigma(X, \mathcal{F})$ there are all the sets $f^{-1}(V)$, where $f \in \mathcal{F}$ and V is a neighborhood of the point $f(x)$ in the topological space $f(X)$. Therefore, all the mappings in the family $\mathcal{F}$ are continuous in $\sigma(X, \mathcal{F})$.

Theorem 1. *$\sigma(X, \mathcal{F})$ is the weakest topology on X in which all the mappings belonging to the family $\mathcal{F}$ are continuous.*

Proof. Let τ be some topology in which all the mappings in the family $\mathcal{F}$ are continuous. Let us show that any set of the form $U_{n,\{f_k\}_{k=1}^n,\{V_k\}_{k=1}^n}(x)$ will be a neighborhood of the point x in the topology τ. This will prove that $\tau \succ \sigma(X, \mathcal{F})$. By hypothesis, all mappings $f_k: X \to f_k(X)$ are continuous in the topology τ. Hence, the sets $f_k^{-1}(V_k)$ are open neighborhoods of the point x in τ. Therefore, the intersection $U_{n,\{f_k\}_{k=1}^n,\{V_k\}_{k=1}^n}(x)$ of such sets is also an open neighborhood of x. $\square$

Definition 1. The topology $\sigma(X, \mathcal{F})$ is called the *topology generated by the family of mappings* $\mathcal{F}$. Another term (justified by the preceding theorem) is that of the *weakest topology in which all the mappings in the family $\mathcal{F}$ are continuous.*

Definition 2. A family of mappings $\mathcal{F}$ is said to *separate the points* of the set X if for any $x_1, x_2 \in X, x_1 \neq x_2$, there exists a mapping $f \in \mathcal{F}$ such that $f(x_1) \neq f(x_2)$.

Theorem 2. *Suppose all the spaces $f(X), f \in \mathcal{F}$, are Hausdorff. For the topology $\sigma(X, \mathcal{F})$ to be Hausdorff it is necessary and sufficient that the family $\mathcal{F}$ separates the points of the set X.*

Proof. Sufficiency. Suppose $\mathcal{F}$ separates the points of the set X. Then for any $x_1, x_2 \in X, x_1 \neq x_2$, there exists an $f \in \mathcal{F}$ such that $f(x_1) \neq f(x_2)$. Since $f(X)$ is a Hausdorff space, there exist disjoint neighborhoods V_1 and V_2 of the points $f(x_1)$ and $f(x_2)$, respectively. The sets $f^{-1}(V_1)$ and $f^{-1}(V_2)$ are the required $\sigma(X, \mathcal{F})$-neighborhoods that separate the points x_1 and x_2.

Necessity. Suppose $\mathcal{F}$ does not separate the points of X. Then there exist points $x_1, x_2 \in X, \ x_1 \neq x_2$, such that $f(x_1) = f(x_2)$ for all $f \in \mathcal{F}$. Pick an arbitrary $\sigma(X, \mathcal{F})$-neighborhood $U_{n,\{f_k\}_{k=1}^n,\{V_k\}_{k=1}^n}(x_1)$ of the point x_1. Since $f_k(x_1) = f_k(x_2)$ for all $k = 1, 2, \ldots, n$, the point x_2 will also lie in $U_{n,\{f_k\}_{k=1}^n,\{V_k\}_{k=1}^n}(x_1)$. Thus, in the described situation $\sigma(X, \mathcal{F})$ not only is not Hausdorff, it even fails the first separation axiom. $\square$

Theorem 3. *For the filter $\mathfrak{F}$ on X to converge in the topology $\sigma(X, \mathcal{F})$ to some $x \in X$, it is necessary and sufficient that $\lim_{\mathfrak{F}} f = f(x)$ for all $f \in \mathcal{F}$.*

Proof. In view of the continuity in the topology $\sigma(X, \mathcal{F})$ of all mappings $f \in \mathcal{F}$, the necessity follows from Theorem 3 of Subsection 16.1.2. Let us prove the sufficiency. Suppose $\lim_{\mathfrak{F}} f = f(x)$ for all $f \in F$. We need to show that every neighborhood of the form $U_{n,\{f_k\}_{k=1}^n,\{V_k\}_{k=1}^n}(x)$ will be an element of the filter $\mathfrak{F}$. By assumption, $\lim_{\mathfrak{F}} f_k = f_k(x)$, and so $f_k^{-1}(V_k) \in \mathfrak{F}$ for all $k = 1, 2, \ldots, n$. Since a filter is stable under taking finite intersections of elements, $U_{n,\{f_k\}_{k=1}^n,\{V_k\}_{k=1}^n}(x) = \bigcap_{k=1}^n f_k^{-1}(V_k) \in \mathfrak{F}$. $\square$

Let Γ be an index set (i.e., a set whose elements will henceforth referred to as indices). Suppose that to each index $\gamma \in \Gamma$ there is assigned a set X_γ. The *Cartesian product* of the sets X_γ with respect to $\gamma \in \Gamma$ is defined to be the set $\prod_{\gamma \in \Gamma} X_\gamma$ consisting of all mappings $x : \Gamma \to \bigcup_{\gamma \in \Gamma} X_\gamma$ with the property that $x(\gamma) \in X_\gamma$ for any $\gamma \in \Gamma$. In the particular case when all sets X_γ are equal to one and the same set X, the product consists of all functions $x : \Gamma \to X$; then the Cartesian product is called the *Cartesian power* and is denoted by X^Γ.

For the values of a function $x \in \prod_{\gamma \in \Gamma} X_\gamma$, instead of $x(\gamma)$ one uses the notation x_γ. In this notation the element $x \in \prod_{\gamma \in \Gamma} X_\gamma$ itself is usually written in the form $x = \{x_\gamma\}_{\gamma \in \Gamma}$ of an indexed set of values.

For any $\alpha \in \Gamma$, the mapping $P_\alpha : \prod_{\gamma \in \Gamma} X_\gamma \to X_\alpha$, acting by the rule $P_\alpha(x) = x_\alpha$, is called a *coordinate projection*.

Definition 3. Suppose all X_γ, $\gamma \in \Gamma$, are topological spaces. The *Tikhonov topology* on $\prod_{\gamma \in \Gamma} X_\gamma$ is the weakest topology in which all coordinate projections P_α, $\alpha \in \Gamma$, are continuous. The Cartesian product $\prod_{\gamma \in \Gamma} X_\gamma$, equipped with the Tikhonov topology, is called the *Tikhonov product*.

We note that, obviously, the coordinate projections separate the points of the product, and so, by Theorem 2, a Tikhonov product of Hausdorff spaces is again a Hausdorff space. Further, Theorem 3 yields the following assertion:

Convergence Criterion in a Tikhonov Product. *A filter $\mathfrak{F}$ on $\prod_{\gamma \in \Gamma} X_\gamma$ converges in the Tikhonov topology to an element $x = \{x_\gamma\}_{\gamma \in \Gamma}$ if and only if $x_\gamma = \lim_{\mathfrak{F}} P_\gamma$ for all $\gamma \in \Gamma$.*

Let us describe the Tikhonov topology explicitly, i.e., describe in more detail the form that the neighborhoods of the topology generated by a family of maps take in this particular case. Let $x \in \prod_{\gamma \in \Gamma} X_\gamma$; let $N \subset \Gamma$ be a finite set of indices, and $V_\gamma \subset X_\gamma$, $\gamma \in N$, be open neighborhoods of the corresponding points x_γ. Define

$$U_{N,\{V_\gamma\}_{\gamma \in N}}(x) = \left\{ y \in \prod_{\gamma \in \Gamma} X_\gamma : y_\alpha \in V_\alpha \text{ for all } \alpha \in N \right\}.$$

Theorem 4. *The sets of the form $U_{N,\{V_\gamma\}_{\gamma \in N}}(x)$ form a basis of neighborhoods of the point $x \in \prod_{\gamma \in \Gamma} X_\gamma$ in the Tikhonov topology.* $\square$

Theorem 5 (Tikhonov's theorem on products of compact spaces). *Let X_γ, $\gamma \in \Gamma$ be compact topological spaces. Then the Tikhonov product $\prod_{\gamma \in \Gamma} X_\gamma$ is also compact.*

Proof. We use criterion (3) of Theorem 3 in Subsection 16.1.3. Let $\mathfrak{A}$ be an ultrafilter on $\prod_{\gamma \in \Gamma} X_\gamma$. Since all the spaces X_γ are compact, for each $\gamma \in \Gamma$ the coordinate projection P_γ has a limit. Denote it by $y_\gamma = \lim_{\mathfrak{A}} P_\gamma$. Then the element $y = \{y_\gamma\}_{\gamma \in \Gamma}$ is the limit of the ultrafilter $\mathfrak{A}$. $\qquad\qquad\square$

Exercises

1. In the case where $\Gamma = \{1, 2\}$, the definition of the Tikhonov product $\prod_{\gamma \in \Gamma} X_\gamma$ coincides with the definition of the product $X_1 \times X_2$ of topological spaces introduced earlier in Subsection 1.2.2.

2. A particular case of the Tikhonov product — the Tikhonov power X^Γ of the topological space X — is the space of all functions $f \colon \Gamma \to X$. Write in explicit form the neighborhoods of a function f in the Tikhonov topology.

3. Prove that a sequence of functions $f_n \in X^\Gamma$ converges in the Tikhonov topology to a function f if and only if $f_n(x) \to f(x)$ for all $x \in X$. This justifies yet another name used for the Tikhonov topology — *topology of pointwise convergence*.

4. For a particular case of the Tikhonov power — the space $[0, 1]^{[0,1]}$ of all functions $f \colon [0, 1] \to [0, 1]$ — write explicitly the neighborhoods of a function f. Prove that the set of all polynomials with rational coefficients is dense in $[0, 1]^{[0,1]}$, i.e., $[0, 1]^{[0,1]}$ is a separable space.

A topological space X is said to be *sequentially compact* if from any sequence of elements in X one can extract a convergent subsequence.

5. The space $[0, 1]^{[0,1]}$, despite being compact, is not sequentially compact (see Exercise 10 in Subsection 3.2.2).

A subset A of a topological space X is said to be *sequentially dense* if for any $x \in X$ there exists a sequence $a_n \in A$ that converges to x. A topological space X is said to be *sequentially separable*, if X contains a countable sequentially dense set.

6. A sequentially separable Hausdorff topological space cannot have cardinality larger than the cardinality of the continuum.

7. The space $[0, 1]^{[0,1]}$, despite its separability, is not sequentially separable.

8. Let G_γ be a topological group. Equip the Tikhonov product $\prod_{\gamma \in \Gamma} G_\gamma$ with the operation $\{x_\gamma\}_{\gamma \in \Gamma} \cdot \{y_\gamma\}_{\gamma \in \Gamma} = \{x_\gamma \cdot y_\gamma\}_{\gamma \in \Gamma}$. Verify that $\prod_{\gamma \in \Gamma} G_\gamma$ is a topological group.

9. Equip the two-point set $\{0, 1\}$ with the discrete topology. Prove that the Tikhonov power $\{0, 1\}^{\mathbb{N}}$ is homeomorphic to the Cantor perfect set.

10. Let X be a fixed set. Identifying each subset $A \subset X$ with its characteristic function $\mathbb{1}_A$, we obtain a natural identification of the family 2^X of all subsets of the space X with the space $\{0, 1\}^X$ of all functions $f : X \to \{0, 1\}$. Since the two-point set is a (discrete) compact space, the space $\{0, 1\}^X = 2^X$ is compact in the Tikhonov topology. Describe explicitly the neighborhoods of the set $A \subset X$ in the Tikhonov topology on 2^X.

11. The topological space $\beta\mathbb{N}$ defined in Exercise 9 of Subsection 16.1.3 is a closed subset of the compact space $2^{2^{\mathbb{N}}}$. Hence, $\beta\mathbb{N}$ is compact as well. The space $\beta\mathbb{N}$ is called the *Stone–Čech compactification* of the natural numbers.

12. Define the operator $T : C(\beta\mathbb{N}) \to \ell_\infty$ by the rule: $T(f)$ is the sequence with coordinates $x_n = f(\mathfrak{A}_n)$, where $\mathfrak{A}_n$ denotes the trivial ultrafilter generated by the point $n \in \mathbb{N}$. Prove that T is a linear bijective isometry. Therefore, the space ℓ_∞ is isometric to the space of continuous functions on a (admittedly rather exotic) compact space.

16.2 Background Material on Topological Vector Spaces

We have already encountered topologies and the corresponding types of convergence on linear spaces of functions with the feature that the convergence cannot be described as convergence with respect to a norm. These were, for instance, pointwise convergence and convergence in measure. Such types of convergence will, with rare exceptions, be the weak and weak* convergence in Banach spaces — the main objects of study in Chap. 17. An adequate language for describing such topologies and convergences is that of topological vector spaces.

16.2.1 Axiomatics and Terminology

Definition 1. A linear space X (real or complex) endowed with a topology τ is called a *topological vector space* if the topology τ is compatible with the linear structure, in the sense that the operations of addition of elements and multiplication of an element by a scalar are jointly continuous in their variables.

To avoid treating the real and complex cases separately each time, we will assume that all spaces are complex, leaving the simpler case of real spaces to the reader for independent study.

Let us explain Definition 1 in more detail. Let X be a topological vector space. Consider the mappings $F: X \times X \to X$ and $G: \mathbb{C} \times X \to X$, acting by the rules $F(x_1, x_2) = x_1 + x_2$ and $G(\lambda, x) = \lambda x$. The compatibility of the topology with the linear structure means that each of the mappings F and G is jointly continuous in its variables. We will use this continuity step by step to deduce geometric properties of neighborhoods in the topology compatible with the linear structure.

Theorem 1. *Let U be an open set in X. Then*

— *for any $x \in X$, the set $U + x$ is open;*

— *for any $\lambda \in \mathbb{C} \setminus \{0\}$, the set λU is open.*

Proof. Fix $x_2 = -x$ and use the continuity of the mapping $F(x_1, x_2) = x_1 + x_2$ in the first variable when the second variable is fixed. The mapping $f(x_1) = x_1 - x$ is continuous in x_1, and $U + x$ is the preimage of the open set U under f. Consequently, $U + x$ is open. The second property is deduced in exactly the same way, using the continuity of the mapping $g(x) = \frac{1}{\lambda}x$. $\square$

It follows from Theorem 1 that the neighborhoods of an arbitrary element $x \in X$ are the sets $U + x$ with U a neighborhood of zero. Accordingly, the topology τ is uniquely determined by the family $\mathfrak{N}_0$ of neighborhoods of zero. For this reason, further properties of the topology τ will be formulated in the language of neighborhoods of zero. Below $\mathbb{C}_r$ will denote the disc of radius r in $\mathbb{C}$ centered at zero: $\mathbb{C}_r = \{\lambda \in \mathbb{C} : |\lambda| \leqslant r\}$.

Let us recall several definitions from Subsection 5.4.2. A subset A of a linear space X is said to be *absorbing* if for any $x \in X$ there exists an $n \in \mathbb{N}$ such that $x \in tA$ for all $t > n$. A subset $A \subset X$ is said to be *balanced* if for any scalar $\lambda \in \mathbb{C}_1$ it holds that $\lambda A \subset A$.

Theorem 2. *The family $\mathfrak{N}_0$ of neighborhoods of zero in the linear space X has the following properties:*

(i) *Any neighborhood of zero is an absorbing set.*

(ii) *Any neighborhood of zero contains a balanced neighborhood of zero.*

(iii) *For any neighborhood $U \in \mathfrak{N}_0$ there exists a balanced neighborhood $V \in \mathfrak{N}_0$ such that $V + V \subset U$.*

Proof. (i) Fix $x \in X$ and use the continuity of the mapping $f(\lambda) = \lambda x$. Since $f(0) = 0$, continuity at the point $\lambda = 0$ means that for any $U \in \mathfrak{N}_0$ there exists an $\varepsilon > 0$ such that $\lambda x \in U$ for all $\lambda \in \mathbb{C}_\varepsilon$. Defining $t = 1/\lambda$, we see that $x \in tU$ for all $t > 1/\varepsilon$.

(ii) Let $U \in \mathfrak{N}_0$. Thanks to the continuity of the mapping $G(\lambda, x) = \lambda x$ at the point $(0, 0)$, there exist an $\varepsilon > 0$ and a neighborhood $W \in \mathfrak{N}_0$ such that $\lambda x \in U$ for all $\lambda \in \mathbb{C}_\varepsilon$ and all $x \in W$. Set $V = \bigcup_{\lambda \in \mathbb{C}_\varepsilon} \lambda W$. Let us show that the set $V \subset U$ provides the requisite balanced neighborhood of zero. On one hand, $V \supset W$, whence $V \in \mathfrak{N}_0$. On the other hand, for any $\lambda_0 \in \mathbb{C}_1$ we have $\lambda_0 \mathbb{C}_\varepsilon \subset \mathbb{C}_\varepsilon$, and so

$$\lambda_0 V = \bigcup_{\lambda \in \mathbb{C}_\varepsilon} \lambda_0 \lambda W = \bigcup_{\mu \in \lambda_0 \mathbb{C}_\varepsilon} \mu W \subset \bigcup_{\mu \in \mathbb{C}_\varepsilon} \mu W = V;$$

this proves that the neighborhood V is balanced.

(iii) Thanks to the continuity of the mapping $F(x_1, x_2) = x_1 + x_2$ at $(0, 0)$, for any neighborhood $U \in \mathfrak{N}_0$ there exist neighborhoods $V_1, V_2 \in \mathfrak{N}_0$ such that $V_1 + V_2 \subset U$. Then based on item (ii) we choose the requisite balanced neighborhood V of zero so that V is contained in the neighborhood $V_1 \cap V_2$. $\square$

We invite the reader to prove the converse result:

Theorem 3. *Suppose the system $\mathfrak{N}_0$ of neighborhoods of zero in a topology τ on the linear space X satisfies the conditions (i)–(iii) in Theorem 2, and for every point $x \in X$ the system $\mathfrak{N}_x$ of neighborhoods of x is obtained from $\mathfrak{N}_0$ by parallel translation by the vector x. Then the topology τ is compatible with the linear structure.* $\square$

Remark 1. In view of the balancedness property, the condition $V + V \subset U$ of item (iii) of Theorem 2 can be rewritten as $V - V \subset U$.

Theorem 4. *For a topological vector space X to be Hausdorff it is necessary and sufficient that the system $\mathfrak{N}_0$ of neighborhoods of zero satisfies the following condition: for any $x \neq 0$ there exists a $U \in \mathfrak{N}_0$ such that $x \notin U$.*

Proof. Suppose $x \neq y$. Then $x - y \neq 0$ and there exists a neighborhood $U \in \mathfrak{N}_0$ which does not contain $x - y$. Pick a neighborhood $V \in \mathfrak{N}_0$ such that $V - V \subset U$. Then the neighborhoods $x + V$ and $y + V$ are disjoint: assuming the contrary, i.e., that there exists a point z which belongs to both $x + V$ and $y + V$, we have $z - x \in V$, $z - y \in V$, and so $x - y = (z - y) - (z - x) \in V - V \subset U$. $\square$

Exercises

1. A balanced set in $\mathbb{C}$ is either the whole set $\mathbb{C}$, or a disc (open or closed) centered at zero, or, finally, consists only of zero.

2. Replacing $\lambda \in \mathbb{C}_1$ by $\lambda \in [-1, 1]$, formulate the analogue of being a balanced set for real spaces. Prove for real spaces the analogue of Theorem 2.

3. Describe the balanced sets in $\mathbb{R}$.

4. Suppose the topology τ on the linear space X is compatible with the linear structure and satisfies the first separation axiom: every point is a closed set. Then the topology τ is Hausdorff.

5. Every topological vector space is also a topological group with respect to addition.

6. Prove that the discrete topology (namely, all sets are open) on $\mathbb{C}$ is compatible with the additive group structure, but not with the linear structure.

Verify that the spaces listed below are topological vector spaces.

7. The space $L_0(\Omega, \Sigma, \mu)$ of measurable functions on a finite measure space, equipped with the topology of convergence in measure (Subsection 3.2.2). The standard neighborhood basis of the function f is provided by the sets of functions $\{g \in L_0(\Omega, \Sigma, \mu) : \mu\{t : |g(t) - f(t)| > \delta\} < \varepsilon\}, \delta, \varepsilon > 0$. In this space, as usual, functions that coincide almost everywhere are identified: without this convention, the space would not be separated.

8. Any normed space with the topology defined by its norm.

9. Any Tikhonov product $\prod_{\gamma \in \Gamma} X_\gamma$ of topological vector spaces X_γ, with the linear operations defined coordinatewise: $a\{x_\gamma\}_{\gamma \in \Gamma} + b\{y_\gamma\}_{\gamma \in \Gamma} = \{ax_\gamma + by_\gamma\}_{\gamma \in \Gamma}$.

10. Any linear subspace of a topological vector space, equipped with the induced topology.

Other natural examples will be given in Subsection 16.3.2. Prove that in a topological vector space:

11. The interior and closure of a convex set are convex.

12. The closure of any linear subspace is a linear subspace.

13. Any neighborhood of zero contains a balanced open neighborhood of zero.

14. Any neighborhood of zero contains a balanced closed neighborhood of zero.

Any metrizable topological vector space satisfies the first countability axiom: every point has a countable neighborhood basis. For Hausdorff topological vector spaces the converse is also true. The reader will obtain the proof by solving the following chain of exercises.

Suppose X is a Hausdorff topological vector space and the family of neighborhoods of zero of the space X has a countable basis. Then:

15. There exists a neighborhoods basis $\{V_n\}$ of zero consisting of balanced open sets that satisfy the condition $V_{n+1} + V_{n+1} \subset V_n, n = 1, 2, \ldots$.

16. Denote by D the set of dyadic rational numbers in the segment $(0, 1]$. For each $r \in D, r < 1$, write its dyadic fraction expansion: $r = \sum_{k=1}^{n(r)} c_k(r)2^{-k}$, where $c_k(r) \in \{0, 1\}$, and $n(r)$ can be arbitrarily large, and define $U(r) = \sum_{k=1}^{n(r)} c_k(r)V_k$. For $r \geqslant 1$, put $U(r) = X$. Then all the sets $U(r)$ are balanced, open, and satisfy $U(1/2^n) = V_n, n = 1, 2, \ldots$, and $U(r) + U(s) \subset U(r + s)$ for all $r, s \in D$.

17. For each $x \in X$, put $\theta(x) = \inf\{r \in D : x \in U(r)\}$. Then the quantity θ is symmetric: $\theta(-x) = \theta(x)$, and satisfies the triangle inequality $\theta(x + y) \leqslant \theta(x) + \theta(y)$ for all $x, y \in X$.

18. The function $\rho(x, y) = \theta(x - y)$ is a metric on X. The topology defined by the metric ρ coincides with the original topology of the space.

16.2.2   Completeness, Precompactness, Compactness

To work successfully with topological vector spaces, we need to define analogues of the basic notions that are used in the setting of normed spaces. Since in general a topological vector space is not metrizable, we need to renounce the language of sequences and use instead the language of neighborhoods and filters befitting our general situation.

Definition 1. A filter $\mathfrak{F}$ on X is called a *Cauchy filter* if for any neighborhood U of zero there exists an element $A \in \mathfrak{F}$ such that $A - A \subset U$. Such an element A is said to be *small of order U*.

Theorem 1. *If the filter $\mathfrak{F}$ has a limit, then $\mathfrak{F}$ is a Cauchy filter.*

Proof. Suppose $\lim \mathfrak{F} = x$ and $U \in \mathfrak{N}_0$. Pick a $V \in \mathfrak{N}_0$ such that $V - V \subset U$. By the definition of the limit, there exists an $A \in \mathfrak{F}$ such that $A \subset x + V$. Then $A - A \subset (x + V) - (x + V) \subset V - V \subset U$. $\qquad\square$

Theorem 2. *Let $\mathfrak{F}$ be a Cauchy filter on a topological vector space X and x a limit point of $\mathfrak{F}$. Then $\lim \mathfrak{F} = x$.*

Proof. Let $x + U$ be an arbitrary neighborhood of the point x, with $U \in \mathfrak{N}_0$. Pick a neighborhood $V \in \mathfrak{N}_0$ with $V + V \subset U$ and a set $A \in \mathfrak{F}$, small of order V: $A - A \subset V$. By the definition of a limit point, the sets A and $x + V$ intersect, i.e., there exists a point $y \in A \cap (x + V)$. Then

$$x + U \supset x + V + V \supset y + V \supset y + A - A \supset y + A - y = A.$$

Hence, the neighborhood $x + U$ contains an element of $\mathfrak{F}$, and so $x + U \in \mathfrak{F}$. $\qquad\square$

Definition 2. A set A in a topological vector space X is said to be *complete*[2] if any Cauchy filter on X that contains A as an element has a limit which belongs to A. In particular, a topological vector space X is said to be *complete* if every Cauchy filter on X has a limit.

---

[2]Here again the already mentioned terminological confusion is widespread. The current term is introduced to generalize the notion of complete metric space. Equally successfully one could have called complete a set whose linear span coincides with the space X (a term used in the theory of linear spaces) or, by analogy with the theory of normed spaces, call a set complete if its linear span is dense in X. We thus obtain identically named notions which however have nothing in common. The relevant meaning must be figured out from the context.

Theorem 3. *Let X be a subspace of a topological vector space E and $A \subset X$ a complete subset of X. Then A is also complete as a subset of the space E.*

Proof. Let $\mathfrak{F}$ be a Cauchy filter in E which contains A as an element. Then, in particular, $X \in \mathfrak{F}$, i.e., the trace $\mathfrak{F}_X$ on X of the filter $\mathfrak{F}$ is a filter. Next, $\mathfrak{F}_X$ is a Cauchy filter on X which contains A as an element. Hence, in view of the completeness of A in X, the filter $\mathfrak{F}_X$ has in X a limit $a \in A$. The same point a is the limit of the filter $\mathfrak{F}$ in E. $\qquad\square$

Theorem 4. *Every complete subset A of a Hausdorff topological vector space X is closed. In particular, if a subspace of a Hausdorff topological vector space is complete in the induced topology, then this subspace is closed.*

Proof. Suppose the point $x \in X$ belongs to the closure of the set A. We need to show that $x \in A$. Consider the family $\mathfrak{D}$ of all intersections $(x + U) \cap A$, where $U \in \mathfrak{N}_0$. All such intersections are non-empty, and $\mathfrak{D}$ obeys all axioms of a filter basis. The filter $\mathfrak{F}$ generated by the basis $\mathfrak{D}$ majorizes the filter $\mathfrak{N}_x$ of all neighborhoods of the point x, and so $\lim \mathfrak{F} = x$. In particular, $\mathfrak{F}$ is a Cauchy filter. By construction, our complete set A is an element of the filter $\mathfrak{F}$. Hence, by Definition 2, $\mathfrak{F}$ must have a limit in A. Since the limit is unique, $x \in A$, as we needed to prove. $\qquad\square$

Definition 3. A set A in a topological vector space X is called *precompact* if for any neighborhood U of zero there exists a finite set $B \subset X$ such that $A \subset B + U$. Such a set B is called, by analogy with an ε-net, a U-*net* of the set A.

Theorem 5. *For a set A of a Hausdorff topological vector space X to be compact it is necessary and sufficient that A be simultaneously precompact and a complete set in X.*

Proof. Necessity. Let A be a compact set and U be an arbitrary open neighborhood of zero in X. The neighborhoods of the form $x + U$ with $x \in A$ form an open cover of the compact set A, hence there exists a finite subcover $x_1 + U, x_2 + U, ..., x_n + U$, with $x_k \in A$. The set $B = \{x_1, x_2, ..., x_n\}$ is a U-net of the set A. This establishes the precompactness of the compact set A. Now let us prove the completeness. Suppose $\mathfrak{F}$ is a Cauchy filter in X which contains A as an element. Then the trace $\mathfrak{F}_A$ on A of the filter $\mathfrak{F}$ is a filter in the compact topological space A, so $\mathfrak{F}_A$ has a limit point $a \in A$. The same point is then a limit point for $\mathfrak{F}$. But a limit point of a Cauchy filter is the limit of that filter. Therefore, $\mathfrak{F}$ has a limit, and $\lim \mathfrak{F} = a \in A$.

Necessity. Let A be a complete precompact set in X. Let us prove that every ultrafilter $\mathfrak{A}$ on A has a limit. Consider the filter $\widetilde{\mathfrak{A}}$, given already not on A, but on the entire space X, for which $\mathfrak{A}$ is a filter basis: $B \in \widetilde{\mathfrak{A}}$ if and only if $B \cap A \in \mathfrak{A}$. Using the ultrafilter criterion (Theorem 2 in Subsection 16.1.3), it is readily verified that $\widetilde{\mathfrak{A}}$ is an ultrafilter. We claim that $\widetilde{\mathfrak{A}}$ is a Cauchy filter. Indeed, let $U \in \mathfrak{N}_0$. Pick a neighborhood $V \in \mathfrak{N}_0$ such that $V - V \subset U$. Let $B = \{x_1, x_2, ..., x_n\}$ be the corresponding V-net of the precompact set A. Since the sets $x_1 + V, x_2 + V, ..., x_n + V$ form a finite open cover of the element A of the ultrafilter $\widetilde{\mathfrak{A}}$, one of these sets, say, $x_j + V$, will be an element of $\widetilde{\mathfrak{A}}$ (Exercise 4 of Subsection 16.1.3). But $x_j + V$ is small of order U:

$$(x_j + V) - (x_j + V) = V - V \subset U.$$

Thus, $\widetilde{\mathfrak{A}}$ is a Cauchy filter, $A \in \widetilde{\mathfrak{A}}$, and A is a complete set in X. Therefore, there exists $\lim \widetilde{\mathfrak{A}} \in A$. The same element will also be the limit in A of the filter $\mathfrak{A}$, the trace of the filter $\widetilde{\mathfrak{A}}$ on A (Exercise 4 of Subsection 16.1.2). $\qquad\square$

Definition 4. Let X be a topological vector space. We say that the neighborhood $U \in \mathfrak{N}_0$ of zero *absorbs* the set $A \subset X$ if there exists a number $t > 0$ such that $A \subset tU$. The set $A \subset X$ is said to be *bounded* if it is absorbed by every neighborhood of zero.

Theorem 6. *The family of bounded subsets of a topological vector space X enjoys the following properties:*

(a) *If $A \subset X$ is bounded, then for any neighborhood $U \in \mathfrak{N}_0$ there exists a number $N > 0$ such that $A \subset tU$ for all $t \geqslant N$.*

(b) *The union of any finite collection of bounded sets is a bounded set.*

(c) *Every finite set is bounded.*

(d) *Every precompact set in X is bounded.*

Proof. (a) Let $V \in \mathfrak{N}_0$ be a balanced neighborhood which is contained in U. Pick $N > 0$ such that $A \subset NV$. Then for every $t \geqslant N$ we have $A \subset NV = t((N/t)V) \subset tV \subset tU$.

(b) Let $A_1, A_2, \ldots, A_n$ be bounded sets, and U be a neighborhood of zero. By (a), for each of the sets A_k there exists a number $N_k \in \mathbb{N}$ such that $A_k \subset tU$ for all $t > N_k$. Put $N = \max_{1 \leqslant k \leqslant n} N_k$. Then for any $t \geqslant N$ all inclusions $A_k \subset tU$ hold simultaneously, that is, $\bigcup_{k=1}^n A_k \subset tU$.

(c) Any single-point set is bounded, since every neighborhood of zero is an absorbing set. It remains to use assertion (b).

(d) Let A be precompact in X and U be a neighborhood of zero. Pick a balanced neighborhood $V \in \mathfrak{N}_0$ such that $V + V \subset U$. By the definition of precompactness, there exists a finite set $B \subset X$ such that $A \subset B + V$. By (c), one can find a number $N > 1$ such that $B \subset NV$. Then $A \subset B + V \subset NV + V \subset N(V + V) \subset NU$. $\square$

Exercises

1. Let $\mathfrak{F}$ be a Cauchy filter in a topological vector space X. Suppose the filter $\mathfrak{F}_1$ majorizes $\mathfrak{F}$ and $x = \lim \mathfrak{F}_1$. Show that $x = \lim \mathfrak{F}$.

A sequence (x_n) of elements of a topological vector space X is called a *Cauchy sequence* if the filter $\mathfrak{F}_{(x_n)}$ generated by the sequence (x_n) is a Cauchy filter. Prove that:

2. (x_n) is a Cauchy sequence if and only if for any $U \in \mathfrak{N}_0$ there exists a number $N \in \mathbb{N}$ such that $x_n - x_m \in U$ for all $n, m \geq N$.

3. (x_n) is a Cauchy sequence if and only if for every $U \in \mathfrak{N}_0$ there exists a number $N \in \mathbb{N}$, such that $x_n - x_N \in U$ for all $n \geq N$.

4. Suppose the topological vector space X has a countable basis of neighborhoods of zero, and every Cauchy sequence in X has a limit. Then X is a complete space.

5. Suppose the complete topological vector space X has a countable basis of neighborhoods of zero U_n, $n \in \mathbb{N}$, and the neighborhoods U_n are chosen so that $U_{n+1} + U_{n+1} \subset U_n$. Pick in each set U_n one element $x_n \in U_n$. Then show that the series $\sum_{n=1}^{\infty} x_n$ converges.

6. Extend Banach's theorem on the inverse operator (if $T : X \to Y$ is linear, bijective, and continuous, then T^{-1} is continuous) to the case where X and Y are complete metrizable topological vector spaces.

7. Prove the completeness of the space $L_0(\Omega, \Sigma, \mu)$ of all measurable functions on a finite measure space, equipped with the topology of convergence in measure.

A metric ρ on a linear space X is said to be *invariant* if $\rho(x, y) = \rho(x - y, 0)$ for any $x, y \in X$. Suppose the topology τ of the topological vector space X is given by an invariant metric ρ. Then:

8. The sequence $(x_n) \subset X$ is Cauchy in the topology τ if and only if it is Cauchy in the metric ρ.

9. The completeness of the topological vector space (X, τ) is equivalent with the completeness of the metric space (X, ρ).

10. The precompactness of a set A in (X, τ) is equivalent to the precompactness of A in the metric ρ.

11. Warning: the boundedness of a set A in (X, τ) **is not equivalent** to the boundedness of A in the metric ρ. More precisely, boundedness in (X, τ) implies ρ-boundedness, but the converse is not true. As an example consider $X = \mathbb{R}$ with the natural topology, and introduce an invariant metric by the formula $\rho(x, y) = \arctan|x - y|$. Then $A = \mathbb{R}$ is a ρ-bounded set, but obviously A is not a bounded subset of the topological vector space $\mathbb{R}$.

Let X_γ, $\gamma \in \Gamma$ be topological vector spaces. We equip the space $X = \prod_{\gamma \in \Gamma} X_\gamma$ with the Tikhonov topology and the coordinatewise-defined linear operations. As usual, denote by $P_\gamma : X \to X_\gamma$, $\gamma \in \Gamma$, the coordinate projectors. Prove that

12. The set $A \subset X$ is bounded if and only if all images $P_\gamma(A) \subset X_\gamma$ are bounded.

13. The set $A \subset X$ is precompact if and only if all the images $P_\gamma(A) \subset X_\gamma$ are precompact.

14. For the closedness and compactness of a set A in the Tikhonov product the analogous criteria are no longer valid. Give examples showing this in the space $X = \mathbb{R} \times \mathbb{R}$.

15. A filter $\mathfrak{F}$ in $X = \prod_{\gamma \in \Gamma} X_\gamma$ is a Cauchy filter if and only if $P_\gamma(F)$ are Cauchy filters in the corresponding spaces X_γ.

16. If all X_γ, $\gamma \in \Gamma$, are complete spaces, then the space $X = \prod_{\gamma \in \Gamma} X_\gamma$ is also complete.

Consider the space $\mathbb{R}^{\mathbb{N}}$, equipped with the Tikhonov product topology. $\mathbb{R}^{\mathbb{N}}$ can be regarded as the space of all infinite numerical sequences $x = (x_1, x_2, \ldots)$. A neighborhood basis of zero is provided by the sets $U_{n,\varepsilon} = \{x \in \mathbb{R}^{\mathbb{N}} : \max_{1 \leqslant k \leqslant n} |x_k| < \varepsilon\}$. Prove that:

17. In $\mathbb{R}^{\mathbb{N}}$ there exists a countable neighborhood basis of zero, i.e., $\mathbb{R}^{\mathbb{N}}$ is metrizable.

18. A metrization of the space $\mathbb{R}^{\mathbb{N}}$ (under another commonly used name $\mathbb{R}^\omega$) was proposed in Exercise 11 of Subsection 1.3.1. Verify that the metric from that exercise generates the Tikhonov product topology on $\mathbb{R}^{\mathbb{N}}$.

19. Convergence in $\mathbb{R}^{\mathbb{N}}$ is equivalent to coordinatewise (or componentwise) convergence.

20. $\mathbb{R}^{\mathbb{N}}$ is a complete topological vector space.

21. A set $A \subset \mathbb{R}^{\mathbb{N}}$ is bounded if and only if there exists an element $b = (b_1, b_2, \ldots) \in (\mathbb{R}^+)^{\mathbb{N}}$ that majorizes all elements of A: for any $a = (a_1, a_2, \ldots) \in A$, the estimate $|a_n| \leqslant b_n$ holds for all $n \in \mathbb{N}$.

22. In $\mathbb{R}^{\mathbb{N}}$ the classes of bounded sets and precompact sets coincide.

23. Regard the sets $\overline{B}_{c_0}$ and $\overline{B}_{\ell_1}$, i.e., the closed unit balls in the spaces c_0 and ℓ_1, as subsets of the space $\mathbb{R}^{\mathbb{N}}$. Are these sets bounded in $\mathbb{R}^{\mathbb{N}}$? Closed in $\mathbb{R}^{\mathbb{N}}$? Precompact in $\mathbb{R}^{\mathbb{N}}$? Compact in $\mathbb{R}^{\mathbb{N}}$?

16.2.3 Linear Operators and Functionals

Throughout this subsection X and E will be topological vector spaces.

Theorem 1. *A linear operator $T : X \to E$ is continuous if and only if it is continuous at the point $x = 0$.*

Proof. A continuous operator is continuous at all points, in particular, at zero. Conversely, suppose the operator T is continuous at zero. Let us show that T is continuous at any point $x_0 \in X$. Let V be an arbitrary neighborhood of the point Tx_0 in E. Then $V - Tx_0$ is a neighborhood of zero in E. By assumption, $T^{-1}(V - Tx_0)$ is a neighborhood of zero in X. Thanks to the linearity of the operator, $T^{-1}(V) \supset T^{-1}(V - Tx_0) + x_0$, i.e., $T^{-1}(V)$ is a neighborhood of x_0. □

Definition 1. The linear operator $T: X \to E$ is said to be *bounded* if the image under T of any bounded subset of the space X is bounded in E.

Theorem 2. *Any continuous linear operator $T: X \to E$ is bounded.*

Proof. Let A be a bounded subset of X. We need to prove that the set $T(A)$ is bounded. Let V be an arbitrary neighborhood of zero in E and U a neighborhood of zero in X such that $T(U) \subset V$. Using the boundedness of A, pick an $N > 0$ such that $A \subset tU$ for all $t > N$. Then $T(A) \subset tT(U) \subset tV$ for all $t > N$. □

As we will show below, it is quite possible that two different topologies $\tau_1 \succ \tau_2$ on X (for instance, the strong and weak topologies of a normed space) generate one and the same system of bounded sets. In this case the identity operator, regarded as acting from (X, τ_2) into (X, τ_1), will be bounded, but discontinuous.

Theorem 3. *Suppose the operator $T: X \to E$ takes some neighborhood U of zero in the space X into a bounded set. Then T is continuous.*

Proof. Suppose $T(U)$ is a bounded set. For any neighborhood V of zero in E, there exists a $t > 0$ such that $T(U) \subset tV$. Then $\frac{1}{t}U \subset T^{-1}(V)$, i.e., $T^{-1}(V)$ is a neighborhood of zero in X. □

Next, we consider continuity conditions for linear functionals.

Theorem 4. *For a non-zero linear functional f on a topological vector space X, the following conditions are equivalent:*

  (i) *f is continuous;*

 (ii) *the kernel of the functional f is closed;*

(iii) *the kernel of the functional f is not dense in X;*

 (iv) *there exists a neighborhood U of zero for which $f(U)$ is a bounded set.*

Proof.
(i) $\Longrightarrow$ (ii). The preimage of any closed set is closed; in particular, $\mathrm{Ker}\, f = f^{-1}(0)$ is a closed set.

(ii) $\Longrightarrow$ (iii). If the kernel is closed and dense in X, then $\mathrm{Ker}\, f = X$, i.e., $f \equiv 0$.

(iii) $\Longrightarrow$ (iv). Suppose $\mathrm{Ker}\, f$ is not dense. Then there exist a point $x \in X$ and a balanced neighborhood U of zero such that $(U + x) \cap \mathrm{Ker}\, f = \emptyset$. This means that the functional cannot take the value $-f(x)$ at any point $y \in U$. Therefore, $f(U)$ is a balanced set of complex numbers which does not coincide with the whole complex plane $((-f(x) \notin f(U))$. It follows that $f(U)$ is a disc centered at zero (in the real case it would be an interval in $\mathbb{R}$ symmetric with respect to zero).

(iv) $\Longrightarrow$ (i). This implication was already established in Theorem 3. $\square$

As in the case of normed spaces, for a topological vector space X we denote by X^* the set of all continuous linear functionals on X.[3] The geometric form of the Hahn–Banach theorem admits a generalization to topological vector spaces.

Theorem 5 (Hahn–Banach separation theorem for topological vector spaces). *Let A and B be disjoint non-empty convex subsets of a real topological vector space X and let A be open. Then there exist a functional $f \in X^* \setminus \{0\}$ and a scalar $\theta \in \mathbb{R}$ such that $f(a) < \theta$ for all $a \in A$ and $f(b) \geqslant \theta$ for all $b \in B$.*

Using the connection between a linear functional and its real part (Subsection 9.1.1), one can obtain a version of the theorem for a complex space, replacing the conditions above by $\operatorname{Re} f(a) < \theta$ for all $a \in A$ and $\operatorname{Re} f(b) \geqslant \theta$ for all $b \in B$.

Proof. As in the case of normed spaces (Subsection 9.3.2), the theorem reduces to the following particular case: Let $A \subset X$ be an open convex neighborhood of zero in X, and let $x_0 \in X \setminus A$. Then there exists a functional $f \in X^* \setminus \{0\}$ such that $f(a) \leqslant f(x_0)$ for all $a \in A$.

In this last case the Minkowski functional φ_A of the set A is a convex functional (Subsection 5.4.2). Consider the subspace $Y = \operatorname{Lin}\{x_0\}$ and a linear functional f on Y with the property that $f(x_0) = \varphi_A(x_0)$. Then on Y the linear functional f is majorized by the convex functional φ_A (see the proof of the lemma in Subsection 9.3.2).

Now using the analytic form of the Hahn–Banach theorem (Subsection 5.4.3) we extend f to the entire space X with preservation of the linearity and the majorization condition $f(x) \leqslant \varphi_A(x)$. By the definition of the Minkowski functional, $\varphi_A(a) \leqslant 1$ for all $a \in A$, whence $f(a) \leqslant \varphi_A(a) \leqslant 1$ on A. Since $x_0 \notin A$, $\varphi_A(x_0) \geqslant 1$. Therefore, $f(a) \leqslant 1 \leqslant \varphi_A(x_0) = f(x_0)$ for all $a \in A$.

Further, since $f(x_0) \geqslant 1$, f is not identically equal to zero. By Lemma 5 in Subsection 9.3.1, which generalizes with no difficulty to topological vector spaces, the strict inequality $f(a) < f(x_0)$ holds for all $a \in A$. This means that the kernel $\operatorname{Ker} f$ does not intersect the non-empty open set $A - x_0$. Hence, $\operatorname{Ker} f$ cannot be dense, and the functional f is continuous. $\square$

Finally, let us generalize to finite-dimensional topological vector spaces properties of finite-dimensional normed spaces already known to us.

Theorem 6. *Let X be a Hausdorff topological vectors space, with $\dim X = n$. Then:*

(a) *Every linear functional on X is continuous.*

---

[3]Often, in textbooks on topological vector spaces, the symbol X^* is used to denote the set of **all** linear functionals on X, while the set of **continuous** linear functionals is denoted by X'. We will do exactly the opposite, in order to preserve the compatibility with the notations from the theory of normed spaces the reader is already familiar with.

(b) *For any topological vector space E, every linear operator $T: X \to E$ is continuous.*

(c) *X is isomorphic to the n-dimensional Hilbert space ℓ_2^n.*

(d) *X is complete.*

Proof. First note that for fixed n the implications (a) $\Longrightarrow$ (b) $\Longrightarrow$ (c) $\Longrightarrow$ (d) hold true. Indeed, (a) $\Longrightarrow$ (b), because if we choose in X a basis $\{x_k\}_{k=1}^n$ with the coordinate functionals $\{f_k\}_{k=1}^n$, the operator T can be represented in the form

$$T(x) = T\left(\sum_{k=1}^n f_k(x)x_k\right) = \sum_{k=1}^n f_k(x)Tx_k.$$

Thus, the calculation of $T(x)$ reduces to calculating the scalars $f_k(x)$ (this action is continuous due to assumption (a)), multiplying by them the constant vectors Tx_k, and summing the resulting products. But by the axioms of a topological vector space, multiplication by a scalar and taking the sum are continuous operations.

(b) $\Longrightarrow$ (c). Both spaces X and ℓ_2^n have the same dimension n, and so there exists a linear bijection $T: X \to \ell_2^n$. Both T and T^{-1} are continuous by condition (b).

Finally, (c) $\Longrightarrow$ (d) thanks to the completeness of the space ℓ_2^n.

The main assertion (a) is proved by induction on n. For $n = 0$ the space X reduces to $\{0\}$, and so the assertion is trivial. Let us perform the step $n \to n + 1$. Suppose $\dim X = n + 1$ and f is a non-zero linear functional on X. Then Ker f is an n-dimensional space. By the induction hypothesis, and using the already established implications (a) $\Longrightarrow$ (b) $\Longrightarrow$ (c) $\Longrightarrow$ (d), we conclude that Ker f is a complete space. Therefore, Ker f is closed in X, and, by Theorem 4, the functional f is continuous. $\square$

Exercises

1. Suppose the space X has a countable neighborhood basis of zero. Then every bounded linear operator $T: X \to E$ is continuous.

2. Let X be a topological vector space, $Y \subset X$ a subspace, and $q: X \to X/Y$ the quotient mapping. Define a topology τ on X/Y as follows: the set $U \subset X/Y$ is declared to be open if $q^{-1}(U)$ is an open set. Verify that:

— the topology τ is compatible with the linear structure;

— τ is the strongest among all topologies on X/Y in which the quotient mapping q is continuous;

— if the subspace Y is closed, then the space X/Y is separated, even when the initial space X is not separated;

— in the case of normed spaces, the topology τ coincides with the topology given by the quotient norm.

3. Prove the following generalization of Theorem 5 from Subsection 11.2.1: if in a Hausdorff topological vector space X there exists a precompact neighborhood of zero, then X is finite-dimensional.

4. On the example of the identity operator in $\mathbb{R}^{\mathbb{N}}$, show that the sufficient condition for continuity proved in Theorem 3 is not necessary.

5. Where in the proof of Theorem 6 of Subsection 16.2.3 was the assumption that the space is separated used? Will the theorem remain valid if the separation assumption is discarded?

16.3 Locally Convex Spaces

16.3.1 Seminorms and Topology

Definition 1. A topological vector space X is said to be *locally convex* if for any neighborhood U of zero there exists a convex neighborhood V of zero such that $V \subset U$. In other words, the space X is locally convex if the neighborhood system $\mathfrak{N}_0$ of zero has a basis consisting of convex sets.

Theorem 1. *Every convex neighborhood U of zero contains a convex balanced open neighborhood of zero. In particular, in a locally convex space there exists a neighborhood basis of zero consisting of convex balanced open sets.*

Proof. Let $V \subset U$ be an open and balanced neighborhood of zero. Then conv $V \subset U$. Let us show that conv V is a convex balanced and open neighborhood of zero. Convexity is obvious. Further, conv $V \supset V$, and hence conv V is a neighborhood of zero. Let us verify that conv V is balanced. Take $\lambda \in \mathbb{C}_1$, i.e., $|\lambda| \leqslant 1$. Then $\lambda V \subset V$ (since V is balanced), and λ conv $V = \text{conv}(\lambda V) \subset \text{conv } V$. Finally, let us verify that conv V is open. Since V is an open set and the operations of multiplication by non-zero scalars and taking the sum of sets leave the class of open sets invariant, all sets of the form $\sum_{k=1}^{n} \lambda_k V$ with $n \in \mathbb{N}$, $\lambda_k > 0$, and $\sum_{k=1}^{n} \lambda_k = 1$, are open. The conclusion follows from the fact that conv V is a union of sets of the form $\sum_{k=1}^{n} \lambda_k V$. $\qquad \square$

Recall (Definition 2 in Subsection 6.1.1) that a function $p \colon X \to \mathbb{R}$ is called a seminorm if $p(x) \geqslant 0$, $p(\lambda x) = |\lambda| p(x)$ for any $x \in X$ and any scalar λ, and $p(x + y) \leqslant p(x) + p(y)$ for all $x, y \in X$. A seminorm differs from a norm by the fact that $p(x)$ may be equal to zero for some non-zero elements $x \in X$. See also Exercises 10–13 in Subsection 6.1.3.

As in the case of a norm, the unit ball of the seminorm p is the set $B_p = \{x \in X : p(x) < 1\}$. The set B_p is convex and balanced. The seminorm p can be recovered from its unit ball by means of the Minkowski functional: $p(x) = \varphi_{B_p}(x)$ (see Subsection 5.4.2).

Theorem 2. *A seminorm p on a topological vector space X is continuous if and only if B_p is a neighborhood of zero.*

Proof. $B_p = p^{-1}(-1, 1)$ is the preimage of an open set. If p is continuous, then this preimage is open. Conversely, suppose that B_p is a neighborhood of zero, and let us show that the seminorm p is continuous. Thus, given any $x \in X$ and any $\varepsilon > 0$, we need to find a neighborhood U of the point x such that $p(U) \subset (p(x) - \varepsilon, p(x) + \varepsilon)$. Such a neighborhood is provided by $U = x + \varepsilon B_p$. Indeed, any point $y \in U$ has the form $y = x + \varepsilon z$, where $p(z) < 1$. Hence, by the triangle inequality, $p(x) - \varepsilon < p(y) < p(x) + \varepsilon$. $\square$

Definition 2. Let G be a family of seminorms on a linear space X. Denote by $\mathfrak{D}_G$ the collection of all finite intersections of sets rB_p, where $p \in G$ and $r > 0$. The *locally convex topology generated by the family of seminorms G* is the topology τ_G on X, in which a neighborhood basis of zero is $\mathfrak{D}_G$, and a neighborhood basis of a point $x \in X$ is, correspondingly, the collection of sets $x + U$ with $U \in \mathfrak{D}_G$.

A family G of seminorms is said to be *non-degenerate* if for any $x \in X \setminus \{0\}$ there exists a $p \in G$ such that $p(x) \neq 0$.

Theorem 3. I. *Let G be a family of seminorms on a linear space X. Then the topology τ_G generated by the family G is compatible with the linear structure and is locally convex.*
II. *The topology τ_G is separated if and only if the family of seminorms G is non-degenerate.*
III. *The topological vector space X is locally convex if and only if its topology is generated by a family of seminorms.*

Proof. I. Since a ball of a seminorm is a convex balanced and absorbing set and these properties are inherited by finite intersections, a neighborhood basis of zero $\mathfrak{D}_G$ consists of convex balanced absorbing sets. Further, for any $U \in \mathfrak{D}_G$ we have $V = (1/2)U \in \mathfrak{D}_G$, and thanks to convexity, $V + V \subset U$. Thus, we have verified conditions (i)–(iii) of Theorem 3 in Subsection 16.2.1. The proof of the fact that the conditions ensuring the existence of the topology given by families of open neighborhoods are satisfied is left to the reader. The compatibility of the topology with the linear structure follows from Theorem 4 of Subsection 16.2.1.

II. The characterization of the separation property follows from Theorem 4 of Subsection 16.2.1.

III. Let X be a locally convex space. By Theorem 1, X has a neighborhood basis $\mathfrak{D}$ of zero that consists of convex balanced open sets. Then as the elements of the sought-for family of seminorms one takes the seminorms whose unit balls are precisely the elements of the basis $\mathfrak{D}$. $\square$

Theorem 4. *Let X be a topological vector space and f be a linear functional on X. Then for f to be continuous it is necessary and sufficient that there exists a continuous seminorm p on X such that $|f(x)| \leqslant p(x)$ for all $x \in X$.*

Proof. Suppose f is continuous. Then $p(x) = |f(x)|$ is the required seminorm. Conversely, suppose that $|f(x)| \leqslant p(x)$, and p is a continuous seminorm. Then f is bounded on the neighborhood B_p of zero.   □

Theorem 5 (Hahn–Banach extension theorem in locally convex spaces). *Let f be a continuous linear functional given on a subspace Y of a locally convex space X. Then f can be extended to the entire space X with preservation of linearity and continuity.*

Proof. By assumption, the set $U = \{y \in Y : |f(y)| < 1\}$ is an open neighborhood of zero in Y. By the definition of the topology induced on a subspace, there exists a neighborhood V of zero in X such that $U \supset V \cap Y$. Since the space X is locally convex, one can take for the neighborhood V the unit ball of some continuous seminorm p given on X. By construction, for any $y \in Y$, if $p(y) < 1$, then $y \in U$ and $|f(y)| < 1$. That is, $|f(y)| \leqslant p(y)$ everywhere on Y.

Now, like for normed spaces, one needs to argue separately for the real and complex cases. If f is a real functional, then by the analytic form of the Hahn–Banach theorem, f can be extended to the entire space X with preservation of the inequality $f(x) \leqslant p(x)$. Replacing x by $-x$, we also obtain the inequality $-f(x) \leqslant p(x)$. Therefore, $|f(x)| \leqslant p(x)$, and the extended functional f is continuous. In the complex case, the extension can be performed so that the condition $\operatorname{Re} f(x) \leqslant p(x)$ is preserved on the entire space X. Taking for x the element $e^{-i\arg f(x)}x$, we arrive again at the inequality $|f(x)| \leqslant p(x)$, which establishes the continuity on X of the functional f.   □

Remark 1. A set of linear functionals $E \subset X'$ will separate the points if and only if for every $x \neq 0$ there exists an $f \in E$ such that $f(x) \neq 0$. Indeed, if the set E separates the points, then, in particular, E separates x from 0. Conversely, if $x \neq y$ are arbitrary points, then $x - y \neq 0$. A functional $f \in E$ for which $f(x - y) \neq 0$ will separate the points x and y.

Corollary 1. *The set X^* of all continuous linear functionals on a Hausdorff locally convex space X separates the points of X.*

Proof. For any $x \neq 0$ there exists a linear functional f on $\operatorname{Lin} \{x\}$ such that $f(x) \neq 0$. It remains to extend f to X by means of the Hahn–Banach theorem.   □

Exercises

1. On the example of the family of seminorms G consisting of a single norm, verify that the locally convex topology τ_G generated by the family of seminorms G

(Definition 2) does not coincide with the topology $\sigma(X, G)$ generated by the family of mappings G (Subsection 16.1.4). Moreover, $\sigma(X, G)$ is not compatible with the linear structure.

2. Let G be a family of seminorms, and F be the family of all functions of the form $f_x(y) = p(x + y)$, with $p \in G$ and $x \in X$. Then $\tau_G = \sigma(X, F)$.

3. A sequence $x_n \in X$ converges to $x \in X$ in the topology τ_G if and only if $p(x_n - x) \to 0$ as $n \to \infty$ for all $p \in G$.

4. Verify that the spaces listed below are indeed separated locally convex spaces, and describe the convergence of sequences in them. Prove the completeness and metrizability of the spaces in the first three examples. Is the fourth space metrizable? Complete?

— The space $\mathcal{H}(D)$ of holomorphic functions in a domain (i.e., connected open subset) $D \subset \mathbb{C}$, equipped with the locally convex topology generated by the family of all seminorms of $p_K(f) = \max_{z \in K} |f(z)|$, where K is a compact subset of D.

— The space $C^\infty[0, 1]$ of all infinitely differentiable functions on $[0, 1]$, equipped with the locally convex topology generated by the family of seminorms $p_n(f) = \max_{t \in [0,1]} |f^{(n)}(t)|$, $n = 0, 1, 2, \ldots$.

— The space $C^\infty(0, +\infty)$ of all infinitely differentiable functions on $(0, +\infty)$, equipped with the topology generated by the family of seminorms $p_n(f) = \max_{t \in (n^{-1}, n)} |f^{(n-1)}(t)|$, $n \in \mathbb{N}$.

— An infinite-dimensional linear space X, equipped with the *strongest locally convex topology*, i.e., the topology generated by the family of **all** seminorms on X.

5. Any Tikhonov product of locally convex spaces is locally convex.

6. Any subspace of a locally convex space is locally convex.

7. Any quotient space of a locally convex space (see Exercise 2 in Subsection 16.2.3) is locally convex.

8. Show that if U is a balanced set and f is a linear functional such that $\operatorname{Re} f(x) \leqslant a$ for all $x \in U$, then also $|f(x)| \leqslant a$ on U.

9. Applying the geometric form of the Hahn–Banach theorem to a set U and an open neighborhood V of the point x_0, prove the following corollary: Let U be a closed, balanced, and convex subset of a Hausdorff locally convex space X and let $x_0 \in X \setminus U$. Then there exists a continuous linear functional f such that $|f(y)| \leqslant 1$ for all $y \in U$ and $|f(x_0)| > 1$.

10. A series $\sum_{k=1}^\infty x_k$ in a locally convex space X is said to be *absolutely convergent* if $\sum_{k=1}^\infty p(x_k) < \infty$ for any continuous seminorm p on X. Prove that in a complete locally convex space every absolutely convergent series converges.

11. The space $L_0[0, 1]$ with the topology of convergence in measure is not a locally convex space. Moreover, any convex closed neighborhood of zero in $L_0[0, 1]$ coincides with the entire space. In particular, the only continuous linear functional on $L_0[0, 1]$ is the functional identically equal to zero.

16.3.2   Weak Topologies

Definition 1. Let X be a linear space, X' its algebraic dual (i.e., the space of all linear functionals on X), and $E \subset X'$ a subset. The *weak topology on X generated by the set of functionals E* is the weakest topology in which all functionals from E are continuous. This topology is a particular case of the topology defined by a family of mappings (Subsection 16.1.4). Accordingly, we denote it by the same symbol $\sigma(X, E)$.

Let us explain this definition in more detail. For any finite collection of functionals $G = \{g_1, g_2, \ldots, g_n\}$ and any $\varepsilon > 0$, define

$$U_{G,\varepsilon} = \bigcap_{g \in G} \{x \in X : |g(x)| < \varepsilon\} = \left\{x \in X : \max_{g \in G} |g(x)| < \varepsilon\right\}.$$

The family of sets $U_{G,\varepsilon}$ with $G = \{g_1, g_2, \ldots, g_n\} \subset E$ and $\varepsilon > 0$ constitutes a neighborhood basis of zero in the topology $\sigma(X, E)$. For an arbitrary element $x_0 \in X$, a neighborhood basis is provided by the sets of the form

$$\bigcap_{g \in G} \{x \in X : |g(x - x_0)| < \varepsilon\} = x_0 + U_{G,\varepsilon}.$$

This shows that $\sigma(X, E)$ is the locally convex topology generated by the family of seminorms $p_G(x) = \max_{g \in G} |g(x)|$, where G runs over all finite subsets of the set E. For this topology to be separated it is necessary and sufficient that the set of functionals E separates the points of the space X.

As we already remarked in Subsection 16.1.4, a filter $\mathfrak{F}$ on X converges in the topology $\sigma(X, E)$ to the element x if and only if $\lim_{\mathfrak{F}} f = f(x)$ for all $f \in E$. In particular, this convergence criterion is also valid for sequences: $x_n \to x$ in the topology $\sigma(X, E)$ if $f(x_n) \to f(x)$ for all $f \in E$.

We begin our more detailed study of weak topologies with a lemma that was proposed earlier as an exercise on the subject "functionals and codimension" (Subsection 5.3.3, Exercise 16). Here, for the reader's convenience, we provide a direct proof.

Lemma 1. *Let f and $\{f_k\}_{k=1}^n$ be linear functionals on X such that* $\mathrm{Ker}\, f \supset \bigcap_{k=1}^n \mathrm{Ker}\, f_k$. *Then $f \in \mathrm{Lin}\{f_k\}_{k=1}^n$.*

Proof. We use induction on n. The induction base is $n = 1$. If $f_1 = 0$, then $\operatorname{Ker} f \supset \operatorname{Ker} f_1 = X$, i.e., $f = 0$. Now let f_1 be a non-zero functional. Then $Y = \operatorname{Ker} f_1$ is a subspace in X. Therefore, there exists a vector $e \in X \setminus Y$, such that $\operatorname{Lin}\{e, Y\} = X$. Let $a = f(e)$ and $b = f_1(e)$. The functional $f - (a/b)f_1$ vanishes on Y as well as at the point e. Hence, $f - (a/b)f_1$ vanishes on the whole space $X = \operatorname{Lin}\{e, Y\}$, i.e., $f \in \operatorname{Lin}\{f_1\}$.

Step $n \to n + 1$. We introduce the subspace $Y = \bigcap_{k=1}^{n} \operatorname{Ker} f_k$. The condition $\operatorname{Ker} f \supset \bigcap_{k=1}^{n+1} \operatorname{Ker} f_k$ may be interpreted as saying that the kernel of the restriction of the functional f to Y contains the kernel of the restriction of the functional f_{n+1} to Y. Therefore (by the case $n = 1$), there exists a scalar α such that $f - \alpha f_{n+1}$ vanishes on the whole space $Y = \bigcap_{k=1}^{n} \operatorname{Ker} f_k$. That is, $\operatorname{Ker}(f - \alpha f_{n+1}) \supset \bigcap_{k=1}^{n} \operatorname{Ker} f_k$. By the induction hypothesis, $f - \alpha f_{n+1} \in \operatorname{Lin}\{f_k\}_{k=1}^{n}$, i.e., $f \in \operatorname{Lin}\{f_k\}_{k=1}^{n+1}$. $\square$

Lemma 2. *Let Y be a subspace of the linear space X, let $f \in X'$, and suppose there exists an $a > 0$ such that $|f(y)| \leqslant a$ for all Y. Then $f(y) = 0$ for all $y \in Y$.*

Proof. Suppose that there exists an $y_0 \in Y$ such that $f(y_0) \neq 0$. Then for the element $y = (2a/f(y_0))y_0 \in Y$ one has $|f(y)| = 2a > a$, a contradiction. $\square$

We are now ready to describe the functionals that are continuous in a weak topology.

Theorem 1. *A functional $f \in X'$ is continuous in the topology $\sigma(X, E)$ if and only if $f \in \operatorname{Lin} E$. In particular, if $E \subset X'$ is a linear subspace, then the set $(X, \sigma(X, E))^*$ of all $\sigma(X, E)$-continuous functionals on X coincides with E.*

Proof. By the definition of the topology $\sigma(X, E)$, all elements of the set E are $\sigma(X, E)$-continuous functionals. Hence, linear combinations of such functionals are also continuous. Conversely, suppose the functional $f \in X'$ is continuous in the topology $\sigma(X, E)$. Then there exist a finite set of functionals $G = \{g_1, g_2, \ldots, g_n\} \subset E$ and an $\varepsilon > 0$ such that on the neighborhood $U_{G,\varepsilon} = \{x \in X : \max_{g \in G} |g(x)| < \varepsilon\}$ all values of the functional f are bounded in modulus by some number $a > 0$. The same number will also bound the values of f on the subspace $Y = \bigcap_{k=1}^{n} \operatorname{Ker} g_k \subset U_{G,\varepsilon}$. By Lemma 2, the functional f vanishes on Y, which in turn means (Lemma 1) that $f \in \operatorname{Lin}\{g_k\}_{k=1}^{n} \subset \operatorname{Lin} E$. $\square$

Exercises

1. Prove the equality of topologies $\sigma(X, \operatorname{Lin} E) = \sigma(X, E)$.

2. The Tikhonov topology (topology of coordinatewise convergence) on $\mathbb{R}^{\mathbb{N}}$ coincides with the weak topology generated by the family $E = \{e_n^*\}_{n \in \mathbb{N}}$ of coordinate functionals. What is $\left(\mathbb{R}^{\mathbb{N}}\right)^*$ equal to?

3. Let $E \subset X'$ be a subspace. Then the topology $\sigma(X, E)$ has a countable neighborhood basis of zero if and only if the linear space E has an at most countable Hamel basis.

4. Suppose that on X there exists a norm that is continuous in the topology $\sigma(X, E)$. Then the space X is finite-dimensional.

5. Every set that is bounded in the topology $\sigma(X, E)$ is precompact in this topology.

6. Kolmogorov's theorem: *If in the topological vector space X there exists a bounded neighborhood U of zero, then the system of neighborhoods of zero has the countable basis $\{(1/n)U\}_{n\in\mathbb{N}}$. In particular, if this bounded neighborhood U is convex, then the topology of the space can be given by a single seminorm (a single norm, if the space is assumed to be separated).*

7. Let X be an infinite-dimensional linear space and let the family of functionals $E \subset X'$ separate points. Then none of the $\sigma(X, E)$-neighborhoods of zero is a $\sigma(X, E)$-bounded set.

8. The space $X = c_0$ is not complete in the topology $\sigma(X, X^*)$.

9. General result: no infinite-dimensional Banach space X is complete in the topology $\sigma(X, X^*)$.

16.3.3 Eidelheit's Interpolation Theorem

Lemma 1. *Let X be a topological vector space and $Y \subset X$ a closed subspace of finite codimension. If the functional $f \in X'$ is discontinuous, then the restriction of f to Y is also discontinuous.*

Proof. We use the properties of quotient spaces of topological vector spaces given in Exercise 2 of Subsection 16.2.3. Suppose, by contradiction, that the restriction of the functional f to Y is continuous. Then $\widetilde{Y} = Y \cap \text{Ker } f$ is a closed subspace of finite codimension. By the definition of the codimension, the quotient space $X/\widetilde{Y}$ is finite-dimensional. Define the functional $\widetilde{f}$ on $X/\widetilde{Y}$ by the rule $\widetilde{f}(q(x)) = f(x)$, where $q : X \to X/\widetilde{Y}$ is the quotient mapping. Since the space $X/\widetilde{Y}$ is finite-dimensional, $\widetilde{f}$ is continuous on $X/\widetilde{Y}$. Therefore, f is continuous, being the composition of $\widetilde{f}$ and the quotient mapping q. $\qquad\square$

Lemma 2. *Let X be a topological vector space, and $f \in X'$ a discontinuous functional. Then for any scalar a the hyperplane $f_{=a} = \{x \in X : f(x) = a\}$ is dense in X.*

Proof. The fact that the kernel of f is dense is guaranteed by Theorem 4 of Subsection 16.2.3. The hyperplane $f_{=a}$ is obtained from $\text{Ker } f$ by parallel translation by any fixed vector $y \in f_{=a}$. $\qquad\square$

Theorem 1. (M. Eidelheit [52]**).** *Let X be a complete locally convex subspace, the topology of which is given by a sequence of seminorms $p_1 \leqslant p_2 \leqslant p_3 \leqslant \cdots$. Suppose that the sequence of linear functionals $f_n \in X^*$ has the following property: for any $n \in \mathbb{N}$, the functional f_n is discontinuous with respect to the seminorm p_n (i.e., discontinuous in the topology generated by the single seminorm p_n), but is continuous with respect to p_{n+1}, and so also continuous with respect to all seminorms p_k with $k > n$. Then for any sequence of scalars a_n there exists an element $x \in X$ such that $f_n(x) = a_n$, $n = 1, 2, \ldots$.*

Proof. We construct the required element $x \in X$ as the sum of a series $\sum_{k=1}^{\infty} x_k$, the elements of which satisfy the following conditions:

(a) $p_n(x_n) \leqslant \dfrac{1}{2^n}$;

(b) $f_n\left(\sum_{k=1}^{n} x_k\right) = a_n$;

(c) $f_n(x_k) = 0$ for $k > n$.

Condition (a) guarantees the absolute convergence of the series $\sum_{k=1}^{\infty} x_k$ (see Exercise 10 in Subsection 16.3.1). Indeed, if p is a continuous seminorm, then its unit ball contains one of the balls of the seminorms p_n. Hence, starting with some m, the estimate $p \leqslant Cp_n$ holds for all $n \geqslant m$. Consequently, $\sum_{k=m}^{\infty} p(x_k) < C \sum_{k=m}^{\infty} p_k(x_k) < \infty$. This shows that the element $x = \sum_{k=1}^{\infty} x_k$ exists. The conditions (b) and (c) ensure that $f_n(x) = a_n$.

Hence, all we need (if not in general in life, at least in the setting of this proof) is to construct a sequence (x_n) with the properties (a)–(c). The construction will be carried out recursively.

The functional f_1 is discontinuous with respect to the seminorm p_1; hence, the hyperplane $X_1 = \{y \in X \colon f_1(y) = a_1\}$ is p_1-dense in X. In particular, X_1 intersects the ball $B_1 = \{y \in X \colon p_1(y) < 1/2\}$. Now as x_1 we take any element of the set $X_1 \cap B_1$.

Next, suppose the vectors $x_1, \ldots, x_{n-1}$ are already constructed; let us construct x_n. Consider the finite-codimensional subspace $Y = \bigcap_{k=1}^{n-1} \operatorname{Ker} f_k$. Since the functionals f_k are p_n-continuous for $k < n$, Y is a p_n-closed subspace. By Lemma 1, the restriction of the functional f_n to Y is discontinuous with respect to the seminorm p_n. Therefore, the hyperplane $X_n = \{y \in Y \colon f_n(y) = a_n - \sum_{k=1}^{n-1} f_n(x_k)\}$ is dense in Y with respect to the seminorm p_{n+1}. It follows that the ball $B_n = \{y \in Y \colon p_n(y) < 2^{-n}\}$ intersects the hyperplane X_n. For x_n we take an arbitrary element of $X_n \cap B_n$. The fact that the vector x_n belongs to B_n, X_n, and Y guarantees the fulfillment of condition (a), (b), and (c), respectively. $\qquad\square$

Let us give a couple of examples which demonstrate how the interpolation theorem just proved applies in problems of mathematical analysis.

Theorem 2. *For any sequence of scalars a_n, $n = 0, 1, 2, \ldots$, there exists an infinitely differentiable function $x(t)$ on the interval $[0, 1]$ such that $x(0) = a_0$, $x'(0) = a_1,\ldots$, $x^{(n)}(0) = a_n,\ldots$.*

Proof. Observe that the natural approach to give the solution in the form of a Taylor series $x(t) = \sum_{n=0}^{\infty} \frac{a_n}{n!} t^n$ fails: if a_n tends rapidly to infinity, then the radius of convergence of the Taylor series will be equal to zero. The interpolation theorem, however, provides a very economical solution to the problem.

In the space $C^{\infty}[0, 1]$ of infinitely differentiable functions on $[0, 1]$, consider the sequence of seminorms $p_0 \leqslant p_1 \leqslant p_2 \leqslant \cdots$ given by

$$p_0 = 0, \quad p_1(x) = \max_{t \in [0,1]} |x(t)|, \quad p_2(x) = \max\{p_1(x), p_1(x')\}, \ldots,$$

$$p_n(x) = \max\{p_1(x), p_1(x'), \ldots, p_1(x^{(n-1)})\}, \ldots,$$

and the sequence of functionals

$$f_0(x) = x(0), \quad f_1(x) = x'(0), \ldots, f_n(x) = x^{(n)}(0), \ldots .$$

The chosen sequence of seminorms gives on $C^{\infty}[0, 1]$ the topology of uniform convergence of functions and all their derivatives. In this topology the space $C^{\infty}[0, 1]$ is complete. Further, $|f_n(x)| \leqslant p_{n+1}(x)$, i.e., the functional f_n is continuous with respect to p_{n+1}. However, there is no constant C such that the inequality $|f_n| \leqslant Cp_n$ is satisfied, as one can readily verify by substituting into the inequality, say, the sequence of functions $x_m(t) = \sin(\pi mt)$. All the conditions of Theorem 1 are satisfied, so it remains only to apply it.   □

Theorem 3. *For any sequence of scalars (a_n), $n = 1, 2, \ldots$, there exists a function $x(z)$ such that $x(n) = a_n$ for all n.*

Proof. In the space $\mathcal{H}(\mathbb{C})$ of entire functions consider the sequence of seminorms $p_1 \leq p_2 \leq \cdots$, $p_n(x) = \max_{|z| \leqslant n-1} |x(z)|$, and the functionals $f_n(x) = x(n)$. This sequence of seminorms gives the topology of uniform convergence on compact sets, in which the space $\mathcal{H}(\mathbb{C})$ is complete. Again, as in the preceding theorem, $|f_n(x)| \leqslant p_{n+1}(x)$, whereas there is no C such that $|f_n| \leqslant Cp_n$ (substitute the functions $x_m(z) = z^m$). And again, the sought-for function $x(z)$ is obtained by applying Theorem 1. □

A slightly more general variant of the interpolation theorem above and its application to the moment problem can be found in B. M. Makarov's work [70].

Exercises

1. Verify the correctness of the definition of the functional $\widetilde{f}$ in the proof of Lemma 1, namely, that if $q(x) = q(y)$, then $f(x) = f(y)$. That is, that $\widetilde{f}(qx) = f(x)$ depends on qx, but not on x.

2. Let $t_1, \ldots, t_j$ be a finite collection of distinct points of the interval $[0, 1]$, and $\{a_{n,k}\}_{n \in \mathbb{N} \cup \{0\}, k \in \{1, \ldots, j\}}$ be numbers. Show that there exists a function $x \in C^{\infty}[0, 1]$ such that $x^{(n)}(t_k) = a_{n,k}$ for all $n \in \mathbb{N} \cup \{0\}$ and $k \in \{1, \ldots, j\}$.

3. Let $z_n \in \mathbb{C}$ be an arbitrary collection of points ("interpolation nodes"). The following conditions are equivalent:

— for any sequence of scalars a_n, $n = 1, 2, \ldots$, there exists an entire function $x(z)$ such that $x(z_n) = a_n$ for all n;

— $z_n \neq z_m$ for $n \neq m$, and $|z_n| \to \infty$ as $n \to \infty$.

4. For any sequence of scalars a_n, $n = 1, 2, \ldots$, and any sequence of indices $m_1 < m_2 < m_3 < \ldots$, there exists a "lacunary" entire function $x(z)$ of the form $x(z) = \sum_{k=1}^{\infty} b_k z^{m_k}$ such that $x(n) = a_n$ for all $n \in \mathbb{N}$.

A sequence x_n of elements of a topological vector space X is said to be ω-*linearly independent* if for any sequence of scalars b_n, the equality $\sum_{k=1}^{\infty} b_k x_k = 0$ implies that all b_n are equal to zero. The Erdős–Straus theorem (P. Erdős, E.G. Straus, 1953)[4] asserts that in a normed space from any linearly independent sequence one can extract an ω-linearly independent subsequence.

5. Suppose that the topological vector space X carries a continuous norm. Then from any linearly independent sequence in X one can extract an ω-linearly independent subsequence.

6. Consider the vectors $x_1 = (1, 2, 3, \ldots, n, \ldots)$, $x_2 = (1^2, 2^2, 3^2, \ldots, n^2, \ldots)$, $x_3 = (1^3, 2^3, 3^3, \ldots, n^3, \ldots), \ldots$ in the space $\mathbb{R}^{\mathbb{N}}$. This sequence is linearly independent, but it contains no ω-linearly independent subsequence. (Use Exercise 4 above.)

7. The Bessaga–Pełczyński theorem (C. Bessaga, A. Pełczyński, 1957). If a complete metrizable locally convex space X admits no continuous norm, then X contains a real subspace isomorphic to $\mathbb{R}^{\mathbb{N}}$.

8. Deduce from the three preceding exercises that for a complete metrizable locally convex space X the following conditions are equivalent:

— from any linearly independent sequence in X one can extract an ω-linearly independent subsequence;

— there exists a continuous norm on X.

---

[4]During the preparation for publication of the second volume of his monograph [40], I. Singer discovered a gap in the original proof of Erdős and Straus. He distributed a letter to other specialists in the theory of bases, asking for an alternative proof of the result. Such proofs were obtained by P. Terenzi and at about the same time by V.I. Gurariĭ, who back then, in 1980, was an active participant in our Kharkiv seminar on the theory of Banach spaces. I have nostalgic memories about those times: in the spring of 1980 I was a third-year student, and this was the first "mature" problem to which I devoted serious thought. The example in Exercise 6 — the fruit of this pondering — was later mentioned by Singer in his monograph. One can imagine how proud I was for discovering this example ...It is amusing that I published this observation only after 10 years and a bit [56].

16.3.4 Precompactness and Boundedness

Definition 1. A topological vector space space X is said to belong to the *Montel class* (or to be a *Montel space*) if any closed bounded set in X is compact.

By Riesz's theorem, a normed space is Montel only if it is finite-dimensional. At the same time, many of the topological vector spaces arising naturally in problems of analysis are Montel, despite being infinite-dimensional. In this subsection we shall give examples of Montel space. The name "Montel class" comes from Montel's theorem, which establishes a compactness criterion in the space $\mathcal{H}(D)$ of holomorphic functions. In modern complex analysis courses this theorem serves as the basis for the proof of Riemann's theorem on the existence of conformal maps (the Riemann mapping theorem).

Definition 2. Let A, B be subsets of the linear space X. We say that the set A is *B-precompact* (and write $A \prec_c B$) if for any $\varepsilon > 0$ there exists a finite set Q such that $A \subset \varepsilon B + Q$.

If X is a normed space, then a subset $A \subset X$ is precompact if and only if $A \prec_c B_X$. A subset A of the topological vector space X is precompact if and only if $A \prec_c U$ for all neighborhoods U of zero in the space X.

Theorem 1. *The relation $\prec_c$ between subsets of a linear space X has the following properties:*

(a) *if $A \prec_c B$ and $A_1 \subset A$, then $A_1 \prec_c B$;*

(b) *if $A \prec_c B$ and $B \subset B_1$, then $A \prec_c B_1$;*

(c) *if $A \prec_c B$ and $t > 0$, then $A \prec_c tB$;*

(d) *if $A_1 \prec_c B$ and $A_2 \prec_c B$, then $A_1 \cup A_2 \prec_c B$;*

(e) *if $A \prec_c B$, Y is a linear subspace, and $T: X \to Y$ is a linear operator, then $T(A) \prec_c T(B)$;*

(f) *if $A_1 \prec_c B$, $A_2 \prec_c B$, and B is a convex set, then $A_1 + A_2 \prec_c B$;*

(g) *if $A \prec_c B$ and $B - B \subset U$, then $A \prec_c U$; moreover, for any $\varepsilon > 0$ there exists a finite set Q such that $Q \subset A$ and $A \subset \varepsilon U + Q$;*

(h) *let $A \prec_c B$, B a convex balanced set, Y a linear space, $T: Y \to X$ a linear operator, and $A \subset T(Y)$. Then $T^{-1}(A) \prec_c T^{-1}(B)$.*

Proof. Properties (a)–(e) are obvious. Let us prove the remaining properties.

(f) Fix $\varepsilon > 0$. Let $Q_1, Q_2 \subset X$ be finite sets for which $A_1 \subset (\varepsilon/2)B + Q_1$, and $A_2 \subset (\varepsilon/2)B + Q_2$. Then $A_1 + A_2 \subset (\varepsilon/2)B + (\varepsilon/2)B + Q_1 + Q_2$. Thanks to convexity, $(1/2)B + (1/2)B \subset B$, and hence $A_1 + A_2 \subset \varepsilon B + Q_1 + Q_2$. It remains to note that the set $Q_1 + Q_2$ is finite.

(g) By hypothesis, there exists a finite set $Q_1 \subset X$ such that $A \subset \varepsilon B + Q_1$. Let us introduce a mapping $f: Q_1 \to A$ with the property that for any $q \in Q_1$, if $q + \varepsilon B$

intersects A, then $f(q) \in (q + \varepsilon B) \cap A$. We claim that $A \subset \varepsilon U + f(Q_1)$, i.e., that $f(Q_1)$ can be taken as the required set Q. Indeed, for every $a_0 \in A$ there exists a $q \in Q_1$ such that $a_0 \in q + \varepsilon B$. For this vector q the sets $q + \varepsilon B$ and A intersect (a_0 is one of the intersection points), and so $f(q) \in (q + \varepsilon B) \cap A$. We have

$$a_0 \in q + \varepsilon B = f(q) + \varepsilon B + (q - f(q)) \subset f(q) + \varepsilon B - \varepsilon B \subset f(q) + \varepsilon U.$$

(h) Since, by property (c), $A \prec_c (1/2)B$ and $(1/2)B - (1/2)B \subset B$, the property (g) proved above says that for any $\varepsilon > 0$ there exists a finite set $Q \subset A$ such that $A \subset \varepsilon B + Q$. Then $Q \subset T(Y)$, and we can construct a mapping $f: Q \to Y$ such that $T(f(q)) = q$ for all $q \in Q$. Let us show that $T^{-1}(A) \subset \varepsilon T^{-1}(B) + f(Q)$. Let $y \in T^{-1}(A)$. Then $T(y) \in A$, and there exist a $b \in B$ and a $q \in Q$ such that $T(y) = q + \varepsilon b$. Since $T(f(q)) = q$, we have $T(y - f(q)) = \varepsilon b$, i.e., $(y - f(q))/\varepsilon \in T^{-1}(B)$. Consequently, $y = \varepsilon(y - f(q))/\varepsilon + f(q) \in \varepsilon T^{-1}(B) + f(Q)$. □

Theorem 2. *Suppose X is a complete topological vector space and for every neighborhood U of zero there exists a neighborhood V of zero such that $V \prec_c U$. Then X belongs to the Montel class.*

Proof. In view of the completeness of the space X, it suffices to prove that every bounded subset $A \subset X$ is precompact (Theorem 5 in Subsection 16.2.2). So, let A be bounded and U be an arbitrary neighborhood of zero. By hypothesis, there exists a neighborhood of zero V with $V \prec_c U$. By the definition of boundedness, $A \subset nV$ for n large enough. By items (a) and (c) in the preceding theorems, $A \subset nV \prec_c nU$, that is, $A \prec_c U$. □

Example 1. The space $\mathcal{H}(D)$ of holomorphic functions on a domain $D \subset \mathbb{C}$ belongs to the Montel class.

To verify this, we use Theorem 2. Let U be an arbitrary neighborhood of zero in $\mathcal{H}(D)$. Recalling the definition of the topology on $\mathcal{H}(D)$ (Exercise 4 in Subsection 16.3.1), we may assume that U is the unit ball of the seminorm $p_K(f) = \max_{z \in K} |f(z)|$, where K is a compact subset of D, and without loss of generality we may assume that K is a finite union of closed disks. Consider a rectifiable contour $\Gamma \subset D$ which includes K in its interior; denote by K_1 the compact set which includes K and has Γ as its boundary, and by V the unit ball of the seminorm p_{K_1}. We show that $V \prec_c U$. Let $\delta = \min\{|z - \zeta| : z \in K, \zeta \in \Gamma\}$, and l be the length of the contour Γ. By the Cauchy integral formula for the derivative, for any function $f \in V$ and any $z \in K$ we have

$$|f'(z)| \leqslant \frac{1}{2\pi} \left| \int_\Gamma \frac{f(\zeta) d\zeta}{(z - \zeta)^2} \right| \leqslant \frac{1}{2\pi} \frac{p_{K_1}(f)}{\delta^2} l \leqslant \frac{l}{2\pi \delta^2}.$$

Thus, the first derivatives of the functions in the family V are uniformly bounded on K. Further, the family V itself is bounded in modulus on K (and even on the larger compact set K_1) by 1. By Arzelà's theorem, V is precompact if regarded as a subset of $C(K)$.

Now consider the operator $T : \mathcal{H}(D) \to C(K)$ which maps each function into the restriction of the function to K. The fact that we just proved can be formulated as follows: the set $T(V)$ is precompact in $C(K)$. In other words, $T(V) \prec_c B_{C(K)}$. Since $T^{-1}(B_{C(K)}) = U$, item (h) of Theorem 1 shows that $V \prec_c U$. $\qquad\square$

Example 2. The space $C^\infty[0, 1]$ belongs to the Montel class.

Recall that the topology of the space $C^\infty[0, 1]$ is given by the family of seminorms $p_n(f) = \max_{t \in [0,1]} |f^{(n)}(t)|, n = 0, 1, 2, \ldots$. Denote the unit ball of the seminorm p_n by B_n. A neighborhood basis of zero is provided by the sets $r U_n$, where $r > 0$, and $U_n = \bigcap_{k=0}^n B_k = \{f \in C^\infty[0, 1] : \max_{k=0,1,\ldots,n} \max_{t \in [0,1]} |f^{(k)}(t)| < 1\}$. By Theorem 2, to justify our example it suffices to show that $U_{n+1} \prec_c U_n$ for all $n = 0, 1, 2, \ldots$. We proceed by induction on n.

$n = 0$. Consider the identity embedding operator $T : C^\infty[0, 1] \to C[0, 1]$. The set $T(U_1)$ (which coincides with U_1) consists of infinitely differentiable functions that obey the conditions $|f(t)| < 1$ and $|f'(t)| < 1$ for all $t \in [0, 1]$. By Arzelà's theorem, $T(U_1)$ is precompact in $C[0, 1]$, i.e., $T(U_1) \prec_c B_{C[0,1]}$. According to item (h) of Theorem 1, $U_1 \prec_c T^{-1}(B_{C[0,1]}) = U_0$.

$n \to n+1$. Suppose $U_{n+1} \prec_c U_n$. Consider the integration operator $G : C^\infty[0, 1] \to C^\infty[0, 1]$, $(Gf)(t) = \int_0^t f(\tau) d\tau$. By item (e) of Theorem 1, $G(U_{n+1}) \prec_c G(U_n)$. Since $G(U_n) \subset U_{n+1}$, we deduce that

$$G(U_{n+1}) \prec_c U_{n+1}. \tag{1}$$

On the other hand, since every function $f \in U_{n+2}$ can be represented as $f(t) = f(0) + \int_0^t f'(\tau) d\tau$, where $f' \in U_{n+1}$, and $|f(0)| < 1$, we have $U_{n+2} \subset A + G(U_{n+1})$, where A consists of constants smaller than 1 in modulus. Condition (1) combined with the obvious condition $A \prec_c U_{n+1}$ (A is a one-dimensional bounded set) allows us to apply assertion (f) of Theorem 2: $U_{n+2} \subset A + G(U_{n+1}) \prec_c U_{n+1}$, as we needed to prove.

Exercises

1. Show that $C^\infty(0, +\infty)$ is a Montel space.

2. Every linear space X equipped with the strongest locally convex topology is a Montel space. Moreover, in such a space every bounded set is finite-dimensional.

3. Any Tikhonov product of Montel spaces is a Montel space.

4. Any closed subspace of a Montel space is itself a Montel space.

Chapter 17
Elements of Duality Theory

17.1 Duality in Locally Convex Spaces

17.1.1 The General Notion of Duality. Polars

Definition 1. Let X, Y be linear spaces over the same field $\mathbb{K} = \mathbb{R}$ or $\mathbb{C}$. A mapping that to each pair of elements $(x, y) \in X \times Y$ assigns a number $\langle x, y \rangle \in \mathbb{K}$ is called a *duality pairing* (or *duality mapping*, or simply a *duality*) if:

(a) $\langle x, y \rangle$ is a bilinear form:

$$\langle a_1 x_1 + a_2 x_2, y \rangle = a_1 \langle x_1, y \rangle + a_2 \langle x_2, y \rangle;$$
$$\langle x, a_1 y_1 + a_2 y_2 \rangle = a_1 \langle x, y_1 \rangle + a_2 \langle x, y_2 \rangle;$$

(b) $\langle x, y \rangle$ satisfies the non-degeneracy condition:

— for every $x \in X \setminus \{0\}$ there exists an $y \in Y$ such that $\langle x, y \rangle \neq 0$, and
— for every $y \in Y \setminus \{0\}$ there exists an $x \in X$ such that $\langle x, y \rangle \neq 0$.

A pair of spaces X, Y with a duality pairing given on them is called a *dual pair*, or a *pair of spaces in duality*.

As in the previous chapter, in order to avoid treating separately the real and complex cases, we will tacitly assume that $\mathbb{K} = \mathbb{C}$. The simpler case of real spaces differs from the complex one only by absence of the symbol Re when one applies the Hahn–Banach separation theorem.

For us the most important example of a dual pair will consist of a locally convex space X and its dual space $Y = X^*$, with the duality pairing given by the evaluation of functionals on elements: $\langle x, y \rangle = y(x)$. To a certain extent this example describes the general situation.

Definition 2. Let (X, Y) be a dual pair. For each element $y \in Y$ we define its action on the elements of the space X by the rule $y(x) = \langle x, y \rangle$. With this definition every

© Springer International Publishing AG, part of Springer Nature 2018
V. Kadets, *A Course in Functional Analysis and Measure Theory*,
Universitext, https://doi.org/10.1007/978-3-319-92004-7_17

element $y \in Y$ becomes a linear functional on X, i.e., $Y \subset X'$. The *weak topology* on X is defined as the topology $\sigma(X, Y)$ introduced in Definition 1 of Subsection 16.3.2. That is, a neighborhood basis of zero in the topology $\sigma(X, Y)$ is given by the family of sets $\{x \in X : \max_{y \in G} |\langle x, y \rangle| < \varepsilon\}$, where $\varepsilon > 0$ and G runs over all finite subsets of the space Y.

Axiom (b) of a dual pair guarantees that the weak topology is separated. By Theorem 1 of Subsection 16.3.2, $(X, \sigma(X, Y))^* = Y$, so that any dual pair can be regarded as a pair of the form (X, X^*). Nevertheless, the general definition of a dual pair has its own merit: in that definition, the spaces X and Y play completely equivalent roles. In particular, we could equally successfully regard the elements of the space X as functionals on Y and consider the weak topology $\sigma(Y, X)$ already on the space Y. This equivalence of the roles played by the two spaces allows one to transfer, by symmetry, the properties of one of the spaces in a dual pair to the other space.

Recall that $\sigma(X, Y)$ is the weakest topology in which all functionals $y \in Y$, i.e., all functionals of the form $y(x) = \langle x, y \rangle$, are continuous. In particular, if X is a locally convex space and $Y = X^*$, then $\sigma(X, Y)$ is weaker (possibly not strictly) than the initial topology of the space X. This remark can be taken as a justification of the term "weak topology".

Let us mention yet another important connection between the original topology of a locally convex space X and the weak topology $\sigma(X, X^*)$.

Theorem 1. *Any convex closed subset of a locally convex space X is also closed in the weak topology $\sigma(X, X^*)$. In particular, every closed linear subspace of a locally convex space X is $\sigma(X, X^*)$-closed.*

Proof. Let $A \subset X$ be a convex closed subset. Let us pick an arbitrary point $x \in X \setminus A$ and show that x is not a $\sigma(X, X^*)$-limit point of the set A. Since A is closed, there exists an open neighborhood U of the point x such that $U \cap A = \emptyset$. Since the space X is locally convex, we may assume that the neighborhood U is convex. By the geometric form of the Hahn–Banach theorem, there exist a functional $f \in X^* \setminus \{0\}$ and a scalar $\theta \in \mathbb{R}$ such that $\operatorname{Re} f(u) < \theta$ for all $u \in U$ and $\operatorname{Re} f(a) \geqslant \theta$ for all $a \in A$. In particular, $\operatorname{Re} f(x) < \theta$. Since f, and together with it $\operatorname{Re} f$ are $\sigma(X, X^*)$-continuous functions, the point x, at which $\operatorname{Re} f(x) < \theta$, cannot be a $\sigma(X, X^*)$-limit point for the set A, on which $\operatorname{Re} f(a) \geqslant \theta$. $\qquad\square$

Definition 3. Let (X, Y) be a dual pair. The *polar* of the set $A \subset X$ is the set $A^\circ \subset Y$, defined by the following rule: $y \in A^\circ$ if $|\langle x, y \rangle| \leqslant 1$ for all $x \in A$. The polar $A^\circ \subset X$ of a set $A \subset Y$ is defined by symmetry.

The *annihilator* of the set $A \subset X$ is the set $A^\perp \subset Y$ consisting of all elements $y \in Y$ such that $\langle x, y \rangle = 0$ for all $x \in A$. Obviously, $A^\perp \subset A^\circ$, and, by Lemma 2 of Subsection 16.3.2, if A is a linear subspace, then $A^\perp = A^\circ$. Moreover, $A^\perp = (\operatorname{Lin} A)^\perp$.

Example 1. Consider the pair (X, X^*), where X is a Banach space. Then $(B_X)^\circ = \overline{B}_{X^*}$. Indeed,

$$f \in \overline{B}_{X^*} \iff \|f\| \leqslant 1 \iff \sup_{x \in B_X} |f(x)| \leqslant 1 \iff f \in (B_X)^\circ.$$

Theorem 2. *Polars possess the following properties:*

(i) *If $A \subset B$, then $A^\circ \supset B^\circ$.*

(ii) *$\{0_X\}^\circ = Y$ and $X^\circ = \{0_Y\}$, where 0_X and 0_Y are the zero elements of the spaces X and Y, respectively.*

(iii) *$(\lambda A)^\circ = \frac{1}{\lambda} A^\circ$ for any $\lambda \neq 0$.*

(iv) *$\left(\bigcup_{A \in \mathfrak{E}} A \right)^\circ = \bigcap_{A \in \mathfrak{E}} A^\circ$ for any family $\mathfrak{E}$ of subsets of the space X. In particular, $(A_1 \cup A_2)^\circ = A_1^\circ \cap A_2^\circ$.*

(v) *For any point $x \in X$, the set $\{x\}^\circ$ is a convex balanced $\sigma(Y, X)$-closed neighborhood of zero.*

(vi) *The polar of any set is a convex balanced $\sigma(Y, X)$-closed set.*

(vii) *The sets of the form A°, where A runs over all finite subsets of the space X, constitute a neighborhood basis of zero in the topology $\sigma(Y, X)$.*

Proof. Properties (i)–(iv) are obvious. The fact that the set

$$\{x\}^0 = \{y \in Y : |\langle x, y \rangle| \leqslant 1\} = \{y \in Y : |x(y)| \leqslant 1\}$$
$$= x^{-1}(\{\lambda \in \mathbb{C} : |\lambda| \leqslant 1\}) \tag{1}$$

is convex and balanced is a consequence of the linearity of x as a functional on Y. Since $\mathbb{C}_1 = \{\lambda \in \mathbb{C} : |\lambda| \leqslant 1\}$ is a closed neighborhood of zero in $\mathbb{C}$, and the functional x is continuous in the $\sigma(Y, X)$-topology, relation (1) means that $\{x\}^\circ$ is a $\sigma(Y, X)$-closed neighborhood of zero. This establishes property (v). Property (vi) follows from (v) thanks to property (iv): $A^\circ = \bigcap_{x \in A} \{x\}^\circ$, and the intersection operation does not destroy the properties of convexity, closedness, and balancedness.

Now let us turn to property (vii). If the subset $A \subset X$ is finite, then $A^\circ = \bigcap_{x \in A} \{x\}^\circ$ is a finite intersection of $\sigma(Y, X)$-neighborhoods. Hence, the polar of a finite set is a weak neighborhood. Further, by definition, every $\sigma(Y, X)$-neighborhood contains a set of the form $U_{G,\varepsilon} \doteq \{y \in Y : \max_{g \in G} |g(x)| < \varepsilon\}$, where $G = \{g_1, g_2, \ldots, g_n\} \subset X$ and $\varepsilon > 0$. For $A = \frac{1}{2\varepsilon} G$ we have $U_{G,\varepsilon} \supset A^\circ$. That is, every $\sigma(Y, X)$-neighborhood contains a set of the form A°, with $A \subset X$ finite. $\square$

Corollary 1. *The annihilator of any set $A \subset X$ is a $\sigma(Y, X)$-closed linear subspace.*

Proof. The linearity is verified directly, while $\sigma(Y, X)$-closedness follows, for example, from property (vi) above and the relations $A^\perp = (\mathrm{Lin}\, A)^\perp = (\mathrm{Lin}\, A)^\circ$. $\square$

Similarly to how with a convex sets one associates its convex hull, and with a subspace its linear hull (span), convex balanced sets lead to the absolute convex hull construction.

Definition 4. Let X be a linear space. An *absolutely convex combination* of a finite collection of elements $\{x_k\}_{k=1}^n \subset X$ is any sum $\sum_{k=1}^n \lambda_k x_k$, where λ_k are scalars with

$\sum_{k=1}^{n} |\lambda_k| \leqslant 1$. The *absolutely convex hull of the set* A in the linear space X is the set aconv A consisting of all absolutely convex combinations of elements of A. The closure of the set aconv A in the topology τ is denoted by τ-$\overline{\text{aconv}} A$ or, if the context makes it clear, simply by $\overline{\text{aconv}} A$.

Exercises

1. Consider the real dual pair $(\mathbb{R}^2, \mathbb{R}^2)$ with the scalar product as the duality pairing. Construct in the plane the polars of the following sets:

- $\{(0, 1)\}$;

- $\{(1, 1)\}$;

- $\{(1, 1), (0, 1)\}$;

- $\{(x_1, x_2) : |x_1| + |x_2| \leqslant 1\}$;

- $\{(x_1, x_2) : |x_1|^2 + |x_2|^2 \leqslant 1\}$.

2. Consider the pair (X, Y), where $X = Y = C[0, 1]$. Which of the expressions listed below give a duality on this pair?

- $\langle f, g \rangle = \int_0^1 f(t)g(t)dt$;

- $\langle f, g \rangle = \int_0^{1/2} f(t)g(t)dt$;

- $\langle f, g \rangle = \int_0^{1/2} f(t)g(t)dt - \int_{1/2}^1 f(t)g(t)dt$;

- $\langle f, g \rangle = \int_0^1 f^2(t)g(t)dt$;

- $\langle f, g \rangle = \int_0^1 f(t^2)g(t)dt$;

- $\langle f, g \rangle = f(0)g(0)$;

- $\langle f, g \rangle = \int_0^1 f(t)g(t)dt + f(0)g(0)$.

3. Let A be a subset of a linear space X. Then aconv A is a convex balanced set.

4. Any convex balanced set containing the set A also contains aconv A.

5. aconv A is equal to the intersection of all convex balanced sets containing A.

6. Let A be a subspace of the linear space X. Then aconv $A = A$.

7. Let A be a subset of a topological space X. Then $\overline{\text{aconv}} A$ is the smallest, with respect to inclusion, closed convex balanced set containing A.

8. Let (X, Y) be a dual pair, $A \subset X$. Then $(\mathrm{aconv}\, A)^\circ = A^\circ$.

9. Consider on X the weak topology $\sigma(X, Y)$. Then the polar of any set coincides with the polar of its closure. Further, $(\overline{\mathrm{aconv}}\, A)^\circ = A^\circ$ for any set $A \subset X$.

10. For any dual pair (X, X^*), where X is a Banach space, describe explicitly the neighborhoods of zero in the topologies $\sigma(X, X^*)$ and $\sigma(X^*, X)$.

11. Is the open unit ball in a Banach space X a $\sigma(X, X^*)$-open set?

12. Is the closed unit ball in a Banach space X a $\sigma(X, X^*)$-closed set?

Note that the number of elements in an absolutely convex combination can be arbitrarily large. Consequently, even for a compact set A the absolutely convex hull is not necessarily closed: $\overline{\mathrm{aconv}}\, A$ will contain, in particular, the infinite sums of the form $\sum_{k=1}^{\infty} \lambda_k x_k$ with $\sum_{k=1}^{\infty} |\lambda_k| \leqslant 1$.

13. On the following example, convince yourself that $\overline{\mathrm{aconv}}\, A$ is not exhausted by the sums $\sum_{k=1}^{\infty} \lambda_k x_k$ with $\sum_{k=1}^{\infty} |\lambda_k| \leqslant 1$. Let $A = \{x_n = e_1 + e_{n+1} : n \in \mathbb{N}\}$, where e_n is an orthonormal system in the Hilbert space H. Then the vector e_1 lies in $\overline{\mathrm{aconv}}\, A$, but cannot be written as $\sum_{k=1}^{\infty} \lambda_k x_k$.

14. In a finite-dimensional space the absolutely convex hull of any closed bounded set is closed.

17.1.2 The Bipolar Theorem

Let (X, Y) be a dual pair, $A \subset X$. Then $A^\circ \subset Y$, and now one can consider the polar of A°.

Definition 1. The set $(A^\circ)^\circ \subset X$ is called the *bipolar* $A \subset X$ and is denoted by $A^{\circ\circ}$.

Theorem 1. *The bipolar $A^{\circ\circ}$ of a set $A \subset X$ coincides with the $\sigma(X, Y)$-closed absolutely convex hull of A.*

Proof. First, we note that $A^{\circ\circ} \supset A$. Indeed, if $x \in A$, then by the definition of the set A°, $|\langle x, y \rangle| \leqslant 1$ for any $y \in A^\circ$. But this means precisely that x belongs to the polar A°.

Further, the bipolar is a particular case of a polar. Therefore, by assertion (vi) in Theorem 2 of Subsection 17.1.1, $A^{\circ\circ}$ is a convex balanced $\sigma(X, Y)$-closed set. Hence, $A^{\circ\circ} \supset \overline{\mathrm{aconv}}\, A$. To prove the opposite inclusion, pick an arbitrary point $x_0 \in X \setminus \overline{\mathrm{aconv}}\, A$, and let us show that $x_0 \notin A^{\circ\circ}$. Indeed, since $x_0 \notin \overline{\mathrm{aconv}}\, A$ and $\overline{\mathrm{aconv}}\, A$ is a convex balanced $\sigma(X, Y)$-closed set, the Hahn–Banach theorem in the form indicated in Exercise 9 in Subsection 16.3.1 says that there exists a $\sigma(X, Y)$-continuous linear functional y on X such that

I. $|y(x)| \leqslant 1$ for all $x \in \overline{\mathrm{aconv}}A$, and

II. $|y(x_0)| > 1$.

But every $\sigma(X, Y)$-continuous linear functional is an element of the space Y. Condition I means that $y \in (\overline{\mathrm{aconv}}A)^\circ \subset A^\circ$. But then condition II means that $x_0 \notin A^{\circ\circ}$, as we needed to show. $\square$

Corollary 1. *If $A \subset X$ is a $\sigma(X, Y)$-closed convex and balanced set, then $A^{\circ\circ} = A$. In particular, $B^{\circ\circ\circ} = B^\circ$ for any set $B \subset Y$.*

Corollary 2. $A^{\perp\perp} = \overline{\mathrm{Lin}}\, A$ *for any set $A \subset X$. If A is a linear subspace, then $A^{\perp\perp} = \overline{A}$. Finally, $B^{\perp\perp\perp} = B^\perp$ for any $B \subset Y$.*

Proof. $A^{\perp\perp} = (A^\perp)^\perp = ((\mathrm{Lin}\, A)^\perp)^\perp = (\mathrm{Lin}\, A)^{\circ\circ} = \overline{\mathrm{Lin}}\, A$. $\square$

Corollary 3. *If $A_1, A_2 \subset X$ are $\sigma(X, Y)$-closed convex and balanced sets, then the equality $A_1 = A_2$ is equivalent to the equality $A_1^\circ = A_2^\circ$. Moreover, if A_1, A_2 are subspaces, then the equality $A_1 = A_2$ is equivalent to the equality $A_1^\perp = A_2^\perp$.*

Proof. If two sets are equal, then their polars are also equal, so the implication $A_1 = A_2 \Longrightarrow A_1^\circ = A_2^\circ$ is obvious with no supplementary restrictions on the sets. Conversely, if $A_1^\circ = A_2^\circ$, then $A_1^{\circ\circ} = A_2^{\circ\circ}$, and it remains to use the bipolar theorem. $\square$

Recall that the spaces figuring in a dual pair (X, Y) play equivalent roles, and all assertions concerning the polars and bipolars of subsets of the space X are valid, upon exchanging the roles of the spaces in the dual pair, for the subsets of the space Y.

Theorem 2. *Let (X, Y) be a dual pair and A be a subset of Y. Then the following conditions are equivalent:*

  (i) *the set of functionals $A \subset Y$ separates the points of the space X;*

 (ii) $A^\perp = \{0\}$;

(iii) $A^{\perp\perp} = Y$;

(iv) *the linear span of the set A is $\sigma(Y, X)$-dense in Y.*

Proof. (i) $\Longrightarrow$ (ii). The inclusion $A^\perp \supset \{0\}$ holds always. If now $x \in X \setminus \{0\}$, then according to (i), there exists an $y \in A$ such that $\langle x, y \rangle \neq 0$. Consequently, $x \notin A^\perp$.

   (ii) $\Longrightarrow$ (i). Let $x \in X \setminus \{0\}$. Then $x \notin A^\perp$, and so there exists an $y \in A$ such that $\langle x, y \rangle \neq 0$.

   (ii) $\Longleftrightarrow$ (iii). Since $A^\perp$ and $\{0\}$ are $\sigma(X, Y)$-closed subspaces, Corollary 3 applies.

   (iii) $\Longleftrightarrow$ (iv). By Corollary 2, $A^{\perp\perp} = \overline{\mathrm{Lin}}\, A$.

Exercises

1. Construct in the plane the bipolars of the sets from Exercise 1 of Subsection 17.1.1.

2. Consider the real dual pair (X, Y), where $X = Y = \mathbb{R}^2$, with the duality pairing $\langle \mathbf{x}, \mathbf{y} \rangle = x_1 y_2 - 2x_2 y_1$. Construct the polars and bipolars of the following subsets of the space X:

— $\{(0, 1)\}$;

— $\{(1, 1)\}$;

— $\{(1, 1), (0, 1)\}$;

— $\{(x_1, x_2) : |x_1| + |x_2| \leqslant 1\}$;

— $\{(x_1, x_2) : |x_1|^2 + |x_2|^2 \leqslant 1\}$.

3. Do the polars of the above sets change if they are regarded as subsets of Y? Do the bipolars change?

4. Consider the real dual pair (X, Y), where $X = C[0, 1]$ and $Y = L_1[0, 1]$, with the duality pairing $\langle f, g \rangle = \int_0^1 f(t)g(t)dt$. Construct the polars and bipolars of the following subsets of the space X:

— $\{f \in C[0, 1] : f(t) > 0 \text{ for all } t \in [0, 1]\}$;

— $\{f \in C[0, 1] : 0 \leqslant f(t) \leqslant 1 \text{ for all } t \in [0, 1]\}$;

— $\{f \in C[0, 1] : f(t) = 0 \text{ for all } t \in [0, 1/2]\}$.

5. Consider the same sets of continuous functions as subsets of the space $Y = L_1[0, 1]$. Construct their polars and bipolars with respect to the duality from the preceding exercise.

17.1.3 The Adjoint Operator

Definition 1. Let X_1, X_2 be linear spaces, and let $T : X_1 \to X_2$ be a linear operator. The *algebraic adjoint* (or *algebraic conjugate*, or *algebraic dual*) of the operator T is the operator $T' : Y' \to X'$ acting as $T'f = f \circ T$.

Further, let (X_1, Y_1), (X_2, Y_2) be dual pairs, and $T : X_1 \to X_2$ a linear operator. We say that an *adjoint operator* $T^* : Y_2 \to Y_1$ exists for T if for any $y \in Y_2$ there exists an element $T^*y \in Y_1$, such that $\langle Tx, y \rangle = \langle x, T^*y \rangle$ for all $x \in X_1$.

Treating the elements of the spaces Y_1 and Y_2 as functionals on the space X_1 and X_2, respectively, we see that $T^*y = y \circ T$, which explains the correctness of the definition and the linearity of the operator T^*. Clearly, an adjoint operator exists for the operator T if and only if $T'(Y_2) \subset Y_1$, and in this case T^* is the restriction of the algebraic adjoint T' to Y_2. For dual pairs (X_1, X_1^*), (X_2, X_2^*), where X_1 and X_2 are Banach spaces, the new definition of adjoint (conjugate, or dual) operator agrees with the already familiar definition of the adjoint (conjugate, or dual) of an operator in Banach spaces.

Let us mention a few simple facts.

Theorem 1. *Let X_1, X_2 be locally convex spaces, and let $T: X_1 \to X_2$ be a continuous linear operator. Then there exists the adjoint operator $T^*: X_2^* \to X_1^*$ for T.*

Proof. Let $f \in X_2^*$. Then the functional $T' f = f \circ T$ is continuous, as the composition of two continuous mappings. Therefore, $T'(X_2^*) \subset X_1^*$. $\square$

Theorem 2. *Let (X_1, Y_1), (X_2, Y_2) be dual pairs, and let $T: X_1 \to X_2$ be a linear operator, and $T^*: Y_2 \to Y_1$ its adjoint. Then for any set $A \subset Y_2$,*

$$T^{-1}(A^\circ) = (T^* A)^\circ. \tag{2}$$

Proof. We have

$$x \in T^{-1}(A^\circ) \iff Tx \in A^\circ \iff \forall y \in A, \ |\langle Tx, y \rangle| \leqslant 1$$

$$\iff \forall y \in A, \ |\langle x, T^* y \rangle| \leqslant 1 \iff \forall z \in T^* A, \ |\langle x, z \rangle| \leqslant 1 \iff x \in (T^* A)^\circ.$$
$\square$

Theorem 3. *Let (X_1, Y_1), (X_2, Y_2) be dual pairs, and let $T: X_1 \to X_2$ be a linear operator. Then the following conditions are equivalent:*

(a) *the operator T has an adjoint;*

(b) *T is weakly continuous, i.e., continuous as an operator acting from $(X_1, \sigma(X_1, Y_1))$ to $(X_2, \sigma(X_2, Y_2))$.*

Proof. (a) $\Longrightarrow$ (b). In view of the linearity, it suffices to verify the continuity of the operator at zero. Recall (item (vii) Theorem 2 in Subsection 17.1.1) that a neighborhood basis of zero in the topology $\sigma(X_2, Y_2)$ is provided by the polars of the finite subsets $A \subset Y_2$. By relation (2), the preimage $T^{-1}(A^\circ)$ of the neighborhood A° is itself the polar $(T^* A)^\circ$ of the finite set $T^* A \subset Y_1$. That is, $T^{-1}(A^\circ)$ is a neighborhood of zero in $\sigma(X_1, Y_1)$.

By Theorem 1 of Subsection 16.3.2, which describes the dual space of a space equipped with the weak topology, the implication (b) $\Longrightarrow$ (a) is a particular case of Theorem 1. $\square$

Theorems 1 and 3 admit the following consequence.

Corollary 1. *Let X_1, X_2 be locally convex spaces, and let $T: X_1 \to X_2$ be a continuous linear operator. Then T is weakly continuous, i.e., it is continuous in the topologies $\sigma(X_1, X_1^*)$ and $\sigma(X_2, X_2^*)$.*

Switching the roles of spaces in dual pairs, we obtain the following statement.

Theorem 4. (I) *Let* (X_1, Y_1), (X_2, Y_2) *be dual pairs, and let* $T: X_1 \to X_2$ *be a weakly continuous linear operator. Then the adjoint operator* $T^*: Y_2 \to Y_1$ *is continuous in the weak topologies* $\sigma(Y_2, X_2)$, $\sigma(Y_1, X_1)$.

(II) *For the operator* $R: Y_2 \to Y_1$ *to be the adjoint of some weakly continuous operator acting from* X_1 *to* X_2 *it is necessary and sufficient that* R *is continuous in the topologies* $\sigma(Y_2, X_2)$, $\sigma(Y_1, X_1)$.

Proof. (I) The formula $\langle Tx, y \rangle = \langle x, T^*y \rangle$ says that the operator T^* has an adjoint: $(T^*)^* = T$. By Theorem 3, applied to T^* instead of T, the operator T^* is weakly continuous.

(II) The necessity of the condition follows from assertion (I). Now suppose R is continuous in the topologies $\sigma(Y_2, X_2)$ and $\sigma(Y_1, X_1)$. Then R has an adjoint $R^*: X_1 \to X_2$. By (I), the operator R^* is weakly continuous. Hence, there exists the adjoint $(R^*)^*: Y_2 \to Y_1$. Since $\langle x, (R^*)^*y \rangle = \langle R^*x, y \rangle = \langle x, Ry \rangle$, we see that $R = (R^*)^*$, and so we have proved that R is an adjoint operator, more precisely, the adjoint of R^*. $\qquad\qquad\square$

Recall that in Subsection 9.4.1 we established the following result. Let X_1, X_2 be Banach spaces, and $T: X_1 \to X_2$ a continuous operator. Then $T^*(X_2^*) \subset (\operatorname{Ker} T)^\perp$. In particular, if the operator T^* is surjective, then T is injective. We are now ready to make this result more precise.

Theorem 5. *Let* X_1, X_2 *be locally convex spaces, and let* $T: X_1 \to X_2$ *be a continuous linear operator. Then:*

(a) $\left(T^*(X_2^*)\right)^\perp = \operatorname{Ker} T$.

(b) $(\operatorname{Ker} T)^\perp$ *coincides with the* $\sigma(X_1^*, X_1)$*-closure of the subspace* $T^*(X_2^*)$.

(c) *The operator* T *is injective if and only if the image of the operator* T^* *is* $\sigma(X_1^*, X_1)$*-dense in* X_1^*.

Proof. (a) Using Theorem 2 and the equality of the polars and annihilators of subspaces, we have

$$(T^*(X_2^*))^\perp = (T^*(X_2^*))^\circ = T^{-1}((X_2^*)^\circ) = T^{-1}(0) = \operatorname{Ker} T.$$

(b) follows from (a) via a direct application of Corollary 2 of the bipolar theorem.

Finally, to prove (c), we remark that injectivity means that $\operatorname{Ker} T = \{0\}$. Both sides of this equality are closed (and hence, by Theorem 1 of Subsection 17.1.1, also weakly closed) subspaces of X_1. By Corollary 3, $\operatorname{Ker} T = \{0\}$ if and only if $(\operatorname{Ker} T)^\perp = X_1^*$, which in view of (b) is equivalent to $T^*(X_2^*)$ being $\sigma(X_1^*, X_1)$-dense in X_1^*. $\qquad\square$

Exercises

1. Apply Theorem 2 of Subsection 17.1.2 to prove Theorem 5.

2. Under the assumptions of Theorem 5, the operator T is injective if and only if the image of the operator T^* separates the points of X_1.

3. Consider the pair of spaces (X, Y), where X is the space of infinitely differentiable functions f on the interval $[0, 1]$ that obey the conditions $f(0) = f(1) = 0$, and $Y = C^\infty[0, 1]$. Equip this pair with the duality pairing $\langle f, g \rangle = \int_0^1 f(t)g(t)dt$. Let $T: X \to X$ be the differentiation operator: $Tf = df/dt$. Does T admit an adjoint operator $T^*: Y \to Y$? Is T a weakly continuous operator?

4. In the preceding exercise, replace Y by $C[0, 1]$. Do the conclusions change?

5. Endow the pair (X, Y) considered in Exercise 3 above with an another duality pairing, given by $\langle f, g \rangle = \int_0^1 f(t)g(t^2)dt$. What is T^* equal to in this case?

17.1.4 Alaoglu's Theorem

Let X be a linear space. The Cartesian power $\mathbb{C}^X$ is the linear space of all complex-valued functions on X, and X' is the space of linear functionals on X. Every functional is a complex-valued function, so X' can be regarded as a subspace of $\mathbb{C}^X$. Equip $\mathbb{C}^X$ with the Tikhonov product topology. Then $\mathbb{C}^X$ becomes a locally convex space with the topology given by the neighborhood basis of zero consisting of the sets $U_{G,\varepsilon} = \{f \in \mathbb{C}^X : \max_{x \in G} |f(x)| < \varepsilon\}$, where G runs over all finite subsets of the space X and $\varepsilon > 0$.

Theorem 1. $\left(X', \sigma(X', X)\right)$ *is a closed subspace of the topological vector space* $\mathbb{C}^X$. *In other words,*

(i) X' *is closed in* $\mathbb{C}^X$, *and*

(ii) *the topology induced by* $\mathbb{C}^X$ *on* X' *coincides with the weak topology* $\sigma(X', X)$.

Proof. (i) For any $x_1, x_2 \in X$ and $a_1, a_2 \in \mathbb{C}$, define the function $F_{x_1, x_2, a_1, a_2}: \mathbb{C}^X \to \mathbb{C}$ by the formula

$$F_{x_1, x_2, a_1, a_2}(f) = f(a_1 x_1 + a_2 x_2) - a_1 f(x_1) - a_2 f(x_2).$$

An element $f \in \mathbb{C}^X$ is a linear functional if and only if $F_{x_1, x_2, a_1, a_2}(f) = 0$ for all $x_1, x_2 \in X$ and all $a_1, a_2 \in \mathbb{C}$. In other words,

$$X' = \bigcap_{x_1, x_2, a_1, a_2} \operatorname{Ker} F_{x_1, x_2, a_1, a_2}.$$

All the functionals F_{x_1,x_2,a_1,a_2} are continuous, being linear combinations of coordinate projectors. Hence, their kernels are closed, and so their intersection X' is also closed.

(ii) It suffices to recall that the neighborhoods of zero in the topology $\sigma(X', X)$ are defined by means of the duality pairing $\langle f, x \rangle = f(x)$ on the dual pair (X', X). In other words, a neighborhood basis of zero in the topology $\sigma(X', X)$ is given by the sets $\{f \in X' : \max_{x \in G} |\langle f, x \rangle| < \varepsilon\} = U_{G,\varepsilon} \cap X'$. □

Theorem 2. *Let U be an absorbing subset of the linear space X. Then the polar $U^\circ \subset X'$ is $\sigma(X', X)$-compact.*

Proof. By Theorem 1, it suffices to establish the compactness of U° as a subset of the Tikhonov power $\mathbb{C}^X$. First, note that the polar U° is $\sigma(X', X)$-closed in X', and X', in turn, is closed in $\mathbb{C}^X$ by Theorem 1. Hence, U° is closed as a subset of the Tikhonov product $\mathbb{C}^X$. Next, for any $x \in X$ denote by $n(x)$ the smallest number $n \in \mathbb{N}$ for which $x \in nU$. Then for any $x \in X$ and any $f \in U^\circ$ one has the estimate $|f(x)| \leqslant n(x)$. This means that $U^\circ \subset \prod_{x \in X} \mathbb{C}_{n(x)}$, where $\mathbb{C}_{n(x)}$ denotes the closed disc in $\mathbb{C}$ centered at zero and of radius $n(x)$. By Tikhonov's theorem on products of compact sets, $\prod_{x \in X} \mathbb{C}_{n(x)}$ is compact $\mathbb{C}^X$. Therefore, U° is a closed subset of a compact set, and as such is compact. □

Corollary 1. *Let U be a neighborhood of zero in the locally convex space X. Then $U^\circ \subset X^*$ is $\sigma(X^*, X)$-compact.*

Proof. Consider first the dual pair (X, X'). The polar of the set U in X' consists of functionals that are bounded on the neighborhood U, i.e., of continuous functionals. Therefore, one can equally well consider the polar U° in the dual pair (X, X') or in the dual pair (X, X^*) — we obtain the same set $U^\circ \subset X^*$. By Theorem 2, U° is $\sigma(X', X)$-compact. It remains to remark that on X^* the topologies $\sigma(X', X)$ and $\sigma(X^*, X)$ coincide. □

Corollary 2 (L. Alaoglu, 1940).[1] *Let X be a Banach space. Then the closed unit ball of the dual Banach space X^* is $\sigma(X^*, X)$-compact.*

Proof. Indeed, $\overline{B}_{X^*} = (B_X)^\circ$. □

Recall that in an infinite-dimensional Banach space with the norm topology balls cannot be compact (Riesz's theorem, Subsection 11.2.1). This restricts considerably the applicability of geometric intuition in the infinite-dimensional case: as it turns out, all arguments relying on the extraction of a convergent subsequence from a bounded sequence are prohibited. However, Alaoglu's theorem gives rise to the hope that this

---

[1] Alaoglu proved this assertion by generalizing results of Banach obtained earlier in the language of pointwise convergent sequences and transfinite sequences of functionals. For this reason, the theorem is often referred to as the Banach–Alaoglu theorem. For locally convex spaces (Corollary 1) the theorem was first obtained by Bourbaki (Nicolas Bourbaki is the collective pseudonym of a group of mainly French mathematicians). Note, however, that this formulation, like Theorem 2, differs only slightly in content and proof from the original variant of Alaoglu's theorem.

prohibition can be partially lifted, at least in dual spaces, and not for the convergence in norm, but in the weaker, $\sigma(X^*, X)$-convergence. A difficulty we still need to face is that Alaoglu's theorem deals with compactness, not sequential compactness; that is, the possibility of extracting convergent subsequences remains under question at this point. This issue will be studied in detail in Sect. 17.2. For the moment, before taking leave from the general theory of duality and passing to the Banach spaces so dear to the author, we will formulate in a series of exercises several additional results, in particular, an important theorem of Mackey and Arens which describes, for a dual pair (X, Y), those topologies on X for which $X^* = Y$. For a detailed exposition the reader is referred to the textbook [37].

Exercises: Topology of Uniform Convergence

Let (X, Y) be a dual pair.

1. For a set $A \subset Y$ to be $\sigma(Y, X)$-bounded it is necessary and sufficient that its polar $A^\circ \subset X$ is absorbing.

2. Let τ be a locally convex topology on X. Then for a functional $y \in Y$ to be τ-continuous it is necessary and sufficient that the set $\{y\}^\circ$ is a neighborhood of zero in the topology τ.

A family $\mathfrak{C}$ of subsets of the space Y is said to be *admissible* if it obeys the following conditions:

— $\{y\} \in \mathfrak{C}$ for any $y \in Y$;

— for any $A \in \mathfrak{C}$ and any scalar λ, the set λA also belongs to $\mathfrak{C}$;

— if $A, B \in \mathfrak{C}$, then there exists a $C \in \mathfrak{C}$ such that $A \cup B \subset C$;

— any element $A \in \mathfrak{C}$ is bounded in the topology $\sigma(Y, X)$.

3. Show that the following families of subsets of a space Y are admissible:

— the family $\mathfrak{Fin}(Y)$ of all finite subsets;

— the family $\mathfrak{Comp}(Y)$ of all absolutely convex $\sigma(Y, X)$-compact subsets;

— the family $\mathfrak{Bound}(Y)$ of all $\sigma(Y, X)$-bounded subsets.

4. Any admissible family $\mathfrak{C}$ satisfies the condition $\mathfrak{Fin}(Y) \subset \mathfrak{C} \subset \mathfrak{Bound}(Y)$.

5. Let τ be a locally convex topology on X with the property that $X^* \supset Y$. Then the family τ° of sets of the form A°, where A runs over all neighborhoods of zero in the topology τ, is admissible.

6. If $\mathfrak{C}$ is an admissible family of subsets of the space Y, then the family of sets A° with $A \in \mathfrak{C}$ constitutes a neighborhood basis of zero in a separated locally convex topology on X.

7. A sequence $x_n \in X$ converges in the topology of the preceding exercise to a point $x \in X$ if and only if $\sup_{y \in A}|\langle x_n - x, y \rangle| \to 0$ as $n \to \infty$ for all $A \in \mathfrak{C}$.

Let $\mathfrak{C}$ be an admissible family of subsets of the space Y. The topology in which a neighborhood basis of zero is formed by all sets A° with $A \in \mathfrak{C}$ is called the *topology of uniform convergence on the sets of the family* $\mathfrak{C}$.

A locally convex topology τ on the space X is said to be *compatible with duality*, if $(X, \tau)^* = Y$.

8. Suppose the topology τ on X is compatible with duality. Then the topology of uniform convergence on the sets of the family τ° from Exercise 5 above coincides with τ.

9. The topology $\sigma(X, Y)$ coincides with the topology of uniform convergence on the sets of the family $\mathfrak{Fin}(Y)$.

The topology of uniform convergence on the sets of family $\mathfrak{Comp}(Y)$ is called the *Mackey topology*, and is denoted by $\tau(X, Y)$.

10. Mackey–Arens theorem. *The topology τ on X is compatible with duality if and only if $\sigma(X, Y) \prec \tau \prec \tau(X, Y)$.*

11. Every $\sigma(X, Y)$-bounded set is also $\tau(X, Y)$-bounded, i.e., all topologies compatible with duality have the same supply of bounded sets.

12. For the dual pair (c_0, ℓ_1) with the duality pairing $\langle x, y \rangle = \sum_{k=1}^{\infty} x_k y_k$, the Mackey topology $\tau(c_0, \ell_1)$ coincides with the norm topology on the space c_0.

13. The identity operator I, regarded as an operator acting from the space c_0 endowed with the topology $\sigma(c_0, \ell_1)$ into c_0 with the norm topology, is an example of a bounded discontinuous linear operator.

The topology of uniform convergence on the sets of the family $\mathfrak{Bound}(Y)$ is called the *strong topology*, and is denoted by $\beta(X, Y)$.

14. $\beta(\ell_1, c_0)$ coincides with the topology of the normed space ℓ_1. On this example convince yourself that the strong topology is not always compatible with duality.

15. Describe the topology $\beta(c_0, \ell_1)$. Prove that $\beta(c_0, \ell_1)$ is compatible with duality.

17.2   Duality in Banach Spaces

In Banach spaces the convergence in norm is called *strong convergence*. In this section we dwell in detail on two weaker forms of convergence: weak and weak* (weak star) convergence.

17.2.1   w*-Convergence

Throughout this subsection we will consider a dual pair (X, X^*), where X is a Banach space.

Definition 1. The topology $\sigma(X^*, X)$ is called the *w*-topology* or, in words, *weak-star topology on the Banach space* X^*. A sequence of functionals $x_n^* \in X^*$ is said to *w*-converge* to a functional $x^* \in X^*$ (and one writes $x_n^* \xrightarrow{w^*} x^*$; the notation $x_n^* \xrightarrow{*}$ x^* is also frequently used) if it converges in the w^*-topology. In detail: $x_n^* \xrightarrow{w^*} x^*$ if $x_n^*(x) \to x^*(x)$ as $n \to \infty$ for all $x \in X$.

As the notation indicates, w^*-convergence is a particular case of the pointwise convergence we are familiar with even for operators, not only for functionals. In particular, the following assertions (Theorems 1 and 2 of Subsection 10.4.2) hold true for w^*-convergence.

Banach–Steinhaus theorem. *If* $x_n^* \xrightarrow{w^*} x^*$, *then* $\sup_{n \in \mathbb{N}} \|x_n^*\| < \infty$.   □

w^*-convergence criterion. *Let* $A \subset X$ *be a dense subset, and* x_n^*, $x^* \in X^*$. *Then the following conditions are equivalent:*

— $x_n^* \xrightarrow{w^*} x^*$;

— $\sup_{n \in \mathbb{N}} \|x_n^*\| < \infty$ *and* $x_n^*(x) \to x^*(x)$ *as* $n \to \infty$, *for all* $x \in A$.   □

Since pointwise convergence on A implies pointwise convergence on $\mathrm{Lin}\, A$, in the last criterion it suffices to require that, instead of the set A itself, its linear span is a dense set.

Based on this general criterion, we next derive w^*-convergence criteria in the sequence spaces we are familiar with.

Theorem 1. *Let* X *be the sequence space* c_0 *or* ℓ_p *with* $1 \leqslant p < \infty$, *and* X^* *be the space* ℓ_1 *or* $\ell_{p'}$ *with* $1 < p' \leqslant \infty$, *respectively. Then for a sequence of elements* $x_n = (x_{n,j})_{j \in \mathbb{N}} \in X^*, n = 1, 2, \ldots,$ *to w*-converge to an element* $y = (y_j)_{j \in \mathbb{N}} \in X^*$, *it is necessary and sufficient that this sequence is bounded in norm and converges to* y *componentwise (coordinatewise):* $x_{n,j} \to y_j$ *as* $n \to \infty$, *for all* $j \in \mathbb{N}$.

Proof. Consider the standard basis of the sequence space X: $e_1 = (1, 0, 0, \ldots), e_2 = (0, 1, 0, 0, \ldots), \ldots$. Since $\langle f, e_j \rangle = f_j$ for any $f = (f_j)_{j \in \mathbb{N}} \in X^*$, componentwise (coordinatewise) convergence is the convergence on the elements e_j of the standard basis. It remains to use the fact that the set $\mathrm{Lin}\{e_j\}_{j\in\mathbb{N}}$ is dense in X and apply the criterion obtained above. □

Since componentwise convergence in ℓ_p does not imply convergence in norm (the sequence of e_j from the proof of Theorem 1 is a typical example), the w^*-convergence in these spaces does not coincide with the norm convergence. This is the case not only in ℓ_p, but also in all infinite-dimensional normed spaces (Josefson [55], Nissenzweig [72]).

In Subsection 6.4.3 (Theorem 2 and Exercise 4) it was shown that the norm on the space X^* is lower semi-continuous with respect to weak* convergence: if $x_n^* \xrightarrow{w^*} x^*$, then $\|x^*\| \leqslant \underline{\lim}_{n\to\infty} \|x_n^*\|$. In fact, the following stronger property holds.

Theorem 2. *The norm on the dual space X^* is lower semi-continuous with respect to the w^*-topology.*

Proof. Recall (Subsection 1.2.4) that a function $f: E \to \mathbb{R}$, defined on a topological space E, is said to be *lower semi-continuous* if for any $a \in \mathbb{R}$ the set $f^{-1}((a, +\infty))$ is open. In our case E is the dual space X^*, equipped with the topology $\sigma(X^*, X)$, and the function of interest is $f(x^*) = \|x^*\|$. Hence, $f^{-1}((a, +\infty))$ is either the entire space X^* (when $a < 0$), which is indeed open, or the set $X^* \setminus a\overline{B}_{X^*}$. Since the ball $\overline{B}_{X^*}$ of the space X^* is $\sigma(X^*, X)$-closed (and even $\sigma(X^*, X)$-compact, thanks to Alaoglu's theorem), $X^* \setminus a\overline{B}_{X^*}$ is a w^*-open set. □

Exercises

1. Let X be a Banach space with a basis $\{e_n\}_1^\infty$. Then, as observed in Subsection 10.5.3, every functional $f \in X^*$ can be identified with the numerical sequence $(f(e_1), f(e_2), \ldots, f(e_n), \ldots)$, and accordingly the space X^* can be identified with the set of all such sequences. Extend Theorem 1 to this case.

2. Let X be a Banach space with a basis $\{e_n\}_1^\infty$. Then the linear span of the set $\{e_n^*\}_1^\infty$ of coordinate functionals is w^*-dense in X^*.

3. On the example of the space $X = \ell_1$ with the standard basis $e_1 = (1, 0, 0, \ldots)$, $e_2 = (0, 1, 0, 0, \ldots), \ldots$, show that the linear span of the set of coordinate functionals is not necessarily norm-dense in X^*.

4. Using the w^*-convergence criterion and the diagonal method, show that from any bounded sequence of functionals $x_n^* \in X^*$ on a separable Banach space X one can extract a w^*-convergent subsequence. Below we shall deduce this result from Alaoglu's theorem and metrizability considerations.

5. Suppose the Banach space E is the dual of some Banach space F, i.e., $E = F^*$. Suppose any finite linked system of closed balls in E has non-empty intersection. Then E has the linked balls property (see the exercises in Subsection 9.3.3 for the definition).

6. The real spaces $L_\infty(\Omega, \Sigma, \mu)$ have the linked balls property, and consequently are 1-injective. This extends the result of Exercise 4 in Subsection 9.3.3.

Recall (see Exercises 2–4 in Subsection 8.4.6 and the comments to them) that any function of bounded variation $F \colon [0, 1] \to \mathbb{R}$ can be regarded as a continuous linear functional on $C[0, 1]$ if one defines its action on elements $f \in C[0, 1]$ by the formula $\langle F, f \rangle = \int_0^1 f\, dF$. Prove the following theorem of Helly.

7. Helly's theorem. *Suppose $F_n \colon [0, 1] \to \mathbb{R}$ is a sequence of functions which converges pointwise to a function F, and the variations of the functions F_n are jointly bounded. Then F also has finite variation, and the functionals on $C[0, 1]$ generated by the functions F_n w^*-converge to the functional generated by the function F.*

Helly's theorem admits the following partial converse.

8. *Suppose the functions $F_n, F \colon [0, 1] \to \mathbb{R}$ of bounded variation are right-continuous everywhere, except possibly at zero, and vanish at zero. Further, suppose that the sequence of functionals on $C[0, 1]$ generated by the functions F_n w^*-converge to the functional generated by the function F. Then the variations of the functions F_n are jointly bounded, and $F_n(t) \to F(t)$ as $n \to \infty$ in each point of continuity t of the function F.*

9. Helly's second theorem. *From any uniformly bounded sequence of functions $F_n \colon [0, 1] \to \mathbb{R}$ with jointly bounded variation one can extract a subsequence that converges pointwise on $[0, 1]$.*

17.2.2 The Second Dual

Let X be a Banach space. Then X^* is also a Banach space, and so one can in turn consider the dual of X^*. This dual space, $(X^*)^*$, is called the *second dual* of X, and is denoted by X^{**}. The elements of the space X^{**} are, by definition, continuous linear functionals on X^*. We are already familiar with a wide class of such functionals from the general duality theory: namely, they are the elements of the initial space X, regarded as functionals on X^*.

Recall that an element $x \in X$ acts on an element $x^* \in X^*$ by the rule $x(x^*) = x^*(x)$. Thus, x can be regarded as a linear functional on X^*, and in this way X becomes a linear subspace of the second dual X^{**}. Furthermore, the formula $\|x\|_X = \sup_{f \in S_{X^*}} |f(x)|$ (Lemma in Subsection 9.4.1), recast as $\|x\|_X = \sup_{f \in S_{X^*}} |x(f)|$, acquires a new meaning: the norm of the element x in the space X is equal to the norm of x as a linear functional on X^*. Accordingly, X can be treated as subspace of the Banach space X^{**}. Of individual interest is the question of when this inclusion becomes an equality. Spaces with this property are called *reflexive* and will be considered in Subsection 17.2.6.

The next theorem, due to Goldstine, is an easy consequence of the bipolar theorem. However, one should note that Goldstine's theorem appeared earlier and was the original result from which, properly speaking, the bipolar theorem was molded.

Theorem 1 (Goldstine). *The closed unit ball of a Banach space X is w^*-dense (i.e., dense in the topology $\sigma(X^{**}, X^*)$) in the closed unit ball of the space X^{**}. The space X is w^*-dense in its second dual X^{**}.*

Proof. Consider the dual pair (X^{**}, X^*). The ball $\overline{B}_X$ is a convex balanced subset of the space X^{**}. By the bipolar theorem, $\overline{B}_X$ is $\sigma(X^{**}, X^*)$-dense in $(\overline{B}_X)^{\circ\circ}$. But $(\overline{B}_X)^\circ = \overline{B}_{X^*}$ and $(\overline{B}_X)^{\circ\circ} = \overline{B}_{X^{**}}$. Therefore, $\overline{B}_X$ is w^*-dense in $\overline{B}_{X^{**}}$.

The fact that X is w^*-dense in X^{**} can be alternatively deduced again from the bipolar theorem in the dual pair (X^{**}, X^*): $X^\perp = \{0\}$, $X^{\perp\perp} = X^{**}$, or by using the fact that the space X^{**} is the union of the balls $n\overline{B}_{X^{**}}$, and each ball $n\overline{B}_{X^{**}}$ contains as a w^*-dense subset the corresponding ball of the space X. $\qquad\square$

Exercises

1. A model example of a triple X, X^*, X^{**} is provided by the spaces c_0, ℓ_1, ℓ_∞. Based on the w^*-convergence criterion in ℓ_∞, prove that the unit ball of c_0 is dense in the $\sigma(\ell_\infty, \ell_1)$-topology in the unit ball of ℓ_∞. That is, prove Goldstine's theorem in this particular case.

2. c_0 is a Banach space, and so, being complete, it must be closed in any ambient space. In particular, c_0 is closed in ℓ_∞. On the other hand, c_0 does not coincide with ℓ_∞, hence it cannot be dense in ℓ_∞. Doesn't this contradict Goldstine's theorem?

3. X^* is also a Banach space, and so one can consider the canonical embedding $X^* \subset X^{***}$. For every element $x^{***} \in X^{***}$, define the element $P(x^{***}) \in X^*$ as the restriction of the functional x^{***} to the subspace $X \subset X^{**}$. Prove that P is a projector and $\|P\| = 1$. That is, for any Banach space the first dual is complemented in the third dual.

4. If $X = X^{**}$, the projector P introduced in the preceding exercise is bijective, i.e., in the present case $X^* = X^{***}$.

5. The ball $\overline{B}_X$ is a complete set in the topology $\sigma(X^{**}, X^*)$ if and only if $\overline{B}_X = \overline{B}_{X^{**}}$, i.e., $X = X^{**}$.

6. The space X is a complete set in the topology $\sigma(X^{**}, X^*)$ if and only if X is finite-dimensional (this fact is connected with the existence of discontinuous linear functionals on X^* in the infinite-dimensional case).

7. Let $A \in L(X, Y)$. Then one can define the second adjoint operator: $A^{**} = (A^*)^*$, $A^{**} \in L(X^{**}, Y^{**})$. Prove that the restriction of the operator A^{**} to X coincides with the original operator A.

8. Using the preceding exercise, prove the converse to Schauder's theorem on the compactness of the adjoint operator (Theorem 3 of Subsection 11.3.2). Namely, show that if the adjoint operator is compact, then so is the original operator.

17.2.3 Weak Convergence in Banach Spaces

Definition 1. The topology $\sigma(X, X^*)$ is called the *weak topology of the Banach space X*. A sequence of elements $x_n \in X$ is said to *converge weakly* to the element $x \in X$ (notation: $x_n \xrightarrow{w} x$, also frequently $x_n^* \xrightarrow{w} x^*$ or simply $x_n^* \rightharpoonup x^*$) if $f(x_n) \to f(x)$ as $n \to \infty$ for any $f \in X^*$ (one says that (x_n) is a *weakly convergent sequence*).

Note that $\sigma(X, X^*)$ is the restriction to X of the topology $\sigma(X^{**}, X^*)$, and the weak convergence of a sequence $x_n \in X$ to the elements $x \in X$ is simultaneously the w^*-convergence of the same sequence in the space X^{**}. Accordingly, the simplest properties of the w^*-convergence mentioned in Subsection 17.2.1 carry over to the weak convergence:

— if $x_n \xrightarrow{w} x$, then $\sup_{n \in \mathbb{N}} \|x_n\| < \infty$;

— if $x_n \xrightarrow{w} x$, then $\|x\| \leqslant \underline{\lim}_{n \to \infty} \|x_n\|$.

The convergence criterion is also preserved, with the roles of the spaces X and X^* switched. More precisely, let $A \subset X^*$ be a subset whose linear span is dense in X^* in the strong topology; let $x_n, x \in X$. Then the following conditions are equivalent:

A. $x_n \xrightarrow{w} x$;

B. $\sup_{n \in \mathbb{N}} \|x_n\| < \infty$ and $x^*(x_n) \to x^*$ for all $x^* \in A$.

From this fact one can readily deduce a weak convergence criterion analogous to Theorem 1 of Subsection 17.2.1 for the class of spaces defined below.

Definition 2. Let X be a Banach space with a basis $\{e_n\}_1^\infty$, and $\{e_n^*\}_1^\infty$ be the corresponding coordinate (evaluation) functionals. The basis $\{e_n\}_1^\infty$ is called a *shrinking basis* if the linear span of the set $\{e_n^*\}_1^\infty$ is dense (in norm) in X^*.

Theorem 1. *Let X be a Banach space with a shrinking basis $\{e_n\}_1^\infty$, and let $x_n, x \in X, x_n = \sum_{j=1}^\infty a_{n,j} e_j, x = \sum_{j=1}^\infty a_j e_j$. Then for the sequence (x_n) to converge weakly to an element x, it is necessary and sufficient that this sequence is bounded in norm and converges componentwise to x: $a_{n,j} \to a_j$ as $n \to \infty$, for all $j \in \mathbb{N}$.*

In particular, "boundedness plus coordinatewise convergence" is a weak convergence criterion in such sequence spaces as c_0 or ℓ_p with $1 < p < \infty$, in which the standard basis is shrinking.

Caution! The standard basis is not a basis in the space ℓ_∞, while in the space ℓ_1 it is a basis, but it is not shrinking (Exercise 3 of Subsection 17.2.1). Consequently, in ℓ_1 and ℓ_∞ boundedness plus componentwise convergence are not sufficient for weak convergence. In ℓ_1 the weak convergence criterion is rather unusual: in this space, according to a theorem of Schur,[2] the weak and strong convergence coincide. In ℓ_∞ there is no conveniently verifiable criterion for weak convergence.

Theorem 2 (**weak convergence criterion in** $C(K)$). *Let K be a compact topological space. For functions $x_n, x \in C(K)$ the following conditions are equivalent:*

(i) $x_n \xrightarrow{w} x$;

(ii) $\sup_{n \in \mathbb{N}} \|x_n\| < \infty$ and $x_n(t) \to x(t)$ as $n \to \infty$ in all points $t \in K$.

Proof. (i) $\Longrightarrow$ (ii). The boundendness of a weakly convergent sequence is a general result. The condition $x_n(t) \to x(t)$ as $n \to \infty$ is simply convergence on the evaluation functional $\delta_t \in C(K)^*$, acting by the rule $\delta_t(f) = f(t)$.

(ii) $\Longrightarrow$ (i). We have to show that $F(x_n) \to F(x)$ as $n \to \infty$ for any $F \in C(K)^*$. Taking into account the general form of a linear functional on $C(K)$, we need to show that $\int_K x_n d\nu \to \int_K x \, d\nu$ for any regular Borel charge ν on K. By hypothesis, we are given a sequence of functions x_n that are uniformly bounded and converge pointwise to the function x. It remains to apply the Lebesgue dominated convergence theorem. $\qquad\square$

The results stated below point to a closer connection between weak and strong convergence than the one observed between w^*-convergence and strong convergence.

Theorem 3. *Let A be a convex subset of the Banach space X. Then the following conditions are equivalent:*

(a) *A is weakly closed;*

(b) *A is weakly sequentially closed, i.e., if $x_n \xrightarrow{w} x$ and $x_n \in A$, then $x \in A$;*

(c) *A is closed in the strong topology.*

Proof. The implication (a) $\Longrightarrow$ (b) is obvious (closedness implies sequential closedness in any topology, not only in the weak one).

(b) $\Longrightarrow$ (c). Let $x_n \in A$ and $\|x_n - x\| \to 0$. Then $x_n \xrightarrow{w} x$ and, by condition (b), $x \in A$.

(c) $\Longrightarrow$ (a). This implication was proved in Theorem 1 of Subsection 17.1.1 not only for Banach spaces, but also for arbitrary locally convex ones. $\qquad\square$

---

[2] See, e.g., [22, p. 293].

Theorem 4 (Mazur). *Suppose the sequence* (x_n) *of elements of the Banach space X converges weakly to an element* $x \in X$. *Then x lies in the strong closure of the convex hull of the sequence* (x_n). *Moreover, there exists a sequence* (y_n) *of convex combinations of the elements* x_n *that converges strongly to x, such that* $y_n \in \mathrm{conv}\{x_k\}_{k=n}^{\infty}$, $n = 1, 2, \ldots$.

Proof. Let us show that for each $n \in \mathbb{N}$ there exists an element $y_n \in \mathrm{conv}\{x_k\}_{k=n}^{\infty}$ such that $\|y_n - x\| < 1/n$. Then (y_n) will be the sought-for sequence. Indeed, let A_n denote the strong closure of the set $\mathrm{conv}\{x_k\}_{k=n}^{\infty}$. By the preceding theorem, A_n is also a weakly closed set. Since A_n contains all x_k with $k \geqslant n$, A_n also contains the weak limit x. Therefore, x lies in the strong closure of the set $\mathrm{conv}\{x_k\}_{k=n}^{\infty}$, and x can be arbitrarily well approximated by elements of this convex hull.   □

Theorem 5. *Let X, Y be Banach spaces. Then for a linear operator* $T : X \to Y$ *the following conditions are equivalent:*

(A) *T is continuous in the weak topologies of the spaces X and Y;*

(B) *T maps any sequence that converges weakly to zero again into one that converges weakly to zero;*

(C) *T is continuous in the strong topologies of the spaces X and Y.*

Proof. The implication (A) $\Longrightarrow$ (B) is obvious.

(B) $\Longrightarrow$ (C). We apply assertion (3) of Theorem 1 of Subsection 6.4.1: T is continuous if and only if it maps sequences that converge to zero into bounded sequences. Suppose $\|x_n\| \to 0$ as $n \to \infty$. Then $x_n \xrightarrow{w} 0$. By condition (B), this means that $Tx_n \xrightarrow{w} 0$. But then $\sup_{n \in \mathbb{N}} \|Tx_n\| < \infty$.

(C) $\Longrightarrow$ (A). Here it suffices to use Corollary 1 in Subsection 17.1.3.

Exercises

1. Let Y be a subspace of the Banach space X. Then the restriction of the topology $\sigma(X, X^*)$ to Y coincides with the topology $\sigma(Y, Y^*)$. In particular, any sequence that converges weakly in Y also converges weakly in X, and any weakly compact subset of Y is also weakly compact in X. Henceforth these facts will be used without further clarifications.

2. On the example of the standard basis verify that the weak convergence criterion established in ℓ_p, $1 < p < \infty$, (boundedness plus componentwise convergence) fails in ℓ_1.

3. Consider the sequence $x_1 = (1, 0, 0, \ldots)$, $x_2 = (1, 1, 0, 0, \ldots)$, $x_3 = (1, 1, 1, 0, \ldots)$, $\ldots$ in ℓ_∞. Verify that this sequence is bounded and converges componentwise to $x = (1, 1, 1, \ldots, 1, \ldots)$. However, Mazur's theorem does not hold for the sequence (x_n), i.e., (x_n) does not converge weakly to x. Thus, the weak convergence criterion "bounded plus componentwise convergence" is not valid in ℓ_∞.

4. Prove the following weak convergence criterion in the space $L_p[0, 1]$, $1 < p < \infty$: $f_n \xrightarrow{w} f$ if and only if $\sup_{n \in \mathbb{N}} \|f_n\| < \infty$ and $\int_a^b f_n(t)\, dt \to \int_a^b f(t)\, dt$ for any subinterval $[a, b] \subset [0, 1]$.

5. Using the example of the sequence $f_n = 2^n\left(\mathbb{1}_{[0,2^{-(n+1)}]} - \mathbb{1}_{[2^{-(n+1)},2^{-n}]}\right)$ convince yourself that in $L_1[0, 1]$ the criterion established in Exercise 4 fails.

6. Prove the following weak convergence criterion in the space $L_1[0, 1]$: $f_n \xrightarrow{w} f$ if and only if $\sup_{n \in \mathbb{N}} \|f_n\| < \infty$ and $\int_A f_n(t)\, dt \to \int_A f(t)\, dt$ for any measurable subset $A \subset [0, 1]$.

7. Let K be a convex weakly compact set in a Banach space that possesses a normal structure (see the exercises in Subsection 15.3.1 for the definition). Show that the Kakutani fixed point theorem remains valid for mappings of the set K.

Let D be a metric space. A mapping $f : D \to D$ is said to be *non-expansive* if for any $x_1, x_2 \in D$ one has $\rho(f(x_1), f(x_2)) \leqslant \rho(x_1, x_2)$.

8. In the setting of Exercise 7, show that every non-expansive mapping $f : K \to K$ has a fixed point.

9. Does there exists a common fixed point for all non-expansive mappings $f : [0, 1] \to [0, 1]$?

17.2.4 Total and Norming Sets. Metrizability Conditions

Definition 1. Let X be a linear space and $Y \subset X$ be a subspace. A set $F \subset X'$ is called a *total set over* Y, if it separates the points of the subspace Y. In detail, F is total over Y if for any $y \in Y \setminus \{0\}$ there exists an $f \in F$ such that $f(y) \neq 0$.

Definition 2. Let X be a Banach space, $Y \subset X$ a linear subspace, and $\theta \in (0, 1]$. A set $F \subset X^*$ is said to be θ-*norming over* Y if

$$\sup_{f \in F \setminus \{0\}} \frac{|f(y)|}{\|f\|} \geqslant \theta \|y\| \tag{1}$$

for all $y \in Y$. The set $F \subset X^*$ is said to be *norming over* Y if there exists a $\theta \in (0, 1)$ such that F is θ-norming over Y.

Obviously, every norming set is total: if $y \neq 0$, then condition (1) means that, in any case, there exists an $f \in F$ such that $f(y) \neq 0$.

In the case when F lies on the unit sphere of the space X^*, condition (1) reads $\sup_{f \in F} |f(y)| \geqslant \theta \|y\|$. In view of the homogeneity in y of this inequality, it suffices to verify it for $y \in S_Y$. Therefore, for $F \subset S_{X^*}$ Definition 2 can be restated as follows: the set F is θ-norming over Y if the inequality

$$\sup_{f \in F} |f(y)| \geqslant \theta \tag{2}$$

holds for all $y \in S_Y$.

Theorem 1. *Let X be a Banach space, $Y \subset X$ a linear subspace, and $G \subset S_{X^*}$ a set that is 1-norming over Y. Further, let $\varepsilon \in (0, 1)$ and suppose that on the unit sphere of the subspace Y there is given an ε-net D. For each element $x \in D$ fix a functional $f_x \in G$ such that $f_x(x) > 1 - \varepsilon$. Then the set $F = \{f_x\}_{x \in D}$ is θ-norming over Y, with $\theta = 1 - 2\varepsilon$.*

Proof. Fix a $y \in S_Y$ and let us show that estimate (2) holds for y. By the definition of an ε-net, there exists an element $x_0 \in D$ such that $\|y - x_0\| < \varepsilon$. We have:

$$\sup_{f \in F} |f(y)| = \sup_{x \in D} |f_x(y)| \geqslant |f_{x_0}(y)| = |f_{x_0}(x_0) - f_{x_0}(y - x_0)|$$

$$\geqslant 1 - \varepsilon - \|y - x_0\| > 1 - 2\varepsilon = \theta. \qquad \square$$

From this theorem, recalling that in a finite-dimensional space the unit sphere contains a finite ε-net, one derives the following

Corollary 1. *Let X be a Banach space, and let $Y \subset X$ be a finite-dimensional subspace. Then for any $\theta \in (0, 1)$ there exists a finite θ-norming set over Y.*

Here is a slightly more complex variant that will prove suitable in the next subsection.

Corollary 1′. *Let E be a Banach space, and let $Y \subset E^{**}$ be a finite-dimensional subspace. Then for any $\theta \in (0, 1)$ there exists a finite θ-norming set over Y consisting of elements of the unit sphere of the space E^*.*

Proof. Apply Theorem 1 with $X = E^{**}$ and $G = S_{E^*}$. $\qquad \square$

For infinite-dimensional separable spaces we obtain the following result.

Corollary 2. *Let X be a Banach space, and $Y \subset X$ a separable subspace. Then there exists a countable 1-norming set over Y.*

Proof. Take as the set D figuring in Theorem 1 a countable dense subset of the unit sphere in the subspace Y and put $G = S_{X^*}$. The set D forms an ε-net in S_Y for all $\varepsilon > 0$. Therefore, the set $F = \{f_x\}_{x \in D}$ consisting of the support functionals ($\|f_x\| = 1 = f_x(X)$) is θ-norming over Y for all $\theta < 1$. But if condition (2) is satisfied for all $\theta < 1$, then it is also satisfied for $\theta = 1$. $\qquad \square$

Recall (Exercises 15–18 in Subsection 16.2.1) that a Hausdorff topological vector space is metrizable if and only if it has a countable neighborhood basis of zero. Further, according to Exercise 3 in Subsection 16.3.2, the topology $\sigma(X, X^*)$ has a countable neighborhood basis of zero if and only if the space X^* has an at most countable Hamel basis. Since a Hamel basis of an infinite-dimensional Banach space is necessarily uncountable (Exercise 4 of Subsection 6.3.2), the weak topology of an infinite-dimensional space, if regarded on the whole space, is not metrizable. The situation changes drastically when instead we consider the weak topology on subsets.

Theorem 2. *Let (X, E) a dual pair, let $B \subset X$ be compact in the topology $\sigma(X, E)$, and let $Y = \mathrm{Lin}\, B$. Denote by $\sigma_B(X, E)$ the topology induced on B by the weak topology $\sigma(X, E)$. Further, suppose there exists a countable total set over Y, $F = \{f_1, f_2, \ldots\} \subset E$. Then there exists a norm p on Y such that the norm topology generated on B by p coincides with $\sigma_B(X, E)$. In particular, on B the weak topology is metrizable.*

Proof. Since B is a weakly compact set and the functionals f_n are continuous in the weak topology $\sigma(X, E)$, each of the f_n's is bounded on B in modulus by some number a_n. With no loss of generality, we may assume that $a_n \leqslant 1$ (otherwise, we multiply f_n by the factor $1/a_n$, without affecting the totalness of the sequence of functionals on Y). For each $y \in Y$ we put $p(y) = \sum_{n=1}^{\infty} 2^{-n} |f_n(y)|$. Let us show that p is the requisite norm. Each of the terms $2^{-n} |f_n(y)|$ is non-negative and satisfies the triangle inequality and the positive homogeneity condition. Therefore, p inherits the same properties. The non-degeneracy ($p(y) = 0 \Longrightarrow y = 0$) follows from the fact that the set F is total.

Now let us compare the topologies involved. Let $x \in B$, $r > 0$. Consider the set $U(x, r) = \{y \in B : p(x - y) < r\}$, i.e., the ball in B of radius r and center x generated by the norm p. Pick an $N \in \mathbb{N}$ such that $2^{-N} < r/2$, and consider the following weak neighborhood of the point x in the set B:

$$V = \left\{ y \in B : \max_{1 \leqslant k \leqslant N} |f_k(x - y)| < r/4 \right\}.$$

If $y \in V$, then

$$p(x - y) = \sum_{n=1}^{\infty} \frac{|f_n(x - y)|}{2^n} \leqslant \sum_{n=1}^{N} \frac{|f_n(x - y)|}{2^n} + \frac{1}{2^N}$$

$$< 2 \max_{1 \leqslant k \leqslant N} |f_k(x - y)| + \frac{r}{2} < r$$

and $y \in U(x, r)$. That is, $V \subset U(x, r)$. This establishes that on B the topology generated by the norm p is weaker than the topology $\sigma_B(X, E)$. Since it is not possible to strictly weaken the topology of a compact set so that it remains separated (see Subsection 1.2.3, second paragraph), we conclude that $\sigma(X, E)$ and p induce on B one and the same topology. $\qquad\square$

Corollary 3. *Let X be a Banach space. Then on any separable weakly compact subset B of X the weak topology $\sigma(X, X^*)$ is metrizable.*

Proof. It suffices to apply Theorem 2 to the dual pair (X, X^*). Here the existence of a countable total set follows from Corollary 2.   □

Corollary 4. *Let X be a separable Banach space. Then the w^*-topology $\sigma(X^*, X)$ is metrizable on the bounded subsets of the space X^*.*

Proof. Consider the dual pair (X^*, X) and take $B = \overline{B}_{X^*}$. By Alaoglu's theorem, B is compact in the $\sigma(X^*, X)$-topology. By assumption, X contains a countable dense set F. This set F is total over X^*. Theorem 2 shows that the w^*-topology on the unit ball of the space X^* is metrizable. To complete the proof, it remains to observe that any bounded subset of X^* lies in some ball of the form $r\overline{B}_{X^*}$.   □

Corollary 5. *Let X be a separable Banach space. Then from any bounded sequence of functionals $x_n^* \in X^*$ one can extract a convergent subsequence.*

Proof. With no loss of generality we can assume that $x_n^* \in \overline{B}_{X^*}$ (otherwise we multiply all x_n^* by the factor $(\sup_n \|x_n^*\|)^{-1}$). By Alaoglu's theorem, $\overline{B}_{X^*}$ is w^*-compact, and by Corollary 4, this compact set is metrizable. It remains to recall that from any sequence of elements of a compact metric space one can extract a convergent subsequence.   □

Exercises

1. Let X be a linear space, and $Y \subset X$ a subspace. Then there exists a finite total set over Y if and only Y is finite-dimensional.

2. Let X be a Banach space, and $\theta \in (0, 1)$. A set $F \subset X^*$ is θ-norming for X if and only if the w^*-closure of the absolute convex hull of F contains $\theta\overline{B}_{X^*}$.

3. For a Banach space X the following conditions are equivalent:

— there exists a countable total set over X;

— there exists an injective continuous linear operator mapping X into ℓ_2.

4. For a Banach space X the following conditions are equivalent:

— there exists a countable norming set over X;

— there exists a bounded below continuous linear operator mapping X into ℓ_∞.

5. For a Banach space X the following conditions are equivalent:

— there exists a countable 1-norming set over X;

— there exists a linear isometric embedding of the space X into ℓ_∞.

6. In particular, every separable Banach space admits an isometric embedding into the space ℓ_∞.

A separable Banach space E is called *universal* if among its subspaces are isometric copies of all separable Banach spaces. The following two exercises give a sketch of the proof of the famous Banach–Mazur theorem on the universality of the space $C[0, 1]$.

7. Let X be a separable Banach space and $\mathcal{K}$ be the Cantor set. Equip the closed ball $\overline{B}_{X^*}$ with the w^*-topology. By Exercise 6 of Subsection 1.4.4 and Alaoglu's theorem, there exists a surjective continuous mapping $F \colon \mathcal{K} \to \overline{B}_{X^*}$. Define the operator $T \colon X \to C(\mathcal{K})$ by the formula $(Tx)(t) = \langle F(t), x \rangle$. Prove that the operator T effects a linear isometric embedding of the space X into $C(\mathcal{K})$. This will establish that the space $C(\mathcal{K})$ of continuous functions on the Cantor set is universal.

8. Prove that $C(\mathcal{K})$, where $\mathcal{K}$ is the Cantor set, admits an isometric embedding into $C[0, 1]$. Deduce from this the universality of the space $C[0, 1]$.

Since $X \subset X^{**}$, one can talk about subsets of the Banach space X that are total or norming over X^*.

9. For a linear subspace Y of a Banach space X the following conditions are equivalent:

— Y is total over X^*;

— Y is a norming set over X^*;

— Y is dense in X.

10. Denote by e_n^* the functional on ℓ_∞ which assigns to each element $x = (x_1, x_2, \ldots)$ of the space ℓ_∞ its n-th coordinate: $e_n^*(x) = x_n$. On the example of the sequence $(e_n^*)_1^\infty$ convince yourself that on non-separable spaces there exists bounded sequences of functionals which contain no w^*-convergent subsequences. In particular, this example shows that, despite Alaoglu's theorem, the unit ball of the dual space is not necessarily w^*-sequentially compact.

17.2.5 The Eberlein–Smulian Theorem

The last exercise in the preceding subsection reminds us that in non-metrizable topological spaces compactness and sequential compactness are, generally speaking, distinct properties. The weak topology of an infinite-dimensional Banach space is not metrizable. It is therefore even more surprising that the weak compactness of a set in a Banach space is equivalent to its weak sequential compactness. This theorem of Eberlein and Smulian consists of two parts, the first of which was proved by V.L. Smulian (in the literature one encounters also the spelling Šmulian or Shmulian) in 1940, and the second by W.F. Eberlein in 1947.

Theorem 1 (Smulian). *Let K be a weakly compact set in a Banach space X. Then from every sequence $x_n \in K$, $n = 1, 2, \ldots$, one can extract a weakly convergent subsequence.*

Proof. Consider the closed linear span Y of the sequence $(x_n)_{n=1}^{\infty}$ and let $\widetilde{K} = K \cap Y$. Since Y is separable, $\widetilde{K}$ is a separable set. Further, any closed linear subspace is a weakly closed set (Theorem 3 of Subsection 17.2.3), hence $\widetilde{K}$, being the intersection of a weakly compact set with a weakly closed set, is a weakly compact set. By Corollary 3 of Subsection 17.2.4, the weak topology $\sigma(X, X^*)$ is metrizable on $\widetilde{K}$. Finally, $x_n \in \widetilde{K}$ by construction, and we know that from any sequence of elements of a metrizable compact set one can extract a convergent (in the present case, weakly convergent, since we are dealing with the weak topology) subsequence. □

Theorem 2 (Eberlein). *Let K be a weakly sequentially compact subset of a Banach space, i.e., such that from every sequence $x_n \in K$, $n = 1, 2, \ldots$, one can extract a weakly convergent subsequence, and the limit of the subsequence again lies in K. Then K is weakly compact.*

Proof. To begin with, note that K is a bounded set. Indeed, assuming that K is not bounded, there would exist a sequence $x_n \in K$ such that $\|x_n\| \to \infty$. But such a sequence cannot contain bounded subsequences, and hence it cannot contain weakly convergent subsequences.

Since any bounded set can be transformed into a subset of the unit ball by multiplying it by a small positive number, we can assume for simplicity that $K \subset \overline{B}_X$. Further, using the embedding $\overline{B}_X \subset \overline{B}_{X^{**}}$, K can be regarded as a subset of the ball $\overline{B}_{X^{**}}$. Since on the space X, and hence also on the set K, the topologies $\sigma(X, X^*)$ and $\sigma(X^{**}, X^*)$ coincide, it suffices to show that K is $\sigma(X^{**}, X^*)$-compact. By Alaoglu's theorem, $\overline{B}_{X^{**}}$ is $\sigma(X^{**}, X^*)$-compact. Hence, to prove that K is $\sigma(X^{**}, X^*)$-compact, it in turn suffices to show that K is $\sigma(X^{**}, X^*)$-closed in $\overline{B}_{X^{**}}$.

So, let $x^{**} \in \overline{B}_{X^{**}}$ be an arbitrary $\sigma(X^{**}, X^*)$-limit point of the set K. We need to verify that $x^{**} \in K$. Recalling the form of the neighborhoods in the topology $\sigma(X^{**}, X^*)$, we see that the condition that x^{**} is a $\sigma(X^{**}, X^*)$-limit point can be expressed as follows:

(A) for any finite set of functionals $D \subset X^*$ and any $\varepsilon > 0$, there exists an element $x \in K$ such that $\max_{y^* \in D} |y^*(x^{**} - x)| < \varepsilon$.

The main idea of the proof is to construct a sequence $x_n \in K$ such that none of its subsequences can converge weakly to a point different from x^{**}. Since by assumption any sequence $x_n \in K$ contains a subsequence that converges to some point of K, this will establish that $x^{**} \in K$.

The requisite sequence $x_n \in K$ will be constructed recursively, using at each step property (A) and Corollary 1' of Subsection 17.2.4. Fix some $\theta \in (0, 1)$ and a sequence $\varepsilon_n \to 0$. Consider $Y_0 = \mathrm{Lin}\,\{x^{**}\}$. By Corollary 1' of Subsection 17.2.4, there exists a finite θ-norming set $D_0 \subset S_{X^*}$ over Y_0. Using property (A), choose a point $x_1 \in K$ such that $\max_{y^* \in D_0} |y^*(x^{**} - x_1)| < \varepsilon_1$. Next, consider $Y_1 = \mathrm{Lin}\,\{x^{**}, x_1\}$. Again, by the same Corollary 1' of Subsection 17.2.4, there exists a θ-norming set $D_1 \subset S_{X^*}$ over Y_1. With no loss of generality we may assume that $D_0 \subset D_1$: otherwise, we can replace D_1 by the union $D_0 \cup D_1$. Using again property (A), we choose $x_2 \in K$ such that $\max_{y^* \in D_1} |y^*(x^{**} - x_2)| < \varepsilon_2$. Continuing this construction, we obtain a sequence of elements $x_n \in K$, a sequence of subspaces $Y_n = \mathrm{Lin}\{x^{**}, x_1, x_2, \ldots, x_n\}$, and a sequence of finite subsets $D_0 \subset D_1 \subset D_2 \subset \cdots$ of the unit sphere of the space X^*, such that D_n is a θ-norming set over Y_n and

$$\max_{y^* \in D_{n-1}} |y^*(x^{**} - x_n)| < \varepsilon_n. \tag{3}$$

Denote $\bigcup_{n=0}^{\infty} D_n$ by D, and the norm-closure of the subspace $\mathrm{Lin}\{x^{**}, x_1, x_2, \ldots\}$ by Y. Suppose that some subsequence (x_{n_j}) of the sequence (x_n) converges weakly to a point $x \in K$. Let us prove that $x = x^{**}$: as we explained above, this will complete the proof of the entire theorem. First we note that, by Mazur's theorem (Theorem 4 in Subsection 17.2.3), $x \in Y$. Hence, $x - x^{**} \in Y$, too. By construction, the set D is θ-norming over all subspaces Y_n. Consequently, D is θ-norming over $\bigcup_{n=1}^{\infty} Y_n$, and hence also over Y, the strong closure of this union. It follows that

$$\|x - x^{**}\| \leqslant \frac{1}{\theta} \sup_{y^* \in D} |y^*(x^{**} - x)|.$$

We claim that the right-hand side of the last inequality is equal to zero. Indeed, for every $y^* \in D$ there exists a number $N \in \mathbb{N}$ such that $y^* \in D_m$ for all $m \geqslant N$. By condition (3), this means that $|y^*(x^{**} - x_m)| < \varepsilon_m$ for all $m > N$. Since x is a weak limit point of the sequence (x_n), this allows us to conclude that $|y^*(x^{**} - x)| = 0$.

$\square$

Exercises

Prove the following two assertions, treated as obvious above:

1. Let $Y_0 \subset Y_1 \subset Y_2 \subset \cdots$ be an increasing chain of subspaces, and D be a θ-norming set over all Y_n's. Then D is also θ-norming over $\bigcup_{n=1}^{\infty} Y_n$.

2. Let D be a θ-norming set over the linear subspace E. Then D is also θ-norming over the strong closure of the subspace E.

The Eberlein–Smulian theorem may create the illusion that in the weak topology of a Banach space all topological properties may be adequately expressed in the language of sequences. The exercises below will help to dispel this illusion.

3. In the space ℓ_2 consider the standard basis $\{e_n\}$. Set $x_n = n^{1/4} e_n$. Prove that:

— 0 is a limit point of the sequence (x_n) in the weak topology;

— the sequence (x_n) contains no bounded subsequence, and hence no weakly convergent subsequence.

The *weak sequential closure* of the set A in the Banach space X is defined to be the set of weak limits of all weakly convergent sequences of elements of A.

4. von Neumann's example. In the space ℓ_2 consider the set A of the vectors $x_{n,m} = e_n + ne_m, n,\ m \in \mathbb{N}, m > n$ (where e_n is, as above, the standard basis). Verify that

— for $m \to \infty$ and fixed n, the sequence $(x_{n,m})$ converges weakly to e_n. That is to say, all the vectors e_n lie in the weak sequential closure of the set A.

— 0 does not belong to the weak sequential closure of A.

— The weak sequential closure of the set A is not a weakly sequentially closed set.

17.2.6 Reflexive Spaces

A Banach space X is said to be *reflexive* if $X = X^{**}$. In a reflexive space, thanks to the equality $X = X^{**}$, the weak topology $\sigma(X, X^*)$ coincides with $\sigma(X^{**}, X^*)$, and together with the usual properties of a weak topology (equivalence of closedness and weak closedness of convex sets, equivalence of continuity and weak continuity of linear operators), it also enjoys the main nice feature of the weak* topology, namely, the compactness of the unit ball. This combination of properties makes reflexive spaces far more convenient in applications.

Theorem 1. *For a Banach space X the following conditions are equivalent:*

(i) *X is reflexive;*

(ii) *the closed ball $\overline{B}_X$ is weakly compact;*

(iii) *from any bounded sequence $x_n \in X$ one can extract a weakly convergent subsequence.*

Proof. (i) $\Longrightarrow$ (ii). If $X = X^{**}$, then $\overline{B}_X = \overline{B}_{X^{**}}$; and, by Alaoglu's theorem, $\overline{B}_{X^{**}}$ is $\sigma(X^{**}, X^*)$-compact.

(ii) $\Longrightarrow$ (i). Since the restriction of the topology $\sigma(X^{**}, X^*)$ to X coincides with the weak topology $\sigma(X, X^*)$, (ii) implies that $\overline{B}_X$ is a $\sigma(X^{**}, X^*)$-compact subset of the space X^{**}. In particular, $\overline{B}_X \subset \overline{B}_{X^{**}}$ is a w^*-closed subset. By Goldstine's theorem (Theorem 1 of Subsection 17.2.2), $\overline{B}_X$ is w^*-dense in $\overline{B}_{X^{**}}$, and hence $\overline{B}_X = \overline{B}_{X^{**}}$. Passing to linear spans, we obtain the needed equality $X = X^{**}$.

Finally, the equivalence (ii) $\Longleftrightarrow$ (iii) follows from the Eberlein–Smulian theorem. $\square$

Theorem 2. *If the Banach space X is reflexive, then its subspaces and quotient spaces are reflexive.*

Proof. Let Y be a subspace of the Banach space X. By our convention, Y is a closed linear subspace. Therefore, Y is weakly closed in X. Since $\overline{B}_X$ is weakly compact, the set $\overline{B}_Y = Y \cap \overline{B}_X$ will also be weakly compact. This establishes the reflexivity of the space Y.

Now let us consider the quotient space X/Y. Recall that the quotient mapping $q: X \to X/Y$, which sends each element of the space X into its equivalence class, is a continuous linear operator. Hence, q is a weakly continuous operator. In particular, $q(\overline{B}_X)$, being the image of a weakly compact set under a weakly continuous map, is a weakly compact subset of the space X/Y. Further, for any $[x] \in \overline{B}_{X/Y}$ there exists a representative $\widetilde{x} \in [x]$ such that $\|\widetilde{x}\| \leqslant \|[x]\| + 1 \leqslant 2$. That is, $\overline{B}_{X/Y} \subset 2q(\overline{B}_X)$. Therefore, the set $\overline{B}_{X/Y}$ is weakly compact, as a weakly closed subset of a weakly compact set. $\qquad\square$

Theorem 3. *The Banach space X is reflexive if and only if its dual X^* is reflexive.*

Proof. Suppose $X = X^{**}$. Passing to dual spaces, we have that $X^* = X^{***}$ (see also Exercises 3 and 4 of Subsection 17.2.2). Thus, the reflexivity of the original space implies the reflexivity of its dual. Conversely, suppose that the space X^* is reflexive. Then, as we have just seen, its dual X^{**} is also reflexive. But X is a subspace of X^{**}, and as such X must be reflexive. $\qquad\square$

Let us list several properties of reflexive spaces which find application in problems of approximation theory and variational calculus.

Theorem 4. *Let $A \neq \emptyset$ be a convex closed subset of a reflexive Banach space X. Then for any $x \in X$, the set A contains a closest point to x.*

Proof. Denote $\rho(x, A)$ by r and consider the sets

$$A_n = \{a \in A : \|a - x\| \leqslant r + (1/n)\} = A \cap \left(x + (r + (1/n))\overline{B}_X\right).$$

Since A is a weakly closed and $\overline{B}_X$ is weakly compact, each of the sets A_n is weakly compact. Any decreasing chain of compact sets (in fact, even any centered family of compact sets) has a non-empty intersection. Every element $y \in \bigcap_n A_n$ lies in A and lies at distance r from x. Thus, y is the sought-for point of A that is closest to x. $\quad\square$

We suggest to the reader to compare Theorem 4 with the theorem on best approximation in a Hilbert space and Exercises 4–6 in Subsection 12.2.1. In particular, under the assumptions of Theorem 4, the uniqueness of the closest point is not guaranteed even in the finite-dimensional case (see Exercise 6 below).

Theorem 5. *For any continuous linear functional f on a reflexive Banach space X there exists an element $x \in S_X$ such that $f(x) = \|f\|$. Therefore, in a reflexive Banach space every linear functional attains the maximum of its modulus on the unit sphere.*

Proof. By Theorem 1 of Subsection 9.2.1, for any point $f \in X^*$ there exists a supporting functional $x \in S_{X^{**}}$ at the point f. For this element one has $f(x) = \|f\|$. It remains to recall that $X = X^{**}$, i.e., x does not just lie somewhere in the second dual space, but, as required, it lies on the unit sphere of the original space X.   □

Let us note that the converse to Theorem 5 also holds: if the Banach space X is not reflexive, then there exists a functional $f \in X^*$ which does not attain on S_X its supremum. The proof of this by far not simple theorem of R.C. James can be found in the first chapter of J. Diestel's monograph [12].

To conclude, let us record which of the spaces known to us are reflexive, and which are not.

— All finite-dimensional spaces are reflexive.

— All spaces L_p and ℓ_p with $1 < p < \infty$ are reflexive (this follows from the theorem on the general form of linear functionals on L_p).

— The space c_0 is not reflexive, since $(c_0)^{**} = \ell_\infty \neq c_0$.

— The space ℓ_1 is not reflexive, since it is the dual of the non-reflexive space c_0.

— The space ℓ_∞ is not reflexive, since it is the dual of the non-reflexive space ℓ_1.

— The space $C(K)$, where K is an infinite compact space, is not reflexive, since it contains a non-reflexive subspace isomorphic to c_0. The reader is invited to verify that for any sequence of functions $f_n \in S_{C(K)}$ with disjoint supports, $\overline{\mathrm{Lin}}\{f_n\}$ is a subspace of $C(K)$ isometric to c_0. In particular, the space $C[0, 1]$ is not reflexive.

— The space $L_1(\Omega, \Sigma, \mu)$, where Ω cannot be decomposed into a finite union of atoms of the measure μ, is not reflexive; indeed, it contains a non-reflexive subspace isomorphic to ℓ_1. (For any sequence of functions $f_n \in S_{L_1(\Omega, \Sigma, \mu)}$ with disjoint supports, $\overline{\mathrm{Lin}}\{f_n\}$ is a subspace of $L_1(\Omega, \Sigma, \mu)$ isometric to ℓ_1). In particular, the space $L_1[0, 1]$ is not reflexive.

Remark 1. The definitions of the spaces L_p and ℓ_p look more complicated and less natural than those of the spaces $C[0, 1]$, $L_1[0, 1]$, or c_0. The list given above sheds light on the reason for the wide utilization of the spaces L_p: their relatively complicated definition is more than compensated by their pleasant features, first and foremost, by their reflexivity.

Exercises

1. In each of the examples of non-reflexive spaces listed above construct explicitly a bounded sequence that contains no weakly convergent subsequence. This will lead to another way of establishing that the spaces in question are not reflexive.

Using the weak compactness of the unit ball, prove the following generalization of Theorem 5.

2. Let $T \in L(X, Y)$, with X reflexive and Y finite-dimensional. Then there exists an element $x \in S_X$ such that $\|T(x)\| = \|T\|$.

3. Provide an example of a continuous linear functional on c_0 which does not attain its supremum on the unit sphere. Give a complete description of such functionals.

4. Solve the analogues of the preceding exercise for the spaces ℓ_1, $L_1[0, 1]$, and $C[0, 1]$.

5. Let A be a weakly compact subset of a Banach space X. Then for any $x \in X$ in A there exists a point closest to x.

6. In the space $\mathbb{R}^2$ equipped with the norm $\|(x_1, x_2)\| = \max\{|x_1|, |x_2|\}$, consider the set $A = \{(1, a) : a \in [-1, 1]\}$. Verify that for $x = 0$ the point in A closest to x is not unique.

7. An operator $A \in L(X, Y)$ is called a *Dunford–Pettis operator* if it maps weakly convergent sequences into norm-convergent sequences. Show that any compact operator is a Dunford–Pettis operator. If the space X is reflexive, then every Dunford–Pettis operator is compact. In non-reflexive spaces (say, in $C[0, 1]$) there exist non-compact Dunford–Pettis operators.

Comments on the Exercises

Subsection 17.2.1

Exercise 7. Denote by A the set of all points where at least one of the functions F_n, F is discontinuous. Consider the space X of all bounded functions on $[0, 1]$ that have a left and a right limit at every point, are continuous at all the points of the set A, and have at most finitely many discontinuity points. Equip X with the norm $\|f\| = \sup_{[0,1]} |f(t)|$. The required relation $\int_0^1 f \, dF_n \to \int_0^1 f \, dF$ is simpler to prove not for $f \in C[0, 1]$, but for the wider class of $f \in X$. To this end we need to show the relation $\int_0^1 f \, dF_n \to \int_0^1 f \, dF$ for $f = \mathbb{1}_{[a,b]}$, then extend it by linearity to the set of all piecewise-constant functions, and use the pointwise convergence criterion for operators (convergence on a dense subset + boundedness in norm).

Exercise 8. Pass to the Borel charges ν_n, for which $F_n(t) = \nu_n([0, t])$ for $t \in (0, 1]$. First deal with the case where ν_n are measures.

Exercise 9. Use the fact that the set of discontinuity points of a function of bounded variation is at most countable, and Exercises 4 and 8. For another argument, see the textbook by A. Kolomogorov and S. Fomin, Chap. VI, § 6.

Subsection 17.2.5

Exercise 3. The example is taken from the paper [51]. The indicated sequence (x_n) will converge to zero with respect to the statistical filter $\mathfrak{F}_s$ (Exercise 7 in Subsection 16.1.1). The following general result holds true [59]: for a sequence of numbers $a_n > 0$ the following conditions are equivalent: (1) there exists a sequence $x_n \in \ell_2$ with $\|x_n\| = a_n$, for which 0 is a weak limit point, and (2) $\sum a_n^{-2} = \infty$.

Chapter 18
The Krein–Milman Theorem and Its Applications

18.1 Extreme Points of Convex Sets

As we remarked earlier, one of the main merits of the functional analysis-based approach to problems of classical analysis is that it reduces problems formulated analytically to problems of a geometric character. The geometric objects that arise in this way lie in infinite-dimensional spaces, but they can be manipulated by using analogies with figures in the plane or in three-dimensional space. To exploit this analogy more freely, to understand when it helps, rather than mislead us, we have studied above many properties of spaces, subspaces, convex sets, compact and weakly compact sets, linear operators, emphasizing each time the coincidences and differences with the finite-dimensional versions of those objects and properties. In the present chapter we add to the already built arsenal of geometric tools yet another one: the study of convex sets by means of their extreme points. Although extreme points are a direct generalization of the vertices of a polygon or polyhedron, in the framework of classical geometry this purely geometric concept was not used for general figures. The study and application of extreme points to problems of geometry (including finite-dimensional ones), functional analysis, mathematical economics, is one of the achievements of the bygone 20th century.

18.1.1 Definitions and Examples

Let A be a convex subset of a linear space X. A point $x \in A$ is called an *extreme point* of the set A if it is not the midpoint of any non-degenerate segment whose endpoints lie in A. The set of extreme points of the set A is denoted by ext A. That is, $x \in \text{ext } A$ if and only if for any $x_1, x_2 \in A$, if $(x_1 + x_2)/2 = x$, then $x_1 = x_2$ (and hence both vectors x_1 and x_2 coincide with x).

Theorem 1. *Let A be a convex subset of a topological vector space X. Then none of the interior points of the set A is an extreme point of A.*

© Springer International Publishing AG, part of Springer Nature 2018
V. Kadets, *A Course in Functional Analysis and Measure Theory*,
Universitext, https://doi.org/10.1007/978-3-319-92004-7_18

Proof. Let $x \in A$ be an interior point. Then there exists a balanced neighborhood U of zero such that $x + U \subset A$. Let $y \in U \setminus \{0\}$. Set $x_1 = x + y$, $x_2 = x - y$. Then $x_1, x_2 \in A$, $(x_1 + x_2)/2 = x$, but $x_1 \neq x_2$. $\qquad \square$

Thus, all extreme points of a set lie on its boundary. Needless to say, this provides rather incomplete information on the positioning of extreme points. Note that ext A depends only on the convex geometry of the set A, but does not depend on the ambient linear space in which A is considered, or on which topology is given on A.

Theorem 2 (examples).

(a) *If A is a convex polygon in the plane, then ext A is the set of vertices of A.*

(b) *If A is a disc, then ext A is its bounding circle.*

(c) *The set of extreme points of the closed unit ball $\overline{B}_H$ of the Hilbert space H is the unit sphere S_H.*

(d) *The closed unit ball $\overline{B}_{c_0}$ of the space c_0 has no extreme points.*

Proof. Assertions (a) and (b) are obvious. Let us address (c). By Theorem 1, ext $\overline{B}_H \subset S_H$. Let us establish the opposite inclusion. Let $x, x_1, x_2 \in S_H$ and put $(x_1 + x_2)/2 = x$. This means that $\|x_1\| = \|x_2\| = 1$ and $\|x_1 + x_2\| = 2$. But then, by the parallelogram equality,

$$\|x_1 - x_2\|^2 = \|x_1 + x_2\|^2 - 2\|x_1\|^2 - 2\|x_2\|^2 = 4 - 2 - 2 = 0,$$

and so $x_1 = x_2$.

(d) Let us show that no point of $\overline{B}_{c_0}$ is an extreme point. Let $a = (a_1, a_2, \ldots) \in \overline{B}_{c_0}$. This means that all the coordinates a_j are not larger in modulus than 1 and $a_j \to 0$ as $j \to \infty$. From the last fact it follows that there exists an n such that $|a_n| < 1/2$. Consider the following vectors x_1 and x_2, all coordinates of which coincide with those of a, except for the nth one, where they differ from a by $\pm 1/2$:

$$x_1 = (a_1, a_2, \ldots, a_{n-1}, a_n + 1/2, a_{n+1}, \ldots)$$

and

$$x_2 = (a_1, a_2, \ldots, a_{n-1}, a_n - 1/2, a_{n+1}, \ldots).$$

Then x_1 and x_2 lie in $\overline{B}_{c_0}$, $(x_1 + x_2)/2 = a$, but $x_1 \neq x_2$. $\qquad \square$

The extreme points of a Cartesian product of convex sets admit a simple description.

Theorem 3. *Let Γ be an index set, X_γ, $\gamma \in \Gamma$, be linear spaces, and $A_\gamma \subset X_\gamma$ be convex sets. Then* ext $\left(\prod_{\gamma \in \Gamma} A_\gamma \right) = \prod_{\gamma \in \Gamma}$ ext A_γ.

Proof. Let $x = (x_\gamma)_{\gamma \in \Gamma} \in \prod_{\gamma \in \Gamma}$ ext A_γ, i.e., $x_\gamma \in$ ext A_γ for all $\gamma \in \Gamma$. Let us show that $x \in$ ext $\left(\prod_{\gamma \in \Gamma} A_\gamma \right)$. Consider elements $y = (y_\gamma)_{\gamma \in \Gamma}$ and $z = (z_\gamma)_{\gamma \in \Gamma}$ in

$\prod_{\gamma \in \Gamma} A_\gamma$ such that $(y + z)/2 = x$. Then $(y_\gamma + z_\gamma)/2 = x_\gamma$ and $y_\gamma, z_\gamma \in A_\gamma$. Since $x_\gamma \in \text{ext}\, A_\gamma$, it follows that $y_\gamma = z_\gamma$ for all $\gamma \in \Gamma$, and so $y = z$. This establishes the inclusion $\text{ext}\left(\prod_{\gamma \in \Gamma} A_\gamma\right) \supset \prod_{\gamma \in \Gamma} \text{ext}\, A_\gamma$.

Now let us establish the opposite inclusion $\text{ext}\left(\prod_{\gamma \in \Gamma} A_\gamma\right) \subset \prod_{\gamma \in \Gamma} \text{ext}\, A_\gamma$. Let $x = \left(x_\gamma\right)_{\gamma \in \Gamma} \in \left(\prod_{\gamma \in \Gamma} A_\gamma\right) \setminus \left(\prod_{\gamma \in \Gamma} \text{ext}\, A_\gamma\right)$. Then there is an index $\gamma_0 \in \Gamma$ for which $x_{\gamma_0} \in A_{\gamma_0} \setminus \text{ext}\, A_{\gamma_0}$. By definition, this means that there exists elements $y_{\gamma_0}, z_{\gamma_0} \in A_{\gamma_0}$, $y_{\gamma_0} \neq z_{\gamma_0}$, such that $(y_{\gamma_0} + z_{\gamma_0})/2 = x_{\gamma_0}$. Now define the elements $y, z \in \prod_{\gamma \in \Gamma} A_\gamma$ as follows: for $\gamma \neq \gamma_0$ put $y_\gamma = z_\gamma = x_\gamma$, while for the index γ_0 take as coordinates precisely the elements y_{γ_0} and z_{γ_0}, respectively. Then $y \neq z$ (they differ in the γ_0 coordinate), but $(y + z)/2 = x$. Therefore, $x \notin \text{ext}\left(\prod_{\gamma \in \Gamma} A_\gamma\right)$. $\square$

An obvious consequence of this theorem is the following descriptions of the extreme points of two important n-dimensional bodies.

Corollary 1. *The extreme points of the n-dimensional cube $[-1, 1]^n$ are precisely the vectors with all coordinates equal to ± 1.* $\square$

Let us recall the notations $\mathbb{C}_1 = \{\lambda \in \mathbb{C} : |\lambda| \leqslant 1\}$ and $\mathbb{T} = \{\lambda \in \mathbb{C} : |\lambda| = 1\}$ for the unit disc and the unit circle.

Corollary 2. *The set of extreme points of the n-dimensional polydisc $(\mathbb{C}_1)^n$ is the skeleton of the polydisc, i.e., the set $\mathbb{T}^n$.* $\square$

Exercises

1. In the real space $C[0, 1]$ the closed unit ball has only two extreme points, the functions $f = 1$ and $g = -1$.

2. In the space $L_1[0, 1]$ the closed unit ball has no extreme points.

3. For $1 < p < \infty$ every element of the unit sphere in the space $L_p[0, 1]$ is an extreme point of the closed unit ball. In other words, $L_p[0, 1]$ is a strictly convex space (see Exercise 6 in Subsection 12.2.1).

4. Using the preceding exercise and the reflexivity, prove the following result: Let $1 < p < \infty$, and let $A \subset L_p[0, 1]$ be a convex closed subset. Then for any $x \in X$ in A there is a unique closest point to x.

5. Let X, Y be linear spaces and $T : X \to Y$ be an injective linear operator. Then for any convex subset $A \subset X$ it holds that $T(\text{ext}\, A) = \text{ext}\, T(A)$.

6. Give an example showing that in the preceding exercise the injectivity assumption cannot be discarded.

7. For any convex compact subset $A \subset \mathbb{R}^2$, the set ext A is closed.

8. Give an example of a convex compact subset $A \subset \mathbb{R}^3$ whose set of extreme points is not closed.

18.1.2 The Krein–Milman Theorem

In this subsection we shall prove the main result of this chapter, namely, the existence of extreme points for any convex compact set.

Definition 1. Let A be a convex subset of a linear space X. The set $B \subset A$ is said to be an *extreme subset of the set A* if it meets the following requirements:

— B is not empty;

— B is convex;

— for any two points $x_1, x_2 \in A$, if $(x_1 + x_2)/2 \in B$, then $x_1, x_2 \in B$.

Obviously, a subset consisting of a single point x will be extreme if and only if x is an extreme point. If A is a triangle in the plane, then an example of an extreme subset B is provided by any side of the triangle.

Lemma 1. *Let X, Y be linear spaces, $T: X \to Y$ be a linear operator, and $A \subset X$ be a convex subset. Then for any extreme subset B of the set $T(A)$, the set $T^{-1}(B) \cap A$ (the complete preimage in A of the set B) is an extreme subset of the original set A. In particular, the complete preimage in A of any extreme point of the set $T(A)$ is an extreme subset of A.*

Proof. Suppose $x_1, x_2 \in A$ and $(x_1 + x_2)/2 \in T^{-1}(B)$. Then $Tx_1, Tx_2 \in T(A)$ and $(Tx_1 + Tx_2)/2 \in B$. Since B is an extreme subset of $T(A)$, this means that $Tx_1, Tx_2 \in B$, and so $x_1, x_2 \in T^{-1}(B)$. $\quad\square$

Lemma 2. *Suppose A is a convex set, B is an extreme subset of A, and C is an extreme subset of B. Then C is an extreme subset of A. In particular, an extreme point of an extreme subset of a set A is an extreme point of A.*

Proof. Suppose $x_1, x_2 \in A$ and $(x_1 + x_2)/2 \in C$. Then, in particular, $(x_1 + x_2)/2 \in B$. Since B is an extreme subset of A, this implies that $x_1, x_2 \in B$. But now recalling that $(x_1 + x_2)/2 \in C$ and C is an extreme subset of B, we conclude that $x_1, x_2 \in C$, as needed. $\quad\square$

Now let us change the setting from arbitrary linear spaces to locally convex topological vector spaces, and from arbitrary convex sets to convex compact sets.

Lemma 3. *Let A be a convex compact set in a topological vector space X, f a continuous real linear functional on X, and $b = \max_{x \in A} f(x)$. Then the set $M(f, A) = \{x \in A : f(x) = b\}$ of points x in which f attains its maximum on A is an extreme subset of A.*

Proof. The set $f(A)$ is an interval $[a, b]$ joining the minimal and maximal values on A of the functional f. Hence, b is an extreme point of the set $f(A)$. By Lemma 1, $M(f, A) = f^{-1}(b) \cap A$ is an extreme subset. $\qquad\square$

Lemma 4. *Let A be a convex compact subset of a topological vector space X and $\mathfrak{M}$ be a centered family of closed extreme subsets of A. Then the intersection $D = \bigcap_{B \in \mathfrak{M}} B$ of all the elements of the family $\mathfrak{M}$ is also a closed extreme subset of A.*

Proof. The compactness of A ensures that the set D is not empty. Convexity and closedness are inherited by intersections of sets, so D is convex and closed. Now let $x_1, x_2 \in A$ and $(x_1 + x_2)/2 \in D$. Then, in particular, $(x_1 + x_2)/2 \in B$ for any $B \in \mathfrak{M}$. Therefore, $x_1, x_2 \in B$ for all $B \in \mathfrak{M}$, whence $x_1, x_2 \in \bigcap_{B \in \mathfrak{M}} B = D$. $\qquad\square$

Lemma 5. *Let A be a convex compact subset of a separated locally convex topological vector space, and suppose A consists of more than one point. Then A contains a closed extreme subset B such that $B \neq A$.*

Proof. Suppose $x_1, x_2 \in A$ and $x_1 \neq x_2$. Since the dual of a separated locally convex space separates points, there exists a real continuous linear functional f such that $f(x_1) \neq f(x_2)$. Hence, f is not identically constant on A, and for the required set B we can take the set $M(f, A)$ from Lemma 3. $\qquad\square$

Theorem 1 (weak formulation of the Krein–Milman theorem).[1] *Every convex compact set K in a separated locally convex space has extreme points.*

Proof. Consider the family $\mathfrak{Ext}(K)$ of all closed extreme subsets of the compact set K. We equip $\mathfrak{Ext}(K)$ with the decreasing order of sets. By Lemma 4, $\mathfrak{Ext}(K)$ is an inductively ordered set. By Zorn's lemma, there exists a minimal with respect to inclusion closed extreme subset A of the compact set K. By Lemma 5, A consists of exactly one point, which is the sought-for extreme point of K. $\qquad\square$

Remark 1. For a convex compact set in a non-locally convex separated topological vector space the assertion of Theorem 1 may fail. A corresponding counterexample was constructed by Roberts [73].

The next result has numerous applications in linear optimization problems and, in particular, in problems of mathematical economics.

Theorem 2. *Let K be a convex compact set in a separated locally convex space X, f a continuous real linear functional on X, and $b = \max_{x \in K} f(x)$. Then there exists a point $x \in \text{ext } K$ in which $f(x) = b$. In other words, when searching for the maximum of a linear functional on a convex compact set, it suffices to consider the values in the extreme points of the compact set under consideration.*

---

[1] Mark Krein and David Milman were Odessa mathematicians. For this reason, in contrast to the theorems of the Lviv school led by Stefan Banach, which became "Ukrainian" only as a result of post-war geopolitical changes, a Ukrainian patriot like me can be proud that the Krein–Milman theorem is "genuinely Ukrainian".

Proof. By Lemma 3, $M(f, K) = \{x \in K : f(x) = b\}$ is an extreme subset of the compact set K, and is closed thanks to the continuity of the functional f. Since the set $M(f, K)$ is convex and compact, it has an extreme point x_0, and in view of the definition of $M(f, K)$, $f(x_0) = b$. It remains to apply Lemma 2: an extreme point of an extreme subset is an extreme point of the original set. □

The application of Theorem 2 becomes especially effective in the case when the set K is a finite-dimensional polyhedron. In this case ext K is a finite set, and the task of calculating the maximum of a linear functional reduces to a finite (admittedly possibly large) item-by-item examination. This examination can be carried out, in particular, by means of the famous *simplex method* of Kantorovich, which these days is presented in every linear programming textbook.

Lemma 6. *Let A and B be convex closed subsets of a locally convex space X. Then the following conditions are equivalent:*

(i) $A = B$;

(ii) $\sup_{x \in A} f(x) = \sup_{x \in B} f(x)$ *for any real linear functional* $f \in X^*$.

Proof. The implication (i) $\Longrightarrow$ (ii) is obvious. Let us prove the converse implication (ii) $\Longrightarrow$ (i). Since the sets A and B play symmetric roles, it suffices to prove that $A \subset B$. Suppose this inclusions does not hold. Then there exists a point $x_0 \in A \setminus B$. Since B is closed, x_0 has a neighborhood U such that $U \cap B = \emptyset$. By the geometric form of the Hahn–Banach theorem, applied to the sets U and B, there exist a continuous real linear functional on X and a constant $a \in \mathbb{R}$ such that $f(x) \leqslant a$ for $x \in B$ and $f(x_0) > a$. Then $\sup_{x \in A} f(x) \geqslant f(x_0) > a \geqslant \sup_{x \in B} f(x)$, which contradicts condition (ii). □

Theorem 3 (Krein–Milman theorem: complete formulation). *Any convex compact set K in a separated locally convex space coincides with the closure of the convex hull of its extreme points.*

Proof. Let $\widetilde{K} = \overline{\mathrm{conv}}\,(\mathrm{ext}\, K)$ and consider an arbitrary continuous real linear functional f on X. Then $\widetilde{K} \subset K$, and so $\sup_{x \in K} f(x) \geqslant \sup_{x \in \widetilde{K}} f(x)$. By Theorem 2, $\sup_{x \in K} f(x) \leqslant \sup_{x \in \mathrm{ext}\, K} f(x) \leqslant \sup_{x \in \widetilde{K}} f(x)$. It remains to apply Lemma 6. □

Thus, we can state that a convex compact set not only has extreme points, but there are "many" such points. For example, if the compact set K is infinite-dimensional, then the set ext K is infinite. Let us give several corollaries.

Corollary 1. *Every convex closed bounded subset of a reflexive space and, in particular, the closed unit ball, has extreme points. If the space is infinite-dimensional, then its closed unit ball has infinitely many extreme points.*

Proof. It suffices to recall that any convex closed bounded subset of a reflexive space is weakly compact. □

This provides another proof of the non-reflexivity of the spaces c_0, $L_1[0, 1]$, and $C[0, 1]$: as we already remarked, in the first two of these spaces the unit ball even has no extreme points, while the unit ball in $C[0, 1]$ has only two extreme points.

If instead of the weak topology we consider the w^*-topology, we obtain another corollary.

Corollary 2. *Let X be a Banach space. Then any convex w^*-closed bounded subset of the space X^* and, in particular, the closed unit ball in X^*, has extreme points. If the space X is infinite-dimensional, then the closed unit ball in the space X^* has infinitely many extreme points.* $\square$

For this reason, the spaces c_0, $L_1[0, 1]$ and $C[0, 1]$ are not just non-reflexive, they actually are not dual to any Banach space (i.e., they are not isometric to any space of the form X^* with X a Banach space).

Exercises

1. Let $A = \mathrm{conv}\, B$. Then $\mathrm{ext}\, A \subset B$.

2. Let $A \subset B$ and $x \in (\mathrm{ext}\, B) \cap A$. Then $x \in \mathrm{ext}\, A$.

3. Let K be a convex compact subset of a strictly convex Banach space. Then a farthest from zero point of the compact set K is an extreme point of K.

4. Let X, Y be Banach spaces, $T \in L(X, Y)$, and K a convex compact set in X. Suppose that $\|Tx\| \leqslant C$ for all $x \in \mathrm{ext}\, K$. Then $\|Tx\| \leqslant C$ for all $x \in K$.

Using the preceding exercise and the description of the extreme points of the n-dimensional cube, prove the following result:

5. Suppose $x_1, \ldots, x_n$ are elements of the Banach space X and the estimate $\left\| \sum_{k=1}^{n} a_k x_k \right\| \leqslant C$ holds for all $a_k = \pm 1$. Then the same estimate holds for all $a_k \in [-1, 1]$.

6. Lindenstrauss–Phelps theorem. *In an infinite-dimensional reflexive Banach space the set of extreme points of the closed unit ball is uncountable.*

The closed unit ball of the space c_0, regarded as a subset of the space $\ell_\infty = \ell_1^*$, is an example of a closed convex and bounded set in a dual space which has no extreme points. Therefore, in Corollary 2 the w^*-closedness assumption cannot be replaced by ordinary closedness. This makes the next result all the more interesting:

7. Let X be a Banach space whose dual X^* is separable. Then any convex closed (in norm) bounded subset of the space X^* has extreme points.

8. None of the spaces c_0, $L_1[0, 1]$ and $C[0, 1]$ can be isomorphically embedded in a separable dual space. In particular, none of these spaces is isomorphic to a dual space.

9. The set of extreme points of a convex metrizable compact subset of a locally convex space is a G_δ-set.

10. For every Banach space X, the identity operator $I \in L(X)$ is an extreme point of the ball $\overline{B}_{L(X)}$.

11. The unit element of any Banach algebra $\mathbf{A}$ is an extreme point of the ball $\overline{B}_\mathbf{A}$.

18.1.3 Weak Integrals and the Krein–Milman Theorem in Integral Form

Let (Ω, Σ, μ) be a finite measure space and X be a locally convex space. A function $f : \Omega \to X$ is said to be *weakly integrable* if for any $x^* \in X^*$ the composition $x^* \circ f$ is an integrable scalar function and there exists an element $x \in X$ such that

$$\int_\Omega x^* \circ f \, d\mu = x^*(x) \tag{1}$$

for all $x^* \in X^*$. In this case the element x is called the *weak integral* of the function f, and is denoted by the symbol $\int_\Omega f \, d\mu$. With this notation formula (1) takes on the form

$$x^* \left(\int_\Omega f \, d\mu_{\scriptscriptstyle\bullet} \right) = \int_\Omega x^* \circ f \, d\mu$$

and can be interpreted as saying that a continuous linear functional can be brought under the integral sign.

The weak integral inherits the simplest properties of the ordinary integral:

— $\int_\Omega a f_1 + b f_2 \, d\mu = a \int_\Omega f_1 \, d\mu + b \int_\Omega f_2 \, d\mu$ (linearity with respect to the function);

— $\int_\Omega f \, d(a\mu_1 + b\mu_2) = a \int_\Omega f \, d\mu_1 + b \int_\Omega f \, d\mu_2$ (linearity with respect to the measure);

— $\int_{\Omega_1 \sqcup \Omega_2} f d\mu = \int_{\Omega_1} f \, d\mu + \int_{\Omega_2} f \, d\mu$ for any disjoint sets $\Omega_1, \Omega_2 \in \Sigma$ (additivity with respect to the integration domain);

Here, in all the three properties, if the integrals on the right-hand side exist, then so does the integral on the left-hand side.

Let us mention that one of the important properties of the Lebesgue integral, namely, that integrability on a set implies integrability on all its measurable subsets, does not hold for the weak integral (see Exercise 1 below). The root of this unpleasant feature is that the weak topology of a space is not necessarily complete. Let us examine on some examples how one calculates the weak integral of a vector-valued function.

Example 1. Let X be the sequence space ℓ_p or c_0, and $e_n^* \in X^*$ be the coordinate (evaluation) functionals. Let $f \colon \Omega \to X$ be a weakly integrable function. For each $t \in \Omega$ denote by $f_n(t)$ the n-th component of the vector $f(t) \colon f(t) = (f_1(t), f_2(t), \ldots)$. Then, by the definition of the weak integral,

$$e_n^* \left(\int_\Omega f \, d\mu \right) = \int_\Omega e_n^* \circ f \, d\mu = \int_\Omega f_n \, d\mu,$$

i.e., $\int_\Omega f \, d\mu$ is the vector with the components $\left(\int_\Omega f_n \, d\mu \right)_{n=1}^\infty$.

Example 2. Let $F \colon \Omega \to C[0, 1]$ be a weakly integrable function. For each $t \in [0, 1]$ and each $\tau \in \Omega$, we define $f(t, \tau) = (F(\tau))(t)$. Using instead of the coordinate functionals the point evaluation functionals, we obtain the following rule for the calculation of the function $\int_\Omega F \, d\mu \in C[0, 1]$:

$$\left(\int_\Omega F \, d\mu \right)(t) = \int_\Omega f(t, \tau) \, d\mu(\tau).$$

As in the scalar case, a function $f \colon \Omega \to X$ is said to be *measurable* if $f^{-1}(A) \in \Sigma$ for any Borel subset A of the space X. Let us mention one useful sufficient condition for weak integrability.

Theorem 1. *Let (Ω, Σ, μ) be a probability space, K a convex compact subset of a separated locally convex space X, and $f \colon \Omega \to K$ a measurable function. Then the function f is weakly integrable and $\int_\Omega f \, d\mu \in K$.*

Proof. Consider the dual pair $((X^*)', X^*)$. Since $K \subset X \subset (X^*)'$, the compact set K can be regarded as a subset of the space $(X^*)'$. Then K will also be compact in the weaker topology $\sigma((X^*)', X^*)$, and so K is a convex $\sigma((X^*)', X^*)$-compact set in $(X^*)'$.

Next, we remark that every functional $x^* \in X^*$ is bounded on K. Consequently, the composition $x^* \circ f$ is a bounded measurable function on Ω, and hence $x^* \circ f$ is an integrable scalar function. Define a linear functional $F \colon X^* \to \mathbb{C}$ by the formula $F(x^*) = \int_\Omega x^* \circ f \, d\mu$. We claim that $F \in K$. Assuming the contrary, there exists an element $x^* \in X^*$ such that $\operatorname{Re} x^*(s) \leqslant 1$ for all $s \in K$ and $\operatorname{Re} x^*(F) > 1$. Then $\operatorname{Re} x^* \circ f \leqslant 1$ everywhere on Ω, and so

$$\operatorname{Re} x^*(F) = \operatorname{Re} F(x^*) = \int_\Omega \operatorname{Re} x^* \circ f \, d\mu \leqslant 1.$$

The contradiction we reached means that $F \in K$. By construction, $x^*(F) = \int_\Omega x^* \circ f \, d\mu$ for all $x^* \in X^*$, i.e., F is the weak integral of the function f. $\qquad \square$

Theorem 2 (Krein–Milman theorem: integral form). *Let K be a convex compact subset of a separated locally convex space X and $x \in K$. Then there exists a regular Borel probability measure μ on $\overline{\text{ext}}\, K$, the closure of the set of extreme points of K, such that*

$$\int_{\overline{\text{ext}}\, K} I \, d\mu = x;$$

here I denotes, as usual, the identity mapping and the integral is understood in the weak sense.

Proof. Recalling the theorem on the general form of linear functionals on the space of continuous functions, we see that the set $M(\overline{\text{ext}}\, K)$ of all regular Borel probability measures on the compact set $\overline{\text{ext}}\, K$ can be regarded as a subset of the space $C(\overline{\text{ext}}\, K)^*$. Moreover, $M(\overline{\text{ext}}\, K)$ is the intersection of the closed unit ball of the space $C(\overline{\text{ext}}\, K)^*$ (i.e., of a convex w^*-compact set) with the w^*-closed set $\{F \in C(\overline{\text{ext}}\, K)^* : F(\mathbb{1}) = 1\}$. As such, $M(\overline{\text{ext}}\, K)$ is a convex w^*-compact subset of $C(\overline{\text{ext}}\, K)^*$.

By the preceding theorem, for each measure $\mu \in M(\overline{\text{ext}}\, K)$ there exists the weak integral $\int_{\overline{\text{ext}}\, K} I \, d\mu$. Consider the operator $T \colon X^* \to C(\overline{\text{ext}}\, K)^*$ which sends each functional $x^* \in X^*$ into its restriction to $\overline{\text{ext}}\, K$. Let us calculate the action of the adjoint operator $T^* \colon C(\overline{\text{ext}}\, K)^* \to X^{**}$ on the elements of the set $M(\overline{\text{ext}}\, K)$. For any measure $\mu \in M(\overline{\text{ext}}\, K)$ and any $x^* \in X^*$ we have

$$\langle T^*\mu, x^* \rangle = \langle \mu, T x^* \rangle = \int_{\overline{\text{ext}}\, K} x^* \, d\mu = x^* \left(\int_{\overline{\text{ext}}\, K} I \, d\mu \right),$$

i.e.,

$$T^*\mu = \int_{\overline{\text{ext}}\, K} I \, d\mu.$$

Therefore, now our task is reduced to proving the equality

$$T^*(M(\overline{\text{ext}}\, K)) = K.$$

The inclusion $T^*(M(\overline{\text{ext}}\, K)) \subset K$ was proved in the preceding theorem. Let us prove the opposite inclusion $T^* \left(M(\overline{\text{ext}}\, K) \right) \supset K$. Denote by δ_x the probability measure supported at the point x. Then for any $x \in \text{ext}\, K$ we have

$$T^*\delta_x = \int_{\overline{\text{ext}}\, K} I \, d\delta_x = x,$$

i.e., $T^*(M(\overline{\text{ext}}\, K)) \supset \text{ext}\, K$. Further, $T^* \left(M(\overline{\text{ext}}\, K) \right)$ is a convex closed set, being the image of the convex w^*-compact set $M(\overline{\text{ext}}\, K)$ under the w^*-continuous operator T^*.

Therefore, $T^*(M(\overline{\operatorname{ext} K})) \supset \overline{\operatorname{conv}}(\operatorname{ext} K)$; but, by Theorem 3 of Subsection 18.1.2, $\overline{\operatorname{conv}}(\operatorname{ext} K) = K$. □

Let us remark that in the metrizable case the measure μ that represents the element x can be selected so that it is supported on the set of extreme points itself, rather than on its closure. Then the integral representation of the element x takes the form $x = \int_{\operatorname{ext} K} I \, d\mu$. The proof of this theorem due to G. Choquet and various generalizations thereof can be found in Phelps' monograph [34].

Another, quite fruitful research direction is connected with the consideration of a narrower set than ext K, namely, the set of strongly exposed points. In terms of such points it was possible to characterize the spaces in which the Radon–Nikodým theorem is valid. Results on this theme, as well as many other interesting branches of the geometry of Banach spaces, can be found in the monographs [3, 12].

Exercises

1. On the interval $[0, 1]$ pick a sequence (Δ_n), $n \in \mathbb{N}$, of pairwise disjoint subintervals. Let e_n, $n \in \mathbb{N}$, denote the standard basis of the space c_0. Define the function $f : [0, 1] \to c_0$ as follows: if the point t lies in none of the intervals Δ_n, put $f(t) = 0$; if t lies in an odd-indexed interval Δ_{2n-1}, put $f(t) = (1/|\Delta_{2n-1}|)e_n$; finally, if t lies in an even-indexed interval Δ_{2n}, put $f(t) = -(1/|\Delta_{2n}|)e_n$. Verify that the function f is weakly integrable on $[0, 1]$ with respect to the Lebesgue measure λ and $\int_{[0,1]} f \, d\lambda = 0$. At the same time, on the subset $\Delta = \bigcup_{n=1}^{\infty} \Delta_{2n-1}$ the function f is not weakly integrable: otherwise, we would have the equality $\int_{\Delta} f \, d\lambda = (1, 1, 1, \ldots)$, but there is no such element in c_0.

2. Prove the following theorem of Carathéodory: *If* $K \subset \mathbb{R}^n$ *is a convex compact set, then every element* $x \in K$ *admits a representation* $x = \sum_{j=1}^{n+1} a_j x_j$, *where* $x_j \in$ ext K, $a_j \geqslant 0$, *and* $\sum_{j=1}^{n+1} a_j = 1$.

3. Based on Choquet's theorem formulated above, prove that if the convex metrizable compact set K in a locally convex space has a countable number of extreme points, then every element $x \in K$ admits a series expansion $x = \sum_{n=1}^{\infty} a_n x_n$, where $x_n \in$ ext K, $a_n \geqslant 0$, and $\sum_{n=1}^{\infty} a_n = 1$.

4. If one removes the requirement that the set of extreme points in countable, the assertion of the preceding exercise may fail. Provide a counterexample.

18.2 Applications

18.2.1 The Connection Between the Properties of the Compact Space K and Those of the Space C(K)

The space $C(K)$ is more convenient to study than the compact space K, because the elements of a function space can be manipulated more freely than the points of a topological space. Indeed, in contrast to the points of the compact space K, functions on K can be added and multiplied by scalars; the topology on $C(K)$ is given by a norm, and so one can speak about Cauchy sequences, completeness, convergence of series, and so on. However, all these advantages would depreciate if in the passage from K to $C(K)$ part of the information about the original compact space is lost. Below we will show that actually no such loss occurs and all the properties of the compact space K can be recovered from the properties of the space $C(K)$.

As usual, we will identify the continuous functionals on $C(K)$ with the regular Borel charges that generate them. In particular, δ_x (the probability measure supported at x) can be regarded as the functional on $C(K)$ which acts as $\langle \delta_x, f \rangle = \int_K f \, d\delta_x = f(x)$, i.e., as the evaluation functional at the point x.

A bit more terminology. The *support* of the regular Borel charge σ is defined to be the support of the measure $|\sigma|$ (see Subsection 8.1.2). As in the case of measures, the support of a charge σ is denoted by $\operatorname{supp} \sigma$. Clearly, $\operatorname{supp} \delta_x = \{x\}$, and if $\operatorname{supp} \sigma = \{x\}$, then $\sigma = \lambda \delta_x$, where λ is a non-zero scalar.

For any Borel-measurable bounded function g on K and any Borel charge σ, we denote by $g \times \sigma$ the Borel charge given by

$$(g \times \sigma)(A) = \int_A g \, d\sigma.$$

The functional defined by the charge $g \times \sigma$ acts by the rule

$$\langle g \times \sigma, f \rangle = \int_K fg \, d\sigma.$$

The operation thus introduced enjoys the natural properties of a product:

— $\mathbb{1} \times \sigma = \sigma$;

— $(g + h) \times \sigma = g \times \sigma + h \times \sigma$;

— $(gh) \times \sigma = (hg) \times \sigma = h \times (g \times \sigma)$;

— $g \times (\nu + \sigma) = g \times \nu + g \times \sigma$.

— Finally, the norm of the charge $g \times \sigma$ is calculated by the formula

$$\|g \times \sigma\| = \int_K |g| \, d|\sigma|.$$

Theorem 1. *The set of extreme points of the closed unit ball of the space $C(K)^*$ coincides with the set of charges of the form $\lambda \delta_x$, where $x \in K$ and $|\lambda| = 1$.*

Proof. First let us show that charges of the form δ_x are extreme points of the set $\overline{B}_{C(K)^*}$. Since the ball is balanced, this will imply that also $\lambda \delta_x \in \operatorname{ext} \overline{B}_{C(K)^*}$ whenever $|\lambda| = 1$. So, assume $\nu_1, \nu_2 \in \overline{B}_{C(K)^*}$ and $(\nu_1 + \nu_2)/2 = \delta_x$. Then $(\nu_1(\{x\}) + \nu_2(\{x\}))/2 = \delta_x(\{x\}) = 1$. Since both numbers $|\nu_1(\{x\})|$, $|\nu_2(\{x\})|$ are not larger than 1, this means that $\nu_1(\{x\}) = \nu_2(\{x\}) = 1$. This in turn means that beyond the point x the charges ν_1 and ν_2 vanish, since otherwise their norms would be strictly larger than 1. That is, $\nu_1 = \nu_2 = \delta_x$.

Now let us show that if the charge $\sigma \in \overline{B}_{C(K)^*}$ is not concentrated at a single point of the compact space K, then it cannot be an extreme point of the unit ball. Indeed, suppose supp σ contains two distinct points $x \neq y$. Surround these points by disjoint neighborhoods U and V. By the definition of the support, the numbers $|\sigma|(U)$ and $|\sigma|(V)$ are different from zero. Let $\varepsilon = \min\{|\sigma|(U), |\sigma|(V)\}$. Consider the function

$$g = \frac{\varepsilon}{|\sigma|(U)} \mathbb{1}_U - \frac{\varepsilon}{|\sigma|(V)} \mathbb{1}_V$$

and the charges

$$\sigma_1 = (1 - g) \times \sigma \quad \text{and} \quad \sigma_2 = (1 + g) \times \sigma.$$

Since $|g| \leqslant 1$, we have $|1 \pm g| = 1 \pm g$. Further, by construction, $\int_K g\, d|\sigma| = 0$. Consequently,

$$\int_K |1 \pm g| d|\sigma| = \int_K d|\sigma| \pm \int_K g\, d|\sigma| = \int_K d|\sigma| = \|\sigma\| \leqslant 1.$$

Hence, $\sigma_1, \sigma_2 \in \overline{B}_{C(K)^*}$. At the same time,

$$\frac{\sigma_1 + \sigma_2}{2} = \sigma$$

and

$$\|\sigma_1 - \sigma_2\| = 2 \int_K |g| d|\sigma| = 4\varepsilon \neq 0;$$

therefore, the charge σ cannot be an extreme point of the unit ball. $\qquad\square$

Suppose we are given a Banach space X and we are told that $X = C(K)$ for some compact space K, but not what this compact space is. Can we recover K from the space X? By the preceding theorem, to this end we need to look at the extreme points of the ball $\overline{B}_{X^*}$.

Let us introduce several definitions and notations. The set ext $\overline{B}_{X^*}$ will be regarded as a subspace of the topological space $(X^*, \sigma(X^*, X))$, i.e., we equip ext $\overline{B}_{X^*}$ with the w^*-topology. Further, we introduce on ext $\overline{B}_{X^*}$ the following equivalence relation: $x^* \sim y^*$ if $x^* = \lambda y^*$ for some scalar λ with $|\lambda| = 1$. The equivalence class of the element $x^* \in$ ext $\overline{B}_{X^*}$ is the pair of points $\pm x^*$ in the real case, and the circle passing through x^*, i.e., $[x^*] = \{\lambda x^* : |\lambda| = 1\}$, in the complex case. We denote the set of equivalence classes into which ext $\overline{B}_{X^*}$ decomposes by $\widetilde{K}(X)$, and denote by q the quotient mapping $q : \text{ext }\overline{B}_{C(K)^*} \to \widetilde{K}(X)$. We equip $\widetilde{K}(X)$ with the strongest topology with respect to which q is w^*-continuous. That is to say, a set $A \subset \widetilde{K}(X)$ is declared to be open in $\widetilde{K}(X)$ if $q^{-1}(A)$ is w^*-open in ext $\overline{B}_{X^*}$. Note that $\widetilde{K}(X)$ is a Hausdorff topological space. Indeed, if $x^*, y^* \in$ ext $\overline{B}_{X^*}$ and $[x^*] \neq [y^*]$, then the functionals x^* and y^* are linearly independent. Hence, the kernel of any of them is not included in the kernel of the other, and so there exists an element $x \in \text{Ker } y^* \setminus \text{Ker } x^*$. Multiplying x by a scalar, one can ensure that $x^*(x) = 1$. Then the points $[x^*], [y^*] \in$

$\widetilde{K}(X)$ are separated by the neighborhoods $U = \{[s^*] \in \widetilde{K} : |s^*(x)| > 1/2\}$ and $V = \{[s^*] \in \widetilde{K} : |s^*(x)| < 1/2\}$.

Theorem 2. *Let $X = C(K)$ for some compact space K. Then K is homeomorphic with the topological space $\widetilde{K}(X)$ constructed above.*

Proof. Define the mapping $\delta \colon K \to \mathrm{ext}\, \overline{B}_{X^*}$ by the formula $\delta(t) = \delta_t$. For any function $f \in C(K)$ we have $\langle \delta(t), f \rangle = f(t)$, which depends continuously on t. Since $\mathrm{ext}\, \overline{B}_{X^*}$ is equipped with the w^*-topology, this means that the mapping δ is continuous. Then, as a composition of continuous mappings, the mapping $j \colon K \to \widetilde{K}(X)$, $j = q \circ \delta$, is also continuous. Since $j(t) = [\delta_t]$, Theorem 1 guarantees that the mapping j is bijective. But any bijective continuous mapping of a compact space onto a Hausdorff space is a homeomorphism. $\square$

Corollary 1. *If for two compact spaces K_1 and K_2 the spaces $C(K_1)$ and $C(K_2)$ are isometric, then the spaces K_1 and K_2 are homeomorphic.* $\square$

Theorem 3. *The space $C(K)$ is separable if and only if the compact space K is metrizable.*

Proof. Suppose $C(K)$ is separable. Then (Corollary 4 in Subsection 17.2.4) the w^*-topology is metrizable on the ball $\overline{B}_{C(K)^*}$. The compact space K is homeomorphic to the subset $\{\delta_t : t \in K\}$ of the ball $\overline{B}_{C(K)^*}$, equipped with the w^*-topology (the homeomorphism is provided by the mapping $t \mapsto \delta_t$). Hence, K is metrizable.

Conversely, suppose K is a compact metric space. Then for each $n \in \mathbb{N}$ there exists a cover of the compact space K by balls $U_{n,1}, U_{n,2}, \ldots, U_{n,m(n)}$ of radius $1/n$. Denote by $\varphi_{n,1}, \varphi_{n,2}, \ldots, \varphi_{n,m(n)}$ a partition of unity subordinate to the cover $U_{n,1}, U_{n,2}, \ldots U_{n,m(n)}$ (see Subsection 15.1.3). Let us prove that the system of elements $\{\varphi_{n,j} : n = 1, \ldots, \infty, \ j = 1, \ldots, m(n)\}$ is complete in $C(K)$. This in turn will establish the desired separability of the space $C(K)$.

Thus, let $f \in C(K)$. For each $\varepsilon > 0$, pick an $n \in \mathbb{N}$ such that for any $t_1, t_2 \in K$, if $\rho(t_1, t_2) < 1/n$, then $|f(t_1) - f(t_2)| < \varepsilon$. Next, in each set $U_{n,j}$ pick one point $t_{n,j}$ and consider the following linear combination f_ε of the functions $\varphi_{n,j}$:

$$f_\varepsilon = f(t_{n,1})\varphi_{n,1} + f(t_{n,2})\varphi_{n,2} + \cdots + f(t_{n,m(n)})\varphi_{n,m(n)}.$$

We claim that $\|f - f_\varepsilon\| < \varepsilon$. Indeed, for any $t \in K$ we have $f(t) = \sum_{j=1}^{m(n)} f(t)\varphi_{n,j}(t)$. Consequently,

$$|f(t) - f_\varepsilon(t)| \leqslant \sum_{j=1}^{m(n)} |f(t) - f(t_{n,j})|\varphi_{n,j}(t).$$

In the last sum, if $\varphi_{n,j}(t) \neq 0$, then $t \in U_{n,j}$, and so $|f(t) - f(t_{n,j})| < \varepsilon$. Continuing the estimate, we conclude that

$$|f(t) - f_\varepsilon(t)| < \sum_{j=1}^{m(n)} \varepsilon \varphi_{n,j}(t) = \varepsilon.$$ $\square$

Remark 1. Jumping ahead, we observe that in the last part of the proof of Theorem 3, the separability of the space $C(K)$ is an easy consequence of the Stone–Weierstrass theorem. But in our opinion the explicit procedure of approximating a function by a partition of unity is instructive in itself.

Exercises

1. Verify that for any Borel-measurable bounded function g on K and any regular Borel charge σ, the charge $g \times \sigma$ is regular (hint: consider first the case $g = \mathbb{1}_A$, then the case of finitely-valued functions, and finally use the approximation of a bounded function by finitely-valued functions).

2. Verify all the properties, listed at the beginning of the subsection, of the operation $g \times \sigma$ of multiplying a regular Borel charge by a bounded Borel function.

3. The fact that the spaces $C(K_1)$ and $C(K_2)$ are isomorphic does not necessarily imply that the compact spaces K_1 and K_2 are homeomorphic. Example: $K_1 = [0, 1]$ and $K_2 = [0, 1] \cup \{2\}$.

18.2.2 The Stone–Weierstrass Theorem

In this subsection we make acquaintance with an exceptionally beautiful, and at the same time, very useful generalization of Weierstrass' theorem on the approximation of functions by polynomials. This generalization, devised by M.H. Stone, is applicable to functions defined not only on an interval, but also on an arbitrary compact space. The proof given below is due to de Branges (L. de Branges, 1959). The application of the same idea of proof to an even more general result, namely Bishop's theorem, can be found in the book [38].

Theorem 1. *Suppose the linear subspace X of the space $C(K)$ has the following properties:*

(a) *$\mathbb{1} \in X$;*

(b) *if f, $g \in X$, then $fg \in X$ (in other words, X is a subalgebra of the algebra $C(K)$);*

(c) *for any function $f \in X$, its complex conjugate $\overline{f}$ also belongs to X;*

(d) *for any $t_1, t_2 \in K$, $t_1 \neq t_2$, there exists a function $f \in X$ such that $f(t_1) \neq f(t_2)$ (i.e., X separates the points of the compact space K).*

Then the subspace X is dense in $C(K)$.

Proof. Suppose the assertion of the theorem is false, i.e., the subspace X is not dense in $C(K)$. Then the annihilator $X^{\perp} \subset C(K)^*$ does not reduce to 0. Recall that $X^{\perp}$ is a w^*-closed subspace in $C(K)^*$ and hence, by Alaoglu's theorem, $\overline{B}_{X^{\perp}} =$

$\overline{B}_{C(K)^*} \cap X^\perp$ is a w^*-compact set. By the Krein–Milman theorem, the ball $\overline{B}_{X^\perp}$ has an extreme point v. Obviously, $v \in S_{X^\perp}$, that is, $\|v\| = 1$. We next study the properties of this regular Borel charge v and show that they are intrinsically contradictory.

We begin with several useful remarks on the properties of the sets X and $X^\perp$:

(i) If $f \in X$, then Re $f \in X$ and Im $f \in X$ (this follows from condition (c) and the formulas Re $f = (f + \bar{f})/2$ and Im $f = (f - \bar{f})/(2i)$).

(ii) If $f \in X$ and $\eta \in X^\perp$, then $f \times \eta \in X^\perp$, where $\times$ is the operation introduced in the preceding subsection. Indeed, for any $g \in X$ the product fg also lies in X, and hence is annihilated by the charge η. We have $\langle f \times \eta, g \rangle = \langle \eta, fg \rangle = 0$, and so $f \times \eta \in X^\perp$.

(iii) If $\eta \in X^\perp$, then supp η contains at least two distinct points. Indeed, if supp $\eta = \{t\}$, then $\eta = a\delta_t$ with $a \in \mathbb{C} \setminus \{0\}$. But then $\langle \eta, \mathbb{1} \rangle = a \neq 0$, i.e., $\eta \notin X^\perp$.

Now let us return to the charge $v \in S_{X^\perp}$, a candidate for the role of an extreme point of the ball $\overline{B}_{X^\perp}$. We use property (iii) above. Let $t_1, t_2 \in$ supp v and $t_1 \neq t_2$. By condition (d), there exists an $f \in X$ such that $f(t_1) \neq f(t_2)$. Then either Re $f(t_1) \neq$ Re $f(t_2)$, or Im $f(t_1) \neq$ Im $f(t_2)$. By property (i), Re f, Im $f \in X$. Therefore, f can be assumed to be a real-valued function: otherwise one can replace it by Re f or by Im f. Further, adding to f a large positive constant one can ensure that that f is positive, and then multiplying by a small positive factor we obtain a function whose values lie in the interval $(0, 1)$. Thus, we proved that there exists a function $f \in X$ such that $f(t_1) \neq f(t_2)$ and $0 < f(t) < 1$ for all $t \in K$.

Let us introduce the auxiliary charges $v_1 = f \times v$ and $v_2 = (1 - f) \times v$. Then

$$\|v_1\| = \int_K f\, d|v|, \quad \|v_2\| = \int_K (1 - f) d|v|,$$

and both numbers are different from zero because, by construction, the functions f and $1 - f$ do not take the value 0. Further,

$$\|v_1\| + \|v_2\| = \int_K d|v| = 1.$$

We have the obvious equality

$$\|v_1\| \frac{v_1}{\|v_1\|} + \|v_2\| \frac{v_2}{\|v_2\|} = v,$$

the geometric meaning of which is as follows: the vector $v \in \overline{B}_{X^\perp}$ is an interior point of the segment that joins the vectors $\frac{v_1}{\|v_1\|} \in \overline{B}_{X^\perp}$ and $\frac{v_2}{\|v_2\|} \in \overline{B}_{X^\perp}$ (the fact that the charges v_1 and v_2 lie in the subspace $X^\perp$ follows from property (ii)). Since by our assumption v is an extreme point of the ball $\overline{B}_{X^\perp}$, the endpoints of the segment must coincide with v: $\frac{v_1}{\|v_1\|} = \frac{v_2}{\|v_2\|} = v$. In particular, $v_1 = \|v_1\| v$, i.e., $(f - \|v_1\|) \times v = 0$. Recalling the formula for the norm of a charge, we get that

$$\int_K \left| f - \|v_1\| \right| d|v| = 0.$$

In view of the continuity of the function f, the last equality means that $f(t) = \|v_1\|$ for all $t \in \text{supp } v$ (Theorem 2 of Subsection 8.1.2). We arrived at a contradiction with the condition $f(t_1) \neq f(t_2)$. $\qquad\square$

Exercises

Deduce from the Stone–Weierstrass theorem that:

1. The set of polynomials is dense in $C(K)$, for any compact subset K of $\mathbb{R}$ (in particular, when $K = [a, b]$). Recall that this fact was used in Subsection 13.1.3 to construct functions of a self-adjoint operator.

2. The set of polynomials in n variables is dense in $C(K)$, where K is any compact subset of $\mathbb{R}^n$.

3. The set of "two-sided" polynomials of the form $\sum_{k=-n}^n a_k z^k$, $n \in \mathbb{N}$, is dense in the space $C(\mathbb{T})$ of continuous functions on the unit circle $\mathbb{T} = \{z \in \mathbb{C} : |z| = 1\}$ (we used this fact earlier to construct functions of an unitary operator).

Consider the half-line $[0, +\infty]$, i.e., the one-point compactification of the half-line $[0, +\infty)$. The neighborhoods of finite points of $[0, +\infty]$ are defined as usual; the neighborhoods of $+\infty$ are the complements of the bounded sets. Show that:

4. In the described topology $[0, +\infty]$ is compact.

5. The space $C[0, +\infty]$ coincides with the space of continuous functions $f(t)$ on $[0, +\infty)$ that have a limit as $t \to +\infty$.

6. The set of exponential functions e^{-at} with $a \in [0, +\infty)$ is a complete system of elements in $C[0, +\infty]$.

18.2.3 Completely Monotone Functions

An infinitely differentiable function f on $[0, +\infty)$ is said to be *completely monotone* if $(-1)^n f^{(n)}(t) \geqslant 0$ for all $n = 0, 1, 2, \ldots$ and all $t \in [0, +\infty)$. In particular, to be completely monotone the function f must be non-negative ($f(t) \geqslant 0$), non-increasing ($(-1)f'(t) \geqslant 0$), and convex ($f''(t) \geqslant 0$). A typical example of a completely monotone function is $f(t) = e^{-t}$. A well-known theorem of S.N. Bernstein[2]

---

[2] Kharkiv is a city that hosted many famous mathematicians. Sergei Natanovich Bernstein not only worked for a period of time in Kharkiv, he spent a major part of his life there, exerting an invaluable influence on the formation of the Kharkiv mathematical school.

asserts that any completely monotone function can be uniquely represented in the form

$$f(x) = \int_0^\infty e^{-tx} d\mu(t), \tag{1}$$

where μ is a finite regular Borel measure on the half-line. In other words, every completely monotone function is in a sense a combination of exponentials. Differentiating under the integral sign one can readily verify that any function of the form (1) is completely monotone, and thus Bernstein's theorem provides a complete description of the class of completely monotone functions.

The representation (1) calls forth a natural association with the Krein–Milman theorem in integral form. The first proof of Bernstein's theorem based on this analogy was proposed by Choquet. Below we provide a sufficiently detailed sketch of this proof, leaving its implementation to the reader. A detailed exposition can be found in the short book [34, Chapter 2].

Theorem 1. *If the function $f : [0, +\infty) \to \mathbb{R}$ admits a representation (1), where μ is a finite regular Borel measure, then this representation is unique.*

Proof. Consider μ as a functional on $C[0, +\infty]$. Formula (1) says that we are given the values of this functional on the exponentials e^{-at}: $\langle \mu, e^{-at} \rangle = f(a)$. By Exercise 6 of Subsection 18.2.2, the set of exponentials e^{-at} with $a \in [0, +\infty)$ is complete in $C[0, +\infty]$. Hence, a continuous functional is uniquely determined by its values on this set. $\qquad\square$

In the space $C^\infty(0, +\infty)$ of infinitely differentiable functions on the open half-line, equipped with the standard topology generated by the seminorms $p_n(f) = \max_{t \in (n^{-1}, n)} |f^{(n-1)}(t)|$, $n \in \mathbb{N}$, consider the set K of all completely monotone functions bounded above by 1. Note that the functions $f \in K$ are defined on the open half-line, but thanks to their monotonicity and boundedness they have limits at 0 and ∞, and consequently can be considered to be defined also at these two points.

Theorem 2. *The set K is convex and compact in $C^\infty(0, +\infty)$.*

Proof. The convexity and closedness are verified directly. Since $C^\infty(0, +\infty)$ is a Montel space (Subsection 16.3.4), to establish the compactness of K it suffices to verify that K is bounded. Now boundedness follows from the following estimate of the n-th derivative of $f \in K$, the proof of which by induction on n is left to the reader:

$$\sup_{a \leqslant t < \infty} |f^{(n)}(t)| \leqslant a^{-n} 2^{n(n+1)/2}$$

for any $a \in (0, 1)$ and any $n = 0, 1, 2, \ldots$. $\qquad\square$

Theorem 3. *Suppose the continuous function $f : (0, +\infty) \to \mathbb{R}$ satisfies for all $x, y \in (0, +\infty)$ the functional equation*

$$f(x + y) = f(x)f(y). \tag{2}$$

Then f is an exponential function of the form $f(x) = a^x$.

Proof. Take for a the value $f(1)$. Substituting into (2) $x = 1$ and $y = 1$, we obtain $f(2) = a^2$. Further, if we fix $x = 1$ and take successively $y = 2, 3, \ldots$, we obtain the equality $f(n) = a^n$. Taking in (2) $x = y = n/2$, we conclude that $f(n/2) = a^{n/2}$. Now taking successively $x = y = n2^{-k}$, we obtain the formula $f(x) = a^x$ for all dyadic-rational numbers. To all the remaining positive real values x the equality $f(x) = a^x$ is extended by continuity. □

Theorem 4. *The set of extreme points of the compact set K introduced above consists of the functions e^{-at}, $a \in [0, +\infty)$, and the null function.*

Proof. Let $f \in \text{ext } K$. Then fix $y > 0$ and consider the auxiliary function $u(x) = f(x + y) - f(x)f(y)$. The reader will be able to verify that the two functions $f_1 = f + u$ and $f_2 = f - u$ lie in K. Since the extreme point f can be written as $f = (f_1 + f_2)/2$, it follows that $u = 0$. This establishes that f satisfies the functional equation (2), and hence that f is an exponential function. But any exponential function that lies in the set K is either 0, or a function of the form e^{-at}.

Now let us show that indeed all the functions indicated in the statement of the theorem lie in ext K. The fact that the functions 0 and 1 lie in ext K follows from the condition $0 \leqslant f(t) \leqslant 1$ that we imposed on all $f \in K$. Further, at least one of the functions $e^{-a_0 t}$ with $0 < a_0 < \infty$ is an extreme point. Otherwise, the set ext K would consist only of the functions 0 and 1 and, by the Krein–Milman theorem, the compact set $K = \overline{\text{conv}}\, \text{ext } K$ would consist only of constants. Now for any $b \in (0, 1)$ the linear operator T which sends each function $f(x)$ into the function $f(bx)$ maps K bijectively onto K. Therefore, the operator T maps extreme points into extreme ones; in particular, the function $e^{-a_0 bt}$ is an extreme point. Since b is arbitrary, this shows that $e^{-at} \in \text{ext } K$ for all $0 < a < \infty$. □

To complete the proof of Bernstein's theorem, consider the bijective mapping $F: [0, +\infty] \to \text{ext } K$ defined by the rule $F(0) = 1$, $F(+\infty) = 0$, and $F(a) = e^{-at}$ for $0 < a < \infty$. The reader will readily verify that the mapping F is continuous. Hence, ext K, being the image of a compact set under a continuous mapping, is a closed set. To conclude, F is a continuous bijective mapping of a compact set into a compact set, and hence a homeomorphism.

Let $f: [0, +\infty) \to \mathbb{R}$ be a completely monotone function. With no loss of generality, one can assume that $f \in K$: this is readily achieved via multiplication by a factor. By the Krein–Milman theorem in integral form, there exists a regular Borel probability measure v on ext K such that

$$f = \int_{\text{ext } K} I \, dv. \tag{3}$$

Define the measure μ on $[0, +\infty]$ to be the preimage of v under the mapping F: $\mu(A) = v(F(A))$. Changing the variables in (3) yields

$$f = \int\limits_{[0,+\infty]} F(t)\,d\mu(t).$$

Since $F(+\infty) = 0$, the point $+\infty$ can be removed from the integration domain:

$$f = \int\limits_{[0,+\infty)} F(t)\,d\mu(t).$$

Finally, applying to both sides of this equality the evaluation functional δ_x at the point x, we obtain the requisite representation (1):

$$f(x) = \langle \delta_x, f \rangle = \int_{[0,+\infty)} \langle \delta_x, F(t) \rangle \, d\mu(t) = \int_{[0,+\infty)} e^{-tx}\,d\mu(t). \qquad \square$$

18.2.4 Lyapunov's Theorem on Vector Measures

We begin this subsection with the "children's" cake-cutting problem. Bart and Todd want to divide a cake in a fair way. The problem is that different parts of the cake have different gastronomical and aesthetical values: some part has marzipan, another candied peel, one carries a chocolate figurine, and so on. An ever bigger issue is the individuality of the children: they may estimate differently the desirability of one and the same piece of the cake. The standard approach to solving this cutting problem goes as follows: Bart cuts the cake into two pieces that from his point of view are equal, and Todd chooses for himself the part that appeals more to him. In this way Bart is convinced that he received exactly one half of the cake, and Todd thinks he received no less than a half. This approach is completely satisfactory as long as Todd does not start bragging that he got a much better part, and Bart is not envious and starts a fight. To avoid such troubles and keep the peace between friends, it is desirable to cut the cakes into two parts such that the parts will be exactly equal from the point of view of Bart, as well as that of Todd. Is this possible? To answer this question, we need a "mature" formulation.[3]

Thus, let Ω be a set (our cake), Σ a σ-algebra of subsets of Ω (the pieces into which one can cut the cake), and μ_1, μ_2 finite countably-additive measures (for each $A \in \Sigma$ the quantity $\mu_1(A)$ (respectively, $\mu_2(A)$) is the "value" that Bart (respectively, Todd) assigns to the cake piece A).[4] Now the problem reads: *is there a set $A \in \Sigma$ such that $\mu_1(A) = \frac{1}{2}\mu_1(\Omega)$ and also $\mu_2(A) = \frac{1}{2}\mu_2(\Omega)$?* These measures μ_1 and μ_2 must also be required to be non-atomic: if some part of the cake cannot be cut into

---

[3] Here simply replacing Bart by Dr. Bartholomew Simpson and Todd by Prof. Todd Flanders is not sufficient to make the formulation "mature".

[4] In principle, μ_1 and μ_2 could also be charges, if some pieces of the cake do not seem appealing to one of the two friends, i.e., have negative value for him.

smaller pieces and both children like very much precisely that part, then the problem is not solvable.

The following theorem of A.A. Lyapunov (1940) shows that the problem has a solution, and in fact not only for two, but also for any finite number of cake lovers. The importance of the theorem is of course not restricted to the fact that it allows a fair cake cutting, regardless of the great importance and applied character of the cake problem. The proof provided below, which uses extreme points, was proposed by Lindenstrauss in 1966.

Theorem 1. *Let $\mu_1, \ldots, \mu_n$ be countably-additive non-atomic real charges on the σ-algebra Σ. Define the vector measure $\mu: \Sigma \to \mathbb{R}^n$ by the formula $\mu(A) = (\mu_1(A), \ldots, \mu_n(A))$. Then the set $\mu(\Sigma)$ of all values of the measure μ is convex and compact in $\mathbb{R}^n$.*

Proof. Consider the scalar-valued measure $\nu = |\mu_1| + \cdots + |\mu_n|$, with respect to which all charges μ_k are absolutely continuous. We use the Radon–Nikodým theorem and denote the derivative $d\mu_k/d\nu$ by g_k. Then $g_k \in L_1(\Omega, \Sigma, \nu)$ and $\mu_k(A) = \int_A g_k \, d\nu$ for all $A \in \Sigma$. Consider the operator $T : L_\infty(\Omega, \Sigma, \nu) \to \mathbb{R}^n$, acting by the rule

$$Tf = \left(\int_\Omega f g_1 \, d\nu, \ldots, \int_\Omega f g_n \, d\nu \right).$$

The set $\mu(\Sigma)$ of all values of the vector measure μ we are interested in coincides with the image under the operator T of the set of functions $\mathbb{1}_A$ with $A \in \Sigma$.

The space $L_\infty(\Omega, \Sigma, \nu)$ will be regarded as the dual to $L_1(\Omega, \Sigma, \nu)$. Then each of the expressions $\int_\Omega f g_k \, d\nu$ is a w^*-continuous with respect to f functional on $L_\infty(\Omega, \Sigma, \nu)$, and therefore the operator T is w^*-continuous. Now in $L_\infty(\Omega, \Sigma, \nu)$ consider the set W of functions f that satisfy the condition $0 \leqslant f \leqslant 1$ ν-a.e. Then W coincides with the closed ball centered at $f = 1/2$ and of radius $1/2$. By Alaoglu's theorem, the set W is w^*-compact. Moreover, W is convex. Thus, $T(W)$ is a convex compact subset of $\mathbb{R}^n$. Let us show that $T(W) = \mu(\Sigma)$. This will complete the proof of the entire theorem.

Since the functions $\mathbb{1}_A$ with $A \in \Sigma$ lie in W, and the values of the measure μ are vectors of the form $T(\mathbb{1}_A)$, we have $\mu(\Sigma) \subset T(W)$. Let us establish the opposite inclusion. Let $x \in T(W)$ be an arbitrary element; $T^{-1}(x)$ is a w^*-closed subset, hence $T^{-1}(x) \cap W$ is a w^*-compact set. Let $f \in \text{ext}\, (T^{-1}(x) \cap W)$. We claim that f takes a.e. the value 0 or 1, i.e., $f = \mathbb{1}_A$ for some set $A \in \Sigma$. Because of the equality $x = T(f) = T(\mathbb{1}_A)$, this will establish the requisite inclusion $T(W) \subset \mu(\Sigma)$.

Consider the set $A = \{t \in \Omega : 0 < f(t) < 1\}$. We need to show that $\nu(A) = 0$. Suppose this is not the case. Define

$$A_n = \left\{ t \in \Omega : \frac{1}{n} < f(t) < 1 - \frac{1}{n} \right\}.$$

By our assumption, the union of the sets A_n is not negligible, so $\nu(A_n) \neq 0$ for some $n \in \mathbb{N}$. Then the subspace $L_\infty(A_n) \subset L_\infty(\Omega, \Sigma, \nu)$ of functions with support in A_n is infinite-dimensional (here in all arguments we use the fact that the measure ν is non-atomic). Since T is a finite-dimensional operator, it cannot be injective on an infinite-dimensional space. Hence, there exists a non-zero element $g \in S_{L_\infty(A_n)}$ such that $Tg = 0$. Then both elements $f \pm \frac{1}{n}g$ lie in $T^{-1}(x) \cap W$, which is impossible because f is an extreme point. $\square$

As the next example will show, the direct extension of Lyapunov's theorem to measures with values in an infinite-dimensional space fails.

Example 1. On the interval $[0, 1]$ define the measure μ with values in $L_2[0, 1]$ by the formula $\mu(A) = \mathbb{1}_A$. This measure is non-atomic and countably additive. At the same time, the set $\mu(\mathfrak{B})$ of all values (the range) of the vector measure μ is not convex: $0, \mathbb{1} \in \mu(\mathfrak{B})$, but the function identically equal to $1/2$ does not belong to $\mu(\mathfrak{B})$.

Using the fact that for any infinite-dimensional Banach space X there exists an injective operator $T: L_2[0, 1] \to X$, one can readily establish the existence of an X-valued non-atomic Borel measure on $[0, 1]$ with non-convex range. Such a measure can be given by the formula $\mu(A) = T(\mathbb{1}_A)$. Nonetheless, infinite-dimensional analogues of the Laypunov theorem do exist, albeit in a weakened form: such generalizations state that the closure $\overline{\mu(\Sigma)}$ of the set of values, rather than the range $\mu(\Sigma)$ itself, is convex.

Definition 1. The Banach space X is said to have the *Lyapunov property* if for any set Ω, any σ-algebra Σ on Ω, and any non-atomic countably-additive measure $\mu: \Sigma \to X$, the set $\overline{\mu(\Sigma)}$ is convex.

The same Example 1 above shows that Hilbert spaces do not have the Lyapunov property. At the same time (see [61]), the spaces c_0 and ℓ_p with $p \in [1, 2) \cup (2, +\infty)$ enjoy the Lyapunov property. Thus, with the Lyapunov property we run into a paradoxical situation: with respect to this property, Hilbert spaces are worse than the (rather badly behaved for other problems and, in particular, non-reflexive) space c_0.

Under additional restrictions on the measure, the community of spaces to which the weaker analogue of the Lyapunov theorem extends widens. For instance, if one considers only measures of bounded variation, then according to a theorem of Uhl (see the last chapter of the book [13], and also the paper [60]), the convexity of the set $\overline{\mu(\Sigma)}$ will hold for non-atomic measures taking values in any space with the Radon–Nikodým property (a class of Banach spaces that includes, in particular, all the reflexive spaces).

Comments on the Exercises

Section 18.1.2

Exercise 6. See [66]. As shown in [48], the set of extreme points of the closed unit ball of a reflexive space is not just uncountable, it cannot even have the "small balls property" (concerning this property, see the exercises of Subsection 11.2.1).

Exercise 7. See [3, Corollary 5.12 and Proposition 5.13].

Exercise 9. See [34, P. 15].

Exercise 10. The solution given here was communicated to us by Dirk Werner. Suppose that for some operator $T \in L(X)$ it holds that $\|I \pm T\| \leqslant 1$. Then $\|I^* \pm T^*\| \leqslant 1$, and for any $x^* \in \text{ext } \overline{B}_{X^*}$ we have $\|x^* \pm T^*x^*\| \leqslant 1$. By the definition of an extreme point, this means that $T^*x^* = 0$. Thus, we have proved that T^* maps into 0 all extreme points of the ball $\overline{B}_{X^*}$, and so $T^* = 0$. Consequently, $T = 0$, too.

Exercise 11. Use the previous exercise and Exercise 6 of Subsection 11.1.1.

Section 18.2.1

Exercise 3. By Milyutin's theorem (see the monograph [33]), if K_1 and K_2 are uncountable metrizable compact spaces, then the spaces $C(K_1)$ and $C(K_2)$ are isomorphic. For the concrete case $K_1 = [0, 1]$ and $K_2 = [0, 1] \cup \{2\}$ the corresponding isomorphism of $C[0, 1]$ and $C(K_2)$ can be given without appealing to the highly non-trivial Milutin's construction. Namely, one finds in $C[0, 1]$ a subspace X isomorphic to c_0 (for any sequence of functions $f_n \in S_{C[0,1]}$ with disjoint supports, $X = \overline{\text{Lin}}\{f_n\}$ is a subspace of $C[0, 1]$ isometric to c_0), then one represents $C[0, 1]$ as a direct sum of the form $X \oplus Y$, writing $C(K_2) = \text{Lin}\{\mathbb{1}_{\{2\}}\} \oplus C[0, 1] = \text{Lin}\{\mathbb{1}_{\{2\}}\} \oplus X \oplus Y$, and finally one proves (using the shift operator) that $\text{Lin}\{\mathbb{1}_{\{2\}}\} \oplus X$ is isomorphic to c_0, and thus is isomorphic to X.

References

1. S. Banach, *Functional Analysis Course* (Ukrainian), Izdat. Rad. Shkola (1948); see also *Theory of Linear Operations*. Translated from the French by F. Jellett. With comments by A. Pełczyński and Cz. Bessaga, vol. 38 (North-Holland Mathematical Library, North-Holland Publishing Co., Amsterdam, 1987)
2. B. Beauzamy, *Introduction to Operator Theory and Invariant Subspaces*, vol. 42 (North-Holland Mathematical Library, North-Holland Publishing Co., Amsterdam, 1988)
3. Y. Benyamini, J. Lindenstrauss, *Geometric Nonlinear Functional Analysis: Vol. 1*, vol. 48 (Colloquium Publications, American Mathematical Society, 2000)
4. Y.M. Berezansky, Z.G. Sheftel, G.F. Us, *Functional Analysis -Vol. I*, Translated from the 1990 Russian original by P.V. Malyshev. Operator Theory: Advances and Applications, vol. 85 (Birkhäuser Verlag, Basel, 1996); Vol. II. Translated from the 1990 Russian original by Peter V. Malyshev. Operator Theory: Advances and Applications, vol. 86 (Birkhäuser, Basel, 1996)
5. W. Blaschke, *Kreis und Kugel (German)* (Chelsea Publishing Co., New York, 1949)
6. R.P. Boas, Jr., *A Primer of Real Functions*, revised and updated by H.P. Boas, 4th ed. The Carus Mathematical Monographs, vol. 13 (The Mathematical Association of America, Washington, 1996)
7. V.G. Boltyanskiĭ, I. Gohberg, Theorems and Problems in Combinatorial Geometry (Russian), Moscow, Izdat. "Nauka" (1965), see also *Results and Problems in Combinatorial Geometry*, Translated from the Russian (Cambridge University Press, Cambridge, 1985)
8. F.F. Bonsall, J. Duncan, *Numerical Ranges of Operators on Normed Spaces and of Elements of Normed Algebras*, London Mathematical Society Lecture Note Series, 2 (Cambridge University Press, London-New York, 1971)
9. F.F. Bonsall, J. Duncan, *Numerical Ranges II*, vol. 10. London Mathematical Society Lecture Notes Series (Cambridge University Press, New York-London, 1973)
10. N. Bourbaki, *General Topology*, Chaps. 1–4. Elements of Mathematics, Translated from the French, Reprint of the 1989 English translation (Springer, Berlin, 1998)
11. M.M. Day, *Normed Linear Spaces*, 3rd edn. Ergebnisse der Mathematik und ihrer Grenzgebiete, Band 21 (Springer, New York-Heidelberg, 1973)
12. J. Diestel, *Geometry of Banach Spaces – Selected Topics*. Lecture Notes in Mathematics, vol. 485 (Springer, Berlin, New York, 1975)
13. J. Diestel, J.J. Uhl, Jr., *Vector Measures*, With a foreword by B.J. Pettis. Mathematical Surveys, vol. 15 (American Mathematical Society, Providence, 1977)
14. N. Dunford, J.T. Schwartz, *Linear Operators. Part I. General Theory*, With the assistance of W.G. Bade, R.G. Bartle. Reprint of the 1958 original. Wiley Classics Library. A Wiley-Interscience Publication (John Wiley & Sons Inc, New York, 1988); Part II. *Spectral Theory*.

© Springer International Publishing AG, part of Springer Nature 2018
V. Kadets, *A Course in Functional Analysis and Measure Theory*,
Universitext, https://doi.org/10.1007/978-3-319-92004-7

Self Adjoint Operators in Hilbert Space, Reprint of the 1963 original. Wiley Classics Library. A Wiley-Interscience Publication (John Wiley & Sons Inc, New York, 1988); Part III *Spectral operators*, Reprint of the 1971 original. Wiley Classics Library. A Wiley-Interscience Publication, (John Wiley & Sons Inc, New York, 1988)

15. R.E. Edwards, *Fourier Series. A Modern Introduction, Vol. 1*, 2nd edn. Graduate Texts in Mathematics, vol. 64 (Springer, New York, Berlin, 1979) ; Vol. 2. Graduate Texts in Mathematics, vol. 85 (Springer, New York, Berlin, 1982)

16. B. Grünbaum, *Studies in Combinatorial Geometry and the Theory of Convex Bodies*, Translated from the English by S. I. Zalgaller. Edited by V. A. Zalgaller and I. M. Jaglom. Mathematics Library. Izdat. "Nauka" (Moscow, 1971); Booklet contains the translations of two articles, *Measures of Symmetry for Convex Sets*, [Convexity (Proc. Sympos. Pure Math.), vol. 7 (American Mathematical Society, Providence, 1963), pp. 233–270; and *Borsuk's problem and related questions*, ibid., pp. 271–284]

17. H. Hadwiger, H. Debrunner, *Combinatorial Geometry in the Plane*. Translated by Victor Klee. With a new chapter and other additional material supplied by the translator, Holt, Rinehart and Winston (New York, 1964)

18. P.R. Halmos, *Measure Theory* (D. Van Nostrand Company Inc, New York, 1950)

19. E. Hewit, K.A. Ross, *Abstract Harmonic Analysis, Volume I, Structure of Topological Groups, Integration theory, Group Representations*, 2nd edn. Grundlehren der Mathematischen Wissenschaften, vol. 115 (Springer, Berlin, New York, 1979)

20. W.B. Johnson, J. Lindenstrauss J. (eds.), *Handbook of the Geometry of Banach Spaces*, vol. 1 (Elsevier Science B.V., Amsterdam, 2001)

21. M.I. Kadets, V.M. Kadets, *Series in Banach spaces. Conditional and Unconditional Convergence*, Operator Theory Advances and Applications, vol. 94 (Birkhäuser, Basel, 1997)

22. L.V. Kantorovich, G.P. Akilov, *Functional Analysis (Russian)*, Moscow, "Nauka", 1984; see also *Functional Analysis* Translated from the Russian by H. L. Silcock, 2nd edn. (Pergamon Press, Oxford-Elmsford, 1982)

23. J.L. Kelley, *General Topology*, Reprint of the 1955 Van Nostrand edn., Graduate Texts in Mathematics, vol. 27 (Springer, New York, Berlin, 1975)

24. A.N. Kolmogorov, S.V. Fomin, *Elements of the Theory of Functions and Functional Analysis* (Russian), with a supplement, "Banach algebras", by V. M. Tikhomirov, 6th edn. "Nauka" (Moscow, 1989); see also *Introductory Real Analysis*, Revised English edition. Translated from the Russian and edited by R.A. Silverman (Prentice-Hall, Inc., Englewood Cliffs, 1970)

25. K. Kuratowski, *Topology*, vol. I. New edition, revised and augmented. Translated from the French by J. Jaworowski (Academic Press, New York-London; Państwowe Wydawnictwo Naukowe, Warsaw, 1966); Vol. II. New edition, revised and augmented. Translated from the French by A. Kirkor (Academic Press, New York-London; Państwowe Wydawnictwo Naukowe Polish Scientific Publishers, Warsaw, 1968)

26. B.Ya. Levin, *Distribution of Zeros of Entire Functions* (Russian) (Tech.–Teor. Lit., Moscow, 1956); see also *Distribution of Zeros of Entire Functions*, Translated from the Russian by R. P. Boas, J. M. Danskin, F. M. Goodspeed, J. Korevaar, A. L. Shields and H. P. Thielman. Revised edition. Translations of Mathematical Monographs, vol. 5 (American Mathematical Society, Providence, 1980)

27. N. Levinson, *Gap and Density Theorems, American Mathematical Society Colloquium Publications*, vol. 26 (American Mathematical Society, New York, 1940)

28. J. Lindenstrauss, L. Tzafriri, *Classical Banach Spaces I: Sequence spaces* and II. *Function spaces*, Ergebnisse der Mathematik und ihrer Grenzgebiete, vols. 92. and 97 (Springer, Berlin, New York, 1977 and 1979)

29. L.A. Lyusternik, V.J. Sobolev, *Elements of Functional Analysis* (Russian), 2nd revised edition, Izdat. "Nauka" (Moscow 1965); see also the authorised third translation from the second extensively enlarged and rewritten Russian edition. International Monographs on Advanced Mathematics & Physics. Hindustan Publishing Corp., Delhi; Halsted Press [John Wiley & Sons, Inc.], New York, 1974

30. V.D. Milman, G. Schechtman, *Asymptotic Theory of Finite-Dimensional Normed Spaces*, vol. 1466, Lecture Notes in Mathematics (Springer, Berlin, 1986)
31. I.P. Natanson, *Theory of Functions of a Real Variable* (Russian), 2nd edition, revised (Gosudarstv. Izdat. Tehn.-Teoret. Lit., Moscow, 1957); see also *Theory of Functions of a Real Variable*, Translated from the Russian by L.F. Boron (Frederick Ungar Publishing Co., New York, 1961)
32. A.L. Paterson, *Amenability, Mathematical Surveys and Monographs* vol. 29 (American Mathematical Society, Providence, 1988)
33. A. Pełczyński, Linear extensions, linear averagings, and their applications to linear topological classification of spaces of continuous functions, Dissertationes Math. Rozprawy Math. **58** (1968)
34. R.R. Phelps, *Lectures on Choquet's Theorem*, vol. 1757, 2nd edn., Lecture Notes in Mathematics (Springer, Berlin, 2001)
35. M. Reed, B. Simon, *Methods of Modern Mathematical Physics I: Functional analysis*, 2nd edn. (Academic Press, Inc. [Harcourt Brace Jovanovich, Publishers], New York, 1980)
36. F. Riesz, B. Sz.-Nagy, in *Functional Analysis*, Translated from the second French edition by L.F. Boron. Reprint of the 1955, original (Dover Books on Advanced Mathematics, Dover Publications Inc, New York, 1990)
37. A.P. Robertson, W. Robertson, in *Topological Vector Spaces*, Reprint of the 2nd edition. Cambridge Tracts in Mathematics, vol. 53 (Cambridge University Press, Cambridge, New York, 1980)
38. W. Rudin, *Functional Analysis*, 2nd edn. International Series in Pure and Applied Mathematics (McGraw-Hill Inc, New York, 1991)
39. I. Singer, *Bases in Banach Spaces*. vol. 1, Die Grundlehren der mathematischen Wissenschaften, Band 154 (Springer, New York-Berlin, 1970)
40. I. Singer, *Bases in Banach Spaces*, vol. 2, Editura Academiei Republicii Socialiste Romania (Bucharest; Springer, Berlin, New York, 1981)
41. A.N. Tikhonov, V.Ya. Arsenin, in *Methods for the Solution of Ill-posed Problems*(Russian), 3rd edn. "Nauka", Moscow, see also Solutions of Ill-Posed Problems (V. H Winston & Sons, Washington, Halsted Press, New York, 1977)
42. E.C. Titchmarsh, *The Theory of Functions*, Reprint of the second (1939) edition (Oxford University Press, Oxford, 1958)
43. S. Wagon, *The Banach–Tarski Paradox*, With a foreword by J. Mycielski. Encyclopedia of Mathematics and its Applications, vol. 24 (Cambridge University Press, Cambridge, 1985)
44. D. Werner, *Funktional Analysis* (German), 7th revised and expanded edition (Springer, Berlin, 2011)

Articles

45. Y.A. Abramovich, C.D. Aliprantis, O. Burkinshaw, The invariant subspace problem: some recent advances, in *Workshop on Measure Theory and Real Analysis*(Italian) (Grado, 1995), Rend. Istit. Mat. Univ. Trieste, **29**, 3–79 (1998), suppl
46. T. Banakh, W.E. Lyantse, Ya.V. Mykytyuk, ∞-convex sets and their applications to the proof of certain classical theorems of functional analysis. Math. Stud. **11**(1), 83–84 (1999)
47. E. Behrends, New proofs of Rosenthal's ℓ^1-theorem and the Josefson-Nissenzweig theorem. Bull. Polish Acad. Sci. Math. **43**(4), 283–295 (1995)
48. E. Behrends, V.M. Kadets, Metric spaces with the small ball property. Studia Math. **148**(3), 275–287 (2001)
49. A. Bezdek, On a generalization of Tarski's plank problem. Discret. Comput. Geom. **38**(2), 189–200 (2007)

50. R.H. Bing, W.W. Bledsoe, R.D. Mauldin, Sets generated by rectangles. Pac. J. Math. **51**(1), 27–36 (1974)

51. J. Connor, M. Ganichev, V. Kadets, A characterization of banach spaces with separable duals via weak statistical convergence. J. Math. Anal. Appl. **244**(1), 251–261 (2000)

52. M. Eidelheit, Zur Theorie der systeme linearer Gleichungen. Stud. Math. **6**(1), 139–148 (1936)

53. P. Enflo, A counterexample to the approximation problem in Banach spaces. Acta Math. **130**, 309–317 (1973)

54. F. Hausdorff, Bemerkung über den Inhalt von Punktmengen. Math. Ann. **75**, 428–433 (1914)

55. B. Josefson, Weak sequential convergence in the dual of a Banach space does not imply norm convergence. Ark. Mat. **13**, 79–89 (1975)

56. V.M. Kadets, Toward a theorem on finding ω-linearly independent sequences, Teor. Funktsiĭ Funktsional. Anal. i Prilozhen. **58**, 78–80 (1992); Translation in J. Math. Sci. (New York) **85**(5), 2201–2202 (1997)

57. V.M. Kadets, Some remarks concerning the Daugavet equation. Quaest. Math. **19**(1–2), 225–235 (1996)

58. V. Kadets, Coverings by convex bodies and inscribed balls. Proc. Am. Math. Soc. **133**(5), 1491–1495 (2005)

59. V.M. Kadets, Weak cluster points of a sequence and coverings by cylinders. Mat. Fiz. Anal. Geom. **11**(2), 161–168 (2004)

60. V.M. Kadets, M.M. Popov, On the Liapunov convexity theorem with applications to sign-embeddings, Ukrain. Mat. Zh. **44**(9), 1192–1200 (1992); Translation in Ukrainian Math. J. **44**(9), 1091–1098 (1992)

61. V.M. Kadets, G. Schechtman, Lyapunov's theorem for ℓ_p-valued measures, Algebra i Analiz **4**(5), 148–154 (1992); Translation in *St. Petersburg Math. J.* **4**(5), 961–966 (1993)

62. V.M. Kadets, R.V. Shvidkoy, G.G. Sirotkin, D. Werner, Banach spaces with the Daugavet property. Trans. Am. Math. Soc. **352**(2), 855–873 (2000)

63. V.M. Kadets, L.M. Tseytlin, On "integration" of non-integrable vector-valued functions. Mat. Fiz. Anal. Geom. **7**(1), 49–65 (2000)

64. V.M. Kadets, M. Martín, R. Payá, Recent progress and open questions on the numerical index of Banach spaces, RACSAM. Rev. R. Acad. Cienc. Exactas Fís. Nat. Ser. A Mat. **100**(1–2), 155–182 (2006)

65. M. Laczkovich, Paradoxical decompositions: a survey of recent results, in *The First European Congress of Mathematics*, vol. 2 (Birkhäuser, Basel, 1992), pp. 159–184

66. J. Lindenstrauss, R.R. Phelps, Extreme point properties of convex bodies in reflexive Banach spaces. Isr. J. Math. **6**, 39–48 (1968)

67. J. Lindenstrauss, L. Tzafriri, On the complemented subspaces problem. Isr. J. Math. **9**, 263–269 (1971)

68. V.I. Lomonosov, Invariant subspaces of the family of operators that commute with a completely continuous operator. (Russian), Funktsional. Anal. i Prilozhen. **7**(3) 55–56 (1973); Translation in Functional analysis and its application **7**, 213–214 (1973)

69. G.G. Lorentz, A contribution to the theory of divergent sequences. Acta Math. **80**, 167–190 (1948)

70. B.M. Makarov, The problem of moments in some functional spaces, (Russian) Dokl. Akad. Nauk SSSR **127**, 957–960 (1959)

71. N.K. Nikolskiĭ, *Invariant subspaces in operator theory and function theory* (Russian), in: "Mathematical analysis", Vol. 12, pp. 199–412, Akad. Nauk SSSR Vsesoyuz. Inst. Nauchn. i Tekhn. Informatsii, Moscow, 1974

72. A. Nissenzweig, w^* sequential convergence. Isr. J. Math. **22**(3–5), 266–272 (1975)

73. J.W. Roberts, A compact convex set with no extreme points. Stud. Math. **60**(3), 255–266 (1977)

74. V.A. Yudin, On Fourier sums in L_p, Trudy Mat. Inst. Steklov **180**, 235–236 (1987); Translation in Proc. Steklov Inst. Math. **180**, 279–280 (1989)

Index

© Springer International Publishing AG, part of Springer Nature 2018
V. Kadets, *A Course in Functional Analysis and Measure Theory*,
Universitext, https://doi.org/10.1007/978-3-319-92004-7

Printed in the United States
By Bookmasters